INTERNATIONAL UNION OF PURE AND APPLIED CHEMISTRY

ANALYTICAL CHEMISTRY DIVISION
COMMISSION ON SOLUBILITY DATA

---

# SOLUBILITY DATA SERIES

Volume 10

## NITROGEN AND AIR

## SOLUBILITY DATA SERIES

Volume 1 H. L. Clever, *Helium and Neon*

Volume 2 H. L. Clever, *Krypton, Xenon and Radon*

Volume 3 M. Salomon, *Silver Azide, Cyanide, Cyanamides, Cyanate, Selenocyanate and Thiocyanate*

Volume 4 H. L. Clever, *Argon*

Volume 5/6 C. L. Young, *Hydrogen and Deuterium*

Volume 7 R. Battino, *Oxygen and Ozone*

Volume 8 C. L. Young, *Oxides of Nitrogen*

Volume 9 W. Hayduk, *Ethane*

Volume 10 R. Battino, *Nitrogen and Air*

Volume 11 B. Scrosati and C. A. Vincent, *Alkali Metal, Alkaline-Earth Metal and Ammonium Halides. Amide Solvents*

Volume 18 O. Popovych, *Tetraphenylborates*

### *Volumes in preparation*

C. L. Young, *Oxides of Sulfur*

H. L. Clever and W. Gerrard, *Hydrogen Halides in Non-Aqueous Solvents*

H. Miyamoto and M. Salomon, *Copper and Silver Halates*

S. Siekierski and J. K. Navratil, *Actinide Nitrates*

E. M. Woolley, *Silver Halides*

Z. Galus and C. Guminski, *Metals and Intermetallic Compounds in Mercury*

J. W. Lorimer, *Alkaline-Earth Metal Sulfates*

A. L. Horvath and F. W. Getzen, *Halogenated Benzenes, Toluenes, Xylenes and Phenols with Water*

H. L. Clever, D. M. Mason and C. L. Young, *Carbon Dioxide*

H. L. Clever and C. L. Young, *Methane*

# SOLUBILITY DATA SERIES

Volume 10

## NITROGEN AND AIR


Volume Editor

RUBINO BATTINO
*Wright State University*
*Dayton, Ohio, USA*

Evaluators

RUBIN BATTINO
*Wright State University*
*Dayton, Ohio, USA*

CHEN-TUNG A. CHEN
*Oregon State University*
*Corvallis, Oregon, USA*

H. LAWRENCE CLEVER
*Emory University*
*Atlanta, Georgia, USA*

COLIN L. YOUNG
*University of Melbourne*
*Parkville, Victoria, Australia*

Compilers

ARDIS L. CRAMER
*Emory University*
*Atlanta, Georgia, USA*

M. ELIZABETH DERRICK
*Valdosta State University*
*Valdosta, Georgia, USA*

J. CHR. GJALDBAEK
*Royal Danish School of Pharmacy*
*Copenhagen, Denmark*

SUSAN A. JOHNSON
*Emory University*
*Atlanta, Georgia, USA*

VLADIMIR KATOVIC
*Wright State University*
*Dayton, Ohio, USA*

TRINA D. KITTREDGE
*Emory University*
*Atlanta, Georgia, USA*

PATRICK L. LONG
*Emory University*
*Atlanta, Georgia, USA*

E. S. THOMSEN
*Royal Danish School of Pharmacy*
*Copenhagen, Denmark*

D. G. T. THORNHILL
*Leicester Polytechnic*
*Leicester, UK*



PERGAMON PRESS

OXFORD · NEW YORK · TORONTO · SYDNEY · PARIS · FRANKFURT

| | |
|---|---|
| U.K. | Pergamon Press Ltd., Headington Hill Hall,<br>Oxford OX3 0BW, England |
| U.S.A. | Pergamon Press Inc., Maxwell House, Fairview Park,<br>Elmsford, New York 10523, U.S.A. |
| CANADA | Pergamon Press Canada Ltd., Suite 104,<br>150 Consumers Rd., Willowdale, Ontario M2J 1P9, Canada |
| AUSTRALIA | Pergamon Press (Aust.) Pty. Ltd., P.O. Box 544,<br>Potts Point, N.S.W. 2011, Australia |
| FRANCE | Pergamon Press SARL, 24 rue des Ecoles,<br>75240 Paris, Cedex 05, France |
| FEDERAL REPUBLIC OF GERMANY | Pergamon Press GmbH, 6242 Kronberg-Taunus,<br>Hammerweg 6, Federal Republic of Germany |



First edition 1982

**Library of Congress Cataloging in Publication Data**

Main entry under title:
Nitrogen and air.
(Solubility data series ; v. 10)
Includes bibliographical references and indexes.
1. Nitrogen—Solubility—Tables. 2. Air—Solubility—Tables. I. Battino, Rubin. II. Series.
QD181.N1N57 1982 546'.711542'0212 82-15046

**British Library Cataloguing in Publication Data**

Nitrogen and air.—(Solubility data series, ISSN 0191-5622; v. 10)
1. Nitrogen—Solubility—Tables
I. Battino, Rubin II. International Union of Pure and Applied Chemistry. *Commission on Solubility Data* III. Series
546'.711'0212 QD181.N1

ISBN 0-08-023922-6

*In order to make this volume available as economically and as rapidly as possible the authors' typescripts have been reproduced in their original forms. This method unfortunately has its typographical limitations but it is hoped that they in no way distract the reader.*

*Printed in Great Britain by A. Wheaton & Co. Ltd., Exeter*

CONTENTS


*Foreword* vii
*Preface* xi

The Solubility of Gases in Liquids
 Introductory Information xiii
 Thermodynamic Considerations xx
The Sechenov Salt Effect Parameter xxix
Representative Values of Factors Needed to Convert
 Sechenov Salt Effect Parameter Forms xxxviii
Conversion Factors xliv

1. Nitrogen solubilities up to 200 kPa
 1.1 Water 1
 1.2 Deuterium oxide 30
 1.3 Seawater 31
 1.4 Salt solutions (aqueous) 45
 1.5 Organic solvents plus water 83
 1.6 Mixed organic solvents 104
 1.7 Hydrocarbons
  Saturated and unsaturated 119
  Cyclic 148
  Aromatic 162
 1.8 Organic compounds containing oxygen 174
 1.9 Organic compounds containing halogen 232
 1.10 Organic compounds containing sulfur 253
 1.11 Organic compounds containing nitrogen 260
 1.12 Organic compounds containing silicon 275
 1.13 Biological fluids 276
 1.14 Miscellaneous fluids 315

2. Nitrogen solubilities above 200 kPa 332
 2.1 Water 333
 2.2 Salt solutions (aqueous) 345
 2.3 Mixed solvents 352
 2.4 Hydrocarbons
  Saturated and unsaturated 361
  Cyclic 439
  Aromatic 443
 2.5 Organic compounds containing oxygen 446
 2.6 Organic compounds containing halogen 455
 2.7 Organic compounds containing nitrogen 462
 2.8 Biological fluids 464
 2.9 Miscellaneous fluids 466

3. Air solubilities 509

System Index 551
Registry Number Index 561
Author Index 565

SOLUBILITY DATA SERIES

*Editor-in-Chief*



INTERNATIONAL UNION OF PURE AND APPLIED CHEMISTRY

IUPAC Secretariat: Bank Court Chambers, 2-3 Pound Way,
Cowley Centre, Oxford OX4 3YF, UK

# FOREWORD

*If the knowledge is*
*undigested or simply wrong,*
*more is not better.*

How to communicate and disseminate numerical data effectively in chemical science and technology has been a problem of serious and growing concern to IUPAC, the International Union of Pure and Applied Chemistry, for the last two decades. The steadily expanding volume of numerical information, the formulation of new interdisciplinary areas in which chemistry is a partner, and the links between these and existing traditional subdisciplines in chemistry, along with an increasing number of users, have been considered as urgent aspects of the information problem in general, and of the numerical data problem in particular.

Among the several numerical data projects initiated and operated by various IUPAC commissions, the *Solubility Data Project* is probably one of the most ambitious ones. It is concerned with preparing a comprehensive critical compilation of data on solubilities in all physical systems, of gases, liquids and solids. Both the basic and applied branches of almost all scientific disciplines require a knowledge of solubilities as a function of solvent, temperature and pressure. Solubility data are basic to the fundamental understanding of processes relevant to agronomy, biology, chemistry, geology and oceanography, medicine and pharmacology, and metallurgy and materials science. Knowledge of solubility is very frequently of great importance to such diverse practical applications as drug dosage and drug solubility in biological fluids, anesthesiology, corrosion by dissolution of metals, properties of glasses, ceramics, concretes and coatings, phase relations in the formation of minerals and alloys, the deposits of minerals and radioactive fission products from ocean waters, the composition of ground waters, and the requirements of oxygen and other gases in life support systems.

The widespread relevance of solubility data to many branches and disciplines of science, medicine, technology and engineering, and the difficulty of recovering solubility data from the literature, lead to the proliferation of published data in an ever increasing number of scientific and technical primary sources. The sheer volume of data has overcome the capacity of the classical secondary and tertiary services to respond effectively.

While the proportion of secondary services of the review article type is generally increasing due to the rapid growth of all forms of primary literature, the review articles become more limited in scope, more specialized. The disturbing phenomenon is that in some disciplines, certainly in chemistry, authors are reluctant to treat even those limited-in-scope reviews exhaustively. There is a trend to preselect the literature, sometimes under the pretext of reducing it to manageable size. The crucial problem with such preselection - as far as numerical data are concerned - is that there is no indication as to whether the material was excluded by design or by a less than thorough literature search. We are equally concerned that most current secondary sources, critical in character as they may be, give scant attention to numerical data.

On the other hand, tertiary sources - handbooks, reference books, and other tabulated and graphical compilations - as they exist today, are comprehensive but, as a rule, uncritical. They usually attempt to cover whole disciplines, thus obviously are superficial in treatment. Since they command a wide market, we believe that their service to advancement of science is at least questionable. Additionally, the change which is taking place in the generation of new and diversified numerical data, and the rate at which this is done, is not reflected in an increased third-level service. The emergence of new tertiary literature sources does not parallel the shift that has occurred in the primary literature.

With the status of current secondary and tertiary services being as briefly stated above, the innovative approach of the *Solubility Data Project* is that its compilation and critical evaluation work involve consolidation and reprocessing services when both activities are based on intellectual and scholarly reworking of information from primary sources. It comprises compact compilation, rationalization and simplification, and the fitting of isolated numerical data into a critically evaluated general framework.

The *Solubility Data Project* has developed a mechanism which involves a number of innovations in exploiting the literature fully, and which contains new elements of a more imaginative approach for transfer of reliable information from primary to secondary/tertiary sources. *The fundamental trend of the Solubility Data Project is toward integration of secondary and tertiary services with the objective of producing in-depth critical analysis and evaluation which are characteristic to secondary services, in a scope as broad as conventional tertiary services.*

Fundamental to the philosophy of the project is the recognition that the basic element of strength is the active participation of career scientists in it. Consolidating primary data, producing a truly critically-evaluated set of numerical data, and synthesizing data in a meaningful relationship are demands considered worthy of the efforts of top scientists. Career scientists, who themselves contribute to science by their involvement in active scientific research, are the backbone of the project. The scholarly work is commissioned to recognized authorities, involving a process of careful selection in the best tradition of IUPAC. This selection in turn is the key to the quality of the output. These top experts are expected to view their specific topics dispassionately, paying equal attention to their own contributions and to those of their peers. They digest literature data into a coherent story by weeding out what is wrong from what is believed to be right. To fulfill this task, the evaluator must cover *all* relevant open literature. No reference is excluded by design and every effort is made to detect every bit of relevant primary source. Poor quality or wrong data are mentioned and explicitly disqualified as such. In fact, it is only when the reliable data are presented alongside the unreliable data that proper justice can be done. The user is bound to have incomparably more confidence in a succinct evaluative commentary and a comprehensive review with a complete bibliography to both good and poor data.

It is the standard practice that any given solute-solvent system consists of two essential parts: I. Critical Evaluation and Recommended Values, and II. Compiled Data Sheets.

The Critical Evaluation part gives the following information:

(i) a verbal text of evaluation which discusses the numerical solubility information appearing in the primary sources located in the literature. The evaluation text concerns primarily the quality of data after consideration of the purity of the materials and their characterization, the experimental method employed and the uncertainties in control of physical parameters, the reproducibility of the data, the agreement of the worker's results on accepted test systems with standard values, and finally, the fitting of data, with suitable statistical tests, to mathematical functions;

(ii) a set of recommended numerical data. Whenever possible, the set of recommended data includes weighted average and standard deviations, and a set of smoothing equations derived from the experimental data endorsed by the evaluator;

(iii) a graphical plot of recommended data.

The compilation part consists of data sheets of the best experimental data in the primary literature. Generally speaking, such independent data sheets are given only to the best and endorsed data covering the known range of experimental parameters. Data sheets based on primary sources where the data are of a lower precision are given only when no better data are available. Experimental data with a precision poorer than considered acceptable are reproduced in the form of data sheets when they are the only known data for a particular system. Such data are considered to be still suitable for some applications, and their presence in the compilation should alert researchers to areas that need more work.

The typical data sheet carries the following information:

(i) components - definition of the system - their names, formulas and Chemical Abstracts registry numbers;

(ii) reference to the primary source where the numerical information is reported. In cases when the primary source is a less common periodical or a report document, published though of limited availability, abstract references are also given;

(iii) experimental variables;

(iv) identification of the compiler;

(v) experimental values as they appear in the primary source. Whenever available, the data may be given both in tabular and graphical form. If auxiliary information is available, the experimental data are converted also to SI units by the compiler.

Under the general heading of Auxiliary Information, the essential experimental details are summarized:

(vi) experimental method used for the generation of data;

(vii) type of apparatus and procedure employed;

(viii) source and purity of materials;

(ix) estimated error;

(x) references relevant to the generation of experimental data as cited in the primary source.

This new approach to numerical data presentation, developed during our four years of existence, has been strongly influenced by the diversity of background of those whom we are supposed to serve. We thus deemed it right to preface the evaluation/compilation sheets in each volume with a detailed discussion of the principles of the accurate determination of relevant solubility data and related thermodynamic information.

Finally, the role of education is more than corollary to the efforts we are seeking. The scientific standards advocated here are necessary to strengthen science and technology, and should be regarded as a major effort in the training and formation of the next generation of scientists and engineers. Specifically, we believe that there is going to be an impact of our project on scientific-communication practices. The quality of consolidation adopted by this program offers down-to-earth guidelines, concrete examples which are bound to make primary publication services more responsive than ever before to the needs of users. The self-regulatory message to scientists of 15 years ago to refrain from unnecessary publication has not achieved much. The literature is still, in 1982, cluttered with poor-quality articles. The Weinberg report (in "Reader in Science Information", Eds. J. Sherrod and A. Hodina, Microcard Editions Books, Indian Head, Inc., 1973, p.292) states that "admonition to authors to restrain themselves from premature, unnecessary publication can have little effect unless the climate of the entire technical and scholarly community encourages restraint..." We think that projects of this kind translate the climate into operational terms by exerting pressure on authors to avoid submitting low-grade material. The type of our output, we hope, will encourage attention to quality as authors will increasingly realize that their work will not be suited for permanent retrievability unless it meets the standards adopted in this project. It should help to dispel confusion in the minds of many authors of what represents a permanently useful bit of information of an archival value, and what does not.

If we succeed in that aim, even partially, we have then done our share in protecting the scientific community from unwanted and irrelevant, wrong numerical information.

A. S. Kertes

# PREFACE

The discovery of nitrogen is usually attributed to Daniel Rutherford in 1772. Priestly, Scheele, and Cavendish also did important experiments on this "phlogisticated" air (in contrast to "dephlogisticated" air or oxygen). Lavoisier named the gas azote from the Greek, meaning without life since it did not support life. Because of its presence in niter or nitre ($KNO_3$) the English name is nitrogen. Nitrogen forms 78.09 per cent of the atmosphere by volume and 75.51 per cent by weight. Despite this great abundance in the atmosphere nitrogen is only present to the extent of 0.0046 weight per cent in the igneous rocks of the crust of the earth. It is not even listed in a table I have on the amounts of elements in the ocean. The great importance of nitrogen is due to its prevalence in air and its "fixed" form as various nitrates.

This particular volume is concerned with the solubility of nitrogen and air in pure liquids, liquid mixtures, aqueous and organic solutions, biological fluids, and miscellaneous fluids and fluid mixtures. High and low pressures are covered in separate sections, the dividing line being a gas partial pressure of about 200 kPa (2 bar). With the higher pressure data there is always a problem of attempting to distinguish between gas solubilities and vapor-liquid equilibria. We have striven to be reasonable in making this distinction. Certain fluids such as molten salts, slags, glasses, metals, and alloys have been excluded from this volume, as well as polymeric materials. The literature, including specialty tabulations and compilations, has been searched through the middle of 1981, although a few more recent papers are included. (The editor would appreciate hearing about articles which were missed.)

The individual data sheets in this volume were prepared using the solubility units presented in the original papers. In general, where a given system was studied by several groups a critical evaluation was prepared. These critical evaluations are meant to be comprehensive, i.e., citing all papers with measurements on that system even though data sheets are not prepared from papers containing data of poor or questionable quality. In these critical evaluations, recommended or tentative smoothed data are usually presented. All cited papers are discussed, albeit briefly in many instances. Related systems are also discussed.

To our knowledge the solubilities of nitrogen and air have been determined only by physical means (if we leave out difference methods in which oxygen and carbon dioxide are analyzed for via chemical methods). The data sheets contain concise descriptions of the methods used. Information is also provided on the source and purity of the components, although we are surprised at how many papers omit this information.

Gas solubilities are presented in the original papers in an almost bewildering number of units. In the data sheets we present the units given in the original papers with explanatory material where necessary. Where it is feasible the compilers have converted these solubilities into the Ostwald coefficient and/or the mole fraction solubility at 101.325 kPa (1 atm) partial pressure of gas. These conversions were carried out for the convenience of the user, and such compiler-generated conversions are noted on the data sheets. Unless otherwise specified the units for the Bunsen coefficients reported herein are $cm^3$ (STP) $cm^{-3}$ $atm^{-1}$.

We are indebted to Dr. William Gerrard for two instances illuminating the necessity of caution in the choice of units to represent gas solubilities. At 295 K if we take the ratio of the solubility of nitrogen in olive oil in units of the Bunsen coefficient to that in water it is 3.5. However, in terms of the mole ratio (number of moles of gas/number of moles of solvent at a specified temperature and pressure) this ratio is 188. Similar ratios at 293 K for nitrogen in several solvents are (solvent, ratio of Bunsen coefficients, ratio of mole ratios): benzene, 7.10, 35.0; methanol, 8.66, 19.4; ethanol, 8.77, 28.4; 1-pentanol, 7.50, 41.7; acetone, 8.94, 36.4; and $CH_3CO_2C_5H_{11}$, 9.44, 83.6. One last example involves the solubility of nitrogen in some biological fluids at 298.5 K. The solvent is given first, the the Bunsen coefficient, then the mole ratio solubility: olive oil, 0.0686, 0.00273; tetralin, 0.0638, 0.000357; oleic acid, 0.085, 0.00111; and ethyl palmitate 0.109, 0.00144. The question of which solvent is "best" for a given gas thus depends on the solubility unit you are using. Caution is obviously required in talking about "solvent power." Solubility units need to be matched to a given model for rational use.

On the data sheets data are reported in the units presented in the original paper. The data are also converted to SI in the tables where this is conveniently feasible, and always in SI in the "VARIABLES" box. With respect to naming components, we have attempted to use IUPAC nomenclature throughout. The Hill system is used for the formulas and the unique Chemical Abstracts Service Registry Numbers are used for all identifiable substances. The Standard Order of Arrangement is used for the sequencing of systems in a given section. Sorting of systems using the table of contents worked unambiguously for most systems. However, some systems had to be assigned arbitrarily by the editor. Common names of substances are used in addition to the systematic names as an aid to the reader.

Error estimates are given where the author(s) give this information. They are also estimated by the compilers in many cases.

Real gas corrections are not necessary for the data on nitrogen solubilities reported in this volume since we found no data of a precision to require the corrections. At STP the molar volume of nitrogen is 22,403.8 $cm^3$ which corresponds to a correction of 0.05 per cent. Recommended second virial coefficients taken from the second edition of Dymond and Smith are shown below along with the per cent error for not applying the correction. Around room temperature the corrections are negligible.

| T/K | $B_{22}/cm^3\ mol^{-1}$ | Per cent Error | T/K | $B_{22}/cm^3\ mol^{-1}$ | Per cent Error |
|---|---|---|---|---|---|
| 100 | -160 | 1.9 | 300 | -4.2 | 0.02 |
| 150 | - 71.5 | 0.58 | 400 | 9.0 | 0.03 |
| 200 | - 35.2 | 0.21 | 500 | 16.9 | 0.04 |
| 250 | - 16.2 | 0.08 | | | |

One of the surprising results of the literature search for this work was the discovery of the relatively small number of measurements for the solubility of air in liquids. Of course, reasonably good values of air solubilities may be calculated from nitrogen and oxygen solubilities since these are additive within at least one per cent. Our literature search unearthed only one paper on nitrogen in heavy water, $D_2O$. This system should be studied further.

In the interest of helping the reader we have included two special papers written specifically for this volume by Drs. H. L. Clever and E. Wilhelm. The paper by Dr. Clever expands on the salt-effect parameters pioneered by Sechenov. Useful equations and interconversions are developed along with a recommended systematic notation. An appendix to this paper includes practical tables for some common salts. The paper by Dr. Wilhelm presents a rigorous thermodynamic treatment for the solubility of gases in liquids. The formalism developed is particularly important for measurements of high precision. We are pleased to include these two papers and express a special thanks to both authors.

The Editor acknowledges the advice and helpful suggestions of his colleagues in IUPAC Commission V.8, the Commission on Solubility Data. He also gratefully acknowledges the work of his research assistant, Mr. Chai-Jing Chou, and the secretaries who worked on this volume - Marty Tucker, Carolyn Dowie, and Lesley Flanagan.

Acknowledgment is made to the National Bureau of Standards for a grant which supported part of this work.

Finally, acknowledgment is made to the following scientists who critically read separate sections of this volume: Drs. R. W. Cargill, C-T. A. Chen, H. L. Clever, C. M. Criss, W. Gerrard, J. Chr. Gjaldbaek, W. Hayduk, D. J. Karl, T. R. Rettich, E. S. Thomsen, R. F. Weiss, and W.-Y. Wen.

Rubin Battino

Dayton, Ohio
April, 1982

# THE SOLUBILITY OF GASES AND LIQUIDS

## Introductory Information

C. L. Young, R. Battino, and H. L. Clever

### INTRODUCTION

The Solubility Data Project aims to make a comprehensive search of the literature for data on the solubility of gases, liquids and solids in liquids. Data of suitable accuracy are compiled into data sheets set out in a uniform format. The data for each system are evaluated and where data of sufficient accuracy are available values are recommended and in some cases a smoothing equation is given to represent the variation of solubility with pressure and/or temperature. A text giving an evaluation and recommended values and the compiled data sheets are published on consecutive pages. The following paper by E. Wilhelm gives a rigorous thermodynamic treatment on the solubility of gases in liquids.

### DEFINITION OF GAS SOLUBILITY

The distinction between vapor-liquid equilibria and the solubility of gases in liquids is arbitrary. It is generally accepted that the equilibrium set up at 300K between a typical gas such as argon and a liquid such as water is gas-liquid solubility whereas the equilibrium set up between hexane and cyclohexane at 350K is an example of vapor-liquid equilibrium. However, the distinction between gas-liquid solubility and vapor-liquid equilibrium is often not so clear. The equilibria set up between methane and propane above the critical temperature of methane and below the critical temperature of propane may be classed as vapor-liquid equilibrium or as gas-liquid solubility depending on the particular range of pressure considered and the particular worker concerned.

The difficulty partly stems from our inability to rigorously distinguish between a gas, a vapor, and a liquid; a subject which has been discussed in numerous textbooks. We have taken a fairly liberal view in these volumes and have included systems which may be regarded, by some workers, as vapor-liquid equilibria.

### UNITS AND QUANTITIES

The solubility of gases in liquids is of interest to a wide range of scientific and technological disciplines and not solely to chemistry. Therefore a variety of ways for reporting gas solubility have been used in the primary literature. Sometimes, because of insufficient available information, it has been necessary to use several quantities in the compiled tables. Where possible, the gas solubility has been quoted as a mole fraction of the gaseous component in the liquid phase. The units of pressure used are bar, pascal, millimeters of mercury, and atmosphere. Temperatures are reported in Kelvins.

### EVALUATION AND COMPILATION

The solubility of comparatively few systems is known with sufficient accuracy to enable a set of recommended values to be presented. This is true both of the measurements near atmospheric pressure and at high pressures. Although a considerable number of systems have been studied by at least two workers, the range of pressures and/or temperatures is often sufficiently different to make meaningful comparison impossible.

Occasionally, it is not clear why two groups of workers obtained very different sets of results at the same temperature and pressure, although both sets of results were obtained by reliable methods and are internally consistent. In such cases, sometimes an incorrect assessment has been given. There are several examples where two or more sets of data have been classified as tentative although the sets are mutually inconsistent.

Many high pressure solubility data have been published in a smoothed form. Such data are particularly difficult to evaluate, and unless specifically discussed by the authors, the estimated error on such values can only be regarded as an "informed guess".

Many of the high pressure solubility data have been obtained in a more general study of high pressure vapor-liquid equilibrium. In such cases a note is included to indicate that additional vapor-liquid equilibrium data are given in the source. Since the evaluation is for the compiled data, it is possible that the solubility data are given a classification which is better than that which would be given for the complete vapor-liquid data (or vice versa). For example, it is difficult to determine coexisting liquid and vapor compositions near the critical point of a mixture using some widely used experimental techniques which yield accurate high pressure solubility data. As another example, conventional methods of analysis may give results with an expected error which would be regarded as sufficiently small for vapor-liquid equilibrium data but an order of magnitude too large for acceptable high pressure gas-liquid solubility.

It is occasionally possible to evaluate data on mixtures of a given substance with a member of a homologous series by considering all the available data for the given substance with other members of the homologous series. In this study the use of such a technique has been limited.

The estimated error is often omitted in the original article and sometimes the errors quoted do not cover all the variables. In order to increase the usefulness of the compiled tables *estimated* errors have been included even when absent from the original article. If the error on *any* variable has been inserted by the compiler, this has been noted.

## PURITY OF MATERIALS

The purity of materials has been quoted in the compiled tables where given in the original publication. The solubility is usually more sensitive to impurities in the gaseous component than to liquid impurities in the liquid component. However, the most important impurities are traces of a gas dissolved in the liquid. Inadequate degassing of the absorbing liquid is probably the most often overlooked serious source of error in gas solubility measurements.

## APPARATUS AND PROCEDURES

In the compiled tables brief mention is made of the apparatus and procedure. There are several reviews on experimental methods of determining gas solubilities and these are given in References 1-7.

## METHODS OF EXPRESSING GAS SOLUBILITIES

Because gas solubilities are important for many different scientific and engineering problems, they have been expressed in a great many ways:

### The Mole Fraction, x(g)

The mole fraction solubility for a binary system is given by:

$$x(g) = \frac{n(g)}{n(g) + n(l)}$$

$$= \frac{W(g)/M(g)}{\{W(g)/M(g)\} + \{W(l)/M(l)\}}$$

here n is the number of moles of a substance (an *amount* of substance), W is the mass of a substance, and M is the molecular mass. To be unambiguous, the partial pressure of the gas (or the total pressure) and the temperature of measurement must be specified.

### The Weight Per Cent Solubility, wt%

For a binary system this is given by

$$\text{wt\%} = 100\, W(g)/\{W(g) + W(l)\}$$

where W is the weight of substance. As in the case of mole fraction, the pressure (partial or total) and the temperature must be specified. The weight per cent solubility is related to the mole fraction solubility by

$$x(g) = \frac{\{wt\%/M(g)\}}{\{wt\%/M(g)\} + \{(100 - wt\%)/M(1)\}}$$

The Weight Solubility, $C_W$

The weight solubility is the number of moles of dissolved gas per gram of solvent when the partial pressure of gas is 1 atmosphere. The weight solubility is related to the mole fraction solubility at one atmosphere partial pressure by

$$x(g) \text{ (partial pressure 1 atm)} = \frac{C_W M(1)}{1 + C_W M(1)}$$

where M(1) is the molecular weight of the solvent.

The Moles Per Unit Volume Solubility, n

Often for multicomponent systems the density of the liquid mixture is not known and the solubility is quoted as moles of gas per unit volume of liquid mixture. This is related to the mole fraction solubility by

$$x(g) = \frac{n\, v^{o}(1)}{1 + n\, v^{o}(1)}$$

where $v^{o}(1)$ is the molar volume of the liquid component.

The Bunsen Coefficient, $\alpha$

The Bunsen coefficient is defined as the volume of gas reduced to 273.15K and 1 atmosphere pressure which is absorbed by unit volume of solvent (at the temperature of measurement) under a partial pressure of 1 atmosphere. If ideal gas behavior and Henry's law are assumed to be obeyed, then

$$\alpha = \frac{V(g)}{V(1)} \frac{273.15}{T}$$

where V(g) is the volume of gas absorbed and V(1) is the original (starting) volume of absorbing solvent. The mole fraction solubility x is related to the Bunsen coefficient by

$$x(g, 1 \text{ atm}) = \frac{\alpha}{\alpha + \frac{273.15}{T} \frac{v^{o}(g)}{v^{o}(1)}}$$

where $v^{o}(g)$ and $v^{o}(1)$ are the molar volumes of gas and solvent at a pressure of one atmosphere. If the gas is ideal,

$$x(g) = \frac{\alpha}{\alpha + \frac{273.15R}{v^{o}(1)}}$$

Real gases do not follow the ideal gas law and it is important to establish the real gas law used for calculating $\alpha$ in the original publication and to make the necessary adjustments when calculating the mole fraction solubility.

The Kuenen Coefficient, S

This is the volume of gas, reduced to 273.15K and 1 atmosphere pressure, dissolved at a partial pressure of gas of 1 atmosphere by 1 gram of solvent.

TABLE 1 Interconversion of parameters used for reporting solubility

$$L = \alpha(T/273.15)$$

$$C_w = \alpha/v_o\rho$$

$$K_H = \frac{17.033 \times 10^6 \rho(\text{soln})}{\alpha\ M(1)} + 760$$

$$L = C_w\ v_{t,gas}\ \rho$$

where $v_o$ is the molal volume of the gas in $cm^3 mol^{-1}$ at 0°C, $\rho$ the density of the solvent at the temperature of the measurement, $\rho_{soln}$ the density of the solution at the temperature of the measurement, and $v_{t,gas}$ the molal volume of the gas ($cm^3 mol^{-1}$) at the temperature of the measurement.

SALT EFFECTS

Salt effect studies have been carried out for many years. The results are often reported as Sechenov (Setchenow) salt effect parameters. There appears to be no common agreement on the units of either the gas solubility, or the electrolyte concentration.

Many of the older papers report the salt effect parameter in a form equivalent to

$$k_{scc}/\text{mol dm}^{-3} = (1/(c_2/\text{mol dm}^{-3}))\ \log\ ((c_1^\circ/\text{mol dm}^{-3})/(c_1/\text{mol dm}^{-3}))$$

where the molar gas solubility ratio, $c_1^\circ/c_1$, is identical to the Bunsen coefficient ratio, $\alpha^\circ/\alpha$, or the Ostwald coefficient ratio, $L^\circ/L$. One can designate the salt effect parameters calculated from the three gas solubility ratios as $k_{scc}$, $k_{sc\alpha}$, $k_{scL}$, respectively, but they are identical, and $k_{scc}/dm^3\ mol^{-1}$ describes all of them. The superzero refers to the solubility in the pure solvent.

Recent statistical mechanical theories favor a molal measure of the electrolyte and gas solubility. Some of the more recent salt effects are reported in the form

$$k_{smm}/\text{kg mol}^{-1} = (1/(m_2/\text{mol kg}^{-1})\ \log\ ((m_1^\circ/\text{mol kg}^{-1})/(m_1/\text{mol kg}^{-1})$$

In this equation the $m_1^\circ/m_1$ ratio is identical to the Kuenen coefficient ratio, $s_1^\circ/s_1$, or the solvomolality ratio referenced to water, $A_{sm}^\circ/A_{sm}$. Thus the salt effect parameters $k_{smm}$, $k_{sms}$, and $k_{smA_{sm}}$ are well represented by the $k_{smm}/\text{kg mol}^{-1}$.

Some experimentalists and theoreticians prefer the gas solubility ratio as a mole fraction ratio, $x_1^\circ/x_1$. It appears that most calculate the mole fraction on the basis of the total number of ions. The salt effect parameters

$$k_{scx}/\text{dm}^3\ \text{mol}^{-1} = (1/(c_2/\text{mol dm}^{-3}))\ \log\ (x_1^\circ/x_1)$$

and

$$k_{smx}/\text{kg mol}^{-1} = (1/(m_2/\text{mol kg}^{-1}))\ \log\ (x_1^\circ/x_1)$$

are both in the literature, but $k_{scx}$ appears to be the more common.

The following conversions were worked out among the various forms of the salt effect parameter from standard definitions of molarity, molality, and mole fraction assuming the gas solubilities are small.

$$k_{smc} = (c_2/m_2)\ k_{scc} = (c_2/m_2)\ k_{scm} + F_{1m}$$

$$k_{scm} = k_{scc} - F_{1c} = (m_2/c_2)\ k_{smc} - F_{1c} = (m_2/c_2)\ k_{smm}$$

$$k_{scx} = (m_2/c_2)\ k_{smx} = (m_2/c_2)\ ksmm + F_{2c}$$

$$k_{smm} = k_{smx} - F_{2m} = (c_2/m_2)\ k_{scx} - F_{2m}$$

$$k_{smx} = (c_2/m_2)\ k_{scx} = (c_2/m_2)\ k_{scc} + F_{3m}$$

$$k_{scc} = k_{scx} - F_{3c} = (m_2/c_2)\ k_{smx} - F_{3c}$$

where

$$F_{1m} = (1/m_2)\ \log\ [(\rho°/\rho)\ (1000 + m_2M_2)/1000]$$

$$F_{1c} = (m_2/c_2)\ F_{1m}$$

$$F_{2m} = (1/m_2)\ \log\ [(1000 + \nu m_3\ M_3)/1000]$$

$$F_{2c} = (m_2/c_2)\ F_{2m}$$

$$F_{3m} = (1/m_2)\ \log\ [(1000\rho + (\nu M_3 - M_2)\ c_2)/1000\rho°)$$

$$F_{3c} = (m_2/c_2)\ F_{3m}$$

The factors $F_{1m}$, $F_{1c}$, $F_{2m}$, $F_{2c}$, $F_{3m}$, and $F_{3c}$ can easily be calculated from aqueous electrolyte data such as weight per cent and density as found in Volume III of the International Critical Tables. The values are small and change nearly linearly with both temperature and molality. The factors normally amount to no more than 10 to 20 per cent of the value of the salt effect parameter.

The symbols in the equations above are defined below:

| Component | Molar Concentration $c/\text{mol dm}^{-3}$ | Molal Concentration $m/\text{mol kg}^{-1}$ | Mole Fraction $x$ | Molecular Weight $M/\text{g mol}^{-1}$ |
|---|---|---|---|---|
| Nonelectrolyte | $c_1^\circ$, $c_1$ | $m_1^\circ$, $m_1$ | $x_1^\circ$, $x_1$ | $M_1$ |
| Electrolyte | $c_2$ | $m_2$ | $x_2$ | $M_2$ |
| Solvent | $c_3$ | $m_3$ | $x_3$ | $M_3$ |

The superscript "°" refers to the nonelectrolyte solubility in the pure solvent. The pure solvent and solution densities are $\rho°/\text{g cm}^{-3}$ and $\rho/\text{g cm}^{-3}$, respectively. They should be the densities of gas saturated solvent (water) and salt solution, but the gas free densities will differ negligibly in the $\rho°/\rho$ ratio. The number of ions per formula of electrolyte is symbolized by $\nu$.

The following table gives estimated errors in $k_{scc}$ for various salt concentrations and a range of random errors in the gas solubility measurement

Error in $k_{scc}/\text{dm}^3\ \text{mol}^{-1}$ [a]

| $c_2/\text{mol dm}^{-3}$ | Random Error in gas solubility Measurement ±2% | ±1% | ±0.5% | ±0.1% | ±0.05% |
|---|---|---|---|---|---|
| 1 | ±18% | ±9% | ±5% | ±1.5% | ±1% |
| 0.1 | ±175% | ±87% | ±43% | ±9% | ±4% |
| 0.05 | ±350% | ±174% | ±87% | ±17% | ±9% |
| 0.01 | ±1750% | ±870% | ±435% | ±87% | ±43% |

[a] Based on a $k_{scc}$ value of 0.100.

AQUAMOLAL OR SOLVOMOLAL, $A_{sm}$ or $m_i^{(s)}$

The term aquamolal was suggested by R. E. Kerwin (9). The unit was first used in connection with $D_2O$ and $H_2O + D_2O$ mixtures. It has since been extended in use to other solvents. The unit represents the numbers of moles of solute per 55.51 moles of solvent. It is represented by

$m_i^{(s)}$/mol kg$^{-1}$ = $(n_1M_2/w_2)(w_2/M_0) = m_i(M_2/M_0)$ where an amount of $n_i$ of solute $i$ is dissolved in a mass $w_2$ of solvent of molar mass $M_2$; $M_0$ is the molar mass of a reference solvent and $m_i$/mol kg$^{-1}$ is the conventional molality in the reference solvent. The reference solvent is normally water.

TEMPERATURE DEPENDENCE OF GAS SOLUBILITY

In a few cases it has been found possible to fit the mole fraction solubility at various temperatures using an equation of the form

$$\ln x = A + B / (T/100K) + C \ln (T/100K) + DT/100K$$

It is then possible to write the thermodynamic functions $\Delta\overline{G}_1^\circ$, $\Delta\overline{H}_1^\circ$, $\Delta\overline{S}_1^\circ$ and $\Delta\overline{C}_{p_1}$ for the transfer of the gas from the vapor phase at 101,325 Pa partial pressure to the (hypothetical) solution phase of unit mole fraction as:

$$\Delta\overline{G}_1^\circ = -RAT - 100\,RB - RCT \ln (T/100) - RDT^2/100$$

$$\Delta\overline{S}_1^\circ = RA + RC \ln (T/100) + RC + 2\,RDT/100$$

$$\Delta\overline{H}_1^\circ = -100\,RB + RCT + RDT^2/100$$

$$\Delta\overline{C}_{p_1}^\circ = RC + 2\,RDT/100$$

In cases where there are solubilities at only a few temperatures it is convenient to use the simpler equations

$$\Delta\overline{G}_1^\circ = -RT \ln x = A + BT$$

in which case $A = \Delta\overline{H}_1^\circ$ and $-B = \Delta\overline{S}_1^\circ$

REFERENCES


1. Battino, R.; Clever, H. L. *Chem. Rev.* 1966, *66*, 395.
2. Clever, H. L.; Battino, R. in *Solutions and Solubilities*, Ed. M. R. J. Dack, J. Wiley & Sons, New York, 1975, Chapter 7.
3. Hildebrand, J. H.; Prausnitz, J. M.; Scott, R. L. *Regular and related Solutions*, Van Nostrand Reinhold, New York, 1970, Chapter 8.
4. Markham, A. E.; Kobe, K. A. *Chem. Rev.* 1941, *63*, 449.
5. Wilhelm, E.; Battino, R. *Chem. Rev.* 1973, *73*, 1.
6. Wilhelm, E.; Battino, R.; Wilcock, R. J. *Chem. Rev.* 1977, *77*, 219.
7. Kertes, A. S.; Levy, O.; Markovits, G. Y. in *Experimental Thermochemistry* Vol. II, Ed. B. Vodar and B. LeNaindre, Butterworth, London, 1974, Chapter 15.
8. Long, F. A.; McDevit, W. F. *Chem. Rev.* 1952, *51*, 119.
9. Kerwin, K. E., Ph.D. Thesis, University of Pittsburgh 1964.


Revised: April 1982 (R.B., H.L.C.)

## The Ostwald Coefficient, L

The Ostwald coefficient, L, is defined as the ratio of the volume of gas absorbed to the volume of the absorbing liquid, all measured at the same temperature:

$$L = \frac{V(g)}{V(l)}$$

If the gas is ideal and Henry's Law is applicable, the Ostwald coefficient is independent of the partial pressure of the gas. It is necessary, in practice, to state the temperature and total pressure for which the Ostwald coefficient is measured. The mole fraction solubility, x(g), is related to the Ostwald coefficient by

$$x(g) = \left[\frac{RT}{P(g)\ L\ v^{o}(l)} + 1\right]^{-1}$$

where P is the partial pressure of gas. The mole fraction solubility will be at a partial pressure of P(g). (See the following paper by E. Wilhelm for a more igorous definition of the Ostwald coefficient.)

## The Absorption Coefficient, β

There are several "absorption coefficients", the most commonly used one being defined as the volume of gas, reduced to 273.15K and 1 atmosphere, absorbed per unit volume of liquid when the total pressure is 1 atmosphere. β is related to the Bunsen coefficient by

$$\beta = \alpha(1 - P(l))$$

where P(l) is the partial pressure of the liquid in atmospheres.

## The Henry's Law Constant

A generally used formulation of Henry's Law may be expressed as

$$P(g) = K_H x(g)$$

where $K_H$ is the Henry's Law constant and x(g) the mole fraction solubility. Other formulations are

$$P(g) = K_2 C(l) \qquad \text{or} \qquad C(g) = K_c C(l)$$

where $K_2$ and $K_c$ are constants, C the concentration, and (l) and (g) refer to the liquid and gas phases. Unfortunately, $K_H$, $K_2$ and $K_c$ are all sometimes referred to as Henry's Law constants. Henry's Law is a *limiting law* but can sometimes be used for converting solubility data from the experimental pressure to a partial gas pressure of 1 atmosphere, provided the mole fraction of the gas in the liquid is small, and that the difference in pressures is small. Great caution must be exercised in using Henry's Law.

## The Mole Ratio, N

The mole ratio, N, is defined by

$$N = n(g)/n(l)$$

Table 1 contains a presentation of the most commonly used inter-conversions not already discussed.

For gas solubilities greater than about 0.01 mole fraciton at a partial pressure of 1 atmosphere there are several additional factors which must be taken into account to unambiguously report gas solubilities. Solution densities or the partial molar volume of gases must be known. Corrections should be made for the possible non-ideality of the gas or the non-applicability of Henry's Law.

# THE SOLUBILITY OF GASES IN LIQUIDS

## Thermodynamic Considerations

*Emmerich Wilhelm*, Institut für Physikalische Chemie,

Universität Wien, Währingerstraße 42, A-1090 Wien, Austria

## 1. INTRODUCTION

Progress on the scope and accuracy of experimental methods for measuring the solubility of gases in liquids as well as advances in the development of estimation procedures have been quite impressive during the last decade or so. Hence, reasonably frequent updating of the introductory material of a data series concerned with this topic, perhaps reflecting shifting emphasis, is indicated. Various aspects of "gas solubility in liquids", including statistical thermodynamical theory, have been treated at different levels of sophistication in several recent review articles, monographs, introductory sections to critical data compilations, and primary publications (1-19), thus documenting unabated interest. Rather than attempting to present a comprehensive state-of-the-art report, we will focus in this article on essentially *classical* thermodynamics associated with the evaluation of Henry coefficients from high-precision measurements of the solubility of pure gases in pure liquids (20-22), that is to say on problems encountered when experimental imprecision is reduced to about 0.05% (or less). Gas solubilities in mixed solvent systems, though of considerable theoretical and practical interest, will not be considered. For the moment, a few representative recent publications, such as refs. (23-28), should suffice to indicate some problems and trends in this area of research.

## 2. THERMODYNAMICS

The thermodynamic analysis is based on the rigorous criterion for vapor(V)-liquid(L) phase equilibrium in PVT-systems of uniform temperature T and pressure P

$$f_i^V(T,P,y_i) = f_i^L(T,P,x_i) \tag{1}$$

that is, on the equality of the fugacities $f_i$ for each component i in all phases. Here, $x_i$ denotes the liquid phase mole fraction and $y_i$ is the vapor phase mole fraction. Let component 2 be the liquid solvent, and component 1 the so-called dissolved gas (not necessarily a supercritical fluid). Introducing the auxiliary functions liquid-phase activity coefficient $\gamma_i$ for which the asymmetric convention is adopted

$$\begin{aligned}\gamma_2 &\to 1 \text{ for } x_2 \to 1\\ \gamma_1 &\to 1 \text{ for } x_1 \to 0\end{aligned} \tag{2}$$

and vapor-phase fugacity coefficient $\phi_i$, the equilibrium condition may be written as

$$y_i P\phi_i(T,P,y_i) = x_i\gamma_i(T,P,x_i)f_i^\circ(T,P) \tag{3}$$

$f_i^\circ(T,P)$ is an appropriate liquid-phase standard-state fugacity at the experimental system temperature and pressure. For the solvent, this is the Poynting-corrected *fugacity of pure solvent* at the same temperature and at orthobaric conditions (indicated by the subscript s), i.e., at saturation pressure $P_{s,2}$ of the solvent:

$$f_2^\circ(T,P) = P_{s,2}\phi_2^\circ(T,P_{s,2})\exp\{\int_{P_{s,2}}^{P}[V_2^{\circ L}(T,P)/RT]dP\} \tag{4}$$

$\phi_2^\circ(T,P_{s,2})$ is the fugacity coefficient of pure saturated solvent vapor, and $V_2^{\circ L}(T,P)$ denotes the molar volume of pure liquid solvent. Provided the pressure range $[P_{s,2},P]$ is not too large and the temperature well below the critical temperature $T_{c,2}$ of the solvent, a common useful approximation is the assumption that $V_2^{oL}$ is independent of pressure and equal to the orthobaric molar liquid volume $V_{s,2}^{oL} = V_2^{oL}(T,P_{s,2})$. However, for large $[P_{s,2},P]$ and in the critical region, where $V_2^{oL}$ changes rapidly with pressure, this simplification will, in general, *not* be satisfactory.

If needed, the pressure dependence of the molar volume of the pure liquid may be accounted for by any suitable equation of state, such as the modified Tait equation (29-31)

$$V^{oL}(T,P) = V^{oL}(T,P_{ref})\{1 + n(P-P_{ref})\beta_T^{oL}(T,P_{ref})\}^{-1/n} , \qquad (5)$$

where $P_{ref}$ is a conveniently chosen reference pressure (for instance $P_{ref} = P_s$), $\beta_T^{oL} = -(V^{oL})^{-1}(\partial V^{oL}/\partial P)_T$ is the corresponding isothermal compressibility and n is a *pressure-independent* parameter. Note, that from eq. (5) it follows that

$$1/\beta_T^{oL}(T,P) = 1/\beta_T^{oL}(T,P_{ref}) + n(P - P_{ref}) \qquad (6)$$

which relation is identical with the linear tangent-modulus equation (30).

The standard-state fugacity for the gaseous component is the *Henry coefficient* $H_{1,2}$ for solute 1 dissolved in solvent 2,

$$f_1^o(T,P) = H_{1,2}(T,P) = H_{1,2}(T,P_{s,2})\exp\{\int_{P_{s,2}}^{P}[V_1^{\infty L}(T,P)/RT]dP\} \qquad (7)$$

where $V_1^{\infty L}(T,P)$ denotes the partial molar volume of dissolved gas at infinite dilution. As compared to the amount of experimental data available on the dependence of $V_2^{oL}$ upon T and P, considerably less is known for $V_1^{\infty L}(T,P)$. Thus for want of a better assumption, it is common practice to assume $V_1^{\infty L}$ to be independent of pressure and equal to its value at $P = P_{s,2}$. We emphasize, however, that for similar reasons as above this simplification will not be satisfactory for large pressure ranges $[P_{s,2},P]$ and in the critical region. The preferred experimental method for determining $V_1^{\infty L}$ is either densimetry or dilatometry (32,33) at very small $x_1$ in conjunction with a suitable extrapolation procedure (see also section 5).

At the saturation pressure of the solvent, $H_{1,2}(T,P_{s,2})$ is rigorously accessible through experimentally determined ratios of fugacity over liquid phase mole fraction according to (2) (at T = const)

$$H_{1,2}(T,P_{s,2}) = \lim_{\substack{x_1\to 0\\ P\to P_{s,2}}}(f_1^L/x_1) = \lim_{\substack{x_1\to 0\\ P\to P_{s,2}}}[y_1P\phi_1(T,P,y_1)/x_1] \qquad (8)$$

since by definition the limiting values of both the activity coefficient $\gamma_1$ and the Poynting correction term are 1 as $x_1 \to 0$ (or $P\to P_{s,2}$). Thus, evaluation of Henry coefficients from measured isothermal sets $\{x_1, y_1, P\}$ according to the extrapolation procedure eq. (8) requires a vapor-phase equation of state for calculating $\phi_1(T,P,y_1)$.

The majority of gas solubility measurements are in the low-pressure domain, say with P not exceeding 1 MPa. Hence for many systems (excluding, for instance, strongly associating fluids) the *virial equation of state*

$$Z = PV/RT = 1 + B/V + C/V^2 + \ldots. \qquad (9)$$

is convenient*), and often yields entirely satisfactory results even when truncated after the term with $V^{-1}$. Here, Z is the compressibility factor, B and C are respectively termed the second and third virial coefficients, and V is the molar volume of the vapor phase. For conciseness, the superscript V denoting vapor phase was omitted. The virial equation has the important advantage over other equations of state of providing theoretically rigorous expressions for the dependence of the mixture virial coefficients on composition. For an n-component mixture

$$B(T,y_1,\ldots,y_n) = \sum_{i=1}^{n}\sum_{j=1}^{n} y_i y_j B_{ij}(T) \tag{10a}$$

$$C(T,y_1,\ldots,y_n) = \sum_{i=1}^{n}\sum_{j=1}^{n}\sum_{k=1}^{n} y_i y_j y_k C_{ijk}(T) \quad \text{etc.} \tag{10b}$$

$B_{ii}$, $C_{iii}$, etc. refer to pure substance i, and $B_{ij} = B_{ji}$, $C_{ijk} = C_{ikj} = C_{jik} = \ldots$ etc., are the cross virial coefficients. As indicated, they are functions of temperature only.

To the order of approximation implied by using virial coefficients up to the third, we obtain for the *fugacity coefficient* of component i in the vapor phase

$$\ell n\phi_i = \left(\frac{2}{V}\right)\sum_{j=1}^{n} y_j B_{ij} + \left(\frac{3}{2}\right)\left(\frac{1}{V^2}\right)\sum_{j=1}^{n}\sum_{k=1}^{n} y_j y_k C_{ijk} - \ell n Z \tag{11}$$

where, for the sake of clarity, the superscript V has again been omitted. For the vast majority of practical applications, eq. (11) provides satisfactory results up to the critical region. Unfortunately, experimental information on third virial coefficients is rather limited (36) and one has to rely heavily on correlation methods (2). This, and the computational convenience in using {T,P} rather than {T,V} as primary variables, leads to a widely used approximation for *low* pressures:

$$Z = 1 + BP/RT, \tag{12}$$

where the second virial coefficient of the mixture is again given by eq. (10a). The corresponding expression for the fugacity coefficient is then

$$\ell n\phi_i = \frac{P}{RT}\left(2\sum_{j=i}^{n} y_j B_{ij} - B\right) \tag{13}$$

For binary mixtures, eqs. (12) and (13) yield the familiar relations.

$$Z = 1 + \frac{P}{RT}(y_1 B_{11} + y_2 B_{22} + y_1 y_2 \delta_{12}) \tag{14}$$

$$\ell n\phi_i = \frac{P}{RT}\,(B_{ii} + y_j^2 \delta_{12}) \tag{15}$$

where the quantity $\delta_{12}$ is defined by $\delta_{12}=2B_{12}-B_{11}-B_{22}$.

A comprehensive critical compilation of virial coefficients of pure gases as well as of cross virial coefficients (including data up to early 1979) has been prepared by Dymond and Smith (36). For recent advances in correlation methods see refs. (34,35,37-40), and the reviews, refs. (2, 12,13,18,41).

---

* Large negative deviations from vapor phase ideality, even at rather low pressures, may occur in mixtures of strongly, that is "chemically", interacting components, such as in mixtures containing carboxylic acids (association equilibria). The virial equation of state is not useful for highly associating fluids, and appropriate chemical theories have been developed (34,35) which will, however, not be considered here.

## 3. VARIATION OF $H_{1,2}$ WITH TEMPERATURE

To date, the most useful approaches to a rational representation of the temperature dependence of Henry coefficients are those introduced by *Clarke* and *Glew* (42), and by *Benson* and *Krause* (20,21). These authors assume that $\ln(H_{1,2}/\text{Pa})$ and the associated standard thermodynamic function changes are well-behaved, continuous and derivable functions of either the thermodynamic temperature T, or the inverse temperature $T^{-1}$. Subsequently, the enthalpy of solution is expanded in a Taylor series about either (a) a suitably chosen reference temperature or (b) an inverse reference temperature. When adopting the former procedure, the experimental results are fitted by (unweighted) least-squares analysis to a Clarke-Glew-type equation (abbreviated as CG)

$$\ell n[H_{1,2}(T,P_{s,2})/\text{Pa}] = A_0 + A_1(T/K)^{-1} + A_2 \ell n(T/K) + A_3(T/K) + A_4(T/K)^2 + \ldots \quad (16)$$

Frequently, for computational convenience $\tau = 10^{-2}T/K$ is used instead of T/K. When the method of Benson and Krause (BK) is used, the results are fitted by a power series in $T^{-1}$

$$\ell n[H_{1,2}(T,P_{s,2})/\text{Pa}] = \sum_{i=0}^{n} a_i (T/K)^{-i} \quad (17)$$

CG-type equations have been used extensively by Wilhelm et al. (13) in their recent critical review on low-pressure gas solubilities in water. We note that the 3-term version of eq. (16) is the well known Valentiner-equation (43).

The number of terms in either series eq. (16) or (17) should be based upon appropriate statistical criteria, such as the $F_\chi$-*test*(44), in conjunction with experience as to the precision attainable with any particular experimental method. While numerical differences between values generated from either the CG or BK equation are in general exceedingly small within the temperature range covered by experiment, care must be exercised when extrapolation to higher temperatures is desired because of increasing divergence of the smoothing functions. Closely related to this problem is Beutier and Renon's recent study (45) on the limiting behavior of $H_{1,2}(T,P_{s,2})$ for $T \to T_{c,2}$ (see also section 5 below).

Both correlating equations (CG and BK) present the temperature variation of the Henry coefficient at saturation pressure $P_{s,2}=P_{s,2}(T)$ of the solvent, that is to say, temperature and pressure variation are coupled through the Clausius-Clapeyron equation:

$$dP_{s,2}/dT = P_{s,2}\Delta H_2^{vap}/RT^2\Delta Z_2 \quad (18)$$

where $\Delta H_2^{vap}$ is the enthalpy of vaporization of pure solvent 2, and $\Delta Z_2 = Z_2^V - Z_2^L$ is the difference of the compressibility factors of pure coexisting gaseous and liquid solvent. For instance, the partial molar enthalpy of solution at $\{T,P_{s,2}\}$, that is the difference between the partial molar enthalpy $H_1^{\infty L}$ of solute 1 at infinite dilution in the liquid phase and the molar enthalpy $H_1^{oV}$ of pure 1 in the perfect-gas reference state, is given by

$$H_1^{\infty L} - H_1^{oV} = -RT^2\left(\frac{\partial\, \ell n[H_{1,2}(T,P)/\text{Pa}]}{\partial T}\right)_{P=P_{s,2}}$$

$$= -RT^2\left(\frac{d\,\ell n[H_{1,2}(T,P_{s,2})/\text{Pa}]}{dT}\right) + TV_1^{\infty L}\left(\frac{dP_{s,2}}{dT}\right) \quad (19)$$

The first term on the right-hand side is obtained from either one of the smoothing equations, and the temperature dependence of $P_{s,2}$ may be calculated from any suitable vapor pressure correlation, for example the Antoine equation. For temperatures well below $T_{c,2}$, the correction term $TV_1^{\infty L}(dP_{s,2}/dT)$ will generally be rather small [see, for instance, ref. (22)];

its magnitude will usually be within the uncertainty limits resulting from experimental imprecision and thus it may be neglected. However, at elevated temperatures it will contribute substantially, predominantly because $dP_{s,2}/dT$ increases rapidly with increasing T, **though** $V_1^{\infty L}$ may also increase significantly. The evaluation of $TV_1^{\infty L}(dP_{s,2}/dT)$ in the temperature region where it is significant requires $V_1^{\infty L}$ data which are often not available and one has to rely upon correlation methods, such as the one advanced by Brelvi and O'Connell (46).

## 4. RELATION OF $H_{1,2}$ TO THE OSTWALD COEFFICIENT

By way of example we will now apply results of preceding sections to establish a rigorous relation between the Henry coefficient $H_{1,2}(T,P_{s,2})$ and the *Ostwald coefficient* L, another widely used practical measure of gas solutility (22). This quantity is of importance in the theory of hydrophobic interaction [see, for example, ref. (47), and in particular ref. (48)].

Let the Ostwald coefficient be defined by

$$L(T,P) = (c_1^L/c_1^V)_{equil} \tag{20}$$

where $c_1 = n_1/v$ with the appropriate superscript is the amount-of-substance concentration (or simply, concentration) of solute 1 in the coexisting liquid phase solution or vapor phase solution, respectively, and $v/(n_1+n_2) = V$ is the corresponding molar volume. Its *limiting value* for vanishingly small concentration is

$$L^{\infty} = \lim_{c_1^L \to 0} (c_1^L/c_1^V)_{equil} \tag{21}$$

Note, that for $c_1^L \to 0$, $x_1 \to 0$, $y_1 \to 0$, $P \to P_{s,2}$ and $V \to V_2^{\circ}(T,P_{s,2})$.
Eq. (20) in conjunction with eqs. (3) and (7) yields, after some algebraic manipulation,

$$L = \frac{RTZ^V\phi_1}{\gamma_1 H_{1,2}(T,P)V_2^L}\,[1 - c_1^L(V_1^L - V_2^L)] \tag{22}$$

where $V_2^L$ and $V_1^L$ denotes, respectively, the partial molar volume of solvent and gas at $\{T,P,x_1\}$ in the liquid phase. Hence, in the limit (22) of vanishingly small concentration $c_1^L$

$$L^{\infty} = \frac{RTZ_2^V(T,P_{s,2})\phi_1^{\infty}(T,P_{s,2})}{H_{1,2}(T,P_{s,2})V_2^{\circ L}(T,P_{s,2})} \tag{23}$$

Here, $Z_2^V(T,P_{s,2})$ is the compressibility factor of pure solvent vapor, and $\phi_1^{\infty}(T,P_{s,2})$ is the fugacity coefficient of gas at infinite dilution in the vapor phase. For this limiting condition "infinitely diluted solution of solute 1 in solvent 2" we obtain through use of eq. (11)

$$L^{\infty} = \left(\frac{RT}{H_{1,2}(T,P_{s,2})V_{s,2}^{\circ L}}\right) \exp\,[2B_{12}/V_{s,2}^{\circ V} + \tfrac{3}{2}C_{122}/(V_{s,2}^{\circ V})^2] \tag{24}$$

At low pressures, the exponential may of course be expanded in a series and truncated after the term with $B_{12}$. Note, that the orthobaric molar volume $V_{s,2}^{\circ V}$ of pure solvent vapor has to be determined by an iterative procedure through use of the virial equation of state eq. (9).

Alternatively, for the low-pressure region, we may use the virial equation in its volume-explicit form eq. (12), which approach leads to (22) (after series expansion of $\phi_1^\infty$)

$$L^\infty = \left(\frac{RT}{H_{1,2}(T,P_{s,2})V_{s,2}^{\circ L}}\right)\left[1 + 2\left(\frac{P_{s,2}}{RT}\right)B_{12} + \left(\frac{P_{s,2}}{RT}\right)^2 B_{22}(2B_{12} - B_{22})\right] \qquad (25)$$

For $P_{s,2} < 0.1$ MPa, differences between results obtained by either eq. (24) (truncated after the $B_{12}$-term) or eq. (25) will usually be negligibly small. The above conversion formulae are exact to the order of approximation implied by the use of the virial equation, and supersede those given earlier (1,13,17). For correlating purposes, the variation of $L^\infty$ (or L) with temperature over limited temperature ranges well below $T_{c,2}$ may either be accounted for by a CG-type or a BK-type equation.

## 5. CONCLUDING REMARKS

Frequently, the available experimental data do not warrant a thermodynamic treatment *in extenso* as discussed in sections 1-3. This may be due to a variety of reasons: limited experimental data, lack of auxiliary data for determining reliably the various correction terms, appreciable experimental uncertainties, or any combination thereof. Several reasonable simplifications are possible and will be presented below.

More often than not, Henry coefficients are *not* obtained through application of the thermodynamically rigorous prescription eq. (9), but through measurement (at constant temperature) of $x_1$ and $y_1$ at only *one* rather low pressure (and hence very small $x_1$). Evaluation according to $H_{1,2} = y_1\phi_1 P/x_1$ then assumes a Poynting correction equal 1, and further that under these specific experimental conditions, deviations of the activity coefficient from its limiting value $\gamma_1=1$ are indistinguishable from experimental scatter. For many typical gas solubility measurements the latter approximation will be entirely reasonable, but it may lead to serious errors for systems with strong (specific) ij-interactions, for elevated pressures and/or for conditions approaching those of the critical region.

As already indicated in section 2, one may often assume that $V_1^{\infty L}$ is independent of pressure and equal to $V_1^{\infty L}(T,P_{s,2})$, whence

$$\int_{P_{s,2}}^{P} \frac{V_1^{\infty L}}{RT}\,dP \approx V_1^{\infty L}(T,P_{s,2})(P-P_{s,2})/RT \qquad (26)$$

Preferably, $V_1^{\infty L}$ should be obtained by either dilatometry or densimetry (see section 2). Some pitfalls associated with its evaluation via the Krichevsky-Kasarnovsky equation have been discussed in detail by Gibbs and Van Ness (49) for ($H_2O + CO_2$). The Poynting-correction corresponding to eq. (26), as a function of pressure, is tabulated below for a specific example ($V_1^{\infty L} = 5\cdot10^{-5}\,m^3\,mol^{-1}$, T = 298.15 K):

| $(P-P_{s,2})$/kPa | $\exp[V_1^{\infty L}(P-P_{s,2})/RT]$ |
|---|---|
| 10 | 1.0002 |
| 100 | 1.0020 |
| 1000 | 1.0204 |
| 10000 | 1.2235 |

It gives an indication of the possible error associated with the approximation $\exp[V_1^{\infty L}(P-P_{s,2})/RT] = 1$.

A particularly simple approximation for calculating vapor-phase fugacity coefficients at low pressures is the *Lewis rule*. It is based on the vapor phase behaving as an ideal mixture:

$$\phi_i(T,P,y_i) = \phi_i^\circ(T,P) \qquad (27)$$

that is to say at constant temperature and pressure, $\phi_i$ is assumed to be independent of the mixture composition and equal to the fugacity coefficient of pure i. From eq. (15) it follows that this can only be achieved if $B_{12} = (B_{11} + B_{22})/2$, a condition which will be rarely fulfilled. Nevertheless, the Lewis fugacity rule usually yields reasonable results when component i is present in large excess and/or when the physical properties of the components are not too different ($\delta_{12}$ fairly small compared to $B_{ii}$).

The approximation $\phi_i = 1$ (the vapor phase behaves as a mixture of ideal gases), together with $\gamma_1=1$ and ignoring the Poynting correction, leads to the most familiar and *simplest* form of Henry's law.

$$P_1 = x_1H_{1,2} \qquad (28)$$

which may be used to obtain a first, relatively rough approximation for the true Henry coefficient $H_{1,2}(T,P_{s,2})$. No general statements as to the magnitude of associated errors are possible.

Prediction of Henry coefficients at elevated temperatures is often required in industrial applications, and when involving aqueous solutions, in the study of geochemical processes (50). Unfortunately, the majority of experimental investigations has been confined to temperatures below the normal boiling point of the solvent [see, for instance, refs. (13, 17)].

However, thermodynamics provides a rigorous expression for the *limiting value* of $H_{1,2}(T,P_{s,2})$ for $T \to T_{c,2}$ (and $P_{s,2} \to P_{c,2}$, the critical pressure of solvent 2), which should be incorporated in any rational wide-temperature range correlation extending up to the critical region. Evidently,

$$\lim_{T \to T_{c,2}} L^\infty = 1 \qquad (29)$$

whence from eq. (23).

$$\lim_{T \to T_{c,2}} H_{1,2}(T,P_{s,2}) = P_{c,2}\phi_1^\infty(T_{c,2},P_{c,2}) \qquad (30)$$

is obtained, which limiting relation is identical with that given by Beutier and Renon (45).

## 6. REFERENCES


1. Battino, R.; Clever, H. L. *Chem. Rev.* 1966, *66*, 395.
2. Prausnitz, J. M. Molecular *Thermodynamics of Fluid Phase Equilibria*. Prentice Hall: Englewood Cliffs, N.J. 1969.
3. Hildebrand, J. H.; Prausnitz, J. M.; Scott, R. L. *Regular and Related Solutions*. Van Nostrand Reinhold: New York 1970.
4. Neff, R. O.; McQuarrie, D. A. *J. Phys. Chem.* 1973, *77*, 413.
5. Klapper, M. H. *Adv. Chem. Phys.* 1973, *23*, 55.
6. Wilhelm, E.; Battino, R. *Chem. Rev.* 1973, *73*, 1.
7. Kertes, A. S.; Levy, O.; Markovits, G. Y. in *Experimental Thermochemistry*, vol. II. B. Vodar and B. Le Naindre: editors. Butterworth: London 1974.
8. Clever, H. L.; Battino, R. in *Solutions and Solubilities*, vol. 8, part 1 of Weissberger: Techniques of Chemistry. M. R. J. Dack: editor. Wiley: New York 1975.

9. Pierotti, R. A. *Chem. Rev.* 1976, *76*, 717.

10. Gerrard, W. *Solubility of Gases in Liquids. A Graphic Approach.* Plenum Press: New York 1976.

11. Landolt-Börnstein, 4. Band:Wärmetechnik, Bestandteil c: Gleichgewicht der Absorption von Gasen in Flüssigkeiten. Teil c 1: *Absorption in Flüssigkeiten von niedrigem dampfdruck.* Springer-Verlag: Berlin 1976.

12. Wilhelm, E. *Fortschr. Verfahrenstechn. A, Grundlagen* 1977, *15*, 21.

13. Wilhelm, E.; Battino, R., Wilcock, R. J. *Chem. Rev.* 1977, *77*, 219.

14. Goldman, S. *J. Chem. Phys.* 1977, *67*, 727.

15. Linford, R. G.; Thornhill, D. G. T. *J. Appl. Chem. Biotechnol.* 1977, *27*, 479.

16. Goldman, S. *Acc. Chem. Res.* 1979, *12*, 409.

17. *Solubility Data Series-IUPAC.* A. S. Kertes: editor. Pergamon Press: Oxford 1979 and later.

18. Wilhelm, E. *Prog. Chem. Eng. A, Fundam. Chem. Eng.* 1980, *18*, 21.

19. Prausnitz, J. M.; Anderson, T. F.; Grens, E. A.; Eckert, C. A.; Hsieh, R.; O'Connell, J. P. *Computer Calculations for Multicomponent Vapor-Liquid and Liquid-Liquid Equilibria.* Prentice Hall: Englewood Cliffs, N.J. 1980.

20. Benson, B. B.; Krause, D. *J. Chem. Phys.* 1976, *64*, 689.

21. Benson, B. B.; Krause, D.; Peterson, M. A. *J. Solution Chem.* 1979, *8*, 655.

22. Rettich, T. R.; Handa, Y. P.; Battino, R.; Wilhelm, E., *J. Phys. Chem.* 1981, *85*, 3230.

23. Tiepel, E. W.; Gubbins, K. E. *Can. J. Chem. Eng.* 1972, *50*, 361.

24. Abrams, D. S. ; Seneci, F.; Chueh, P. L.; Prausnitz, J. M. *Ind. Eng. Chem., Fundam.* 1975, *14*, 52.

25. Van Ness, H. C.; Abbott, M. M. *AIChE J.* 1979, *25*, 645.

26. O' Connell, J. P. *Fluid Phase Equil.* 1981, *6*, 21.

27. Mathias, P. M.; O' Connell, J. P. *Chem. Eng. Sci.* 1981,*36*, 1123.

28. Brandani,V.; Prausnitz, J. M. *Fluid Phase Equil.* 1981, *7*, 259.

29. Macdonald, J. R. *Rev. Mod. Phys.* 1966,*38*,669.

30. Hayward, A. T. J. *Brit. J. Appl. Phys.* 1967,*18*,965.

31. Wilhelm, E. *J. Chem. Phys.* 1975, *63*, 3379.

32. Moore, J. C.; Battino, R.; Rettich, T. R.; Handa, Y. P.; Wilhelm, E. *J. Chem. Eng. Data* 1982, *27*,22.

33. Handa, Y. P.; D'Arcy, P. J.; Benson, G. C. Private communication.

34. Nothnagel, K.-H.; Abrams, D. S.; Prausnitz, J. M. *Ind. Eng. Chem., Process Des. Develop.* 1973, *12*, 25.

35. Hayden, J. G.; O' Connell, J. P. *Ind. Eng. Chem., Process Des. Develop.* 1975, *14*, 221.

36. Dymond, J. H.; Smith, E. B. *The Virial Coefficients of Pure Gases and Mixtures. A Critical Compilation.* Clarendon Press: Oxford 1980.

37. Tsonopoulos, C. *AIChe J.* 1974, *20*, 263; 1975, *21*, 827; 1978, *24*, 1112.

38. Tarakad, R. R.; Danner, R. P. *AIChE J.* 1977, *23*, 685.

39. Pesuit, D. R. *Ind. Eng. Chem., Fundam.* 1978, *17*, 235.

40. Kohler, F.; Fischer, J.; Wilhelm, E. *J. Molec. Structure*, 1982, in press.

41. Reid, R. C.; Prausnitz, J. M.; Sherwood, T. K. *The Properties of Gases and Liquids*, 3rd edition. McGraw-Hill: New York 1977.

42. Clarke, E. C. W.; Glew, D. N. *Trans. Faraday Soc.* 1966, *62*, 539. For a condensed treatment see Bolton, P. D. *J. Chem. Educ.* 1970, *47*, 638.

43. Valentiner, S. *Z. Phys.* 1927, *42*, 253.

44. Bevington, P. R. *Data Reduction and Error Analysis for the Physical Sciences*. McGraw-Hill: New York 1969.

45. Beutier, D.; Renon, H. *AIChE J.* 1978, *24*, 1122.

46. Brelvi, S. W.; O' Connell, J. P. *AIChE J.* 1972, *18*, 1239.

47. Ben-Naim, A. *Water and Aqueous Solutions. Introduction to a Molecular Theory*. Plenum Press: New York 1974.

48. Ben-Naim, A. *Hydrophobic Interactions*. Plenum Press: New York 1980.

49. Gibbs, R. E.; Van Ness, H. C. *Ind. Eng. Chem., Fundam.* 1971, *10*, 312.

50. Potter II, R. W.; Clynne, M. A. *J. Solution Chem.* 1978, *7*, 837.


April 1982

| The Sechenov Salt Effect Parameter. | H. Lawrence Clever<br>Department of Chemistry<br>Emory University<br>Atlanta, GA 30322 USA<br><br>1982, March |
|---|---|

# THE SECHENOV SALT EFFECT PARAMETER

## Forms of Expression and Interconversion

In the following discussion the components are indexed so that:

the nonelectrolyte (gas) is component 1,

the electrolyte is component 2, and

the solvent (water) is component 3.

The symbols and units are:

| Component | Molar Concentration $c$/mol dm$^{-3}$ | Molal Concentration $m$/mol kg$^{-1}$ | Mole Fraction $x$ | Molecular Mass $M$/g mol$^{-1}$ |
|---|---|---|---|---|
| Nonelectrolyte | $c_1^0$, $c_1$ | $m_1^0$, $m_1$ | $x_1^0$, $x_1$ | $M_1$ |
| Electrolyte | $c_2$ | $m_2$ | $x_2$ | $M_2$ |
| Solvent | $c_3$ | $m_3$ | $x_3$ | $M_3$ |

The superscript "$^0$" refers to the nonelectrolyte solubility in the pure solvent. The pure solvent and solution densities are $\rho^0$/g cm$^{-3}$ and $\rho$/g cm$^{-3}$, respectively. They should be the densities of gas saturated solvent (water) and salt solution, but the gas free densities will differ negligibly in the $\rho^0/\rho$ ratio. The number of ions per formula of electrolyte is symbolized by $\nu$, and the electrolyte apparent molar volume is symbolized by $\phi_{v2}$.

The salt effect parameter is defined in any one of several forms,

$$k_{scc} \ / \ dm^3mol^{-1} = (1/c_2) \log (c_1^0/c_1)$$

which numerically is the same as

$$k_{sc\alpha} \ / \ dm^3mol^{-1} = (1/c_2) \log (\alpha_1^0/\alpha_1)$$

and

$$k_{scL} \ / \ dm^3mol^{-1} = (1/c_2) \log (L_1^0/L_1)$$

where $\alpha/cm^3 (STP)\ cm^{-3}\ atm^{-1}$ is the Bunsen coefficient, and $L/cm^3\ cm^{-3}$ is the Ostwald coefficient. The Sechenov parameter subscripts have the following significance: the first letter, s, refers to the Sechenov salt effect, the second letter the electrolyte measure of concentration, and the third letter the gas solubility unit of measure.

The ratios $c_1^0/c_1 = \alpha_1^0/\alpha_1 = L_1^0/L_1$ since $c_1 = 10^3\alpha_1/RT$ with T/K = 273.15, and $c_1 = L_{c,\text{ gas phase}} = 10^3\ L\ p/RT$ with T/K the temperature of the gas solubility measurement.

Another frequently used form is

$$k_{smm} / kg\ mol^{-1} = (1/m_2) \log (m_1^0/m_1)$$

which is equivalent to

$$k_{sms} / kg\ mol^{-1} = (1/m_2) \log (s_1^0/s_1)$$

where $s/cm^3 (STP)\ g^{-1}\ atm^{-1}$ is the Kuenen coefficient and $m_1 = 10^3\ s_1/RT$ with T/K = 273.15.

A mole fraction nonelectrolyte solubility ratio is also used. It may be combined with either a molar or a molal electrolyte concentration.

$$k_{scx} / dm^3\ mol^{-1} = (1/c_2) \log (x_1^0/x_1)$$

and

$$k_{smx} / kg\ mol^{-1} = (1/m_2) \log (x_1^0/x_1)$$

Care should be taken to define how the mole fraction is calculated. Some calculate the electrolyte contribution to the total number of moles on the basis of the electrolyte formula and others on the basis of the total number of moles of ions. Our definition shall use the ions from the electrolyte assuming 100 per cent dissociation.

Masterton and co-workers (1) have applied scaled particle theory and used the salt effect parameter in the form $k_{scx} / dm^3\ mol^{-1}$. Others (2) favor the $k_{smm} / kg\ mol^{-1}$ form for statistical mechanical approaches aimed at understanding salt effects.

Friedman (3) gives convincing arguments for using a form consistent with definitions of thermodynamics. The form he uses is

$$k_{1,2} = -(\delta \ln m_1/\delta m_2)_{\mu_1}$$

However, if $k_{1,2}$ is independent of $m_2$, then $k_{smm} = k_{1,2}/2.303$ since

$$k_{1,2} \int_0^{m_2} \delta m_2 = -\int_{m_1^0}^{m_1} \delta \ln m_1$$

$$k_{1,2}\, m_2 = -(\ln m_1 - \ln m_1^0) = \ln (m_1^0/m_1)$$

Ben-Naim (4) calculates Gibbs energy of transfer between pure water and electrolyte solution. His work implies the form $k_{smc}$ of the salt effect parameter.

Most of the current workers in the area seem to favor either mole fraction or molality for the unit to express the nonelectrolyte solubility. However, there are arguments that suggest a volume concentration is better, especially when large pressures and/or solubilities are involved.

Some workers have reported the nonelectrolyte solubility as moles per kg of solution. We do not believe this is a satisfactory unit and it is not discussed further.

Krestov and co-workers (5) have started to use solvomolality as a unit. This is the number of moles per 55.51 mole of solvent. For water ($H_2O$) this is the same as molality, and for other solvents ($D_2O$, methanol, etc.) it is proportional to the molality. Thus its use will result in a $k_{smm}$/kg mol$^{-1}$ value. (Solvomolality, $m_i^{(s)} = m_i(M_3/M_{H_2O})$).

I. Molarity and Molality

$$k_{scc} = \frac{1}{c_2} \log \frac{c_1^0}{c_1} = \frac{1}{c_2} \log \frac{\left[\dfrac{m_1^0}{(1000 + m_1^0 M_1) / 1000\rho^0}\right]}{\left[\dfrac{m_1}{(1000 + m_1 M_1 + m_2 M_2) / 1000\rho}\right]}$$

$$= \frac{1}{c_2} \log \left(\frac{m_1^0}{m_1}\right)\left(\frac{\rho^0}{\rho}\right)\left(\frac{1000 + m_1 M_1 + m_2 M_2}{1000 + m_1^0 M_1}\right)$$

$$= \frac{1}{c_2} \log \left(\frac{m_1^0}{m_1}\right) + \frac{1}{c_2} \log \left(\frac{\rho^0}{\rho}\right)\left(\frac{1000 + m_1 M_1 + m_2 M_2}{1000 + m_1^0 M_1}\right)$$

$$k_{scc} = k_{scm} + \frac{1}{c_2} \log \left(\frac{\rho^0}{\rho}\right)\left(\frac{1000 + m_1 M_1 + m_2 M_2}{1000 + m_1^0 M_1}\right)$$

and in addition

$$k_{smc} = \frac{1}{m_2} \log \frac{c_1^0}{c_1}$$

$$= \frac{c_2}{m_2} k_{scc} = \frac{c_2}{m_2} k_{scm} + \frac{c_2}{m_2}\left\{\frac{1}{c_2} \log \left(\frac{\rho^0}{\rho}\right)\left(\frac{1000 + m_1 M_1 + m_2 M_2}{1000 + m_1^0 M_1}\right)\right\}$$

$$= k_{smm} + \frac{c_2}{m_2}\left\{\frac{1}{c_2} \log \left(\frac{\rho^0}{\rho}\right)\left(\frac{1000 + m_1 M_1 + m_2 M_2}{1000 + m_1^0 M_1}\right)\right\}$$

The terms $m_1 M_1$ and $m_i^0 M_1$ are negligibly small except for the very soluble gases. Thus for most cases

$$k_{smc} = \frac{c_2}{m_2} k_{scc} \simeq k_{smm} + \frac{c_2}{m_2}\left\{\frac{1}{c_2} \log \left(\frac{\rho^0}{\rho}\right)\left(\frac{1000 + m_2 M_2}{1000}\right)\right\}$$

$$\simeq \frac{c_2}{m_2} k_{scm} + \frac{c_2}{m_2}\left\{\frac{1}{c_2} \log \left(\frac{\rho^0}{\rho}\right)\left(\frac{1000 + m_2 M_2}{1000}\right)\right\}$$

II. Mole Fraction and Molality

$$x_1^0 = \frac{m_1^0}{\frac{1000}{M_3} + m_1^0} = \frac{m_1^0 M_3}{1000 + m_1^0 M_3}$$

$$x_1 = \frac{m_1}{\frac{1000}{M_3} + m_1 + \nu m_2} = \frac{m_1 M_3}{1000 + m_1 M_3 + \nu m_2 M_3}$$

$$\frac{x_1^0}{x_1} = \frac{m_1^0}{m_1} \cdot \frac{(1000 + m_1 M_3 + \nu m_2 M_3)}{(1000 + m_1^0 M_3)}$$

$$\frac{1}{m_2} \log \frac{x_1^0}{x_1} = \frac{1}{m_2} \log \frac{m_1^0}{m_1} + \frac{1}{m_2} \log\left(\frac{(1000 + m_1 M_3 + \nu m_2 M_3)}{(1000 + m_1^0 M_3)}\right)$$

or

$$k_{smx} = k_{smm} + \frac{1}{m_2} \log\left(\frac{(1000 + m_1 M_3 + \nu m_2 M_3)}{(1000 + m_1^0 M_3)}\right)$$

and

$$k_{scx} = \frac{1}{c_2} \log \frac{x_1^0}{x_1} = \frac{m_2}{c_2} k_{smx}$$

$$= \frac{m_2}{c_2} k_{smm} + \frac{m_2}{c_2} \left\{ \frac{1}{m_2} \log \left( \frac{(1000 + m_1 M_3 + \nu m_2 M_3)}{(1000 + m_1^0 M_3)} \right) \right\}$$

where $c_2$ is the electrolyte molar concentration of solution molality $m_2$.

For systems in which $m_1^0$ and $m_1$ are small, as is the usual case for gas solubilities, the terms $m_1^0 M_3$ and $m_1 M_3$ may be omitted from the sums in the log term.

For a 1-1 electrolyte at 1 mol kg$^{-1}$ in which $m_1^0$ and $m_1$ are sufficiently small to ignore in the sums in the log term the equations simplify to

$$k_{scx} = \frac{1}{c_2} k_{smx} = \frac{1}{c_2} k_{smm} + \frac{1}{c_2} \log \left\{ \frac{(1000 + (2)\ (18.0152))}{1000} \right\}$$

$$= \frac{1}{c_2} k_{smm} + \frac{1}{c_2} \frac{\ln (1 + 0.0360)}{2.303}$$

and using the first term of the expansion ln (1+x) =

$x - (1/2)\ x^2 + (1/3)\ x^3 - (1/4)\ x^4 + \cdots \qquad (-1 < x < 1)$

$$= \frac{1}{c_2} k_{smm} + \frac{1}{c_2} \cdot \frac{0.0360}{2.303}$$

which is a form given by Masterton, Bolocofsky, and Lee (1). In this equation $c_2$ is the electrolyte molar concentration of the 1 mol kg$^{-1}$ 1-1 electrolyte.

For any electrolyte molality, $m_2$.

$$k_{scx} = \frac{1}{c_2} \log \frac{x_1^0}{x_1} = \frac{m_2}{c_2} k_{smx} = \frac{m_2}{c_2} k_{smm} + \frac{m_2}{c_2} \left\{ \frac{1}{m_2} \log \left( \frac{(1000 + 2m_2\ 18.015)}{1000} \right) \right\}$$

$$= \frac{m_2}{c_2} k_{smm} + \frac{m_2}{c_2} \frac{1}{m_2} \frac{\ln (1 + 0.0360 m_2)}{2.303}$$

$$= \frac{m_2}{c_2} k_{smm} + \frac{m_2}{c_2} \cdot \frac{0.0360}{2.303}$$

III. Mole Fraction and Molarity

$$x_1^0 = \frac{c_1^0}{\left\{\dfrac{1000\ \rho^0 - c_1^0\ M_1}{M_3}\right\} + c_1^0} = \frac{c_1^0\ M_3}{1000\ \rho^0 + (M_3 - M_1)\ c_1^0}$$

$$x_1 = \frac{c_1}{\left\{\dfrac{1000\ \rho - c_1\ M_1 - c_2\ M_2}{M_3}\right\} + c_1 + \nu c_2}$$

$$= \frac{c_1\ M_3}{1000\ \rho + (M_3 - M_1)\ c_1 + (\nu M_3 - M_2)\ c_2}$$

The densities $\rho^0$ and $\rho$ should be the densities of gas saturated water and salt solution. However, use of gas-free water and gas-free salt solution densities makes a negligible difference (0.3 % in $\rho$ for a gas mole fraction of $10^{-3}$).

$$\frac{x_1^0}{x_1} = \frac{c_1^0}{c_1} \cdot \frac{(1000\rho + (M_3 - M_1)\ c_1 + (\nu M_3 - M_2)\ c_2)}{(1000\rho^0 + (M_3 - M_1)\ c_1^0)}$$

and

$$k_{scx} = \frac{1}{c_2} \log \frac{x_1^0}{x_1}$$

$$= \frac{1}{c_2} \log \frac{c_1^0}{c_1} + \frac{1}{c_2} \log \left\{\frac{(1000\rho + (M_3 - M_1)\ c_1 + (\nu M_3 - M_2)\ c_2)}{(1000\rho^0 + (M_3 - M_1)\ c_1^0)}\right\}$$

or

$$k_{scx} = k_{scc} + \frac{1}{c_2} \log \left\{\frac{(1000\rho + (M_3 - M_1)\ c_1 + (\nu M_3 - M_2)\ c_2)}{1000\rho^0 + (M_3 - M_1)\ c_1^0}\right\}$$

and

$$k_{smx} = \frac{1}{m_2} \log \frac{x_1^0}{x_1} = \frac{c_2}{m_2} \left\{\frac{1}{c_2} \log \frac{x_1^0}{x_1}\right\}$$

$$= \frac{c_2}{m_2} \left\{\frac{1}{c_2} \log \frac{c_1^0}{c_1} + \frac{1}{c_2} \log \left\{\frac{(1000\rho + (M_3 - M_1)\ c_1 + (\nu M_3 - M_2)\ c_2)}{(1000\rho^0 + (M_3 - M_1)\ c_1^0)}\right\}\right\}$$

or

$$k_{smx} = \frac{c_2}{m_2} k_{scx}$$

$$= \frac{c_2}{m_2} k_{scc} + \frac{c_2}{m_2} \left\{\frac{1}{c_2} \log \left\{\frac{(1000\rho + (M_3 - M_1)\ c_1 + (\nu M_3 - M_2)\ c_2)}{(1000\rho^0 + (M_3 - M_1)\ c_1^0)}\right\}\right\}$$

For systems in which $c_1^0$ and $c_1$ are small enough to ignore in the sums

$$k_{smx} = \frac{c_2}{m_2} k_{scx} = \frac{c_2}{m_2} k_{scc} + \frac{c_2}{m_2}\left[\frac{1}{c_2} \log\left(\frac{(1000\rho + (\nu M_3 - M_2)\, c_2)}{(1000\rho^0)}\right)\right]$$

and for a 1-1 electrolyte at a concentration of one mol dm$^{-3}$

$$k_{smx} = \frac{1}{m_2} k_{scx} = \frac{1}{m_2} k_{scc} + \frac{1}{m_2} \log\left(\frac{(1000\rho + (2M_3 - M_2)\, c_2)}{(1000\rho^0)}\right)$$

where $m_2$ is the molality of the one molar solution.

An alternate approach is to use the apparent molar volume, $\phi_{v2}$, of the electrolyte. Thus

$$x_1^0 = \frac{c_1^0}{\dfrac{1000}{M_3} + c_1^0} = \frac{c_1^0\, M_3}{1000 + c_1^0\, M_3}$$

$$x_1 = \frac{c_1}{\dfrac{1000 - \phi_{v2}}{M_3} + c_1 + \nu c_2} = \frac{c_1\, M_3}{1000 - \phi_{v2} + c_1\, M_3 + \nu c_2\, M_3}$$

$$\frac{x_1^0}{x_1} = \frac{c_1^0}{c_1} \cdot \frac{(1000 - \phi_{v2} + c_1\, M_3 + \nu c_2\, M_3)}{(1000 + c_1^0\, M_3)}$$

$$\frac{1}{c_2} \log \frac{x_1^0}{x_1} = \frac{1}{c_2} \log \frac{c_1^0}{c_1} + \frac{1}{c_2} \log\left(\frac{(1000 - \phi_{v2} + c_1\, M_3 + \nu c_2\, M_3)}{(1000 + c_1^0\, M_3)}\right)$$

The equation can be simplified for a one mol dm$^{-3}$ 1-1 electrolyte solution in which the nonelectrolyte solubilities, $c_1^0$ and $c_1$ are small enough to be negligible in the sums to

$$k_{scx} = k_{scc} + \log\left(\frac{(1000 - \phi_{v2} + 2M_3)}{1000}\right)$$

$$k_{scx} = k_{scc} + \log\left[1 + \frac{(2)\ (18.0152)}{1000} - \frac{\phi_{v2}}{1000}\right]$$

$$= k_{scc} + \frac{\ln\left(1 + (0.0360 - \phi_{v2} / 1000)\right)}{2.303}$$

expanding the ln term as ln $(1+x) \cong x$ where $x = (0.0360 - \phi_{v2} / 1000)$ gives

$$k_{scx} \simeq k_{scc} + \frac{(0.0360 - \phi_{v2} / 1000)}{2.303}$$

This is a form given by Masterton *et al.* (1).

## Summary

The equations can be put in the following form for use.

From Part I

$$k_{smc} = \frac{c_2}{m_2} k_{scc} = \frac{c_2}{m_2} k_{scm} + F_{1m} = k_{smm} + F_{1m}$$

or

$$k_{scm} = k_{scc} - F_{1c} = \frac{m_2}{c_2} k_{smc} - F_{1c} = \frac{m_2}{c_2} k_{smm}$$

where

$$F_{1m} = \frac{1}{m_2} \log\left(\frac{\rho^0}{\rho}\right)\left(\frac{(1000 + m_1 M_1 + m_2 M_2)}{(1000 + m_1^0 M_1)}\right)$$

$$F_{1c} = \frac{1}{c_2} \log\left(\frac{\rho^0}{\rho}\right)\left(\frac{(1000 + m_1 M_1 + m_2 M_2)}{(1000 + m_1^0 M_1)}\right)$$

For all but very soluble gases the $m_1 M_1$ and $m_1^0 M_1$ terms may be neglected in the $F_{1m}$ and $F_{1c}$ terms.

From Part II

$$k_{scx} = \frac{m_2}{c_2} k_{smx} = \frac{m_2}{c_2} k_{smm} + F_{2c}$$

or

$$k_{smm} = k_{smx} - F_{2m} = \frac{c_2}{m_2} k_{scx} - F_{2m}$$

where

$$F_{2m} = \frac{1}{m_2} \log\left(\frac{(1000 + m_1 M_3 + \nu m_2 M_3)}{(1000 + m_1^0 M_3)}\right)$$

$$F_{2c} = \frac{1}{c_2} \log\left(\frac{(1000 + m_1 M_3 + \nu m_2 M_3)}{(1000 + m_1^0 M_3)}\right)$$

Again for all but the very soluble gases the $m_1 M_3$ and $m_1^0 M_3$ terms can be neglected in $F_{2m}$ and $F_{2c}$.

From Part III

$$k_{smx} = \frac{c_2}{m_2} k_{scx} = \frac{c_2}{m_2} k_{scc} + F_{3m}$$

or

$$k_{scc} = k_{scx} - F_{3c} = \frac{m_2}{c_2} k_{smx} - F_{3c}$$

where

$$F_{3m} = \frac{1}{m_2} \log \left[ \frac{(1000\rho + (M_3 - M_1)\ c_1 + (\nu M_3 - M_2)\ c_2)}{(1000\rho^0 + (M_3 - M_1)\ c_1^0)} \right]$$

and

$$F_{3c} = \frac{1}{c_2} \log \left[ \frac{(1000\rho + (M_3 - M_1)\ c_1 + (\nu M_3 - M_2)\ c_2)}{(1000\rho^0 + (M_3 - M_1)\ c_1^0)} \right]$$

and again the terms in $c_1$ and $c_1^0$ in the $F_{3m}$ and $F_{3c}$ factors can normally be neglected.

Acknowledgement:

The author wishes to thank Professors J. E. Desnoyers, H. L. Friedman and W. L. Masterton who provided detailed comments and advice on salt effect parameters.

References:

(1) Masterton, W. L.; Bolocofsky, D.; Lee, T. P.
*J. Phys. Chem.* 1971, *75*, 2809-2815.

(2) Perron, G.; Joly, D.; Desnoyers, J. E.; Avedikian, L.; Morel, J. P. *Can. J. Chem.* 1978, *56*, 552-559.

(3) Krishnan, C. V.; Friedman, H. L.
*J. Solution Chem.* 1974, *3*, 727-744.

(4) Ben-Naim, A.
*J. Phys. Chem.* 1965, *69*, 3240-45, 3245-50, 3250-53.
(M. Egel-Thal is co-author of the third paper).

(5) Abrosimov, V. K.; Badelin, V. G.; Krestov, G. A.
*Izv. Vyssh. Uchebn. Zaved., Khim. Khim. Tekhnol.*
1976, *19*, 1381-1386; 1977, *20*, 1864-1866.

| Represenative values of factors needed to convert Sechenov salt effect parameter forms. | H. Lawrence Clever<br>Department of Chemistry<br>Emory University<br>Atlanta, GA 30322 USA<br><br>1982, March |
|---|---|

# CONVERSION AMONG THE VARIOUS FORMS OF THE SECHENOV SALT EFFECT PARAMETER

## Factors for Selected Electrolytes

93[1]Magnesium Sulfate; $MgSO_4$; [7487-88-9]

| Wt % | $c_2$/ mol dm$^{-3}$ | $m_2$/ mol kg$^{-1}$ | $F_{1c}$ | $F_{1m}$ | $F_{2c}$ | $F_{2m}$ | $F_{3c}$ | $F_{3m}$ |
|---|---|---|---|---|---|---|---|---|
| | 283.15 K | (10 °C) | | | | | | |
| 8 | 0.7205 | 0.7224 | 0.0014 | 0.0014 | 0.0155 | 0.0154 | 0.0141 | 0.0140 |
| 10 | 0.9189 | 0.9231 | 0.0020 | 0.0020 | 0.0155 | 0.0154 | 0.0135 | 0.0134 |
| 12 | 1.1251 | 1.1329 | 0.0025 | 0.0025 | 0.0154 | 0.0153 | 0.0129 | 0.0128 |
| 14 | 1.3394 | 1.3524 | 0.0031 | 0.0030 | 0.0154 | 0.0153 | 0.0124 | 0.0123 |
| | 303.15 K | (30 °C) | | | | | | |
| 2 | 0.1688 | 0.1695 | 0.0004 | 0.0004 | 0.0157 | 0.0156 | 0.0152 | 0.0152 |
| 4 | 0.3443 | 0.3462 | 0.0011 | 0.0011 | 0.0156 | 0.0155 | 0.0145 | 0.0144 |
| 6 | 0.5269 | 0.5303 | 0.0017 | 0.0017 | 0.0156 | 0.0155 | 0.0139 | 0.0138 |
| 8 | 0.7166 | 0.7224 | 0.0023 | 0.0022 | 0.0156 | 0.0154 | 0.0133 | 0.0132 |
| 10 | 0.9138 | 0.9231 | 0.0027 | 0.0027 | 0.0155 | 0.0154 | 0.0128 | 0.0127 |
| 12 | 1.1186 | 1.1329 | 0.0032 | 0.0032 | 0.0155 | 0.0153 | 0.0123 | 0.0121 |
| 14 | 1.3314 | 1.3524 | 0.0037 | 0.0036 | 0.0155 | 0.0153 | 0.0118 | 0.0116 |
| 16 | 1.5524 | 1.5824 | 0.0041 | 0.0041 | 0.0155 | 0.0152 | 0.0114 | 0.0112 |
| 18 | 1.7819 | 1.8236 | 0.0046 | 0.0045 | 0.0155 | 0.0152 | 0.0109 | 0.0107 |
| 20 | 2.0203 | 2.0769 | 0.0050 | 0.0049 | 0.0155 | 0.0151 | 0.0105 | 0.0102 |
| 22 | 2.2678 | 2.3432 | 0.0054 | 0.0053 | 0.0155 | 0.0150 | 0.0101 | 0.0098 |
| 24 | 2.5246 | 2.6235 | 0.0059 | 0.0056 | 0.0155 | 0.0150 | 0.0097 | 0.0093 |
| 26 | 2.7912 | 2.9189 | 0.0063 | 0.0060 | 0.0155 | 0.0149 | 0.0093 | 0.0089 |
| | 323.15 K | (50 °C) | | | | | | |
| 8 | 0.7111 | 0.7224 | 0.0027 | 0.0027 | 0.0157 | 0.0154 | 0.0130 | 0.0128 |
| 10 | 0.9068 | 0.9231 | 0.0031 | 0.0031 | 0.0157 | 0.0154 | 0.0125 | 0.0123 |
| 12 | 1.1101 | 1.1329 | 0.0035 | 0.0035 | 0.0157 | 0.0153 | 0.0121 | 0.0119 |
| 14 | 1.3214 | 1.3524 | 0.0039 | 0.0038 | 0.0156 | 0.0153 | 0.0117 | 0.0114 |

96[1]Barium Chloride; $BaCl_2$; [10361-37-2]

| Wt % | $c_2$/ mol dm$^{-3}$ | $m_2$/ mol kg$^{-1}$ | $F_{1c}$ | $F_{1m}$ | $F_{2c}$ | $F_{2m}$ | $F_{3c}$ | $F_{3m}$ |
|---|---|---|---|---|---|---|---|---|
| | 283.15 K | (10 °C) | | | | | | |
| 14 | 0.7647 | 0.7817 | 0.0124 | 0.0121 | 0.0235 | 0.0230 | 0.0111 | 0.0109 |
| 16 | 0.8911 | 0.9147 | 0.0125 | 0.0122 | 0.0235 | 0.0229 | 0.0110 | 0.0107 |
| 18 | 1.0226 | 1.0541 | 0.0127 | 0.0124 | 0.0235 | 0.0228 | 0.0108 | 0.0105 |
| 20 | 1.1593 | 1.2005 | 0.0130 | 0.0125 | 0.0235 | 0.0227 | 0.0106 | 0.0102 |

[1] Number preceding electrolyte is the standard order of arrangement number.

96 Barium Chloride; $BaCl_2$; [10361-37-2] (Continued)

| Wt % | $c_2$/ mol dm$^{-3}$ | $m_2$/ mol kg$^{-1}$ | $F_{1c}$ | $F_{1m}$ | $F_{2c}$ | $F_{2m}$ | $F_{3c}$ | $F_{3m}$ |
|---|---|---|---|---|---|---|---|---|
| | 298.15 K | (25 °C) | | | | | | |
| 2 | 0.0975 | 0.0980 | 0.0118 | 0.0118 | 0.0235 | 0.0234 | 0.0117 | 0.0116 |
| 4 | 0.1984 | 0.2001 | 0.0122 | 0.0121 | 0.0235 | 0.0233 | 0.0113 | 0.0112 |
| 6 | 0.3029 | 0.3065 | 0.0126 | 0.0125 | 0.0235 | 0.0233 | 0.0109 | 0.0108 |
| 8 | 0.4113 | 0.4176 | 0.0129 | 0.0127 | 0.0236 | 0.0232 | 0.0107 | 0.0105 |
| 10 | 0.5237 | 0.5336 | 0.0131 | 0.0128 | 0.0236 | 0.0231 | 0.0105 | 0.0103 |
| 12 | 0.6403 | 0.6548 | 0.0132 | 0.0129 | 0.0236 | 0.0231 | 0.0103 | 0.0101 |
| 14 | 0.7614 | 0.7817 | 0.0134 | 0.0130 | 0.0236 | 0.0230 | 0.0102 | 0.0100 |
| 16 | 0.8871 | 0.9147 | 0.0135 | 0.0131 | 0.0236 | 0.0229 | 0.0101 | 0.0098 |
| 18 | 1.0177 | 1.0541 | 0.0137 | 0.0132 | 0.0236 | 0.0228 | 0.0099 | 0.0096 |
| 20 | 1.1534 | 1.2005 | 0.0139 | 0.0134 | 0.0237 | 0.0227 | 0.0097 | 0.0093 |
| 22 | 1.2947 | 1.3544 | 0.0141 | 0.0135 | 0.0237 | 0.0227 | 0.0096 | 0.0091 |
| 24 | 1.4415 | 1.5164 | 0.0144 | 0.0137 | 0.0237 | 0.0226 | 0.0094 | 0.0089 |
| 26 | 1.5942 | 1.6872 | 0.0146 | 0.0138 | 0.0238 | 0.0225 | 0.0091 | 0.0086 |
| | 313.15 K | (40 °C) | | | | | | |
| 14 | 0.7569 | 0.7817 | 0.0140 | 0.0136 | 0.0237 | 0.0230 | 0.0097 | 0.0094 |
| 16 | 0.8819 | 0.9147 | 0.0141 | 0.0136 | 0.0238 | 0.0229 | 0.0096 | 0.0093 |
| 18 | 1.0117 | 1.0541 | 0.0143 | 0.0137 | 0.0238 | 0.0228 | 0.0095 | 0.0091 |
| 20 | 1.1465 | 1.2005 | 0.0145 | 0.0138 | 0.0238 | 0.0227 | 0.0093 | 0.0089 |

98 Lithium Chloride; LiCl; [7447-41-8]

| Wt % | $c_2$/ mol dm$^{-3}$ | $m_2$/ mol kg$^{-1}$ | $F_{1c}$ | $F_{1m}$ | $F_{2c}$ | $F_{2m}$ | $F_{3c}$ | $F_{3m}$ |
|---|---|---|---|---|---|---|---|---|
| | 283.15 K | (10 °C) | | | | | | |
| 2 | 0.4773 | 0.4814 | 0.0075 | 0.0075 | 0.0156 | 0.0155 | 0.0081 | 0.0080 |
| 4 | 0.9656 | 0.9828 | 0.0078 | 0.0077 | 0.0157 | 0.0154 | 0.0078 | 0.0077 |
| 6 | 1.4650 | 1.5056 | 0.0080 | 0.0078 | 0.0157 | 0.0152 | 0.0076 | 0.0074 |
| 8 | 1.9752 | 2.0512 | 0.0082 | 0.0079 | 0.0157 | 0.0151 | 0.0074 | 0.0072 |
| | 303.15 K | (30 °C) | | | | | | |
| 1 | 0.2362 | 0.2383 | 0.0077 | 0.0076 | 0.0157 | 0.0156 | 0.0080 | 0.0079 |
| 2 | 0.4752 | 0.4814 | 0.0079 | 0.0078 | 0.0157 | 0.0155 | 0.0078 | 0.0077 |
| 4 | 0.9612 | 0.9828 | 0.0081 | 0.0079 | 0.0157 | 0.0154 | 0.0076 | 0.0076 |
| 6 | 1.4579 | 1.5056 | 0.0083 | 0.0080 | 0.0157 | 0.0152 | 0.0074 | 0.0072 |
| 8 | 1.9654 | 2.0512 | 0.0085 | 0.0081 | 0.0158 | 0.0151 | 0.0073 | 0.0070 |
| 10 | 2.4838 | 2.6209 | 0.0086 | 0.0082 | 0.0158 | 0.0150 | 0.0071 | 0.0068 |
| 12 | 3.0131 | 3.2166 | 0.0088 | 0.0082 | 0.0158 | 0.0148 | 0.0070 | 0.0066 |
| 14 | 3.5536 | 3.8400 | 0.0089 | 0.0083 | 0.0158 | 0.0147 | 0.0069 | 0.0064 |
| 16 | 4.1056 | 4.4930 | 0.0091 | 0.0083 | 0.0159 | 0.0145 | 0.0068 | 0.0062 |
| 18 | 4.6693 | 5.1779 | 0.0092 | 0.0083 | 0.0159 | 0.0143 | 0.0067 | 0.0060 |
| 20 | 5.2451 | 5.8971 | 0.0093 | 0.0083 | 0.0159 | 0.0142 | 0.0066 | 0.0059 |
| | 323.15 K | (50 °C) | | | | | | |
| 2 | 0.4716 | 0.4814 | 0.0085 | 0.0083 | 0.0158 | 0.0155 | 0.0073 | 0.0072 |
| 4 | 0.9541 | 0.9828 | 0.0084 | 0.0081 | 0.0158 | 0.0154 | 0.0075 | 0.0072 |
| 6 | 1.4474 | 1.5056 | 0.0086 | 0.0081 | 0.0158 | 0.0152 | 0.0074 | 0.0071 |
| 8 | 1.9516 | 2.0512 | 0.0086 | 0.0082 | 0.0159 | 0.0151 | 0.0073 | 0.0069 |

99 (1) Sodium Chloride; NaCl; [7447-41-8]

| Wt % | $c_2$/ mol dm$^{-3}$ | $m_2$/ mol kg$^{-1}$ | $F_{1c}$ | $F_{1m}$ | $F_{2c}$ | $F_{2m}$ | $F_{3c}$ | $F_{3m}$ |
|---|---|---|---|---|---|---|---|---|
| | 283.15 K | (10 °C) | | | | | | |
| 2 | 0.3472 | 0.3492 | 0.0070 | 0.0069 | 0.0156 | 0.0155 | 0.0087 | 0.0086 |
| 4 | 0.7044 | 0.7130 | 0.0072 | 0.0072 | 0.0156 | 0.0154 | 0.0084 | 0.0083 |
| 6 | 1.0720 | 1.0922 | 0.0075 | 0.0073 | 0.0156 | 0.0153 | 0.0082 | 0.0080 |
| 8 | 1.4498 | 1.4880 | 0.0077 | 0.0075 | 0.0156 | 0.0152 | 0.0079 | 0.0077 |
| | 298.15 K | (25 °C) | | | | | | |
| 1 | 0.1718 | 0.1728 | 0.0076 | 0.0076 | 0.0157 | 0.0156 | 0.0081 | 0.0080 |
| 2 | 0.3460 | 0.3492 | 0.0078 | 0.0077 | 0.0157 | 0.0155 | 0.0079 | 0.0079 |
| 4 | 0.7018 | 0.7130 | 0.0080 | 0.0078 | 0.0157 | 0.0154 | 0.0077 | 0.0076 |
| 6 | 1.0674 | 1.0922 | 0.0082 | 0.0080 | 0.0157 | 0.0153 | 0.0075 | 0.0074 |
| 8 | 1.4430 | 1.4880 | 0.0083 | 0.0081 | 0.0157 | 0.0152 | 0.0074 | 0.0072 |
| 10 | 1.8289 | 1.9013 | 0.0085 | 0.0082 | 0.0157 | 0.0151 | 0.0072 | 0.0069 |
| 12 | 2.2252 | 2.3334 | 0.0087 | 0.0083 | 0.0158 | 0.0150 | 0.0071 | 0.0067 |
| 14 | 2.6321 | 2.7856 | 0.0089 | 0.0084 | 0.0158 | 0.0149 | 0.0069 | 0.0065 |
| 16 | 3.0500 | 3.2594 | 0.0090 | 0.0085 | 0.0158 | 0.0148 | 0.0068 | 0.0063 |
| 18 | 3.4791 | 3.7562 | 0.0092 | 0.0085 | 0.0158 | 0.0147 | 0.0066 | 0.0062 |
| 20 | 3.9197 | 4.2779 | 0.0094 | 0.0086 | 0.0159 | 0.0146 | 0.0065 | 0.0060 |
| | 313.15 K | (40 °C) | | | | | | |
| 2 | 0.3443 | 0.3492 | 0.0082 | 0.0081 | 0.0158 | 0.0155 | 0.0076 | 0.0075 |
| 4 | 0.6980 | 0.7130 | 0.0084 | 0.0082 | 0.0158 | 0.0154 | 0.0074 | 0.0073 |
| 6 | 1.0614 | 1.0922 | 0.0085 | 0.0083 | 0.0158 | 0.0153 | 0.0073 | 0.0071 |
| 8 | 1.4347 | 1.4880 | 0.0087 | 0.0084 | 0.0158 | 0.0152 | 0.0071 | 0.0069 |

99 (2) Sodium Sulfate; $Na_2SO_4$; [7757-82-6]

| Wt % | $c_2$/ mol dm$^{-3}$ | $m_2$/ mol kg$^{-1}$ | $F_{1c}$ | $F_{1m}$ | $F_{2c}$ | $F_{2m}$ | $F_{3c}$ | $F_{3m}$ |
|---|---|---|---|---|---|---|---|---|
| | 283.15 K | (10 °C) | | | | | | |
| 10 | 0.7708 | 0.7823 | 0.0082 | 0.0080 | 0.0233 | 0.0230 | 0.0152 | 0.0149 |
| 12 | 0.9416 | 0.9600 | 0.0088 | 0.0087 | 0.0233 | 0.0229 | 0.0145 | 0.0142 |
| 14 | 1.1182 | 1.1461 | 0.0095 | 0.0092 | 0.0233 | 0.0228 | 0.0139 | 0.0136 |
| 16 | 1.3008 | 1.3410 | 0.0101 | 0.0098 | 0.0234 | 0.0227 | 0.0133 | 0.0129 |
| | 303.15 K | (30 °C) | | | | | | |
| 1 | 0.0707 | 0.0711 | 0.0068 | 0.0067 | 0.0235 | 0.0234 | 0.0168 | 0.0167 |
| 2 | 0.1427 | 0.1437 | 0.0074 | 0.0074 | 0.0235 | 0.0234 | 0.0161 | 0.0160 |
| 4 | 0.2905 | 0.2933 | 0.0081 | 0.0081 | 0.0235 | 0.0233 | 0.0154 | 0.0152 |
| 6 | 0.4434 | 0.4494 | 0.0088 | 0.0087 | 0.0235 | 0.0232 | 0.0147 | 0.0145 |
| 8 | 0.6016 | 0.6122 | 0.0094 | 0.0093 | 0.0235 | 0.0231 | 0.0141 | 0.0138 |
| 10 | 0.7653 | 0.7822 | 0.0100 | 0.0098 | 0.0235 | 0.0230 | 0.0135 | 0.0132 |
| 12 | 0.9346 | 0.9600 | 0.0105 | 0.0102 | 0.0235 | 0.0229 | 0.0130 | 0.0127 |
| 14 | 1.1095 | 1.1461 | 0.0110 | 0.0106 | 0.0235 | 0.0228 | 0.0125 | 0.0121 |
| 16 | 1.2904 | 1.3410 | 0.0115 | 0.0110 | 0.0235 | 0.0227 | 0.0121 | 0.0116 |
| 18 | 1.4775 | 1.5454 | 0.0119 | 0.0114 | 0.0236 | 0.0225 | 0.0116 | 0.0111 |
| 20 | 1.6706 | 1.7601 | 0.0124 | 0.0118 | 0.0236 | 0.0224 | 0.0112 | 0.0106 |
| 22 | 1.8704 | 1.9857 | 0.0129 | 0.0121 | 0.0237 | 0.0223 | 0.0108 | 0.0102 |
| 24 | 2.0769 | 2.2232 | 0.0133 | 0.0124 | 0.0237 | 0.0222 | 0.0104 | 0.0097 |

99 (2) Sodium Sulfate; $Na_2SO_4$; [7757-82-6] (Continued)

| Wt % | $c_2$/ mol dm$^{-3}$ | $m_2$/ mol kg$^{-1}$ | $F_{1c}$ | $F_{1m}$ | $F_{2c}$ | $F_{2m}$ | $F_{3c}$ | $F_{3m}$ |
|---|---|---|---|---|---|---|---|---|
| | 323.15 K | (50 °C) | | | | | | |
| 10 | 0.7591 | 0.7823 | 0.0107 | 0.0104 | 0.0237 | 0.0230 | 0.0130 | 0.0126 |
| 12 | 0.9269 | 0.9600 | 0.0112 | 0.0108 | 0.0237 | 0.0229 | 0.0125 | 0.0121 |
| 14 | 1.1001 | 1.1461 | 0.0117 | 0.0112 | 0.0237 | 0.0228 | 0.0120 | 0.0115 |
| 16 | 1.2792 | 1.3410 | 0.0122 | 0.0116 | 0.0238 | 0.0227 | 0.0116 | 0.0110 |

100 (1) Potassium Hydroxide; KOH; [1310-58-3]

| Wt % | $c_2$/ mol dm$^{-3}$ | $m_2$/ mol kg$^{-1}$ | $F_{1c}$ | $F_{1m}$ | $F_{2c}$ | $F_{2m}$ | $F_{3c}$ | $F_{3m}$ |
|---|---|---|---|---|---|---|---|---|
| | 288.15 K | (15 °C) | | | | | | |
| 1 | 0.1797 | 0.1800 | 0.0021 | 0.0021 | 0.0156 | 0.0156 | 0.0135 | 0.0135 |
| 2 | 0.3627 | 0.3637 | 0.0023 | 0.0023 | 0.0156 | 0.0155 | 0.0132 | 0.0132 |
| 3 | 0.5489 | 0.5512 | 0.0025 | 0.0025 | 0.0156 | 0.0155 | 0.0130 | 0.0130 |
| 4 | 0.7385 | 0.7426 | 0.0027 | 0.0027 | 0.0155 | 0.0154 | 0.0128 | 0.0127 |
| 5 | 0.9314 | 0.9380 | 0.0029 | 0.0029 | 0.0155 | 0.0154 | 0.0126 | 0.0125 |
| 6 | 1.1275 | 1.1376 | 0.0031 | 0.0031 | 0.0155 | 0.0153 | 0.0124 | 0.0123 |
| 8 | 1.5299 | 1.5498 | 0.0034 | 0.0034 | 0.0154 | 0.0152 | 0.0120 | 0.0119 |
| 10 | 1.9458 | 1.9803 | 0.0037 | 0.0036 | 0.0154 | 0.0151 | 0.0117 | 0.0115 |
| 12 | 2.3756 | 2.4303 | 0.0040 | 0.0039 | 0.0153 | 0.0150 | 0.0114 | 0.0111 |
| 14 | 2.8192 | 2.9013 | 0.0043 | 0.0042 | 0.0153 | 0.0149 | 0.0110 | 0.0107 |
| 16 | 3.2773 | 3.3947 | 0.0045 | 0.0044 | 0.0153 | 0.0148 | 0.0107 | 0.0104 |
| 20 | 4.2360 | 4.4556 | 0.0051 | 0.0048 | 0.0153 | 0.0145 | 0.0102 | 0.0097 |
| 24 | 5.2547 | 5.6281 | 0.0056 | 0.0052 | 0.0153 | 0.0142 | 0.0097 | 0.0090 |
| 28 | 6.3351 | 6.9309 | 0.0061 | 0.0056 | 0.0153 | 0.0140 | 0.0092 | 0.0084 |
| 32 | 7.4808 | 8.3870 | 0.0066 | 0.0059 | 0.0153 | 0.0137 | 0.0087 | 0.0078 |
| 36 | 8.6931 | 10.0251 | 0.0071 | 0.0061 | 0.0154 | 0.0134 | 0.0083 | 0.0072 |
| 40 | 9.9741 | 11.8815 | 0.0076 | 0.0064 | 0.0155 | 0.0130 | 0.0079 | 0.0067 |
| 44 | 11.3259 | 14.0033 | 0.0081 | 0.0066 | 0.0157 | 0.0127 | 0.0076 | 0.0061 |
| 48 | 12.7525 | 16.4514 | 0.0086 | 0.0067 | 0.0159 | 0.0123 | 0.0072 | 0.0056 |

100 (2) Potassium Chloride; KCl; [7447-40-7]

| Wt % | $c_2$/ mol dm$^{-3}$ | $m_2$/ mol kg$^{-1}$ | $F_{1c}$ | $F_{1m}$ | $F_{2c}$ | $F_{2m}$ | $F_{3c}$ | $F_{3m}$ |
|---|---|---|---|---|---|---|---|---|
| | 283.15 K | (10 °C) | | | | | | |
| 4 | 0.5504 | 0.5589 | 0.0118 | 0.0116 | 0.0157 | 0.0155 | 0.0040 | 0.0039 |
| 6 | 0.8363 | 0.8561 | 0.0120 | 0.0117 | 0.0158 | 0.0154 | 0.0038 | 0.0037 |
| 8 | 1.1294 | 1.1663 | 0.0122 | 0.0119 | 0.0158 | 0.0153 | 0.0036 | 0.0035 |
| 10 | 1.4300 | 1.4903 | 0.0125 | 0.0120 | 0.0159 | 0.0152 | 0.0034 | 0.0033 |
| | 303.15 K | (30 °C) | | | | | | |
| 1 | 0.1344 | 0.1355 | 0.0120 | 0.0119 | 0.0157 | 0.0156 | 0.0037 | 0.0037 |
| 2 | 0.2705 | 0.2737 | 0.0122 | 0.0120 | 0.0158 | 0.0156 | 0.0036 | 0.0035 |
| 4 | 0.5477 | 0.5588 | 0.0124 | 0.0122 | 0.0158 | 0.0155 | 0.0034 | 0.0033 |
| 6 | 0.8319 | 0.8561 | 0.0127 | 0.0123 | 0.0159 | 0.0154 | 0.0032 | 0.0031 |
| 8 | 1.1231 | 1.1663 | 0.0129 | 0.0124 | 0.0159 | 0.0153 | 0.0030 | 0.0029 |
| 10 | 1.4215 | 1.4902 | 0.0131 | 0.0125 | 0.0160 | 0.0152 | 0.0029 | 0.0028 |
| 12 | 1.7272 | 1.8289 | 0.0133 | 0.0125 | 0.0160 | 0.0151 | 0.0028 | 0.0026 |

100 (2) Potassium Chloride; KCl; [7447-40-7] (Continued)

| Wt % | $c_2$/ mol dm$^{-3}$ | $m_2$/ mol kg$^{-1}$ | $F_{1c}$ | $F_{1m}$ | $F_{2c}$ | $F_{2m}$ | $F_{3c}$ | $F_{3m}$ |
|---|---|---|---|---|---|---|---|---|
| | 303.15 K | (30 °C) (Continued) | | | | | | |
| 14 | 2.0405 | 2.1833 | 0.0135 | 0.0126 | 0.0161 | 0.0151 | 0.0026 | 0.0025 |
| 16 | 2.3614 | 2.5547 | 0.0137 | 0.0126 | 0.0162 | 0.0150 | 0.0025 | 0.0023 |
| 18 | 2.6902 | 2.9441 | 0.0139 | 0.0127 | 0.0163 | 0.0149 | 0.0024 | 0.0022 |
| 20 | 3.0271 | 3.3530 | 0.0140 | 0.0127 | 0.0164 | 0.0148 | 0.0023 | 0.0021 |
| 22 | 3.3724 | 3.7829 | 0.0142 | 0.0127 | 0.0165 | 0.0147 | 0.0022 | 0.0020 |
| 24 | 3.7262 | 4.2354 | 0.0144 | 0.0127 | 0.0166 | 0.0146 | 0.0021 | 0.0019 |
| 26 | 4.0888 | 4.7123 | 0.0146 | 0.0127 | 0.0167 | 0.0145 | 0.0020 | 0.0018 |
| | 323.15 K | (50 °C) | | | | | | |
| 4 | 0.5435 | 0.5589 | 0.0133 | 0.0130 | 0.0159 | 0.0155 | 0.0026 | 0.0025 |
| 6 | 0.8253 | 0.8561 | 0.0134 | 0.0129 | 0.0160 | 0.0154 | 0.0026 | 0.0025 |
| 8 | 1.1141 | 1.1663 | 0.0135 | 0.0129 | 0.0160 | 0.0153 | 0.0026 | 0.0025 |
| 10 | 1.4100 | 1.4903 | 0.0136 | 0.0129 | 0.0161 | 0.0152 | 0.0025 | 0.0024 |

100 (3) Potassium Iodide; KI; [7681-11-0]

| Wt % | $c_2$/ mol dm$^{-3}$ | $m_2$/ mol kg$^{-1}$ | $F_{1c}$ | $F_{1m}$ | $F_{2c}$ | $F_{2m}$ | $F_{3c}$ | $F_{3m}$ |
|---|---|---|---|---|---|---|---|---|
| | 283.15 K | (10 °C) | | | | | | |
| 12 | 0.7922 | 0.8214 | 0.0197 | 0.0190 | 0.0160 | 0.0154 | -0.0037 | -0.0036 |
| 14 | 0.9391 | 0.9806 | 0.0199 | 0.0190 | 0.0161 | 0.0154 | -0.0038 | -0.0037 |
| 16 | 1.0908 | 1.1474 | 0.0200 | 0.0190 | 0.0161 | 0.0153 | -0.0039 | -0.0037 |
| 18 | 1.2474 | 1.3223 | 0.0202 | 0.0190 | 0.0162 | 0.0153 | -0.0040 | -0.0038 |
| | 303.15 K | (30 °C) | | | | | | |
| 1 | 0.0604 | 0.0608 | 0.0201 | 0.0200 | 0.0157 | 0.0156 | -0.0044 | -0.0043 |
| 2 | 0.1217 | 0.1229 | 0.0202 | 0.0200 | 0.0158 | 0.0156 | -0.0044 | -0.0044 |
| 4 | 0.2470 | 0.2510 | 0.0203 | 0.0200 | 0.0158 | 0.0156 | -0.0045 | -0.0044 |
| 6 | 0.3761 | 0.3845 | 0.0205 | 0.0200 | 0.0159 | 0.0155 | -0.0046 | -0.0045 |
| 8 | 0.5091 | 0.5238 | 0.0206 | 0.0200 | 0.0159 | 0.0155 | -0.0047 | -0.0045 |
| 10 | 0.6461 | 0.6693 | 0.0207 | 0.0200 | 0.0160 | 0.0155 | -0.0047 | -0.0046 |
| 12 | 0.7875 | 0.8214 | 0.0209 | 0.0200 | 0.0161 | 0.0154 | -0.0048 | -0.0046 |
| 14 | 0.9332 | 0.9806 | 0.0210 | 0.0200 | 0.0162 | 0.0154 | -0.0049 | -0.0046 |
| 16 | 1.0837 | 1.1474 | 0.0211 | 0.0200 | 0.0162 | 0.0153 | -0.0049 | -0.0046 |
| 18 | 1.2390 | 1.3223 | 0.0213 | 0.0200 | 0.0163 | 0.0153 | -0.0050 | -0.0047 |
| 20 | 1.3994 | 1.5059 | 0.0214 | 0.0199 | 0.0164 | 0.0152 | -0.0050 | -0.0047 |
| 24 | 1.7366 | 1.9022 | 0.0217 | 0.0198 | 0.0166 | 0.0151 | -0.0051 | -0.0048 |
| 28 | 2.0974 | 2.3426 | 0.0220 | 0.0197 | 0.0168 | 0.0150 | -0.0052 | -0.0047 |
| 35 | 2.7933 | 3.2436 | 0.0226 | 0.0194 | 0.0172 | 0.0148 | -0.0054 | -0.0046 |
| 45 | 3.9582 | 4.9285 | 0.0236 | 0.0189 | 0.0179 | 0.0144 | -0.0056 | -0.0045 |
| 55 | 5.3825 | 7.3623 | 0.0249 | 0.0182 | 0.0190 | 0.0139 | -0.0059 | -0.0043 |
| | 323.15 K | (50 °C) | | | | | | |
| 12 | 0.7808 | 0.8214 | 0.0219 | 0.0208 | 0.0162 | 0.0154 | -0.0057 | -0.0054 |
| 14 | 0.9252 | 0.9806 | 0.0220 | 0.0207 | 0.0163 | 0.0154 | -0.0057 | -0.0054 |
| 16 | 1.0743 | 1.1474 | 0.0221 | 0.0207 | 0.0164 | 0.0153 | -0.0057 | -0.0053 |
| 18 | 1.2281 | 1.3223 | 0.0222 | 0.0206 | 0.0165 | 0.0153 | -0.0057 | -0.0053 |

The calculations above are based on weight per cent and aqueous electrolyte density data that appear in *International Critical Tables*, McGraw-Hill Book Co., New York, 1928, Volume III.

| I. M. Sechenov [J. Setchenow] | H. Lawrence Clever<br>Department of Chemistry<br>Emory University<br>Atlanta, GA 30322<br><br>1982, March |
|---|---|

Ivan Mikhaylovich Sechenov (J. Setchenow, J. Setchenoff)

born 1829, August 1, Teply Stan, Simbirsk guberniya [now Sechenovo, Arzamas Oblast], Russia

died 1905, November 2, Moscow, Russia

Research interests: physiology, physical chemistry, psychology.

Selected Works, A. A. Subkov, Editor, Moscow, 1935.

A selection of Sechenov's publications. Those written originally in German are republished in German, the originals in Russian are republished in English. Many of the papers are on psychology.

There are earlier collections of Sechenov's papers. An 1889 one in French, and one *ca.* 1907 in Russian.

Sechenov papers on the salt effect and on the solubility of carbon dioxide in various electrolyte and nonelectrolyte aqueous solutions.

1. *Pflug. Arch.* 1874, *8*, 1 - 39.
2. *Akad. Nauk SSSR, Mem. Acad. Imp. Sci., St. Petersbourg* 1877, *22*,102-
3. *Ibid.* 1879, *26*,
4. *Ibid.* 1886, *34*,
5. *Ibid.* 1887, *35*, 1 - 59.
6. *Nouv. Mem. Soc. Imp. Nat. Moscow* 1889, *15*,203 - 274.
7. *Z. Phys. Chem., Stoechiom. Verwandtsschaftsl.* 1889, *4*, 117 - 125.
8. *Ibid.* 1891, *8*, 657 - 660.
9. *Ann. Chim. Phys.* [*6*] 1892, *25*, 226 - 270.

Paper 7. is a long abstract of paper 6.
Paper 9. and paper 5. contain some data in common.
I do not have copies of papers 3. and 4. Their citations are not certain. All of the other papers are in German except 9. which is in French.

The Sechenov Salt Effect Equation.

$y = \alpha\ f(x)$ — $\alpha$ is the Bunsen coefficient in pure water, y is the Bunsen coefficient in the salt solution, and $x$ is the dilution of the salt solution.

Sechenov tried several $f(x)$ on his data. They included

$\sqrt[x]{m}$ where $m < 1$,

$p^{k/x}$ where $p < 1$, and

$e^{-k/x}$. He decided the last function was the best.

Thus, he gave the equation as $y = \alpha e^{-k/x}$ or $k = -x \ln \frac{y}{\alpha} = x \ln \frac{\alpha}{y}$ .

Our present designation would be $k_{scc}/\text{dm}\ \text{mol}^{-1} = (1/c) \log (\alpha^0/\alpha)$.

Sechenov often gave weight and volume data from which the electrolyte concentration can be calculated. Sometimes he did not give the information to calculate the concentration. He would prepare a saturated solution and dilute the solution. He usually checked his equation by taking the Bunsen coefficient in water and in the most concentrated electrolyte solution to predict the Bunsen coefficient in the electrolyte solutions at each dilution. The agreement bewteen the calculated and the experimental Bunsen coefficients was satisfactory. The gas solubility apparatus used by Sechenov used a large solvent volume (up to *ca.* 500 $cm^3$).His solubility values agree within a few per cent of modern values.

## APPENDIX I. Conversion Factors k and $k^{-1}$

| Non-SI Unit | k<br>1 (non-SI Unit) = k (SI Unit) | | $k^{-1}$<br>1 (SI Unit) = $k^{-1}$ (non-SI Unit) | |
|---|---|---|---|---|
| **LENGTH** | | | SI Unit, m | |
| Å (angstrom) | $1 \times 10^{-10}$ | (*) | $1 \times 10^{10}$ | (*) |
| cm (centimeter) | $1 \times 10^{-2}$ | (*) | $1 \times 10^{2}$ | (*) |
| in (inch) | $254 \times 10^{-4}$ | (*) | $3\ 937\ 008 \times 10^{-5}$ | |
| ft (foot) | $3\ 048 \times 10^{-4}$ | (*) | $3\ 280\ 840 \times 10^{-6}$ | |
| **AREA** | | | SI Unit, $m^2$ | |
| $cm^2$ | $1 \times 10^{-4}$ | (*) | $1 \times 10^{4}$ | (*) |
| $in^2$ | $64\ 516 \times 10^{-8}$ | (*) | $1\ 550\ 003 \times 10^{-3}$ | |
| $ft^2$ | $9\ 290\ 304 \times 10^{-8}$ | (*) | $1\ 076\ 391 \times 10^{-5}$ | |
| **VOLUME** | | | SI Unit, $m^3$ | |
| $cm^3$ | $1 \times 10^{-6}$ | (*) | $1 \times 10^{6}$ | (*) |
| $in^3$ | $16\ 387\ 064 \times 10^{-12}$ | (*) | $6\ 102\ 374 \times 10^{-2}$ | |
| $ft^3$ | $2\ 831\ 685 \times 10^{-8}$ | | $3\ 531\ 467 \times 10^{-5}$ | |
| l (litre) | $1 \times 10^{-3}$ | (*) | $1 \times 10^{3}$ | (*) |
| UKgal (UK gallon) | $45\ 461 \times 10^{-7}$ | | $21\ 997 \times 10^{-2}$ | |
| USgal (US gallon) | $37\ 854 \times 10^{-7}$ | | $26\ 417 \times 10^{-2}$ | |
| **MASS** | | | SI Unit, kg | |
| g (gram) | $1 \times 10^{-3}$ | (*) | $1 \times 10^{3}$ | (*) |
| t (tonne) | $1 \times 10^{3}$ | (*) | $1 \times 10^{-3}$ | (*) |
| lb (pound) | $45\ 359\ 237 \times 10^{-8}$ | (*) | $2\ 204\ 623 \times 10^{-6}$ | |
| **DENSITY** | | | SI Unit, $kg\ m^{-3}$ | |
| $g\ cm^{-3}$ | $1 \times 10^{3}$ | (*) | $1 \times 10^{-3}$ | (*) |
| $g\ l^{-1}$ | 1 | (*) | 1 | (*) |
| $lb\ in^{-3}$ | $2\ 767\ 991 \times 10^{-2}$ | | $3\ 612\ 728 \times 10^{-11}$ | |
| $lb\ ft^{-3}$ | $1\ 601\ 847 \times 10^{-5}$ | | $6\ 242\ 795 \times 10^{-8}$ | |
| $lb\ UKgal^{-1}$ | $99\ 776 \times 10^{-3}$ | | $100\ 224 \times 10^{-7}$ | |
| $lb\ USgal^{-1}$ | $1\ 198\ 264 \times 10^{-4}$ | | $8\ 345\ 406 \times 10^{-9}$ | |
| **PRESSURE** | | | SI Unit, Pa (pascal, $kg\ m^{-1} s^{-2}$) | |
| $dyn\ cm^{-2}$ | $1 \times 10^{-1}$ | (*) | $1 \times 10$ | (*) |
| at ($kgf\ cm^{-2}$) | $980\ 665 \times 10^{-1}$ | (*) | $1\ 019\ 716 \times 10^{-11}$ | |
| atm (atmosphere) | 101 325 | (*) | $9\ 869\ 233 \times 10^{-12}$ | |
| bar | $1 \times 10^{5}$ | (*) | $1 \times 10^{-5}$ | (*) |
| $lbf\ in^{-2}$ (p.s.i.) | $6\ 894\ 757 \times 10^{-3}$ | | $1\ 450\ 377 \times 10^{-10}$ | |
| $lbf\ ft^{-2}$ | $47\ 880 \times 10^{-3}$ | | $20\ 886 \times 10^{-6}$ | |
| inHg (inch of mercury) | $3\ 386\ 388 \times 10^{-3}$ | | $2\ 952\ 999 \times 10^{-10}$ | |
| mmHg (millimeter of mercury, torr) | $1\ 333\ 224 \times 10^{-4}$ | | $7\ 500\ 617 \times 10^{-9}$ | |

## APPENDIX I. Conversion Factors k and $k^{-1}$

| Non-SI Unit | k<br>1 (non-SI Unit) =<br>k (SI Unit) | | $k^{-1}$<br>1 (SI Unit) =<br>$k^{-1}$ (non-SI Unit) | |
|---|---|---|---|---|
| ENERGY | | | Unit, J (joule, kg $m^2s^{-2}$) | |
| erg | $1 \times 10^{-7}$ | (*) | $1 \times 10^{7}$ | (*) |
| $cal_{IT}$ (I.T. calorie) | $41\ 868 \times 10^{-4}$ | (*) | $2\ 388\ 459 \times 10^{-7}$ | |
| $cal_{th}$ (thermochemical calorie) | $4\ 184 \times 10^{-3}$ | (*) | $2\ 390\ 057 \times 10^{-7}$ | |
| kW h (kilowatt hour) | $36 \times 10^{5}$ | (*) | $2\ 777\ 778 \times 10^{-13}$ | |
| 1 atm | $101\ 325 \times 10^{-3}$ | (*) | $9\ 869\ 233 \times 10^{-9}$ | |
| ft lbf | $1\ 355\ 818 \times 10^{-6}$ | | $7\ 375\ 622 \times 10^{-7}$ | |
| hp h (horse power hour) | 2 684 519 | | $3\ 725\ 062 \times 10^{-13}$ | |
| Btu (British thermal unit) | $1\ 055\ 056 \times 10^{-3}$ | | $9\ 478\ 172 \times 10^{-10}$ | |

An asterisk (*) denotes an exact relationship

| COMPONENTS: | EVALUATOR: |
|---|---|
| (1) Nitrogen; $N_2$; [7727-37-9]<br>(2) Water; $H_2O$; [7732-18-5] | Rubin Battino<br>Department of Chemistry<br>Wright State University<br>Dayton, Ohio 45435<br>U.S.A.<br>December 1981 |

CRITICAL EVALUATION:

Nitrogen is one of the atmospheric gases that has been studied by many workers over a long period of time. We selected the work of nine groups to use for the smoothing equation. In the process of fitting the data to the smoothing equation any data points which differed from the smooth equation by about two standard deviations or more were rejected. The 74 points used for the final smoothing were obtained as follows: (reference number-number of data points from that reference): 1-6; 2-28; 3-10; 4-1; 5-18; 6-1; 7-1; 8-1; 9-8. The fitting equation used was

$$\ln x_1 = A + B/(T/100K) + C \ln (T/100K) + DT/100K \quad (1)$$

Using T/100K as the variable rather than T/K gives coefficients of approximately equal magnitude. The best fit for the 74 points was

$$\ln x_1 = -67.38765 + 86.32129/(T/100K) + 24.79808 \ln (T/100K) \quad (2)$$

where $x_1$ is the mole fraction solubility at 101,325 Pa partial pressure of gas. The fit in $\ln x_1$ gave a standard deviation of 0.72% taken at the middle of the temperature range. Table 1 gives smoothed values of the mole fraction solubility and the Ostwald coefficient at 5K intervals.

Table 1 also gives the thermodynamic functions $\Delta\bar{G}_1^\circ$, $\Delta\bar{H}_1^\circ$, $\Delta\bar{S}_1^\circ$, and $\Delta\bar{C}_{p_1}^\circ$ for the transfer of the gas from the vapor phase at 101,325 Pa partial gas pressure to the (hypothetical) solution of unit mole fraction. These thermodynamic properties were calculated from the smoothing equation according to the following equations:

$$\Delta\bar{G}_1^\circ = -RAT - 100RB - RCT \ln (T/100) - RDT^2/100 \quad (3)$$

$$\Delta\bar{S}_1^\circ = RA + RC \ln (T/100) + RC + 2RDT/100 \quad (4)$$

$$\Delta\bar{H}_1^\circ = -100RB + RCT + RDT^2/100 \quad (5)$$

$$\Delta\bar{C}_{p_1}^\circ = RC + 2RDT/100 \quad (6)$$

The experimental results of twenty-one other groups were not used for the smoothing equation for a variety of reasons. Although Douglas' (10) work is in general of high quality, the nitrogen solubility data appeared to vary erratically around the smoothing equation. Winkler's values (11) were mostly low. Van Slyke's two values (12) were close and high so both were discarded. Müller's results (13) were about 3% low. Just's values (14) were 1-3% high. Ikels single measurement (15) was 3% low. Orcutt and Seevers' single value (16) was 3% high. Behnke and Yarbrough's single value was 2% high (17). Stoddard's single value (18) was 8% low. Grollman's single value was 6% high (19). Morrison and Billett's single value (20) was 4% low. Brasted and Hirayama's single value (21) was 12% high. Drucker and Moles' single value (22) was 2% low. Christoff's single value was 10% low (23). Bunsen's five experimental values (24) were about 15% low. Braun's five values (25) ranged from 8% high to 2% low. Adeney and Becker's results (26) ranged from 1% high to about 6% low. Power and Stegall's single value (27) was 1.1% low. Wilcock and Battino's value at 298.15 K (28) was 1.4% low. The Wilcock and Battino paper also showed that the solubility of one gas is not affected by the presence of another within their experimental precision, and at the solubilities concerned. Cassuto's values (29) were off by more than 6%. The early measurements by Petterson and Sonden (30) were very low.

Weiss (31) analyzed the data then current on nitrogen solubilities in water and seawater and provided smoothing equations and smoothed data. Klots and Benson (32) determined the ratio of the solubility of $^{28}N_2/^{29}N_2$ in water in the range 275 to 300 K and found the ratio of the Henry's law

continued on following page

| COMPONENTS: | EVALUATOR: |
| --- | --- |
| (1) Nitrogen; $N_2$; [7727-37-9]<br>(2) Water; $H_2O$; [7732-18-5] | Rubin Battino<br>Department of Chemistry<br>Wright State University<br>Dayton, Ohio 45435<br>U.S.A.<br><br>December 1981 |

CRITICAL EVALUATION:

continued

constants (mole fraction and atm units) to be 1.00085 ± 0.00010 in this temperature range. Note that the heavier isotope is more soluble.

Figure 1 shows the temperature dependence of the solubility of nitrogen in water as mole fraction versus T/K. The curve was obtained from the smoothing equation. The solubility expressed as the Ostwald coefficient shows a minimum at about 338K, but over the temperature range of the data used for the smoothing equation no minimum was evident, although the trend to a minimum was apparent.

Table 1. Smoothed values of the solubility of nitrogen in water and thermodynamic functions[a] at 5K intervals using equation 2 at 101,325 Pa partial pressure of gas.

| T/K | $10^5 x_1$ [b] | $10^2 L$ [c] | $\Delta\bar{G}_1^\circ$/kJ mol$^{-1}$ | $\Delta\bar{H}_1^\circ$/kJ mol$^{-1}$ | $\Delta\bar{S}_1^\circ$/J mol$^{-1}$K$^{-1}$ |
| --- | --- | --- | --- | --- | --- |
| 273.15 | 1.908 | 2.373 | 24.68 | -15.45 | -147 |
| 278.15 | 1.695 | 2.147 | 25.41 | -14.42 | -143 |
| 283.15 | 1.524 | 1.965 | 26.11 | -13.39 | -140 |
| 288.15 | 1.386 | 1.818 | 26.80 | -12.36 | -136 |
| 293.15 | 1.274 | 1.698 | 27.47 | -11.33 | -132 |
| 298.15 | 1.183 | 1.601 | 28.12 | -10.30 | -129 |
| 303.15 | 1.108 | 1.523 | 28.76 | -9.27 | -125 |
| 308.15 | 1.047 | 1.461 | 29.38 | -8.24 | -122 |
| 313.15 | 0.9981 | 1.413 | 29.98 | -7.21 | -119 |
| 318.15 | 0.9585 | 1.376 | 30.57 | -6.17 | -115 |
| 323.15 | 0.9273 | 1.349 | 31.14 | -5.14 | -112 |
| 328.15 | 0.9033 | 1.331 | 31.69 | -4.11 | -109 |
| 333.15 | 0.8855 | 1.321 | 32.23 | -3.08 | -106 |
| 338.15 | 0.8735 | 1.319 | 32.75 | -2.05 | -103 |
| 343.15 | 0.8666 | 1.324 | 33.26 | -1.02 | -100 |
| 348.15 | 0.8644 | 1.336 | 33.75 | 0.01 | - 97 |

[a] $\Delta\bar{C}_{p_1}^\circ$ was independent of temperature and had the value 206 J mol$^{-1}$ K$^{-1}$.

[b] Mole fraction solubility at 101,325 Pa partial pressure of gas.

[c] Ostwald coefficient.

In a review D. M. Himmelblau (33) reported on the partial molal heats and entropies of solution for gases ($O_2$, $N_2$, $H_2$, He, Xe, and $CH_4$) dissolved in water from the freezing to near the critical point. He also gave constants for a general solubility equation for these gases. Recently Clever and Han (34) summarized work on the solubility of hydrogen nitrogen, oxygen, and the noble gases in water from 350-600K. They give smoothing equations and critically evaluate the literature for these systems at high temperatures.

Schulze and Prausnitz (35) have correlated gas solubility data for aqueous systems at high temperatures within the framework of scaled scaled particle theory.

continued on following page

| COMPONENTS: | EVALUATOR: |
| --- | --- |
| (1) Nitrogen; $N_2$; [7727-37-9]<br>(2) Water; $H_2O$; [7732-18-5] | Rubin Battino<br>Department of Chemistry<br>Wright State University<br>Dayton, Ohio 45435<br>U.S.A.<br>December 1981 |

CRITICAL EVALUATION:

continued

Note: The critical evaluation in the following section on nitrogen solubility in seawater which was prepared by Dr. C-T. A. Chen does not use the solubility data of Klots and Benson (2) for the pure water part of his smoothing equation. In that sense the pure water solubilities recommended in this part and the following one are not compatible. Differences are small and indicate different ways of judging data.

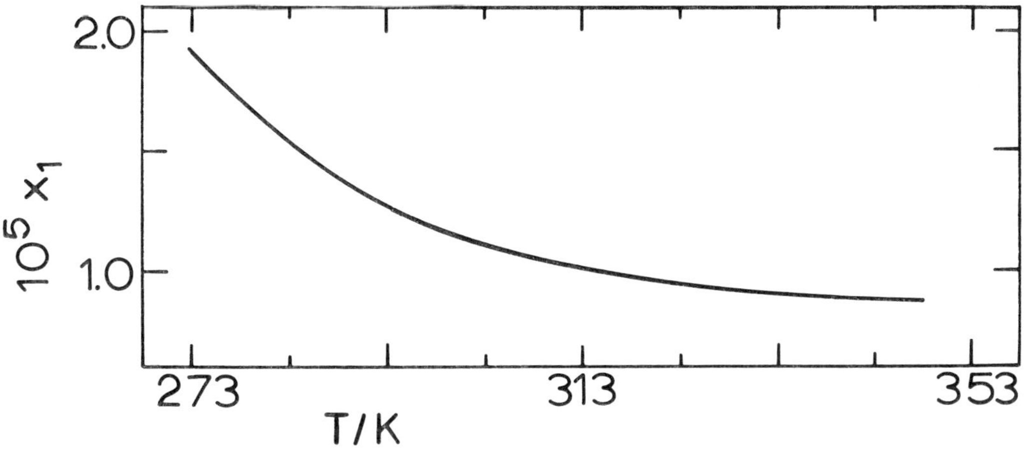


References

1. Farhi, L. E.; Edwards, A. W. T.; Homma, T. *J. Appl. Physiol.* 1963, *18*, 97-106.
2. Klots, C. E.; Benson, B. B. *Sears Foundation: J. Marine Res.* 1963, *21*, 48-57.
3. Morrison, T. J.; Billett, F. *J. Chem. Soc.* 1952, 3819-22.
4. Bohr, C. *Z. Physik. Chem.* 1910, *71*, 47-50.
5. Fox, C. J. J. *Trans. Faraday Soc.* 1909, *5*, 68-87.
6. Hawkins, J. A.; Shilling, C. W. *J. Biol. Chem.* 1936, *113*, 273-8.
7. Hüfner, G. *Z. Physik. Chem.* 1907, *57*, 611-24.
8. Clever, H. L.; Battino, R.; Saylor, J. H.; Gross, P. M. *J. Phys. Chem.* 1957, *61*, 1078-82.
9. Murray, C. N.; Riley, J. P.; Wilson, T. R. S. *Deep-Sea Research* 1969, *16*, 297-310.
10. Douglas, E. *J. Phys. Chem.* 1964, *68*, 169-74.
11. Winkler, L. W. *Ber.* 1891, *24*, 3602-10.
12. Van Slyke, D. D.; Dillon, R. T.; Margaria, R. *J. Biol. Chem.* 1934, *105*, 571-96.
13. Müller, C. *Z. Physik. Chem.* 1912, *81*, 483-503.
14. Just, G. *Z. Physik. Chem.* 1901, *37*, 342-67.
15. Ikels, K. G. *DDC, Report No. SAM-TDR-64-28* 1964.
16. Orcutt, F. S.; Seevers, M. H. *J. Biol. Chem.* 1937, *117*, 501-7.
17. Behnke, A. R.; Yarbrough, O. D. *U.S. Naval Medical Bulletin* 1938, *36*, 542-59.

continued on following page

| COMPONENTS: | EVALUATOR: |
|---|---|
| (1) Nitrogen; $N_2$; [7727-37-9]<br>(2) Water; $H_2O$; [7732-18-5] | Rubin Battino<br>Department of Chemistry<br>Wright State University<br>Dayton, Ohio 45435<br>U.S.A.<br><br>December 1981 |

CRITICAL EVALUATION:

continued


18. Stoddard, J. L. *J. Biol. Chem.* 1926-27, *71*, 629-92.
19. Grollman, A. *J. Biol. Chem.* 1929, *82*, 317-25.
20. Morrison, T. J.; Billett, F. *J. Chem. Soc.* 1948, 2033-5.
21. Brasted, R. C.; Hirayama, C. *J. Phys. Chem.* 1958, *62*, 125-6.
22. Drucker, K.; Moles. E. *Z. Physik. Chem.* 1910, *75*, 405-36.
23. Christoff, A. *Z. Physik. Chem.* 1906, *55*, 622-34.
24. Bunsen, R. *Ann.* 1855, *93*, 1, 496; *Phil. Mag.* 1855, 9, 116, 181.
25. Braun, L. *Z. Physik. Chem.* 1900, *33*, 721-41.
26. Adeney, W. E.; Becker, H. G. *Royal Dublin Soc. Scientific Pro.* 1919, *15*, 609-28.
27. Power, G. G.; Stegall, H. *J. Appl. Physiology* 1970, *29* 145-9.
28. Wilcock, R. J. Battino, R. *Nature* 1974, *252*, 614-5.
29. Cassuto, L. *Physik. Z.* 1904, *5*, 233-6.
30. Petterson, O.; Sonden, K. *Ber.* 1889, 22, 1434-9, 1439-46.
31. Weiss, R. F. *Deep-Sea Research* 1970, *17*, 721-35.
32. Klots, C. E.; Benson, B. B. *J. Chem. Phys.* 1963, *38*, 890-2.
33. Himmelblau, D. M. *J. Phys. Chem.* 1959, *63*, 1803-8.
34. Clever, H. L.; Han, C. H. "Thermodynamics of Aqueous Systems With Industrial Applications," Ed. S. A. Newman, ACS Symposium Series 133, Amer. Chem. Soc., Washington D.C. 1980, 513-36.
35. Schulze, G.; Prausnitz, J. M. *I & EC Fundam.* 1981, *20*, 175-7.

**COMPONENTS:**

(1) Nitrogen; $N_2$; [7727-37-9]

(2) Water; $H_2O$; [7732-18-5]

**ORIGINAL MEASUREMENTS:**

Winkler, L. W.

*Ber.* 1891, *24*, 3602-10.

**VARIABLES:**

T/K = 273-353

**PREPARED BY:**

R. Battino

**EXPERIMENTAL VALUES:**

| t/°C | $T^a$/K | $10^5 x_1$ [a,b] | $10^2 L$ [a,c] | $10^2 \alpha$ [d] / $cm^3$ (STP) $cm^{-3}$ $atm^{-1}$ |
|---|---|---|---|---|
| 0.08 | 273.23 | 1.884 | 2.344 | 2.3435 |
| 10.01 | 283.16 | 1.492 | 1.924 | 1.8564 |
| 20.01 | 293.16 | 1.242 | 1.655 | 1.5420 |
| 29.97 | 303.12 | 1.082 | 1.487 | 1.3400 |
| 40.00 | 313.15 | 0.958 | 1.356 | 1.1825 |
| 50.00 | 323.15 | 0.884 | 1.286 | 1.087 |
| 60.02 | 333.17 | 0.836 | 1.247 | 1.022 |
| 69.99 | 343.14 | 0.802 | 1.226 | 0.976 |
| 79.99 | 353.14 | 0.791 | 1.237 | 0.957 |

[a] Calculated by compiler.

[b] Mole fraction solubility at 101.325 kPa partial pressure of gas.

[c] Ostwald coefficient.

[d] Bunsen coefficient. Values reported in this table are averages of three to eight measurements at each temperature.

**AUXILIARY INFORMATION**

**METHOD/APPARATUS/PROCEDURE:**

Used Bunsen's absorptiometer and method.

**SOURCE AND PURITY OF MATERIALS:**

(1) From air by passing it through a heated to glowing copper tube.
(2) No details given.

**ESTIMATED ERROR:**

$\delta\alpha/\alpha = \pm 0.03$, compiler's estimate.

**REFERENCES:**

COMPONENTS:

(1) Nitrogen; $N_2$; [7727-37-9]

(2) Water; $H_2O$; [7732-18-5]

ORIGINAL MEASUREMENTS:

Hüfner, G.

*Z. Physik. Chem.* 1907, 57, 611-24.

VARIABLES:

T/K = 293

PREPARED BY:

R. Battino

EXPERIMENTAL VALUES:

| T/K | P/mmHg | P/kPa | $10^5 x_1$ [a] | $10^2 \alpha$ [b] |
|---|---|---|---|---|
| 293.38 | 695.5 | 92.73 | | 1.590 |
| 293.28 | 689.2 | 91.89 | | 1.552 |
| 293.30 | 648.6 | 86.47 | | 1.552 |
| 293.35 | 665.3 | 88.70 | | 1.565 |
| 293.33 | | | 1.260[c] | 1.565[c] |

[a] Mole fraction solubility (calculated by compiler) at 101.325 Pa partial pressure of gas.

[b] Bunsen coefficient.

[c] Average of four previous values.

AUXILIARY INFORMATION

METHOD/APPARATUS/PROCEDURE:

Apparatus described in reference 1. An absorption type using gas burets and an absorption flask. Solubilities also determined in several other solvents and solutions.

SOURCE AND PURITY OF MATERIALS:

(1) No details given.

(2) "Pure"

ESTIMATED ERROR:

$\delta\alpha/\alpha = \pm 0.02$, compiler's estimate

REFERENCES:

1. Hüfner, G. *Archiv. G. Anatomie und Physiologie, Physiolog. Abtlg.* 1894, S., 191.

| COMPONENTS: | ORIGINAL MEASUREMENTS: |
|---|---|
| (1) Nitrogen; $N_2$; [7727-37-9]<br>(2) Water; $H_2O$; [7732-18-5] | Fox, C. J. J.<br>*Trans. Faraday Soc.* 1909, 5, 68-87. |
| VARIABLES: | PREPARED BY: |
| T/K = 274-325 | R. Battino |

EXPERIMENTAL VALUES:

| $T^a$/K | Mole Fraction[b] $10^5 x_1$ | Ostwald Coeff.[c] $10^2 L$ | Bunsen Coeff.[d] $10^2 \alpha$ |
|---|---|---|---|
| 273.68 | 1.866 | 2.326 | 2.321 |
| 279.15 | 1.653 | 2.101 | 2.056 |
| 283.15 | 1.560 | 1.959 | 1.890 |
| 293.75 | 1.269 | 1.695 | 1.576 |
| 298.05 | 1.187 | 1.607 | 1.473 |
| 303.03 | 1.116 | 1.534 | 1.383 |
| 308.14 | 1.061 | 1.480 | 1.312 |
| 314.64 | 0.998 | 1.418 | 1.231 |
| 319.94 | 0.945 | 1.362 | 1.163 |
| 323.04 | 0.922 | 1.340 | 1.133 |
| 273.54[e] | 1.874 | 2.334 | 2.331 |
| 277.55 | 1.710 | 2.162 | 2.128 |
| 280.55 | 1.601 | 2.046 | 1.992 |
| 284.05 | 1.494 | 1.932 | 1.858 |
| 285.95 | 1.444 | 1.880 | 1.796 |
| 287.40 | 1.404 | 1.836 | 1.745 |

continued on following page

AUXILIARY INFORMATION

METHOD/APPARATUS/PROCEDURE:

A modified form of the Estreicher (1) apparatus was used. Diagram and procedure given in paper. The water is degassed by boiling using a double flask system. The quantity of water is determined by weighing. The dry gas is introduced from a gas buret, the flask shaken and a final reading of the volume of gas absorbed is made after there are no changes. A manometer is used to take readings at different pressures. Several readings at different temperatures are taken on one loading of the absorption flask.

SOURCE AND PURITY OF MATERIALS:

(1) Nitrogen - obtained by several time passing air to and fro over warm white phosphorus, soda lime, and $P_2O_5$.

(2) Water - "distilled water".

ESTIMATED ERROR:

$\delta\alpha/\alpha = \pm 0.015$, estimated by compiler.

REFERENCES:

(1) Estreicher, T. *Z. Physik. Chem.* 1899, *31*, 176.

| COMPONENTS: | ORIGINAL MEASUREMENTS: |
|---|---|
| (1) Nitrogen; $N_2$; [7727-37-9] | Fox, C. J. J. |
| (2) Water; $H_2O$; [7732-18-5] | *Trans. Faraday Soc.* 1909, 5, 68-87. |

EXPERIMENTAL VALUES:

continued

| $T^a$/K | Mole Fraction[b] $10^5 x_1$ | Ostwald Coeff.[c] $10^2 L$ | Bunsen Coeff.[d] $10^2 \alpha$ |
|---|---|---|---|
| 292.15 | 1.266 | 1.682 | 1.573 |
| 296.45 | 1.216 | 1.638 | 1.509 |
| 303.05 | 1.117 | 1.535 | 1.384 |
| 308.91 | 1.047 | 1.465 | 1.295 |
| 313.43 | 1.008 | 1.427 | 1.244 |
| 318.88 | 0.957 | 1.376 | 1.179 |
| 321.85 | 0.930 | 1.348 | 1.144 |
| 325.16 | 0.900 | 1.315 | 1.105 |

[a] Original temperature reported as °C to 0.01°C.

[b] Mole fraction solubility at 101325 Pa (1atm) partial pressure of gas calculated by compiler.

[c] Ostwald coefficients calculated by compiler.

[d] Bunsen coefficients reported in original paper.

[e] Second series of measurements.

[f] Smoothing equation of author given as $1000\alpha = 22.998 - 0.5298t + 0.009196t^2 - 0.00006779t^3$ with t in °C.

[g] Author also determined the solubility of nitrogen in seawater from 0 to 28°C and 0 to 20 per mille Cl.

| COMPONENTS: | ORIGINAL MEASUREMENTS: |
|---|---|
| (1) Nitrogen; $N_2$; [7727-37-9]<br>(2) Water; $H_2O$; [7732-18-5] | Bohr, C.<br>*Z. Physik. Chem.* 1910, *71*, 47-50. |
| VARIABLES:<br>T/K = 294 | PREPARED BY:<br>R. Battino |

EXPERIMENTAL VALUES:

| $T^a/K$ | $10^5x_1{}^b$ | $10^2L^c$ | $10^2\alpha^d$ |
|---|---|---|---|
| 294.35 | 1.256 | 1.681 | 1.560 |

[a] Temperature reported to 0.1°C.

[b] Mole fraction solubility at 101.325 kPa (1 atm) partial pressure of gas calculated by compiler.

[c] Ostwald coefficient calculated by compiler.

[d] Bunsen coefficient.

AUXILIARY INFORMATION

METHOD/APPARATUS/PROCEDURE:

The absorption measuring apparatus is described in reference (1). This paper also reports solubilities in sulfuric acid solutions.

SOURCE AND PURITY OF MATERIALS:

(1) Nitrogen - no comment by author.

(2) Water - no comment by author.

ESTIMATED ERROR:

$\delta\alpha/\alpha = \pm 0.03$, estimate by author.

REFERENCES:

(1) Bohr, C. Sauerstoffaufnahme des Blutfarbstoffes. Kopenhagen 1895. Jolin, Archiv F. (Anat. u.) Physiologie 1889, 265.

COMPONENTS:

(1) Nitrogen; $N_2$; [7727-37-9]

(2) Water; $H_2O$; [7732-18-5]

ORIGINAL MEASUREMENTS:

Müller, C.

*Z. Physik. Chem.* 1912, *81*, 483-503.

VARIABLES:

T/K = 289-290

PREPARED BY:

R. Battino

EXPERIMENTAL VALUES:

| $T^a$/K | Mole Fraction[b] $10^5 x_1$ | Ostwald Coeff.[c] $10^2 L$ | Bunsen Coeff.[d] $10^2 \alpha$ |
|---|---|---|---|
| 289.35 | 1.344 | 1.769 | 1.670 |
| 290.35 | 1.304 | 1.724 | 1.622 |

[a] Temperature reported to 0.1°C.

[b] Mole fraction solubility at 101.325 kPa (1 atm) partial pressure of gas calculated by compiler.

[c] Ostwald coefficient calculated by compiler.

[d] Bunsen coefficient.

AUXILIARY INFORMATION

METHOD/APPARATUS/PROCEDURE:

Water is degassed by pumping under vacuum. The absorption vessel is about 600 $cm^3$ and is separately thermostatted from the buret gas measuring system. Both are connected by a copper capillary tube. Gas uptake is read on the gas burets. The absorption flask volume and all liquid volumes were calibrated using mercury. The original paper contains details of operation and a drawing. Solubilities were also measured in aqueous solutions of sucrose, chloral hydrate, and glycerol.

SOURCE AND PURITY OF MATERIALS:

(1) Nitrogen - from heating potassium nitrite and ammonium nitrite. Washed with potassium hydroxide solution, ferrous sulfate solution, and sulfuric acid; and finally passed through a glowing copper tube.

(2) Water - no comment by author.

ESTIMATED ERROR

$\delta\alpha/\alpha = \pm 0.01$, estimate by compiler.

REFERENCES:

| COMPONENTS: | ORIGINAL MEASUREMENTS: |
|---|---|
| (1) Nitrogen; $N_2$; [7727-37-9]<br>(2) Water; $H_2O$; [7732-18-5] | Adeney, W. E.; Becker, H. G.<br>*Royal Dublin Soc. Scientific Pro.* 1919, *15*, 609-28. |
| VARIABLES: | PREPARED BY: |
| T/K = 277-308 | R. Battino |

EXPERIMENTAL VALUES:

| t/°C | $T^a$/K | $10^5 x_1$ [a,b] | $10^2 L$ [a,c] | $10^2 \alpha$ [d]/$cm^3$ (STP) $cm^{-3}$ $atm^{-1}$ |
|---|---|---|---|---|
| 3.5 | 276.7 | 1.771 | 2.232 | 2.203 |
| 11.2 | 284.4 | 1.463 | 1.895 | 1.820 |
| 15.0 | 288.2 | 1.369 | 1.795 | 1.701 |
| 20.1 | 293.3 | 1.247 | 1.663 | 1.549 |
| 24.8 | 296.0 | 1.173 | 1.588 | 1.456 |
| 30.4 | 303.6 | 1.067 | 1.469 | 1.322 |
| 35.1 | 308.3 | 0.986 | 1.377 | 1.220 |

[a] Calculated by compiler.

[b] Mole fraction solubility at 101.325kPa partial pressure of gas.

[c] Ostwald coefficient.

[d] Bunsen coefficient.

AUXILIARY INFORMATION

METHOD/APPARATUS/PROCEDURE:

A manometric system using water-jacketed burets was used.

SOURCE AND PURITY OF MATERIALS:

(1) Prepared from potassium bichromate, ammonium sulphate, and sodium nitrite.

(2) Distilled.

ESTIMATED ERROR:

$\delta\alpha/\alpha = \pm 0.01$, compiler's estimate
$\delta T/K = \pm 0.1$

REFERENCES:

**COMPONENTS:**

(1) Nitrogen; $N_2$; [7727-37-9]

(2) Water; $H_2O$; [7732-18-5]

**ORIGINAL MEASUREMENTS:**

Frolich, P.K.; Tauch, E.J.; Hogan, J.J.; Peer, A.A.

*Ind. Eng. Chem.* 1931, *23*, 548-50.

**VARIABLES:**

T/K = 298.15
P/MPa (absolute) = 0.1-19

**PREPARED BY:**

C-T. A. Chen.

**EXPERIMENTAL VALUES:**

| $P^a$/atm | $S^b$ |
|---|---|
| 20 (2.03 MPa) | 0.28 |
| 60 (6.08 MPa) | 0.83 |
| 100 (10.13 MPa) | 1.39 |
| 140 (14.18 MPa) | 1.95 |
| 190 (19.25 MPa) | 2.64 |

[a] Absolute pressure.

[b] Volume of nitrogen at 298.15 K and 1 atm per volume of water; estimated from figure 3 of the original paper.

AUXILIARY INFORMATION

**METHOD/APPARATUS/PROCEDURE:**

Water was introduced into a steel cylinder of 2 liters capacity. Nitrogen was then forced in under pressure and the cylinder agitated in a water bath maintained at 298.15 K. After the pressure had become constant, some water was withdrawn. The water and the nitrogen separating from it on release of pressure were collected over mercury in a buret and the two volumes were read directly. Corrections were made for the vapor pressure of the liquid and the atmospheric solubility of the gas. Pressures were measured with a Bourdon Gauge.

**SOURCE AND PURITY OF MATERIALS:**

No details given

**ESTIMATED ERROR:**

$\delta S/S = \pm 0.05$, compiler's estimate

**REFERENCES:**

**COMPONENTS:**

(1) Nitrogen; $N_2$; [7727-37-9]

(2) Water; $H_2O$; [7732-18-5]

**ORIGINAL MEASUREMENTS:**

Goodman, J. B.; Krase, N. W.

*Ind. Eng. Chem.* 1931, *23*, 401-4

**VARIABLES:**

T/K = 273-443
P/MPa (hydrostatic) = 10.13 - 30.39

**PREPARED BY:**

C-T. A. Chen

**EXPERIMENTAL VALUES:**

| $p^a$/atm | $S^b$ | | | | | | |
|---|---|---|---|---|---|---|---|
| | 273.15K[c] | 298.15K | 323.15K | 353.15K | 373.15K | 417.15K | 442.15K |
| 100(10.13MPa) | 1.46 | 1.07 | 1.003 | 0.934 | 0.954 | 1.025 | 1.08 |
| 125(12.66MPa) | 1.76 | 1.44 | 1.24 | 1.15 | 1.17 | 1.30 | 1.52 |
| 200(20.26MPa) | 3.19 | 2.76 | 2.49 | 2.27 | 2.25 | 2.68 | 3.29 |
| 300(30.39MPa) | 3.60 | 3.25 | 2.99 | 2.86 | 2.91 | 3.46 | 3.83 |

[a] Hydrostatic pressure.

[b] $cm^3$ of nitrogen at STP per gram of water.

[c] Calculated by compiler.

**AUXILIARY INFORMATION**

**METHOD/APPARATUS/PROCEDURE:**

Nitrogen was forced into a 61 cm long, 6.3 cm inside-diameter steel cylinder, which contained water saturated with nitrogen at room temperature and the desired pressure. The water was then transferred to a thermostated steel pipet where it was stirred at the desired temperature and pressure. After equilibrium, a sample was withdrawn into a weighed glass trap. The nitrogen released was then measured over mercury in a buret. The pressure was measured by a Bourdon gauge.

**SOURCE AND PURITY OF MATERIALS:**

Compressed nitrogen from unspecified commercial source was drawn over heated copper wire at 450°C. Water and carbon dioxide were removed by passing nitrogen through soda lime.

**ESTIMATED ERROR:**

**REFERENCES:**

COMPONENTS:

(1) Nitrogen; $N_2$; [7727-37-9]

(2) Water; $H_2O$; [7732-18-5]

ORIGINAL MEASUREMENTS:

Wiebe, R.; Gaddy, V. L.; Heins, C.Jr.

*J. Am. Chem. Soc.* 1933, *55*, 947-53.

VARIABLES:

T/K = 298-373
P/MPa (hydrostatic) = 0-101

PREPARED BY:

C-T. A. Chen

EXPERIMENTAL VALUES:

$cm^3$ $N_2$ at STP per gram of water

| $P^a$/atm | $T/K^{b,c}$ 298.15 | $T/K^b$ = 323.15 | $T/K^b$ = 348.15 | $T/K^b$ = 373.15 |
|---|---|---|---|---|
| 25 (2.530 MPa) | 0.348 | 0.273 | 0.254 | 0.266 |
| 50 (5.071 MPa) | 0.674 | 0.533 | 0.494 | 0.516 |
| 100 (10.13 MPa) | 1.264 | 1.011 | 0.946 | 0.986 |
| 200 (20.27 MPa) | 2.257 | 1.830 | 1.732 | 1.822 |
| 300 (30.40 MPa) | 3.061 | 2.534 | 2.413 | 2.546 |
| 500 (50.66 MPa) | 4.441 | 3.720 | 3.583 | 3.799 |
| 800 (81.06 MPa) | 6.134 | 5.221 | 5.062 | 5.365 |
| 1000 (101.3 MPa) | 7.15 | 6.123 | 5.934 | 6.256 |

[a] Hydrostatic pressure.

[b] Calculated by compiler.

[c] Data include some values previously published (Wiebe, et al., 1932)

AUXILIARY INFORMATION

METHOD/APPARATUS/PROCEDURE:

Two cylinders of 1000 cc and 300 cc capacity, respectively were filled with water to three-fourth capacity. The nitrogen was bubbled through the larger and the smaller cylinder successively for several hours before water samples were taken from the smaller cylinder for analysis. The larger cylinder was used as a reservoir. The experiments were approached from both under and supersaturation.

SOURCE AND PURITY OF MATERIALS:

(1) 99.9% pure. Impurities are argon and traces of oxygen.

ESTIMATED ERROR:

$\delta T/K = \pm 0.05$
$\delta S/S = \pm 0.005$, authors' estimate

REFERENCES:

1. Wiebe, R.; Gaddy, V.L.; Heins, C. Jr. *Ind. Eng. Chem.* 1932, *24*, 927.

**COMPONENTS:**

(1) Nitrogen; $N_2$; [7727-37-9]

(2) Water, $H_2O$; [7732-18-5]

**ORIGINAL MEASUREMENTS:**

Van Slyke, D. D.; Dillon, R. T.; Margaria, R.

*J. Biol. Chem.* 1934, *105*, 571-96.

**VARIABLES:**

T/K = 298-311

**PREPARED BY:**

R. Battino

**EXPERIMENTAL VALUES:**

| T/°C | $T^a$/K | $P^b_{N_2}$/mmHg | $P^a_{N_2}$/kPa | $10^2\alpha^c$/$cm^3$(STP) $cm^{-3}$ $atm^{-1}$ | $10^2\alpha°^d$/$cm^3$(STP) $g^{-1}$ $H_2O$ |
|---|---|---|---|---|---|
| 25 | 298.15 | 580 | 77.3 | 1.469 | 1.473 |
| 25 | 298.15 | 580 | 77.3 | 1.475 | 1.479 |
| 25 | 298.15 | 587 | 78.3 | 1.483 | 1.487 |
| 25 | 298.15 | 587 | 78.3 | 1.483 | 1.487 |
| 25 | 298.15 | 589 | 78.5 | 1.480 | 1.484 |
| 25 | 298.15 | 589 | 78.5 | 1.486 | 1.490 |
| 25 | 298.15 | 589 | 78.5 | 1.485 | 1.489 |
| 25 | 298.15 | 589 | 78.5 | 1.486 | 1.489 |
| 38 | 311.15 | 564.5 | 75.26 | 1.272 | 1.281 |

[a] Calculated by compiler.

[b] Partial pressure of nitrogen in air or gas mixture.

[c] Bunsen coefficient.

[d] Solubility in $cm^3$ (STP) of nitrogen per gram of water.

**AUXILIARY INFORMATION**

**METHOD/APPARATUS/PROCEDURE:**

The water was saturated with gas using the rotating double tonometer method of Austin, et al. (1). 2-2.5 h was allowed for saturation. The amount of dissolved nitrogen was determined using the method of Van Slyke and Neill (2).

**SOURCE AND PURITY OF MATERIALS:**

(1) From air.

(2) "Freshly distilled."

**ESTIMATED ERROR:**

$\delta T/K = \pm 0.04$
$\delta\alpha/\alpha = \pm 0.01$, compiler's estimate

**REFERENCES:**

1. Austin, J.H.; Cullen, G.E.; Hastings, A.B.; McLean, F.C.; Peters, J.D.; Van Slyke, D.D. *J. Biol. Chem.* 1922, *54*, 134.
2. Van Slyke, D.D.; Neil, J.M. *J. Biol. Chem.* 1924, *61*, 523.

**COMPONENTS:**

(1) Nitrogen; $N_2$; [7727-37-9]

(2) Water; $H_2O$; [7732-18-5]

**ORIGINAL MEASUREMENTS:**

Hawkins, J. A.; Shilling, C. W.

*J. Biol. Chem.* 1936, *113*, 273-8.

**VARIABLES:**

T/K = 311
P/MPa = 0.1-0.6

**PREPARED BY:**

R.Battino

**EXPERIMENTAL VALUES:**

| T/K | P/atm | P/MPa | $10^5 x_1$ [a] | $10^2 \alpha$ [b] |
|---|---|---|---|---|
| 311.15 | 0.939 | 0.0951 | | 1.281 |
| 311.15 | 0.945 | 0.0958 | | 1.259 |
| 311.15 | 1.955 | 0.1981 | | 1.299 |
| 311.15 | 3.921 | 0.3973 | | 1.261 |
| 311.15 | 5.958 | 0.6037 | | 1.246 |
| 311.15 | 6.042 | 0.6122 | | 1.270 |
| 311.15 | | | 1.028 [c] | 1.269 [c] |

[a] Mole fraction solubility (calculated by compiler) at 101.325 Pa partial pressure of gas.

[b] Bunsen coefficient. Each value is average of duplicate runs.

[c] Average value.

**AUXILIARY INFORMATION**

**METHOD/APPARATUS/PROCEDURE:**

Procedure followed is that of Austin, et al., (1). This is the Van Slyke approach. Also determined nitrogen solubility in ox and dog blood.

**SOURCE AND PURITY OF MATERIALS:**

No details given.

**ESTIMATED ERROR:**

$\delta\alpha/\alpha$ = ±0.015, compiler's estimate.

**REFERENCES:**

1. Austin, J.H.; Cullin, C.E.; Hastings, A.B.; MacLean, F.C.; Peters, J.P.; Van Slyke, D.D. *J. Biol. Chem.* 1922, *54*, 134.

| COMPONENTS: | ORIGINAL MEASUREMENTS: |
|---|---|
| (1) Nitrogen; $N_2$; [7727-37-9]<br>(2) Water; $H_2O$; [7732-18-5] | Morrison, T. J.; Billett, F.<br>*J. Chem. Soc.* 1952, 3819-22. |
| VARIABLES:<br>T/K = 286-346 | PREPARED BY:<br>R. Battino |

EXPERIMENTAL VALUES:

| T/K[a] | Mole Fraction[b] $10^5 x_1$ | Ostwald Coeff.[c] $10^2 L$ | $S_o$[d] |
|---|---|---|---|
| 285.65 | 1.477 | 1.921 | 18.37 |
| 288.65 | 1.385 | 1.820 | 17.22 |
| 293.45 | 1.280 | 1.707 | 15.89 |
| 298.65 | 1.190 | 1.614 | 14.76 |
| 303.15 | 1.112 | 1.528 | 13.77 |
| 308.75 | 1.030 | 1.440 | 12.74 |
| 314.75 | 0.9702 | 1.379 | 11.97 |
| 319.55 | 0.9364 | 1.349 | 11.53 |
| 324.15 | 0.9156 | 1.335 | 11.25 |
| 328.65 | 0.9029 | 1.332 | 11.07 |
| 336.95 | 0.8781 | 1.322 | 10.72 |
| 345.65 | 0.8677 | 1.334 | 10.54 |

[a] Temperature reported to 0.1°C.

[b] Mole fraction solubility at 101.325 Pa (1 atm) partial pressure of gas calculated by compiler.

[c] Ostwald coefficient calculated by compiler.

[d] $S_o$ is the solubility in $cm^3$ at standard temperature and pressure per 1000 g of water at a total gas pressure of 1 atm.

AUXILIARY INFORMATION

METHOD/APPARATUS/PROCEDURE:

Solvent is degassed by thorough boiling. It then flows in a thin film down a glass spiral and into a buret system containing the gas to be dissolved. Equilibrium is rapidly attained. Flow rates may be varied over a wide range without affecting the solubility. The solvent serves as its own indicator in the gas buret and volumes read on the burets are used to calculate the solubility The method is described in reference (1). The present paper describes an improved degassing section based on the vapor-pump principle.

SOURCE AND PURITY OF MATERIALS:

(1) Nitrogen - from a cylinder and from ammonium nitrite gave the same results.
(2) Water - no comment by authors.

ESTIMATED ERROR:

$\delta S_o/S_o$ = ±0.005, estimate of reproducibility by authors.

REFERENCES:

(1) Morrison, T. J.; Billett, F. *J. Chem. Soc.* 1948, 2033-5.

**COMPONENTS:**

(1) Nitrogen; $N_2$; [7727-37-9]

(2) Water; $H_2O$; [7732-18-5]

**ORIGINAL MEASUREMENTS:**

Suciu, S. N.; Zoss, L. M.; Sibbitt, W. L.

*Am. Soc. Mech. Eng. Paper No.* 53-A-64, Presented at the Annual Meeting, New York, 1953.

**VARIABLES:**

**PREPARED BY:**

H. L. Clever

**EXPERIMENTAL VALUES:**

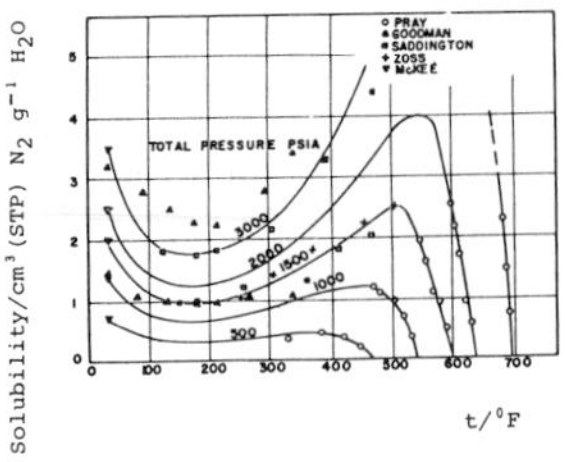


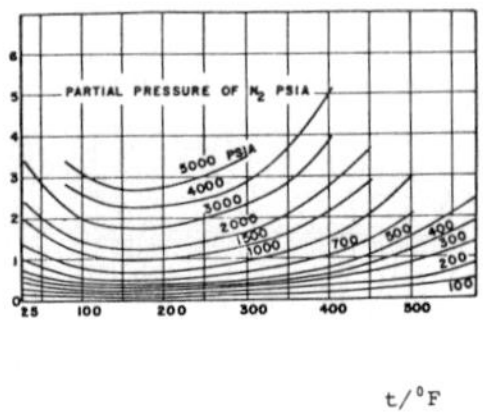


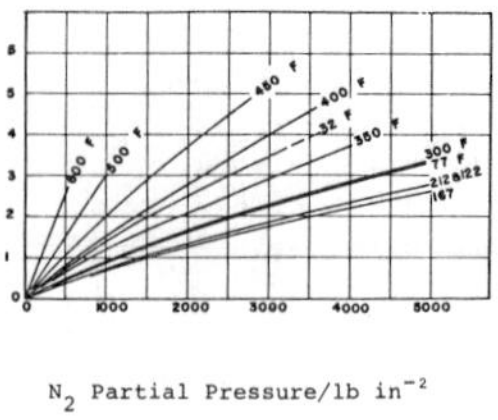


The authors prepared the figures above. They show the solubility of nitrogen in $cm^3$(STP) $g^{-1}$ $H_2O$ as a function of temperature at several total pressures (left); as a function of temperature at a dozen nitrogen partial pressures (middle), and as a function of nitrogen partial pressure at 11 different temperatures (right). They used data from:

Pray, H. A.; Schweickert, C. E.; Minnick, B. H. Battelle Memorial Institute Reprot BMI- T-25; *Ind. Eng. Chem.* 1952, *44*, 1146 - 1151.

Goodman, J. B.; Krase, N. W. *Ind. Eng. Chem.* 1931, *23*, 401.

Saddington, A. W.; Krase, N. W. *J. Am. Chem. Soc.* 1934, *56*, 353 - 361.

Zoss, L. M. Ph. D. thesis, Purdue University, 1952.

McKee, O. G. Ph. D. thesis, Purdue University, 1953.

Suciu, S. N. Ph. D. thesis, Purdue University, 1951.

AUXILIARY INFORMATION

**METHOD/APPARATUS/PROCEDURE:**

**SOURCE AND PURITY OF MATERIALS:**

**ESTIMATED ERROR:**

**REFERENCES:**

| COMPONENTS: | ORIGINAL MEASUREMENTS: |
|---|---|
| (1) Nitrogen; $N_2$; [7727-37-9]<br>(2) Water; $H_2O$; [7732-18-5] | Clever, H. L.; Battino, R.; Saylor, J. H.; Gross, D.M.<br>*J. Phys. Chem.* 1957, *61*, 1078-82. |
| VARIABLES: | PREPARED BY: |
| T/K = 298 | R. Battino |

EXPERIMENTAL VALUES:

| T/K | $10^5 x_1$ [a] | $10^2 \beta$ [b] |
|---|---|---|
| 298.15 | 1.185 | 1.47 |

[a] Mole fraction solubility at 101.325 Pa partial pressure of gas (calculated by compiler).

[b] Bunsen coefficient.

AUXILIARY INFORMATION

METHOD/APPARATUS/PROCEDURE:

A modification of the apparatus of Morrison and Billett (1) was used. The degassed liquid flows in a thin film through the gas down a 6 ft. absorption spiral. Gas uptake is measured on a gas buret where the solvent is the confining medium. See reference 1 for details.

SOURCE AND PURITY OF MATERIALS:

No details given.

ESTIMATED ERROR:

$\delta\alpha/\alpha = \pm 0.01$, compiler's estimate.

REFERENCES:

1. Morrison, T. J.; Billett, F. *J. Chem. Soc.* 1948, 2033; *ibid.*, 1952, 3819.

**COMPONENTS:**

(1) Nitrogen; $N_2$; [7727-37-9]

(2) Water; $H_2O$; [7732-18-5]

**ORIGINAL MEASUREMENTS:**

Farhi, L. E.; Edwards, A. W. T.; Homma, T.

*J. Appl. Physiol.* 1963, *18*, 97-106.

**VARIABLES:**

T/K = 276-310

**PREPARED BY:**

R. Battino

**EXPERIMENTAL VALUES:**

| t/°C | $T^a$/K | $10^5 x_1$ [a,b] | $10^2 L$ [a,c] | $10^2 \alpha$ [d] |
|---|---|---|---|---|
| 3.1 | 276.25 | 1.783 | 2.243 | 2.218 |
| 26.8 | 299.95 | 1.153 | 1.570 | 1.430 |
| 27.15 | 300.30 | 1.150 | 1.567 | 1.425 |
| 32.0 | 305.15 | 1.080 | 1.494 | 1.337 |
| 32.05 | 305.20 | 1.079 | 1.493 | 1.336 |
| 37.05 | 310.20 | 1.027 | 1.442 | 1.270 |

[a] Calculated by compiler.

[b] Mole fraction solubility at 101.325 kPa partial pressure of gas.

[c] Ostwald coefficient.

[d] Bunsen coefficient in units of $cm^3$ (STP) $cm^{-3}$ $atm^{-1}$.

AUXILIARY INFORMATION

**METHOD/APPARATUS/PROCEDURE:**

The Van Slyke method was used in conjunction with gas chromatography.

**SOURCE AND PURITY OF MATERIALS:**

(1) From air.

(2) Distilled.

**ESTIMATED ERROR:**

$\delta\alpha/\alpha = \pm 0.01$, compiler's estimate
$\delta T/K = \pm 0.05$.

**REFERENCES:**

| COMPONENTS: | ORIGINAL MEASUREMENTS: |
|---|---|
| (1) Nitrogen; $N_2$; [7727-37-9]<br>(2) Water; $H_2O$; [7732-18-5] | Klots, C. E.; Benson, B. B.<br>*J. Marine Res. (Sears Found. Marine Res.)* 1963, *21*, 48-57. |
| VARIABLES: | PREPARED BY: |
| T/K = 275-300 | R. Battino |

EXPERIMENTAL VALUES:

| $T^a$/K | Mole Fraction[b] $10^5x_1$ | Ostwald Coeff.[c] $10^2L$ | Bunsen Coeff.[d] $10^2\alpha$ | Henry's Law[e] $10^{-4}K$ |
|---|---|---|---|---|
| 275.01 | 1.846 | 2.313 | 2.297 | 5.417 |
| 275.33 | 1.833 | 2.299 | 2.281 | 5.453 |
| 275.36 | 1.829 | 2.293 | 2.275 | 5.467 |
| 275.86 | 1.809 | 2.273 | 2.251 | 5.528 |
| 276.51 | 1.776 | 2.237 | 2.210 | 5.627 |
| 276.90 | 1.753 | 2.210 | 2.181 | 5.704 |
| 276.98 | 1.749 | 2.207 | 2.176 | 5.718 |
| 277.96 | 1.706 | 2.160 | 2.123 | 5.861 |
| 278.40 | 1.686 | 2.138 | 2.098 | 5.929 |
| 279.04 | 1.671 | 2.124 | 2.079 | 5.984 |
| 279.07 | 1.664 | 2.115 | 2.070 | 6.010 |
| 280.03 | 1.635 | 2.085 | 2.034 | 6.114 |
| 280.81 | 1.603 | 2.050 | 1.994 | 6.238 |
| 282.08 | 1.562 | 2.007 | 1.943 | 6.400 |
| 282.38 | 1.552 | 1.996 | 1.931 | 6.441 |
| 283.25 | 1.519 | 1.959 | 1.889 | 6.583 |
| 284.18 | 1.492 | 1.930 | 1.855 | 6.704 |
| 285.40 | 1.462 | 1.900 | 1.818 | 6.840 |
| 285.92 | 1.445 | 1.881 | 1.797 | 6.918 |
| 287.07 | 1.415 | 1.849 | 1.759 | 7.067 |
| 288.08 | 1.389 | 1.821 | 1.727 | 7.199 |
| 289.08 | 1.367 | 1.798 | 1.699 | 7.315 |
| 289.99 | 1.344 | 1.773 | 1.670 | 7.440 |
| 290.97 | 1.318 | 1.745 | 1.638 | 7.587 |
| 292.64 | 1.287 | 1.712 | 1.598 | 7.776 |

continued on following page

AUXILIARY INFORMATION

| METHOD/APPARATUS/PROCEDURE: | SOURCE AND PURITY OF MATERIALS: |
|---|---|
| Two bulbs of ca. 300 $cm^3$ capacity are connected by two stopcocks. Degassed water is introduced into one bulb. The volumes of both are known to better than 0.05%. The stopcocks between the bulbs are opened, about one-half of the water is moved into the gas bulb and the apparatus agitated in a thermostat for up to 48 hours. The bulbs are placed upright and the saturated water and gas phases isolated. The undissolved gas is removed (via a Toepler pump) and its pressure, volume and temperature measured in the dry state to determine the number of moles of gas undissolved. The moles of dissolved gas are similarly determined. The constancy of the gas mass in a given experiment was about 0.05%. Solubilities are calculated from the known volumes and quantities with appropriate corrections for gas non-ideality. | (1) Nitrogen - "Tank" grade with about 0.1% each oxygen and argon.<br>(2) Water - "distilled"<br><br>ESTIMATED ERROR:<br>Precision $\delta\alpha/\alpha$ = ±0.002, Author's estimate.<br><br>REFERENCES: |

COMPONENTS:

(1) Nitrogen; $N_2$; [7727-37-9]

(2) Water; $H_2O$; [7732-18-5]

ORIGINAL MEASUREMENTS:

Klots, C. E.; Benson, B. B.

*J. Marine Res. (Sears Found. Marine Res.)* 1963, *21*, 48-57.

EXPERIMENTAL VALUES:

continued

| $T^a/K$ | Mole Fraction[b] $10^5 x_1$ | Ostwald Coeff.[c] $10^2 L$ | Bunsen Coeff.[d] $10^2 \alpha$ | Henry's Law[e] $10^{-4} K$ |
|---|---|---|---|---|
| 293.33 | 1.271 | 1.696 | 1.579 | 7.864 |
| 294.41 | 1.246 | 1.667 | 1.547 | 8.028 |
| 295.27 | 1.232 | 1.653 | 1.530 | 8.113 |
| 296.29 | 1.207 | 1.625 | 1.498 | 8.288 |
| 297.20 | 1.197 | 1.616 | 1.485 | 8.354 |
| 298.22 | 1.183 | 1.603 | 1.468 | 8.453 |
| 299.25 | 1.164 | 1.582 | 1.444 | 8.584 |
| 300.16 | 1.149 | 1.565 | 1.424 | 8.711 |

| $T^a/K$ | $\alpha(Ar)/\alpha(N_2)$[f] |
|---|---|
| 275.50 | 2.234 |
| 276.17 | 2.232 |
| 279.91 | 2.222 |
| 282.10 | 2.209 |
| 283.71 | 2.200 |
| 284.40 | 2.199 |
| 286.34 | 2.190 |
| 288.16 | 2.176 |
| 290.76 | 2.168 |
| 292.45 | 2.160 |
| 293.37 | 2.158 |
| 295.54 | 2.150 |
| 297.53 | 2.143 |
| 298.74 | 2.137 |
| 299.87 | 2.137 |

| $T^a/K$ | $\alpha(Ar)/\alpha(N_2)$[g] |
|---|---|
| 276.34 | 2.238 |
| 285.71 | 2.207 |
| 294.97 | 2.147 |
| 300.43 | 2.132 |

[a] Calculated by compiler. Temperatures reported to 0.01°C.

[b] Mole fraction solubility at 101,325 Pa (1 atm) partial pressure of gas calculated by compiler.

[c] Ostwald coefficient calculated by compiler.

[d] Bunsen coefficient - data originally reported as.

[e] Henry's Law constant - data originally reported as. Defined as $P_i = K_i x_i$ where $P_i$ is the fugacity of nitrogen in atmospheres, and $x_i$ is the mole fraction solubility.

[f] Ratio of argon solubility to nitrogen solubility expressed as Bunsen coefficients from original mixture of roughly 50:50 composition. Determined manometrically.

[g] Same as in (f), but determined mass spectrometrically. The precision in the solubility ratios is about 0.2%.

**COMPONENTS:**

(1) Nitrogen; $N_2$; [7727-37-9]

(2) Water; $H_2O$; [7732-18-5]

**ORIGINAL MEASUREMENTS:**

Douglas, E.

*J. Phys. Chem.* 1964, *68*, 169-74.

**VARIABLES:**

T/K = 276-303
P/kPa = 101.325

**PREPARED BY:**

R. Battino

**EXPERIMENTAL VALUES:**

| $T^a$/K | Mole Fraction[b] $10^5 x_1$ | Ostwald Coeff.[c] $10^2 L$ | Bunsen Coeff.[d] $10^2 \alpha$ | No.Pts.[e] | Range in[f] $10^2 \alpha$ |
|---|---|---|---|---|---|
| 276.25 | 1.762 | 2.217 | 2.192 | 3 | 2.187-2.197 |
| 280.98 | 1.575 | 2.015 | 1.959 | 10 | 1.951-1.966 |
| 284.83 | 1.460 | 1.894 | 1.816 | 5 | 1.812-1.817 |
| 293.10 | 1.254 | 1.672 | 1.558 | 10 | 1.550-1.565 |
| 302.65 | 1.093 | 1.500 | 1.354 | 10 | 1.348-1.363 |

[a] Temperature reported to 0.01°C.

[b] Mole fraction solubility at 101.325kPa (1 atm) partial pressure of gas calculated by compiler.

[c] Ostwald coefficient calculated by compiler.

[d] Bunsen coefficient.

[e] Number of replicate measurements at that temperature. Value reported as Bunsen coefficient is an average of these measurements.

[f] The range in the replicate measurements.

AUXILIARY INFORMATION

**METHOD/APPARATUS/PROCEDURE:**

The apparatus and procedures are based on the Scholander microgasometric apparatus (1). This version uses one arm with an expanded solvent reservoir. Degassed water is introduced into this reservoir and sealed over mercury. The water is then tipped into the absorption chamber. Gas is introduced and the entire apparatus is shaken in a water thermostat (controlled to 0.01°C). Gas and liquid volumes are read on a micrometer syringe by displacement of mercury.

**SOURCE AND PURITY OF MATERIALS:**

(1) Nitrogen - no comment by author.

(2) Water - double distilled.

**ESTIMATED ERROR:**

$\delta\alpha/\alpha$ = ±0.003, estimate by author.
$\delta T$/K = ±0.02, estimate by author.

**REFERENCES:**

(1) Scholander, P.F. J. Biol. Chem. 1947, 167, 235.

COMPONENTS:

(1) Nitrogen; $N_2$; [7727-37-9]

(2) Water; $H_2O$; [7732-18-5]

ORIGINAL MEASUREMENTS:

Enns, T.; Scholander, P. F.; Bradstreet, E. D.

*J. Phys. Chem.* 1965, *69*, 389-91.

VARIABLES:

T/K = 298
P/MPa (hydrostatic) = 0-10

PREPARED BY:

C-T.A. Chen.

EXPERIMENTAL VALUES:

| $P^a$/atm | $P^b$/MPa | $P^c$/mm Hg | $P^c$/mm Hg | $P^c$/mm Hg |
|---|---|---|---|---|
| | | $T^b$/K = 298.15 | | |
| 0 | 0 | 733 | 705 | 732 |
| 34 | 3.45 | 773 | 744 | 769 |
| 68 | 6.89 | 803 | 777 | 806.5 |
| 102 | 10.34 | 843 | 811 | 844 |

[a] Hydrostatic pressure.

[b] Calculated by compiler.

[c] The values in the table are the nitrogen equilibrium pressures in units of mm Hg. The table shows the effect of the hydrostatic pressure on the nitrogen equilibrium pressure.

AUXILIARY INFORMATION

METHOD/APPARATUS/PROCEDURE:

Water was gas-extracted before the experiment. Nitrogen at the desired initial pressure was dissolved in water at a constant temperature. The equilibrium vessel was shaken for 0.5 to 1 hr. After that, the water was transferred to a syringe which has a teflon tube attached to it. The pressure developed within the tube was then read on a mercury manometer.

SOURCE AND PURITY OF MATERIALS:

No details given.

ESTIMATED ERROR:

REFERENCES:

**COMPONENTS:**

(1) Nitrogen; $N_2$; [7727-37-9]

(2) Water; $H_2O$; [7732-18-5]

**ORIGINAL MEASUREMENTS:**

Murray, C. N.; Riley, J. P.; Wilson, T. R. S.

*Deep-Sea Research* 1969, *16*, 297-310.

**VARIABLES:**

T/K = 274-304
P/kPa = 101.325

**PREPARED BY:**

R. Battino

**EXPERIMENTAL VALUES:**

| $T^a$/K | Mole fraction[b] $10^5 x_1$ | Ostwald Coeff.[c] $10^2 L$ | Bunsen Coeff.[d] $10^2 \alpha$ |
|---|---|---|---|
| 273.73 | 1.875 | 2.337 | 2.332 |
| 275.92 | 1.782 | 2.199 | 2.217 |
| 278.10 | 1.686 | 2.136 | 2.098 |
| 280.02 | 1.615 | 2.060 | 2.009 |
| 282.30 | 1.538 | 1.977 | 1.913 |
| 284.23 | 1.477 | 1.912 | 1.837 |
| 288.03 | 1.375 | 1.802 | 1.709 |
| 293.33 | 1.254 | 1.672 | 1.557 |
| 297.52 | 1.175 | 1.588 | 1.458 |
| 303.86 | 1.075 | 1.482 | 1.332 |

[a] Temperature reported to 0.01°C.

[b] Mole fraction solubility at 101.325 kPa (1 atm) partial pressure of gas calculated by compiler.

[c] Ostwald coefficient calculated by compiler.

[d] Bunsen coefficient.

**AUXILIARY INFORMATION**

**METHOD/APPARATUS/PROCEDURE:**

The apparatus was a modification of the design by Ben-Naim and Baer (1). Degassed water is forced by centrifugal action up two outside tubes of the absorption vessel and down a central tube contacting the gas along the way. Gas uptake is read on a gas buret system. The whole apparatus is thermostatted to 0.012°C. The degassing apparatus is a modification of the design of Battino and Evans (2) utilizing a magnetically driven pump.

**SOURCE AND PURITY OF MATERIALS:**

(1) Nitrogen - British Oxygen Company white spot grade (99.9%). Maximum levels of impurity were 0.001% $O_2$ and 0.005% Ar.

(2) Water - distilled

**ESTIMATED ERROR:**

$\delta T = \pm 0.02$K, estimate by compiler.
$\delta\alpha/\alpha = \pm 0.0019$, estimate of authors

**REFERENCES:**

(1) Ben-Naim, A.; Baer, S. *Trans. Faraday Soc.* 1963, *59*, 2735-8.

(2) Battino, R.; Evans, F. D. *Anal. Chem.* 1966, *38*, 1627-9.

COMPONENTS:

(1) Nitrogen; $N_2$; [7727-37-9]

(2) Water; $H_2O$; [7732-18-5]

ORIGINAL MEASUREMENTS:

Power, G.G.; Stegall, H.

*J. Appl.Physiology*, 1970, *29*, 145-9.

VARIABLES:

T/K = 310

PREPARED BY:

C.L. Young.

EXPERIMENTAL VALUES:

| T/K | Bunsen coefficient, $\alpha/cm^3(STP)\ cm^{-3}atm^{-1}$ | S.D.* | No. of measurements |
|---|---|---|---|
| 310.15 | 0.01257 | 0.00009 | 4 |

* Standard deviation.

AUXILIARY INFORMATION

METHOD/APPARATUS/PROCEDURE:

5 to 12 $cm^3$ samples placed in a stirrer cell and gas, saturated with water vapor passed through water for 30-60 mins. Samples of saturated liquid withdrawn and transferred to Van Slyke apparatus. Dissolved gas removed under reduced pressure.

SOURCE AND PURITY OF MATERIALS:

(1) Matheson sample, purity better than 99.7 mole per cent.

ESTIMATED ERROR:

$\delta T/K = \pm 0.1$

REFERENCES:

**COMPONENTS:**

(1) Nitrogen; $N_2$; [7727-37-9]

(2) Water; $H_2O$; [7732-18-5]

**ORIGINAL MEASUREMENTS:**

Spink, T. J.

M.S. Thesis, Oregan State University, Corvallis, OR, 1971.

**VARIABLES:**

T/K = 285-299
P/MPa (hydrostatic) = 0-0.20

**PREPARED BY:**

C-T. A. Chen

**EXPERIMENTAL VALUES:**

| t/°C | $T^a$/K | $P^b$=0 (0 MPa) | | P=1 atm (0.1 MPa) | | P=2 atm (0.2 MPa) | |
|---|---|---|---|---|---|---|---|
| | | $10^6\ x_1$ | $10^3\beta^c$ | $10^6\ x_1$ | $10^3\beta^c$ | $10^6\ x_1$ | $10^3\beta^c$ |
| | | Mixed System | | | | | |
| 12.3 | 285.45 | 18.78 | 22.95 | 33.88 | 21.08 | 56.30 | 23.35 |
| 19.6 | 292.75 | 18.08 | 22.50 | 34.27 | 21.32 | 53.32 | 22.11 |
| 25.5 | 298.65 | 15.96 | 19.95 | 33.01 | 20.54 | 48.60 | 20.16 |
| | | Unmixed System | | | | | |
| 25.5 | 298.65 | | 15.73 | | | | |

a Calculated by compiler.

b Hydrostatic pressure, in atm.

c Bunsen coefficient.

AUXILIARY INFORMATION

**METHOD/APPARATUS/PROCEDURE:**

The saturation chamber was a six inch piece of Pyrex glass pipe, four inches in diameter. The nitrogen entered tangentially at the bottom and exitted from the top where a manometer was connected for pressure measurement. Water samples were drawn after five days and the dissolved nitrogen concentration measured by a Varian Aerograph 2100 chromatograph. A different experiment was also performed where the gas entered the chamber from the top of the chamber and no mixing was allowed.

**SOURCE AND PURITY OF MATERIALS:**

(1) Matheson pre-purified grade, >99.997% purity.

(2) Doubly distilled.

**ESTIMATED ERROR:**

$\delta x_1/x_1$ = ±0.2, compiler's estimate.

**REFERENCES:**

| COMPONENTS: | ORIGINAL MEASUREMENTS: |
|---|---|
| (1) Nitrogen; $N_2$; [7727-37-9]<br>(2) Water; $H_2O$; [7732-18-5] | Wilcock, R. J.; Battino, R.<br>*Nature* 1974, *252*, 614-5. |
| **VARIABLES:** | **PREPARED BY:** |
| T/K = 298 | R. Battino |

EXPERIMENTAL VALUES:

| T/K | $10^5x_1(N_2)$ [a] | $10^5x_1(O_2/N_2)$ [b] |
|---|---|---|
| 298.15 | 1.166 | 1.775 |
| 298.15 | 1.166 | 1.779 |
| 298.15 | | 1.757 |

[a] Mole fraction solubility at 101.325kPa partial pressure of gas.

[b] Mole fraction, but for a mixture of 49.5 mol per cent oxygen and 50.5 mol per cent nitrogen.

AUXILIARY INFORMATION

METHOD/APPARATUS/PROCEDURE:

Degassed water is flowed slowly through an absorption spiral containing the gas. As the gas dissolves the uptake is determined on burets. Measurements are made at constant pressure. Details are given in reference (1).

SOURCE AND PURITY OF MATERIALS:

(1) No details given.

(2) Distilled.

ESTIMATED ERROR:

$\delta x_1/x_1 = \pm 0.01$, compiler's estimate
$\delta T/K = \pm 0.05$, compiler's estimate.

REFERENCES:

(1) Battino, R.; Evans, F. D.; Danforth, W. F. *J. Am. Oil Chem. Soc.* 1968, *45*, 830.

**COMPONENTS:**

(1) Nitrogen; $N_2$; [7727-37-9]

(2) Water; $H_2O$; [7732-18-5]

**ORIGINAL MEASUREMENTS:**

Cosgrove, B. A.; Walkley, J.

*J. Chromatogr.* 1981, *216*, 161 - 7.

**VARIABLES:**

$T$/K = 278.15 - 313.15

$p_1$/kPa = 101.325

**PREPARED BY:**

H. L. Clever

**EXPERIMENTAL VALUES:**

| Temperature $T$/K | Mol Fraction $10^4 x_1$ |
|---|---|
| 278.15 | 0.1692 |
| 283.15 | 0.1513 |
| 288.15 | 0.1372 |
| 293.15 | 0.1275 |
| 298.15 | 0.1175 |
| 303.15 | 0.1116 |
| 308.15 | 0.1062 |
| 313.15 | 0.09986 |

**AUXILIARY INFORMATION**

**METHOD/APPARATUS/PROCEDURE:**

A 20 ml volume of degassed solvent (sublimation technique) is transferred to a previously evacuated ($10^{-4}$mmHg) saturation cell immersed in an insulated controlled (± 0.01 K) water bath. The gas is dispersed through the constantly stirred solution at 1 atm by a course, fritted glass disc. Saturation is obtained within a few hours. Prior to analysis the solution is allowed to sit under 1 atm gas pressure for 1 hour.

A saturated sample is withdrawn from the saturation cell using a greaseless, gas tight (2.500 ± 0.001 ml) Gilmont syringe. A 0.250 ml sample is injected to "wet" the frit. It is stripped and then four 0.500 ml samples are injected sequentially in to the cell. The stripped gas is dried before entering the column. The gas is analyzed on a dual filamnet conductivity detector. Calibrations with pure gas are made before and after each series of runs.

**SOURCE AND PURITY OF MATERIALS:**

(1) Nitrogen. No information.

(2) Water. No information.

**ESTIMATED ERROR:**

$\delta x_1/x_1$ = ± 0.015 (compiler)

**REFERENCES:**

**COMPONENTS:**

(1) Nitrogen; $N_2$; [7727-37-9]

(2) Water-$d_2$; $D_2O$; [7789-20-0]

**ORIGINAL MEASUREMENTS:**

Cosgrove, B. A.; Walkley, J.

*J. Chromatogr.* 1981, *216*, 161 - 7.

**VARIABLES:**

$T$/K = 278.15 - 313.15

$p_1$/kPa = 101.325

**PREPARED BY:**

H. L. Clever

**EXPERIMENTAL VALUES:**

| Temperature $T$/K | Mol Fraction $10^4x_1$ |
|---|---|
| 278.15 | 0.1862 |
| 283.15 | 0.1691 |
| 288.15 | 0.1561 |
| 293.15 | 0.1467 |
| 298.15 | 0.1335 |
| 303.15 | 0.1249 |
| 308.15 | 0.1170 |
| 313.15 | 0.1069 |

**AUXILIARY INFORMATION**

**METHOD/APPARATUS/PROCEDURE:**

A 20 ml volume of degassed solvent (sublimation technique) is transferred to a previously evacuated ($10^{-4}$ mmHg) saturation cell immersed in an insulated controlled water bath (± 0.01 K). The gas is dispersed through the constantly stirred solution at 1 atm by a course, fritted glass disc. Saturation is obtained within a few hours. Prior to analysis the solution is allowed to sit under 1 atm gas pressure for 1 hour.

A saturated sample is withdrawn from the saturation cell using a greaseless, gas tight (2.500 ± 0.001 ml) Gilmont syringe.A 0.250 ml sample is injected to "wet" the frit. It is stripped and then four 0.500 ml samples are injected sequentially into the cell. The stripped gas is dried before entering the column. The gas is analyzed on a dual filament conductivity detector. Calibrations with pure gas are made before and after each series of runs.

**SOURCE AND PURITY OF MATERIALS:**

(1) Nitrogen. No information

(2) Water-$d_2$. No information.

**ESTIMATED ERROR:**

$\delta x_1/x_1$ = ± 0.015 (compiler)

**REFERENCES:**

| COMPONENTS: | EVALUATOR: |
|---|---|
| (1) Nitrogen; $N_2$; [7727-37-9]<br>(2) Seawater | Chen-Tung A. Chen<br>School of Oceanography<br>Oregon State University<br>Corvallis, OR 97331<br>U.S.A.<br>February 1982 |

CRITICAL EVALUATION:

Nitrogen solubility measurements covering a wide range of salinity and temperature have been made by Törnoe (1), Dittmar (2), Hamberg (3), Fox (4), Rakestraw and Emmel (5), Benson and Parker (6), Douglas (7,8), and Murray, Riley and Wilson (9). The first three are only of historical interest because of the comparatively crude techniques used (9,10). The results of Fox (4) and Rakestraw and Emmel (5) are more precise but apparently suffer from systematic errors (9). Benson and Parker (6) did not measure the nitrogen solubility directly but only reported the ratio of solubility of nitrogen to argon or oxygen. Because of these limitations, these three are cited here but are excluded from further analysis.

The first accurate nitrogen solubility measurements on water and seawater were carried out by Douglas (7, 8) and later confirmed by Murray, Riley and Wilson (9). Since the data of these workers are similar in precision and have been shown to agree well (9,11) their results were combined to form the data base for this study. The following equation is used to represent the Bunsen solubility coefficient with a standard deviation of $5.4 \times 10^{-5}$ units (approximately 0.4%):

$$\ln C = 59.7745 - 76.7685/(T/100\ K) - 88.327 \ln (T/100\ K) + 19.5287\ (T/100\ K) + S\left[(7.1485 \times 10^{-3} - 3.9793 \times 10^{-2}/(T/100\ K)\right] \quad (1)$$

where T is the absolute temperature and S is the salinity in parts per thousand.

It should be pointed out that neither Douglas (7, 8) nor Murray, Riley and Wilson (9) corrected their data for the gas dissolution effect on the volume of the aqueous phase (11). Although their original experimental data have been compiled in this study, corrected values were used in the least squares fit. The corrections were made according to the following equation (Weiss, personal communication, 1979):

$$\Delta = \frac{273.15\Phi_v}{22404\ T} \quad (2)$$

In the present study, the partial molal volume of nitrogen ($\Phi_v$) in water is taken as 33.3 $cm^3$/mol (12), and the volume of 1 mole of nitrogen is taken as 22404 ml at STP.

For practical oceanographic purposes, it is convenient to use an equation for the nitrogen solubilities from water-saturated air at 1 atm total pressure in ml(STP)/l as a function of salinity and temperature (11, 13, 14). In order to take full advantage of the precision of the data, it is also desirable to express the solubility unit in terms of μmol/kg, since it is pressure and temperature independent (11, 13, 14). Consequently, the combined data sets have also been fit to the following equations (15, 16):

$$\ln C\ (ml/l) = -29.1410 + 53.3161/(T/100\ K) + 7.499 \ln (T/100\ K) + 1.8298\ (T/100\ K) + S\ (7.365 \times 10^{-3} - 4.038 \times 10^{-2}/(T/100\ K)) \pm 0.4\% \quad (3)$$

continued on following page

| COMPONENTS: | EVALUATOR: |
|---|---|
| (1) Nitrogen; $N_2$; [7727-37-9]<br>(2) Seawater | Chen-Tung A. Chen<br>School of Oceanography<br>Oregon State University<br>Corvallis, OR 97331<br>U.S.A.<br>February 1982 |

CRITICAL EVALUATION:

continued

$$\ln C\ (\mu mol/kg) = -29.2710 + 58.6753/(T/100\ K) + 10.3401 \ln (T/100\ K) + 1.5045\ (T/100\ K) + S\ (7.116 \times 10^{-3} - 4.186 \times 10^{-2}/(T/100\ K)) \pm 0.4\% \quad (4)$$

The vapor pressures for pure water and seawater have been calculated using the formulas given by Besley and Bottomley (17, for pure water) and by Robinson (18, for seawater). The densities have been calculated using the equation of Millero, Gonzalez and Ward (19).

Limited information is available on the pressure effect of nitrogen solubility in pure water (12, 20-24), but nothing is available for seawater. It is the evaluator's opinion that, within the precision of these measurements, the pure water data can be applied to seawater. Since no seawater data are available, the more recent pure water data are compiled for reference but should be used for seawater only with caution.

References


1. Tőrnoe, H. *The Norwegian North Atlantic Expedition* 1876-78, *V.24*, Grondahl Christiania, 1880.
2. Dittmar, W. "Report on the Scientific Results of the Exploring Voyage of H.M.S. Challenger. Physics and Chemistry", *V.1* H. M. Stationery Office. London, 1884.
3. Hamberg, A. *J. Prakt. Chem.* 1886, *33*, 433-41.
4. Fox, C. J. J. *Trans. Faraday Soc.*, 1909, *5*, 68-87.
5. Rakestraw, N. W.; Emmel, V. M. *J. Phys. Chem.* 1938, *42*, 1211-5.
6. Benson, B. B.; Parker, P. D. M. *J. Phys. Chem.* 1961, *65*, 1489-96.
7. Douglas, E. *J. Phys. Chem.* 1964, *68*, 169-74.
8. Douglas, E. *J. Phys. Chem.* 1965, *69*, 2608-10.
9. Murray, C. N.; Riley, J. P.; Wilson, T. R. S. *Deep-Sea Res.* 1969, *16*, 297-310.
10. Coste, J. H. *J. Phys. Chem.* 1927, *31*, 81-7.
11. Weiss, R. E. *Deep-Sea Res.* 1970, *17*, 721-35.
12. Enns, T.; Scholander, P. F.; Bradstreet, E. D. *J. Phys. Chem.* 1965, *69*, 389-91.
13. Craig, H.; Weiss, R. F. *Earth Planet. Sci. Lett.* 1968, *5*, 175-83.
14. Kester, D. in "Chemical Oceanography" V. 1, 2nd edition, J. P. Riley and G. Skirrow, eds., Academic Press, 1975.
15. Chen, C. T. in "Solubility Data Series. V. 4. Argon", H. L. Clever, ed., Pergamon Press, 1980.
16. Chen, C. T. in "Solubility Data Series V.7, Oxygen and Ozone", R. Battino, ed., Pergamon Press, 1981.


continued on following page

| COMPONENTS: | EVALUATOR: |
|---|---|
| (1) Nitrogen; $N_2$; [7727-37-9]<br>(2) Seawater | Chen-Tung A. Chen<br>School of Oceanography<br>Oregon State University<br>Corvallis, OR 97331<br>U.S.A.<br>February 1982 |

CRITICAL EVALUATION:

continued


17. Besley, L.; Bottomley, G. A. *J. Chem. Thermodyn.* 1973, *5*, 397-410.

18. Robinson, R. A. *J. Mar. Biol. Ass. U.K.* 1954, *33*,449-55.

19. Millero, F. J.; Gonzalez, A.; Ward, G. K. *J.Mar. Res.* 1976, *34*., 61-93.

20. Goodman, J. B.; Norman, W. K., *Ind. Eng. Chem*, 1931, *23* , 401-4.

21. Frolich, P. K.; Tauch, E. J.; Hogan, J. J.; Peer, A. A. *Ind. Eng. Chem.*, 1931, *23* , 548-50.

22. Wiebe, R.; Gaddy, V. L.; Heins, C. Jr., *Ind. Eng. Chem.*, 1932, *24*, 927.

23. Wiebe, R.; Gaddy, V. L.; Heins, C. Jr. *J. Am. Chem. Soc.* 1933, *55*, 947-53.

24. Spink, T. J. M. S. thesis, Oregon State Univ., 1971.

| COMPONENTS: | ORIGINAL MEASUREMENTS: |
|---|---|
| (1) Nitrogen; $N_2$; [7727-37-9]<br>(2) Seawater | Fox, C.J.J<br>*Trans. Faraday Soc.* 1909, *5*, 68-87. |
| **VARIABLES:** | **PREPARED BY:** |
| T/K = 273-301<br>Chlorinity/‰ = 0-20 | C-T.A. Chen |

EXPERIMENTAL VALUES:

| $T^a$/K | Chlorinity (‰) | $S^b$ (cc/1000cc) | $T^a$/K | Chlorinity (‰) | S (cc/1000cc) |
|---|---|---|---|---|---|
| | 0 | 18.64 | | 0 | 13.45 |
| 273.15 | 4 | 17.77 | | 4 | 12.94 |
| | 8 | 16.90 | 289.15 | 8 | 12.44 |
| | 12 | 16.03 | | 12 | 11.93 |
| | 16 | 15.18 | | 16 | 11.73 |
| | 20 | 14.31 | | 20 | 10.92 |
| | 0 | 17.02 | | 0 | 12.59 |
| | 4 | 16.27 | | 4 | 12.15 |
| 277.15 | 8 | 15.51 | 293.15 | 8 | 11.70 |
| | 12 | 14.75 | | 12 | 11.25 |
| | 16 | 14.00 | | 16 | 10.81 |
| | 20 | 13.24 | | 20 | 10.36 |
| | 0 | 15.63 | | 0 | 11.86 |
| | 4 | 14.98 | | 4 | 11.46 |
| | 8 | 14.32 | | 8 | 11.07 |
| 281.15 | 12 | 13.66 | 297.15 | 12 | 10.67 |
| | 16 | 13.00 | | 16 | 10.27 |
| | 20 | 12.34 | | 20 | 9.87 |

continued on following page

AUXILIARY INFORMATION

METHOD/APPARATUS/PROCEDURE:

A modified form of Estreicher's (1) method. The seawater was acidified before boiling and weighing. The dry air was introduced from a burette to a glass bulb containing the gas-free seawater. The bulb was subsequently shaken and the resulting decrease of air in the burette measured until there were no changes.

SOURCE AND PURITY OF MATERIALS:

Unspecified air and seawater

ESTIMATED ERROR:

$\delta S/S = \pm 0.004$, compiler's estimate

REFERENCES:

1. Estreicher, T.Z. *Physik. Chem.* 1899, *31*, 176

| COMPONENTS: | ORIGINAL MEASUREMENTS: |
|---|---|
| (1) Nitrogen; $N_2$; [7727-37-9] | Fox, C. J. J. |
| (2) Seawater | *Trans. Faraday Soc.* 1909, 5, 68-87. |

EXPERIMENTAL VALUES:

continued

| $T^a$/K | Chlorinity (‰) | $S^b$ (cc/1000cc) | $T^a$/K | Chlorinity (‰) | S (cc/1000cc) |
|---|---|---|---|---|---|
| | 0 | 14.45 | | 0 | 11.25 |
| | 4 | 13.88 | | 4 | 10.89 |
| | 8 | 13.30 | 301.15 | 8 | 10.52 |
| 285.15 | 12 | 12.72 | | 12 | 10.16 |
| | 16 | 12.15 | | 16 | 9.80 |
| | 20 | 11.57 | | 20 | 9.44 |

[a] Original temperature reported in °C to 0.01°C.

[b] Number of cc of nitrogen (containing Argon) absorbed by 1000cc of seawater from a free dry atmosphere at 760 mmHg total pressure.

**COMPONENTS:**

(1) Nitrogen; $N_2$; [7727-37-9]

(2) Seawater

**ORIGINAL MEASUREMENTS:**

Rakestraw, N. W.; Emmel, V. M.

*J. Phys. Chem.* 1938, *42*, 1211-5.

**VARIABLES:**

T/K = 275-298
Chlorinity/‰ = 16-20

**PREPARED BY:**

R. Battino

**EXPERIMENTAL VALUES:**

| t/°C | $T^a$/K | $10^2\alpha^b/cm^3(STP)cm^{-3}\ atm^{-1}$ | | |
|---|---|---|---|---|
| Chlorinity/‰ | | 16.03 | 18.00 | 19.99 |
| 2.11 | 275.26 | 1.422 | 1.381 | 1.350 |
| 7.08 | 280.23 | 1.268 | 1.240 | 1.208 |
| 12.15 | 285.30 | 1.153 | 1.127 | 1.095 |
| 18.03 | 291.18 | 1.033 | 1.005 | 0.989 |
| 25.05 | 298.20 | 0.919 | 0.900 | 0.881 |

[a] Calculated by compiler.

[b] Bunsen coefficient.

[c] Chlorinity in grams of chlorine per kg of seawater.

AUXILIARY INFORMATION

**METHOD/APPARATUS/PROCEDURE:**

Duplicate samples were saturated with air and then "analyzed for nitrogen" by a method unspecified.

**SOURCE AND PURITY OF MATERIALS:**

(1) From air.

(2) "Natura" seawater, diluted as needed.

**ESTIMATED ERROR:**

$\delta\alpha/\alpha = \pm 0.01$, compiler's estimate.

**REFERENCES:**

**COMPONENTS:**

(1) Nitrogen; $N_2$; [7727-37-9]

(2) Seawater
Nitrogen/Argon and
Nitrogen/Oxygen ratios.

**ORIGINAL MEASUREMENTS:**

Benson, B.B.; Parker, P.D.M.

*J. Phys. Chem.* 1961, *65*, 1489-96.

**VARIABLES:**

T/K = 273 - 303

Concentration

**PREPARED BY:**

R. Battino

**EXPERIMENTAL VALUES:**

| Salinity/‰ | t/°C | $T^a$/K | $N_2/Ar^b$ | $N_2/O_2{}^b$ |
|---|---|---|---|---|
| 0.066 | 0.0 | 273.2 | 37.92 | 1.836 |
| 0.066 | 5.0 | 278.2 | 38.29 | 1.825 |
| 0.48 | 10.0 | 283.2 | 38.42 | 1.826 |
| 0.066 | 15.0 | 288.2 | 38.66 | 1.859 |
| 0.084 | 20.0 | 293.2 | 39.03 | 1.887 |
| 0.11 | 25.0 | 298.2 | 39.43 | 1.941 |
| 0.102 | 30.0 | 303.2 | 39.71 | 1.987 |
| 32.59 | 0.0 | 273.2 | 37.78 | 1.833 |
| 32.565 | 5.0 | 278.2 | 37.78 | 1.803 |
| 32.60 | 10.0 | 283.2 | 38.00 | 1.830 |
| 32.65 | 15.0 | 288.2 | 37.51[c] | 1.821[c] |
| 32.58 | 20.0 | 293.2 | 38.47 | 1.907 |
| 32.65 | 25.0 | 298.2 | 38.78 | 1.934 |
| 32.97 | 30.0 | 303.2 | 39.25 | 1.954 |
| 34.70 | 0.0 | 273.2 | 37.52 | 1.845 |
| 34.70 | 5.0 | 278.2 | 37.64 | 1.816 |
| 34.71 | 10.0 | 283.2 | 38.05 | 1.828 |
| 34.67 | 15.0 | 288.2 | 37.96 | 1.833 |
| 34.695 | 20.0 | 293.2 | 38.28 | 1.858 |
| 34.65 | 25.0 | 298.2 | 38.89 | 1.947 |
| 35.11 | 30.0 | 303.2 | 39.23 | 1.926 |
| 36.12 | 0.0 | 273.2 | 37.74 | 1.860 |

continued on following page

**AUXILIARY INFORMATION**

**METHOD/APPARATUS/PROCEDURE:**

Four flasks containing distilled water and water of salinities ca. 32, 34, and 36‰ were thermostatted in contact with the atmosphere for over sixty hours. Samples were removed for salinity determinations. The dissolved gas was extracted from other samples and the Na/Ar and $Na/O_2$ ratios were determined on a mass spectrometer tuned to be sensitive to the ratios. Details of the procedure are given in the paper.

**SOURCE AND PURITY OF MATERIALS:**

(1) From air of compositions: 78.08% $N_2$; 20.94% $O_2$, 0.93% Ar, 0.032% $CO_2$, traces of rare gases.

(2) Distilled, sea water, or sea water diluted with distilled water.

**ESTIMATED ERROR:**

$\delta(N_2/Ar)/(N_2/Ar) = \pm 0.015$, authors' estimate.

$\delta T/K = \pm 0.15$

**REFERENCES:**

| COMPONENTS: | ORIGINAL MEASUREMENTS: |
|---|---|
| (1) Nitrogen; $N_2$; [7727-37-9]<br>(2) Seawater<br>Nitrogen/Argon and<br>Nitrogen/Oxygen ratios. | Benson, B.B.; Parker. P.D.M.<br>*J. Phys. Chem.* 1961, *65*, 1489-96. |

CRITICAL EVALUATION: continued

| Salinity/‰ | t/°C | $T^a$/K | $N_2/Ar^b$ | $N_2/O_2{}^b$ |
|---|---|---|---|---|
| 36.14 | 5.0 | 278.2 | 37.72 | 1.821 |
| 36.17 | 10.0 | 283.2 | 37.80 | 1.803 |
| 36.16 | 15.0 | 288.2 | 37.96 | 1.830 |
| 36.17 | 20.0 | 293.2 | 38.45 | 1.870 |
| 36.23 | 25.0 | 298.2 | 38.73 | 1.914 |
| 36.61 | 30.0 | 303.2 | 39.24 | 1.918 |

a Calculated by compiler.

b $N_2/Ar$ and $N_2/O_2$ are the ratios of the concentrations of each species found in sea water saturated with air. To obtain the ratio of the Bunsen coefficients, $\alpha$, use

$$\alpha_{N_2}/\alpha_{O_2} = (20.94/78.08)(N_2/O_2) \text{ and}$$

$$\alpha_{N_2}/\alpha_{Ar} = (0.93/78.08)(N_2/Ar).$$

c This point rejected by the authors.

d The authors give the following least squares smoothed equations:

(1) $N_2/O_2 = 1.800 + 0.00533\ (t/°C) \pm 1.3\%$ for pure water.

(2) $N_2/O_2 = 1.807 + 0.00396\ (t/°C) \pm 1.5\%$ for salinities of 34.5‰.

(3) $N_2/Ar = 37.90 + 0.0590\ (t/°C) \pm 0.2$ to 1.5% for pure water.

(4) $N_2/Ar = 37.48 + 0.0522\ (t/°C) \pm 0.46\%$ for salinities of 34.5‰.

| COMPONENTS: | ORIGINAL MEASUREMENTS: |
|---|---|
| (1) Nitrogen; $N_2$; [7727-37-9] | Douglas, E. |
| (2) Seawater | *J. Phys. Chem.* 1965, *69*, 2608-10. |

EXPERIMENTAL VALUES:

continued

| t/°C | $T^a$/K | $10^2\alpha^c/cm^3$(STP) $cm^{-3}$ $atm^{-1}$ | | |
|---|---|---|---|---|
| | | Cl‰ = 20.985 | | |
| 2.10 | 275.25 | 1.689 | 1.692 | 1.698 |
| 5.92 | 279.07 | 1.566 | 1.563 | 1.567 |
| 10.15 | 283.30 | 1.442 | 1.436 | 1.442 |
| 15.05 | 288.20 | 1.319 | 1.320 | 1.320 |
| 19.41 | 292.56 | 1.227 | 1.222 | 1.231 |
| 24.88 | 298.03 | 1.140 | 1.139 | 1.139 |
| 29.99 | 303.14 | 1.076 | 1.078 | 1.071 |

[a] Calculated by compiler.

[b] Chlorinity is in grams of chlorine per kilogram of seawater.

[c] Bunsen coefficient.

COMPONENTS:

(1) Nitrogen; $N_2$; [7727-37-9]

(2) Seawater

ORIGINAL MEASUREMENTS:

Douglas, E.

*J. Phys. Chem.* 1965, *69*, 2608-10.

VARIABLES:

T/K = 275-303
Chlorinity/‰ = 15-21

PREPARED BY:

R. Battino

EXPERIMENTAL VALUES:

| t/°C | $T^a$/K | $10^2\alpha^c$/cm$^3$(STP) cm$^{-3}$ atm$^{-1}$ | | |
|---|---|---|---|---|
| | | Cl ‰ = 15.376 | | |
| 1.50 | 274.65 | 1.857 | 1.858 | 1.852 |
| 6.46 | 279.61 | 1.661 | 1.667 | 1.670 |
| 10.00 | 283.15 | 1.544 | 1.550 | 1.551 |
| 14.81 | 287.96 | 1.416 | 1.422 | 1.417 |
| 19.99 | 293.14 | 1.299 | 1.303 | 1.304 |
| 25.08 | 298.23 | 1.209 | 1.210 | 1.203 |
| 29.83 | 302.98 | 1.137 | 1.144 | 1.139 |
| | | Cl ‰ = 18.604 | | |
| 2.17 | 275.32 | 1.746 | 1.748 | 1.746 |
| 6.80 | 279.95 | 1.588 | 1.589 | 1.592 |
| 10.25 | 283.40 | 1.485 | 1.476 | 1.485 |
| 14.51 | 287.66 | 1.373 | 1.367 | 1.370 |
| 19.41 | 292.56 | 1.264 | 1.269 | 1.268 |
| 25.27 | 298.42 | 1.161 | 1.168 | 1.165 |
| 29.72 | 302.87 | 1.109 | 1.111 | 1.107 |

continued on following page

AUXILIARY INFORMATION

METHOD/APPARATUS/PROCEDURE:

Used a modification of the Scholander microgasometric apparatus described in an earlier paper (1) of the author's. Gas-free seawater is equilibrated with the gas. All volumes are determined on a micrometer buret. The apparatus takes about 8 cm$^3$ of seawater.

SOURCE AND PURITY OF MATERIALS:

(1) Mass spectrometer analysis showed 0.04% $O_2$ and less than 0.03% Ar.

(2) Gas-free, millipore filtered, stored at 4°C.

ESTIMATED ERROR:

$\delta\alpha/\alpha$ = ±0.005, author's estimate
$\delta$Cl % to ±0.05‰, author's estimate
$\delta T$/K = ±0.01, compiler's estimate

REFERENCES:

(1) Douglas, E. *J. Phys. Chem.* 1964, *68*, 169-74.

COMPONENTS:

(1) Nitrogen; $N_2$; [7727-37-9]

(2) Seawater

ORIGINAL MEASUREMENTS:

Murray, C.N.; Riley, J.P.; Wilson, T.R.S.

*Deep-Sea Research* 1969, *16*, 297-310.

VARIABLES:

T/K = 274-304
Salinity/‰ = 0-40

PREPARED BY:

R. Battino

EXPERIMENTAL VALUES:

| Salinity /‰ | Chlorinity /‰ | $10^2\alpha^a$ | Salinity /‰ | Chlorinity /‰ | $10^2\alpha^a$ |
|---|---|---|---|---|---|
| 0.58°C | (273.73 K) | | 6.87°C | (280.02 K) | |
| 0.00 | 0.00 | 2.332 | 0.00 | 0.00 | 2.009 |
| 3.45 | 1.91 | 2.277 | 6.10 | 3.38 | 1.927 |
| 5.80 | 3.21 | 2.230 | 15.45 | 8.55 | 1.801 |
| 16.26 | 9.00 | 2.083 | 23.90 | 13.23 | 1.700 |
| 37.70 | 20.87 | 1.766 | 34.00 | 18.82 | 1.587 |
| 2.77°C | (275.92 K) | | 9.15°C | (282.30 K) | |
| 0.00 | 0.00 | 2.217 | 0.00 | 0.00 | 1.913 |
| 8.65 | 4.79 | 2.078 | 10.01 | 5.45 | 1.784 |
| 18.80 | 10.41 | 1.938 | 13.55 | 7.50 | 1.739 |
| 26.42 | 14.62 | 1.837 | 20.73 | 11.48 | 1.663 |
| 33.20 | 18.38 | 1.731 | 31.15 | 17.24 | 1.546 |
| 4.95°C | (278.10 K) | | 11.08°C | (284.23 K) | |
| 0.00 | 0.00 | 2.098 | 0.00 | 0.00 | 1.837 |
| 11.36 | 6.29 | 1.934 | 7.65 | 4.24 | 1.746 |
| 16.25 | 9.00 | 1.873 | 9.60 | 5.31 | 1.722 |
| 26.58 | 14.71 | 1.740 | 21.50 | 11.90 | 1.598 |
| 32.55 | 18.02 | 1.662 | 26.60 | 14.72 | 1.540 |
| | | | 34.25 | 18.96 | 1.456 |

continued on following page

AUXILIARY INFORMATION

METHOD/APPARATUS/PROCEDURE:

The Ben-Naim and Baer (1) apparatus was used. In this a vortex caused by high-speed stirring forces liquid up two side-arms to return via a central arm. Equilibrium is rapid. Gas uptake (in the wet state) is measured on a gas buret system. The paper contains drawings and a detailed description of the apparatus and procedure.

SOURCE AND PURITY OF MATERIALS:

(1) British Oxygen Company, "white spot" grade, 99.9%

(2) Surface seawater

ESTIMATED ERROR:

$\delta T/K = \pm 0.02$
$\delta S‰ = \pm 0.02$
$\delta\alpha/\alpha = \pm 0.002$

REFERENCES:

1. Ben-Naim, A.; Baer, S. *Trans. Faraday Soc.* 1963, *59*, 2735-8.

| COMPONENTS: | ORIGINAL MEASUREMENTS: |
|---|---|
| (1) Nitrogen; $N_2$; [7727-37-9] | Murray, C.N.; Riley, J.P.; Wilson, T.R.S. |
| (2) Seawater | *Deep-Sea Research* 1969, *16*, 297-310. |

EXPERIMENTAL VALUES:

continued

| Salinity /‰ | Chlorinity /‰ | $10^2\alpha$ [a] | Salinity /‰ | Chlorinity ‰ | $10^2\alpha$ [a] |
|---|---|---|---|---|---|
| 14.88°C (288.03 K) | | | 24.37°C (297.52 K) | | |
| 0.00 | 0.00 | 1.709 | 0.00 | 0.00 | 1.458 |
| 8.35 | 4.62 | 1.604 | 14.15 | 7.83 | 1.329 |
| 23.90 | 13.23 | 1.455 | 21.50 | 11.90 | 1.266 |
| 26.75 | 14.81 | 1.424 | 31.85 | 17.63 | 1.189 |
| 34.05 | 18.83 | 1.365 | 35.81 | 19.82 | 1.158 |
| 20.18°C (293.33 K) | | | 30.71°C (303.86 K) | | |
| 0.00 | 0.00 | 1.557 | 0.00 | 0.00 | 1.332 |
| 9.94 | 5.50 | 1.451 | 11.65 | 6.45 | 1.242 |
| 21.89 | 12.12 | 1.351 | 23.95 | 13.26 | 1.147 |
| 23.52 | 13.02 | 1.335 | 35.20 | 19.49 | 1.074 |
| 36.28 | 20.08 | 1.232 | 40.05 | 22.17 | 1.050 |

[a] Bunsen coefficient, $cm^3$ (STP) $cm^{-3}$ $atm^{-1}$.

The table below gives the authors' smoothed values of the Bunsen coefficient as a function of temperature and chlorinity.

| T/K | t/°C | Chlorinity(°/₀₀) 0 | 2 | 4 | 6 | 8 | 10 | 12 | 14 | 16 | 18 | 20 |
|---|---|---|---|---|---|---|---|---|---|---|---|---|
| 271.15 | -1 | --- | --- | --- | --- | --- | --- | --- | --- | --- | --- | 19.00 |
| 272.15 | -2 | --- | --- | --- | --- | --- | --- | 20.75 | 20.20 | 19.64 | 19.09 | 18.54 |
| 273.15 | 0 | 23.69 | 23.05 | 22.48 | 21.92 | 21.36 | 20.82 | 20.27 | 19.73 | 19.18 | 18.64 | 18.10 |
| 274.15 | 1 | 23.10 | 22.47 | 21.91 | 21.37 | 20.83 | 20.31 | 19.78 | 19.25 | 18.73 | 18.21 | 17.68 |
| 275.15 | 2 | 22.54 | 21.91 | 21.37 | 20.85 | 20.33 | 19.82 | 19.31 | 18.80 | 18.29 | 17.79 | 17.29 |
| 276.15 | 3 | 22.00 | 21.38 | 20.86 | 20.35 | 19.85 | 19.35 | 18.86 | 18.37 | 17.88 | 17.39 | 16.91 |
| 277.15 | 4 | 21.48 | 20.88 | 20.37 | 19.87 | 19.38 | 18.91 | 18.43 | 17.95 | 17.48 | 17.01 | 16.54 |
| 278.15 | 5 | 20.99 | 20.39 | 19.90 | 19.42 | 18.94 | 18.48 | 18.02 | 17.56 | 17.10 | 16.65 | 16.19 |
| 279.15 | 6 | 20.52 | 19.93 | 19.45 | 18.98 | 18.52 | 18.07 | 17.62 | 17.18 | 16.73 | 16.30 | 15.86 |
| 280.15 | 7 | 20.06 | 19.49 | 19.02 | 18.56 | 18.12 | 17.68 | 17.25 | 16.82 | 16.39 | 15.96 | 15.53 |
| 281.15 | 8 | 19.63 | 19.06 | 18.60 | 18.16 | 17.73 | 17.31 | 16.89 | 16.47 | 16.05 | 15.64 | 15.22 |
| 282.15 | 9 | 19.22 | 18.66 | 18.21 | 17.78 | 17.36 | 16.95 | 16.54 | 16.13 | 15.73 | 15.33 | 14.93 |
| 283.15 | 10 | 18.82 | 18.27 | 17.83 | 17.42 | 17.01 | 16.61 | 16.21 | 15.81 | 15.42 | 15.03 | 14.64 |
| 284.15 | 11 | 18.44 | 17.90 | 17.47 | 17.07 | 16.67 | 16.28 | 15.89 | 15.51 | 15.13 | 14.75 | 14.37 |
| 285.15 | 12 | 18.07 | 17.54 | 17.13 | 16.73 | 16.34 | 15.96 | 15.59 | 15.21 | 14.84 | 14.48 | 14.11 |
| 286.15 | 13 | 17.72 | 17.20 | 16.79 | 16.41 | 16.03 | 15.66 | 15.30 | 14.93 | 14.57 | 14.22 | 13.86 |
| 287.15 | 14 | 17.38 | 16.87 | 16.48 | 16.10 | 15.73 | 15.37 | 15.02 | 14.66 | 14.31 | 13.96 | 13.62 |
| 288.15 | 15 | 17.06 | 16.56 | 16.17 | 15.80 | 15.44 | 15.09 | 14.75 | 14.40 | 14.06 | 13.72 | 13.38 |
| 289.15 | 16 | 16.75 | 16.26 | 15.88 | 15.52 | 15.17 | 14.83 | 14.49 | 14.15 | 13.82 | 13.49 | 13.16 |
| 290.15 | 17 | 16.45 | 15.97 | 15.60 | 15.25 | 14.91 | 14.57 | 14.24 | 13.91 | 13.59 | 13.27 | 12.94 |
| 291.15 | 18 | 16.17 | 15.69 | 15.33 | 14.99 | 14.65 | 14.33 | 14.00 | 13.68 | 13.37 | 13.05 | 12.74 |
| 292.15 | 19 | 15.89 | 15.43 | 15.07 | 14.74 | 14.41 | 14.09 | 13.78 | 13.46 | 13.15 | 12.85 | 12.54 |
| 293.15 | 20 | 15.63 | 15.17 | 14.82 | 14.50 | 14.18 | 13.87 | 13.56 | 13.25 | 12.95 | 12.65 | 12.35 |
| 294.15 | 21 | 15.37 | 14.93 | 14.59 | 14.27 | 13.95 | 13.65 | 13.35 | 13.05 | 12.75 | 12.46 | 12.16 |
| 295.15 | 22 | 15.13 | 14.69 | 14.36 | 14.05 | 13.74 | 13.44 | 13.14 | 12.85 | 12.56 | 12.27 | 11.99 |

continued on following page

| COMPONENTS: | ORIGINAL MEASUREMENTS: |
|---|---|
| (1) Nitrogen; $N_2$; [7727-37-9]<br>(2) Seawater | Murray, C.N.; Riley, J.P.; Wilson, T.R.S.<br>*Deep-Sea Research* 1969, *16*, 297-310. |

EXPERIMENTAL VALUES: continued

| T/K | t/°C | Chlorinity (°/₀₀) 0 | 2 | 4 | 6 | 8 | 10 | 12 | 14 | 16 | 18 | 20 |
|---|---|---|---|---|---|---|---|---|---|---|---|---|
| 296.15 | 23 | 14.90 | 14.47 | 14.14 | 13.83 | 13.53 | 13.24 | 12.95 | 12.66 | 12.38 | 12.10 | 11.82 |
| 297.15 | 24 | 14.67 | 14.25 | 13.93 | 13.63 | 13.33 | 13.05 | 12.76 | 12.48 | 12.20 | 11.93 | 11.65 |
| 298.15 | 25 | 14.46 | 14.04 | 13.73 | 13.43 | 13.14 | 12.86 | 12.58 | 12.31 | 12.04 | 11.77 | 11.50 |
| 299.15 | 26 | 14.25 | 13.84 | 13.54 | 13.25 | 12.96 | 12.69 | 12.41 | 12.14 | 11.87 | 11.61 | 11.34 |
| 300.15 | 27 | 14.05 | 13.65 | 13.35 | 13.06 | 12.79 | 12.52 | 12.25 | 11.98 | 11.72 | 11.46 | 11.20 |
| 301.15 | 28 | 13.85 | 13.47 | 13.17 | 12.89 | 12.62 | 12.35 | 12.09 | 11.83 | 11.57 | 11.31 | 11.06 |
| 302.15 | 29 | 13.67 | 13.29 | 13.00 | 12.72 | 12.46 | 12.19 | 11.94 | 11.68 | 11.42 | 11.17 | 10.92 |
| 303.15 | 30 | 13.49 | 13.12 | 12.83 | 12.56 | 12.30 | 12.04 | 11.79 | 11.54 | 11.29 | 11.04 | 10.79 |
| 304.15 | 31 | 13.32 | 12.96 | 12.68 | 12.41 | 12.15 | 11.90 | 11.65 | 11.40 | 11.15 | 10.91 | 10.67 |
| 305.15 | 32 | 13.16 | 12.80 | 12.52 | 12.26 | 12.01 | 11.76 | 11.51 | 11.27 | 11.03 | 10.79 | 10.55 |
| 306.15 | 33 | 13.00 | 12.65 | 12.38 | 12.12 | 11.87 | 11.63 | 11.38 | 11.14 | 10.90 | 10.67 | 10.43 |
| 307.15 | 34 | 12.85 | 12.50 | 12.24 | 11.99 | 11.74 | 11.50 | 11.26 | 11.02 | 10.78 | 10.55 | 10.32 |

The following table gives the authors' smoothed values of nitrogen solubility in units of $cm^3$ (STP)/$Lm^3$ from an atmosphere of 78.08% $N_2$ and 100% relative humidity.

| T/K | t/°C | Chlorinity (°/₀₀) 0 | 2 | 4 | 6 | 8 | 10 | 12 | 14 | 16 | 18 | 20 |
|---|---|---|---|---|---|---|---|---|---|---|---|---|
| 271.15 | -2 | --- | --- | --- | --- | --- | --- | --- | --- | --- | --- | 14.74 |
| 272.15 | -1 | --- | --- | --- | --- | --- | --- | 16.13 | 15.69 | 15.26 | 14.82 | 14.39 |
| 273.15 | 0 | 18.39 | 17.89 | 17.45 | 17.01 | 16.58 | 16.16 | 15.73 | 15.31 | 14.89 | 14.47 | 14.05 |
| 274.15 | 1 | 17.92 | 17.43 | 17.00 | 16.58 | 16.16 | 15.76 | 15.35 | 14.93 | 14.53 | 14.13 | 13.72 |
| 275.15 | 2 | 17.48 | 16.99 | 16.57 | 16.17 | 15.76 | 15.37 | 14.97 | 14.58 | 14.18 | 13.79 | 13.41 |
| 276.15 | 3 | 17.05 | 16.57 | 16.17 | 15.77 | 15.38 | 15.00 | 14.62 | 14.24 | 13.86 | 13.48 | 13.11 |
| 277.15 | 4 | 16.64 | 16.17 | 15.78 | 15.39 | 15.01 | 14.65 | 14.28 | 13.90 | 13.54 | 13.18 | 12.81 |
| 278.15 | 5 | 16.25 | 15.78 | 15.41 | 15.03 | 14.66 | 14.31 | 13.95 | 13.59 | 13.24 | 12.89 | 12.53 |
| 279.15 | 6 | 15.88 | 15.42 | 15.05 | 14.68 | 14.33 | 13.98 | 13.63 | 13.29 | 12.94 | 12.61 | 12.27 |
| 280.15 | 7 | 15.51 | 15.07 | 14.71 | 14.35 | 14.01 | 13.67 | 13.34 | 13.00 | 12.67 | 12.34 | 12.01 |
| 281.15 | 8 | 15.17 | 14.73 | 14.37 | 14.03 | 13.70 | 13.37 | 13.05 | 12.72 | 12.40 | 12.08 | 11.76 |
| 282.15 | 9 | 14.84 | 14.41 | 14.06 | 13.73 | 13.40 | 13.09 | 12.77 | 12.45 | 12.14 | 11.84 | 11.53 |
| 283.15 | 10 | 14.52 | 14.09 | 13.75 | 13.44 | 13.12 | 12.81 | 12.50 | 12.20 | 11.90 | 11.59 | 11.29 |
| 284.15 | 11 | 14.21 | 13.80 | 13.47 | 13.16 | 12.85 | 12.55 | 12.25 | 11.95 | 11.66 | 11.37 | 11.08 |
| 285.15 | 12 | 13.91 | 13.51 | 13.19 | 12.88 | 12.58 | 12.29 | 12.01 | 11.71 | 11.43 | 11.15 | 10.87 |
| 286.15 | 13 | 13.63 | 13.23 | 12.92 | 12.62 | 12.33 | 12.05 | 11.77 | 11.49 | 11.21 | 10.94 | 10.66 |
| 287.15 | 14 | 13.36 | 12.97 | 12.67 | 12.37 | 12.09 | 11.81 | 11.54 | 11.27 | 11.00 | 10.73 | 10.47 |
| 288.15 | 15 | 13.10 | 12.71 | 12.41 | 12.13 | 11.85 | 11.59 | 11.32 | 11.06 | 10.79 | 10.53 | 10.27 |
| 289.15 | 16 | 12.84 | 12.47 | 12.18 | 11.90 | 11.63 | 11.37 | 11.11 | 10.85 | 10.60 | 10.35 | 10.09 |
| 290.15 | 17 | 12.60 | 12.23 | 11.95 | 11.68 | 11.42 | 11.16 | 10.91 | 10.65 | 10.41 | 10.16 | 9.91 |
| 291.15 | 18 | 12.37 | 12.00 | 11.73 | 11.47 | 11.21 | 10.96 | 10.71 | 10.46 | 10.23 | 9.98 | 9.75 |
| 292.15 | 19 | 12.14 | 11.79 | 11.51 | 11.26 | 11.01 | 10.76 | 10.53 | 10.28 | 10.05 | 9.82 | 9.58 |
| 293.15 | 20 | 11.92 | 11.57 | 11.31 | 11.06 | 10.82 | 10.58 | 10.34 | 10.11 | 9.88 | 9.65 | 9.42 |
| 294.15 | 21 | 11.71 | 11.37 | 11.11 | 10.87 | 10.63 | 10.40 | 10.17 | 9.94 | 9.71 | 9.49 | 9.26 |
| 295.15 | 22 | 11.51 | 11.17 | 10.92 | 10.69 | 10.45 | 10.22 | 9.99 | 9.77 | 9.55 | 9.33 | 9.12 |
| 296.15 | 23 | 11.31 | 10.99 | 10.74 | 10.50 | 10.27 | 10.05 | 9.83 | 9.61 | 9.40 | 9.19 | 8.97 |
| 297.15 | 24 | 11.12 | 10.80 | 10.56 | 10.33 | 10.10 | 9.90 | 9.67 | 9.46 | 9.25 | 9.04 | 8.83 |
| 298.15 | 25 | 10.94 | 10.62 | 10.39 | 10.16 | 9.94 | 9.73 | 9.52 | 9.31 | 9.11 | 8.90 | 8.70 |
| 299.15 | 26 | 10.76 | 10.45 | 10.22 | 10.00 | 9.78 | 9.58 | 9.37 | 9.17 | 8.96 | 8.77 | 8.56 |
| 300.15 | 27 | 10.59 | 10.28 | 10.06 | 9.84 | 9.64 | 9.43 | 9.23 | 9.03 | 8.83 | 8.63 | 8.44 |

continued on following page

| COMPONENTS: | ORIGINAL MEASUREMENTS: |
|---|---|
| (1) Nitrogen; $N_2$; [7727-37-9] | Murray, C.N.; Riley, J.P.; Wilson, T.R.S. |
| (2) Seawater | *Deep-Sea Research* 1969, *16*, 297-310. |

EXPERIMENTAL VALUES:

continued

| | | Chlorinity (‰) | | | | | | | | | | |
|---|---|---|---|---|---|---|---|---|---|---|---|---|
| T/K | t/°C | 0 | 2 | 4 | 6 | 8 | 10 | 12 | 14 | 16 | 18 | 20 |
| 301.15 | 28 | 10.41 | 10.13 | 9.90 | 9.69 | 9.49 | 9.28 | 9.09 | 8.89 | 8.70 | 8.50 | 8.31 |
| 302.15 | 29 | 10.25 | 9.97 | 9.75 | 9.54 | 9.35 | 9.14 | 8.96 | 8.76 | 8.57 | 8.38 | 8.19 |
| 303.15 | 30 | 10.09 | 9.82 | 9.60 | 9.40 | 9.20 | 9.01 | 8.82 | 8.63 | 8.45 | 8.26 | 8.07 |
| 304.15 | 31 | 9.94 | 9.67 | 9.46 | 9.26 | 9.07 | 8.88 | 8.69 | 8.51 | 8.32 | 8.14 | 7.96 |
| 305.15 | 32 | 9.79 | 9.53 | 9.32 | 9.12 | 8.94 | 8.75 | 8.57 | 8.39 | 8.21 | 8.03 | 7.85 |
| 306.15 | 33 | 9.65 | 9.39 | 9.19 | 8.99 | 8.81 | 8.63 | 8.45 | 8.27 | 8.09 | 7.92 | 7.74 |
| 307.15 | 34 | 9.51 | 9.25 | 9.06 | 8.87 | 8.69 | 8.51 | 8.33 | 8.15 | 7.98 | 7.81 | 7.64 |

| COMPONENTS: | EVALUATOR: |
|---|---|
| (1) Nitrogen; $N_2$; [7727-37-9]<br>(2) Electrolyte<br>(3) Water; $H_2O$; [7732-18-5] | H. Lawrence Clever<br>Department of Chemistry<br>Emory University<br>Atlanta, GA 30322 USA<br>1981, December |

CRITICAL EVALUATION:

An Evaluation of the Solubility of Nitrogen in Aqueous Electrolyte Solutions.

Not enough workers have measured the solubility of nitrogen in any one aqueous electrolyte system over common ranges of concentration and temperature to recommend solubility values. Most of the available data are classed as tentative.

In order to have a common basis for comparison, where possible the solubility data have been converted to Sechenov (Setschenow) salt effect parameters at a nitrogen partial pressure of 101.325 kPa in the form

$$k_{sc\alpha}/\mathrm{dm^3\ mol^{-1}} = (1/(c_2/\mathrm{mol\ dm^{-3}}))\log(\alpha°/\alpha)$$

where $c_2$ is the electrolyte concentration in mol $dm^{-3}$, and $\alpha°$ and $\alpha$ are the Bunsen coefficients in pure water and electrolyte solution, respectively. The Ostwald coefficient ratio, $L°/L$, will give the same value, but the salt effect parameter is symbolized, $k_{scL}$. Both ratios are equivalent to a molar gas solubility ratio, $c_1°/c_1$, thus

$$k_{scc} = k_{sc\alpha} = k_{scL} = (1/(c_2/\mathrm{mol\ dm^{-3}}))\ \log\ (c°_{1,sat}/c_{1,sat})$$

Other forms of the salt effect parameter will be found on the data sheets that follow this discussion. They include

$$k_{sms}/\mathrm{kg\ mol^{-1}} = (1/(m_2/\mathrm{mol\ kg^{-1}}))\log(s°/s)$$

$$k_{scx}/\mathrm{dm^3\ mol^{-1}} = (1/(c_2/\mathrm{mol\ dm^{-3}}))\log(x°/x)$$

$$k_{smx}/\mathrm{kg\ mol^{-1}} = (1/(m_2/\mathrm{mol\ kg^{-1}}))\log(x°/x)$$

where $m_2$ is the electrolyte molality, $s°/s$ is the Kuenen coefficient ratio, and $x°/x$ is the mole fraction gas solubility ratio usually calculated with respect to all ions in the solution. The Kuenen coefficient, $s$, is referenced to 1 g of water whether it is for pure water or the electrolyte solution. Thus the $s°/s$ ratio is equal to a gas molality ratio $m_1°/m_1$, and $k_{smm} = k_{sms}$. The density data required to convert the salt effect parameter from one form to another were taken from the *International Critical Tables*, McGraw-Hill Co., Inc., 1928, Vol. III.

The various Sechenov constants can be interconverted by use of the relationships

$$k_{scc} = \frac{m_2}{c_2}\ k_{smm} + \frac{1}{c_2}\ \log\left[\left(\frac{\rho°}{\rho}\right)\left(\frac{1000 + m_2M_2}{1000}\right)\right] \qquad (1)$$

$$k_{scx} = \frac{m_2}{c_2}\ k_{smx} = \frac{m_2}{c_2}\ k_{smm} + \frac{m_2}{c_2}\left[\frac{1}{m_2}\left[\log\left(\frac{1000 + \nu M_3 m_2}{1000}\right)\right]\right] \qquad (2)$$

and

$$k_{smx} = \frac{c_2}{m_2}\ k_{scx} = \frac{c_2}{m_2}\ k_{scc} + \frac{c_2}{m_2}\left[\frac{1}{c_2}\left[\log\left(\frac{1000 + (\nu M_3 - M_2)c_2}{1000\rho°}\right)\right]\right] \qquad (3)$$

where $\rho$ and $\rho°$ are the densities of gas saturated electrolyte solution and pure water respectively, and $\nu$ is the number of ions in the electrolyte formula. For very soluble gases the log terms above may need additional terms containing the gas concentrations in either molality or molarity.

| COMPONENTS: | EVALUATOR: |
|---|---|
| (1) Nitrogen; $N_2$; [7727-37-9]<br>(2) Electrolyte<br>(3) Water; $H_2O$; [7732-18-5] | H. Lawrence Clever<br>Department of Chemistry<br>Emory University<br>Atlanta, GA 30322 USA<br><br>1981, December |

CRITICAL EVALUATION:

The activity coefficient of the dissolved nitrogen, $y_1$, is a function of the concentration of all solute species, which in the present systems are the electrolyte and the nonelectrolyte. At a given temperature log $y_1$ can be represented as a power series in $c_1$ and $c_2$

$$\log y_1 = \sum_{nm} k_{nm} c_1^n c_2^m \tag{4}$$

If it is assumed that for low concentrations of both nonelectrolyte, $c_1$, and electrolyte, $c_2$, the only important terms are the linear ones,

$$\log y_1 = k_1 c_1 + k_2 c_2 \tag{5}$$

The expression has been experimentally verified for moderately dilute solutions in which there is no chemical interaction between the solute species.

The measurements of the nonelectrolyte solubility in pure water, $c^\circ_{1,sat}$, and in a salt solution, $c_{1,sat}$, give directly the activity coefficient of the nonelectrolyte. Long and McDevit (8) show that

$$\log(y_1/y_1^\circ) = \log(c^\circ_{1,sat}/c_{1,sat}) = k_1(c_{1,sat} - c^\circ_{1,sat}) + k_2 c_2 \tag{6}$$

If the nonelectrolyte solubility values are low, as they generally are for a gas, the term in $k_1$ can be neglected, even though $k_1$ is similar in magnitude to $k_2$. Thus for low nonelectrolyte concentration

$$\log(y_1/y_1^\circ) = \log(c^\circ_{1,sat}/c_{1,sat}) = k_2 c_2 \tag{7}$$

The solubility data which are evaluated in this section do not always meet the requirements set forth above. Thus, the experimental Setschenow salt effect parameters, $k_s$, may not be equivalent to the theoretically important salt effect parameter, $k_2$.

Although for a given system the experimental values of $k_{scc}$, $k_{sc\alpha}$, and $k_{scL}$ will have the same magnitude and units, there may be a unit problem when one tries to use equation (3). The constant $k_2$ will have inverse $c_2$ units, $dm^3\ mol^{-1}$, and $k_1$ will have inverse $c_1$ units. If the gas solubility is expressed as $c_1/mol\ dm^{-3}$, $k_1$ units will be $dm^3\ mol^{-1}$, but if the Bunsen or Ostwald coefficients are used for $c_{1sat}$ and $c^\circ_{1sat}$ in equation 3, then $k_1$ will have units which are the inverse of the Bunsen or Ostwald coefficients, $cm^3\ atm\ (cm^3(STP))^{-1}$ and $cm^3 cm^{-3}$ respectively.

A plot of log $(\alpha^\circ/\alpha)$ *vs.* $c_2$ is usually linear over moderate concentrations of the electrolyte. However, curvature of the plot is often seen at above moderate concentrations, and in extreme cases one observes regions of both salting-out and salting-in over different concentration regions of the same isotherm.

The salt effect parameters, $k_{sc\alpha}/dm^3\ mol^{-1}$, are presented by several methods in the evaluation.

| COMPONENTS: | EVALUATOR: |
|---|---|
| (1) Nitrogen; $N_2$; [7727-37-9]<br>(2) Electrolyte<br>(3) Water; $H_2O$; [7732-18-5] | H. Lawrence Clever<br>Department of Chemistry<br>Emory University<br>Atlanta, GA 30322 USA<br>1981, December |

CRITICAL EVALUATION:

In the first method $k_{sc\alpha}$ values are calculated for each experimental determination, and a graph is prepared of $k_{sc\alpha}$ *vs.* $c_2$. If the plot is linear and of zero slope, $k_{sc\alpha}$ is taken to be independent of concentration. In such cases an average value of $k_{sc\alpha}$ is given. If the plot is linear, but of non-zero slope, $k_{sc\alpha}$ is fitted to a linear equation in $c_2$.

In the second method a graph is prepared of $\log(\alpha°/\alpha)$ *vs.* $c_2$. A linear plot shows no concentration dependence of the salt effect parameter and the slope is $k_{sc\alpha}$. Recently some workers have fitted such plots that show curvature to a function

$$\log(\alpha°/\alpha) = k°_{sc\alpha}\, c_2/(1 + k'_{sc\alpha}\, c_2) \qquad (8)$$

In a sense the first method, in which the salt effect parameter is given by a linear function of concentration, $k_{sc\alpha} = a + bc_2$, is equivalent to

$$\log(\alpha°/\alpha) = (a + bc_2)c_2 = ac_2 + bc_2^2 \qquad (9)$$

however, the constants should be redetermined from the $\log(\alpha°/\alpha)$ and $c_2$ data.

Data on the solubility of nitrogen in aqueous electrolyte solutions are meager and, with a few exceptions, of poor quality. Solubility data are reported by more than one laboratory for only three systems. They are aqueous sulfuric acid, sodium chloride, and sodium sulfate. Overall there are data on about 25 systems.

The systems are given in the order of the standard arrangement for inorganic compounds used by the U. S. National Bureau of Standards publications. The number before each system is the standard order number for the electrolyte cation.

The general policy in evaluating salt effects on gas solubility has been to convert all salt effect data into sechenov salt effect parameters of the form $k_{scc}/dm^3\ mol^{-1}$ for purposes of comparison. The salt effect parameter is defined in terms of the ratio of the solubility in pure water to the solubility in the salt solution. When available, the author's solubility in pure water is used, rather than Battino's recommended values (this volume, pp. 1-3), in the belief that the author's systematic errors may partially cancel in the ratio. However, the choice of the nitrogen solubility in pure water is important and can change the value of $k_{scc}$ especially when the electrolyte concentration is small. Further comments on this point will be made in several of the evaluations.

2 Nitrogen + Sulfuric acid [7664-93-9] + Water

Both Christoff (2) and Bohr (3) report Ostwald coefficients of nitrogen in aqueous sulfuric acid solutions at temperatures near 293.15 K. The salt effect parameter, $k_{scc}$, decreases as the sulfuric acid concentration increases.

| COMPONENTS: | EVALUATOR: |
|---|---|
| (1) Nitrogen; $N_2$; [7727-37-9]<br>(2) Electrolyte<br>(3) Water; $H_2O$; [7732-18-5] | H. Lawrence Clever<br>Department of Chemistry<br>Emory University<br>Atlanta, GA 30322 USA<br>1981, December |

CRITICAL EVALUATION:

| $T$/K | Sulfuric Acid $c_2$/mol dm$^{-3}$ | Salt Effect Parameter, $k_{scc}$/dm$^3$ mol$^{-1}$ Bohr | Christoff |
|---|---|---|---|
| 294.03 | 0.0 | -- | -- |
| | 2.45 | 0.096 | -- |
| | 4.45 | 0.075 | -- |
| 293.15 | 4.63 | -- | 0.056 |
| 293.35 | 5.35 | 0.070 | -- |
| 293.15 | 9.52 | -- | 0.024 |
| 294.25 | 10.15 | 0.050 | -- |
| 294.65 | 12.4 | 0.041 | -- |
| 293.95 | 14.8 | 0.033 | -- |
| 294.05 | 17.15 | 0.011 | -- |
| 294.25 | 17.9 | 0.005 | -- |
| 293.15 | 17.9 | -- | -0.002 |

At 17.9 mol dm$^{-3}$ sulfuric acid Bohr's data indicates a small salting out, and Christoff's a small salting in. Both sets of data are classed as tentative with no reason to prefer one set over the other.

29 Nitrogen + Hydrated aluminum oxide [1333-84-2] + Water
41 Nitrogen + Hydrated ferric oxide [12259-21-1] + Water

The data of Shkol'nikova (11) are classed as tentative. She did not report a nitrogen in water solubility value. No salt effect parameters were calculated for the colloidal solutions.

93 Nitrogen + Magnesium sulfate [7487-88-9] + Water

Smith, Kelemen and Nagy (13) measured the solubility of nitrogen in 0, 1.25, and 2.49 mol dm$^{-3}$ magnesium sulfate solutions at 303.15 K and a total pressure range of 1.5 to 7.0 MPa. The evaluator calculated values of $k_{scc}$ assuming Henry's law was obeyed at each electrolyte concentration. The nitrogen solubility was reported as mole fraction which was calculated with respect to the number of moles of electrolyte (formula) rather than the number of moles of ions. The salt effect parameters change little with concentration. The data are classed as tentative.

| $T$/K | Magnesium Sulfate $c_2$/mol dm$^{-3}$ | Salt Effect Parameter $k_{scc}$/dm$^3$mol$^{-1}$ | Standard Deviation $\sigma$/dm$^3$ mol$^{-1}$ |
|---|---|---|---|
| 303.15 | 1.25 | 0.222 | 0.012 |
| | 2.49 | 0.228 | 0.011 |

94 Nitrogen + Calcium chloride [10043-53-4] + Water

Smith, Kelemen, and Nagy (13) measured the solubility of nitrogen in 0, 0.50, 1.75, 3.50, and 5.60 mol dm$^{-3}$ calcium chloride solutions at 303.15 K and a total pressure range of 12 to 73 bar. The evaluator calculated $k_{scx}$ and $k_{scc}$ values at four or five pressures at each concentration assuming Henry's law is obeyed. The nitrogen solubility was reported as mole fraction which was calculated with respect to moles of calcium chloride ($CaCl_2$) rather than respect to the moles of ions ($Ca^{2+}$, $Cl^-$). The salt effect parameter shows a definite decrease as the electrolyte concentration increases. The results are classed as tentative.

| COMPONENTS: | EVALUATOR: |
|---|---|
| (1) Nitrogen; $N_2$; [7727-37-9]<br>(2) Electrolyte<br>(3) Water; $H_2O$; [7732-18-5] | H. Lawrence Clever<br>Department of Chemistry<br>Emory University<br>Atlanta, GA 30322 USA<br><br>1981, December |

CRITICAL EVALUATION:

| $T$/K | Calcium Chloride $c_2$/mol dm$^{-3}$ | Salt Effect Parameter $k_{scc}$/dm$^3$ mol$^{-1}$ | Standard Deviation $\sigma$/dm$^3$ mol$^{-1}$ |
|---|---|---|---|
| 303.15 | 0.50 | 0.212 | 0.020 |
| | 1.75 | 0.177 | 0.008 |
| | 3.50 | 0.157 | 0.008 |
| | 5.60 | 0.149 | 0.026 |

96 Nitrogen + Barium chloride [10361-37-2] + Water

Braun (1) measured the solubility of nitrogen in water and in six solutions of barium chloride at molalities between 0.165 and 0.7707 mol kg$^{-1}$ at five degree intervals from 278.2 to 298.2 K. The salt effect parameters calculated from the solubility data decreased with concentration and increased with temperature. The solubility parameters were greatest at the lower barium chloride concentrations and leveled off to values about 2/3 as large at the four higher concentrations. The variation of the salt effect parameter with concentration depends to some extent on the value of the solubility of nitrogen in water. The effect described above is not as pronounced when Battino's recommended nitrogen in water solubility is used instead of Barun's value. The increase of the parameter with temperature is contrary to normal behavior, thus the data are classed as doubtful. Average values of the parameter at barium chloride molalities from 0.35 to 0.77 mol kg$^{-1}$ are below.

| $T$/K | 278.2 | 283.2 | 288.2 | 293.2 | 298.2 |
|---|---|---|---|---|---|
| $k_{scc}$/dm$^3$ mol$^{-1}$ | 0.31 | 0.31 | 0.32 | 0.34 | 0.36 |

98 Nitrogen + Lithium chloride [7447-41-8] + Water

Morrison and Billett (9) measured the solubility of nitrogen in water and in one mol kg$^{-1}$ lithium chloride solution at four temperatures. Only the salt effect parameter based on the two solubility measurements was reported by the authors. They are classed as tentative. Values of $k_{smm}$, $k_{smx}$, and $k_{scc}$ are given below.

| $T$/K | $k_{smm}$/kg mol$^{-1}$ | $k_{smx}$/kg mol$^{-1}$ | $k_{scc}$/dm$^3$ mol$^{-1}$ |
|---|---|---|---|
| 285.75 | 0.125 | 0.140 | 0.135 |
| 303.15 | 0.095 | 0.110 | 0.105 |
| 322.55 | 0.084 | 0.099 | 0.095 |
| 344.85 | 0.092 | 0.107 | 0.104 |

99(1) Nitrogen + Sodium **Hydroxide** [1310-73-3] + Water

Levina and Stisbarovskaya (7) measured the solubility of nitorgen in 0, 0.5, and 4 mol dm$^{-3}$ sodium hydroxide at ten temperatures between 273 and 513 K at a total pressure of 9.81 MPa. The authors calculated the solubility of nitrogen at a nitrogen partial pressure at 9.81 MPa.

The salt effect parameters, $k_{scm}$, calculated from the data by the evaluator show considerable scatter. No effort was made to calculate $k_{scc}$ values since it is believed the uncertainty in the data is larger than the conversion factor. The data are classed as tentative in the 275 to 323 K range and doubtful at the high temperatures.

| COMPONENTS: | EVALUATOR: |
|---|---|
| (1) Nitrogen; $N_2$; [7727-37-9]<br>(2) Electrolyte<br>(3) Water; $H_2O$; [7732-18-5] | H. Lawrence Clever<br>Department of Chemistry<br>Emory University<br>Atlanta, GA 30322 USA<br>1981, December |

CRITICAL EVALUATION:

| $T$/K | 273 | 298 | 308 | 323 |
|---|---|---|---|---|
| $k_{scm}$/dm$^3$mol$^{-1}$ | 0.20 | 0.22 | 0.19 | 0.22 |

| $T$/K | 348 | 373 | 398 | 423 | 473 | 513 |
|---|---|---|---|---|---|---|
| $k_{scm}$/dm$^3$mol$^{-1}$ | 0.16 | 0.16 | 0.19 | 0.25 | 0.20 | 0.14 |

99(2)Nitrogen + Sodium chloride [7647-14-5] + Water

There are seven papers on the solubility of nitrogen in aqueous sodium chloride solution. Four of the papers (1, 4, 9, 12) report measurements of a nitrogen partial pressure near 101.325 kPa (1 atm), and three (13, 14, 15) report solubility measurements at nitrogen partial pressures up to 7 MPa. Although there is some scatter in the calculated salt effect parameters from the data in the seven papers, the values are of similar magnitude.

The salt effect parameter appears to decrease as the sodium chloride concentration increases. However, this trend depends to some extent on the value of the gas solubility in pure water chosen to calculate the solubility ratio. The choice of the value is more critical for dilute electrolyte solutions than concentrated solutions.

Braun (1) measured the solubility of nitrogen in 12 aqueous sodium chloride solutions over the 0.12 to 2.27 mol dm$^{-3}$ concentration range at five degree intervals over the 278.2 to 298.2 K temperature range. The salt effect parameters calculated from the data decrease in magnitude as the sodium chloride concentration increases and increase in magnitude as the temperature increases. Both the concentration and temperature trends lead the evaluator to have doubts about the reliability of the data. At sodium chloride concentrations near 0.12 mol dm$^{-3}$ the salt effect parameter is about 100 per cent greater than the values in the 1.0 to 2.2 mol dm$^{-3}$ concentration range. The increase in $k_{scc}$ magnitude with temperature is definitely contrary to the normal behavior in this temperature range. Braun's value for the solubility of nitrogen in pure water is greater than the recommended value at 298.2 K and less than the recommended value at the other temperatures. The data are classed as doubtful.

Van Slyke, Dillon, and Margaria (4) measured the solubility of nitrogen in water and in 0.155 mol dm$^{-3}$ sodium chloride solution at 298.15 and 311.15 K. The salt effect parameters calculated from their data are sensitive to the nitrogen in water solubility values used. Values of $k_{scc}$ are given for both the Van Slyke *et al.* and the Battino recommended value of the nitrogen + water solubility. The data are classed as tentative, but they were not used in the smoothed data fit.

Morrison and Billett (9) measured the solubility of nitrogen in water and in one mol kg$^{-1}$ sodium chloride solution at four temperatures between 285.75 and 344.85 K at atmospheric pressure. Their data are classed as tentative.

Mishnina, Avdeeva, and Bozhovakaya (12) made some measurements of their own, but also used the measurements of Morrison and Billett (9) to prepare a table of smoothed values of the Bunsen coefficient and the salt effect parameter for sodium chloride solutions up to saturation at five degree intervals from 283.15 to 393.15 K. They apparently assume the salt effect parameter is independent of the sodium chloride concentration. Since their data are already smoothed the comparison of their data with the data of others should be used with caution. The data are classed as tentative.

| COMPONENTS: | EVALUATOR: |
|---|---|
| (1) Nitrogen; $N_2$; [7727-37-9]<br>(2) Electrolyte<br>(3) Water; $H_2O$; [7732-18-5] | H. Lawrence Clever<br>Department of Chemistry<br>Emory University<br>Atlanta, GA 30322 USA<br>1981, December |

CRITICAL EVALUATION:

| $T$/K | Salt Effect Parameter, $k_{scc}$/$dm^3$ $mol^{-1}$ | | | | | |
|---|---|---|---|---|---|---|
| | Braun[1] | Van Slyke, Dillon, Margaria | Morrison, Billett | Mishnina, Avdeeva, Bozhovakaya | Smith, *et al.* | Smoothed Values[6] |
| 278.15 | 0.158 | | | | | -- |
| 283.15 | 0.166 | | | 0.160 | | 0.161 |
| 285.75 | -- | | 0.164 | -- | | -- |
| 288.15 | 0.168 | | | 0.148 | | 0.150 |
| 293.15 | 0.185 | | | 0.137 | | 0.141 |
| 298.15 | 0.295 | 0.114[2]/0.144[3] | | 0.129 | | 0.134 |
| 303.15 | | | 0.132 | 0.121 | 0.124[4] | 0.127 |
| 308.15 | | | | 0.117 | | 0.122 |
| 311.15 | | 0.090[2]/0.117[3] | | -- | | -- |
| 313.15 | | | | 0.113 | | 0.118 |
| 318.15 | | | | 0.109 | | 0.115 |
| 322.55 | | | 0.118 | -- | | -- |
| 323.15 | | | | 0.106 | 0.122[5] | 0.113 |
| 324.65 | | | | -- | | -- |
| 328.15 | | | | 0.106 | | 0.111 |
| 333.15 | | | | 0.106 | | 0.110 |
| 338.15 | | | | 0.106 | | 0.109 |
| 343.15 | | | | 0.106 | | 0.109 |
| 353.15 | | | | | | 0.110 |
| 344.85 | | | 0.119 | | | -- |
| 363.15 | | | | | | 0.113 |
| 373.15 | | | | | | 0.118 |
| 375.65 | | | | | 0.116[5] | -- |
| 398.15 | | | | | 0.132[5] | -- |

[1] Average value of the salt effect parameter from Braun's data at sodium chloride concentrations between 1.17 and 2.27 mol $dm^{-3}$. The standard deviations ranged from 0.006 to 0.009. The salt effect parameters based on Braun's data over the 0.12 to 0.75 mol $dm^{-3}$ range were up to 100 per cent larger. The larger values were rejected. The use of Battino's recommended nitrogen in water solubilities increased the 298.15 K $k_{scc}$ value about 10 per cent and decreased the values at the other temperatures by 5 to 10 per cent.

[2] Values calculated using $\alpha°$ values of Van Slyke *et al.*

[3] Values calculated using $\alpha°$ values based on Battino's recommended values (*This volume*, pp. 1-3.).

[4] Smith, Kelemen, and Nagy (13).

[5] O'Sullivan and Smith (15).

[6] The smoothed values are based on a linear regression that weighted the Morrison and Billett (9) and the Smith *et al.* (13, 15) values two times and the Misnina *et al.* (12) values one time.

| COMPONENTS: | EVALUATOR: |
|---|---|
| (1) Nitrogen; $N_2$; [7727-37-9]<br>(2) Electrolyte<br>(3) Water; $H_2O$; [7732-18-5] | H. Lawrence Clever<br>Department of Chemistry<br>Emory University<br>Atlanta, GA 30322 USA<br>1981, December |

CRITICAL EVALUATION:

Smith, Kelemen and Nagy (13), and O'Sullivan and Smith (14, 15) measured the solubility of nitrogen in water and in aqueous sodium chloride solution up to concentrations of 5.27 mol $kg^{-1}$ and temperatures from 303.15 to 398.15 K under nitrogen pressures up to near 7 MPa. They reported the nitrogen solubility as mole fraction, but it should be noted that they calculated the mole fraction on the basis of the formula (NaCl) not the moles of the individual ions ($Na^+$, $Cl^-$). The evaluator calculated salt effect parameters from their data assuming Henry's law behavior. The results agree with those of Morrison and Billett (9), and Mishnina *et al.* (12) at one molar sodium chloride. At 303.15 K the salt effect parameters from the data of Smith *et al.* (13, 15) decrease with the salt concentration. At other temperatures the parameter does not show a consistent change with concentration. See table below.

| Sodium Chloride $c_2$/mol $dm^{-3}$ | $k_{scc}$/$dm^3$ $mol^{-1}$ | | | |
|---|---|---|---|---|
| | 303.15 K | 324.65 K | 375.65 K | 398.15 K |
| 1.00 | 0.124 | 0.122 | 0.116 | 0.132 |
| 2.7 | 0.113 | -- | -- | -- |
| 4 | -- | 0.120 | 0.124 | 0.126 |
| 5.47 | 0.098 | -- | -- | -- |

The smoothed values of $k_{scc}$ shown in the earlier comparison table were obtained by a linear regression that weighted the Morrison and Billett and the Smith *et al.* values two times, and the Mišhnina *et al.* values one time. The equation is

$$\ln k_{scc} = -41.900 + 70.099/(T/100\ K) + 20.492 \ln (T/100\ K).$$

The Smith *et al.* value at 398.15 K was omitted.

99(3)Nitrogen + Sodium sulfite [10579-83-6] + Water
99(4)Nitrogen + Sodium sulfate [7757-82-6] + Water

Yasunishi (18) measured the solubility of nitrogen in water and in up to ten* concentrations of sodium sulfite and sodium sulfate between 0 and 1.0 mol $dm^{-3}$ at 298.15 K. He made vapor pressure and solvent dilation corrections. The salt effect parameters calculated from his data show little variation with electrolyte concentration. The values were averaged. His results appear to be reliable and they are classed as tentative.

Smith, Kelemen, and Nagy (13) measured the solubility of nitrogen in 0, 0.50, 1.25, and 2.10 mol $dm^{-3}$ sodium sulfate solution over the total pressure range of 1.5 to 7.0 MPa. The evaluator calculated the salt effect parameters assuming Henry's law is obeyed at each salt concentration. The nitrogen solubility was reported as mole fraction which was calculated with respect to the number of moles of electrolyte ($Na_2SO_4$), not the number of moles of ions ($Na^+$, $SO_4^{2-}$). The salt effect parameters, converted to $k_{scc}$ values, decrease in magnitude as the sodium sulfate concentration increases. The value at 0.50 mol $dm^{-3}$ is about 4 per cent less than Yasumishi's value at 298.15 K which is a reasonable decrease for the five degree increase in temperature. The other values are significantly lower in value.

Both sets of data are classed as tentative, but the Yasunishi data are preferred.

*different

| COMPONENTS: | EVALUATOR: |
|---|---|
| (1) Nitrogen; $N_2$; [7727-37-9]<br>(2) Electrolyte<br>(3) Water; $H_2O$; [7732-18-5] | H. Lawrence Clever<br>Department of Chemistry<br>Emory University<br>Atlanta, GA 30322 USA<br><br>1981, December |

CRITICAL EVALUATION:

| $T/K$ | Electrolyte | Number of Determinations[1] | Concentration Range $c_2$/mol dm$^{-3}$ | Salt Effect Parameter $k_{scc}$/dm$^3$mol$^{-1}$ |
|---|---|---|---|---|
| 298.15 | $Na_2SO_3$ | 11 | 0 - 1.04 | 0.383 ± 0.012[2] |
| 298.15 | $Na_2SO_4$ | 8 | 0 - 1.08 | 0.353 ± 0.005[2] |
| 303.15 | $Na_2SO_4$ | 2 | 0 - 0.5 | 0.339 |
| | | 2 | 0 - 1.25 | 0.30 |
| | | 2 | 0 - 2.10 | 0.24 |

[1] Includes measurement in pure water.

[2] Average ± standard deviation.

99(5)Nitrogen + Sulfuric acid [7664-93-9]
+ Sodium sulfate [7757-82-6] + Water

The single solubility value of Kobe and Kenton (6) is classed as tentative. A data sheet on a gas mixture of nitrogen, oxygen, and carbon dioxide is also classed as tentative.

99(6)Nitrogen + Sodium sulfite [10579-83-6]
+ Sodium sulfate [7757-82-6] + Water

Yasunishi (18), see 99(3) and (4) above, measured the solubility of nitrogen in a mixed electrolyte solution of varying concentration and ratio of sulfite and sulfate ion. The salt effect parameters for the mixtures fall between the values reported above for the single salt solutions. They are classed as tentative.

99(7)Nitrogen + Sodium bicarbonate [144-55-8] + Water
99(8)Nitrogen + Sodium carbonate [497-19-8] + Water

Van Slyke, Dillon, and Margaria (4) measured the solubility of nitrogen in water and in solutions of sodium bicarbonate and sodium carbonate of one concentration. The salt effect parameter values were calculated using the author's value of the solubility of nitrogen in water. The use of Battino's recommended nitrogen + water solubility value would decrease the $k_{scc}$ values by 18 and 16 per cent respectively. (Values are 0.139 and 0.312).

| $T/K$ | Electrolyte | Number of Determinations | Concentration $c_2$/mol dm$^{-3}$ | Salt Effect Parameter[1] $k_{scc}$/dm$^3$ mol$^{-1}$ |
|---|---|---|---|---|
| 298.15 | $NaHCO_3$ | 4 | 0.153 | 0.170 ± 0.007 |
| | $Na_2CO_3$ | 2 | 0.077 | 0.373 ± 0.006 |

[1] Average ± standard deviation.

| COMPONENTS: | EVALUATOR: |
|---|---|
| (1) Nitrogen; $N_2$; [7727-37-9]<br>(2) Electrolyte<br>(3) Water; $H_2O$; [7732-18-5] | H. Lawrence Clever<br>Department of Chemistry<br>Emory University<br>Atlanta, GA 30322 USA<br><br>1981, December |

CRITICAL EVALUATION:

99(9) Nitrogen + Acetic acid [64-19-7] + Sodium acetate [127-09-3] + Metal (2+) acetate + Water

Brasted and Hirayama (10) measured the solubility of nitrogen in water and in aqueous mixtures containing 1.0 mol $dm^{-3}$ acetic acid, 0.50 mol $dm^{-3}$ sodium acetate, and several concentrations of a transition metal (2+) acetate. There is some confusion in the paper over the Bunsen and Ostwald coefficient definitions, but the solubility data are clearly given as Ostwald coefficients.

It is unfortunate that the authors did not measure the nitrogen solubility in just the acetic acid and sodium acetate mixture. Such a solubility value would have been helpful in understanding the influence of the metal (2+) acetate on the nitrogen solubility. The salt effect parameter was calculated by the evaluator on the basis of both total electrolyte molarity and on ionic strength. The trends in the salt effect parameter values show no systematic pattern with either molarity or ionic strength.

| $T$/K | Electrolyte[1] M(2+) Acetate | M(2+) Acetate Concentration[2] $c_4$/mol $dm^{-3}$ | Salt Effect Parameter $k_{scc}$/$dm^3$ $mol^{-1}$ | Salt Effect Parameter $k_{sIc}$/$dm^3$ $mol^{-1}$ |
|---|---|---|---|---|
| 298.15 | 33. $Zn(C_2H_3O_2)_2$ | 0.15 | 0.169 | 0.116 |
| | 35. $Hg(C_2H_3O_2)_2$ | 0.05 | 0.083 | 0.070 |
| | | 0.15 | 0.113 | 0.077 |
| | 36. $Cu(C_2H_3O_2)_2$ | 0.05 | 0.169 | 0.143 |
| | | 0.157 | 0.123 | 0.084 |
| | 37. $Ni(C_2H_3O_2)_2$ | 0.051 | 0.148 | 0.126 |
| | | 0.15 | 0.098 | 0.067 |
| | 40. $Co(C_2H_3O_2)_2$ | 0.033 | 0.187 | 0.166 |
| | | 0.10 | 0.155 | 0.116 |
| | 48. $Mn(C_2H_3O_2)_2$ | 0.05 | 0.145 | 0.123 |
| | | 0.15 | 0.123 | 0.084 |

[1] The number before the formula is the standard order number.

[2] The other concentrations are acetic acid, $c_2$/mol $dm^{-3}$ = 1.00, and sodium acetate, $c_3$/mol $dm^{-3}$ = 0.50. The acetic acid concentration was not used in the calculation of the $k_{scc}$ and $k_{sIc}$ values.

100(1) Nitrogen + Potassium chloride [7447-40-7] + Water

Bikov (5) measured the solubility of nitrogen in 0, 0.5, 1.0, and 2.0 mol $dm^{-3}$ aqueous potassium chloride solution per gram of solvent at a total pressure of 9.81 MPa at eight temperatures between 273 and 513 K. The author calculated the solubility per g of solution at a nitrogen partial pressure of 9.81 MPa apparently by subtracting the water vapor pressure from the total pressure and assuming Henry's law is obeyed.

| COMPONENTS: | EVALUATOR: |
|---|---|
| (1) Nitrogen; $N_2$; [7727-37-9]<br>(2) Electrolyte<br>(3) Water; $H_2O$; [7732-18-5] | H. Lawrence Clever<br>Department of Chemistry<br>Emory University<br>Atlanta, GA 30322 USA<br><br>1981, December |

CRITICAL EVALUATION:

The salt effect parameters, $k_{scm}$, calculated from the data by the evaluator showed scatter with respect to both the potassium chloride concentration and the temperature. The $k_{scm}$ values are the average values for the three KCl solutions at each temperature. The data are classed as doubtful.

| $T$/K | 273 | 298 | 323 | 348 | 373 | 423 | 473 | 513 |
|---|---|---|---|---|---|---|---|---|
| $k_{scm}/dm^3\ mol^{-1}$ | 0.13 | 0.16 | 0.10 | 0.09 | 0.09 | 0.13 | 0.17 | 0.12 |

100(2)Nitrogen + Potassium iodide [7681-11-0] + Water

Morrison and Billett (9) measured the solubility of nitrogen in water and in one mol $kg^{-1}$ potassium iodide solution at four temperatures between 285.75 and 344.85 K. The authors reported the solubility in water and the solubility parameter, $k_{smm}$, but not the solubility in the solution. The data are classed as tentative.

| $T$/K | Salt Effect Parameter | | |
|---|---|---|---|
| | $k_{smm}/kg\ mol^{-1}$ | $k_{smx}/kg\ mol^{-1}$ | $k_{scc}/dm^3\ mol^{-1}$ |
| 285.75 | 0.134 | 0.149 | 0.160 |
| 303.15 | 0.100 | 0.115 | 0.126 |
| 322.55 | 0.086 | 0.101 | 0.113 |
| 344.85 | 0.084 | 0.099 | 0.112 |

100(3)Nitrogen + Potassium carbonate [584-08-7] + Water

Schröder (17) measured the solubility of nitrogen in 0, 0.5, 1, 2, 4, and 6 mol $dm^{-3}$ aqueous potassium carbonate solutions at six temperatures between 323 and 403 K at nitrogen partial pressures between 1.01 and 4.05 MPa (10 and 40 atm). Most of the results are presented in graphs, but some results are tabulated. Results from the present work were combined with results from an earlier Schröder paper (16) to calculate several $k_{scc}$ values at the nitrogen partial pressure of 4.05 MPa.

| $T$/K | Potassium Carbonate $c_2$/mol $dm^{-3}$ | Salt Effect Parameter $k_{scc}/dm^3\ mol^{-1}$ |
|---|---|---|
| 323.15 | 0.5 | 0.272 |
| | 1 | 0.263 |
| | 2 | 0.240 |
| 363.15 | 0.5 | 0.477 |
| | 1 | 0.456 |
| | 2 | 0.171, 0.329[1] |
| | 4 | 0.209 |
| | 6 | 0.143 |

[1] Average 0.250.

The values at 363.15 for the 0.5 and 1 mol $dm^{-3}$ potassium carbonate are classed as doubtful. The other values are classed as tentative.

| COMPONENTS: | EVALUATOR: |
|---|---|
| (1) Nitrogen; $N_2$; [7727-37-9]<br>(2) Electrolyte<br>(3) Water; $H_2O$; [7732-18-5] | H. Lawrence Clever<br>Department of Chemistry<br>Emory University<br>Atlanta, GA 30322 USA<br><br>1981, December |

CRITICAL EVALUATION:

References:


1. Braun, L. *Z. Phys. Chem.* 1900, *33*, 721.
2. Christoff, A. *Z. Phys. Chem.* 1906, *55*, 622.
3. Bohr, C. *Z. Physik. Chem.* 1910, *71*, 47.
4. Van Slyke, D. D.; Dillon, R. T.; Margaria, R. *J. Biol. Chem.* 1934, *105*, 571.
5. Bikov, M. M. *Tr. Voronezhsk. Gos. Univ.* 1937, *9*, 29.
6. Kobe, K. A.; Kenton, F. H. *Ind. Eng. Chem., Anal. Ed.* 1938, *10*, 76.
7. Levina, M. I.; Stsibarovskaya, N. P. *Russ. J. Phys. Chem.* 1939, *12*, 653.
8. Long, F. A.; McDevit, W. F. *Chem. Rev.* 1952, *51*, 119.
9. Morrison, T. J.; Billett, F. *J. Chem. Soc.* 1952, 3819.
10. Brasted, R. C.; Harayama, C. *J. Phys. Chem.* 1958, *62*, 125.
11. Shkol'nikova, R. I. *Uch. Zap. Leningr. Gos. Univ., Ser. Khim. Nauk.* 1959, Nr. 18, 64. *Chem. Abstr.* 1961, *55*, 25443b.
12. Mishnina, T. A.; Avdeeva, O. I.; Bozhovakaya, T. K. *Materialy Vses. Nauchn. Issled. Geol. Inst.* 1961, *46*, 93.
13. Smith, N. O.; Kelemen, S.; Nagy, B. *Geochim. Cosmochim. Acta* 1962, *26*, 921.
14. O'Sullivan, T. D.; Smith, N. O. *Geochim. Cosmochim. Acta* 1966, *30*, 617.
15. O'Sullivan, T. D.; Smith, N. O. *J. Phys. Chem.* 1970, *74*, 1460. *(Corrn.-J. Phys. Chem.* 1970, *74*, 4612).
16. Schröder, W. *Z. Naturforsch,* 1969, 24b, 500.
17. Schröder, W. *Z. Phys. Chem. (Wiesbaden)* 1971, *75*, 248.
18. Yasunishi, A. *J. Chem. Eng. Jpn.* 1977, *10*, 89.

**COMPONENTS:**

(1) Nitrogen; $N_2$; [7727-37-9]

(2) Sulfuric acid; $H_2SO_4$; [7664-93-9]

(3) Water; $H_2O$; [7732-18-5]

**ORIGINAL MEASUREMENTS:**

Christoff, A.

*Z. Phys. Chem.* 1906, *55*, 622-34.

**VARIABLES:**

T/K = 293.15
P/kPa = Atmospheric
$H_2SO_4$/wt % = 0 - 95.6

**PREPARED BY:**

M. E. Derrick
H. L. Clever

**EXPERIMENTAL VALUES:**

| T/K | $H_2SO_4$/wt % | $m_{H_2SO_4}$/mol kg$^{-1}$ | Ostwald Coefficient $L \times 10^2$ |
|---|---|---|---|
| 293.15 | 0 | 0.0 | 1.537 |
| | 35.82 | 5.690 | 0.8447 |
| | 61.62 | 16.37 | 0.6144 |
| | 95.6 | 222. | 1.672 |

**AUXILIARY INFORMATION**

**METHOD/APPARATUS/PROCEDURE:**

The apparatus was an Ostwald type (1) with a lead capillary tube through which the gas flows to the absorption flask, the gas buret, and the gas-ometer.

The acid solution was degassed by boiling under reflux. The author estimates a one per cent change in the acid concentration due to the degassing procedure. The absorption vessel was filled with solvent, the gas was introduced, and the system shaken until equilibrium was reached.

**SOURCE AND PURITY OF MATERIALS:**

(1) Nitrogen. Oxygen was removed from air by passage over copper in a combustion furnace.

(2) Sulfuric acid. Merck (Darmstadt). Specific gravity 1.271, 1.523, and 1.839 for 35.82, 61.61, and 95.6 wt per cent acid, respectively.

(3) Water. Distilled.

**ESTIMATED ERROR:**

$\delta T/K$ = 0.02 for acid
0.5 for gas

Barometric fluctuations are stated to be negligible.

**REFERENCES:**

1. Ostwald, W.
*Lehrbuch der allgem. Chemie* (2 Aufl.), *1*, 615.

COMPONENTS:

(1) Nitrogen; $N_2$; [7727-37-9]

(2) Sulfuric Acid; $H_2SO_4$; [7664-93-9]

(3) Water; $H_2O$; [7732-18-5]

ORIGINAL MEASUREMENTS:

Bohr, C.

*Z. Physik. Chem.* 1910, *71*, 47-50.

VARIABLES:

T/K = 294
P/kPa = 101
C/N = 0-36

PREPARED BY:

R. Battino

EXPERIMENTAL VALUES:

| $T^a$/K | Normality[b] | Ostwald Coeff.[c] $10^2L$ | Bunsen Coeff.[d] $10^2\alpha$ |
|---|---|---|---|
| 294.05 | 0.0 | 1.68 | 1.56 |
| 294.05 | 4.9 | 0.98 | 0.91 |
| 294.05 | 8.9 | 0.78 | 0.72 |
| 294.35 | 10.7 | 0.71 | 0.66 |
| 294.25 | 20.3 | 0.53 | 0.49 |
| 294.65 | 24.8 | 0.52 | 0.48 |
| 293.95 | 29.6 | 0.55 | 0.51 |
| 294.05 | 34.3 | 1.08 | 1.00 |
| 294.25 | 35.8 | 1.39 | 1.29 |

[a] Temperature reported to 0.1°C.

[b] Normality of sulfuric acid solution in g-equiv per liter.

[c] Ostwald coefficient calculated by compiler.

[d] Bunsen coefficient.

AUXILIARY INFORMATION

METHOD/APPARATUS/PROCEDURE:

The absorption measuring apperatus is described in reference (1).

SOURCE AND PURITY OF MATERIALS:

No details given.

ESTIMATED ERROR:

$\delta\alpha/\alpha$ = ±0.03, estimate by author.

Normality to 0.1, estimate by author.

REFERENCES:

(1) Bohr, C. *Sauerstoffaufnahme des Blutfarbstoffes*, Kopenhagen 1895.
Jolin, *Archiv. F. (Anat. U.) Physiologie* 1889, 265.

**COMPONENTS:**

(1) Nitrogen; $N_2$; [7727-37-9]

(2) Nitric acid; $HNO_3$; [7697-37-2]

(3) Water; $H_2O$; [7732-18-5]

**ORIGINAL MEASUREMENTS:**

Shapka, A. V.; Yudina, L. V.; Zhuravleva, A.N.; Vorona, N.Ya; Drushlyak, O.G.

*Vestn. Khar'kov Politekhn in-ta* 1974, (98), 44-5.

*Chem. Abstr.* 1975, *83*, 137686m.

**VARIABLES:**

T/K = 290
P/MPa = 1.1-4.2

**PREPARED BY:**

V. Katovic

**EXPERIMENTAL VALUES:**

t/°C = 17 (290 K)

| $C_{HNO_3}$/% | P/atm 11 | P/atm 21 | P/atm 31 | P/atm 41 |
|---|---|---|---|---|
| | $S^a$ | $S^a$ | $S^a$ | $S^a$ |
| 0 | 0.161 | 0.290 | 0.45 | 0.5 |
| | | 0.280 | 0.465 | 0.439 |
| 18 | 0.155 | 0.300 | 0.404 | 0.643 |
| | 0.165 | 0.320 | 0.425 | |
| | | 0.323 | 0.440 | |
| 42 | 0.139 | 0.292 | 0.40 | 0.45 |
| 50 | 0.115 | 0.230 | 0.326 | 0.43 |
| | | | 0.342 | 0.45 |
| 67 | 0.154 | 0.225 | 0.365 | 0.44 |
| | 0.16 | 0.300 | 0.41 | 0.46 |
| | | 0.25 | 0.37 | 0.495 |
| | | 0.31 | 0.45 | 0.53 |
| | (1.11)[b] | (2.13) | (3.14) | (4.15) |

[a] Solubility of nitrogen in ℓ $N_2$ (STP)/ℓ solution.
[b] Values in parentheses are the pressure in MPa.

**AUXILIARY INFORMATION**

**METHOD/APPARATUS/PROCEDURE:**

Nitrogen at 60 atm was brought into contact with a known volume of nitric acid solution in a thermostated absorption vessel. The system was equilibrated for 30-40 minutes with stirring during which time the pressure was lowered to the desired value. A sample of the saturated solution was withdrawn and analyzed for the dissolved nitrogen content.

**SOURCE AND PURITY OF MATERIALS:**

No details given.

**ESTIMATED ERROR:**

$\delta S/S = \pm 0.03$, compiler's estimate.

**REFERENCES:**

COMPONENTS:

(1) Nitrogen; $N_2$; [7727-37-9]

(2) Hydrated aluminum oxide; $Al_2O_3 \cdot xH_2O$; [1333-84-2]

(3) Water; $H_2O$; [7732-18-5]

ORIGINAL MEASUREMENTS:

Shkol'nikova, R. I.

*Uch. Zap. Leningr. Gos. Univ., Ser. Khim. Nauk.* 1959, Nr. 18, 64-86.

*Chem. Abstr.* 1961, *55*, 25443b.

VARIABLES:

T/K = 293.15 - 313.15
P/kPa = 101.325
$Al_2O_3$/wt % = 0.1, 0.35

PREPARED BY:

A. L. Cramer
H. L. Clever

EXPERIMENTAL VALUES:

| T/K | $Al_2O_3$/Wt % | Bunsen Coefficient $\alpha \times 10^3$ |
|---|---|---|
| 293.15 | 0.1 | 15.8 |
| | 0.35 | 12.6 |
| 303.15 | 0.1 | 12.9 |
| | 0.35 | 10.3 |
| 313.15 | 0.1 | 10.0 |
| | 0.35 | 8.2 |

The enthalpy of solution of nitrogen was calculated from the temperature coefficient of the Bunsen coefficient. It is 3060, 4180, and 4460 cal $mol^{-1}$ in water, and 0.1 and 0.35 wt % $Al_2O_3$ respectively.

AUXILIARY INFORMATION

METHOD/APPARATUS/PROCEDURE:

The apparatus and procedure of Lannung were modified (1).

SOURCE AND PURITY OF MATERIALS:

(1) Nitrogen. Source not given. Stated to be near 100 per cent.

(2) Aluminium oxide. No information.

(3) Water. No information.

ESTIMATED ERROR:

REFERENCES:

1. Lannung, A. *J. Am. Chem. Soc.* 1930, *52*, 68

**COMPONENTS:**

(1) Nitrogen; $N_2$; [7727-37-9]

(2) Hydrated ferric oxide; $Fe_2O_3 \cdot xH_2O$; [12259-21-1]

(3) Water; $H_2O$· [7732-18-5]

**ORIGINAL MEASUREMENTS:**

Shkol'nikova, R. I.

*Uch. Zap. Leningr. Gos. Univ., Ser. Khim. Nauk.* 1959, Nr. 18, 64-86.

*Chem. Abstr.* 1961, *55*, 25443b.

**VARIABLES:**

T/K = 293.15 - 313.15
P/kPa = 101.325
$Fe_2O_3$/wt % = 0.1 - 0.8

**PREPARED BY:**

A. L. Cramer
H. L. Clever

**EXPERIMENTAL VALUES:**

| T/K | $Fe_2O_3$/Wt % | Bunsen Coefficient $\alpha \times 10^3$ |
|---|---|---|
| 293.15 | 0.1 | 16.0 |
| | 0.5 | 13.4 |
| | 0.8 | 13.0 |
| 303.15 | 0.1 | 13.1 |
| | 0.5 | 11.1 |
| | 0.8 | 10.7 |
| 313.15 | 0.1 | 11.0 |
| | 0.5 | 9.9 |
| | 0.8 | 8.6 |

The enthalpy of solution of nitrogen was calculated from the temperature coefficient of the Bunsen coefficient. It is 3060, 3420, 3570, and 3760 cal $mol^{-1}$ in water, and 0.1, 0.5, and 0.8 wt % $Fe_2O_3$ respectively.

AUXILIARY INFORMATION

**METHOD/APPARATUS/PROCEDURE:**

The apparatus and procedure of Lannung were modified (1).

**SOURCE AND PURITY OF MATERIALS:**

(1) Nitrogen. Source not given. Stated to be near 100 per cent.

(2) Ferric oxide. No information.

(3) Water. No information.

**ESTIMATED ERROR:**

**REFERENCES:**

1. Lannung, A.
*J. Am. Chem. Soc.* 1930, *52*, 68-80.

**COMPONENTS:**

(1) Nitrogen; $N_2$; [7727-37-9]

(2) Barium chloride; $BaCl_2$; [10361-37-2]

(3) Water; $H_2O$; [7732-18-5]

**ORIGINAL MEASUREMENTS:**

Braun, L.

*Z. Phys. Chem.* 1900, *33*, 721-41.

**VARIABLES:**

T/K = 278-298
Concentration

**PREPARED BY:**

C.L. Young

**EXPERIMENTAL VALUES:**

| T/K | Wt. of salt per 100g soln./g | Bunsen coefficient, α |
|---|---|---|
| 278.2 | 13.830 | 0.01270 |
| | 11.927 | 0.01368 |
| | 6.903 | 0.01598 |
| | 6.738 | 0.01628 |
| | 3.870 | 0.01802 |
| | 3.325 | 0.01826 |
| | 0.0 | 0.02173 |
| 283.2 | 13.830 | 0.01166 |
| | 11.927 | 0.01249 |
| | 6.903 | 0.01474 |
| | 6.738 | 0.01494 |
| | 3.870 | 0.01660 |
| | 3.325 | 0.01681 |
| | 0.0 | 0.02003 |
| 288.2 | 13.830 | 0.01036 |
| | 11.927 | 0.01139 |
| | 6.903 | 0.01317 |
| | 6.738 | 0.01340 |
| | 3.870 | 0.01480 |
| | 3.325 | 0.01502 |
| | 0.0 | 0.01789 |
| 293.2 | 13.830 | 0.00923 |
| | 11.927 | 0.00976 |
| | 6.903 | 0.01184 |
| | 6.738 | 0.01182 |
| | 3.870 | 0.01323 |
| | 3.325 | 0.01346 |

AUXILIARY INFORMATION

**METHOD/APPARATUS/PROCEDURE:**

Ostwald method, using gas buret and pipet. Measurement of volume of gas before and after absorption. Vapor pressure of water was allowed for assuming Raoult's law. Solution was degassed. Concentration of salt solution estimated by titration with silver nitrate solution.

**SOURCE AND PURITY OF MATERIALS:**

No details given.

**ESTIMATED ERROR:**

$\delta\alpha/\alpha = \pm 0.03$

**REFERENCES:**

| COMPONENTS: | ORIGINAL MEASUREMENTS: |
|---|---|
| (1) Nitrogen; $N_2$; [7727-37-9]<br>(2) Barium chloride; $BaCl_2$; [10361-37-2]<br>(3) Water; $H_2O$; [7732-18-5] | Braun, L.<br>*Z. Phys. Chem.* 1900, *33*, 721-41. |

EXPERIMENTAL VALUES:

| T/K | Wt. of salt per 100g soln. /g | Bunsen coefficient, $\alpha$ |
|---|---|---|
| 293.2 | 0.0 | 0.01621 |
| 298.2 | 13.830 | 0.00783 |
| | 11.927 | 0.00855 |
| | 6.903 | 0.01044 |
| | 6.738 | 0.01036 |
| | 3.870 | 0.01137 |
| | 3.325 | 0.01190 |
| | 0.0 | 0.01432 |

COMPONENTS:

(1) Nitrogen; $N_2$; [7727-37-9]

(2) Lithium Chloride; LiCl; [7447-41-8]

(3) Water; $H_2O$; [7732-18-5]

ORIGINAL MEASUREMENTS:

Morrison, T. J.; Billett, F.

*J. Chem. Soc.* 1952, 3819-22.

VARIABLES:

T/K = 285.75 - 344.85

P/kPa = 101.325

PREPARED BY:

H. L. Clever, R. Battino

EXPERIMENTAL VALUES:

| T/K | 1/(T/K) | Salt Effect Parameters $(1/m)\log(S°/S)$[1] | Salt Effect Parameters $(1/m)\log(x°/x)$ |
|---|---|---|---|
| 285.75 | 0.0035 | 0.125 | 0.140 |
| 303.15 | 0.0033 | 0.095 | 0.110 |
| 322.55 | 0.0031 | 0.084 | 0.099 |
| 344.85 | 0.0029 | 0.092 | 0.107 |

[1] The authors used $(1/c)\log(S°/S)$ with c defined as g eq salt per kg of water. For the 1-1 electrolyte the compiler changed the c to an m for $m_{LiCl}/\text{mol kg}^{-1}$. The nitrogen solubility S is $\text{cm}^3$(STP) $\text{kg}^{-1}$.

The salt effect parameters were calculated from two measurements. The solubility of nitrogen in water, S°, and in the one molal salt solution, S. Only the solubility of the nitrogen in water, and the value of the salt effect parameter are given in the paper. The solubility values in the salt solution are not given.

The compiler calculated the values of the salt effect parameter using the mole fraction gas solubility ratio.

AUXILIARY INFORMATION

METHOD/APPARATUS/PROCEDURE:

The degassed solvent flows in a thin film down an absorption helix containing the nitrogen gas plus solvent vapor at a total pressure of one atmosphere. The volume of gas absorbed is measured in an attached buret system (1).

SOURCE AND PURITY OF MATERIALS:

(1) Nitrogen. From a cylinder and from ammonium nitrite gave the same results.

(2) Lithium Chloride. "AnalaR" material.

(3) Water. No information given.

ESTIMATED ERROR:

$\delta k = 0.010$

REFERENCES:

1. Morrison, T. J.; Billett, F. *J. Chem. Soc.* 1948, 2033-5.

**COMPONENTS:**

(1) Nitrogen; $N_2$; [7727-37-9]

(2) Sodium hydroxide; NaOH; [1310-73-2]

(3) Water; $H_2O$; [7732-18-5]

**ORIGINAL MEASUREMENTS:**

Levina, M. I.; Stsibarovskaya, N. P.

*Russ. J. Phys. Chem.* 1939, 12, 653-9.

**VARIABLES:**

T/K = 273-533
P/MPa = 9.8
Concentration

**PREPARED BY:**

V. Katovic

**EXPERIMENTAL VALUES:**

| t/°C | T[a]/K | $H_2O$ | Oxygen Solubility[b] NaOH Solution/N 0.5 | 2 | 4 |
|---|---|---|---|---|---|
| Total pressure at 100 kg/cm$^2$ (9.81 MPa) | | | | | |
| 0 | 273 | 1.24 | 1.03 | 0.41 | 0.18 |
| 25 | 298 | 1.01 | 0.76 | 0.37 | 0.16 |
| 35 | 308 | 0.88 | 0.65 | 0.35 | 0.14 |
| 50 | 323 | 0.78 | 0.58 | 0.26 | 0.15 |
| 75 | 348 | 0.64 | 0.55 | 0.25 | 0.17 |
| 100 | 373 | 0.71 | 0.59 | 0.31 | 0.20 |
| 125 | 398 | 0.77 | 0.62 | 0.33 | -- |
| 150 | 423 | 0.99 | 0.80 | 0.44 | 0.24 |
| 200 | 473 | 1.46 | 1.07 | 0.62 | 0.37 |
| 240 | 513 | 1.66 | 1.44 | 0.84 | 0.51 |

[a] Calculated by compiler.

[b] Solubility in cm$^3$(STP) per 1 gram of solvent at total pressure of 100 kg/cm$^2$ (9.81 MPa).

continued on following page

**AUXILIARY INFORMATION**

**METHOD/APPARATUS/PROCEDURE:**

Solubility determined by an apparatus described in the paper, by measuring the volume of liquid and air. Solubility of nitrogen was calculated from measured volume of air and oxygen (determined by pyrogallol).

**SOURCE AND PURITY OF MATERIALS:**

(1) From air.

(2,3) No details given.

**ESTIMATED ERROR:**

$\delta$ solubility/solubility = ±0.05, compiler's estimate.

**REFERENCES:**

COMPONENTS:

(1) Nitrogen; $N_2$; [7727-37-9]

(2) Sodium hydroxide; NaOH; [1310-73-2]

(3) Water; $H_2O$; [7732-18-5]

ORIGINAL MEASUREMENTS:

Levina, M. I.; Stsibarovskaya, N. P.

*Russ. J. Phys. Chem.* 1939, *12*, 653-9.

EXPERIMENTAL VALUES:

continued

| t/°C | $T^a$/K | Solubility Coefficient[b] $H_2O$ | NaOH Solution/N 0.5 | 2 | 4 |
|---|---|---|---|---|---|
| 0 | 273 | 1.58 | 1.31 | 0.52 | 0.23 |
| 25 | 298 | 1.28 | 0.96 | 0.47 | 0.20 |
| 35 | 308 | 1.11 | 0.82 | 0.44 | 0.18 |
| 50 | 323 | 0.99 | 0.74 | 0.33 | 0.19 |
| 75 | 348 | 0.81 | 0.70 | 0.32 | 0.21 |
| 100 | 373 | 0.91 | 0.75 | 0.39 | 0.25 |
| 125 | 398 | 1.01 | 0.80 | 0.43 | -- |
| 150 | 423 | 1.32 | 0.58 | 0.31 | -- |
| 200 | 473 | 2.20 | 1.61 | 0.93 | 0.54 |
| 240 | 513 | 3.20 | 2.76 | 1.55 | 0.92 |

[a] Calculated by compiler.

[b] Solubility coefficient of $cm^3$(STP) per 1 gram of a solution at a partial nitrogen pressure of 100 $kg/cm^2$ (9.81 MPa).

COMPONENTS:

(1) Nitrogen; $N_2$; [7727-37-9]

(2) Sodium chloride; NaCl; [7647-14-5]

(3) Water; $H_2O$; [7732-18-5]

ORIGINAL MEASUREMENTS:

Braun, L.

*Z. Phys. Chem.* 1900, *33*, 721- 41.

VARIABLES:

T/K = 278-298
Concentration

PREPARED BY:

C.L. Young

EXPERIMENTAL VALUES:

| T/K | Wt. of salt per 100g soln./g | Bunsen coefficient, $\alpha$ |
|---|---|---|
| 278.2 | 11.732 | 0.01016 |
| | 10.945 | 0.01052 |
| | 8.135 | 0.01266 |
| | 8.033 | 0.01248 |
| | 6.595 | 0.01380 |
| | 6.400 | 0.01375 |
| | 4.196 | 0.01579 |
| | 3.880 | 0.01615 |
| | 2.120 | 0.01795 |
| | 2.100 | 0.01805 |
| | 0.686 | 0.01994 |
| | 0.671 | 0.02000 |
| | 0.0 | 0.02173 |
| 283.2 | 11.732 | 0.00930 |
| | 10.945 | 0.00912 |
| | 8.135 | 0.01131 |
| | 8.033 | 0.01121 |
| | 6.595 | 0.01252 |
| | 6.400 | 0.01259 |
| | 4.196 | 0.01451 |
| | 3.880 | 0.01475 |
| | 2.120 | 0.01638 |
| | 2.100 | 0.01656 |
| | 0.686 | 0.01833 |
| | 0.671 | 0.01845 |
| | 0.0 | 0.02003 |

AUXILIARY INFORMATION

METHOD/APPARATUS/PROCEDURE:

Ostwald method, using gas buret and pipet. Measurement of volume of gas before and after absorption. Vapor pressure of water was allowed for assuming Raoult's law. Solution was degassed. Concentration of salt solution estimated by titration with silver nitrate solution.

SOURCE AND PURITY OF MATERIALS:

No details given.

ESTIMATED ERROR:

$\delta\alpha/\alpha = \pm 0.03$.

REFERENCES:

| COMPONENTS: | ORIGINAL MEASUREMENTS |
|---|---|
| (1) Nitrogen; $N_2$; [7727-37-9]<br>(2) Sodium chloride; NaCl; [7647-14-5]<br>(3) Water; $H_2O$; [7732-18-5] | Braun, L.<br>*Z. Phys. Chem.* 1900, *33*, 721-41. |

EXPERIMENTAL VALUES:

| T/K | Wt. of salt per 100g soln./g | Bunsen coefficient, $\alpha$ |
|---|---|---|
| 288.2 | 11.732 | 0.00810 |
| | 10.945 | 0.00824 |
| | 8.135 | 0.01014 |
| | 8.033 | 0.00995 |
| | 6.595 | 0.01120 |
| | 6.400 | 0.01134 |
| | 4.196 | 0.01294 |
| | 3.880 | 0.01316 |
| | 2.120 | 0.01469 |
| | 2.100 | 0.01467 |
| | 0.686 | 0.01640 |
| | 0.671 | 0.01642 |
| | 0.0 | 0.01789 |
| 293.2 | 11.732 | 0.00657 |
| | 10.945 | 0.00703 |
| | 8.135 | 0.00872 |
| | 8.033 | 0.00871 |
| | 6.595 | 0.00972 |
| | 6.400 | 0.00975 |
| | 4.196 | 0.01151 |
| | 3.880 | 0.01168 |
| | 2.120 | 0.01311 |
| | 2.100 | 0.01314 |
| | 0.686 | 0.01477 |
| | 0.671 | 0.01484 |
| | 0.0 | 0.01621 |
| 298.2 | 11.732 | 0.00470 |
| | 10.945 | 0.00565 |
| | 8.135 | 0.00749 |
| | 8.033 | 0.00729 |
| | 6.595 | 0.00802 |
| | 6.400 | 0.00826 |
| | 4.196 | 0.00990 |
| | 3.880 | 0.01005 |
| | 2.120 | 0.01131 |
| | 2.100 | 0.01133 |
| | 0.686 | 0.01295 |
| | 0.671 | 0.01304 |
| | 0.0 | 0.01432 |

| COMPONENTS: | ORIGINAL MEASUREMENTS: |
|---|---|
| (1) Nitrogen; $N_2$; [7727-37-9]<br>(2) Sodium Chloride; NaCl; [7647-14-5]<br>(3) Water; $H_2O$; [7732-18-5] | Morrison, T. J.; Billett, F.<br>*J. Chem. Soc.* 1952, 3819-22. |
| VARIABLES:<br>T/K = 285.75 - 344.85<br>P/kPa = 101.325 | PREPARED BY:<br>H. L. Clever, R. Battino |

EXPERIMENTAL VALUES:

| T/K | 1/(T/K) | Salt Effect Parameters | |
|---|---|---|---|
| | | (1/m)log(S°/S)[1] | (1/m)log($x$°/$x$) |
| 285.75 | 0.0035 | 0.154 | 0.169 |
| 303.15 | 0.0033 | 0.121 | 0.136 |
| 322.55 | 0.0031 | 0.106 | 0.121 |
| 344.85 | 0.0029 | 0.106 | 0.121 |

[1] The authors used (1/c)log(S°/S) with c defined as g eq salt per kg of water. For the 1-1 electrolyte the compiler changed the c to an m for $m_{NaCl}$/mol $kg^{-1}$. The nitrogen solubility S is $cm^3$(STP) $kg^{-1}$.

The salt effect parameters were calculated from two measurements. The solubility of nitrogen in water, S°, and in the one molal salt solution, S. Only the solubility of the nitrogen in water, and the value of the salt effect parameter are given in the paper. The solubility values in the salt solution are not given.

The compiler calculated the values of the salt effect parameter using the mole fraction gas solubility ratio.

AUXILIARY INFORMATION

METHOD/APPARATUS/PROCEDURE:

The degassed solvent flows in a thin film down an absorption helix containing the nitrogen gas plus solvent vapor at a total pressure of one atmosphere. The volume of gas absorbed is measured in an attached buret system (1).

SOURCE AND PURITY OF MATERIALS:

(1) Nitrogen. From a cylinder and from ammonium nitrite gave the same results.

(2) Sodium Chloride. "AnalaR" material.

(3) Water. No information given.

ESTIMATED ERROR:

$\delta k = 0.010$

REFERENCES:

1. Morrison, T. J.; Billett, F. *J. Chem. Soc.* 1948, 2033-5.

| COMPONENTS: | ORIGINAL MEASUREMNTS: |
|---|---|
| (1) Nitrogen; $N_2$; [7727-37-9]<br>(2) Sodium chloride; NaCl; [7647-14-5]<br>(3) Water; $H_2O$; [7732-18-5] | Mishnina, T. A.; Avdeeva, O. I. Bozhovakaya, T. K.<br>*Materialy Vses. Nauchn. Issled. Geol. Inst.* 1961, *46*, 93-110. |
| VARIABLES:<br>T/K = 283.15 - 343.15<br>P/kPa = 101.325<br>$c_2$/mol dm$^{-3}$ = 0 - 5.4 | PREPARED BY:<br>A. L. Cramer<br>H. L. Clever |

EXPERIMENTAL VALUES:

| NaCl/g eq dm$^{-3}$ | Bunsen Coefficient, $\alpha \times 10^3$ | | | | | | | | | | | | Salt Effect Parameter |
|---|---|---|---|---|---|---|---|---|---|---|---|---|---|
| T/K | 0.0 | 0.5 | 1.0 | 1.5 | 2.0 | 2.5 | 3.0 | 3.5 | 4.0 | 4.5 | 5.0 | 5.4 | $k_s$ |
| 283.15 | 19.5 | 16.2 | 13.5 | 11.2 | 9.3 | 7.7 | 6.4 | 5.3 | 4.5 | 3.8 | 3.1 | 2.7 | 0.160 |
| 288.75 | 17.4 | 14.7 | 12.4 | 10.4 | 8.8 | 7.4 | 6.3 | 5.3 | 4.4 | 3.7 | 3.2 | 2.8 | 0.148 |
| 293.15 | 16.0 | 13.7 | 11.7 | 10.0 | 8.5 | 7.3 | 6.2 | 5.3 | 4.5 | 3.9 | 3.3 | 2.9 | 0.137 |
| 298.15 | 14.8 | 12.8 | 11.0 | 9.5 | 8.1 | 7.0 | 6.1 | 5.2 | 4.5 | 3.9 | 3.4 | 3.0 | 0.129 |
| 303.15 | 13.8 | 12.0 | 10.4 | 9.1 | 7.9 | 6.9 | 6.0 | 5.2 | 4.5 | 3.9 | 3.4 | 3.1 | 0.121 |
| 308.15 | 12.8 | 11.2 | 9.8 | 8.5 | 7.5 | 6.5 | 5.7 | 5.0 | 4.4 | 3.8 | 3.3 | 3.0 | 0.117 |
| 313.15 | 12.2 | 10.7 | 9.4 | 8.3 | 7.2 | 6.4 | 5.6 | 4.9 | 4.3 | 3.8 | 3.3 | 3.0 | 0.113 |
| 318.15 | 11.5 | 10.1 | 9.0 | 7.9 | 7.0 | 6.1 | 5.4 | 4.8 | 4.2 | 3.7 | 3.3 | 3.0 | 0.109 |
| 323.15 | 11.3 | 10.0 | 8.8 | 7.8 | 6.9 | 6.1 | 5.4 | 4.8 | 4.3 | 3.8 | 3.3 | 3.0 | 0.106 |
| 328.15 | 11.1 | 9.8 | 8.7 | 7.7 | 6.8 | 6.0 | 5.3 | 4.7 | 4.2 | 3.7 | 3.3 | 3.0 | 0.106 |
| 333.15 | 10.8 | 9.6 | 8.5 | 7.5 | 6.6 | 5.9 | 5.2 | 4.6 | 4.1 | 3.6 | 3.2 | 2.9 | 0.106 |
| 338.15 | 10.7 | 9.5 | 8.4 | 7.4 | 6.6 | 5.8 | 5.1 | 4.6 | 4.0 | 3.6 | 3.2 | 2.9 | 0.106 |
| 343.15 | 10.6 | 9.4 | 8.3 | 7.4 | 6.5 | 5.8 | 5.1 | 4.5 | 3.9 | 3.5 | 3.1 | 2.8 | 0.106 |

The table of smoothed Bunsen coefficients of nitrogen dissolved in aqueous sodium chloride solutions was prepared by the authors. The complete source of data for the table is not clear. A 1958 report of A. A. Cherepinnikov, mentioned in the paper, was not available to us.

**COMPONENTS:**

(1) Nitrogen; $N_2$; [7727-37-9]

(2) Sodium salts; see below.

(3) Water; $H_2O$; [7732-18-5]

**ORIGINAL MEASUREMENTS:**

Van Slyke, D.D.; Dillon, R.T.; Margaria, R.

*J. Biol. Chem.* 1934, *105*, 571-96.

**VARIABLES:**

T/K = 298-311
Concentration

**PREPARED BY:**

R. Battino

**EXPERIMENTAL VALUES:**

At 25°C (298.15K) Except For Indicated Point

| Liquid Phase | $P_{N_2}$ [a]/mmHg | $P_{N_2}$ [b]/kPa | $10^2\alpha$ [c]/ $cm^3$ (STP) $cm^3$ $atm^{-1}$ | $10^2\alpha°$ [d]/ $cm^3$ (STP) $g^{-1}$ $H_2O$ |
|---|---|---|---|---|
| 0.155 M NaCl [e] | 582 | 77.6 | 1.408 | 1.416 |
| 0.155 M NaCl | 582 | 77.6 | 1.410 | 1.418 |
| 0.155 M NaCl | 590 | 78.7 | 1.406 | 1.414 |
| 0.155 M NaCl | 590 | 78.7 | 1.410 | 1.418 |
| 0.155 M NaCl [f] | 564.5 | 75.26 | 1.220 | |
| 0.077 M $Na_2CO_3$ [g] | 587 | 78.3 | 1.387 | 1.391 |
| 0.077 M $Na_2CO_3$ | 587 | 78.3 | 1.389 | 1.393 |
| 0.153 M $NaHCO_3$ [h] | 555 | 74.0 | 1.396 [h] | 1.404 |
| 0.153 M $NaHCO_3$ | 555 | 74.0 | 1.402 [h] | 1.410 |
| 0.153 M $NaHCO_3$ | 555 | 74.0 | 1.394 [h] | 1.402 |
| 0.153 M $NaHCO_3$ | 555 | 74.0 | 1.396 [h] | 1.404 |

[a] Nitrogen partial pressure. Solubilities calculated at 101 kPa $P_{N_2}$.
[b] Calculated by compiler.
[c] Bunsen coefficient.
[d] Solubility in $cm^3$ (STP) nitrogen per gram of water.
[e] Sodium chloride; NaCl; [7647-14-5].
[f] At 38°C (311.15 K).
[g] Sodium carbonate; $Na_2CO_3$; [497-19-8].
[h] Sodium bicarbonate; $NaHCO_3$; [144-55-8]. Solution also contained 0.003 M $Na_2CO_3$. Values have been corrected by adding 0.00004.

**AUXILIARY INFORMATION**

**METHOD/APPARATUS/PROCEDURE:**

The solutions were saturated with gas using the rotating double tonometer method of Austin, et al. (1). 2-2.5 h were allowed for saturation. The amount of dissolved nitrogen was determined using the method of Van Slyke and Neil (2).

**SOURCE AND PURITY OF MATERIALS:**

(1) From air.

(2) Reagent grade.

(3) "Freshly Distilled".

**ESTIMATED ERROR:**

$\delta T/K = \pm 0.04$
$\delta\alpha/\alpha = \pm 0.01$, compiler's estimate.

**REFERENCES:**

1. Austin, J.H.; Cullen, G.E.; Hastings, A.B.; McLean, F.C.; Peters, J.D.; Van Slyke, D.D. *J. Biol. Chem.* 1922, *54*, 134.
2. Van Slyke, D.D.; Neil, J.M. *J. Biol. Chem.* 1924, *61*, 523.

COMPONENTS:

(1) Nitrogen; $N_2$; [7727-37-9]

(2) Sulfurous acid, disodium salt, (Sodium sulfite); $Na_2SO_3$; [10579-83-6]

(3) Water; $H_2O$; [7732-18-5]

ORIGINAL MEASUREMENTS:

Yasunishi, A.

*J. Chem. Eng. Jpn.* 1977, *10*, 89-94

VARIABLES:

T/K = 298
Concentration

PREPARED BY:

C.L. Young

EXPERIMENTAL VALUES:

Pressure = $1.01325 \times 10^5$ Pa

| T/K | Conc of salt soln. /mol $l^{-1}$ | Ostwald coefficient, *L* |
|---|---|---|
| 298.15 | 0.0 | 0.01619 |
| | 0.102 | 0.01474 |
| | 0.106 | 0.01470 |
| | 0.209 | 0.01338 |
| | 0.318 | 0.01239 |
| | 0.472 | 0.01073 |
| | 0.636 | 0.00931 |
| | 0.778 | 0.00808 |
| | 0.785 | 0.00830 |
| | 0.938 | 0.00720 |
| | 1.043 | 0.00650 |

AUXILIARY INFORMATION

METHOD/APPARATUS/PROCEDURE:

Volumetric apparatus with multibulb buret and magnetic stirrer. Amount of solution and gas absorbed determined volumetrically. Density and composition of solution determined after absorption. Details in source.

SOURCE AND PURITY OF MATERIALS:

(1) Purity 99.99 mole per cent

(2) Guaranteed reagent

(3) Degassed

ESTIMATED ERROR:

$\delta T/K = \pm 0.05$; $\delta L = \pm 1\%$

(estimated by compiler)

REFERENCES:

| COMPONENTS: | ORIGINAL MEASUREMENTS: |
|---|---|
| (1) Nitrogen; $N_2$; [7727-37-9]<br>(2) Sulfuric acid, disodium salt, (Sodium sulfate); $Na_2SO_4$; [7757-82-6]<br>(3) Water; $H_2O$; [7732-18-5] | Yasunishi, A.<br>*J. Chem. Eng. Jpn.* 1977, *40*, 89-94. |
| VARIABLES:<br>T/K = 298<br>Concentration | PREPARED BY:<br>C.L. Young |

EXPERIMENTAL VALUES:

Pressure = $1.01325 \times 10^5$ Pa

| T/K | Conc of salt soln. /mol $l^{-1}$ | Ostwald coefficient, *L* |
|---|---|---|
| 298.15 | 0.0 | 0.01619 |
| | 0.101 | 0.01491 |
| | 0.201 | 0.01377 |
| | 0.303 | 0.01262 |
| | 0.516 | 0.01075 |
| | 0.764 | 0.00871 |
| | 0.912 | 0.00761 |
| | 1.079 | 0.00670 |

AUXILIARY INFORMATION

METHOD /APPARATUS/PROCEDURE:

Volumetric apparatus with multibulb buret and magnetic stirrer. Amount of solution and gas absorbed determined volumetrically. Density and composition of solution determined after absorption. Details in source.

SOURCE AND PURITY OF MATERIALS:

(1) Purity 99.99 mole per cent

(2) Guaranteed reagent

(3) Degassed

ESTIMATED ERROR:

$\delta T/K = \pm 0.05$; $\delta L = \pm 1\%$

(estimated by compiler)

REFERENCES:

**COMPONENTS:**

(1) Nitrogen; $N_2$; [7727-37-9]

(2) Sodium sulfate; $Na_2SO_4$; [7757-82-6]

(3) Sulfuric acid; $H_2SO_4$; [7664-93-9]

(4) Water; $H_2O$; [7732-18-5]

**ORIGINAL MEASUREMENTS:**

Kobe, K. A.; Kenton, F. H.

*Ind. Eng. Chem., Anal. Ed.* 1938, *10*, 76-7.

**VARIABLES:**

T/K = 298.15

$N_2$ P/kPa = 101.325

**PREPARED BY:**

P. L. Long
H. L. Clever

**EXPERIMENTAL VALUES:**

| Temperature t/°C | T/K | Solvent Volume cm³ | Absorbed Oxygen Volume cm³ | Bunsen Coefficient $\alpha \times 10^2$ | Ostwald Coefficient $L \times 10^2$ |
|---|---|---|---|---|---|
| 25 | 298.15 | 99.54 | 0.50 | | |
| | | 99.54 | 0.48 | 0.45 av. | 0.49 av. |

The solvent is a mixture of 800 g $H_2O$
200 g $Na_2SO_4$ (anhydrous)
40 cm³ $H_2SO_4$ (36 N)

Thus the solution molality is:

$m_{Na_2SO_4}$/mol kg$^{-1}$ 1.76

$m_{H_2SO_4}$/mol kg$^{-1}$ 0.90

**AUXILIARY INFORMATION**

**METHOD/APPARATUS/PROCEDURE:**

The apparatus was described in detail in an earlier paper (1). The apparatus consists of a gas buret, a pressure compensator, a 200 cm³ absorption bulb, and a mercury leveling bulb. The absorption bulb is attached to a shaking machine.

The solvent and the gas are placed in an absorption bulb. The bulb is shaken until equilibrium is reached. The remaining gas is returned to the buret. The difference in the final and initial volumes is taken as the gas absorbed.

The authors also measured the solvent vapor pressure.

**SOURCE AND PURITY OF MATERIALS:**

(1) Nitrogen. Source not given. Purity stated to be 99$^+$ per cent.

(2 & 3) Sodium sulfate and sulfuric acid. Sources not given. Analytical grades.

(4) Water. Distilled.

**ESTIMATED ERROR:**

$\delta\alpha$/cm³ = 0.001 (authors)

**REFERENCES:**

1. Kobe, K. A.; Williams, J. S. *Ind. Eng. Chem., Anal. Ed.* 1935, *7*, 37.

| COMPONENTS: | ORIGINAL MEASUREMENTS: |
|---|---|
| (1) Nitrogen; $N_2$; [7727-37-9]<br>(2) Sulfuric acid, disodium salt, (Sodium sulfate); $Na_2SO_4$; [7757-82-6]<br>(3) Sulfurous acid, disodium salt (Sodium sulfite); $Na_2SO_4$; [10579-83-6]<br>(4) Water; $H_2O$; [7732-18-5] | Yasunishi, A.<br><br>*J. Chem. Eng. Jpn.* <u>1977</u>, *10*, 89-94 |
| **VARIABLES:** | **PREPARED BY:** |
| T/K = 298<br>Concentration | C.L. Young |

EXPERIMENTAL VALUES:

Pressure = $1.01325 \times 10^5$ Pa

| T/K | Total conc of salt/mol $l^{-1}$ | Mole fraction of sodium sulfite in mixed salt. | Ostwald coefficient, $L$ |
|---|---|---|---|
| 298.15 | 0.305 | 0.253 | 0.01244 |
| | 0.300 | 0.502 | 0.01246 |
| | 0.303 | 0.713 | 0.01251 |
| | 0.598 | 0.504 | 0.00973 |
| | 0.864 | 0.258 | 0.00796 |
| | 0.883 | 0.508 | 0.00767 |

AUXILIARY INFORMATION

METHOD/APPARATUS/PROCEDURE:

Volumetric apparatus with multibulb buret and magnetic stirrer. Amount of solution and gas absorbed determined volumetrically. Density and composition of solution determined after absorption. Details in source.

SOURCE AND PURITY OF MATERIALS:

(1) Purity 99.99 mole per cent

(2) and (3) Guaranteed reagents

(4) Degassed

ESTIMATED ERROR:

$\delta T/K = \pm 0.05$; $\delta L = \pm 1\%$

(estimated by compiler)

REFERENCES:

**COMPONENTS:**

(1) Nitrogen; $N_2$; [7727-37-9]

(2) Oxygen; $O_2$; [7782-44-7]

(3) Carbon dioxide; $CO_2$; [124-38-9]

(4) Sulfuric acid; $H_2SO_4$; [7664-93-9]

(5) Sodium sulfate; $Na_2SO_4$; [7757-82-6]

(6) Water; $H_2O$; [7732-18-5]

**ORIGINAL MEASUREMENTS:**

Kobe, K. A.; Kenton, F. H.

*Ind. Eng. Chem., Anal. Ed.* 1938, *10*, 76-7.

**VARIABLES:**

$T$/K = 298.15

$p_1$/kPa = 101.325

**PREPARED BY:**

P. L. Long
H. L. Clever

**EXPERIMENTAL VALUES:**

| Temperature $t$/°C | $T$/K | Gas Composition volume per cent | Solvent Volume $V/cm^3$ | $N_2$, $O_2$ and $CO_2$ Volume Absorbed $v_1/cm^3$ | Bunsen Coefficient $\alpha/cm^3(STP)cm^{-3}atm^{-1}$ | Ostwald Coefficient $L/cm^3cm^{-3}$ |
|---|---|---|---|---|---|---|
| 25 | 298.15 | 6.1% $O_2$ | 49.54 | 1.53 | 0.0284 | 0.0310 |
| | | 14.5% $CO_2$ | 49.54 | 1.54 | | |
| | | 79.4% $N_2$ | | | | |

The solvent is a mixture of 800 g $H_2O$
200 g $Na_2SO_4$ (anhydrous)
40 ml $H_2SO_4$ (Conc., 36 normal)

Thus the molality of the solution is

$$m_4/\text{mol kg}^{-1} = 0.90\ (H_2SO_4)$$

$$m_5/\text{mol kg}^{-1} = 1.76\ (Na_2SO_4)$$

**AUXILIARY INFORMATION**

**METHOD/APPARATUS/PROCEDURE:**

The apparatus is described in detail in an earlier paper (1). The apparatus consists of a gas buret, a pressure compensator, and a 200 $cm^3$ absorption bulb and mercury leveling bulb. The absorption bulb is attached to a shaking mechanism.

The solvent and the gas are placed in the absorption bulb. The bulb is shaken until equilibrium is reached. The remaining gas is returned to the buret. The difference in the final and initial volumes is taken as the volume of gas absorbed.

**SOURCE AND PURITY OF MATERIALS:**

(1,2,3) Nitrogen, Oxygen, and Carbon dioxide. Source not given. Purity stated to be 99+ per cent.

(4,5) Sulfuric acid and sodium sulfate. Sources not given. Analytical grade.

(6) Water. Distilled.

**ESTIMATED ERROR:**

$\delta\alpha/cm^3 = \pm 0.005$ (authors)

**REFERENCES:**

1. Kobe, K. A.; Williams, J. S. *Ind. Eng. Chem., Anal. Ed.* 1935, *7*, 37.

| COMPONENTS: | ORIGINAL MEASUREMENTS: |
|---|---|
| (1) Nitrogen; $N_2$; [7727-37-9]<br>(2) Acetic acid; $C_2H_4O_2$; [64-19-7]<br>(3) Sodium acetate; $C_2H_4O_2 \cdot Ha$; [127-09-3]<br>(4) Metal (2+) acetates<br>(5) Water; $H_2O$; [7732-18-5] | Brasted, R. C.; Hirayama, C.<br>*J. Phys. Chem.* 1958, *62*, 125-6. |
| **VARIABLES:** | **PREPARED BY:** |
| T/K = 298.15<br>P/kPa = 101.325 | P. L. Long<br>H. L. Clever |

EXPERIMENTAL VALUES: All metal acetate solutions are one mol $dm^{-3}$ acetic acid.

| Temperature t/°C | T/K | $c_{M(C_2H_4O_2)_2}$/mol $dm^-$ | $c_{NaC_2H_4O_2}$/mol $dm^{-3}$ | Ostwald Coefficient $L \times 10^2$ |
|---|---|---|---|---|
| 25.00 | 298.15 | 0 | 0 | 1.61 |
| | | Cobalt (2+) acetate; $Co(C_2H_4O_2)_2$; [71-48-7] | | |
| | | 0.033 | 0.50 | 1.28 |
| | | 0.10 | 0.50 | 1.30 |
| | | Copper (2+) acetate; $Cu(C_2H_4O_2)_2$; [142-71-2] | | |
| | | 0.050 | 0.50 | 1.30 |
| | | 0.157 | 0.52 | 1.33 |
| | | Manganese (2+) acetate; $Mn(C_2H_4O_2)_2$; [638-38-0] | | |
| | | 0.050 | 0.50 | 1.34 |
| | | 0.15 | 0.50 | 1.34 |
| | | Mercury (2+) acetate; $Hg(C_2H_4O_2)_2$; [1600-27-7] | | |
| | | 0.05 | 0.50 | 1.45 |
| | | 0.15 | 0.50 | 1.36 |
| | | Nickel (2+) acetate; $Ni(C_2H_4O_2)_2$; [373-02-4] | | |
| | | 0.051 | 0.51 | 1.33 |
| | | 0.15 | 0.50 | 1.39 |
| | | Zinc (2+) acetate; $Zn(C_2H_4O_2)_2$; [557-34-6] | | |
| | | 0.15 | 0.50 | 1.25 |

AUXILIARY INFORMATION

METHOD/APPARATUS/PROCEDURE:

[1] The authors quote (with an error carried along) from Lannung (1), the definition of the Bunsen coefficient However, they call it an Ostwald coefficient. They compare their value of the solubility of $N_2$ in $H_2O$ with a handbook Bunsen coefficient. They state they do not understand why their value, $1.61 \times 10^{-2}$, is high. Their value in water agrees well with the recommended Ostwald coefficient. The evaluator has labelled the result as Ostwald coefficient.

The apparatus was not described. The authors did measure the solution vapor pressure in a static vapor pressure apparatus.

All solutions were one molar in acetic acid. The first entry in the table is for pure water.

SOURCE AND PURITY OF MATERIALS:

The chemicals were of reagent grade. They were used without further purification. The source of the chemicals was not given.

ESTIMATED ERROR:

$\delta L/L = 0.01$ (authors)

REFERENCES:

1. Lannung, A.
*J. Am. Chem. Soc.* 1930, *52*, 68-80.

COMPONENTS:

(1) Nitrogen: $N_2$; [7727-37-9]

(2) Potassium Chloride; KCl; [7447-40-7]

(3) Water; $H_2O$; [7732-18-5]

ORIGINAL MEASUREMENTS:

Bikov, M. M.

*Tr. Voronezhsk. Gos. Univ.* 1937, 9, 29-57.

VARIABLES:

T/K = 273-510
P/MPa = 9.81
Concentration

PREPARED BY:

V. Katovic

EXPERIMENTAL VALUES:

| t/°C | $T^a$/K | Nitrogen Solubility[b] $H_2O$ | KCl Solution/N 0.5 | 1.0 | 2.0 |
|---|---|---|---|---|---|
| Total Pressure of 100 $kg/cm^2$ (9.81 MPa) | | | | | |
| 0 | 273 | 1.254 | 1.215 | 0.974 | 0.629 |
| 25 | 298 | 1.013 | 0.834 | 0.696 | 0.478 |
| 50 | 323 | 0.746 | 0.664 | 0.598 | 0.461 |
| 75 | 348 | 0.640 | 0.629 | 0.543 | 0.404 |
| 100 | 373 | 0.715 | 0.656 | 0.579 | 0.432 |
| 150 | 423 | 0.994 | 0.885 | 0.685 | 0.566 |
| 200 | 473 | 1.460 | 1.188 | 0.975 | 0.771 |
| 240 | 513 | 1.664 | 1.661 | 1.342 | 0.938 |

[a] Calculated by compiler.

[b] Solubility in $cm^3$(STP) per 1 gram of solvent.

continued on following page

AUXILIARY INFORMATION

METHOD/APPARATUS/PROCEDURE:

Details of the method and a drawing of the apparatus are given in the paper.

SOURCE AND PURITY OF MATERIALS:

(1) From air.

(2) No details given.

(3) Distilled.

ESTIMATED ERROR:

$\delta T/K = \pm 0.2$
$\delta S/s = \pm 0.01$ compiler's estimate for solubility.

REFERENCES:

| COMPONENTS: | ORIGINAL MEASUREMENTS: |
|---|---|
| (1) Nitrogen; $N_2$; [7727-37-9]<br>(2) Potassium Chloride; KCl; [7447-40-7]<br>(3) Water; $H_2O$; [7732-18-5] | Bikov, M. M.<br>*Tr. Voronezhsk. Gos. Univ.* 1937, 9, 29-57. |

EXPERIMENTAL VALUES:

continued

| t/°C | $T^a$/K | Solubility Coefficient[b] $H_2O$ | KCl Solution/N 0.5 | 1.0 | 2.0 |
|---|---|---|---|---|---|
| 0 | 273 | 1.587 | 1.538 | 1.233 | 0.796 |
| 25 | 298 | 1.282 | 1.056 | 0.881 | 0.605 |
| 50 | 323 | 0.944 | 0.841 | 0.758 | 0.587 |
| 75 | 348 | 0.814 | 0.799 | 0.690 | 0.514 |
| 100 | 373 | 0.914 | 0.839 | 0.741 | 0.552 |
| 150 | 423 | 1.323 | 0.176[c] | 0.910 | 0.750 |
| 200 | 473 | 2.196 | 1.776 | 1.456 | 1.142 |
| 240 | 513 | 3.195 | 3.166 | 2.537 | 1.746 |

[a] Calculated by compiler.

[b] The solubility coefficient in $cm^3$ (STP) of nitrogen dissolved per 1 gram of solution at a nitrogen partial pressure of 100 kg/$cm^2$ (9.81 MPa).

[c] Probably a typographical error.

COMPONENTS:

(1) Nitrogen; $N_2$; [7727-37-9]

(2) Potassium Iodide; KI; [7681-11-0]

(3) Water; $H_2O$; [7732-18-5]

ORIGINAL MEASUREMENTS:

Morrison, T. J.; Billett, F.

*J. Chem. Soc.* 1952, 3819-22.

VARIABLES:

T/K = 285.75 - 344.85
P/kPa = 101.325

PREPARED BY:

H. L. Clever, R. Battino

EXPERIMENTAL VALUES:

| T/K | 1/(T/K) | Salt Effect Parameters $(1/m)\log(S°/S)$[1] | $(1/m)\log(x°/x)$ |
|---|---|---|---|
| 285.75 | 0.0035 | 0.134 | 0.149 |
| 303.15 | 0.0033 | 0.100 | 0.115 |
| 322.55 | 0.0031 | 0.086 | 0.101 |
| 344.85 | 0.0029 | 0.084 | 0.099 |

[1] The authors used $(1/c)\log(S°/S)$ with c defined as g eq salt per kg of water. For the 1-1 electrolyte the compiler changed the c to an m for $m_{KI}$ /mol $kg^{-1}$. The nitrogen solubility S is $cm^3$(STP) $kg^{-1}$.

The salt effect parameters were calculated from two measurements. The solubility of nitrogen in water, S°, and in the one molal salt solution, S. Only the solubility of the nitrogen in water, and the value of the salt effect parameter are given in the paper. The solubility values in the salt solution are not given.

The compiler calculated the values of the salt effect parameter using the mole fraction gas solubility ratio.

AUXILIARY INFORMATION

METHOD/APPARATUS/PROCEDURE:

The degassed solvent flows in a thin film down an absorption helix containing the nitrogen gas plus solvent vapor at a total pressure of one atmosphere. The volume of gas absorbed is measured in an attached buret system (1).

SOURCE AND PURITY OF MATERIALS:

(1) Nitrogen. From a cylinder and from ammonium nitrite gave the same results.

(2) Potassium Iodide, "AnalaR" material.

(3) Water. No information given.

ESTIMATED ERROR:

$\delta k = 0.010$

REFERENCES:

1. Morrison, T. J.; Billett, F. *J. Chem. Soc.* 1948, 2033-5.

| COMPONENTS: | ORIGINAL MEASUREMENTS: |
|---|---|
| (1) Nitrogen; $N_2$; [7727-37-9]<br>(2) Potassium carbonate; $K_2CO_3$; [584-08-7]<br>(3) Water; $H_2O$; [7732-18-5] | Schröder, W.<br>*Z. Phys. Chem. (Wiesbaden)* 1971, *75*, 248 - 64. |
| VARIABLES:<br>$T$/K = 323.15 - 403.15<br>$p_1$/MPa = 0.709 - 7.701 (7-76 atm)<br>$K_2CO_3$ $c_2$/mol dm$^{-3}$ = 0.5 - 6 | PREPARED BY:<br>H. L. Clever |

EXPERIMENTAL VALUES:

| Temperature $t/^{\circ}C$ | Temperature $T$/K | Potassium Carbonate $c_2$/mol dm$^{-3}$ | Nitrogen Pressure $p_1$/atm | Bunsen Coefficient $\alpha$/m$^3$(STP) m$^{-3}$ atm$^{-1}$ |
|---|---|---|---|---|
| 90 | 363.15 | 2 | 10 | 0.0099[2] |
| | | | 14.5 | 0.0072[2] |
| | | | 23 | 0.0056[2] |
| | | | 40 | 0.0056[2], 0.0027[3] |
| | | 4 | 11.5 | 0.0076[2] |
| | | | 16 | 0.0057[2] |
| | | | 21.5 | 0.0042[2] |
| | | | 29 | 0.0033[2] |
| | | | 40 | 0.0018[3] |
| | | | 49 | 0.0022[2] |
| | | 6 | 12 | 0.0060[2] |
| | | | 14 | 0.0052[2] |
| | | | 21 | 0.0036[2] |
| | | | 33 | 0.0021[2] |
| | | | 42 | 0.0017[2] |
| 110 | 383.15 | 1 | 10 | 0.0145[1] |
| | | | 13.5 | 0.0120[1] |
| | | | 17 | 0.0105[1] |
| | | | 25.5 | 0.0088[1] |
| | | | 41 | 0.0075[1] |
| 130 | 403.15 | 1 | 15 | 0.0109[1] |

[1] [2] Author's "break points"from Tables 2 and 3, respectively.
[3] Values from Author's Table 4. [4] Author's earlier paper (1).

AUXILIARY INFORMATION

METHOD/APPARATUS/PROCEDURE:

The apparatus is described in an earlier paper (1). It is a bubbling-type autoclave based on the design of Wiebe *et al.* (2).In general the upper pressure limit was 70 - 80 atm.

The nitrogen partial pressure was obtained by subtraction of the solvent vapor pressure from the total pressure.

The values of the solubility of nitrogen in water, footnote [4], were read by the compiler from diagrams in the author's earlier paper (1). They were added to give a basis of comparison for the salt effect.

SOURCE AND PURITY OF MATERIALS:

No information.

ESTIMATED ERROR:

REFERENCES:

1. Schröder, W. *Z. Naturforsch.* 1969, *24b*, 500.
2. Wiebe, R. ; Gaddy, V. L.; Heins, C. Jr. *J. Am. Chem. Soc.* 1933, *55*, 947.

| COMPONENTS: | ORIGINAL MEASUREMENTS: |
|---|---|
| (1) Nitrogen; $N_2$; [7727-37-9]<br>(2) Potassium carbonate; $K_2CO_3$; [584-08-7]<br>(2) Water; $H_2O$; [7732-18-5] | Schröder, W.<br>*Z. Phys. Chem. (Wiesbaden)* 1971, *75*, 248 - 64. |

EXPERIMENTAL VALUES:

The author measured the solubility of nitrogen in aqueous potassium carbonate as a function of temperature, pressure, and salt concentration.

The ranges covered by the experiments are summarized in the table below. Numerical data were not given. Isotherms of the Bunsen coefficient, $\alpha$, *vs.* pressure, $p_1$/atm, were displayed for the three systems for which the pressure range and number of experimental points are given in the table.

| Potassium Carbonate $c_2$/mol dm$^{-3}$ | Isotherm $T$/K | Pressure Range $p_1$/atm | Number of Experimental Values |
|---|---|---|---|
| 0.5 | 323.15 | | |
| 1 | 323.15 | | |
| | 343.15 | | |
| | 363.15 | | |
| | 383.15 | 7 - 52 | 60 |
| | 403.15 | 8 - 28 | 33 |
| 2 | 363.15 | | |
| 4 | 363.15 | 7 - 76 | 61 |
| 6 | 363.15 | | |

The author states that the isotherms can be broken in a maximum of 6 linear regions. He tabulated the Bunsen coefficient and pressure at the "break points" (knickpunkte) where linear regions met. The tabulated values plus some other values from this and from an earlier of the author's publications are given in the table below. The values are not experimental values, but the values read from the diagrams by the author and in some cases the compiler. The values reproduce the isotherms reasonably well.

| Temperature $t/^{\circ}C$ | $T$/K | Potassium Carbonate $c_2$/mol dm$^{-3}$ | Nitrogen Pressure $p_1$/atm | Bunsen Coefficient $\alpha$/m$^3$(STP) m$^{-3}$ atm$^{-1}$ |
|---|---|---|---|---|
| 50 | 323.15 | 0.0 | 10 | 0.0187[4] |
| | | | 15 | 0.0161[4] |
| | | | 20 | 0.0146[4] |
| | | | 30 | 0.0136[4] |
| | | | 40 | 0.0130[3] [4] |
| | | 0.5 | 40 | 0.0095[3] |
| | | 1 | 14.5 | 0.0102[1] |
| | | | 30 | 0.0075[1] |
| | | | 40 | 0.0071[3] |
| | | 2 | 40 | 0.0043[3] |
| 70 | 343.15 | 1 | 13 | 0.0117[1] |
| | | | 23 | 0.0086[1] |
| | | | 38 | 0.0070[1] |
| 90 | 363.15 | 0.0 | 10 | 0.0174[4] |
| | | | 15 | 0.0158[4] |
| | | | 20 | 0.0144[4] |
| | | | 30 | 0.0130[4] |
| | | | 40 | 0.0123[3] [4] |
| | | 0.5 | 40 | 0.0071[3] |
| | | 1 | 10 | 0.0142[1] [2] |
| | | | 16 | 0.0108[1] [2] |
| | | | 26 | 0.0085[1] [2] |
| | | | 38 | 0.0072[1] [2] |
| | | | 40 | 0.0043[3] |

Continued on next page.

| COMPONENTS: | EVALUATOR: |
|---|---|
| (1) Nitrogen; $N_2$; [7727-37-9]<br>(2) Polar Organic Compounds<br>(3) Water; $H_2O$; [7732-18-5] | H. Lawrence Clever<br>Department of Chemistry<br>Emory University<br>Atlanta, GA 30322 USA<br>1981, December |

CRITICAL EVALUATION:

The Solubility of Nitrogen in Mixtures of Polar Organic Compounds + Water

In this section ten nitrogen + polar organic compound + water systems are evaluated. The evaluation is difficult because all but two of the systems were studied at only one laboratory. The two systems studied at more than one laboratory were not studied at the same temperature or composition. Thus direct comparison of results from various laboratories is not possible.

Comparison of the nitrogen solubility in the pure components with the results of other workers can be made, especially for the nitrogen solubility in water, where we have Battino's recommended values to use (this volume, pp 1-3). Most of the papers report nitrogen in water Ostwald coefficients that are within 2 to 3 per cent of the recommended values. Tokunaga's (8) nitrogen in water solubility values are 4 to 5 per cent larger than the recommended values. Kretschmer, Nowakowska, and Wiebe (6) did not report a nitrogen + water solubility.

The solubility studies carried out before 1912 (1-5) used an Ostwald type apparatus. Just (2) gives a detailed description of the apparatus and the technique of the measurement. Both Kretschmer *et al.* (6) and Tokunaga (8) used modifications of the Horiuti apparatus. Kretschmer *et al.* either measured or used literature solvent vapor pressures and applied a correction for the dilation of the solvent due to the dissolved nitrogen. Tokunaga calculated the mixed solvent vapor pressure assuming Raoult's law and he did not correct for dilation, assuming the correction to be negligible.

Nitrogen + Methanol [67-56-1] + Water

Tokunaga (8) measured the solubility of nitrogen in the pure liquids and in thirteen methanol + water mixtures at 293.15 and 313.15 K. He made extra measurements in the 0 to 0.1 methanol mole fraction region where he showed there is a maximum in the solubility *vs.* mole fraction curve at 0.03-0.06 mole fraction methanol. The temperature coefficient of solubility changes from negative to positive as the composition changes from pure water to pure methanol. Similar behavior was reported earlier for argon in aqueous methanol (Solubility Series, Volume 4). Tokunaga's value of the solubility of nitrogen in water is 4 to 5 per cent greater than the recommended value and his nitrogen solubility in methanol is lower than the values reported by most others. The deviation towards high solubility values was also found in an evaluation of his work on oxygen as solute (Solubility Series, Volume 7). The data are classed as tentative.

Nitrogen + Ethanol [64-17-5] + Water

Just (2) measured the solubility of nitrogen in the two pure components and in two mixtures at 298.15 K. Kretschmer *et al.* (6) measured the solubility in ethanol and one mixture of 95 volume per cent ethanol at 298.15 K. The solubilities in the pure components agree well with accepted values and both sets of data are classed as tentative. Tokunaga (8) measured the solubility of nitrogen in the pure components and thirteen ethanol + water mixtures at 293.15 and 313.15 K. The features observed for the aqueous methanol system were also observed for aqueous ethanol. Tokunaga's nitrogen in water solubility values were discussed in the aqueous methanol system above. His solubilities in ethanol appear greater at 293.15 K and less at 313.15 K than the general run of the data of others. The data are classed as tentative.

| COMPONENTS: | EVALUATOR: |
|---|---|
| (1) Nitrogen; $N_2$; [7727-37-9]<br>(2) Polar Organic Compounds<br>(3) Water; $H_2O$; [7732-18-5] | H. Lawrence Clever<br>Department of Chemistry<br>Emory University<br>Atlanta, GA 30322 USA<br><br>1981, December |

CRITICAL EVALUATION:

Nitrogen + 1-Propanol [71-32-8] + Water
Nitrogen + 2-Propanol [67-63-0] + Water

Tokunaga (8) measured the solubility of nitrogen in water, 1-propanol, 2-propanol and up to twelve alcohol + water mixtures at 293.15 and 313.15 K. The maximum at low alcohol concentrations and the sign change in the temperature coefficient of solubility observed in the aqueous methanol and ethanol solutions are also observed in these systems. The data are classed as tentative.

Nitrogen + 1,2,3-Propanetriol [56-81-5] + Water

Drucker and Moles (4) measured the solubility of nitrogen in water and in five 1,2,3-propanetriol + water mixtures at 298.15 K. Müller (5) measured the solubility of nitrogen in water and in seven mixtures near 288.15 K. The values of the nitrogen solubility in water are 2.5 and 1.6 per cent, respectively, lower than the recommended value. The two data sets differ in that Müller made more measurements at the large 1,2,3-propanetriol compositions where he observed a minimum in the solubility *vs.* solvent composition curve at about 80 weight per cent alcohol. Both sets of data are classed as tentative.

Nitrogen + Propanoic acid [79-09-4] + Water

Braun (1) measured the solubility of nitrogen in water and in eight acid + water mixtures at concentrations up to 1.7 mol $kg^{-1}$ at five degree intervals between 278.2 and 298.2 K. Braun's values for the solubility of nitrogen in water range from 3.1 per cent greater at 278.2 to 2.4 per cent less at 298.2 K than the recommended values. The data are classed as tentative, but should be used with some caution. See the discussion of Braun's data in the aqueous electrolyte section under barium chloride and sodium chloride.

Nitrogen + 2-Methylpropanoic acid [79-31-2] + Water

Drucker and Moles (4) measured the solubility of nitrogen in water and up to six acid + water mixtures at 296.17 and 302.17 K. The data are classed as tentative.

Nitrogen + 2,2,2-Trichloro-1,1-ethanetriol [302-17-0] + Water

Müller (5) measured the solubility of nitrogen in water and in seven chloral hydrate + water mixtures at temperatures near 288.15 K. In addition to the original data he gives a table of solubility values corrected to 288.15 K. He observed a minimum in the solubility *vs.* solvent composition curve at about 56 weight per cent chloral hydrate. The data are classed as tentative.

Nitrogen + Acetamide [60-35-5] + Water

Hüfner (3) measured the solubility of nitrogen in water and in one mol $dm^{-3}$ acetamide solution near 293 K. His value for the solubility of nitrogen in water agrees well with the recommended value. The data are classed as tentative.

| COMPONENTS: | EVALUATOR: |
|---|---|
| (1) Nitrogen; $N_2$; [7727-37-9]<br>(2) Polar Organic Compounds<br>(3) Water; $H_2O$; [7732-18-5] | H. Lawrence Clever<br>Department of Chemistry<br>Emory University<br>Atlanta, GA 30322 USA<br><br>1981, December |

CRITICAL EVALUATION:

Nitrogen + Detergent + Water

Enns, Scholander, and Bradstreet (7) studied the effect of hydrostatic pressure up to about 10 MPa on the equilibrium pressure of nitrogen dissolved in water and in a solution of an unidentified detergent + water which had a surface tension equal to 30 per cent that of water. The effect of the detergent and the resulting reduced surface tension was negligible. The detergent concentration was not given, but it may be assumed that it was small. Thus, it appears that the bulk properties of water which influence solubility are not effected by the small amounts of detergent.

References:

1. Braun, L. *Z. Phys. Chem., Stoechiom. Verwandtschaftsl.* 1900, *33*, 721.

2. Just, G. *Z. Phys. Chem., Stoechiom. Verwandtschaftsl.* 1901, *37*, 342.

3. Hüfner, G. *Z. Phys. Chem., Stoechiom. Verwandtschaftsl.* 1907, *57*, 611.

4. Drucker, K.; Moles, E. *Z. Phys. Chem., Stoechiom. Verwandtschaftsl.* 1910, *75*, 405.

5. Müller, C., *Z. Phys. Chem., Stoechiom. Verwandtschaftsl.* 1912, *81*, 483.

6. Kretschmer, C. B.; Nowakowska, J.; Wiebe, R. *Ind. Eng. Chem.* 1946, *38*, 506.

7. Enns, T.; Scholander, P. F.; Bradstreet, E. D. *J. Phys. Chem.* 1965, *69*, 389.

8. Tokunaga, J. *J. Chem. Eng. Data* 1975, *20*, 41.

| COMPONENTS: | ORIGINAL MEASUREMENTS: |
|---|---|
| (1) Nitrogen; $N_2$; [7727-37-9]<br>(2) Methanol; $CH_4O$; [67-56-1]<br>(3) Water; $H_2O$; [7732-18-5] | Tokunaga, J.<br>*J. Chem. Eng. Data.* 1975, *20*, 41-6. |
| VARIABLES:<br>T/K = 293-313<br>Composition | PREPARED BY:<br>C.L. Young |

EXPERIMENTAL VALUES:

| T/K | Mole fraction of alcohol, $x_{CH_4O}$ | Ostwald coefficient,* $L$ | Henry's Constant /atm | Mole fraction of nitrogen in liquid, ** $x_{N_2}$ |
|---|---|---|---|---|
| 293.15 | 0.0000 | 0.0179 | 74500 | 0.0000134 |
| | 0.0274 | 0.0188 | 68800 | 0.0000145 |
| | 0.0514 | 0.0197 | 64100 | 0.0000156 |
| | 0.0670 | 0.0198 | 62700 | 0.0000159 |
| | 0.0967 | 0.0183 | 66000 | 0.0000152 |
| | 0.1653 | 0.0202 | 56000 | 0.0000179 |
| | 0.2084 | 0.0217 | 50200 | 0.0000199 |
| | 0.2931 | 0.0261 | 38800 | 0.0000258 |
| | 0.4050 | 0.0318 | 28900 | 0.0000346 |
| | 0.5038 | 0.0448 | 19000 | 0.0000526 |
| | 0.6088 | 0.0598 | 13100 | 0.0000763 |
| | 0.6784 | 0.0723 | 10300 | 0.0000971 |
| | 0.7630 | 0.0910 | 7670 | 0.0000130 |
| | 0.8973 | 0.118 | 5390 | 0.0000186 |
| | 1.0000 | 0.138 | 4310 | 0.0000232 |
| 313.15 | 0.0000 | 0.0147 | 96600 | 0.0000104 |
| | 0.0225 | 0.0168 | 82700 | 0.0000121 |
| | 0.0478 | 0.0164 | 82300 | 0.0000122 |
| | 0.0664 | 0.0177 | 73500 | 0.0000136 |
| | 0.1009 | 0.0186 | 68800 | 0.0000145 |
| | 0.2036 | 0.0219 | 53300 | 0.0000188 |
| | 0.2971 | 0.0279 | 38400 | 0.0000260 |
| | 0.4278 | 0.0433 | 22200 | 0.0000450 |

AUXILIARY INFORMATION

METHOD /APPARATUS/PROCEDURE:

Volumetric apparatus with multibulb buret and magnetic stirrer. Amount of solution and gas absorbed determined volumetrically. Partial pressure determined from total pressure and vapor pressure of solvent solution. Details in source.

SOURCE AND PURITY OF MATERIALS:

(1) High purity grade sample, purity 99.99 mole per cent.

(2/3) Guaranteed reagent, further purified by distillation.

Solution degassed by boiling.

ESTIMATED ERROR:

$\delta T/K = \pm 0.5$; $\delta x_{N_2} = \pm 1-2\%$.

REFERENCES:

COMPONENTS:

(1) Nitrogen; $N_2$; [7727-37-9]

(2) Methanol; $CH_4O$; [67-56-1]

(3) Water; $H_2O$ [7732-18-5]

ORIGINAL MEASUREMENTS:

Tokunaga, J.

*J. Chem. Eng. Data.* 1975, *20*, 41-6.

EXPERIMENTAL VALUES :

| T/K | Mole fraction of alcohol, $x_{CH_4O}$ | Ostwald coefficient,* $L$ | Henry's Constant /atm | Mole fraction of nitrogen in liquid,** $x_{N_2}$ |
|---|---|---|---|---|
| 313.15 | 0.5200 | 0.0534 | 16700 | 0.0000599 |
| | 0.6712 | 0.0805 | 9850 | 0.000102 |
| | 0.7074 | 0.0867 | 8910 | 0.000112 |
| | 0.7897 | 0.101 | 7190 | 0.000139 |
| | 0.8957 | 0.117 | 5790 | 0.000173 |
| | 1.0000 | 0.146 | 4250 | 0.000235 |

* at $1.01325 \times 10^5$ Pa partial pressure

** at $1.01325 \times 10^5$ Pa partial pressure (calculated from the reciprocal of Henry's Constant).

**COMPONENTS:**

(1) Nitrogen; $N_2$; [7727-37-9]

(2) Ethanol; $C_2H_6O$; [64-17-5]

(3) Water; $H_2O$; [7732-18-5]

**ORIGINAL MEASUREMENTS:**

Just, G.

*Z. Phys. Chem.* 1901, *37*, 342-67.

**VARIABLES:**

T/K = 298.15
P/kPa = 101.325
Ethanol/vol % = 0 - 99.8

**PREPARED BY:**

M. E. Derrick
H. L. Clever

**EXPERIMENTAL VALUES:**

| T/K | Ethanol/ volume per cent | Bunsen Coefficient $\alpha$ | Ostwald Coefficient L |
|---|---|---|---|
| 298.15 | 0 | 0.01497 | 0.01634 |
| | 20 | 0.01407 | 0.01536 |
| | 33 | 0.01575 | 0.01719 |
| | 99.8 | 0.1312 | 0.1432 |

The author measured the Ostwald coefficient at a pressure of about 746 mmHg. The compiler assumed the Ostwald coefficient to be independent of pressure, and calculated the mole fraction and Bunsen coefficient values at 101.325 kPa (1 atm) partial pressure of the gas.

**AUXILIARY INFORMATION**

**METHOD/APPARATUS/PROCEDURE:**

An Ostwald apparatus as modified by Timofejew (1), and Steiner (2) is used. The apparatus consists of a gas buret, an absorption flask, and a mercury manometer. The system is thermostated with a water jacket.

The gas is introduced into the degassed liquid. The gas volume absorbed is determined by the gas buret. The solvent volume is determined at the end of the experiment by pouring the solvent into a graduated flask.

**SOURCE AND PURITY OF MATERIALS:**

(1) Nitrogen. Prepared in aqueous solution by the reaction of sodium nitrite with ammonium nitrate in the presence of calcium dichromate.

(2) Ethanol. No information

(3) Water. Distilled.

**ESTIMATED ERROR:**

$\delta L/L = 0.03$ (compiler)

**REFERENCES:**

1. Timofejew, W.
*Z. Physik. Chem.* 1890, *6*, 141-52.

2. Steiner,
*Ann. Phys. (Leipzig)* 1894, *52*, 275.

**COMPONENTS:**

(1) Nitrogen; $N_2$; [7727-37-9]

(2) Ethyl Alcohol (Ethanol); $C_2H_6O$; [64-17-5]; (95% by volume)

(3) Water; $H_2O$; [7732-18-5]

**ORIGINAL MEASUREMENTS:**

Kretschmer, C. B.; Nowakowska, J.; Wiebe, R.

*Ind. Eng. Chem.* 1946, *38*, 506-9.

**VARIABLES:**

T/K = 248.15 - 323.15
P/kPa = 101.325
Ethanol/vol % = 95

**PREPARED BY:**

P. L. Long
H. L. Clever

**EXPERIMENTAL VALUES:**

| T/K | Bunsen Coefficient $\alpha$ | Ostwald Coefficient L |
|---|---|---|
| 248.15 | 0.1039 | 0.0944 |
| 273.15 | 0.1053 | 0.1053 |
| 298.15 | 0.1063 | 0.1160 |
| 323.15 | 0.1090 | 0.1290 |

The Ostwald coefficients were measured at a total pressure of 101.325 kPa (1 atm). The compiler calculated the Bunsen coefficient value for an oxygen partial pressure of 101.325 kPa (1 atm) assuming the Ostwald coefficient was independent of pressure.

The solvent is 95 volume percent ethanol, 5 volume per cent water.

**AUXILIARY INFORMATION**

**METHOD/APPARATUS/PROCEDURE:**

The apparatus was similar to that of Horiuti (1), it consisted of a gas buret, manometer, and absorption pipet equipped with a magnetic stirrer. All volumes were calibrated by mercury displacement.

The solvent was partly degassed by boiling under reflux. About 40 $cm^3$ of solvent was admitted into the absorption pipet, and pumped on to complete the degassing. Dry gas, measured in the buret, was admitted, final equilibrium was established after 1 to 3 hours stirring. During the solution process the total pressure was kept at one atm by additions of dry gas.

The solubility value was corrected for the increase of solvent volume with gas absorption

**SOURCE AND PURITY OF MATERIALS:**

(1) Nitrogen. Commercial, but source not given. Passed through alkaline pyrogallol to remove $O_2$, passed through Drierite to dry.

(2) Ethanol. The alcohol was distilled made 95 per cent by volume by the addition of water. $\rho_4^{25}$ /g $cm^{-3}$ = 0.80724.

(3) Water.

**ESTIMATED ERROR:**

Gas buret $\delta V/cm^3$ = 0.005
Average deviation from the mean 0.3%, maximum deviation 0.9% (authors).

**REFERENCES:**

1. Horiuti, J. *J. Sci Papers Inst. Phys. Chem. Res. (Tokyo)* 1931, *17*, 125-256.

| COMPONENTS: | ORIGINAL MEASUREMENTS: |
|---|---|
| (1) Nitrogen; $N_2$; [7727-37-9]<br>(2) Ethanol; $C_2H_6O$; [64-17-5]<br>(3) Water; $H_2O$; [7732-18-5] | Tokunaga, J.<br>*J. Chem. Eng. Data.* 1975, *20*, 41-6. |
| **VARIABLES:** | **PREPARED BY:** |
| T/K = 293-313 | C.L. Young |

EXPERIMENTAL VALUES:

| T/K | Mole fraction of alcohol, $x_{C_2H_6O}$ | Ostwald coefficient*, $L$ | Henry's Constant /atm | Mole fraction of nitrogen in liquid,** $x_{N_2}$ |
|---|---|---|---|---|
| 293.15 | 0.0000 | 0.0179 | 74500 | 0.0000134 |
| | 0.0070 | 0.0193 | 68100 | 0.0000147 |
| | 0.0199 | 0.0200 | 64100 | 0.0000156 |
| | 0.0558 | 0.0212 | 56700 | 0.0000176 |
| | 0.0815 | 0.0214 | 53800 | 0.0000186 |
| | 0.1015 | 0.0217 | 51300 | 0.0000195 |
| | 0.1558 | 0.0240 | 42600 | 0.0000235 |
| | 0.2757 | 0.0363 | 23600 | 0.0000424 |
| | 0.3206 | 0.0421 | 19100 | 0.0000524 |
| | 0.4628 | 0.0622 | 10900 | 0.0000917 |
| | 0.5952 | 0.0846 | 6930 | 0.000144 |
| | 0.7130 | 0.102 | 5150 | 0.000194 |
| | 0.8746 | 0.126 | 3620 | 0.000276 |
| | 0.9140 | 0.134 | 3320 | 0.000301 |
| | 1.0000 | 0.151 | 2730 | 0.000366 |
| 313.15 | 0.0000 | 0.0147 | 96600 | 0.0000104 |
| | 0.0195 | 0.0163 | 84100 | 0.0000119 |
| | 0.0385 | 0.0161 | 82000 | 0.0000122 |
| | 0.0509 | 0.0159 | 80200 | 0.0000125 |
| | 0.0574 | 0.0159 | 80200 | 0.0000125 |
| | 0.0734 | 0.0170 | 73100 | 0.0000137 |
| | 0.1087 | 0.0186 | 63100 | 0.0000158 |
| | 0.2237 | 0.0278 | 35200 | 0.0000284 |

AUXILIARY INFORMATION

METHOD /APPARATUS/PROCEDURE:

Volumetric apparatus with multibulb buret and magnetic stirrer. Amount of solution and gas absorbed determined volumetrically. Partial pressure determined from total pressure and vapor pressure of solvent solution. Details in source.

SOURCE AND PURITY OF MATERIALS:

(1) High purity grade sample, purity 99.99 mole per cent.

(2/3) Guaranteed reagent, further purified by distillation. Solution degassed by boiling.

ESTIMATED ERROR:

$\delta T/K = \pm 0.5$; $\delta x_{N_2} = \pm 1\text{-}2\%$.

REFERENCES:

COMPONENTS :

(1) Nitrogen; $N_2$; [7727-37-9]

(2) Ethanol; $C_2H_6O$; [64-17-5]

(3) Water; $H_2O$; [7732-18-5]

ORIGINAL MEASUREMENTS :

Tokunaga, J.

*J. Chem. Eng. Data.* 1975, *20*, 41-6.

EXPERIMENTAL VALUES :

| T/K | Mole fraction of alcohol, $x_{C_2H_6O}$ | Ostwald coefficient,* $L$ | Henry's Constant /atm | Mole fraction of nitrogen in liquid,** $x_{N_2}$ |
|---|---|---|---|---|
| 313.15 | 0.3120 | 0.0432 | 20000 | 0.0000500 |
| | 0.4162 | 0.0554 | 13700 | 0.0000730 |
| | 0.5061 | 0.0707 | 9680 | 0.000103 |
| | 0.6062 | 0.0852 | 7230 | 0.000138 |
| | 0.6946 | 0.100 | 5650 | 0.000177 |
| | 0.7751 | 0.115 | 4570 | 0.000219 |
| | 0.8814 | 0.136 | 3490 | 0.000287 |
| | 0.9702 | 0.151 | 2920 | 0.000343 |
| | 1.0000 | 0.156 | 2760 | 0.000362 |

* at $1.01325 \times 10^5$ Pa partial pressure.

** at $1.01325 \times 10^5$ Pa partial pressure (calculated from the reciprocal of Henry's Constant).

| COMPONENTS: | ORIGINAL MEASUREMENTS: |
|---|---|
| (1) Nitrogen; $N_2$; [7727-37-9]<br>(2) 2,2,2,-trichloro-1.1-ethanediol (chloral hydrate); $C_2H_3Cl0_2$; [302-17-0]<br>(3) Water; $H_2O$ [7732-18-5] | Müller, C.<br>*Z. Physik. Chem.* 1912, *81*, 483-503. |
| VARIABLES: | PREPARED BY: |
| T/K = 288-290<br>P/kPa = 101 | R. Battino |

EXPERIMENTAL VALUES:

| $T^a$/K | Weight Percentage[b] Chloral hydrate | Ostwald Coeff.[c] $10^2L$ | Bunsen Coeff.[d] $10^2\alpha$ |
|---|---|---|---|
| 288.75 | 15.8 | 1.664 | 1.574 |
| 288.55 | 28.2 | 1.498 | 1.418 |
| 289.55 | 37.25 | 1.365 | 1.288 |
| 290.05 | 47. | 1.378 | 1.260 |
| 290.15 | 56.52 | 1.307 | 1.230 |
| 288.45 | 71.5 | 1.494 | 1.415 |
| 287.95 | 78.8 | 1.525 | 1.447 |

[a] Temperature reported to 0.1°C.

[b] 2,2,2 trichloro-1,1-ethanediol.

[c] Ostwald coefficient calculated by compiler.

[d] Bunsen coefficient.

AUXILIARY INFORMATION

METHOD/APPARATUS/PROCEDURE:

Water is degassed by pumping under vacuum. The absorption vessel is about 600 $cm^3$ and is separately thermostatted from the buret gas measuring system. Both are connected by a copper capillary tube. Gas uptake is read on the gas burets. The absorption flask volume and all liquid volumes were calibrated using mercury. The original paper contains details of operation and a drawing of the apparatus. Solubilities were also measured in pure water and sucrose.

SOURCE AND PURITY OF MATERIALS:

(1) Nitrogen - from heating potassium nitrite and ammonium nitrite. Washed with potassium hydroxide solution and sulfuric acid; and finally passed through a glowing copper tube.
(2) Chloral hydrate - from Merck or from Münster.
(3) Water - no comment by author.

ESTIMATED ERROR:

$\delta\alpha/\alpha = \pm 0.01$ estimated by compiler.

REFERENCES:

**COMPONENTS:**

(1) Nitrogen; $N_2$; [7727-37-9]

(2) Acetamide; $C_2H_5NO$ or $CH_3CONH_2$; [60-35-5]

(3) Water; $H_2O$; [7732-18-5]

**ORIGINAL MEASUREMENTS:**

Hüfner, G.

*Z. Physik. Chem.* 1907, *57*, 611-24.

**VARIABLES:**

T/K = 293.35, 293.40

$N_2$ P/kPa = 101.325

**PREPARED BY:**

H. L. Clever

**EXPERIMENTAL VALUES:**

| Temperature t/°C | T/K | $c_{C_2H_5NO}$/mol dm$^{-3}$ | Gas Pressure/ mmHg | Absorbed Gas Volume[1] | Bunse Coefficient $\alpha \times 10^2$ |
|---|---|---|---|---|---|
| 20.18 | 293.33 | 0.0 | --- | -- | 1.565[2] |
| 20.20 | 293.35 | 1.0 | 672.3 | 5.37 | 1.481 |
| 20.25 | 293.40 | 1.0 | 680.8 | 5.40 | 1.470 |

[1] Volume of gas reduced to 273.15 K and 760 mmHg absorbed in 490.94 cm$^3$ solution at the stated pressure.

[2] Average of four runs, see Hüfner's $N_2$ + $H_2O$ data sheet.

AUXILIARY INFORMATION

**METHOD/APPARATUS/PROCEDURE:**

The apparatus, described in reference (1), was the Bunsen absorption type. It consisted of an absorption flask, which holds 409.94 cm$^3$ of solvent, and gas burets.

The final partial pressure of the gas ranged between 648.6 and 720.7 mmHg. The author reported both the volume of gas absorbed at the final pressure in 409.94 cm$^3$ of solvent reduced to 273.15 K and 760 mmHg, and the Bunsen coefficient, which was calculated assuming Henry's law.

**SOURCE AND PURITY OF MATERIALS:**

(1) Nitrogen. Source not given. Stated to be argon free.

(2) Acetamide. Source not given.

(3) Water. Pure.

**ESTIMATED ERROR:**

$\delta T/K = 0.01$

**REFERENCES:**

1. Hüfner, G. *Archiv. F. Anatomie und Physiologie, Physiolog. Abtly.* 1894, *5*, 191.

| COMPONENTS: | ORIGINAL MEASUREMENTS: |
|---|---|
| (1) Nitrogen; $N_2$; [7727-37-9]<br>(2) Propanoic acid, (Propionic acid); $C_3H_6O_2$; [79-09-4]<br>(3) Water; $H_2O$; [7732-18-5] | Braun, L.<br>*Z. Phys. Chem.* 1900, *33*, 721-41. |
| VARIABLES:<br>T/K = 278-298<br>Concentration | PREPARED BY:<br>C.L. Young |

EXPERIMENTAL VALUES:

| T/K | Wt. of acid per 100g soln./g | Bunsen coefficient, α |
|---|---|---|
| 278.2 | 11.220 | 0.01951 |
| | 11.023 | 0.01977 |
| | 9.537 | 0.02040 |
| | 9.155 | 0.02026 |
| | 6.066 | 0.02077 |
| | 5.891 | 0.02089 |
| | 4.081 | 0.02095 |
| | 3.816 | 0.02087 |
| | 0.0 | 0.02173 |
| 283.2 | 11.220 | 0.01779 |
| | 11.023 | 0.01800 |
| | 9.537 | 0.01823 |
| | 9.155 | 0.01845 |
| | 6.066 | 0.01855 |
| | 5.891 | 0.01872 |
| | 4.081 | 0.01919 |
| | 3.816 | 0.01915 |
| | 0.0 | 0.02003 |
| 288.2 | 11.220 | 0.01593 |
| | 11.023 | 0.01585 |
| | 9.537 | 0.01634 |
| | 9.155 | 0.01630 |
| | 6.066 | 0.01637 |
| | 5.891 | 0.01648 |
| | 4.081 | 0.01688 |

AUXILIARY INFORMATION

| METHOD/APPARATUS/PROCEDURE: | SOURCE AND PURITY OF MATERIALS: |
|---|---|
| Ostwald method, using gas buret and pipet. Measurement of volume of gas before and after absorption. Vapor pressure of water was allowed for assuming Raoult's law. Solution was degassed. Concentration of salt solution estimated by titration with silver nitrate solution. | No details given |
| | ESTIMATED ERROR:<br>$\delta\alpha/\alpha = \pm 0.03$ |
| | REFERENCES: |

COMPONENTS:

(1) Nitrogen; $N_2$; [7727-37-9]

(2) Propanoic acid, (Propionic acid); $C_3H_6O_2$; [79-09-4]

(3) Water; $H_2O$; [7732-18-5]

ORIGINAL MEASUREMENTS:

Braun, L.

*Z. Phys. Chem.* 1900, *33*, 721-41.

EXPERIMENTAL VALUES:

| T/K | Wt. of acid per 100g soln./g | Bunsen coefficient, $\alpha$ |
|---|---|---|
| 288.2 | 3.816 | 0.01674 |
| | 0.0 | 0.01789 |
| 293.2 | 11.220 | 0.01463 |
| | 11.023 | 0.01447 |
| | 9.537 | 0.01471 |
| | 9.155 | 0.01469 |
| | 6.066 | 0.01476 |
| | 5.891 | 0.01480 |
| | 4.081 | 0.01541 |
| | 3.816 | 0.01547 |
| | 0.0 | 0.01621 |
| 298.2 | 11.220 | 0.01301 |
| | 11.023 | 0.01295 |
| | 9.537 | 0.01336 |
| | 9.155 | 0.01329 |
| | 6.066 | 0.01335 |
| | 5.891 | 0.01338 |
| | 4.081 | 0.01365 |
| | 3.816 | 0.01371 |
| | 0.0 | 0.01432 |

COMPONENTS:

(1) Nitrogen; $N_2$; [7727-37-9]

(2) 1-Propanol; $C_3H_8O$; [71-32-8]

(3) Water; $H_2O$; [7732-18-5]

ORIGINAL MEASUREMENTS:

Tokunaga, J.

*J. Chem. Eng. Data.* 1975, *20*, 41-6.

VARIABLES:

T/K = 293-313

Composition

PREPARED BY:

C.L. Young

EXPERIMENTAL VALUES:

| T/K | Mole fraction of alcohol, $x_{C_3H_8O}$ | Ostwald coefficient,* $L$ | Henry's Constant /atm | Mole fraction of nitrogen in liquid,** $x_{N_2}$ |
|---|---|---|---|---|
| 293.15 | 0.0000 | 0.0179 | 74500 | 0.0000134 |
| | 0.0247 | 0.0210 | 59300 | 0.0000169 |
| | 0.0349 | 0.0180 | 67400 | 0.0000148 |
| | 0.0758 | 0.0202 | 54300 | 0.0000184 |
| | 0.0987 | 0.0234 | 44400 | 0.0000225 |
| | 0.1997 | 0.0371 | 22500 | 0.0000444 |
| | 0.3020 | 0.0541 | 12900 | 0.0000775 |
| | 0.4045 | 0.0636 | 9390 | 0.000106 |
| | 0.5014 | 0.0712 | 7380 | 0.000136 |
| | 0.6058 | 0.0846 | 5490 | 0.000182 |
| | 0.7775 | 0.101 | 3880 | 0.000258 |
| | 0.8640 | 0.109 | 3310 | 0.000302 |
| | 1.0000 | 0.123 | 2650 | 0.000377 |
| 313.15 | 0.0000 | 0.0147 | 96600 | 0.0000104 |
| | 0.0247 | 0.0170 | 77700 | 0.0000129 |
| | 0.0480 | 0.0163 | 76400 | 0.0000131 |
| | 0.0717 | 0.0205 | 57300 | 0.0000175 |
| | 0.1008 | 0.0283 | 38600 | 0.0000259 |
| | 0.1395 | 0.0365 | 27300 | 0.0000366 |
| | 0.2903 | 0.0616 | 12100 | 0.0000826 |
| | 0.4107 | 0.0798 | 7770 | 0.000129 |
| | 0.4955 | 0.0882 | 6280 | 0.000159 |
| | 0.6135 | 0.0989 | 4880 | 0.000205 |
| | 0.6993 | 0.108 | 4110 | 0.000243 |

AUXILIARY INFORMATION

METHOD /APPARATUS/PROCEDURE:

Volumetric apparatus with multibulb buret and magnetic stirrer. Amount of solution and gas absorbed determined volumetrically. Partial pressure determined from total pressure and vapor pressure of solvent solution. Details in source.

SOURCE AND PURITY OF MATERIALS:

(1) High purity grade sample, purity 99.99 mole per cent.

(2/3) Guaranteed reagent, further purified by distillation. Solution degassed by boiling.

ESTIMATED ERROR:

$\delta T/K = \pm 0.5$; $\delta x_{N_2} = \pm 1\text{-}2\%$.

REFERENCES:

| COMPONENTS | ORIGINAL MEASUREMENTS |
|---|---|
| (1) Nitrogen; $N_2$; [7727-37-9]<br>(2) 1-Propanol; $C_3H_8O$; [71-23-8]<br>(3) Water; $H_2O$; [7732-18-5] | Tokunaga, J.<br>*J. Chem. Eng. Data.* 1975, *20*, 41-6. |

EXPERIMENTAL VALUES :

| T/K | Mole fraction of alcohol, $x_{C_3H_8O}$ | Ostwald coefficient,* $L$ | Henry's Constant /atm | Mole fraction of nitrogen in liquid,** $x_{N_2}$ |
|---|---|---|---|---|
| 313.15 | 0.8251 | 0.118 | 3310 | 0.000302 |
| | 0.8761 | 0.122 | 3050 | 0.000328 |
| | 1.0000 | 0.132 | 2570 | 0.000389 |

* at $1.01325 \times 10^5$ Pa partial pressure

** at $1.01325 \times 10^5$ Pa partial pressure (calculated from the reciprocal of Henry's Constant).

COMPONENTS:

(1) Nitrogen; $N_2$; [7727-37-9]

(2) 2-Propanol; $C_3H_8O$; [67-63-0]

(3) Water; $H_2O$; [7732-18-5]

ORIGINAL MEASUREMENTS:

Tokunaga, J.

*J. Chem. Eng. Data.* 1975, *20*, 41-6.

VARIABLES:

T/K = 293-313

Composition

PREPARED BY:

C.L. Young

EXPERIMENTAL VALUES:

| T/K | Mole fraction of alcohol, $x_{C_3H_8O}$ | Ostwald coefficient,* $L$ | Henry's Constant /atm | Mole fraction of nitrogen in liquid,** $x_{N_2}$ |
|---|---|---|---|---|
| 293.15 | 0.0000 | 0.0179 | 74500 | 0.0000134 |
| | 0.0709 | 0.0168 | 65900 | 0.0000152 |
| | 0.1585 | 0.0246 | 37100 | 0.0000270 |
| | 0.3288 | 0.0511 | 13000 | 0.0000769 |
| | 0.4000 | 0.0636 | 9350 | 0.000107 |
| | 0.5351 | 0.0773 | 6410 | 0.000156 |
| | 0.6396 | 0.0939 | 4680 | 0.000214 |
| | 0.7125 | 0.0986 | 4130 | 0.000242 |
| | 0.8147 | 0.117 | 3160 | 0.000316 |
| | 0.9031 | 0.119 | 2860 | 0.000350 |
| | 1.0000 | 0.133 | 2390 | 0.000418 |
| 313.15 | 0.0000 | 0.0147 | 96600 | 0.0000104 |
| | 0.0480 | 0.0136 | 91300 | 0.0000110 |
| | 0.1046 | 0.0205 | 52700 | 0.0000190 |
| | 0.1620 | 0.0299 | 31700 | 0.0000315 |
| | 0.2200 | 0.0395 | 21400 | 0.0000467 |
| | 0.3249 | 0.0586 | 12000 | 0.0000833 |
| | 0.4174 | 0.0732 | 8330 | 0.000120 |
| | 0.5300 | 0.0921 | 5700 | 0.000175 |
| | 0.6031 | 0.0990 | 4860 | 0.000206 |

AUXILIARY INFORMATION

METHOD /APPARATUS/PROCEDURE:

Volumetric apparatus with multibulb buret and magnetic stirrer. Amount of solution and gas absorbed determined volumetrically. Partial pressure determined from total pressure and vapor pressure of solvent solution. Details in source.

SOURCE AND PURITY OF MATERIALS:

(1) High purity grade sample, purity 99.99 mole per cent.

(2/3) Guaranteed reagent, further purified by distillation. Solution degassed by boiling.

ESTIMATED ERROR:

$\delta T/K = \pm 0.5$; $\delta x_{N_2} = \pm 1\text{-}2\%$.

REFERENCES:

COMPONENTS:

(1) Nitrogen; $N_2$; [7727-37-9]

(2) 2-Propanol; $C_3H_8O$; [67-63-0]

(3) Water; $H_2O$; [7732-18-5]

ORIGINAL MEASUREMENTS:

Tokunaga, J.

*J. Chem. Eng. Data.* 1975, *20*, 41-6.

EXPERIMENTAL VALUES :

| T/K | Mole fraction of alcohol, $x_{C_3H_8O}$ | Ostwald coefficient,* $L$ | Henry's Constant /atm | Mole fraction of nitrogen in liquid,** $x_{N_2}$ |
|---|---|---|---|---|
| 313.15 | 0.7743 | 0.117 | 3420 | 0.000292 |
| | 0.8487 | 0.126 | 2970 | 0.000337 |
| | 0.9535 | 0.140 | 2450 | 0.000408 |
| | 1.0000 | 0.146 | 2210 | 0.000452 |

* at $1.01325 \times 10^5$ Pa partial pressure.

** at $1.01325 \times 10^5$ Pa partial pressure (calculated from the reciprocal of Henry's Constant).

| COMPONENTS: | ORIGINAL MEASUREMENTS: |
|---|---|
| (1) Nitrogen; $N_2$; [7727-37-9]<br>(2) 1,2,3-Propanetriol (Glycerol); $C_3H_8O_3$; [56-81-5]<br>(3) Water; $H_2O$; [7732-18-5] | Drucker, K.; Moles, E.<br>*Z. Physik. Chem.* 1910, *75*, 405- 36. |
| **VARIABLES:**<br>T/K = 298<br>Wt. fraction of 1,2,3-Propanetriol | **PREPARED BY:**<br>C. L. Young |

EXPERIMENTAL VALUES:

| T/K | Wt. fraction of 1,2,3-Propanetriol | *p*/mmHg | *p*/kPa | Ostwald coefficient, *L* | Ratio of density of soln./density of $H_2O$ |
|---|---|---|---|---|---|
| 298.2 | 0.0 | 760 | 101 | 0.0156 | 1.0 |
| | 0.16 | 598.4 | 79.8 | 0.0103 | 1.0392 |
| | | 915.5 | 122.1 | 0.0103 | |
| | 0.297 | 556.5 | 74.2 | 0.0067 | 1.0744 |
| | | 846.5 | 112.9 | 0.0068 | |
| | 0.489 | 617.7 | 82.4 | 0.0052 | 1.1263 |
| | | 859.8 | 114.6 | 0.0051 | |
| | 0.745 | 588.5 | 78.5 | 0.0025 | 1.1931 |
| | 0.841 | 637.3 | 85.0 | 0.0024 | 1.2213 |
| | | 757.0 | 100.9 | 0.0024 | |

AUXILIARY INFORMATION

METHOD/APPARATUS/PROCEDURE:

Volumetric method using an Ostwald type gas buret and pipet. Solvent degassed before being admitted to buret. Density of solution determined using a Sprengel pyknometer.

SOURCE AND PURITY OF MATERIALS:

(1) From cylinder, passed through concentrated alkali, alkaline pyrogallol and concentrated sulfuric acid then over hot copper.

(2) Pure sample, purity about 98 mole per cent.

(3) Distilled.

ESTIMATED ERROR:

$\delta L/L$ =±1-3% for dil. soln.;

±5-7% for conc. soln.

REFERENCES:

| COMPONENTS: | ORIGINAL MEASUREMENTS: |
|---|---|
| (1) Nitrogen; $N_2$; [7727-37-9]<br>(2) 1,2,3-propanetriol(glycerol); $C_3H_8O_3$; [56-81-5]<br>(3) Water; $H_2O$; [7732-18-5] | Müller, C.<br>*Z. Physik. Chem.* 1912, *81*, 483-503. |
| **VARIABLES:** | **PREPARED BY:** |
| T/K = 287-291<br>P/kPa = 101.325 | R. Battino |

EXPERIMENTAL VALUES:

| $T^a$/K | Weight Percentage Glycerol[b] | Ostwald Coeff.[c] $10^2L$ | Bunsen Coeff.[d] $10^2\alpha$ |
|---|---|---|---|
| 289.25 | 25. | 1.313 | 1.240 |
| 288.75 | 42.2 | 1.053 | 0.996 |
| 287.85 | 51.5 | 0.800 | 0.759 |
| 288.05 | 58.0 | 0.741 | 0.703 |
| 289.05 | 80.25 | 0.550 | 0.520 |
| 289.35 | 90. | 0.604 | 0.570 |
| 291.15 | 95. | 0.616 | 0.578 |

[a] Temperature reported to 0.1°C.

[b] 1,2,3-propanetriol.

[c] Ostwald coefficient calculated by compiler.

[d] Bunsen coefficient.

AUXILIARY INFORMATION

METHOD/APPARATUS/PROCEDURE:

Water is degassed by pumping under vacuum. The absorption vessel is about 600 $cm^3$ and is separately thermostatted from the buret gas measuring system. Both are connected by a copper capillary tube. Gas uptake is read on the gas burets. The absorption flask volume and all liquid volumes were calibrated using mercury. The original paper contains details of operation and a drawing of the apparatus. Solubilities were also measured in pure water, sucrose, and chloral hydrate.

SOURCE AND PURITY OF MATERIALS:

(1) Nitrogen - from heating potassium nitrite and ammonium nitrite. Washed with potassium hydroxide solution, ferrous sulfate solution and sulfuric acid; and finally passed through a glowing copper tube.

(2) Glycerol - from Merck.

(3) Water - no comment by author.

ESTIMATED ERROR:

$\delta\alpha/\alpha$ = ±0.01, estimated by compiler.

REFERENCES:

**COMPONENTS:**

(1) Nitrogen; $N_2$; [7727-37-9]

(2) 1,2,3-Propanetriol (glycerine); $C_3H_8O_3$; [56-81-5]

(3) Water; $H_2O$; [7732-18-5]

**ORIGINAL MEASUREMENTS:**

Hammel, A. von

*Z. Physik. Chem.* 1915, *90*, 121-5.

**VARIABLES:**

T/K = 288

Concentration

**PREPARED BY:**

R. Battino

**EXPERIMENTAL VALUES:**

15°C (288 K)

| wt. % (2) | $10^2L^a$ | wt. % (2) | $10^2L^a$ |
|---|---|---|---|
| 0 | 1.707 | 72.8 | 0.552 |
| 0 | 1.708 | 74.7 | 0.597 |
| 15.7 | 1.425 | 77.0 | 0.527 |
| 15.7 | 1.376 | 85.1 | 0.482 |
| 29.9 | 1.087 | 87.3 | 0.492 |
| 46.6 | 0.840 | 88.5 | 0.536 |
| 57.6 | 0.698 | 99.25 | 0.524 |
| 67.1 | 0.635 | | |

[a] Ostwald coefficient.

## AUXILIARY INFORMATION

**METHOD/APPARATUS/PROCEDURE:**

The Ostwald apparatus as modified by Geffcken (1) was used.

**SOURCE AND PURITY OF MATERIALS:**

No details given.

**ESTIMATED ERROR:**

$\delta L/L = \pm 0.03$, Compiler's estimate

**REFERENCES:**

1. Geffcken, G. *Z. Physik. Chem.* 1904, *49*, 257-302.

COMPONENTS:

(1) Nitrogen; $N_2$; [7727-37-9]

(2) 2-Methylpropanoic acid; $C_4H_8O_2$; [79-31-2]

(3) Water; $H_2O$; [7732-18-5]

ORIGINAL MEASUREMENTS:

Drucker, K.; Moles, E.

*Z. Physik. Chem.* 1910, *75*, 405-36.

VARIABLES:

T/K = 296-302
P/kPa = 31-116

PREPARED BY:

C. L. Young

EXPERIMENTAL VALUES:

Solvent contained 37.5% component 2.

| T/K | *p*/mmHg | *p*/kPa | Ostwald Coefficient, *L* | Ratio of density of soln./density of $H_2O$ at 298.15 K |
|---|---|---|---|---|
| 296.17 | 246.2 | 32.82 | 0.0393 | 0.9984 |
| | 492.2 | 65.62 | 0.0393 | |
| | 563.6 | 75.14 | 0.0393 | |
| | 836.3 | 111.5 | 0.0400 | |
| | 867.3 | 115.6 | 0.0401 | |
| 302.17 | 231 | 30.80 | 0.0373 | |
| | 468.4 | 62.45 | 0.0384 | |
| | 480.7 | 64.09 | 0.0383 | |
| | 536 | 71.46 | 0.0385 | |
| | 656 | 87.46 | 0.0384 | |
| | 720 | 95.99 | 0.0386 | |

AUXILIARY INFORMATION

METHOD/APPARATUS/PROCEDURE:

Volumetric method using an Ostwald type gas buret and pipet. Solvent degassed before being admitted to pipet. Density of solution determined using a Sprengel pyknometer.

SOURCE AND PURITY OF MATERIALS:

(1) From cylinder, passed through concentrated alkali, alkaline pyrogallol, concentrated sulfuric acid and over hot copper.

(2) From Kahlbaum, distilled four times.

(3) Distilled.

ESTIMATED ERROR:

$\delta L/L$ = ±1-3% for dil. soln.;

±5-7% for conc. soln.

REFERENCES:

| COMPONENTS: | EVALUATOR: |
|---|---|
| (1) Nitrogen; $N_2$; [7727-37-9]<br>(2, 3) Binary Component Organic Solvent | H. Lawrence Clever<br>Department of Chemistry<br>Emory University<br>Atlanta, GA 30322 USA<br>1981, December |

CRITICAL EVALUATION:

The Solubility of Nitrogen in Binary Organic Solvents

The solubility of nitrogen in binary organic solvents is reported in three papers (1, 2, 3). Seven systems were studied. No system was studied by more than one laboratory. Thus, no direct comparison of the data from different sources can be made and the data are all classed as tentative.

Kretschmer, Nowakowska, and Wiebe (1) measured the solubility of nitrogen at just one mole fraction and in the pure components at three temperatures. Thus, their data are of only limited value in giving information on the solubility of nitrogen in mixed solvents. Wiebe and co-workers are well known for reliable work. They used the technique and apparatus of the type first described by Horiuti. They either measured or used literature solvent vapor pressures and they corrected for dilation. Their results are classed as tentative.

Nitta, Katayama and co-workers (2, 3) also used a modified Horiuti apparatus and technique. They assumed Raoult's law valid to calculate the mixed solvent vapor pressures. They considered the solvent dilation correction too small to be significant. The binary solvents were stated to change composition less than 0.7 per cent on degassing and through the solubility procedure. The mole fraction solubilities are for a nitrogen partial pressure of 101.325 kPa (1 atm). Their work appears to have been carried out carefully and the data are classed as tentative.

The systems studied are listed below.

1. Nitrogen + Cyclohexane [110-82-7] + 2,2,4-Trimethylpentane [540-84-1]

   Nitta, Tatsuishi, and Katayama (2) measured the solubility of nitrogen in nine mixtures and the two pure components at 298.15 K.

2. Nitrogen + Ethanol [64-17-5] + 2,2,4-Trimethylpentane [540-84-1]

   Kretschmer, Nowakowska, and Wiebe (1) measured the solubility of nitrogen in the pure components and one mixture 0.739 mole fraction ethanol at 273.15, 298.15, and 323.15 K.

3. Nitrogen + 1-Propanol [71-23-8] + 2,2,4-Trimethylpentane [540-84-1]

   Nitta, Tatsuishi, and Katayama (2) measured the solubility of nitrogen in the pure components and nine mixtures at 298.15 K.

4. Nitrogen + Ethanol [64-17-5] + Benzene [71-43-2]

   Nitta, Fujio, and Katayama (3) measured the solubility of nitrogen in the pure components and nine mixtures at 298.15 K.

| COMPONENTS: | EVALUATOR: |
|---|---|
| (1) Nitrogen; $N_2$; [7727-37-9]<br>(2, 3) Binary Component Organic Solvent | H. Lawrence Clever<br>Department of Chemistry<br>Emory University<br>Atlanta, GA 30322 USA<br>1981, December |

CRITICAL EVALUATION:

5. Nitrogen + Ethanol [64-17-5] + 1,1'-Oxybisethane [60-29-7]

Nitta, Fujio, and Katayama (3) measured the solubility of nitrogen in the pure components and nine mixtures at 273.15 K.

6. Nitrogen + Ethanol [64-17-5] + 2-Propanone [67-64-1]

Kretschmer, Nowakowska, and Wiebe (1) measured the solubility of nitrogen in the two pure components and one mixture of 0.558 mole fraction ethanol at temperatures of 273.15, 298.15 and 323.15 K.

7. Nitrogen + Ethanol [64-17-5] + Acetic acid, ethyl ester [141-78-6]

Nitta, Fujio, and Katayama (3) measured the solubility of nitrogen in the two pure components and nine mixtures at 298.15 K.

References:

1. Kretschmer, C. B.; Nowakowska, J.; Wiebe, R. *Ind. Eng. Chem.* 1946, *38*, 506.

2. Nitta, T.; Tatsuishi, A.; Katayama, T. *J. Chem. Eng. Japan* 1973, *6*, 475.

3. Nitta, T.; Fujio, J.; Katayama, T. *J. Chem. Eng. Data* 1978, *23*, 157.

**COMPONENTS:**

(1) Nitrogen; $N_2$; [7727-37-9]

(2) Ethanol or Ethyl Alcohol; $C_2H_6O$; [64-17-5]

(3) 2-Propanone; $C_3H_6O$; [67-64-1]

**ORIGINAL MEASUREMENTS:**

Kretschmer, C. B.; Nowakowska, J.; Wiebe, R.

*Ind. Eng. Chem.* 1946, *38*, 506-9.

**VARIABLES:**

$T$/K = 273.15 - 323.15
$p$/kPa = 101.325
$C_2H_6O/x_2$ = 0.558

**PREPARED BY:**

D. G. T. Thornhill

**EXPERIMENTAL VALUES:**

| $T$/K | Ethanol Mol Fraction $x_2$ | Nitrogen Mol Fraction $10^4 x_1$ | Nitrogen Ostwald Coefficient $L_1$/cm$^3$ cm$^{-3}$ |
|---|---|---|---|
| 273.15 | 0.558 | 4.05 | 0.1427 |
| 298.15 | 0.558 | 4.30 | 0.1609 |
| 323.15 | 0.558 | 4.65 | 0.1829 |

Smoothing equation:

$\Delta G_1^\circ$/J mol$^{-1}$ = $-RT \ln x_1$/J mol$^{-1}$ = A + BT/K.

Coefficients in the smoothing equation, A = $\Delta H_1^\circ$/J mol$^{-1}$ and -B = $S_1^\circ$/J mol$^{-1}$ K$^{-1}$, and standard deviation, $\sigma_d$, were determined by least-squares analysis.

$\sigma_d = \left(\Sigma(\Delta G_1^\circ(\text{calc}) - \Delta G_1^\circ)^2/(N-1)\right)^{1/2}$, N being the number of experimental values.

$\Delta H_1^\circ$/J mol$^{-1}$ = 2042 $\Delta S_1^\circ$/J mol$^{-1}$ K$^{-1}$ = -57.5 $\sigma_d$/J mol$^{-1}$ = 22

**AUXILIARY INFORMATION**

**METHOD/APPARATUS/PROCEDURE:**

Saturation of degassed solvent with dry gas. Solubilities were calculated from appropriate determinations of gas pressures and volumes, and of solvent volumes and vapour pressures.
The apparatus was similar to that of Horiuti (1). Solvent was initially degassed by boiling it under reflux. Then it was transferred to the evacuated absorption vessel, and degassing completed by pumping on it under vacuum. Mercury was now added to the vessel until it was completely filled. An accurately measured amount of gas was then added to the vessel from a buret until the mercury was displaced to a calibrated mark. The volume of displaced mercury equalled the gas phase volume. Subtraction of this latter value from the total volume of the absorption vessel yielded the volume of solvent. Gas and solvent were mixed by agitation with a stirrer, and intermittently fresh aliquots of gas were added to the vessel. During dissolution of gas care was taken to avoid diffusion of wet gas and solvent vapour from the absorption vessel to the buret. Final equilibrium was achieved after 1-3 hours. Corrections were made for increased solvent volume with gas absorption.

**SOURCE AND PURITY OF MATERIALS:**

(1) Nitrogen. Commercial nitrogen was freed of oxygen by passage through alkaline pyrogallol. Dried thoroughly by passage through $CaSO_4$.

(2) Ethanol. Fractionated and dried with Mg. $\rho_2$(298.15 K)/kg m$^{-3}$ = 785.08.

(3) 2-Propanone. Treated with $Ag_2O$, dried with $CaSO_4$ ("Drierite"), and fractionated. $\rho_3$(298.15 K)/kg m$^{-3}$ = 784.90. Boiling point (760 mmHg)/K = 329.25 - 329.29.

**ESTIMATED ERROR:** Average deviation of individual measurements from mean was 0.3%; maximum deviation was 0.9%. $\delta x_1/x_1 = \pm$ 0.005 (evaluator).

**REFERENCES:**

1. Horiuti, J. *Sci. Papers Inst. Phys. Chem. Research* (Tokyo) 1931, *17*, 125.

| COMPONENTS: | ORIGINAL MEASUREMENTS: |
|---|---|
| (1) Nitrogen; $N_2$; [7727-37-9]<br>(2) Ethanol; $C_2H_6O$; [64-17-5]<br>(3) Acetic acid, ethyl ester, (Ethyl acetate); $C_4H_8O_2$; [141-78-6] | Nitta, T.; Fujio, J.; Katayama, T.;<br>*J. Chem. Eng. Data.* 1978, *23*, 157-9. |
| VARIABLES: | PREPARED BY: |
| T/K = 313<br>Composition | C.L. Young, D. G. T. Thornhill |

EXPERIMENTAL VALUES:

| T/K | Mole fraction of of alcohol in mixture (before adsorption of nitrogen), $x_2$ | Ostwald coefficient, $L$ | Henry's Constant /atm | Mole fraction of nitrogen [a] $x_{N_2}$ |
|---|---|---|---|---|
| 313.15 | 0.0 | 0.2059 | 1242 | 0.0008052 |
| | 0.1000 | 0.2052 | 1297 | 0.0007710 |
| | 0.2000 | 0.2010 | 1382 | 0.0007236 |
| | 0.3000 | 0.1983 | 1465 | 0.0006826 |
| | 0.4000 | 0.1953 | 1559 | 0.0006414 |
| | 0.5000 | 0.1908 | 1677 | 0.0005963 |
| | 0.6000 | 0.1863 | 1810 | 0.0005525 |
| | 0.7000 | 0.1801 | 1979 | 0.0005053 |
| | 0.8000 | 0.1727 | 2190 | 0.0004566 |
| | 0.9000 | 0.1653 | 2437 | 0.0004103 |
| | 1.0 | 0.1584 | 2767 | 0.0003614 |

[a] Calculated by compiler. The equation $x_{N_2} = 8.15 \times 10^{-4} - 4.46 \times 10^{-4} x_2$ reproduces the nitrogen solubility with an average deviation of 0.7%.

AUXILIARY INFORMATION

METHOD/APPARATUS/PROCEDURE:

Saturation of degassed solvent with dry gas. Solubilities were calculated from approximate determinations of gas pressures and volumes. The apparatus was similar to those described by Horiuti (1), Kretschmer et al., (2) and Tokunaga et al., (3). Initially mixed solvent of known composition was degassed by pumping on it while it boiled under reflux. Degassed solvent was then transferred to the absorption vessel and degassed mercury added until the vessel was completely filled. An accurately measured amount of gas was now added to the vessel from a buret, and the mercury displaced to a calibrated mark. The volume of displaced mercury equalled of this latter value from the total volume of the absorption vessel yielded

continued on following page

SOURCE AND PURITY OF MATERIALS:

(1) Osaka Oxygen super-pure grade, purity 99.99 mole per cent.

(2) Spectral grade from Nakarai.

(3) Merck Uvasol grade.

ESTIMATED ERROR:

$\delta T/K = \pm 0.1$; $\delta x_{N_2}/x_{N_2} = \pm 0.008$

REFERENCES:

(1) Horiuti, J. *Sci. Papers Phys. Chem. Research (Tokyo)* 1931, *17*, 125.

(2) Nitta, T.; Tatsuishi, A.; Katayama, T. *J. Chem. Eng. Jap.* 1973, *6*, 475.

continued on following page

| COMPONENTS: | ORIGINAL MEASUREMENTS: |
|---|---|
| (1) Nitrogen; $N_2$; [7727-37-9]<br>(2) Ethanol; $C_2H_6O$; [64-17-5]<br>(3) Acetic acid, ethyl ester, (Ethyl acetate); $C_4H_8O_2$; [141-78-6] | Nitta, T.; Fujio, J.; Katayama, T.<br>*J. Chem. Eng. Data.* 1978, *23*, 157-9. |

METHOD/APPARATUS/PROCEDURE:

continued

the volume of solvent. Gas and solvent were mixed by agitation with stirrer bar, and intermittently fresh aliquots of gas were added to the vessel. During dissolution of gas care was taken to avoid diffusion of wet gas and solvent vapour from the absorption vessel to the buret. Final equilibrium was achieved after 1-1½ hours. Corrections were made for increased solvent volume with gas absorption.

REFERENCES:

continued

3. Katayama, T.; Nitta, T.
*J. Chem. Eng. Data* 1976, *21*, 194.

COMPONENTS:

(1) Nitrogen; $N_2$; [7727-37-9]

(2) Ethanol; $C_2H_6O$; [64-17-5]

(3) 1,1'-oxybisethane (diethyl ether); $C_4H_{10}O$; [60-29-7]

ORIGINAL MEASUREMENTS:

Nitta, T.; Fujio, J.; Katayama, T.

*J. Chem. Eng. Data* 1978, *23*, 157-9 .

VARIABLES:

T/K = 273
Composition

PREPARED BY:

C.L. Young , D. G. T. Thornhill

EXPERIMENTAL VALUES:

| T/K | Mole fraction of alcohol in mixture (before adsorption of nitrogen) , $x_2$ | Ostwald coefficient, $L$ | Henry's Constant /atm | Mole fraction[a] of nitrogen $x_{N_2}$ |
|---|---|---|---|---|
| 273.15 | 0.0 | 0.2668 | 835.1 | 0.001197 |
| | 0.1000 | 0.2562 | 909.0 | 0.001100 |
| | 0.2000 | 0.2429 | 1004 | 0.0009960 |
| | 0.3000 | 0.2330 | 1099 | 0.0009099 |
| | 0.4000 | 0.2180 | 1236 | 0.0008091 |
| | 0.50000 | 0.2035 | 1397 | 0.0007185 |
| | 0.6000 | 0.1923 | 1563 | 0.0006398 |
| | 0.7000 | 0.1778 | 1797 | 0.0005565 |
| | 0.8000 | 0.1637 | 2081 | 0.0004805 |
| | 0.9000 | 0.1500 | 2431 | 0.0004114 |
| | 1.0 | 0.1378 | 2847 | 0.0003512 |

[a] Calculated by compiler. The equation $x_{N_2} = 1.170 \times 10^{-3} - 8.55 \times 10^{-4} x_2$ reproduces the nitrogen solubility with an average deviation of 2.7%.

AUXILIARY INFORMATION

METHOD /APPARATUS/PROCEDURE :

Saturation of degassed solvent with dry gas. Solubilities were calculated from approximate determinations of gas pressures and volumes. The apparatus was similar to those described by Horiuti (1), Kretschmer et al., (2) and Tokunaga et al., (3). Initially mixed solvent of known composition was degassed by pumping on it while it boiled under reflux. Degassed solvent was then transferred to the absorption vessel and degassed mercury added until the vessel was completely filled. An accurately measured amount of gas was not added to the vessel from a buret, and the mercury displaced to a calibrated mark. The volume of displaced mercury equalled of this latter value from the total volume of the absorption vessel yielded

continued on following page

SOURCE AND PURITY OF MATERIALS:

(1) Osaka Oxygen super-pure grade, purity 99.99 mole per cent.

(2) Spectral grade from Nakarai.

(3) Nakarai sample, dried and fractionated.

ESTIMATED ERROR:

$\delta T/K = \pm 0.1$; $\delta x_{N_2}/x_2 = \pm 0.008$

REFERENCES:


1. Horiuti, J. *Sci. Papers Phys. Chem. Research (Tokyo)* 1931, *17*, 125.
2. Nitta, T.; Tatsuishi, A.; Katayama, T. *J. Chem. Eng. Jap.* 1973, *6*, 475.


continued on following page

| COMPONENTS: | ORIGINAL MEASUREMENTS: |
|---|---|
| (1) Nitrogen; $N_2$; [7727-37-9]<br>(2) Ethanol; $C_2H_6O$; [64-17-5]<br>(3) 1,1'-oxybisethane (diethyl ether); $C_4H_{10}O$; [60-29-7] | Nitta, T.; Fujio, J.; Katayama, T.<br>*J. Chem. Eng. Data* 1978, *23*, 157-9. |

METHOD/APPARATUS/PROCEDURE:

continued

the volume of solvent. Gas and solvent were mixed by agitation with stirrer bar, and intermittently fresh aliquots of gas were added to the vessel. During dissolution of gas care was taken to avoid diffusion of wet gas and solvent vapour from the absorption vessel to the buret. Final equilibrium was achieved after 1-1½ hours. Corrections were made for increased solvent volume with gas absorption.

REFERENCES:

continued

3. Katayama, T.; Nitta, T.
*J. Chem. Eng. Data* 1976, *21*, 194.

COMPONENTS:

(1) Nitrogen; $N_2$; [7727-37-9]

(2) Ethanol; $C_2H_6O$; [64-17-5]

(3) Benzene; $C_6H_6$; [71-43-2]

ORIGINAL MEASUREMENTS:

Nitta, T.; Fujio, J.; Katayama, T.

*J. Chem. Eng. Data.* 1978, *23*, 157-9.

VARIABLES:

T/K = 298

Composition

PREPARED BY:

C.L. Young, D. G. T. Thornhill

EXPERIMENTAL VALUES:

| T/K | Mole fraction of alcohol in mixture (before adsorption of nitrogen), $x_2$ | Ostwald coefficient, $L$ | Henry's Constant /atm | Mole fraction[a] of nitrogen $x_{N_2}$ |
|---|---|---|---|---|
| 298.15 | 0.0 | 0.1238 | 2211 | 0.0004523 |
| | 0.1000 | 0.1262 | 2245 | 0.0004454 |
| | 0.2000 | 0.1284 | 2287 | 0.0004373 |
| | 0.3000 | 0.1309 | 2330 | 0.0004292 |
| | 0.4000 | 0.1317 | 2408 | 0.0004153 |
| | 0.5000 | 0.1330 | 2488 | 0.0004019 |
| | 0.6000 | 0.1346 | 2562 | 0.0003903 |
| | 0.7000 | 0.1379 | 2615 | 0.0003824 |
| | 0.8000 | 0.1409 | 2683 | 0.0003727 |
| | 0.9000 | 0.1443 | 2750 | 0.0003636 |
| | 1.0 | 0.1494 | 2793 | 0.0003580 |

[a] Calculated by compiler. The equation $x_{N_2} = 4.54 \times 10^{-4} - 1.00 \times 10^{-4} x_2$ reproduces the nitrogen solubility with an average deviation of 0.6%.

AUXILIARY INFORMATION

METHOD /APPARATUS/PROCEDURE

Saturation of degassed solvent with dry gas. Solubilities were calculated from approximate determinations of gas pressures and volumes. The apparatus was similar to those described by Horiuti (1), Kretschmer et al., (2) and Tokunaga et al., (3). Initially mixed solvent of known composition was degassed by pumping on it while it boiled under reflux. Degassed solvent was then transferred to the absorption vessel and degassed mercury added until the vessel was completely filled. An accurately measured amount of gas was not added to the vessel from a buret, and the mercury displaced to a calibrated mark. The volume of displaced mercury equalled of this latter value from the total volume of the absorption vessel yielded

continued on following page

SOURCE AND PURITY OF MATERIALS:

(1) Osaka Oxygen super-pure grade, purity 99.99 mole per cent.

(2) Spectral grade from Nakarai.

(3) Merck Uvasol grade.

ESTIMATED ERROR:

$\delta T/K = \pm 0.1$; $\delta x_{N_2}/x_{N_2} = \pm 0.008$

REFERENCES:

(1) Horiuti, J. *Sci. Papers Phys. Chem. Research (Tokyo)* 1931, *17*, 125.

(2) Nitta, T.; Tatsuishi, A.; Katayama, T. *J. Chem. Eng. Jap.* 1973, *6*, 475.

| COMPONENTS: | ORIGINAL MEASUREMENTS: |
|---|---|
| (1) Nitrogen; $N_2$; [7727-37-9] | Nitta, T.; Fujio, J.; Katayama, T. |
| (2) Ethanol; $C_2H_6O$; [64-17-5] | *J. Chem. Eng. Data.* 1978, *23*, 157-9. |
| (3) Benzene; $C_6H_6$; [71-43-2] | |

METHOD/APPARATUS/PROCEDURE

continued

the volume of solvent. Gas and solvent were mixed by agitation with stirrer bar, and intermittently fresh aliquots of gas were added to the vessel. During dissolution of gas care was taken to avoid diffusion of wet gas and solvent vapour from the absorption vessel to the buret. Final equilibrium was achieved after 1-1½ hours. Corrections were made for increased solvent volume with gas absorption.

REFERENCES:

continued

3. Katayama, T.; Nitta, T.
*J. Chem. Eng. Data* 1976, *21*, 194.

| COMPONENTS: | ORIGINAL MEASUREMENTS: |
|---|---|
| (1) Nitrogen; $N_2$; [7727-37-9]<br>(2) Ethanol or Ethyl Alcohol; $C_2H_6O$; [64-17-5]<br>(3) 2,2,4-Trimethylpentane; $C_8H_{18}$; [540-84-1] | Kretschmer, C. B.; Nowakowska, J.; Wiebe, R.<br>*Ind. Eng. Chem.* 1946, *38*, 506-9. |
| VARIABLES: $T$/K = 273.15 - 323.15<br>$p$/kPa = 101.325<br>$C_2H_6O/x_2$ = 0.739 | PREPARED BY:<br>D. G. T. Thornhill |

EXPERIMENTAL VALUES:

| $T$/K | Ethanol Mol Fraction $x_2$ | Nitrogen Mol Fraction $10^4 x_1$ | Nitrogen Ostwald Coefficient $L_1$/cm$^3$ cm$^{-3}$ |
|---|---|---|---|
| 273.15 | 0.739 | 6.71 | 0.1785 |
| 298.15 | 0.739 | 6.87 | 0.1939 |
| 323.15 | 0.739 | 7.06 | 0.2100 |

Smoothing equation:

$\Delta G_1^\circ$/J mol$^{-1}$ = $-$ RT ln$x_1$/J mol$^{-1}$ = A + BT/K.

Coefficients in the smoothing equation, A = $\Delta H_1^\circ$/J mol$^{-1}$ and

$-$B = $\Delta S_1^\circ$/J mol$^{-1}$ K$^{-1}$, and standard deviation, $\sigma_d$, were determined

by least-squares analysis.

$\sigma_d = \left(\Sigma(\Delta G_1^\circ(\text{calc}) - \Delta G_1^\circ)^2/(N-1)\right)^{1/2}$, N being the number of experimental values.

$\Delta H_1^\circ$/J mol$^{-1}$ = 749

$\Delta S_1^\circ$/J mol$^{-1}$ K$^{-1}$ = $-$58.0

$\sigma_d$/J mol$^{-1}$ = 9

AUXILIARY INFORMATION

METHOD/APPARATUS/PROCEDURE:

Saturation of degassed solvent with dry gas. Solubilities were calculated from appropriate determinations of gas pressures and volumes, and of solvent volumes and vapour pressures.
The apparatus was similar to that of Horiuti (1). Solvent was initially degassed by boiling it under reflux. Then it was transferred to the evacuated absorption vessel, and degassing completed by pumping on it under vacuum. Mercury was now added to the vessel until it was completely filled. An accurately measured amount of gas was then added to the vessel from a buret until the mercury was displaced to a calibrated mark. The volume of displaced mercury equalled the gas phase volume. Subtraction of this latter value from the total volume of the absorption vessel yielded the volume of solvent. Gas and solvent were mixed by agitation with a stirrer, and intermittently fresh aliquots of gas were added to the vessel. During dissolution of gas care was taken to avoid diffusion of wet gas and solvent vapour from the absorption vessel to the buret. Final equilibrium was achieved after 1-3 hours. Corrections were made for increased solvent volume with gas absorption.

SOURCE AND PURITY OF MATERIALS:

(1) Nitrogen. Commercial nitrogen was freed of oxygen by passage through alkaline pyrogallol. Dried thoroughly by passage through $CaSO_4$.

(2) Ethanol. Fractionated and dried with Mg.
$\rho_2$(298.15 K)/kg m$^{-3}$ = 785.08.

(3) 2,2,4-Trimethypentane. Certified material dried with $CaSO_4$, and fractionated.
$\rho_3$(298.15 K)/kg m$^{-3}$ = 687.74.
Boiling point (760 mmHg)/K=372.39.

ESTIMATED ERROR: Average deviation of individual measurements from mean was 0.3%; maximum deviation was 0.9%.
$\delta x_1/x_1$ = ± 0.005 (evaluator).

REFERENCES:

1. Horiuti, J.
*Sci. Papers Inst. Phys. Chem. Research* (Tokyo) 1931, *17*, 125.

COMPONENTS:

(1) Nitrogen; $N_2$; [7727-37-9]

(2) 2-Propanone or acetone; $C_3H_6O$; [67-64-1]

(3) Ethanol; $C_2H_6O$; [64-17-5]

ORIGINAL MEASUREMENTS:

Kretschmer, C. B.; Nowakowska, J.; Wiebe, R.

*Ind. Eng. Chem.* 1946, *38*, 506-9.

VARIABLES:

T/K = 273.15 - 323.15
P/kPa = 101.325
Ethanol/vol % = 50

PREPARED BY:

P. L. Long
H. L. Clever

EXPERIMENTAL VALUES:

| T/K | Bunsen Coefficient $\alpha$ | Ostwald Coefficient L |
|---|---|---|
| 273.15 | 0.1427 | 0.1427 |
| 298.15 | 0.1474 | 0.1609 |
| 323.15 | 0.1546 | 0.1829 |

The Ostwald coefficients were measured at a total pressure of 101.325 kPa (1 atm). The compiler calculated the Bunsen coefficient value for an oxygen partial pressure of 101.325 kPa (1 atm) assuming the Ostwald coefficient was independent of pressure.

The solvent is acetone in ethanol, 50 per cent by volume.

AUXILIARY INFORMATION

METHOD/APPARATUS/PROCEDURE:

The apparatus was similar to that of Horiuti (1), it consisted of a gas buret, manometer, and absorption pipet equipped with a magnetic stirrer. All volumes were calibrated by mercury displacement.

The solvent was partly degassed by boiling under reflux. About 40 $cm^3$ of solvent was admitted into the absorption pipet, and pumped on to complete the degassing. Dry gas, measured in the buret, was admitted, final equilibrium was established after 1 to 3 hours stirring. During the solution process the total pressure was kept at one atm by additions of dry gas.

The solubility value was corrected for the increase of solvent volume with gas absorption

SOURCE AND PURITY OF MATERIALS:

(1) Nitrogen. Commercial, but source not given. Passed through alkaline pyrogallol to remove $O_2$, passed through Drierite to dry.

(2) 2-Propanone. Source not given. Treated with $Ag_2O$, dried over $CaSO_4$, distilled, b.p. (760 mmHg)/°C 56.10-56.14, density, $\rho_4^{25}$/g $cm^{-3}$ 0.78490

(3) Ethanol. Source not given. Dried with Mg. Density, $\rho$/g $cm^{-3}$ 0.78508.

ESTIMATED ERROR:

Gas buret $\delta V/cm^3$ = 0.005
Average deviation from the mean 0.3%, maximum deviation 0.9% (authors).

REFERENCES:

1. Horiuti, J. *J. Sci Papers Inst. Phys. Chem. Res. (Tokyo)* 1931, *17*, 125-256.

| COMPONENTS: | ORIGINAL MEASUREMENTS: |
|---|---|
| (1) Nitrogen; $N_2$; [7727-37-9]<br>(2) 1-Propanol; $C_3H_8O$; [71-23-8]<br>(3) Pentane, 2,2,4-trimethyl-, $C_8H_{18}$; [540-84-1] | Nitta, T.; Tatsuishi. A.; Katayama, T.<br>*J. Chem. Eng. Japan* 1973, *6*, 475-80. |
| VARIABLES:<br>T/K = 298<br>Composition | PREPARED BY:<br>C.L. Young, D. G. T. Thornhill |

EXPERIMENTAL VALUES:

| T/K | Mole fraction of 1-propanol in soln. before saturation. $x_2$ | Ostwald coefficient, $L$ | Henry's Law Constant /atm | Mole fraction of nitrogen[a] $x_{N_2}$ |
|---|---|---|---|---|
| 298.15 | 0.0000 | 0.2255 | 653.8 | 0.001530 |
| | 0.1000 | 0.2207 | 706.2 | 0.001416 |
| | 0.2000 | 0.2157 | 766.7 | 0.001304 |
| | 0.3000 | 0.2110 | 834.6 | 0.001198 |
| | 0.4000 | 0.2052 | 918.7 | 0.001088 |
| | 0.5000 | 0.1973 | 1028 | 0.0009727 |
| | 0.6000 | 0.1874 | 1170 | 0.0008547 |
| | 0.7000 | 0.1753 | 1363 | 0.0007336 |
| | 0.8000 | 0.1622 | 1616 | 0.0006188 |
| | 0.9000 | 0.1474 | 1971 | 0.0005073 |
| | 1.0000 | 0.1312 | 2482 | 0.0004029 |

[a] Calculated by compiler. The equation $x_{N_2} = 1.534 \times 10^{-3} - 1.135 \times 10^{-3} x_2$ reproduces the nitrogen solubility with an average deviation of 0.6%.

AUXILIARY INFORMATION

METHOD /APPARATUS/PROCEDURE

Saturation of degassed solvent with dry gas. Solubilities were calculated from approximate determinations of gas pressures and volumes. The apparatus was similar to those described by Horiuti (1), Kretschmer et al., (2) and Tokunaga et al., (3). Initially mixed solvent of known composition was degassed by pumping on it while it boiled under reflux. Degassed solvent was then transferred to the absorption vessel and degassed mercury added until the vessel was completely filled. An accurately measured amount of gas was not added to the vessel from a buret, and the mercury displaced to a calibrated mark. The volume of displaces mercury equalled of this latter value from the total volume of the absorption vessel yielded the

continued on following page

SOURCE AND PURITY OF MATERIALS:

(1) Osaka Oxygen super-pure grade of purity 99.9995 mole per cent.

(2) Nakarai Chemical sample dried and fractionated, purity about 99.9 mole per cent.

(3) Nakarai Chemical sample dried and fractionated, purity better than 99.98 mole per cent

ESTIMATED ERROR:

$\delta T/K = \pm 0.05$; $\delta x/x = \pm 1\%$ (estimated by compiler.)

REFERENCES:

(1) Horiuti, J. *Sci. Papers Inst. Phys. Chem. Research (Tokyo)* 1931, *17*, 125.

(2) Kretschmer, C. B.; Nowakowska, J.; Wiebe, R. *Ind. Eng. Chem.* 1946, *38*, 506.

continued on following page

| COMPONENTS: | ORIGINAL MEASUREMENTS: |
|---|---|
| (1) Nitrogen; $N_2$; [7727-37-9]<br>(2) 1-Propanol; $C_3H_8O$; [71-23-8]<br>(3) Pentane, 2,2,4-trimethyl-, $C_8H_{18}$; [540-84-1] | Nitta, T.; Tatsuishi, A.; Katayama, T.<br>*J. Chem. Eng. Japan* 1973, *6*, 475-80. |

METHOD/APPARATUS/PROCEDURE:

continued

volume of solvent. Gas and solvent were mixed by agitation with stirrer bar, and intermittently fresh aliquots of gas were added to the vessel. During dissolution of gas care was taken to avoid diffusion of wet gas and solvent vapour from the absorption vessel to the buret. Final equilibrium was achieved after 1-1½ hours. Corrections were made for increased solvent volume with gas absorption.

REFERENCES:

continued

3. Tokunaga, J.; Nitta, T.; Katayama, T. *Kagaku Kogaku* 1969, *33*, 775.

COMPONENTS:

(1) Nitrogen; $N_2$; [7727-37-9]

(2) Cyclohexane; $C_6H_{12}$; [110-82-7]

(3) Pentane, 2,2,4-trimethyl-; $C_8H_{18}$; [540-84-1]

ORIGINAL MEASUREMENTS:

Nitta, T.; Tatsuishi, A.; Katayama, T.

*J. Chem. Eng. Japan* 1973, *6*, 475-80.

VARIABLES:

T/K = 298

Composition

PREPARED BY:

C.L. Young, D. G. T. Thornhill

EXPERIMENTAL VALUES:

| T/K | Mole fraction of cyclohexane in soln. before saturation, $x_2$ | Ostwald coefficient, $L$ | Henry's Law Constant /atm | Mole fraction[a] of nitrogen, $x_{N_2}$ |
|---|---|---|---|---|
| 298.15 | 0.0000 | 0.2255 | 653.8 | 0.001530 |
| | 0.1000 | 0.2216 | 689.0 | 0.001451 |
| | 0.2000 | 0.2168 | 730.4 | 0.001369 |
| | 0.3000 | 0.2131 | 771.6 | 0.001296 |
| | 0.4000 | 0.2084 | 820.2 | 0.001219 |
| | 0.5000 | 0.2040 | 873.1 | 0.001145 |
| | 0.6000 | 0.1984 | 936.8 | 0.001067 |
| | 0.7000 | 0.1923 | 1010 | 0.000990 |
| | 0.8000 | 0.1857 | 1096 | 0.0009124 |
| | 0.9000 | 0.1780 | 1201 | 0.0008326 |
| | 1.0000 | 0.1705 | 1320 | 0.0007575 |

[a] Calculated by compiler. The equation $x_1 = 1.528 \times 10^{-3} - 7.70 \times 10^{-4} x_2$ reproduces the nitrogen solubility with an average deviation of 0.1%.

AUXILIARY INFORMATION

METHOD /APPARATUS/PROCEDURE

Saturation of degassed solvent with dry gas. Solubilities were calculated from approximate determinations of gas pressures and volumes. The apparatus was similar to those described by Horiuti (1), Kretschmer et al., (2) and Tokunaga et al., (3).Initially mixed solvent of known composition was degassed by pumping on it while it boiled under reflux. Degassed solvent was then transferred to the absorption vessel and degassed mercury added until the vessel was completely filled. An accurately measured amount of gas was now added to the vessel from a buret, and the mercury displaced to a calibrated mark. The volume of displaced mercury equalled of this latter value from the total volume of the absorption vessel yielded

continued on following page

SOURCE AND PURITY OF MATERIALS:

(1) Osaka Oxygen super-pure grade of purity 99.9995 mole per cent.

(2) Wako Pure Chemical sample, purity better than 99.9 mole per cent.

(3) Nakarai Chemical sample dried and fractionated, purity better than 99.98 mole per cent.

ESTIMATED ERROR:

$\delta T/K = \pm 0.05$; $\delta x/x = \pm 1\%$

(estimated by compiler).

REFERENCES:

1. Horiuti, J. *Sci. Papers Inst. Phys. Chem. Research (Tokyo)* 1931, *17*, 125.

2. Kretschmer, C.B.; Nowakowska, J. Wiebe, R. *Ind. Eng. Chem.* 1946, *38*, 506.

continued on following page

| COMPONENTS: | ORIGINAL MEASUREMENTS: |
|---|---|
| (1) Nitrogen; $N_2$; [7727-37-9]<br>(2) Cyclohexane; $C_6H_{12}$; [110-82-7]<br>(3) Pentane, 2,2,4-trimethyl-; $C_8H_{18}$; [540-84-1] | Nitta, T.; Tatsuishi, A.; Katayma, T.<br>*J. Chem. Eng. Japan* 1973, *6*, 475-80. |

METHOD/APPARATUS/PROCEDURE

continued

the volume of solvent. Gas and solvent were mixed by agitation with stirrer bar, and intermittently fresh aliquots of gas were added to the vessel. During dissolution of gas care was taken to avoid diffusion of wet gas and solvent vapour from the absorption vessel to the buret. Final equilibrium was achieved after 1-1½ hours. Corrections were made for increased solvent volume with gas absorption.

REFERENCES:

continued

3. Tokunaga, J.; Nitta, T.; Katayama, T. *Kagaku Kogaku* 1969, *33*, 775.

| COMPONENTS: | EVALUATOR: |
|---|---|
| (1) Nitrogen; $N_2$; [7727-37-9]<br>(2) n-Alkanes and 2,2,4-Trimethylpentane | Rubin Battino<br>Department of Chemistry<br>Wright State University<br>Dayton, Ohio 45435<br>U.S.A.<br>December 1981 |

CRITICAL EVALUATION:

The solubility of nitrogen in saturated hydrocarbons has been studied by a number of workers at pressures below two bars. (For critical evaluations of these systems at higher pressures see the high pressure section.) The figure below shows the data of Makranczy, et al. (1)-○; Gjaldbaek and Hildebrand (2)-✕; Patyi, et al. (3)-●; Thomsen and Gjaldbaek (4)-□; Wilcock, et al. (5)-△; Katayama and Nitta (6)-■; and Tremper and Prausnitz (7)-▽ for the n-alkanes. For carbon numbers six through ten the Makranczy, et al., values are significantly higher than those of other workers. (The data of Guerry (8) and Ijams (9) were very low and/or erratic and were not graphed.) Obviously, these systems await a definitive study.

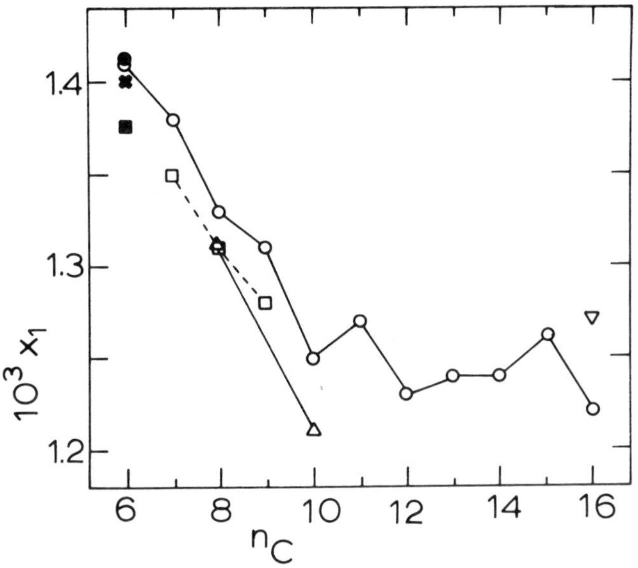


The data for the individual n-alkanes and 2,2,4-trimethylpentane (isooctane)will be discussed individually below.

Nitrogen in n-Hexane

The solubility of nitrogen in n-hexane was determined by five groups (1,2,3,6,8) but Guerry's datum (8) was much too low to be considered. We used the data of the other four groups, rejecting the 313.15 K value from reference 1 to end up with nine points in the temperature range 213.15 to 298.15 K. These data were smoothed to yield

$$\ln x_1 = -6.39422 - 0.553875/(T/100K) \qquad (1)$$

The standard deviation in $\ln x_1$ was 0.016. Smoothed recommended values at 10 K intervals are given below.

| T/K | $10^3 x_1$ | T/K | $10^3 x_1$ |
|---|---|---|---|
| 213.15 | 1.29 | 263.15 | 1.35 |
| 223.15 | 1.30 | 273.15 | 1.36 |
| 233.15 | 1.32 | 283.15 | 1.37 |
| 243.15 | 1.33 | 293.15 | 1.38 |
| 253.15 | 1.34 | 298.15 | 1.39 |

continued on following page

| COMPONENTS: | EVALUATOR: |
|---|---|
| (1) Nitrogen; $N_2$; [7727-37-9]<br>(2) n-Alkanes and 2,2,4-Trimethylpentane | Rubin Battino<br>Department of Chemistry<br>Wright State University<br>Dayton, Ohio 45435<br>U.S.A.<br>December 1981 |

CRITICAL EVALUATION:

continued

The thermodynamic properties are $\Delta\bar{H}°=461$ J $mol^{-1}$ and $\Delta\bar{S}°=-53$ J $K^{-1}$ $mol^{-1}$ and are independent of temperature.

Nitrogen in n-Heptane

The solubility of nitrogen in n-heptane was determined by four groups (1,4,8,9) but we selected the work of Thomsen and Gjaldbaek (4) as being the most reliable. The smoothed data from their paper in the range 298 to 308 K are given by:

$$\ln x_1 = -6.04777 - 1.66951/(T/100K) \qquad (2)$$

The standard deviation in $\ln x_1$ was 0.0049. The smoothed mole fraction at 298.15 K is $1.35 \times 10^{-3}$. The thermodynamic properties are $\Delta\bar{H}°=1388$ J $mol^{-1}$ and $\Delta\bar{S}°=-50$ J $K^{-1}$ $mol^{-1}$.

Nitrogen in n-Octane

This system was studied by four workers (1,4,5,9) and the results obtained by Thomsen and Gjaldbaek (4) and Wilcock et al., (5) were selected as the most reliable. Smoothing these results in the range 298 to 308K yielded:

$$\ln x_1 = -5.98284 - 1.94827/(T/100K) \qquad (3)$$

The standard deviation in $\ln x_1$ was 0.0053. $\Delta\bar{H}°=1620$ J $mol^{-1}$ and $\Delta\bar{S}°=-49$ J $K^{-1}$ $mol^{-1}$.

Nitrogen in n-Nonane

The Thomsen and Gjaldbaek (4) data were the only ones which we considered reliable, but they only determined the solubility at 298.15K. The average of the values is $x_1 = 1.28 \times 10^{-3}$. The Makranczy, et al., (1) values show too great a temperature dependence when compared with other workers and are to be treated as tentative until further checks are made.

Nitrogen in n-Decane

Wilcock, et al.'s values (5) are the most reliable for this system. The other two sets of results (1,9) were not used. Smoothing of Wilcock, et al's three points yields:

$$\ln x_1 = -6.82876 + 0.340428/(T/100K) \qquad (4)$$

This is for the range 283 to 313K. The standard deviation in $\ln x_1$ is 0.00052. Smoothed values at 5K intervals follow. $\Delta\bar{H}°=-283$ J $mol^{-1}$ and $\Delta\bar{S}°=-56.8$ J $K^{-1}$ $mol^{-1}$.

continued on following page

| COMPONENTS: | EVALUATOR: |
|---|---|
| (1) Nitrogen; $N_2$; [7727-37-9]<br>(2) n-Alkanes and 2,2,4-Trimethylpentane | Rubin Battino<br>Department of Chemistry<br>Wright State University<br>Dayton, Ohio 45435<br>U.S.A.<br>December 1981 |

CRITICAL EVALUATION:

continued

| T/K | $10^3x_1$ | T/K | $10^3x_1$ |
|---|---|---|---|
| 283.15 | 1.220 | 303.15 | 1.211 |
| 283.15 | 1.218 | 308.15 | 1.209 |
| 293.15 | 1.215 | 313.15 | 1.206 |
| 298.15 | 1.213 | | |

Nitrogen in 2,2,4-Trimethylpentane

For this system we used the results of Kretschmer, et al., (10) and Baldwin and Daniel (11) rejecting Ijams single value (9). Over the temperature range 248 to 323 K the smoothing equation is:

$$\ln x_1 = -6.30691 - 0.503860/(T/100K) \qquad (5)$$

with a standard deviation of 0.014 in $\ln x_1$. Smoothed values at 10K intervals follow.

| T/K | $10^3x_1$ | T/K | $10^3x_1$ |
|---|---|---|---|
| 248.15 | 1.49 | 298.15 | 1.54 |
| 258.15 | 1.50 | 308.15 | 1.55 |
| 268.15 | 1.51 | 318.15 | 1.56 |
| 278.15 | 1.52 | 328.15 | 1.56 |
| 288.15 | 1.53 | | |

$\Delta\bar{H}°=-419$ J mol$^{-1}$ and $\Delta\bar{S}°=-52$ J K$^{-1}$ mol$^{-1}$.

Nitrogen in n-Hexadecane

The data reported by Tremper and Prausnitz (7) are considered the most reliable here; that of Makranczy, et. al., (1) not being used. The smoothing equation is:

$$\ln x_1 = -5.54557 - 3.45285/(T/100K) \qquad (6)$$

with a standard deviation of 0.019 in $\ln x_1$. Smoothed values are at 25K intervals for the range 300 to 475 K. $\Delta\bar{H}°=2871$ J mol$^{-1}$ and $\Delta\bar{S}°=-46$ J K$^{-1}$ mol$^{-1}$.

| T/K | $10^3x_1$ | T/K | $10^3x_1$ |
|---|---|---|---|
| 300 | 1.24 | 400 | 1.65 |
| 325 | 1.35 | 425 | 1.73 |
| 350 | 1.46 | 450 | 1.81 |
| 375 | 1.55 | 475 | 1.89 |

References

1. Makranczy, J.; Megyery-Balog, Mrs. K.; Rusz, L.; Patyi, L. *Hung. J. Ind. Chem.* 1976, *4*, 269-80.

2. Gjaldbaek, J. C.; Hildebrand, J. H. *J. Am. Chem. Soc.* 1949, *71*, 3147-50.

3. Patyi, L.; Furmer, I. E.; Makranczy, J.; Sadilenko, A. S.; Stepanova, Z. G.; Berengarten, M. G. *Zh. Prikl. Khim.* 1978, *51*, 1296-1300.

continued on following page

| COMPONENTS: | EVALUATOR: |
|---|---|
| (1) Nitrogen; $N_2$; [7727-37-9]<br>(2) n-Alkanes and 2,2,4-Trimethylpentane | Rubin Battino<br>Department of Chemistry<br>Wright State University<br>Dayton, Ohio 45435<br>U.S.A.<br>December 1981 |

CRITICAL EVALUATION:

continued


4. Thomsen, E. S.; Gjaldbaek, J. C. *Acta. Chem. Scand.* 1963, *17*, 127-33.

5. Wilcock, R. J.; Battino, R.; Danforth, W. F.; Wilhelm, E. *J. Chem. Thermodyn.* 1978, *10*, 817-22.

6. Katayama, T.; Nitta, T. *J. Chem. Eng. Data* 1976, *21*, 194-6.

7. Tremper, K. K.; Prausnitz, J. M. *J. Chem. Eng. Data* 1976, *21*, 295-9.

8. Guerry, D. Jr. Ph. D. Thesis 1944, Vanderbilt University, Nashville, TN.

9. Ijams, C. C. Ph. D. Thesis 1941, Vanderbilt University, Nashville, TN.

10. Kretschmer, C. B.; Nowakowska, J.; Wiebe, R. *Ind. Eng. Chem.* 1946, *38*, 506-9.

11. Baldwin, R. R.; Daniel, S. G. *J. Appl. Chem.* 1952, *2*, 161-5; *J. Inst. Petrol., London* 1953, *39*, 105-24.

| COMPONENTS: | ORIGINAL MEASUREMENTS: |
|---|---|
| (1) Nitrogen; $N_2$; [7727-37-9]<br>(2) Pentane; $C_5H_{12}$; [109-66-0] | Makranczy, J.; Megyery-Balog, Mrs. K; Rusz, L.; Patyi, L.<br>*Hung. J. Ind. Chem.* 1976, *4*(2), 269-280. |
| VARIABLES: | PREPARED BY: |
| T/K = 298.15 -<br>Total P/kPa = 101.325 | S. A. Johnson<br>H. L. Clever |

EXPERIMENTAL VALUES:

| T/K | Mol Fraction $X_1 \times 10^3$ | Bunsen Coefficient $\alpha$ | Ostwald Coefficient L |
|---|---|---|---|
| 298.15 | 1.45 | 0.280 | 0.306 |
| 313.15 | 1.23 | 0.232 | 0.266 |

Smoothed Data: $\Delta G^o/J\ mol^{-1} = -RT \ln X_1 = -8{,}516 + 82.91T$

Both the data and the equation are classed as tentative.

Mol Fractions and Bunsen Coefficients were calculated by the compiler at a nitrogen partial pressure of 101.325 kPa (1 atm) with the assumption that the Ostwald coefficient is independent of pressure.

AUXILIARY INFORMATION

METHOD/APPARATUS/PROCEDURE:

Volumetric method, described in detail in reference (1).

SOURCE AND PURITY OF MATERIALS:

Analytical grade reagents of Hungarian and foreign origin were used (both liquids and gases). No further information.

ESTIMATED ERROR:

$\delta X_1/X_1 = 0.03$

REFERENCES:

1. Bodor, E.; Bor, Gy.; Mohai, B.; Sipos, G. *Veszpremi Vegyip. Egy. Kozl.* 1957, *1*, 55. *Chem. Abstr.* 1961, *55*, 3175h.

COMPONENTS:

(1) Nitrogen; $N_2$; [7727-37-9]

(2) Hexane; $C_6H_{14}$; [110-34-3]

ORIGINAL MEASUREMENTS:

Gjaldbaek, J. C.; Hildebrand, J. H.

*J. Am. Chem. Soc.* 1949, *71*, 3147-50.

VARIABLES:

T/K = 298.10 - 298.45
Total P/kPa = 101.325

PREPARED BY:

J. Chr. Gjaldbaek

EXPERIMENTAL VALUES:

| T/K | Mol Fraction $X_1 \times 10^3$ | Bunsen Coefficient $\alpha$ | Ostwald Coefficient L |
|---|---|---|---|
| 298.10 | 1.40 | 0.241 | 0.263 |
| 298.45 | 1.39 | 0.237 | 0.259 |

One value from the paper was rejected.

The mole fraction and Ostwald solubility values were calculated by the compiler.

AUXILIARY INFORMATION

METHOD/APPARATUS/PROCEDURE:

A calibrated all-glass combined manometer and bulb was enclosed in an air thermostat and shaken until equilibrium. Mercury was used for calibration and as the confining liquid. The solvents were degassed in the apparatus. Details are in references 1 and 2.

The absorbed volume of gas was calculated from the initial and final amounts, both saturated with solvent vapor. The amount of solvent was determined by the weight of displaced mercury.

The saturation of the liquid with the gas was carried out close to atmospheric pressure. The solubility values were reported for one atmosphere gas pressure assuming Henry's law is obeyed.

SOURCE AND PURITY OF MATERIALS:

(1) Nitrogen. Stuart Oxygen Co. 99.996 percent $N_2$.

(2) Hexane. Phillips Petroleum Co. Research grade.

ESTIMATED ERROR:

$\delta T/K = 0.05$
$\delta X_1/X_1 = 0.015$

REFERENCES:

1. Lannung, A. *J. Am. Chem. Soc.* 1930, *52*, 68.

2. Gjaldbaek, J. C. *Acta Chem. Scand.* 1952, *6*, 623

COMPONENTS:

(1) Nitrogen; $N_2$; [7727-37-9]

(2) Hexane; $C_6H_{14}$; [110-54-3]

ORIGINAL MEASUREMENTS:

Katayama, T.; Nitta, T.

*J. Chem. Eng. Data.* 1976, *21*, 194-6.

VARIABLES:

T/K = 213-298

PREPARED BY:

C.L. Young

EXPERIMENTAL VALUES:

| T/K | Ostwald coefficient, L | Henry's Constant /atm | Mole fraction* of nitrogen in liquid, $x_{N_2}$ |
|---|---|---|---|
| 298.15 | 0.256 | 727 | 0.001376 |
| 273.15 | 0.235 | 749 | 0.001335 |
| 253.15 | 0.219 | 764 | 0.001309 |
| 233.15 | 0.206 | 766 | 0.001305 |
| 213.15 | 0.195 | 758 | 0.001319 |

* at a partial pressure of 1 atmosphere

AUXILIARY INFORMATION

METHOD /APPARATUS/PROCEDURE:

Volumetric apparatus with equilibrium cell of approximately 0.08 litres. Magnetic stirrer. Solvent carefully degassed. Equilibrium established in 1-1.5 hours. Details in ref. (1).

SOURCE AND PURITY OF MATERIALS:

(1) Osaka Oxygen Co. sample, purity 99.9995 mole per cent.

(2) Merck Uvasol spectral grade sample.

ESTIMATED ERROR:

$\delta T/K = \pm 0.12$ at $T = 213.15$K, $\pm 0.05$ at other temperatures; $\delta x_{N_2} = \pm 1\%$

REFERENCES:

1. Nitta, T.; Tatsuishi. A.; Katayama, T. *J. Chem. Eng. Jpn.* 1973, *6*, 475.

**COMPONENTS:**

(1) Nitrogen; $N_2$; [7727-37-9]

(2) Hexane; $C_6H_{14}$; [110-54-3]

**ORIGINAL MEASUREMENTS:**

Makranczy, J.; Megyery-Balog, Mrs.K; Rusz, L.; Patyi, L.

*Hung. J. Ind. Chem.* 1976, *4*(2), 269-280.

**VARIABLES:**

T/K = 298.15 - 313.15
Total P/kPa = 101.325

**PREPARED BY:**

S. A. Johnson
H. L. Clever

**EXPERIMENTAL VALUES:**

| T/K | Mol Fraction $X_1 \times 10^3$ | Bunsen Coefficient $\alpha$ | Ostwald Coefficient L |
|---|---|---|---|
| 298.15 | 1.41 | 0.240 | 0.262 |
| 313.15 | 1.17 | 0.195 | 0.223 |

Smoothed Data: $\Delta G^o/J\ mol^{-1} = -RT \ln X_1 = -9,655 + 86.96T$

Both the data and the equation are classed as tentative.

Mol Fractions and Bunsen Coefficients were calculated by the compiler at a nitrogen partial pressure of 101.325 kPa (1 atm) with the assumption that the Ostwald coefficient is independent of pressure.

**AUXILIARY INFORMATION**

**METHOD/APPARATUS/PROCEDURE:**

Volumetric method, described in detail in reference (1).

**SOURCE AND PURITY OF MATERIALS:**

Analytical grade reagents of Hungarian and foreign origin were used (both liquids and gases). No further information.

**ESTIMATED ERROR:**

$\delta X_1/X_1 = 0.03$

**REFERENCES:**

1. Bodor, E.; Bor, Gy.; Mohai, B.; Sipos, G. *Veszpremi Vegyip. Egy. Kozl.* 1957, *1*, 55. *Chem. Abstr.* 1961, *55*, 3175h.

| COMPONENTS: | ORIGINAL MEASUREMENTS: |
|---|---|
| (1) Nitrogen; $N_2$; [7727-37-9]<br><br>(2) Hexane; $C_6H_{14}$; [110-54-3] | Patyi, L; Furmer, I. E.; Makranczy, J.; Sadilenko, A. S.; Stepanova, Z. G.; Berengarten, M. G.<br>*Zh. Prikl. Khim.* 1978, *51*, 1296-1300. |
| VARIABLES:<br>T/K = 298 | PREPARED BY:<br>C. L. Young |

EXPERIMENTAL VALUES:

| T/K | $\alpha^*$ | Mole fraction of nitrogen at a partial pressure of 101.325 kPa |
|---|---|---|
| 298.15 | 0.241 | 0.001411 |

* volume of gas (measured at 101.325 kPa and 273.15 K) dissolved by one volume of hexane.

AUXILIARY INFORMATION

METHOD/APPARATUS/PROCEDURE:

Volumetric method. Pressure measured when known amounts of gas were added, in increments, to a known amount of liquid in a vessel of known dimensions. Corrections were made for the partial pressure of solvent. Details in ref. (1).

SOURCE AND PURITY OF MATERIALS:

Purity better than 99 mole per cent as determined by gas chromatography.

ESTIMATED ERROR:

$\delta T/K = \pm 0.1$; $\delta\alpha = \pm 4\%$ or less.

REFERENCES:

1. Bodor, E.; Bor, G. J.; Mohai, B.; Sipos, G. *Veszpremi. Vegyip. Egy. Kozl.* 1957, *1*, 55.

**COMPONENTS:**

(1) Nitrogen; $N_2$; [7727-37-9]

(2) Alkanes; $C_6H_{14}$ and $C_7H_{16}$

**ORIGINAL MEASUREMENTS:**

Guerry, D. Jr.

Ph.D. thesis, 1944
Vanderbilt University
Nashville, TN

**VARIABLES:**

T/K = 293.15, 298.15
P/kPa = 101.325

**PREPARED BY:**

H. L. Clever

**EXPERIMENTAL VALUES:**

| T/K | Mol Fraction $x_1 \times 10^4$ | Bunsen Coefficient $\alpha$ | Ostwald Coefficient L |
|---|---|---|---|
| Hexane; $C_6H_{14}$; [110-54-3] | | | |
| 293.15 | 12.6 | 0.217 | 0.233 |
| 298.15 | 12.5 | 0.213 | 0.232 |
| Heptane; $C_7H_{16}$; [142-82-5] | | | |
| 293.15 | 12.3 | 0.189 | 0.203 |
| 298.15 | 12.4 | 0.189 | 0.206 |

The Ostwald coefficients were calculated by the compiler.

**AUXILIARY INFORMATION**

**METHOD/APPARATUS/PROCEDURE:**

A Van Slyke-Neill Manometric Apparatus manufactured by the Eimer and Amend Co. was used.

The procedure of Van Slyke (1) for pure liquids was modified (2) so that small solvent samples (2 $cm^3$) could be used with almost complete recovery of the sample.

An improved temperature control system was used.

**SOURCE AND PURITY OF MATERIALS:**

Hexane. Eastman Kodak Co.
B.p. (760.3 mmHg) t/°C 68.85 - 68.90.

Heptane. B.p. (753.9 mmHg) t/°C 98.27 - 98.28 (corr.).

**SOURCE AND PURITY OF MATERIALS:**

(1) Nitrogen. Ohio Chemical Co. Oil pumped, 99.8 per cent pure.

(2) Alkanes. Distilled from sodium in air. In addition to the solubility data the thesis contains data of the refractive index, density, vapor pressure, and b.p.

**ESTIMATED ERROR:**

$\delta T/K = 0.05$

**REFERENCES:**


1. Van Slyke, D. D. *J. Biol. Chem.* 1939, *130*, 545.

2. Ijams, C. C. Ph.D. thesis, 1941 Vanderbilt University

**COMPONENTS:**

(1) Nitrogen; $N_2$; [7727-37-9]

(2) Heptane; $C_7H_{16}$; [142-82-5]

**ORIGINAL MEASUREMENTS:**

Thomsen, E. S.; Gjaldbaek, J. C.

*Acta Chem. Scand.* 1963, *17*, 127-33.

**VARIABLES:**

T/K = 298.15 - 308.35
Total P/kPa = 101.325

**PREPARED BY:**

E. S. Thomsen

**EXPERIMENTAL VALUES:**

| T/K | Mol Fraction $X_1 \times 10^3$ | Bunsen Coefficient $\alpha$ | Ostwald Coefficient L |
|---|---|---|---|
| 298.15 | 1.35 | 0.205 | 0.224 |
| 298.15 | 1.36 | 0.207 | 0.226 |
| 298.25 | 1.34 | 0.204 | 0.223 |
| 298.25 | 1.35 | 0.205 | 0.224 |
| 308.15 | 1.38 | 0.207 | 0.223 |
| 308.35 | 1.37 | 0.206 | 0.232 |

The mole fraction and Ostwald solubility values were calculated by the compiler.

Smoothed Data: $\Delta G^o/J\ mol^{-1} = -RT\ ln\ X_1 = 1398.1 + 50.250\ T$

Std. dev. $G^o$ = 13.3, Coef. Corr. = 0.9938

$\Delta H^o/J\ mol^{-1} = 1398.1$, $\Delta S^o/J\ K^{-1}\ mol^{-1} = -50.250$.

| T/K | Mol Fraction $X_1 \times 10^3$ | $\Delta G^o/J\ mol^{-1}$ |
|---|---|---|
| 298.15 | 1.35 | 16,380 |
| 303.15 | 1.36 | 16,631 |
| 308.15 | 1.37 | 16,882 |

**AUXILIARY INFORMATION**

**METHOD/APPARATUS/PROCEDURE:**

A calibrated all-glass combined manometer and bulb was enclosed in an air thermostat and shaken until equilibrium. Mercury was used for calibration and as the confining liquid. The solvents were degassed in the apparatus. Details are in references 1 and 2.

The absorbed volume of gas was calculated from the initial and final amounts, both saturated with solvent vapor. The amount of solvent was determined by the weight of displaced mercury.

The saturation of the liquid with the gas was carried out close to atmospheric pressure. The solubility values were reported for one atmosphere gas pressure assuming Henry's law is obeyed.

**SOURCE AND PURITY OF MATERIALS:**

(1) Nitrogen. Contained 0.2 percent Ar, but no $O_2$.

(2) Heptane. Merck Co. Distillation range 0.09 K.

**ESTIMATED ERROR:**

$\delta T/K = 0.05$
$\delta X_1/X_1 = 0.015$

**REFERENCES:**

1. Lannung, A. *J. Am. Chem. Soc.* 1930, *52*, 68.

2. Gjaldbaek, J. C. *Acta Chem. Scand.* 1952, *6*, 623.

| COMPONENTS: | ORIGINAL MEASUREMENTS: |
|---|---|
| (1) Nitrogen; $N_2$; [7727-37-9]<br>(2) Heptane; $C_7H_{16}$; [142-82-5] | Makranczy, J.; Megyery-Balog, Mrs. K.; Rusz, L.; Patyi, L.<br><br>*Hung. J. Ind. Chem.* 1976, *4*(2), 269-80. |
| **VARIABLES:** | **PREPARED BY:** |
| T/K = 298.15 - 313.15<br>Total P/kPa = 101.325 | S. A. Johnson<br>H. L. Clever |

EXPERIMENTAL VALUES:

| T/K | Mol Fraction $X_1 \times 10^3$ | Bunsen Coefficient $\alpha$ | Ostwald Coefficient $L$ |
|---|---|---|---|
| 298.15 | 1.38 | 0.210 | 0.229 |
| 313.15 | 1.05 | 0.157 | 0.180 |

Smoothed Data: $\Delta G^{\circ}/J\ mol^{-1} = -RT \ln X_1 = -14,143 + 102.19T$

Both the data and the equation are classed as tentative.

Mol Fractions and Bunsen Coefficients were calculated by the compiler at a nitrogen partial pressure of 101.325 kPa (1 atm) with the assumption that the Ostwald coefficient is independent of pressure.

AUXILIARY INFORMATION

METHOD/APPARATUS/PROCEDURE:

Volumetric method, described in detail in reference (1).

SOURCE AND PURITY OF MATERIALS:

Analytical grade reagents of Hungarian and foreign origin were used (both liquids and gases). No further information.

ESTIMATED ERROR:

$\delta X_1/X_1 = 0.03$

REFERENCES:

1. Bodor, E.; Bor, Gy.; Mohai, B.; Sipos, G. *Veszpremi Vegyip. Egy. Kozl.* 1957, *1*, 55. *Chem. Abstr.* 1961, *55*, 3175h.

**COMPONENTS:**

(1) Nitrogen; $N_2$; [7727-37-9]

(2) Alkanes; $C_7H_{16}$, $C_8H_{18}$, and $C_{10}H_{22}$

**ORIGINAL MEASUREMENTS:**

Ijams, C. C.

Ph.D. thesis, 1941
Vanderbilt University
Nashville, TN

**VARIABLES:**

T/K = 298.15
P/kPa = 101.325

**PREPARED BY:**

H. L. Clever

**EXPERIMENTAL VALUES:**

| T/K | Mol Fraction $x_1 \times 10^3$ | Bunsen Coefficient $\alpha$ | Ostwald Coefficient L |
|---|---|---|---|
| Heptane; $C_7H_{16}$; [142-82-5] | | | |
| 298.15 | 1.22 | 0.186 | 0.203 |
| Octane; $C_8H_{18}$; [111-65-9] | | | |
| 298.15 | 1.14 | 0.157 | 0.171 |
| 2,2,4-Trimethylpentane or isooctane; $C_8H_{18}$; [540-84-1] | | | |
| 298.15 | 1.48 | 0.200 | 0.218 |
| Decane; $C_{10}H_{22}$; [124-18-5] | | | |
| 298.15 | 1.23 | 0.141 | 0.154 |

The mole fraction and Ostwald coefficient values were calculated by the compiler.

AUXILIARY INFORMATION

**METHOD/APPARATUS/PROCEDURE:**

A Van Slyke-Neill Manometric Apparatus manufactured by the Eimer and Amend Co. was used.

The procedure of Van Slyke (1) for pure liquids was modified so that small solvent sample (2 $cm^3$) could be used with almost 100 per cent recovery of the solvent.

An improved temperature control system was used.

**SOURCE AND PURITY OF MATERIALS:**

(1) Nitrogen. Air Reduction Co. Purity 99.5 per cent.

(2) Alkanes. Purified. Vapor pressure and boiling point also reported in the thesis.

**ESTIMATED ERROR:**

$\delta T/K = 0.05$

**REFERENCES:**

1. Van Slyke, D. D. *J. Biol. Chem.* 1939, *130*, 545.

COMPONENTS:

(1) Nitrogen; $N_2$; [7727-37-9]

(2) Octane; $C_8H_{18}$; [111-65-9]

ORIGINAL MEASUREMENTS:

Thomsen, E. S.; Gjaldbaek, J. C.

*Acta Chem. Scand.* 1963, *17*, 127-33.

VARIABLES:

T/K = 298.15 - 308.05
Total P/kPa = 101.325

PREPARED BY:

E. S. Thomsen

EXPERIMENTAL VALUES:

| T/K | Mol Fraction $X_1 \times 10^3$ | Bunsen Coefficient $\alpha$ | Ostwald Coefficient L |
|---|---|---|---|
| 298.15 | 1.31 | 0.180 | 0.196 |
| 298.25 | 1.29 | 0.177 | 0.193 |
| 298.25 | 1.32 | 0.181 | 0.198 |
| 298.25 | 1.32 | 0.181 | 0.198 |
| 298.45 | 1.30 | 0.178 | 0.194 |
| 308.05 | 1.34 | 0.182 | 0.205 |

The mole fraction and Ostwald solubility values were calculated by the compiler.

Smoothed Data: $\Delta G^\circ/J\ mol^{-1} = -RT \ln X_1 = 1961.0 + 48.628\ T$

$\Delta H^\circ/J\ mol^{-1} = 1961.0$, $\Delta S^\circ/J\ K^{-1}\ mol^{-1} = -48.628$

| T/K | Mol Fraction $X_1 \times 10^3$ | $\Delta G^\circ/J$ mol |
|---|---|---|
| 298.15 | 1.31 | 16,459 |
| 303.15 | 1.32 | 16,702 |
| 308.15 | 1.34 | 16,946 |

AUXILIARY INFORMATION

METHOD/APPARATUS/PROCEDURE:

A calibrated all-glass combined manometer and bulb was enclosed in an air thermostat and shaken until equilibrium. Mercury was used for calibration and as the confining liquid. The solvents were degassed in the apparatus. Details are in references 1 and 2.

The absorbed volume of gas was calculated from the initial and final amounts, both saturated with solvent vapor. The amount of solvent was determined by the weight of displaced mercury.

The saturation of the liquid with the gas was carried out close to atmospheric pressure. The solubility values were reported for one atmosphere gas pressure assuming Henry's law is obeyed.

SOURCE AND PURITY OF MATERIALS:

(1) Nitrogen. Contained 0.2 percent Ar, but no $O_2$.

(2) Octane. British Drug House. Fractionated, distillation range 0.02 K. About 1 percent impurity.

ESTIMATED ERROR:

$\delta T/K = 0.05$
$\delta X_1/X_1 = 0.015$

REFERENCES:

1. Lannung, A. *J. Am. Chem. Soc.* 1930, *52*, 68.

2. Gjaldbaek, J. C. *Acta Chem. Scand.* 1952, *6*, 623.

| COMPONENTS: | ORIGINAL MEASUREMENTS: |
|---|---|
| (1) Nitrogen; $N_2$; [7727-37-9]<br>(2) Octane; $C_8H_{18}$; [111-65-9] | Makranczy, J.; Megyery-Balog, Mrs.K.; Rusz, L.; Patyi, L.<br>*Hung. J. Ind. Chem.* 1976, *4*(2),269-80. |
| VARIABLES: | PREPARED BY: |
| T/K = 298.15 - 313.15<br>Total P/kPa = 101.325 | S. A. Johnson<br>H. L. Clever |

EXPERIMENTAL VALUES:

| T/K | Mol Fraction $X_1 \times 10^3$ | Bunsen Coefficient $\alpha$ | Ostwald Coefficient $L$ |
|---|---|---|---|
| 298.15 | 1.33 | 0.182 | 0.199 |
| 313.15 | 1.04 | 0.140 | 0.161 |

Smoothed Data: $\Delta G^o/J\ mol^{-1} = -RT\ ln\ X_1 = -12,728 + 97.75T$

Both the data and the equation are classed as tentative.

Mol Fractions and Bunsen Coefficients were calculated by the compiler. at a nitrogen partial pressure of 101.325 kPa (1 atm) with the assumption that the Ostwald coefficient is independent of pressure.

AUXILIARY INFORMATION

METHOD/APPARATUS/PROCEDURE:

Volumetric method, described in detail in reference (1).

SOURCE AND PURITY OF MATERIALS:

Analytical grade reagents of Hungarian and foreign origin were used (both liquids and gases). No further information.

ESTIMATED ERROR:

$\delta X_1/X_1 = 0.03$

REFERENCES:

1. Bodor, E.; Bor, Gy.; Mohai, B.; Sipos, G. *Veszpremi Vegyip. Egy. Kozl.* 1957, *1*, 55. *Chem. Abstr.* 1961, *55*, 3175h.

| COMPONENTS: | ORIGINAL MEASUREMENTS: |
|---|---|
| (1) Nitrogen; $N_2$; [7727-37-9]<br>(2) Octane; $C_8H_{18}$; [111-65-9] | Wilcock, R. J.; Battino, R.; Danforth, W. F.; Wilhelm, E.<br>*J. Chem. Thermodyn.* 1978, *10*, 817-22. |
| VARIABLES:<br>T/K = 298.25<br>P/kPa = 101.325 | PREPARED BY:<br>A. L. Cramer |

EXPERIMENTAL VALUES:

| T/K | Mol Fraction $X_1 \times 10^3$ | Bunsen Coefficient $\alpha$ | Ostwald Coefficient L |
|---|---|---|---|
| 298.25 | 1.311 | 0.1800 | 0.1965 |

AUXILIARY INFORMATION

METHOD/APPARATUS/PROCEDURE:

The solubility apparatus is based on the design of Morrison and Billett (1) and the version used is described by Battino, Evans, and Danforth (2). The degassing apparatus is that described by Battino, Banzhof, Bogan, and Wilhelm (3).

Degassing. Up to 500 $cm^3$ of solvent is placed in a flask of such size that the liquid is about 4 cm deep. The liquid is rapidly stirred, and vacuum is intermittently applied through a liquid $N_2$ trap until the permanent gas residual pressure drops to 5 microns.

Solubility Determination. The degassed solvent is passed in a thin film down a glass helical tube containing solute gas plus the solvent vapor at a total pressure of one atm. The volume of gas absorbed is found by difference between the initial and final volumes in the buret system. The solvent is collected in a tared flask and weighed.

SOURCE AND PURITY OF MATERIALS:

(1) Nitrogen. Matheson Co., Inc. 99.999 per cent.

(2) Octane. Phillips Petroleum Co. 99 mol %, distilled, density $\rho_{298.15} = 0.6988$.

ESTIMATED ERROR:

$\delta T/K = 0.02$
$\delta P/mmHg = 0.5$
$\delta X_1/X_1 = 0.01$

REFERENCES:

1. Morrison, T. J.; Billett, F. *J. Chem. Soc.* 1948, 2033.
2. Battino, R.; Evans, F. D.; Danforth, W. F. *J. Am. Oil Chem. Soc.* 1968, *45*, 830.
3. Battino, R.; Banzhof, M.; Bogan, M.; Wilhelm, E. *Anal. Chem.* 1971, *43*, 806.

| COMPONENTS: | ORIGINAL MEASUREMENTS: |
|---|---|
| (1) Nitrogen; $N_2$; [7727-37-9]<br>(2) 2,2,4-Trimethylpentane or iso-octane; $C_8H_{18}$; [540-84-1] | Kretschmer, C. B.; Nowakowaska, J.; Wiebe, R.<br>*Ind. Eng. Chem.* 1946, *38*, 506-9. |
| VARIABLES:<br>T/K = 248.15 - 323.15<br>P/kPa = 101.325 | PREPARED BY:<br>P. L. Long<br>H. L. Clever |

EXPERIMENTAL VALUES:

| T/K | Mol Fraction $x_1 \times 10^3$ | Bunsen Coefficient $\alpha$ | Ostwald Coefficient $L$ |
|---|---|---|---|
| 248.15 | 1.496 | 0.2139 | 0.1943 |
| 273.15 | 1.503 | 0.2083 | 0.2088 |
| 298.15 | 1.533 | 0.2069 | 0.2258 |
| 323.15 | 1.583 | 0.2073 | 0.2452 |

The Ostwald coefficients were measured at a total pressure of 101.325 kPa (1 atm). The compiler calculated the Bunsen coefficient and mole fraction values for an oxygen partial pressure of 101.325 kPa (1 atm) assuming the Ostwald coefficient was independent of pressure.

Smoothed Data: $\ln x_1 = -6.2753 - 0.5891/(T/100\ \text{K})$

Standard error about the regression line = $1.84 \times 10^{-5}$

| T/K | Mol Fraction $x_1 \times 10^3$ |
|---|---|
| 253.15 | 1.491 |
| 263.15 | 1.505 |
| 273.15 | 1.517 |
| 283.15 | 1.529 |
| 293.15 | 1.540 |
| 298.15 | 1.545 |
| 303.15 | 1.550 |
| 313.15 | 1.559 |
| 323.15 | 1.569 |

AUXILIARY INFORMATION

METHOD/APPARATUS/PROCEDURE:

The apparatus was similar to that of Horiuti (1), it consisted of a gas buret, manometer, and absorption pipet equipped with a magnetic stirrer. All volumes were calibrated by mercury displacement.

The solvent was partly degassed by boiling under reflux. About 40 $cm^3$ of solvent was admitted into the absorption pipet, and pumped on to complete the degassing. Dry gas, measured in the buret, was admitted, final equilibrium was established after 1 to 3 hours stirring. During the solution process the total pressure was kept at one atm by additions of dry gas.

The solubility value was corrected for the increase of solvent volume with gas absorption, and for the nitrogen impurity in the oxygen.

SOURCE AND PURITY OF MATERIALS:

(1) Nitrogen. Commercial, but source not given. Passed through alkaline pyrogallol to remove $O_2$, passed through Drierite to dry.

(2) 2,2,4-Trimethylpentane. "Certified material," source not given. Dried over $CaSO_4$, distilled, b.p. (760 mmHg)/°C 99.24, density, $\rho_4^{25}/\text{g cm}^{-3}$ 0.68774.

ESTIMATED ERROR:

Gas buret $\delta V/\text{cm}^3 = 0.005$
Average deviation from the mean 0.3%, maximum deviation 0.9% (authors).

REFERENCES:

1. Horiuti, J.
*J. Sci Papers Inst. Phys. Chem. Res. (Tokyo)* 1931, *17*, 125.

COMPONENTS:

(1) Nitrogen; $N_2$; [7727-37-9]

(2) 2,2,4-Trimethylpentane (Isooctane); $C_8H_{18}$; [540-84-1]

ORIGINAL MEASUREMENTS:

Baldwin, R. R.; Daniel, S. G.;

*J. Appl. Chem.* 1952, *2* (Apr.), 161-5.

*J. Inst. Petrol.* London 1953, *39*, 105-124.

VARIABLES:

T/K = 293.15

P/kPa = 101.325

PREPARED BY:

P. L. Long
H. L. Clever

EXPERIMENTAL VALUES:

| T/K | Mol Fraction $x_1 \times 10^3$ | Bunsen Coefficient $\alpha$ | Ostwald Coefficient L |
|---|---|---|---|
| 293.15 | 1.55 | 0.211 | 0.226 |

The mole fraction solubility was calculated by the compiler.

AUXILIARY INFORMATION

METHOD/APPARATUS/PROCEDURE:

The method consists of two operations: Saturation of the liquid with the gas under consideration; and determination of the amount of dissolved gas in the liquid saturated with the gas. All gases were passed through a liquid air trap to remove water and $CO_2$ vapors before passage through the liquid. All gases were passed through a long coil, thermostated, before the saturation process.

Oil or similar liquid is deaerated by passing it through a continuously evacuated vessel. Gas is then passed through the liquid to saturation. Liquid is then degassed, and transfered to a storage burette, disconnected from the degassing apparatus, thermostated, and saturated with the desired gas. Burette is now connected to another degassing apparatus, portions of the oil degassed and gas collected.

SOURCE AND PURITY OF MATERIALS:

(1) Nitrogen. From commercial cylinders, and gas analysis on a Bone and Wheeler apparatus shows it to be at least 99.5% pure; no source given.

(2) 2,2,4-Trimethylpentane. No information given.

ESTIMATED ERROR:

$\delta P/P = 0.01$
$\delta \alpha/\alpha = 0.01$

REFERENCES:

| COMPONENTS: | ORIGINAL MEASUREMENTS: |
|---|---|
| (1) Nitrogen; $N_2$; [7727-37-9]<br>(2) Nonane; $C_9H_{20}$; [111-84-2] | Thomsen, E. S.; Gjaldbaek, J. C.<br>*Acta Chem. Scand.* 1963, *17*, 127-33. |

| VARIABLES: | PREPARED BY: |
|---|---|
| T/K = 298.15 - 298.25<br>Total P/kPa = 101.325 | E. S. Thomsen |

EXPERIMENTAL VALUES:

| T/K | Mol Fraction $X_1 \times 10^3$ | Bunsen Coefficient $\alpha$ | Ostwald Coefficient L |
|---|---|---|---|
| 298.15 | 1.28 | 0.160 | 0.175 |
| 298.15 | 1.28 | 0.160 | 0.175 |
| 298.15 | 1.29 | 0.161 | 0.176 |
| 298.25 | 1.28 | 0.160 | 0.175 |

The mole fraction and Ostwald solubility values were calculated by the compiler.

AUXILIARY INFORMATION

METHOD/APPARATUS/PROCEDURE:

A calibrated all-glass combined manometer and bulb was enclosed in an air thermostat and shaken until equilibrium. Mercury was used for calibration and as the confining liquid. The solvents were degassed in the apparatus. Details are in references 1 and 2.

The absorbed volume of gas was calculated from the initial and final amounts, both saturated with solvent vapor. The amount of solvent was determined by the weight of displaced mercury.

The saturation of the liquid with the gas was carried out close to atmospheric pressure. The solubility values were reported for one atmosphere gas pressure assuming Henry's law is obeyed.

SOURCE AND PURITY OF MATERIALS:

(1) Nitrogen. Contained 0.2 percent Ar, but no $O_2$.

(2) Nonane. Fluka "purum". Fractionated, distillation range 0.08 K.

ESTIMATED ERROR:

$\delta T/K = 0.05$

$\delta X_1/X_1 = 0.015$

REFERENCES:


1. Lannung, A. *J. Am. Chem. Soc.* 1930, *52*, 68.
2. Gjaldbaek, J. C. *Acta Chem. Scand.* 1952, *6*, 623.

**COMPONENTS:**

(1) Nitrogen; $N_2$; [7727-37-9]

(2) Nonane; $C_9H_{20}$; [111-84-2]

**ORIGINAL MEASUREMENTS:**

Makranczy, J.; Megyery-Balog, Mrs.K; Rusz, L.; Patyi, L.

*Hung. J. Ind. Chem.* 1976, *4*(2), 269-80.

**VARIABLES:**

T/K = 298.15 - 313.15
Total P/kPa = 101.325

**PREPARED BY:**

S. A. Johnson
H. L. Clever

**EXPERIMENTAL VALUES:**

| T/K | Mol Fraction $X_1 \times 10^3$ | Bunsen Coefficient $\alpha$ | Ostwald Coefficient L |
|---|---|---|---|
| 298.15 | 1.31 | 0.163 | 0.178 |
| 313.15 | 1.03 | 0.126 | 0.145 |

Smoothed Data: $\Delta G^o/J\ mol^{-1} = -RT \ln X_1 = -12,446 + 96.93T$

Both the data and the equation are classed as tentative.

Mol Fractions and Bunsen Coefficients were calculated by the compiler at a nitrogen partial pressure of 101.325 kPa (1 atm) with the assumption that the Ostwald coefficient is independent of pressure.

**AUXILIARY INFORMATION**

**METHOD/APPARATUS/PROCEDURE:**

Volumetric method, described in detail in reference (1).

**SOURCE AND PURITY OF MATERIALS:**

Analytical grade reagents of Hungarian and foreign origin were used (both liquids and gases). No further information.

**ESTIMATED ERROR:**

$\delta X_1/X_1 = 0.03$

**REFERENCES:**

1. Bodor, E.; Bor, Gy.; Mohai, B.; Sipos, G. *Veszpremi Vegyip. Egy. Kozl.* 1957, *1*, 55. *Chem. Abstr.* 1961, *55*, 3175h.

COMPONENTS:

(1) Nitrogen; $N_2$; [7727-37-9]

(2) Decane; $C_{10}H_{22}$; [124-18-5]

ORIGINAL MEASUREMENTS:

Makranczy, J.; Megyery-Balog, Mrs.K; Rusz, L.; Patyi, L.

*Hung. J. Ind. Chem.* 1976, *4*(2), 269-80.

VARIABLES:

T/K = 298.15 - 313.15
Total P/kPa = 101.325

PREPARED BY:

S. A. Johnson
H. L. Clever

EXPERIMENTAL VALUES:

| T/K | Mol Fraction $X_1 \times 10^3$ | Bunsen Coefficient $\alpha$ | Ostwald Coefficient L |
|---|---|---|---|
| 298.15 | 1.25 | 0.143 | 0.156 |
| 313.15 | 1.02 | 0.114 | 0.131 |

Smoothed Data: $\Delta G^o/J\ mol^{-1} = -RT \ln X_1 = -10{,}525 + 90.88T$

Both the data and the equation are classed as tentative.

Mol Fractions and Bunsen Coefficients were calculated by the compiler at a nitrogen partial pressure of 101.325 kPa (1 atm) with the assumption that the Ostwald coefficient is independent of pressure.

AUXILIARY INFORMATION

METHOD/APPARATUS/PROCEDURE:

Volumetric method, described in detail in reference (1).

SOURCE AND PURITY OF MATERIALS:

Analytical grade reagents of Hungarian and foreign origin were used (both liquids and gases). No further information.

ESTIMATED ERROR:

$\delta X_1/X_1 = 0.03$

REFERENCES:

1. Bodor, E.; Bor, Gy.; Mohai, B.; Sipos, G.
*Veszpremi Vegyip. Egy. Kozl.* 1957, *1*, 55.
*Chem. Abstr.* 1961, *55*, 3175h.

**COMPONENTS:**

(1) Nitrogen; $N_2$; [7727-37-9]

(2) Decane; $C_{10}H_{22}$; [124-18-5]

**ORIGINAL MEASUREMENTS:**

Wilcock, R. J.; Battino, R.; Danforth, W. F.; Wilhelm, E.

*J. Chem. Thermodyn.* 1978, *10*, 817-22.

**VARIABLES:**

T/K = 283.18 - 313.19
P/kPa = 101.325

**PREPARED BY:**

H. L. Clever
A. L. Cramer

**EXPERIMENTAL VALUES:**

| T/K | Mol Fraction $X_1 \times 10^3$ | Bunsen Coefficient $\alpha$ | Ostwald Coefficient L |
|---|---|---|---|
| 283.18 | 1.220 | 0.1419 | 0.1471 |
| 298.11 | 1.214 | 0.1391 | 0.1518 |
| 313.19 | 1.206 | 0.1360 | 0.1559 |

The Bunsen coefficients were calculated by the compiler.

The solubility values were adjusted to 101.325 kPa (1 atm) assuming Henry's law.

Smoothed Data: $\Delta G^o/J\ mol^{-1} = -RT\ ln\ X_1 = -284.6 + 56.782\ T$

Std. Dev. $\Delta G^o = 1.6$, Coef. Corr. = 0.9999

$\Delta H^o/J\ mol^{-1} = -284.6$, $\Delta S^o/J\ K^{-1}\ mol^{-1} = -56.782$

| T/K | Mol Fraction $X_1 \times 10^3$ | $\Delta G^o/J\ mol^{-1}$ |
|---|---|---|
| 278.15 | 1.22 | 15,509 |
| 288.15 | 1.22 | 16,077 |
| 298.15 | 1.21 | 16,665 |
| 308.15 | 1.21 | 17,213 |
| 318.15 | 1.20 | 17,780 |

AUXILIARY INFORMATION

**METHOD/APPARATUS/PROCEDURE:**

The solubility apparatus is based on the design of Morrison and Billett (1) and the version used is described by Battino, Evans, and Danforth (2). The degassing apparatus is that described by Battino, Banzhof, Bogan, and Wilhelm (3).

Degassing. Up to 500 $cm^3$ of solvent is placed in a flask of such size that the liquid is about 4 cm deep. The liquid is rapidly stirred, and vacuum is intermittently applied through a liquid $N_2$ trap until the permanent gas residual pressure drops to 5 microns.

Solubility Determination. The degassed solvent is passed in a thin film down a glass helical tube containing solute gas plus the solvent vapor at a total pressure of one atm. The volume of gas absorbed is found by difference between the initial and final volumes in the buret system. The solvent is collected in a tared flask and weighed.

**SOURCE AND PURITY OF MATERIALS:**

(1) Nitrogen. Matheson Co., Inc. 99.999 per cent.

(2) Decane. Phillips Petroleum Co. 99 mol % distilled, density, $\rho_{298.15} = 0.7264$.

**ESTIMATED ERROR:**

$\delta T/K = 0.02$
$\delta P/mmHg = 0.5$
$\delta X_1/X_1 = 0.01$

**REFERENCES:**

1. Morrison, T. J.; Billett, F. *J. Chem. Soc.* 1948, 2033.
2. Battino, R.; Evans, F. D.; Danforth, W. F. *J. Am. Oil Chem. Soc.* 1968, *45*, 830.
3. Battino, R.; Banzhof, M.; Bogan, M.; Wilhelm, E. *Anal. Chem.* 1971, *43*, 806.

COMPONENTS:

(1) Nitrogen; $N_2$; [7727-37-9]

(2) Undecane; $C_{11}H_{24}$; [1120-21-4]

ORIGINAL MEASUREMENTS:

Makranczy, J.; Megyery-Balog, Mrs. K; Rusz, L.; Patyi, L.

*Hung. J. Ind. Chem.* 1976, *4*, 269-80.

VARIABLES:

T/K = 298.15 - 313.15
Total P/kPa = 101.325

PREPARED BY:

S. A. Johnson
H. L. Clever

EXPERIMENTAL VALUES:

| T/K | Mol Fraction $X_1 \times 10^3$ | Bunsen Coefficient $\alpha$ | Ostwald Coefficient L |
|---|---|---|---|
| 298.15 | 1.27 | 0.134 | 0.146 |
| 313.15 | 1.02 | 0.106 | 0.122 |

Smoothed Data: $\Delta G^{\circ}/J\ mol^{-1} = -RT \ln X_1 = -11{,}346 + 93.50T$

Both the data and the equation are classed as tentative.

Mol Fractions and Bunsen Coefficients were calculated by the compiler at a nitrogen partial pressure of 101.325 kPa (1 atm) with the assumption that the Oswald coefficient is independent of pressure.

AUXILIARY INFORMATION

METHOD/APPARATUS/PROCEDURE:

Volumetric method, described in detail in reference (1).

SOURCE AND PURITY OF MATERIALS:

Analytical grade reagents of Hungarian and foreign origin were used (both liquids and gases). No further information.

ESTIMATED ERROR:

$\delta X_1/X_1 = 0.03$

REFERENCES:

1. Bodor, E.; Bor, Gy.; Mohai, B.; Sipos, G. *Veszpremi Vegyip. Egy. Kozl.* 1957, *1*, 55. *Chem. Abstr.* 1961, *55*, 3175h.

| COMPONENTS: | ORIGINAL MEASUREMENTS: |
|---|---|
| (1) Nitrogen, $N_2$; [7727-37-9]<br>(2) Dodecane; $C_{12}H_{26}$; [112-40-3] | Makranczy, J.; Megyery-Balog, Mrs. K; Rusz, L.; Patyi, L.<br>*Hung. J. Ind. Chem.* 1976, *4*(2), 269-80. |
| VARIABLES: | PREPARED BY: |
| T/K = 298.15 - 313.15<br>Total P/kPa = 101.325 | S. A. Johnson<br>H. L. Clever |

EXPERIMENTAL VALUES:

| T/K | Mol Fraction $X_1 \times 10^3$ | Bunsen Coefficient $\alpha$ | Ostwald Coefficient L |
|---|---|---|---|
| 298.15 | 1.23 | 0.121 | 0.132 |
| 313.15 | 0.95 | 0.092 | 0.105 |

Smoothed Data: $\Delta G^\circ/J\ mol^{-1} = -RT \ln X_1 = -13{,}367 + 100.55T$

Both the data and the equation are classed as tentative.

Mol Fractions and Bunsen Coefficients were calculated by the compiler at a nitrogen partial pressure of 101.325 kPa (1 atm) with the assumption that the Ostwald coefficient is independent of pressure.

AUXILIARY INFORMATION

METHOD/APPARATUS/PROCEDURE:

Volumetric method, described in detail in reference (1).

SOURCE AND PURITY OF MATERIALS:

Analytical grade reagents of Hungarian and foreign origin were used (both liquids and gases). No further information.

ESTIMATED ERROR:

$\delta X_1/X_1 = 0.03$

REFERENCES:

1. Bodor, E.; Bor, Gy.; Mohai, B.; Sipos, G. *Veszpremi Vegyip. Egy. Kozl.* 1957, *1*, 55. *Chem. Abstr.* 1961, *55*, 3175h.

**COMPONENTS:**

(1) Nitrogen; $N_2$; [7727-37-9]

(2) Tridecane; $C_{13}H_{28}$; [629-50-5]

**ORIGINAL MEASUREMENTS:**

Makranczy, J.; Megyery-Balog, Mrs. K; Rusz, L.; Patyi, L.

*Hung. J. Ind. Chem.* 1976, *4*(2), 269-80.

**VARIABLES:**

T/K = 298.15 - 313.15
Total P/kPa = 101.325

**PREPARED BY:**

S. A. Johnson
H. L. Clever

**EXPERIMENTAL VALUES:**

| T/K | Mol Fraction $X_1 \times 10^3$ | Bunsen Coefficient $\alpha$ | Ostwald Coefficient $L$ |
|---|---|---|---|
| 298.15 | 1.24 | 0.114 | 0.124 |
| 313.15 | 0.97 | 0.087 | 0.100 |

Smoothed Data: $\Delta G^{\circ}/J\ mol^{-1} = -RT \ln X_1 = -12,708 + 98.27T$

Both the data and the equation are classed as tentative.

Mol Fractions and Bunsen Coefficients were calculated by the compiler at a nitrogen partial pressure of 101.325 kPa (1 atm) with the assumption that the Ostwald coefficient is independent of pressure.

**AUXILIARY INFORMATION**

**METHOD/APPARATUS/PROCEDURE:**

Volumetric method, described in detail in reference (1).

**SOURCE AND PURITY OF MATERIALS:**

Analytical grade reagents of Hungarian and foreign origin were used (both liquids and gases). No further information.

**ESTIMATED ERROR:**

$\delta X_1/X_1 = 0.03$

**REFERENCES:**

1. Bodor, E.; Bor, Gy.; Mohai, B.; Sipos, G. *Veszpremi Vegyip. Egy. Kozl.* 1957, *1*, 55. *Chem. Abstr.* 1961, *55*, 3175h.

COMPONENTS:

(1) Nitrogen; $N_2$; [7727-37-9]

(2) Tetradecane; $C_{14}H_{30}$; [629-59-4]

ORIGINAL MEASUREMENTS:

Makranczy, J.; Megyery-Balog, Mrs. K; Rusz, L.; Patyi, L.

*Hung. J. Ind. Chem.* 1976, *4* (2), 269-80.

VARIABLES:

T/K = 298.15 - 313.15
Total P/kPa = 101.325

PREPARED BY:

S. A. Johnson
H. L. Clever

EXPERIMENTAL VALUES:

| T/K | Mol Fraction $X_1 \times 10^3$ | Bunsen Coefficient $\alpha$ | Ostwald Coefficient $L$ |
|---|---|---|---|
| 298.15 | 1.24 | 0.106 | 0.116 |
| 313.15 | 0.95 | 0.080 | 0.092 |

Smoothed Data: $\Delta G^o/J\ mol^{-1} = -RT \ln X_1 = -13,785 + 101.88T$

Both the data and the equation are classed as tentative.

Mol Fractions and Bunsen Coefficients were calculated by the compiler for a nitrogen partial pressure of 101.325 kPa (1 atm) with the assumption that the Ostwald coefficient was independent of pressure.

AUXILIARY INFORMATION

METHOD/APPARATUS/PROCEDURE:

Volumetric method, described in detail in reference (1).

SOURCE AND PURITY OF MATERIALS:

Analytical grade reagents of Hungarian and foreign origin were used (both liquids and gases). No further information.

ESTIMATED ERROR:

$\delta X_1/X_1 = 0.03$

REFERENCES:

1. Bodor, E.; Bor, Gy.; Mohai, B.; Sipos, G. *Veszpremi Vegyip. Egy. Kozl.* 1957, *1*, 55. *Chem. Abstr.* 1961, *55*, 3175h.

| COMPONENTS: | ORIGINAL MEASUREMENTS: |
|---|---|
| (1) Nitrogen; $N_2$; [7727-37-9]<br>(2) Pentadecane; $C_{15}H_{32}$; [629-62-9] | Makranczy, J.; Megyery-Balog, Mrs. K; Rusz, L.; Patyi, L.<br>*Hung. J. Ind. Chem.* 1976, *4*(2), 269-80. |
| **VARIABLES:** | **PREPARED BY:** |
| T/K = 298.15 - 313.15<br>Total P/kPa = 101.325 | S. A. Johnson<br>H. L. Clever |

EXPERIMENTAL VALUES:

| T/K | Mol Fraction $X_1 \times 10^3$ | Bunsen Coefficient $\alpha$ | Ostwald Coefficient L |
|---|---|---|---|
| 298.15 | 1.26 | 0.102 | 0.111 |
| 313.15 | 0.93 | 0.074 | 0.085 |

Smoothed Data: $\Delta G^o/J\ mol^{-1} = -RT \ln X_1 = -15{,}717 + 108.23T$

Both the data and the equation are classed as tentative.

Mol Fractions and Bunsen Coefficients were calculated by the compiler at a nitrogen partial pressure of 101.325 kPa (1 atm) with the assumption that the Ostwald coefficient was independent of pressure.

AUXILIARY INFORMATION

METHOD/APPARATUS/PROCEDURE:

Volumetric method, described in detail in reference (1).

SOURCE AND PURITY OF MATERIALS:

Analytical grade reagents of Hungarian and foreign origin were used (both liquids and gases). No further information.

ESTIMATED ERROR:

$\delta X_1/X_1 = 0.03$

REFERENCES:

1. Bodor, E.; Bor, Gy.; Mohai, B.; Sipos, G. *Veszpremi Vegyip. Egy. Kozl.* 1957, *1*, 55. *Chem. Abstr.* 1961, *55*, 3175h.

| COMPONENTS: | ORIGINAL MEASUREMENTS: |
|---|---|
| (1) Nitrogen; $N_2$; [7727-37-9]<br>(2) Hexadecane; $C_{16}H_{34}$; [544-76-3] | Makranczy, J.; Megyery-Balog, Mrs. K; Rusz, L.; Patyi, L.<br>*Hung. J. Ind. Chem.* 1976, *4*(2), 269-80. |
| VARIABLES:<br>T/K = 298.15 - 313.15<br>Total P/kPa = 101.325 | PREPARED BY:<br>S. A. Johnson<br>H. L. Clever |

EXPERIMENTAL VALUES:

| T/K | Mol Fraction $X_1 \times 10^3$ | Bunsen Coefficient $\alpha$ | Ostwald Coefficient L |
|---|---|---|---|
| 298.15 | 1.22 | 0.093 | 0.102 |
| 313.15 | 0.94 | 0.071 | 0.081 |

Smoothed Data: $\Delta G^o/J\ mol^{-1} = -RT \ln X_1 = -13.492 + 101.03T$

The data and the equation are classed as tentative.

Mol Fractions and Bunsen Coefficients were calculated by the compiler at a nitrogen partial pressure of 101.325 kPa (1 atm) with the assumption that the Ostwald coefficient was independent of pressure.

AUXILIARY INFORMATION

METHOD/APPARATUS/PROCEDURE:

Volumetric method, described in detail in reference (1).

SOURCE AND PURITY OF MATERIALS:

Analytical grade reagents of Hungarian and foreign origin were used (both liquids and gases). No further information.

ESTIMATED ERROR:

$\delta X_1/X_1 = 0.03$

REFERENCES:

1. Bodor, E.; Bor, Gy.; Mohai, B.; Sipos, G.
*Veszpremi Vegyip. Egy. Kozl.* 1957, *1*, 55.
*Chem. Abstr.* 1961, *55*, 3175h.

| COMPONENTS: | ORIGINAL MEASUREMENTS: |
|---|---|
| (1) Nitrogen; $N_2$; [7727-37-9]<br>(2) Hexadecane; $C_{16}H_{34}$; [544-76-3] | Tremper, K.K.; Prausnitz, J.M.<br>*J. Chem.Eng. Data* 1976, *21*, 295-9. |
| VARIABLES: | PREPARED BY: |
| T/K = 300-475 | C.L. Young |

EXPERIMENTAL VALUES:

| T/K | Henry's Constant[a] /atm | Mole fraction[b] of nitrogen at 1 atm partial pressure, $x_{N_2}$ |
|---|---|---|
| 300 | 790.0 | 0.00127 |
| 325 | 743.0 | 0.00135 |
| 350 | 701.0 | 0.00143 |
| 375 | 661.0 | 0.00151 |
| 400 | 618.0 | 0.00162 |
| 425 | 576.0 | 0.00174 |
| 450 | 546.0 | 0.00183 |
| 475 | 517.0 | 0.00193 |

a. Authors stated measurements were made at several pressures and values of solubility used were all within the Henry's-Law region.

b. Calculated by compiler assuming linear relationship between mole fraction and pressure.

AUXILIARY INFORMATION

METHOD/APPARATUS/PROCEDURE

Volumetric apparatus similar to that described by Dymond and Hildebrand (1). Pressure measured with a null detector and precision gauge. Details in ref. (2).

SOURCE AND PURITY OF MATERIALS:

Solvent degassed, no other details given.

ESTIMATED ERROR:

$\delta T/K = \pm 0.1$; $\delta x_{N_2} = \pm 1\%$.

REFERENCES:

1. Dymond, J.; Hildebrand, J.H. *Ind. Eng. Chem. Fundam.* 1967, *6*, 130.
2. Cukor, P.M.; Prausnitz, J.M. *Ind. Eng. Chem. Fundam.* 1971, *10*, 638.

| COMPONENTS: | EVALUATOR: |
|---|---|
| (1) Nitrogen, $N_2$; [7727-37-9]<br>(2) Cyclohexane; $C_6H_{12}$; [110-82-7] | Rubin Battino<br>Department of Chemistry<br>Wright State University<br>Dayton, OH 45435<br>U.S.A.<br>December 1981 |

CRITICAL EVALUATION:

There are seven groups that determined the solubility of nitrogen in cyclohexane (1-7). Guerry's two points (7) are much too low when compared with more reliable measurements. The single points obtained by Gjaldbaek and Hildebrand (1) and Patyi, et al. (5), are in good agreement but again are lower than the recommended values. The results obtained by Wild, et al. (6), are a special case inasmuch as their results were obtained via two methods (which show good internal agreement), at partial pressures of nitrogen up to 1.9 MPa, and over a wide temperature range - 300 to 443 K. However, the mole fraction solubilities calculated at 101 kPa partial pressure of nitrogen are almost double the results obtained by other workers and also show an opposite slope with respect to temperature dependence. We classify the Wild, et al. (6), values as tentative over the temperature range they covered.

For smoothing we used the data from three papers (2,3,4) covering a temperature range of 283 to 307 K at a nitrogen partial pressure of 101 kPa. The smoothing equation is:

$$\ln x_1 = -6.14235 - 3.06839/(T/100K) \qquad (1)$$

with a standard deviation of 0.0041 in $\ln x_1$. $\Delta\bar{H}° = 2552$ J mol$^{-1}$ and $\Delta\bar{S}° = -51$ J K$^{-1}$ mol$^{-1}$. The smoothed mole fractions which are recommended values at 5K intervals follow.

| T/K | $10^4\ x_1$ | T/K | $10^4\ x_1$ |
|---|---|---|---|
| 283.15 | 7.27 | 298.15 | 7.68 |
| 288.15 | 7.41 | 303.15 | 7.81 |
| 293.15 | 7.55 | 308.15 | 7.94 |

There is not sufficient comparative work reported on other cyclic hydrocarbons to warrant further comment.


References

1. Gjaldbaek, J. C.; Hildebrand, J. H. *J. Am. Chem. Soc.* 1949, *71*, 3147-50.
2. Dymond, J. H. *J. Phys. Chem.* 1967, *71*, 1829-31.
3. Dymond, J. H.; Hildebrand, J. H. *Ind. Eng. Chem., Fundam.* 1967, *6*, 130-1.
4. Wilhelm, E.; Battino, R. *J. Chem. Thermodyn.* 1973, *5*, 117-20.
5. Patyi, L.; Furmer, I. E.; Makranczy, J.; Sadilenko, A. S.; Stepanova, Z. G.; Berengarten, M. G. *Zh. Prikl. Khim.* 1978, *51*, 1296-1300.
6. Wild, J. D.; Sridhar, J.; Potter, O. E. *Chem. Eng. J.* 1978, *15*, 209-14.
7. Guerry, D. Jr. Ph.D. Thesis 1944, Vanderbilt University, Nashville, TN.

COMPONENTS:

(1) Nitrogen; $N_2$; [7727-37-9]

(2) Cyclohexane; $C_6H_{12}$; [110-82-7]

ORIGINAL MEASUREMENTS:

Gjaldbaek, J. C.; Hildebrand, J. H.

*J. Am. Chem. Soc.* 1949, *71*, 3147-50.

VARIABLES:

T/K = 298.13
Total P/kPa = 101.325

PREPARED BY:

J. Chr. Gjaldbaek

EXPERIMENTAL VALUES:

| T/K | Mol Fraction $x_1 \times 10^3$ | Bunsen Coefficient $\alpha$ | Ostwald Coefficient L |
|---|---|---|---|
| 298.13 | 0.755 | 0.156 | 0.170 |
| 298.13 | 0.752 | 0.155 | 0.169 |

The mole fraction and Ostwald solubility values were calculated by the compiler.

AUXILIARY INFORMATION

METHOD/APPARATUS/PROCEDURE:

A calibrated all-glass combined manometer and bulb was enclosed in an air thermostat and shaken until equilibrium. Mercury was used for calibration and as the confining liquid. The solvents were degassed in the apparatus. Details are in references 1 and 2.

The absorped volume of gas was calculated from the initial and final amounts, both saturated with solvent vapor. The amount of solvent was determined by the weight of displaced mercury.

The saturation of the liquid with the gas was carried out close to atmospheric pressure. The solubility values were reported for one atmosphere gas pressure assuming Henry's law is obeyed.

SOURCE AND PURITY OF MATERIALS:

(1) Nitrogen. Stuart Oxygen Co. 99.996 percent $N_2$.

(2) Cyclohexane. Paragon Testing Laboratories. Boiling point 80.63 - 80.66°C at 755.7 mmHg.

ESTIMATED ERROR:

$\delta T/K = 0.05$
$\delta x_1/x_1 = 0.015$

REFERENCES:

1. Lannung, A. *J. Am. Chem. Soc.* 1930, *52*, 68.

2. Gjaldbaek, J. C. *Acta Chem. Scand.* 1952, *6*, 623.

**COMPONENTS:**
(1) Nitrogen; $N_2$; [7727-37-9]
(2) Cyclohexane; $C_6H_{12}$; [110-82-7]

**ORIGINAL MEASUREMENTS:**
Dymond, J. H.
*J. Phys. Chem.* 1967, *71*, 1829-31.

**VARIABLES:**
T/K = 291.00 - 306.52
P/kPa = 101.325

**PREPARED BY:**
M. E. Derrick
H. L. Clever

**EXPERIMENTAL VALUES:**

| Temperature t/°C | T/K | Mol Fraction $x_1 \times 10^4$ | Bunsen Coefficient $\alpha$ | Ostwald Coefficient L |
|---|---|---|---|---|
| 17.85 | 291.00 | 7.52 | 0.156 | 0.166 |
| 25.45 | 298.60 | 7.70 | 0.159 | 0.174 |
| 30.70 | 303.85 | 7.87 | 0.161 | 0.179 |
| 33.37 | 306.52 | 7.89 | 0.161 | 0.181 |

The Bunsen and Ostwald coefficients were calculated by the compiler.

Smoothed Data: $\ln x_1 = -6.196 - 2.8998/(T/100)$

| T/K | Mol Fraction $x_1 \times 10^4$ |
|---|---|
| 293.15 | 7.57 |
| 298.15 | 7.70 |
| 303.15 | 7.82 |
| 308.15 | 7.95 |

AUXILIARY INFORMATION

**METHOD/APPARATUS/PROCEDURE:**
The liquid is saturated with the gas at a gas partial pressure of 1 atm.

The apparatus is that described by Dymond and Hildebrand (1). The apparatus uses an all-glass pumping system to spray slugs of degassed solvent into the gas. The amount of gas dissolved is calculated from the initial and final gas pressure.

**SOURCE AND PURITY OF MATERIALS:**
(1) Nitrogen. General Dynamics Corp. Dried.

(2) Cyclohexane. Matheson, Coleman and Bell chromatoquality reagent. Dried and fractionally frozen. m.p. 6.45°C.

**ESTIMATED ERROR:**

$\delta x_1/x_1 = 0.01$

**REFERENCES:**
1. Dymond, J.; Hildebrand, J. H. *Ind. Eng. Chem. Fundam.* 1967, *6*, 130.

**COMPONENTS:**

(1) Nitrogen; $N_2$; [7727-37-9]

(2) Cyclohexane; $C_6H_{12}$; [110-82-7]

**ORIGINAL MEASUREMENTS:**

Dymond, J.; Hildebrand, J. H.

*Ind. Eng. Chem., Fundam.* 1967, *6*, 130-1.

**VARIABLES:**

T/K = 298.15
Total P/kPa = 101.325

**PREPARED BY:**

A. L. Cramer
H. L. Clever

**EXPERIMENTAL VALUES:**

| T/K | Mol Fraction $x_1 \times 10^4$ | Bunsen Coefficient $\alpha$ | Ostwald Coefficient L |
|---|---|---|---|
| 298.15 | 7.68 | 0.158 | 0.173 |

The Bunsen and Ostwald coefficients were calculated by the compiler.

AUXILIARY INFORMATION

**METHOD/APPARATUS/PROCEDURE:**

The apparatus consists of an all-glass buret system and a two bulb absorption vessel.

A magnetic pump forces solvent from the lower bulb over the wall of the upper bulb which contains the gas. The solvent flow back to the lower bulb. Pumping is continued until there is no further pressure change.

The amount of gas absorbed is calculated from the initial and final gas pressure.

**SOURCE AND PURITY OF MATERIALS:**

No details given.

**ESTIMATED ERROR:**

$\delta T/K = \pm 0.05$
$\delta x_1/x_1$ ±0.01 or less.

**REFERENCES:**

COMPONENTS:

(1) Nitrogen; $N_2$; [7727-37-9]

(2) Cyclohexane; $C_6H_{12}$; [110-82-7]

ORIGINAL MEASUREMENTS:

Wilhelm, E.; Battino, R.

*J. Chem. Thermodyn.* 1973, *5*, 117-20.

VARIABLES:

T/K = 283.39 - 297.59
P/kPa = 101.325

PREPARED BY:

H. L. Clever

EXPERIMENTAL VALUES:

| T/K | Mol Fraction $X_1 \times 10^3$ | Bunsen Coefficient $\alpha$ | Ostwald Coefficient L |
|---|---|---|---|
| 283.39 | 0.726 | 0.152 | 0.158 |
| 283.53 | 0.730 | 0.153 | 0.159 |
| 297.59 | 0.760 | 0.157 | 0.171 |

The solubility values were adjusted to a nitrogen partial pressure of 101.325 kPa (1 atm) by Henry's law.

Smoothed Data: $\Delta G^o/J\ mol^{-1} = -RT \ln X_1 = 2135.2 + 52.541\ T$

Std. Dev. $\Delta G^o$ = 3.5, Coef. Corr. = 0.9999

$\Delta H^o/J\ mol^{-1} = 2135.2$, $\Delta S^o/J\ K^{-1}\ mol^{-1} = -52.541$

| T/K | Mol Fraction $X_1 \times 10^3$ | $\Delta G^o/J\ mol^{-1}$ |
|---|---|---|
| 283.15 | 0.727 | 17,012 |
| 288.15 | 0.739 | 17,275 |
| 293.15 | 0.750 | 17,538 |
| 298.15 | 0.761 | 17,880 |

AUXILIARY INFORMATION

METHOD/APPARATUS/PROCEDURE:

The solubility apparatus is based on the design of Morrison and Billett (1) and the version used is described by Battino, Evans, and Danforth (2). The degassing apparatus is that described by Battino, Banzhof, Bogan, and Wilhelm (3).

Degassing. Up to 500 $cm^3$ of solvent is placed in a flask of such size that the liquid is about 4 cm deep. The liquid is rapidly stirred, and vacuum is intermittently applied through a liquid $N_2$ trap until the permanent gas residual pressure drops to 5 microns.

Solubility Determination. The degassed solvent is passed in a thin film down a glass spiral tube containing solute gas plus the solvent vapor at a total pressure of one atm. The volume of gas absorbed is found by difference between the initial and final volumes in the buret system. The solvent is collected in a tared flask and weighed.

SOURCE AND PURITY OF MATERIALS:

(1) Nitrogen. Matheson Co., Inc. Minimum purity 99.99 mole per cent.

(2) Cyclohexane. Phillips Petroleum Co. Pure Grade. Distilled.

ESTIMATED ERROR:

$\delta T/K = 0.03$
$\delta P/mmHg = 0.5$
$\delta X_1/X_1 = 0.005$

REFERENCES:

1. Morrison, T. J.; Billett, F. *J. Chem. Soc.* 1948, 2033.
2. Battino, R.; Evans, F. D.; Danforth, W. F. *J. Am. Oil Chem. Soc.* 1968, *45*, 830.
3. Battino, R.; Banzhof, M.; Bogan, M.; Wilhelm, E. *Anal. Chem.* 1971 *43*, 806.

**COMPONENTS:**

(1) Nitrogen; $N_2$; [7727-37-9]

(2) Cyclohexane; $C_6H_{12}$; [110-82-7]

**ORIGINAL MEASUREMENTS:**

Patyi, L.; Furmer, I. E.; Makranczy, J.; Sadilenko, A. S.; Stepanova, Z. G.; Berengarten, M. G.

*Zh. Prikl. Khim.* 1978, *51*, 1296-1300.

**VARIABLES:**

T/K = 298

**PREPARED BY:**

C. L. Young

**EXPERIMENTAL VALUES:**

| T/K | $\alpha^*$ | Mole fraction of nitrogen at a partial pressure of 101.325 kPa |
|---|---|---|
| 298.15 | 0.155 | 0.000755 |

* volume of gas (measured at 101.325 kPa and 273.15 K) dissolved by one volume of cyclohexane.

**AUXILIARY INFORMATION**

**METHOD/APPARATUS/PROCEDURE:**

Volumetric method. Pressure measured when known amounts of gas were added, in increments, to a known amount of liquid in a vessel of known dimensions. Corrections were made for the partial pressure of solvent. Details in ref. (1).

**SOURCE AND PURITY OF MATERIALS:**

Purity better than 99 mole per cent as determined by gas chromatography.

**ESTIMATED ERROR:**

$\delta T/K = \pm 0.1$; $\delta\alpha = \pm 4\%$ or less.

**REFERENCES:**

1. Bodor, E.; Bor, G. J.; Mohai, B.; Sipos, G. *Veszpremi. Vegyip. Egy. Kozl.* 1957, *1*, 55.

**COMPONENTS:**

(1) Nitrogen; $N_2$; [7727-37-9]

(2) Cyclic hydrocarbons; $C_6H_{10}$ and $C_6H_{12}$

**ORIGINAL MEASUREMENTS:**

Guerry, D. Jr.

Ph.D. thesis, 1944
Vanderbilt University
Nashville, TN

**VARIABLES:**

T/K = 293.15, 298.15
P/kPa = 101.325

**PREPARED BY:**

H. L. Clever

**EXPERIMENTAL VALUES:**

| T/K | Mol Fraction $x_1 \times 10^4$ | Bunsen Coefficient $\alpha$ | Ostwald Coefficient $L$ |
|---|---|---|---|
| Cyclohexene; $C_6H_{10}$; [110-83-8] | | | |
| 293.15 | 6.10 | 0.135 | 0.145 |
| 298.15 | 6.22 | 0.137 | 0.150 |
| Cyclohexane; $C_6H_{12}$; [110-82-7] | | | |
| 293.15 | 7.22 | 0.150 | 0.161 |
| 298.15 | 7.22 | 0.149 | 0.163 |

The Ostwald coefficients were calculated by the compiler.

AUXILIARY INFORMATION

**METHOD/APPARATUS/PROCEDURE:**

A Van Slyke-Neill Manometric Apparatus manufactured by the Eimer and Amend Co. was used.

The procedure of Van Slyke (1) for pure liquids was modified (2) so that small solvent samples (2 $cm^3$) could be used with almost complete recovery of the sample.

An improved temperature control system was used.

**SOURCE AND PURITY OF MATERIALS:**

(1) Nitrogen. Ohio Chemical Co. Oil pumped, 99.8 per cent pure.

(2) Hydrocarbons. Both were Eastman Kodak Co. products. They were purified by standard methods, and distilled from Na in a nitrogen atm.

**SOURCE AND PURITY OF MATERIALS:**

Cyclohexene. B.p. (756.6 mmHg) t/°C 82.35 - 82.50 (corr.).

Cyclohexane. B.p. (760.7 mmHg) t/°C 80.90 (corr.).

Data on density, refractive index and vapor pressure are in the thesis.

**ESTIMATED ERROR:**

$\delta T/K = 0.05$

**REFERENCES:**


1. Van Slyke, D. D. *J. Biol. Chem.* 1939, *130*, 545.
2. Ijams, C. C. Ph.D. thesis, 1941 Vanderbilt University

**COMPONENTS:**

(1) Nitrogen; $N_2$; [7727-37-9]

(2) Methylcyclohexane; $C_7H_{14}$; [108-87-2]

**ORIGINAL MEASUREMENTS:**

Field, L. R.; Wilhelm, E.; Battino, R.

*J. Chem. Thermodyn.* 1974, *6*, 237-43.

**VARIABLES:**

T/K = 284.29 - 313.27
P/kPa = 101.325

**PREPARED BY:**

H. L. Clever

**EXPERIMENTAL VALUES:**

| T/K | Mol Fraction $X_1 \times 10^3$ | Bunsen Coefficient $\alpha$ | Ostwald Coefficient L |
|---|---|---|---|
| 284.29 | 0.849 | 0.1508 | 0.1569 |
| 298.18 | 0.946 | 0.1653 | 0.1804 |
| 313.27 | 0.988 | 0.1697 | 0.1946 |

The gas solubility values were adjusted to a nitrogen partial pressure of 101.325 kPa (1 atm) by Henry's law.

The Bunsen coefficients were calculated by the compiler.

Smoothed Data: $\Delta G^\circ/J\ mol^{-1} = -RT\ ln\ X_1 = 3824.9 + 45.244\ T$

Std. Dev. $\Delta G^\circ = 45.6$, Coef. Corr. = 0.9977

$\Delta H^\circ/J\ mol^{-1} = 3824.9$, $\Delta S^\circ/J\ K^{-1}\ mol^{-1} = -45.244$

| T/K | Mol Fraction $X_1 \times 10^3$ | $\Delta G^\circ/J\ mol^{-1}$ |
|---|---|---|
| 283.15 | 0.853 | 16,636 |
| 293.15 | 0.902 | 17,088 |
| 303.15 | 0.950 | 17,541 |
| 313.15 | 0.997 | 17,993 |

AUXILIARY INFORMATION

**METHOD/APPARATUS/PROCEDURE:**

The solubility apparatus is based on the design of Morrison and Billett (1) and the version used is described by Battino, Evans, and Danforth (2). The degassing apparatus is that described by Battino, Banzhof, Bogan, and Wilhelm (3).

Degassing. Up to 500 $cm^3$ of solvent is placed in a flask of such size that the liquid is about 4 cm deep. The liquid is rapidly stirred, and vacuum is applied intermittently through a liquid $N_2$ trap until the permanent gas residual pressure drops to 5 microns.

Solubility Determination. The degassed solvent is passed in a thin film down a glass spiral tube containing solute gas plus the solvent vapor at a total pressure of one atm. The volume of gas absorbed is found by difference between the initial and final volumes in the buret system. The solvent is collected in a tared flask and weighed.

**SOURCE AND PURITY OF MATERIALS:**

(1) Nitrogen. Matheson Co., Inc. Minimum purity 99.99 mole per cent.

(2) Methylcyclohexane. Phillips Petroleum Co. Pure Grade. Distilled.

**ESTIMATED ERROR:**

$\delta T/K = 0.03$
$\delta P/mmHg = 0.5$
$\delta X_1/X_1 = 0.005$

**REFERENCES:**


1. Morrison, T. J.; Billett, F. *J. Chem. Soc.* 1948, 2033.
2. Battino, R.; Evans, F. D.; Danforth, W. F. *J. Am. Oil Chem. Soc.* 1968, *45*, 830.
3. Battino, R.; Banzhof, M.; Bogan, M.; Wilhelm, E. *Anal. Chem.* 1971, *43*, 806.

**COMPONENTS:**

(1) Nitrogen; $N_2$; [7727-37-9]

(2) Cyclooctane; $C_8H_{16}$; [292-64-8]

**ORIGINAL MEASUREMENTS:**

Wilcock, R. J.; Battino, R.; Wilhelm, E.

*J. Chem. Thermodyn.* 1977, *9*, 111-5.

**VARIABLES:**

T/K = 289.04 - 313.52
P/kPa = 101.325

**PREPARED BY:**

H. L. Clever
A. L. Cramer

**EXPERIMENTAL VALUES:**

| T/K | Mol Fraction $X_1 \times 10^3$ | Bunsen Coefficient $\alpha$ | Ostwald Coefficient $L$ |
|---|---|---|---|
| 289.04 | 0.570 | 0.0955 | 0.1011 |
| 298.26 | 0.598 | 0.0994 | 0.1085 |
| 313.52 | 0.660 | 0.1081 | 0.1241 |

The Bunsen coefficients were calculated by the compiler.

The solubility values were adjusted to a nitrogen partial pressure of 101.325 kPa (1 atm) by Henry's law.

Smoothed Data: $\Delta G^\circ/J\ mol^{-1} = -RT \ln X_1 = 4564.0 + 46.351\ T$

Std. Dev. $\Delta G^\circ = 14.3$, Coef. Corr. = 0.9999

$\Delta H^\circ/J\ mol^{-1} = 4564.0$, $\Delta S^\circ/J\ K^{-1}\ mol^{-1} = -46.351$

| T/K | Mol Fraction $X_1 \times 10^3$ | $\Delta G^\circ/J\ mol^{-1}$ |
|---|---|---|
| 288.15 | 0.564 | 17,920 |
| 293.15 | 0.583 | 18,152 |
| 298.15 | 0.602 | 18,384 |
| 303.15 | 0.620 | 18,615 |
| 308.15 | 0.639 | 18,847 |
| 313.15 | 0.657 | 19,079 |

**AUXILIARY INFORMATION**

**METHOD/APPARATUS/PROCEDURE:**

The solubility apparatus is based on the design of Morrison and Billett (1) and the version used is described by Battino, Evans, and Danforth (2). The degassing apparatus is that described by Battino, Banzhof, Bogan, and Wilhelm (3).

Degassing. Up to 500 $cm^3$ of solvent is placed in a flask of such size that the liquid is about 4 cm deep. The liquid is rapidly stirred, and vacuum is applied intermittently through a liquid $N_2$ trap until the permanent gas residual pressure drops to 5 microns.

Solubility Determination. The degassed solvent is passed in a thin film down a glass spiral tube containing solute gas plus the solvent vapor at a total pressure of one atm. The volume of gas absorbed is found by difference between the initial and final volumes in the buret system. The solvent is collected in a tared flask and weighed.

**SOURCE AND PURITY OF MATERIALS:**

(1) Nitrogen. Matheson Co., Inc. minimum purity 99.99 mol per cent.

(2) Cyclooctane. Chemical Samples Co. 99 mol percent, distilled, refractive index (NaD, 298.15 K) 1.4562.

**ESTIMATED ERROR:**

$\delta T/K = 0.03$
$\delta P/mmHg = 0.5$
$\delta X_1/X_1 = 0.005$

**REFERENCES:**


1. Morrison, T. J.; Billett, F. *J. Chem. Soc.* 1948, 2033.

2. Battino, R.; Evans, F. D.; Danforth, W. F. *J. Am. Oil Chem. Soc.* 1968, *45*, 830.

3. Battino, R.; Banzhof, M.; Bogan, M.; Wilhelm, E. *Anal. Chem.* 1971, *43*, 806.

**COMPONENTS:**

(1) Nitrogen; $N_2$; [7727-37-9]

(2) *cis*-Dimethylcyclohexane; $C_8H_{16}$; [2207-01-4]

**ORIGINAL MEASUREMENTS:**

Geller, E. B.; Battino, R.; Wilhelm, E.

*J. Chem. Thermodyn.* 1976, *8*, 197-202.

**VARIABLES:**

T/K = 297.97 - 312.97
P/kPa = 101.325

**PREPARED BY:**

H. L. Clever
A. L. Cramer

**EXPERIMENTAL VALUES:**

| T/K | Mol Fraction $X_1 \times 10^3$ | Bunsen Coefficient $\alpha$ | Ostwald Coefficient L |
|---|---|---|---|
| 297.97 | 0.884 | 0.1401 | 0.1528 |
| 312.97 | 0.923 | 0.1440 | 0.1650 |

The Bunsen coefficient was calculated by the compiler.

The solubility values were adjusted to a nitrogen partial pressure of 101.325 kPa (1 atm) by Henry's law.

Smoothed Data: The equation is based on two pair of experimental values and should be used with caution.

$\Delta G^o/J\ mol^{-1} = -RT \ln X_1 = 2223.5 + 50.996\ T$

$\Delta H^o/J\ mol^{-1} = 2223.5$, $\Delta S^o/J\ K^{-1}\ mol^{-1} = -50.996$

| T/K | Mol Fraction $X_1 \times 10^3$ | $\Delta G^o/J\ mol^{-1}$ |
|---|---|---|
| 298.15 | 0.885 | 17,428 |
| 303.15 | 0.898 | 17,683 |
| 308.15 | 0.911 | 17,938 |
| 313.15 | 0.923 | 18,193 |

**AUXILIARY INFORMATION**

**METHOD/APPARATUS/PROCEDURE:**

The solubility apparatus is based on the design of Morrison and Billett (1) and the version used is described by Battino, Evans, and Danforth (2). The degassing apparatus is that described by Battino, Banzhof, Bogan, and Wilhelm (3).

Degassing. Up to 500 $cm^3$ of solvent is placed in a flask of such size that the liquid is about 4 cm deep. The liquid is rapidly stirred, and vacuum is intermittently applied through a liquid $N_2$ trap until the permanent gas residual pressure drops to 5 microns.

Solubility Determination. The degassed solvent is passed in a thin film down a glass spiral tube containing solute gas plus the solvent vapor at a total pressure of one atm. The volume of gas absorbed is found by difference between the initial and final volumes in the buret system. The solvent is collected in a tared flask and weighed.

**SOURCE AND PURITY OF MATERIALS:**

(1) Nitrogen. Either Air Products & Chemicals, Inc. or Matheson Co., Inc. 99 mol per cent or better.

(2) *cis*-1,2-Dimethylcyclohexane. Chemical Samples Co. Fractionally distilled and stored in dark. Refractive index (NaD, 298.15 K) 1.4337.

**ESTIMATED ERROR:**

$\delta T/K = 0.03$
$\delta P/mmHg = 0.5$
$\delta X_1/X_1 = 0.005$

**REFERENCES:**


1. Morrison, T. J.; Billett, F. *J. Chem. Soc.* 1948, 2033.

2. Battino, R.; Evans, F. D.; Danforth, W. F. *J. Am. Oil Chem. Soc.* 1968, *45*, 830.

3. Battino, R.; Banzhof, M.; Bogan, M.; Wilhelm, E. *Anal. Chem.* 1971, *43*, 806.

**COMPONENTS:**

(1) Nitrogen; $N_2$; [7727-37-9]

(2) *trans*-1,2-Dimethylcyclohexane; $C_8H_{16}$; [6876-23-9]

**ORIGINAL MEASUREMENTS:**

Geller, E. B.; Battino, R.; Wilhelm, E.

*J. Chem. Thermodyn.* 1976, *8*, 197-202.

**VARIABLES:**

T/K = 298.25 - 313.01
P/kPa = 101.325

**PREPARED BY:**

H. L. Clever
A. L. Cramer

**EXPERIMENTAL VALUES:**

| T/K | Mol Fraction $x_1 \times 10^3$ | Bunsen Coefficient $\alpha$ | Ostwald Coefficient L |
|---|---|---|---|
| 298.25 | 1.001 | 0.1545 | 0.1687 |
| 313.01 | 1.016 | 0.1544 | 0.1770 |

The Bunsen coefficients were calculated by the compiler.

The solubility values were adjusted to a nitrogen partial pressure of 101.325 kPa (1 atm) by Henry's law.

Smoothed Data: The Gibbs energy equation is based on only two pair of points and should be used with caution.

$\Delta G^o/J\ mol^{-1} = -RT \ln X_1 = 785.05 + 54.793\ T$

$\Delta H^o/J\ mol^{-1} = 785.05$, $\Delta S^o/J\ K^{-1}\ mol^{-1} = -54.793$

AUXILIARY INFORMATION

**METHOD/APPARATUS/PROCEDURE:**

The solubility apparatus is based on the design of Morrison and Billett (1) and the version used is described by Battino, Evans, and Danforth (2). The degassing apparatus is that described by Battino, Banzhof, Bogan, and Wilhelm (3).

Degassing. Up to 500 $cm^3$ of solvent is placed in a flask of such size that the liquid is about 4 cm deep. The liquid is rapidly stirred, and vacuum is intermittently applied through a liquid $N_2$ trap until the permanent gas residual pressure drops to 5 microns.

Solubility Determination. The degassed solvent is passed in a thin film down a glass helical tube containing solute gas plus the solvent vapor at a total pressure of one atm. The volume of gas absorbed is found by difference between the initial and final volumes in the buret system. The solvent is collected in a tared flask and weighed.

**SOURCE AND PURITY OF MATERIALS:**

(1) Nitrogen. Either Air Products & Chemicals Inc., or Matheson Co., Inc. 99 mol per cent or better.

(2) *trans*-1,2-Dimethylcyclohexane. Chemical samples Co. Fractionally distilled and stored in dark. Refractive index (NaD, 298.15 K) 1.4248.

**ESTIMATED ERROR:**

$\delta T/K = 0.03$
$\delta P/mmHg = 0.5$
$\delta X_1/X_1 = 0.005$

**REFERENCES:**

1. Morrison, T. J.; Billett, F. *J. Chem. Soc.* 1948, 2033.
2. Battino, R.; Evans, F. D.; Danforth, W. F. *J. Am. Oil Chem. Soc.* 1968, *45*, 830.
3. Battino, R.; Banzhof, M.; Bogan, M.; Wilhelm, E. *Anal. Chem.* 1971, *43*, 806.

| COMPONENTS: | ORIGINAL MEASUREMENTS: |
|---|---|
| (1) Nitrogen; $N_2$; [7727-37-9]<br>(2) *cis*-1,4-Dimethylcyclohexane, 70 mol %; $C_8H_{16}$; [624-29-3]<br>(3) *trans*-1,4-Dimethylcyclohexane, 30 mol %; $C_8H_{16}$; [2207-04-7] | Geller, E. B.; Battino, R.; Wilhelm, E.<br>*J. Chem. Thermodyn.* 1976, *8*, 197-202. |
| **VARIABLES:** | **PREPARED BY:** |
| T/K = 298.17 - 313.15<br>P/kPa = 101.325 | H. L. Clever<br>A. L. Cramer |

EXPERIMENTAL VALUES:

| T/K | Mol Fraction $X_1 \times 10^3$ | Bunsen Coefficient $\alpha$ | Ostwald Coefficient L |
|---|---|---|---|
| 298.17 | 0.997 | 0.1539 | 0.1681 |
| 313.15 | 1.029 | 0.1565 | 0.1794 |

The Bunsen coefficients were calculated by the compiler.

The solubility values were adjusted to a nitrogen partial pressure of 101.325 kPa (1 atm) by Henry's law.

Smoothed Data: The Gibbs energy equation is based on only two pair of experimental values. It should be used with caution.

$$\Delta G^o/J\ mol^{-1} = -RT\ ln\ X_1 = 1689.1 + 51.802\ T$$

$$\Delta H^o/J\ mol^{-1} = 1689.1,\ \Delta S^o/J\ K^{-1}\ mol^{-1} = -51.802$$

AUXILIARY INFORMATION

METHOD/APPARATUS/PROCEDURE:

The solubility apparatus is based on the design of Morrison and Billett (1) and the version used is described by Battino, Evans, and Danforth (2). The degassing apparatus is that described by Battino, Banzhof, Bogan, and Wilhelm (3).

Degassing. Up to 500 $cm^3$ of solvent is placed in a flask of such size that the liquid is about 4 cm deep. The liquid is rapidly stirred, and vacuum is intermittently applied through a liquid $N_2$ trap until the permanent gas residual pressure drops to 5 microns.

Solubility Determination. The degassed solvent is passed in a thin film down a glass spiral tube containing solute gas plus the solvent vapor at a total pressure of one atm. The volume of gas absorbed is found by difference between the initial and final volumes in the buret system. The solvent is collected in a tared flask and weighed.

SOURCE AND PURITY OF MATERIALS:

(1) Nitrogen. Either Air Products & Chemicals, Inc., or Matheson Co., Inc. 99 mol per cent or better.

(2) *cis*-1,4-Dimethylcyclohexane.

(3) *trans*-1,4-Dimethylcyclohexane. Chemical Samples Co. The binary mixture used as received. Composition determined by refractive index by authors.

ESTIMATED ERROR:

$\delta T/K = 0.03$
$\delta P/mmHg = 0.5$
$\delta X_1/X_1 = 0.005$

REFERENCES:


1. Morrison, T. J.; Billett, F. *J. Chem. Soc.* 1948, 2033.
2. Battino, R.; Evans, F. D.; Danforth, W. F. *J. Am. Oil Chem. Soc.* 1968, *45*, 830.
3. Battino, R.; Banzhof, M.; Bogan, M.; Wilhelm, E. *Anal. Chem.* 1971, *43*, 806.

| COMPONENTS: | ORIGINAL MEASUREMENTS: |
|---|---|
| (1) Nitrogen; $N_2$; [7727-37-9]<br>(2) *cis*-1,3-Dimethylcyclohexane, 59 mol %; $C_8H_{16}$; [638-04-0]<br>(3) *trans*-1,3-Dimethylcyclohexane, 41 mol %; $C_8H_{16}$; [2207-03-6] | Geller, E. B.; Battino, R.; Wilhelm, E.<br>*J. Chem. Thermodyn.* 1976, *8*, 197-202. |
| **VARIABLES:** | **PREPARED BY:** |
| T/K = 298.07 - 313.01<br>P/kPa = 101.325 | H. L. Clever<br>A. L. Cramer |

EXPERIMENTAL VALUES:

| T/K | Mol Fraction $X_1 \times 10^3$ | Bunsen Coefficient $\alpha$ | Ostwald Coefficient L |
|---|---|---|---|
| 298.07 | 1.005 | 0.1547 | 0.1688 |
| 313.01 | 1.043 | 0.1580 | 0.1811 |

The Bunsen coefficients were calculated by the compiler.

The solubility values were adjusted to a nitrogen partial pressure of 101.325 kPa (1 atm) by Henry's law.

Smoothed Data: The Gibbs energy equation is based on only two pair of experimental values and should be used with caution.

$$\Delta G^o/\text{J mol}^{-1} = -RT \ln X_1 = 1924.2 + 50.937\ T$$

$$\Delta H^o/\text{J mol}^{-1} = 1924.2,\ \Delta S^o/\text{J K}^{-1}\ \text{mol}^{-1} = -50.937$$

AUXILIARY INFORMATION

METHOD/APPARATUS/PROCEDURE:

The solubility apparatus is based on the design of Morrison and Billett (1) and the version used is described by Battino, Evans, and Danforth (2). The degassing apparatus is that described by Battino, Banzhof, Bogan, and Wilhelm (3).

Degassing. Up to 500 $cm^3$ of solvent is placed in a flask of such size that the liquid is about 4 cm deep. The liquid is rapidly stirred, and vacuum is applied intermittently through a liquid $N_2$ trap until the permanent gas residual pressure drops to 5 microns.

Solubility Determination. The degassed solvent is passed in a thin film down a glass helical tube containing solute gas plus the solvent vapor at a total pressure of one atm. The volume of gas absorbed is found by difference between the initial and final volumes in the buret system. The solvent is collected in a tared flask and weighed.

SOURCE AND PURITY OF MATERIALS:

(1) Nitrogen. Either Air Products & Chemicals, Inc., or Matheson Co., Inc. 99 mol per cent or better.

(2) *cis*-1,4-Dimethylcyclohexane.

(3) *trans*-1,4-Dimethylcyclohexane. Chemical Samples Co. Binary mixture used as received. Authors analyzed mixture by refractive index.

ESTIMATED ERROR:

$\delta T/K = 0.03$
$\delta P/mmHg = 0.5$
$\delta X_1/X_1 = 0.005$

REFERENCES:

1. Morrison, T. J.; Billett, F. *J. Chem. Soc.* 1948, 2033.
2. Battino, R.; Evans, F. D.; Danforth, W. F. *J. Am. Oil Chem. Soc.* 1968, *45*, 830.
3. Battino, R.; Banzhof, M.; Bogan, M.; Wilhelm, E. *Anal. Chem.* 1971, *43*, 806.

| COMPONENTS: | ORIGINAL MEASUREMENTS: |
|---|---|
| (1) Nitrogen; $N_2$; [7727-37-9]<br>(2) 1,1' - Bicyclohexyl; $C_{12}H_{22}$; [92-51-3] | Tremper, K.K.; Prausnitz, J.M.<br>*J. Chem. Eng. Data* 1976, *21*, 295-9. |
| VARIABLES:<br>T/K = 300-475 | PREPARED BY:<br>C.L. Young |

EXPERIMENTAL VALUES:

| T/K | Henry's Constant[a] /atm | Mole fraction[b] of nitrogen at 1 atm partial pressure, $x_{N_2}$ |
|---|---|---|
| 300 | 1345.0 | 0.0007435 |
| 325 | 1274.0 | 0.0007849 |
| 350 | 1200.0 | 0.0008333 |
| 375 | 1123.0 | 0.0008905 |
| 400 | 1041.0 | 0.0009606 |
| 425 | 945.0 | 0.001058 |
| 450 | 829.0 | 0.001206 |
| 475 | 716.0 | 0.001397 |

a. Authors stated measurements were made at several pressures and values of solubility used were all within the Henry's-Law region.

b. Calculated by compiler assuming linear relationship between mole fraction and pressure

AUXILIARY INFORMATION

METHOD/APPARATUS/PROCEDURE.

Volumetric apparatus similar to that described by Dymond and Hildebrand (1). Pressure measured with a null detector and precision gauge. Details in ref. (2).

SOURCE AND PURITY OF MATERIALS:

Solvent degassed, no other details given.

ESTIMATED ERROR:

$\delta T/K = \pm 0.1$; $\delta x_{N_2} = \pm 1\%$.

REFERENCES:

1. Dymond, J.; Hildebrand, J.H. *Ind. Eng. Chem. Fundam.* 1967, *6*, 130.
2. Cukor, P.M.; Prausnitz, J.M. *Ind. Eng. Chem. Fundam.* 1971, *10*, 638.

| COMPONENTS: | EVALUATOR: |
|---|---|
| (1) Nitrogen; $N_2$; [7727-37-9]<br>(2) Benzene; $C_6H_6$; [71-43-2]<br>(2') Toluene; $C_7H_8$; [108-88-3] | Rubin Battino<br>Department of Chemistry<br>Wright State University<br>Dayton, Ohio 45435<br>U.S.A.<br><br>December 1981 |

CRITICAL EVALUATION:

Benzene

Five papers (1-5) report on the solubility of nitrogen in benzene. Just's two points (1) were significantly lower and Patyi, et al.'s, single point (5) significantly higher than the other three papers' values. Thus, the recommended values were obtained from smoothing the data in references 2,3, and 4. This yielded

$$\ln x_1 = -6.05445 - 4.95673/(T/100K) \quad (1)$$

with a standard deviation in ln $x_1$ of 0.0090. $\Delta\bar{H}° = 4121$ J mol$^{-1}$, $\Delta\bar{S}° = -50$ J K$^{-1}$ mol$^{-1}$. Smoothed values at 5K intervals follow.

| T/K | $10^4\ x_1$ | T/K | $10^4\ x_1$ |
|---|---|---|---|
| 278.15 | 3.95 | 308.15 | 4.70 |
| 283.15 | 4.08 | 313.15 | 4.82 |
| 288.15 | 4.20 | 318.15 | 4.94 |
| 293.15 | 4.33 | 323.15 | 5.06 |
| 298.15 | 4.45 | 328.15 | 5.18 |
| 303.15 | 4.58 | 333.15 | 5.30 |

Toluene

The nitrogen/toluene system was studied by Just (1) and Field, et al., (2). Both sets of data were combined for smoothing but due to the variation in the data the smoothed values may only considered to be tentative. The smoothing equation is

$$\ln x_1 = -6.27576 - 3.71675/(T/100K) \quad (2)$$

with a standard deviation in ln $x_1$ of 0.030. $\Delta\bar{H}° = 3090$ J mol$^{-1}$ and $\Delta\bar{S}° = -52$ J K$^{-1}$ mol$^{-1}$. Smoothed values follow.

| T/K | $10^4\ x_1$ | T/K | $10^4\ x_1$ |
|---|---|---|---|
| 283.15 | 5.06 | 303.15 | 5.52 |
| 288.15 | 5.18 | 308.15 | 5.63 |
| 293.15 | 5.29 | 313.15 | 5.74 |
| 298.15 | 5.41 | | |

References

1. Just, G. *Z. Phys. Chem.* 1901, *37*, 342-67.
2. Horiuti, J. *Sci. Pap. Inst. Phys. Chem. Res. (Jpn)* 1931/32, *17*, 125-256.
3. Gjaldbaek, J.C.; Hildebrand, J.H. *J. Am. Chem. Soc.* 1949, *71*, 3147-50.
4. Byrne, J.E.; Battino, R.; Danforth, W.F. *J. Chem. Thermodyn.* 1974, *6*, 245-50.
5. Patyi, L.; Furmer, I. E.; Makranczy, J.; Sadilenko, A. S.; Stepanova, Z. G.; Berengarten, M. G. *Zh. Prikl. Khim.* 1978, *51*, 1296-1300.
6. Field, L. R.; Wilhelm, E.; Battino, R. *J. Chem. Thermodyn.* 1974, *6*, 237-43.

| COMPONENTS: | ORIGINAL MEASUREMENTS: |
|---|---|
| (1) Nitrogen; $N_2$; [7727-37-9]<br>(2) Benzene; $C_6H_6$; [71-43-2] | Just, G.<br>*Z. Phys. Chem.* 1901, *37*, 342-67. |

| VARIABLES: | PREPARED BY: |
|---|---|
| T/K = 293-298<br>P/kPa = 101.325 | M. E. Derrick<br>H. L. Clever |

EXPERIMENTAL VALUES:

| T/K | Mol Fraction $x_1 \times 10^4$ | Bunsen Coefficient $\alpha$ | Ostwald Coefficient L |
|---|---|---|---|
| 293.15 | 4.12 | 0.1038 | 0.1114 |
| 298.15 | 4.24 | 0.1062 | 0.1159 |

The author measured the Ostwald coefficient at a pressure of about 746 mmHg. The compiler assumed the Ostwald coefficient to be independent of pressure, and calculated the mole fraction and Bunsen coefficient values at 101.325 kPa (1 atm) partial pressure of the gas.

AUXILIARY INFORMATION

METHOD/APPARATUS/PROCEDURE:

An Ostwald apparatus as modified by Timofejew (1), and Steiner (2) is used. The apparatus consists of a gas buret, an absorption flask, and a mercury manometer. The system is thermostated with a water jacket.

The gas is introduced into the degassed liquid. The gas volume absorbed is determined by the gas buret. The solvent volume is determined at the end of the experiment by pouring the solvent into a graduated flask.

SOURCE AND PURITY OF MATERIALS:

(1) Nitrogen. Prepared by the reaction of sodium nitrite and ammonium nitrate in aqueous solution in the presence of calcium dichromate.

(2) Benzene. No information.

ESTIMATED ERROR:

$\delta L/L = 0.03$ (compiler)

REFERENCES:

1. Timofejew, W. *Z. Physik. Chem.* 1890, *6*, 141.

2. Steiner, *Ann. Phys. (Leipzig)* 1894, *52*, 275.

COMPONENTS:

(1) Nitrogen; $N_2$; [7727-37-9]

(2) Benzene; $C_6H_6$; [71-43-2]

ORIGINAL MEASUREMENTS:

Horiuti, J.

*Sci. Pap. Inst. Phys. Chem. Res. (Jpn)* 1931/32, *17*, 125-256.

VARIABLES:

T/K = 280.25 - 333.15

Total P/kPa = 101.325

PREPARED BY:

M. E. Derrick
H. L. Clever

EXPERIMENTAL VALUES:

| T/K | Mol Fraction $X_1 \times 10^3$ | Bunsen Coefficient $\alpha$ | Ostwald Coefficient L |
|---|---|---|---|
| 280.25 | 0.4045 | 0.1036 | 0.1063 |
| 293.15 | 0.4292 | 0.1083 | 0.1162 |
| 313.15 | 0.4799 | 0.1182 | 0.1355 |
| 333.15 | 0.5374 | 0.1291 | 0.1575 |

Smoothed Data: $\ln X_1 = -17.9801 + 12.1852/(T/100) + 5.646 \ln (T/100)$

Standard error about the regression line = $1.55 \times 10^{-6}$

| T/K | Mol Fraction $X_1 \times 10^3$ |
|---|---|
| 283.15 | 0.4096 |
| 293.15 | 0.4303 |
| 303.15 | 0.4534 |

| T/K | Mol Fraction $X_1 \times 10^3$ |
|---|---|
| 313.15 | 0.4789 |
| 323.15 | 0.5071 |
| 333.15 | 0.5378 |

If the gas and solution are assumed to be ideal the fitted equation gives the following thermodynamic values.

| T/K | $\Delta G^o$/kJ mol$^{-1}$ | $\Delta H^o$/J mol$^{-1}$ | $\Delta S^o$/J K$^{-1}$ mol$^{-1}$ | $\Delta C_p^o$/J K$^{-1}$ mol$^{-1}$ |
|---|---|---|---|---|
| 283.15 | 18.363 | 3161 | -53.69 | 46.95 |
| 298.15 | 19.150 | 3865 | -51.27 | 46.95 |

The Ostwald coefficient was measured at a total pressure of one atm. The Bunsen coefficient and the mole fraction solubility values were calculated by the compiler for a gas pressure of one atm assuming the gas to behave ideally and the Ostwald coefficient to be independent of pressure.

AUXILIARY INFORMATION

METHOD/APPARATUS/PROCEDURE:

The apparatus consists of a gas buret, a solvent reservoir, and an absorption pipet. The volume of the pipet was determined at various miniscus heights by weighing a quantity of water. The miniscus height was read with a cathetometer.

The dry gas introduced into the degassed solvent. The gas and solvent are mixed with a magnetic stirrer until saturation. Care taken to prevent solvent vapor from mixing with the solute gas in the gas buret. The volume of gas was determined from a gas buret reading, the volume of the solvent was determined from the miniscus height in the absorption pipet.

SOURCE AND PURITY OF MATERIALS:

(1) Nitrogen. Prepared in the laboratory by heating sodium azide (Kahlbaum, recrystalized and dried). Gas dried over phosphorous pentoxide.

(2) Benzene. Merck. Extra pure, free from sulfur sample refluxed with sodium amalgam. Normal boiling point 80.18$^o$C.

ESTIMATED ERROR:

$\delta T/K = 0.05$

COMPONENTS:

(1) Nitrogen; $N_2$; [7727-37-9]

(2) Benzene; $C_6H_6$; [71-43-2]

ORIGINAL MEASUREMENTS:

Gjaldbaek, J. C.; Hildebrand, J. H.

*J. Am. Chem. Soc.* 1949, *71*, 3147-50.

VARIABLES:

T/K = 298.15
Total P/kPa = 101.325

PREPARED BY:

J. Chr. Gjaldbaek

EXPERIMENTAL VALUES:

| T/K | Mol Fraction $X_1 \times 10^3$ | Bunsen Coefficient $\alpha$ | Ostwald Coefficient L |
|---|---|---|---|
| 298.15 | 0.448 | 0.124 | 0.135 |

The mole fraction and Ostwald solubility values were calculated by the compiler.

AUXILIARY INFORMATION

METHOD/APPARATUS/PROCEDURE:

A calibrated all-glass combined manometer and bulb was enclosed in an air thermostat and shaken until equilibrium. Mercury was used for calibration and as the confining liquid. The solvents were degassed in the apparatus. Details are in references 1 and 2.

The absorbed volume of gas was calculated from the initial and final amounts, both saturated with solvent vapor. The amount of solvent was determined by the weight of displaced mercury.

The saturation of the liquid with the gas was carried out close to atmospheric pressure. The solubility values were reported for one atmosphere gas pressure assuming Henry's law is obeyed.

SOURCE AND PURITY OF MATERIALS:

(1) Nitrogen. Stuart Oxygen Co. 99.996 percent $N_2$.

(2) Benzene. Reagent grade.

ESTIMATED ERROR:

$\delta T/K = 0.05$
$\delta X_1/X_1 = 0.015$

REFERENCES:

1. Lannung, A. *J. Am. Chem. Soc.* 1930, *52*, 68.

2. Gjaldbaek, J. C. *Acta Chem. Scand.* 1952, *6*, 623.

COMPONENTS:

(1) Nitrogen; $N_2$; [7727-37-9]

(2) Benzene; $C_6H_6$; [71-43-2]

ORIGINAL MEASUREMENTS:

Yen, L. C.; McKetta, J. J.

*J. Chem. Eng. Data* 1962, *7*, 288-9.

VARIABLES:

T/K = 301

PREPARED BY:

R. Battino

EXPERIMENTAL VALUES:

t/°C (301 K)

| Run | $\alpha^a$/cm$^3$ (STP) cm$^{-3}$ atm$^{-1}$ |
|---|---|
| N-2 | 0.118 |
| N-3 | 0.115 |
| N-4 | 0.116 |
| N-6 | 0.113 |
| N-7 | 0.118 |
| N-8 | 0.113 |

[a] Bunsen coefficient. Average value is 0.115.

AUXILIARY INFORMATION

METHOD/APPARATUS/PROCEDURE:

Used a modified version of the solubility apparatus of Markham and Kobe (1).

SOURCE AND PURITY OF MATERIALS:

(1) No details given.

(2) J. T. Baker Chemical Co., Baker analyzed grade, distilled.

ESTIMATED ERROR:

$\delta\alpha/\alpha = \pm 0.016$

REFERENCES:

1. Markham, A. E.; Kobe, K. A. *J. Am. Chem. Soc.* 1941, *63*, 449.

**COMPONENTS:**

(1) Nitrogen; $N_2$; [7727-37-9]

(2) Benzene; $C_6H_6$; [71-43-2]

**ORIGINAL MEASUREMENTS:**

Byrne, J. E.; Battino, R.; Danforth, W. F.

*J. Chem. Thermodyn.* 1974, *6*, 245-50.

**VARIABLES:**

T/K = 310.64, 310.69
P/kPa = 101.325

**PREPARED BY:**

H. L. Clever

**EXPERIMENTAL VALUES:**

| T/K | Mol Fraction $X_1 \times 10^3$ | Bunsen Coefficient $\alpha$ | Ostwald Coefficient L |
|---|---|---|---|
| 310.64 | 0.471 | 0.1163 | 0.1323 |
| 310.69 | 0.473 | 0.1168 | 0.1328 |

The Bunsen coefficients and mol fraction solubilities were calculated by the compiler.

**AUXILIARY INFORMATION**

**METHOD/APPARATUS/PROCEDURE:**

The solubility apparatus is based on the design of Morrison and Billett (1) and the version used is described by Battino, Evans, and Danforth (2). The degassing apparatus is that described by Battino, Banzhof, Bogan, and Wilhelm (3).

Degassing. Up to 500 $cm^3$ of solvent is placed in a flask of such size that the liquid is about 4 cm deep. The liquid is rapidly stirred, and vacuum is intermittently applied through a liquid $N_2$ trap until the permanent gas residual pressure drops to 5 microns.

Solubility Determination. The degassed solvent is passed in a thin film down a glass spiral tube containing solute gas plus the solvent vapor at a total pressure of one atm. The volume of gas absorbed is found by difference between the initial and final volumes in the buret system. The solvent is collected in a tared flask and weighed.

**SOURCE AND PURITY OF MATERIALS:**

(1) Nitrogen. Matheson Co., Inc. Minimum purity 99.99 mole per cent.

(2) Benzene. Baker Analyzed Reagent Grade. Thiophene free, 99.9 mol per cent.

**ESTIMATED ERROR:**

$\delta T/K = 0.03$
$\delta P/mmHg = 0.5$
$\delta X_1/X_1 = 0.005$

**REFERENCES:**

1. Morrison, T. J.; Billett, F. *J. Chem. Soc.* 1948, 2033.
2. Battino, R.; Evans, F. D.; Danforth, W. F. *J. Am. Oil Chem. Soc.* 1968, *45*, 830.
3. Battino, R.; Banzhof, M.; Bogan, M.; Wilhelm, E. *Anal. Chem.* 1971, *43*, 806.

| COMPONENTS: | ORIGINAL MEASUREMENTS: |
|---|---|
| (1) Nitrogen; $N_2$; [7727-37-9]<br><br>(2) Benzene; $C_6H_6$; [71-43-2] | Patyi, L.; Furmer, I. E.; Makranczy, J; Sadilenko, A. S.; Stepanova, Z. G.; Berengarten, M.G.<br>*Zh. Prikl. Khim.* 1978, *51*, 1296-1300. |
| VARIABLES: | PREPARED BY: |
| T/K = 298 | C. L. Young |

EXPERIMENTAL VALUES:

| T/K | $\alpha^*$ | Mole fraction of nitrogen at a partial pressure of 101.325 kPa |
|---|---|---|
| 298.15 | 0.124 | 0.000495 |

* volume of gas (measured at 101.325 kPa and 273.15 K) dissolved by one volume of benzene.

AUXILIARY INFORMATION

METHOD/APPARATUS/PROCEDURE:

Volumetric method. Pressure measured when known amounts of gas were added, in increments, to a known amount of liquid in a vessel of known dimensions. Corrections were made for the partial pressure of solvent. Details in ref. (1).

SOURCE AND PURITY OF MATERIALS:

Purity better than 99 mole per cent as determined by gas chromatography.

ESTIMATED ERROR:

$\delta T/K = \pm 0.1$; $\delta\alpha = \pm 4\%$ or less.

REFERENCES:

1. Bodor, E.; Bor, G. J.; Mohai, B.; Sipos, G. *Veszpremi. Vegyip. Egy. Kozl.* 1957, *1*, 55.

**COMPONENTS:**

(1) Nitrogen; $N_2$; [7727-37-9]

(2) Methyl benzene or toluene; $C_7H_8$; [108-88-3]

**ORIGINAL MEASUREMENTS:**

Just, G.

*Z. Phys. Chem.* 1901, *37*, 342-67.

**VARIABLES:**

T/K = 293.15, 298.15
P/kPa = 101.325

**PREPARED BY:**

M. E. Derrick
H. L. Clever

**EXPERIMENTAL VALUES:**

| T/K | Mol Fraction $x_1 \times 10^4$ | Bunsen Coefficient $\alpha$ | Ostwald Coefficient L |
|---|---|---|---|
| 293.15 | 5.24 | 0.1105 | 0.1186 |
| 298.15 | 5.39 | 0.1131 | 0.1235 |

The author measured the Ostwald coefficient at a pressure of about 746 mmHg. The compiler assumed the Ostwald coefficient to be independent of pressure, and calculated the mole fraction and Bunsen coefficient values at 101.325 kPa (1 atm) partial pressure of the gas.

**AUXILIARY INFORMATION**

**METHOD/APPARATUS/PROCEDURE:**

An Ostwald apparatus as modified by Timofejew (1), and Steiner (2) is used. The apparatus consists of a gas buret, an absorption flask, and a mercury manometer. The system is thermostated with a water jacket.

The gas is introduced into the degassed liquid. The gas volume absorbed is determined by the gas buret. The solvent volume is determined at the end of the experiment by pouring the solvent into a graduated flask.

**SOURCE AND PURITY OF MATERIALS:**

(1) Nitrogen. Prepared by the reaction of sodium nitrite and ammonium nitrate in aqueous solution in the presence of calcium dichromate.

(2) Methyl benzene. No information.

**ESTIMATED ERROR:**

$\delta L/L = 0.03$ (compiler)

**REFERENCES:**


1. Timofejew, W.
   *Z. Physik. Chem.* 1890, *6*, 141.

2. Steiner,
   *Ann. Phys. (Leipzig)* 1894, *52*, 275.

**COMPONENTS:**

(1) Nitrogen; $N_2$; [7727-37-9]

(2) Methylbenzene or Toluene; $C_7H_8$; [108-88-3]

**ORIGINAL MEASUREMENTS:**

Field, L. R.; Wilhelm, E.; Battino, R.

*J. Chem. Thermodyn.* 1974, *6*, 237-43.

**VARIABLES:**

T/K = 283.39 - 313.28
P/kPa = 101.325

**PREPARED BY:**

H. L. Clever

**EXPERIMENTAL VALUES:**

| T/K | Mol Fraction $X_1 \times 10^3$ | Bunsen Coefficient $\alpha$ | Ostwald Coefficient L |
|---|---|---|---|
| 283.39 | 0.497 | 0.1061 | 0.1101 |
| 298.29 | 0.574 | 0.1204 | 0.1315 |
| 313.28 | 0.560 | 0.1158 | 0.1328 |

The gas solubility values were adjusted to a nitrogen partial pressure of 101.325 kPa (1 atm) by Henry's law.

The Bunsen coefficients were calculated by the compiler.

Smoothed Data: $\Delta G^\circ/\text{J mol}^{-1} = -RT \ln X_1 = 2875.7 + 52.860\ T$

Std. Dev. $\Delta G^\circ = 116.8$, Coef. Corr. = 0.9894

$\Delta H^\circ/\text{J mol}^{-1} = 2875.7$, $\Delta S^\circ/\text{J K}^{-1}\text{ mol}^{-1} = -52.860$

| T/K | Mol Fraction $X_1 \times 10^3$ | $\Delta G^\circ/\text{J mol}^{-1}$ |
|---|---|---|
| 283.15 | 0.511 | 17,843 |
| 293.15 | 0.533 | 18,372 |
| 303.15 | 0.554 | 18,900 |
| 313.15 | 0.574 | 19,429 |

**AUXILIARY INFORMATION**

**METHOD/APPARATUS/PROCEDURE:**

The solubility apparatus is based on the design of Morrison and Billett (1) and the version used is described by Battino, Evans, and Danforth (2). The degassing apparatus is that described by Battino, Banzhof, Bogan, and Wilhelm (3).

Degassing. Up to 500 $cm^3$ of solvent is placed in a flask of such size that the liquid is about 4 cm deep. The liquid is rapidly stirred, and vacuum is intermittently applied through a liquid $N_2$ trap until the permanent gas residual pressure drops to 5 microns.

Solubility Determination. The degassed solvent is passed in a thin film down a glass spiral tube containing solute gas plus the solvent vapor at a total pressure of one atm. The volume of gas absorbed is found by difference between the initial and final volumes in the buret system. The solvent is collected in a tared flask and weighed.

**SOURCE AND PURITY OF MATERIALS:**

(1) Nitrogen. Matheson Co., Inc. Minimum purity 99.99 mole per cent.

(2) Toluene. Phillips Petroleum Pure Grade. Distilled.

**ESTIMATED ERROR:**

$\delta T/K = 0.03$
$\delta P/\text{mmHg} = 0.5$
$\delta X_1/X_1 = 0.005$

**REFERENCES:**


1. Morrison, T. J.; Billett, F. *J. Chem. Soc.* 1948, 2033.
2. Battino, R.; Evans, F. D.; Danforth, W. F. *J. Am. Oil Chem. Soc.* 1968, *45*, 830.
3. Battino, R.; Banzhof, M.; Bogan, M.; Wilhelm, E. *Anal. Chem.* 1971, *43*, 806.

COMPONENTS:

(1) Nitrogen; $N_2$; [7727-37-9]

(2) Dimethylbenzene or xylene; $C_8H_{10}$; [1330-20-7]

ORIGINAL MEASUREMENTS:

Just, G.

*Z. Phys. Chem.* 1901, *37*, 342-67.

VARIABLES:

T/K = 293.15, 298.15
P/kPa = 101.325

PREPARED BY:

M. E. Derrick
H. L. Clever

EXPERIMENTAL VALUES:

| T/K | Mol Fraction $x_1 \times 10^4$ | Bunsen Coefficient $\alpha$ | Ostwald Coefficient L |
|---|---|---|---|
| 293.15 | 6.03 | 0.1104 | 0.1185 |
| 298.15 | 6.12 | 0.1115 | 0.1217 |

The author measured the Ostwald coefficient at a pressure of about 746 mmHg. The compiler assumed the Ostwald coefficient to be independent of pressure, and calculated the mole fraction and Bunsen coefficient values at 101.325 kPa (1 atm) partial pressure of the gas.

The compiler assumed the dimethylbenzene was a mixture of 40 per cent 1,3-dimethylbenzene, and 20 per cent each of 1,2-dimethylbenzene, 1,4-dimethylbenzene, and ethylbenzene. See Riddick, J. A.; Bunger, W. B. Organic Solvents, 3rd Ed., Wiley-Interscience, New York, 1970, p. 614.

AUXILIARY INFORMATION

METHOD/APPARATUS/PROCEDURE:

An Ostwald apparatus as modified by Timofejew (1), and Steiner (2) is used. The apparatus consists of a gas buret, an absorption flask, and a mercury manometer. The system is thermostated with a water jacket.

The gas is introduced into the degassed liquid. The gas volume absorbed is determined by the gas buret. The solvent volume is determined at the end of the experiment by pouring the solvent into a graduated flask.

SOURCE AND PURITY OF MATERIALS:

(1) Nitrogen. Prepared by the reaction of sodium nitrite and ammonium nitrate in aqueous solution in the presence of calcium dichromate.

(2) Dimethylbenzene. No information given.

ESTIMATED ERROR:

$\delta L/L = 0.03$ (compiler)

REFERENCES:

1. Timofejew, W. *Z. Physik. Chem.* 1890, *6*, 141.

2. Steiner, *Ann. Phys. (Leipzig)* 1894, *52*, 275.

| COMPONENTS: | ORIGINAL MEASUREMENTS: |
|---|---|
| (1) Nitrogen; $N_2$; [7727-37-9]<br>(2) Naphthalene, 1-methyl; $C_{11}H_{10}$; [90-12-0] | Tremper, K.K.; Prausnitz, J.M.<br>*J. Chem. Eng. Data* 1976, *21*, 295-9. |
| VARIABLES: | PREPARED BY: |
| T/K = 300-475 | C.L. Young |

EXPERIMENTAL VALUES:

| T/K | Henry's Constant[a] /atm | Mole fraction[b] of nitrogen at 1 atm partial pressure, $x_{N_2}$ |
|---|---|---|
| 300 | 2970.0 | 0.0003367 |
| 325 | 2729.0 | 0.0003664 |
| 350 | 2481.0 | 0.0004031 |
| 375 | 2222.0 | 0.0004500 |
| 400 | 1949.0 | 0.0005131 |
| 425 | 1655.0 | 0.0006042 |
| 450 | 1338.0 | 0.0007474 |
| 475 | 1108.0 | 0.0009025 |

a. Authors stated measurements were made at several pressures and values of solubility used were all within the Henry's-Law region.

b. Calculated by compiler assuming linear relationship between mole fraction and pressure.

AUXILIARY INFORMATION

METHOD/APPARATUS/PROCEDURE

Volumetric apparatus similar to that described by Dymond and Hildebrand (1). Pressure measured with a null detector and precision gauge. Details in ref. (2).

SOURCE AND PURITY OF MATERIALS:

Solvent degassed, no other details given.

ESTIMATED ERROR:

$\delta T/K = \pm 0.1$; $\delta x_{N_2} = \pm 1\%$.

REFERENCES:

1. Dymond, J.; Hildebrand, J.H. *Ind. Eng. Chem. Fundam.* 1967, *6*, 130.
2. Cukor, P.M.; Prausnitz, J.M. *Ind. Eng. Chem. Fundam.* 1971, *10*, 638.

COMPONENTS:

(1) Nitrogen; $N_2$; [7727-37-9]

(2) Benzene, 1, 1'-methylenebis-
(Diphenylmethane); $C_{13}H_{12}$;
[101-81-5]

ORIGINAL MEASUREMENTS:

Tremper, K.K.; Prausnitz, J.M.

*J. Chem. Eng. Data* 1976, *21*, 295-9.

VARIABLES:

T/K = 300-475

PREPARED BY:

C.L. Young

EXPERIMENTAL VALUES:

| T/K | Henry's Constant[a] /atm | Mole fraction[b] of nitrogen at 1 atm partial pressure, $x_{N_2}$ |
|---|---|---|
| 300 | 2937.0 | 0.0003405 |
| 325 | 2570.0 | 0.0003891 |
| 350 | 2304.0 | 0.0004340 |
| 375 | 2081.0 | 0.0004805 |
| 400 | 1890.0 | 0.0005291 |
| 425 | 1717.0 | 0.0005824 |
| 450 | 1553.0 | 0.0006439 |
| 475 | 1421.0 | 0.0007037 |

a. Authors stated measurements were made at several pressures and values of solubility used were all within the Henry's-Law region.

b. Calculated by compiler assuming linear relationship between mole fraction and pressure.

AUXILIARY INFORMATION

METHOD/APPARATUS/PROCEDURE

Volumetric apparatus similar to that described by Dymond and Hildebrand (1). Pressure measured with a null detector and precision gauge. Details in ref. (2).

SOURCE AND PURITY OF MATERIALS:

Solvent degassed, no details given.

ESTIMATED ERROR:

$\delta T/K = \pm 0.1$; $\delta x_{N_2} = \pm 1\%$.

REFERENCES:

1. Dymond, J.; Hildebrand, J.H.
*Ind. Eng. Chem. Fundam.* 1967, *6*, 130.
2. Cukor, P.M.; Prausnitz, J.M.
*Ind. Eng. Chem. Fundam.* 1971, *10*, 638.

| COMPONENTS: | EVALUATOR: |
|---|---|
| (1) Nitrogen; $N_2$; [7727-37-9]<br>(2) Alcohols | Rubin Battino<br>Department of Chemistry<br>Wright State University<br>Dayton, Ohio 45435<br>U.S.A.<br>December 1981 |

CRITICAL EVALUATION:

In this section we will do a critical evaluation of the 1-alkanols from $C_1$ to $C_{12}$, 2-propanol, and 2-methyl-1-propanol (isobutanol) First we will discuss the data for the individual alcohols and then discuss the data at 298.15K for the 1-alkanols.

Methanol

The nitrogen/methanol system was studied by five groups (1-5). Just's data (1) were rejected for being too low and Makranczy, et al's, datum (5) for being too high. The remaining three sets of data were smoothed to yield.

$$\ln x_1 = -8.17081 - 6.97967 \times 10^{-2}/(T/100K) \quad (1)$$

in the range 213 to 323 K with a standard deviation of 0.019 in $\ln x_1$. Smoothed recommended values at 10 K intervals follow. $\Delta\bar{H}° = 5803$ J mol$^{-1}$ and $\Delta\bar{S}° = 68$ J K$^{-1}$ mol$^{-1}$.

| T/K | $10^4\ x_1$ | T/K | $10^4\ x_1$ | T/K | $10^4\ x_1$ |
|---|---|---|---|---|---|
| 213.15 | 2.74 | 253.15 | 2.75 | 293.15 | 2.76 |
| 223.15 | 2.74 | 263.15 | 2.75 | 298.15 | 2.76 |
| 233.15 | 2.74 | 273.15 | 2.76 | 303.15 | 2.76 |
| 243.15 | 2.75 | 283.15 | 2.76 | 313.15 | 2.77 |
| | | | | 323.15 | 2.77 |

Ethanol

This system was studied by six groups (1-6) with Just's (1) and Metschl's (6) values being too low and Makranczy, et al.'s (5), value being too high. The smoothing equation obtained from references 2,3, and 4 gave

$$\ln x_1 = -7.95312 + 4.43802 \times 10^{-2}/(T/100K) \quad (2)$$

in the range 213 to 323 K with a standard deviation in $\ln x_1$ of 0.016. Smoothed recommended values at 10 K intervals follow. $\Delta\bar{H}° = -37$ J mol$^{-1}$ $\Delta\bar{S}° = -66$ J K$^{-1}$ mol$^{-1}$.

| T/K | $10^4\ x_1$ | T/K | $10^4\ x_1$ | T/K | $10^4\ x_1$ |
|---|---|---|---|---|---|
| 213.15 | 3.59 | 253.15 | 3.58 | 293.15 | 3.57 |
| 223.15 | 3.59 | 263.15 | 3.58 | 298.15 | 3.57 |
| 233.15 | 3.58 | 273.15 | 3.57 | 303.15 | 3.57 |
| 243.15 | 3.58 | 283.15 | 3.57 | 313.15 | 3.57 |
| | | | | 323.15 | 3.56 |

1-Propanol

The nitrogen/1-propanol system was studied by four groups (3,4,5,7). Again, Makranczy, et al.'s (5), value was too high to use. Also the 308 K value of Boyer and Bircher (3) was anomalously high. The remaining data from references 3,4, and 7 gave

$$\ln x_1 = -7.90598 + 0.257433/(T/100K) \quad (3)$$

in the range 213 to 298 K with a standard deviation of 0.015 in $\ln x_1$. Smoothed values at 10K intervals follow. $\Delta\bar{H}° = 214$ J mol$^{-1}$ and $\Delta\bar{S}° = -62$ J K$^{-1}$ mol$^{-1}$.

continued on following page

| COMPONENTS: | EVALUATOR: |
|---|---|
| (1) Nitrogen; $N_2$; [7727-37-9]<br>(2) Alcohols | Rubin Battino<br>Department of Chemistry<br>Wright State University<br>Dayton, Ohio 45435<br>U.S.A.<br><br>December 1981 |

CRITICAL EVALUATION:

continued

| T/K | $10^4\ x_1$ | T/K | $10^4\ x_1$ | T/K | $10^4\ x_1$ |
|---|---|---|---|---|---|
| 213.15 | 4.16 | 243.15 | 4.10 | 273.15 | 4.05 |
| 223.15 | 4.14 | 253.15 | 4.08 | 283.15 | 4.04 |
| 233.15 | 4.12 | 263.15 | 4.06 | 293.15 | 4.02 |
| | | | | 298.15 | 4.02 |

2-Propanol

This system was studied by Kretschmer, et al. (2), and Katayama and Nitta (4). Smoothing their values in the range 213 to 323 K gave

$$\ln x_1 = -7.63455 - 0.110188/(T/100K) \quad (4)$$

with a standard deviation of 0.026 in ln $x_1$. The table below gives smoothed recommended values for this system. $\Delta\bar{H}° = 92$ J mol$^{-1}$ and $\Delta\bar{S}° = -63$ J K$^{-1}$ mol$^{-1}$.

| T/K | $10^4\ x_1$ | T/K | $10^4\ x_1$ | T/K | $10^4\ x_1$ |
|---|---|---|---|---|---|
| 213.15 | 4.59 | 253.15 | 4.63 | 293.15 | 4.66 |
| 223.15 | 4.60 | 263.15 | 4.64 | 298.15 | 4.66 |
| 233.15 | 4.61 | 273.15 | 4.74 | 303.15 | 4.66 |
| 243.15 | 4.62 | 283.15 | 4.75 | 313.15 | 4.67 |
| | | | | 323.15 | 4.67 |

1-Butanol

Four groups (2-5) studied this system and we used all of their data to obtain the following smoothing equation:

$$\ln x_1 = -7.64505 - 0.111660/(T/100K) \quad (5)$$

The standard deviation in ln $x_1$ is 0.026 and $\Delta\bar{H}° = 93$ J mol$^{-1}$ and $\Delta\bar{S}° = -64$ J K$^{-1}$ mol$^{-1}$. Smoothed values at 10 K intervals follow for the range 213 to 323 K.

| T/K | $10^4\ x_1$ | T/K | $10^4\ x_1$ | T/K | $10^4\ x_1$ |
|---|---|---|---|---|---|
| 213.15 | 4.54 | 253.15 | 4.58 | 293.15 | 4.61 |
| 223.15 | 4.55 | 263.15 | 4.59 | 298.15 | 4.61 |
| 233.15 | 4.56 | 273.15 | 4.59 | 303.15 | 4.61 |
| 243.15 | 4.57 | 283.15 | 4.60 | 313.15 | 4.62 |
| | | | | 323.15 | 4.62 |

2-Methyl-1-Propanol (Isobutanol)

This system was only studied by Battino, et al. (8), in the range 274 to 328 K. The smoothing equation is:

$$\ln x_1 = -7.69430 + 0.179798/(T/100K) \quad (6)$$

with a standard deviation of 0.014 in ln $x_1$. Smoothed recommended values at 5 K intervals follow. $\Delta\bar{H}_1° = -149$ J mol$^{-1}$ and $\Delta\bar{S}_1° = -64$ K$^{-1}$ mol$^{-1}$.

continued on following page

| COMPONENTS: | EVALUATOR: |
|---|---|
| (1) Nitrogen; $N_2$; [7727-37-9]<br>(2) Alcohols | Rubin Battino<br>Department of Chemistry<br>Wright State University<br>Dayton, Ohio 45435<br>U.S.A.<br><br>December 1981 |

CRITICAL EVALUATION:

continued

| T/K | $10^4\ x_1$ | T/K | $10^4\ x_1$ | T/K | $10^4\ x_1$ |
|---|---|---|---|---|---|
| 273.15 | 4.86 | 293.15 | 4.84 | 313.15 | 4.82 |
| 278.15 | 4.86 | 298.15 | 4.84 | 318.15 | 4.82 |
| 283.15 | 4.85 | 303.15 | 4.83 | 323.15 | 4.81 |
| 288.15 | 4.85 | 308.15 | 4.83 | 328.15 | 4.81 |

1-Pentanol

The nitrogen/1-pentanol system was studied by four groups (1,3,5,7) but with widely varying results. Thus, the smoothed values based on a least squares fit of the data must be regarded as tentative pending a more definitive study. The smoothing equation in the range 293 to 308K is:

$$\ln x_1 = -6.0966 - 4.3302/(T/100K) \quad (7)$$

with a standard deviation of 0.047 in ln $x_1$. Smoothed values at 5K intervals follow.

| T/K | $10^4\ x_1$ | T/K | $10^4\ x_1$ |
|---|---|---|---|
| 293.15 | 5.14 | 303.15 | 5.39 |
| 298.15 | 5.27 | 308.15 | 5.52 |

1-Octanol

This system was investigated by four groups (3,5,9,10). Makranczy, et al.'s, value (5) was significantly high and was not used. Smoothing of the remaining values in the range 283 to 314 K gave

$$\ln x_1 = -6.18842 - 3.49654/(T/100K) \quad (8)$$

with a standard deviation in ln $x_1$ of 0.044. This is a rather large error and the smoothed results should be considered tentative.

| T/K | $10^4\ x_1$ | T/K | $10^4\ x_1$ |
|---|---|---|---|
| 283.15 | 5.97 | 298.15 | 6.35 |
| 288.15 | 6.10 | 303.15 | 6.48 |
| 293.15 | 6.23 | 308.15 | 6.60 |
| | | 313.15 | 6.72 |

1-Decanol

Makranczy, et al. (5), and Wilcock, et al. (10), both studied this system, but the former's result was judged to be too high. A least squares fitting of Wilcock, et al.'s, three points in the range 283 to 313 K gave

$$\ln x_1 = -6.65033 - 1.98173/(T/100K) \quad (9)$$

with a standard deviation of 0.0029 in ln $x_1$. Smoothed recommended values follow. $\Delta\bar{H}° = 1648$ J mol$^{-1}$ and $\Delta\bar{S}° = -55$ J K$^{-1}$ mol$^{-1}$ for this system.

continued on following page

| COMPONENTS: | EVALUATOR: |
|---|---|
| (1) Nitrogen; $N_2$; [7727-37-9]<br>(2) Alcohols | Rubin Battino<br>Department of Chemistry<br>Wright State University<br>Dayton, Ohio 45435<br>U.S.A. December 1981 |

CRITICAL EVALUATION: continued

| T/K | $10^4\ x_1$ | T/K | $10^4\ x_1$ |
|---|---|---|---|
| 283.15 | 6.42 | 298.15 | 6.65 |
| 288.15 | 6.50 | 303.15 | 6.73 |
| 293.15 | 6.58 | 308.15 | 6.80 |
| | | 313.15 | 6.87 |

1-Alkanols

The figure below shows the results for Boyer and Bircher (3 ) □ , and Makranczy, et al. (5 ), ○.
At a glance it is obvious that the Makranczy, et al. (5 ), values appear to be always significantly higher than the results for other workers. We do not know the reason for this. As an aid we list below the figure the recommended values at 298.15 K for the 1-alkanols up through 1-decanol. A least-squares

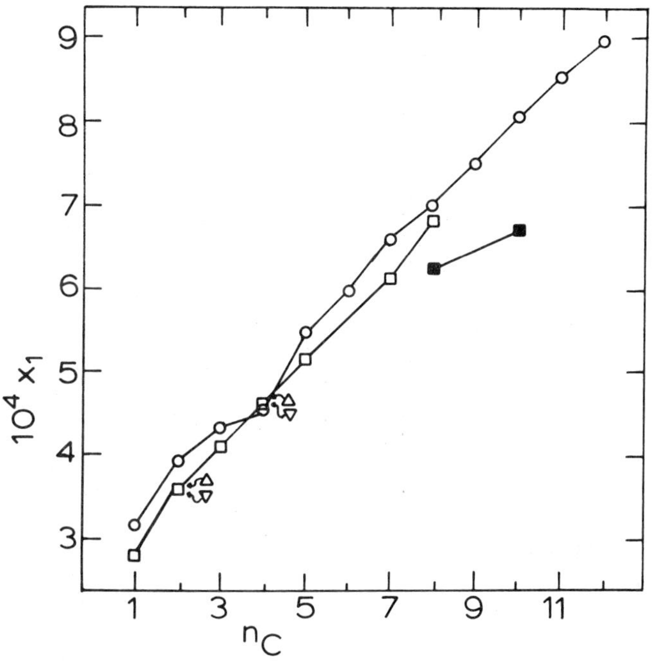


For the above figure the solid squares are for Wilcock et al. (10), the triangles for Katayama and Nitta (4), and the inverted triangles for Kretschmer at al. (2).

continued on following page

| COMPONENTS: | EVALUATOR: |
|---|---|
| (1) Nitrogen; $N_2$; [7727-37-9]<br>(2) Alcohols | Rubin Battino<br>Department of Chemistry<br>Wright State University<br>Dayton, Ohio 45435<br>U.S.A. December 1981 |

CRITICAL EVALUATION:

continued

298.15 K

| Solvent | $10^4\ x_1$ |
|---|---|
| methanol | 2.76 |
| ethanol | 3.57 |
| 1-propanol | 4.02 |
| 1-butanol | 4.61 |
| 1-pentanol | 5.27 |
| 1-hexanol | 5.84 |
| 1-heptanol | 6.10 |
| 1-octanol | 6.35 |
| 1-decanol | 6.65 |

fitting of the mole fraction solubility versus carbon number, n, yields:

$$x_1 = 2.2725 \times 10^{-4} + 4.4885 \times 10^{-5}\ n \qquad (10)$$

with a standard deviation in $x_1$ of $3.0 \times 10^{-5}$. (The smoothed values from equation (10) are obviously different from the recommended values listed in the table.)


References

1. Just, G. *Z. Phys. Chem.* 1901, *37*, 342-67.
2. Kretschmer, C. B.; Nowakowska, J.; Wiebe, R. *Ind. Eng. Chem.* 1946, *38*, 506-9.
3. Boyer, F. L.; Bircher, L. J. *J. Phys. Chem.* 1960, *64*, 1330-1.
4. Katayama, T.; Nitta, T. *J. Chem. Eng. Data* 1976, *21*, 194-6.
5. Makranczy, J.; Rusz, L.; Balog-Megyery, K. *Hung. J. Ind. Chem.* 1979, *7*, 41-6.
6. Metschl, J. *J. Phys. Chem.* 1924, *28*, 417-37.
7. Gjaldbaek, J.C.; Niemann, H. *Acta Chem. Scand.* 1958, *12*, 1015-23.
8. Battino, R.; Evans, F. D.; Danforth, W. F.; Wilhelm, E. *J. Chem. Thermodyn.* 1971, *3*, 743-51.
9. Ijams, C. C. Ph.D. Thesis 1941, Vanderbilt University, Nashville, TN.
10. Wilcock, R. J.; Battino, R.; Danforth, W. F.; Wilhelm, E. *J. Chem. Thermodyn.* 1978, *10*, 817-22.

| COMPONENTS: | ORIGINAL MEASUREMENTS: |
|---|---|
| (1) Nitrogen; $N_2$; [7727-37-9]<br>(2) Methanol; $CH_4O$; [67-56-1] | Just, G.<br>*Z. Phys. Chem.* 1901, *37*, 342-67. |
| VARIABLES:<br>T/K = 293-298<br>P/kPa = 101.325 | PREPARED BY:<br>M. E. Derrick<br>H. L. Clever |

EXPERIMENTAL VALUES:

| T/K | Mol Fraction $x_1 \times 10^4$ | Bunsen Coefficient $\alpha$ | Ostwald Coefficient L |
|---|---|---|---|
| 293.15 | 2.27 | 0.1256 | 0.1348 |
| 298.15 | 2.36 | 0.1296 | 0.1415 |

The author measured the Ostwald coefficient at a pressure of about 746 mmHg. The compiler assumed the Ostwald coefficient to be independent of pressure, and calculated the mole fraction and Bunsen coefficient values at 101.325 kPa (1 atm) partial pressure of the gas.

AUXILIARY INFORMATION

METHOD/APPARATUS/PROCEDURE:

An Ostwald apparatus as modified by Timofejew (1), and Steiner (2) is used. The apparatus consists of a gas buret, an absorption flask, and a mercury manometer. The system is thermostated with a water jacket.

The gas is introduced into the degassed liquid. The gas volume absorbed is determined by the gas buret. The solvent volume is determined at the end of the experiment by pouring the solvent into a graduated flask.

SOURCE AND PURITY OF MATERIALS:

(1) Nitrogen. Prepared by the reaction of sodium nitrite and ammonium nitrate in aqueous solution in the presence of calcium dichromate.

(2) Methanol. No information.

ESTIMATED ERROR:

$\delta L/L = 0.03$ (compiler)

REFERENCES:

1. Timofejew, W. *Z. Physik. Chem.* 1890, *6*, 141.

2. Steiner, *Ann. Phys. (Leipzig)* 1894, *52*, 275.

**COMPONENTS:**

(1) Nitrogen; $N_2$; [7727-37-9]

(2) Methanol; $CH_4O$; [67-56-1]

**ORIGINAL MEASUREMENTS:**

Kretschmer, C. B.; Nowakowska, J.; Wiebe, R.

*Ind. Eng. Chem.* 1946, *38*, 506-9.

**VARIABLES:**

T/K = 248.15 - 323.15
P/kPa = 101.325

**PREPARED BY:**

P. L. Long
H. L. Clever

**EXPERIMENTAL VALUES:**

| T/K | Mol Fraction $x_1 \times 10^3$ | Bunsen Coefficient $\alpha$ | Ostwald Coefficient L |
|---|---|---|---|
| 248.15 | 0.2708 | 0.1580 | 0.1435 |
| 273.15 | 0.2703 | 0.1532 | 0.1532 |
| 298.15 | 0.2739 | 0.1507 | 0.1645 |
| 323.15 | 0.2795 | 0.1492 | 0.1765 |

The Ostwald coefficients were measured at a total pressure of 101.325 kPa (1 atm). The compiler calculated the Bunsen coefficient and mole fraction values for an oxygen partial pressure of 101.325 kPa (1 atm) assuming the Ostwald coefficient was independent of pressure.

Smoothed Data: $\ln x_1 = -8.0863 - 0.3326/(T/100\ \mathrm{K})$

Standard error about the regression line = $2.52 \times 10^{-6}$

| T/K | Mol Fraction $x_1 \times 10^3$ |
|---|---|
| 253.15 | 0.270 |
| 263.15 | 0.271 |
| 273.15 | 0.272 |
| 283.15 | 0.274 |
| 293.15 | 0.275 |
| 298.15 | 0.275 |
| 303.15 | 0.276 |
| 313.15 | 0.277 |
| 323.15 | 0.278 |

## AUXILIARY INFORMATION

**METHOD/APPARATUS/PROCEDURE:**

The apparatus was similar to that of Horiuti (1), it consisted of a gas buret, manometer, and absorption pipet equipped with a magnetic stirrer. All volumes were calibrated by mercury displacement.

The solvent was partly degassed by boiling under reflux. About 40 $cm^3$ of solvent was admitted into the absorption pipet, and pumped on to complete the degassing. Dry gas, measured in the buret, was admitted, final equilibrium was established after 1 to 3 hours stirring. During the solution process the total pressure was kept at one atm by additions of dry gas.

The solubility value was corrected for the increase of solvent volume with gas absorption, and for the nitrogen impurity in the oxygen.

**SOURCE AND PURITY OF MATERIALS:**

(1) Nitrogen. Commercial, but source not given. Passed through alkaline pyrogallol to remove $O_2$, passed through Drierite to dry.

(2) Methanol. Source not given. Distilled, dried with metallic Al, b.p. (760 mmHg)/°C 64.50, density, $\rho_4^{25}/\mathrm{g\ cm^{-3}}$ 0.78650.

**ESTIMATED ERROR:**

Gas buret $\delta V/cm^3 = 0.005$
Average deviation from the mean 0.3%, maximum deviation 0.9% (authors).

**REFERENCES:**

1. Horiuti, J.
*J. Sci Papers Inst. Phys. Chem. Res. (Tokyo)* 1931, *17*, 125.

| COMPONENTS: | ORIGINAL MEASUREMENTS: |
|---|---|
| (1) Nitrogen; $N_2$; [7727-37-9]<br>(2) Methanol; $CH_4O$; [67-56-1] | Boyer, F. L.; Bircher, L. J.<br>*J. Phys. Chem.* 1960, *64*, 1330-1. |
| VARIABLES:<br>T/K = 298.15<br>P/kPa = 101.325 | PREPARED BY:<br>M. E. Derrick |

EXPERIMENTAL VALUES:

| T/K | Mol Fraction $X_1 \times 10^3$ | Bunsen Coefficient $\alpha$ | Ostwald Coefficient L |
|---|---|---|---|
| 298.15 | 0.279 | 0.153 | 0.167 ± 0.003 |

The Bunsen coefficient was calculated by the compiler.

The observed mole fraction solubility was taken from Boyer's dissertation (2).

The Dissertation gives the equations $\log X_1 = -3.594 + 0.449 \log C$ at 298.15 K where C is the number of carbon atoms in the alcohol. The mole fraction solubility given in the original paper was calculated from the 298.15 K equation.

AUXILIARY INFORMATION

METHOD/APPARATUS/PROCEDURE:

Commercial Van Slyke blood gas apparatus (E. H. Sargent Co.) was modified by the authors for their use.

The total pressure of gas and solvent vapor in the solution chamber was adjusted to a pressure of one atm. The pressure was maintained at one atm during the solution process. The saturated solution was transferred to a bulb below the lower stopcock of the extraction vessel and sealed off. The gas and solvent vapor were then brought to volume over mercury. See (1) for details of the extraction procedure.

SOURCE AND PURITY OF MATERIALS:

(1) Nitrogen. Air Reduction Co. 99.7 per cent nitrogen.

(2) Methanol. Source not given. Treated by standard methods to remove aldehydes and ketones, then dried and distilled.

ESTIMATED ERROR:

$\delta T/K = 0.01$

$\delta L/cm^3 = 0.003$

REFERENCES:

1. Peters, J. P.; Van Slyke, D. D. "Quantitative Clinical Chemistry" Williams and Wilkins Co., Baltimore, MD, 1932, Volume II.
2. Boyer, F. L. Ph.D. Dissertation Vanderbilt U., 1959, Nashville, TN

**COMPONENTS:**

(1) Nitrogen; $N_2$; [7727-37-9]

(2) Methanol; $CH_4O$; [67-56-1]

**ORIGINAL MEASUREMENTS:**

Katayama, T.; Nitta, T.

*J. Chem. Eng. Data* 1976, *21*, 194-6.

**VARIABLES:**

T/K = 213-298

**PREPARED BY:**

C.L. Young

**EXPERIMENTAL VALUES:**

| T/K | Ostwald coefficient, L | Henry's Constant /atm | Mole fraction* of nitrogen in liquid, $x_{N_2}$ |
|---|---|---|---|
| 298.15 | 0.164 | 3660 | 0.000273 |
| 273.15 | 0.152 | 3720 | 0.000269 |
| 253.15 | 0.145 | 3700 | 0.000270 |
| 233.15 | 0.139 | 3640 | 0.000275 |
| 213.15 | 0.134 | 3520 | 0.000284 |

* at a partial pressure of 1 atmosphere

**AUXILIARY INFORMATION**

**METHOD /APPARATUS/PROCEDURE:**

Volumetric apparatus with equilibrium cell of approximately 0.08 litres. Magnetic stirrer. Solvent carefully degassed. Equilibrium established in 1-1.5 hours. Details in ref. (1).

**SOURCE AND PURITY OF MATERIALS:**

(1) Osaka Oxygen Co. sample, purity 99.9995 mole per cent.

(2) Nakarai Chemicals sample, purity, 99.9 mole per cent

**ESTIMATED ERROR:**

$\delta T/K = \pm 0.12$ at T= 213.15K, ±0.05 at other temperatures; $\delta x_{N_2} = \pm 1\%$

**REFERENCES:**

1. Nitta, T.; Tatsuishi, A.; Katayama, T. *J. Chem. Eng. Jpn.* 1973 *6*, 475.

| COMPONENTS: | ORIGINAL MEASUREMENTS: |
|---|---|
| (1) Nitrogen; $N_2$; [7727-37-9]<br>(2) Methanol; $CH_4O$; [67-56-1] | Makranczy, J.; Rusz, L.; Balog-Megyery, K.<br>*Hung. J. Ind. Chem.* 1979, *7*, 41-6. |
| VARIABLES: | PREPARED BY: |
| T/K = 298 | C.L. Young |

EXPERIMENTAL VALUES:

| T/K | $P^+$/kPa | Ostwald coefficient | Mole fraction of nitrogen*, $x_{N_2}$ |
|---|---|---|---|
| 298.15 | 101.3 | 0.186 | 0.0003096 |

* calculated by compiler

\+ partial pressure of nitrogen.

AUXILIARY INFORMATION

| METHOD/APPARATUS/PROCEDURE: | SOURCE AND PURITY OF MATERIALS: |
|---|---|
| Volumetric method. The apparatus of Bodor, Bor, Mohai, and Sipos (1) was used. | No details given. |

ESTIMATED ERROR:

$\delta x_{N_2} = \pm 3\%$.

REFERENCES:

1. Bodor, E.; Bor, Gy.; Mohai, B.; Sipos, G. *Veszpremi Vegyip. Egy. Kozl.* 1957, *1*, 55, *Chem. Abstr.* 1961, *55*, 3175h

| COMPONENTS: | ORIGINAL MEASUREMENTS: |
|---|---|
| (1) Nitrogen; $N_2$; [7727-37-9]<br>(2) 1,2-Ethanediol or Ethylene glycol; $C_2H_5O_2$; [107-21-1] | Gjaldbaek, J. C.; Niemann, H.<br>*Acta Chem. Scand.* 1958, *12*, 1015-23. |
| **VARIABLES:** | **PREPARED BY:** |
| T/K = 298.16<br>Total P/kPa = 101.325 | J. Chr. Gjaldbaek |

EXPERIMENTAL VALUES:

| T/K | Mol Fraction $X_1 \times 10^3$ | Bunsen Coefficient $\alpha$ | Ostwald Coefficient L |
|---|---|---|---|
| 298.16 | 0.0342 | 0.01374 | 0.01500 |
| 298.17 | 0.0356 | 0.01428 | 0.01559 |

The mole fraction and Ostwald solubility values were calculated by the compiler.

AUXILIARY INFORMATION

METHOD/APPARATUS/PROCEDURE:

A calibrated all-glass combined manometer and bulb was enclosed in an air thermostat and shaken until equilibrium. Mercury was used for calibration and as the confining liquid. The solvents were degassed in the apparatus. Details are in references 1 and 2.

The absorbed volume of gas was calculated from the initial and final amounts, both saturated with solvent vapor. The amount of solvent was determined by the weight of displaced mercury.

The saturation of the liquid with the gas was carried out close to atmospheric pressure. The solubility values were reported for one atmosphere gas pressure assuming Henry's law is obeyed.

SOURCE AND PURITY OF MATERIALS:

(1) Nitrogen. Dansk Ilt-og Brint Fabrik. Contained less than 0.13 percent oxygen.

(2) 1,2-Ethanediol. Fractionated in a column at low pressure. Normal boiling point 197.30 - 197.42°C, refractive index at 20°C, $n_D$ = 1.4320.

ESTIMATED ERROR:

$\delta T/K = 0.05$
$\delta X_1/X_1 = 0.015$

REFERENCES:

1. Lannung, A. *J. Am. Chem. Soc.* 1930, *52*, 68.

2. Gjaldbaek, J. C. *Acta Chem. Scand.* 1952, *6*, 623.

**COMPONENTS:**

(1) Nitrogen; $N_2$; [7727-37-9]

(2) Ethanol; $C_2H_6O$; [64-17-5]

**ORIGINAL MEASUREMENTS:**

Just, G.

*Z. Phys. Chem.* 1901, *37*, 342 - 367.

**VARIABLES:**

T/K = 293.15, 298.15
P/kPa = 101.325

**PREPARED BY:**

M. E. Derrick
H. L. Clever

**EXPERIMENTAL VALUES:**

| T/K | Mol Fraction $x_1 \times 10^4$ | Bunsen Coefficient $\alpha$ | Ostwald Coefficient L |
|---|---|---|---|
| 293.15 | 3.40 | 0.1304 | 0.1400 |
| 298.15 | 3.44 | 0.1312 | 0.1432 |

The author measured the Ostwald coefficient at a pressure of about 746 mmHg. The compiler assumed the Ostwald coefficient to be independent of pressure, and calculated the mole fraction and Bunsen coefficient values at 101.325 kPa (1 atm) partial pressure of the gas.

**AUXILIARY INFORMATION**

**METHOD/APPARATUS/PROCEDURE:**

An Ostwald apparatus as modified by Timofejew (1), and Steiner (2) is used. The apparatus consists of a gas buret, an absorption flask, and a mercury manometer. The system is thermostated with a water jacket.

The gas is introduced into the degassed liquid. The gas volume absorbed is determined by the gas buret. The solvent volume is determined at the end of the experiment by pouring the solvent into a graduated flask.

**SOURCE AND PURITY OF MATERIALS:**

(1) Nitrogen. Prepared by the reaction of sodium nitrite and ammonium nitrate in aqueous solution in the presence of calcium dichromate.

(2) Ethanol. No information.

**ESTIMATED ERROR:**

$\delta L/L = 0.03$ (compiler)

**REFERENCES:**


1. Timofejew, W. *Z. Physik. Chem.* 1890, *6*, 141.

2. Steiner, *Ann. Phys. (Leipzig)* 1894, *52*, 275.

**COMPONENTS:**

(1) Nitrogen; $N_2$; [7727-37-9]

(2) Ethanol; $C_2H_6O$; [64-17-5]

**ORIGINAL MEASUREMENTS:**

Metschl, J.

*J. Phys. Chem.* 1924, *28*, 417-37.

**VARIABLES:**

T/K = 298.15
P/kPa = 101.325

**PREPARED BY:**

M. E. Derrick
H. L. Clever

**EXPERIMENTAL VALUES:**

| T/K | Mol Fraction $x_1 \times 10^4$ | Bunsen Coefficient $\alpha$ | Ostwald Coefficient L |
|---|---|---|---|
| 298.15 | 3.40 | 0.130 | 0.142 |

The compiler calculated the mole fraction and Ostwald coefficient values.

**AUXILIARY INFORMATION**

**METHOD/APPARATUS/PROCEDURE:**

The degassed solvent is saturated with nitrogen at 6 to 10 different pressures between 1.5 and 4.5 atm.

The pressure is reduced to one atm, and the "supersaturated gas" is shaken out of the solvent and measured.

The slope of a plot of the volume of gas reduced to standard conditions shaken out of 100 $cm^3$ of solvent *vs.* the pressure of saturation is nearly identical to the solubility-pressure graph (Henry's law). The slope of the curve is used to obtain the Bunsen coefficient. From the method of calculation the value represents the volume of gas shaken out of the solvent when the pressure is reduced from 2 to 1 atm.

**SOURCE AND PURITY OF MATERIALS:**

(1) Nitrogen. Prepared by the reaction of sodium nitrite and ammonium nitrate in the presence of calcium dichromate in aqueous solution.

(2) Ethanol. Treated with metallic sodium and distilled. Specific gravity 0.7992 at 15°C. About 98 per cent pure.

**ESTIMATED ERROR:**

**REFERENCES:**

| COMPONENTS: | ORIGINAL MEASUREMENTS: |
|---|---|
| (1) Nitrogen; $N_2$; [7727-37-9]<br>(2) Ethanol; $C_2H_6O$; [64-17-5] | Kretschmer, C. B.; Nowakowska, J.; Wiebe, R.<br>*Ind. Eng. Chem.* 1946, *38*, 506-9. |
| VARIABLES:<br>T/K = 248.15 - 323.15<br>P/kPa = 101.325 | PREPARED BY:<br>P. L. Long<br>H. L. Clever |

EXPERIMENTAL VALUES:

| T/K | Mol Fraction $x_1 \times 10^3$ | Bunsen Coefficient $\alpha$ | Ostwald Coefficient L |
|---|---|---|---|
| 248.15 | 0.3498 | 0.1409 | 0.1280 |
| 273.15 | 0.3545 | 0.1391 | 0.1391 |
| 298.15 | 0.3572 | 0.1364 | 0.1489 |
| 323.15 | 0.3655 | 0.1358 | 0.1606 |

The Ostwald coefficients were measured at a total pressure of 101.325 kPa (1 atm). The compiler calculated the Bunsen coefficient and mole fraction values for an oxygen partial pressure of 101.325 kPa (1 atm) assuming the Ostwald coefficient was independent of pressure.

Smoothed Data: $\ln x_1 = -7.7832 - 0.4397/(T/100\ K)$

Standard error about the regression line = $2.18 \times 10^{-6}$

| T/K | Mol Fraction $x_1 \times 10^3$ |
|---|---|
| 253.15 | 0.350 |
| 263.15 | 0.3525 |
| 273.15 | 0.355 |
| 283.15 | 0.357 |
| 293.15 | 0.359 |
| 298.15 | 0.3595 |
| 303.15 | 0.360 |
| 313.15 | 0.362 |
| 323.15 | 0.364 |

AUXILIARY INFORMATION

METHOD/APPARATUS/PROCEDURE:

The apparatus was similar to that of Horiuti (1), it consisted of a gas buret, manometer, and absorption pipet equipped with a magnetic stirrer. All volumes were calibrated by mercury displacement.

The solvent was partly degassed by boiling under reflux. About 40 $cm^3$ of solvent was admitted into the absorption pipet, and pumped on to complete the degassing. Dry gas, measured in the buret, was admitted, final equilibrium was established after 1 to 3 hours stirring. During the solution process the total pressure was kept at one atm by additions of dry gas.

The solubility value was corrected for the increase of solvent volume with gas absorption, and for the nitrogen impurity in the oxygen.

SOURCE AND PURITY OF MATERIALS:

(1) Nitrogen. Commercial, but source not given. Passed through alkaline pyrogallol to remove $O_2$, passed through Drierite to dry.

(2) Ethanol. Source not given. Dried with Mg. Density, $\rho/g\ cm^{-3}$ 0.78508.

ESTIMATED ERROR:

Gas buret $\delta V/cm^3 = 0.005$
Average deviation from the mean 0.3%, maximum deviation 0.9% (authors).

REFERENCES:

1. Horiuti, J.
*J. Sci Papers Inst. Phys. Chem. Res. (Tokyo)* 1931, *17*, 125.

| COMPONENTS: | ORIGINAL MEASUREMENTS: |
|---|---|
| (1) Nitrogen; $N_2$; [7727-37-9]<br>(2) Ethanol; $C_2H_6O$; [64-17-5] | Boyer, F. L.; Bircher, L. J.<br>*J. Phys. Chem.* 1960, *64*, 1330-1. |

| VARIABLES: | PREPARED BY: |
|---|---|
| T/K = 298.15<br>P/kPa = 101.325 | M. E. Derrick |

EXPERIMENTAL VALUES:

| T/K | Mol Fraction $X_1 \times 10^3$ | Bunsen Coefficient $\alpha$ | Ostwald Coefficient L |
|---|---|---|---|
| 298.15 | 0.359 | 0.136 | 0.149 $\pm$ 0.002 |

The Bunsen coefficient was calculated by the compiler.

The observed mole fraction solubility was taken from Boyer's dissertation (2).

The Dissertation gives the equations log $X_1$ = -3.594 + 0.449 log C at 298.15 K where C is the number of carbon atoms in the alcohol. The mole fraction solubility given in the original paper was calculated from the 298.15 K equation.

AUXILIARY INFORMATION

METHOD/APPARATUS/PROCEDURE:

Commercial Van Slyke blood gas apparatus (E. H. Sargent Co.) was modified by the authors for their use.

The total pressure of gas and solvent vapor in the solution chamber was adjusted to a pressure of one atm. The pressure was maintained at one atm during the solution process. The saturated solution was transferred to a bulb below the lower stopcock of the extraction vessel and sealed off. The gas and solvent vapor were then brought to volume over mercury. See (1) for details of the extraction procedure.

SOURCE AND PURITY OF MATERIALS:

(1) Nitrogen. Air Reduction Co. 99.7 per cent nitrogen.

(2) Ethanol. Source not given. Treated by standard methods to remove aldehydes and ketones, then dried and distilled.

ESTIMATED ERROR:

$\delta T/K = 0.01$

$\delta L/cm^3 = 0.002$

REFERENCES:

1. Peters, J. P.; Van Slyke, D. D. "Quantitative Clinical Chemistry" Williams and Wilkins Co., Baltimore, M.D., 1932, Volume II.
2. Boyer, F. L. Ph.D. Dissertation Vanderbilt U., 1959, Nashville,TN

| COMPONENTS: | ORIGINAL MEASUREMENTS: |
|---|---|
| (1) Nitrogen; $N_2$; [7727-37-9]<br>(2) Ethanol; $C_2H_6O$; [64-17-5] | Katayama, T.; Nitta, T.<br>*J. Chem.Eng. Data* 1976, *21*, 194-6. |
| VARIABLES: | PREPARED BY: |
| T/K = 213-298 | C.L. Young |

EXPERIMENTAL VALUES:

| T/K | Ostwald coefficient, L | Henry's Constant /atm | Mole fraction* of nitrogen in liquid, $x_{N_2}$ |
|---|---|---|---|
| 298.15 | 0.149 | 2800 | 0.000357 |
| 273.15 | 0.138 | 2840 | 0.000352 |
| 253.15 | 0.131 | 2830 | 0.000353 |
| 233.15 | 0.125 | 2800 | 0.000357 |
| 213.15 | 0.121 | 2700 | 0.000370 |

* at partial pressure of 1 atmosphere

AUXILIARY INFORMATION

METHOD /APPARATUS/PROCEDURE:

Volumetric apparatus with equilibrium cell of approximately 0.08 litres. Magnetic stirrer. Solvent carefully degassed. Equilibrium established in 1-1.5 hours. Details in ref. (1).

SOURCE AND PURITY OF MATERIALS:

(1) Osaka Oxygen Co. sample, purity 99.9995 mole per cent.

(2) Nakarai Chemicals sample, purity 99.9 mole per cent.

ESTIMATED ERROR:

$\delta T/K = \pm 0.12$ at $T = 213.15K$, $\pm 0.05$ at other temperatures; $\delta x_{N_2} = \pm 1\%$

REFERENCES:

1. Nitta, T.; Tatsuishi, A.; Katayama, T. *J. Chem.Eng.Jpn.* 1973 *6*, 475.

| COMPONENTS: | ORIGINAL MEASUREMENTS: |
|---|---|
| (1) Nitrogen; $N_2$; [7727-37-9]<br>(2) Ethanol; $C_2H_6O$; [64-17-5] | Makranczy, J.; Rusz, L.; Balog-Megyery, K.<br>*Hung. J. Ind. Chem.* 1979, *7*, 41-6. |
| VARIABLES: | PREPARED BY: |
| T/K = 298 | C.L. Young |

EXPERIMENTAL VALUES:

| T/K | $P^+$/kPa | Ostwald coefficient | Mole fraction of nitrogen*, $x_{N_2}$ |
|---|---|---|---|
| 298.15 | 101.3 | 0.164 | 0.000393 |

* calculated by compiler

\+ partial pressure of nitrogen

AUXILIARY INFORMATION

METHOD/APPARATUS/PROCEDURE:

Volumetric method. The apparatus of Bodor, Bor, Mohai, and Sipos (1) was used.

SOURCE AND PURITY OF MATERIALS:

No details given.

ESTIMATED ERROR:

$\delta x_{N_2} = \pm 3\%$.

REFERENCES:

1. Bodor, E.; Bor, Gy.; Mohai, B.; Sipos, G.
*Veszpremi Vegyip. Egy. Kozl.* 1957, *1*, 55.
*Chem. Abstr.* 1961, *55*, 3175h

**COMPONENTS:**

(1) Nitrogen; $N_2$; [7727-37-9]

(2) 1-Propanol; $C_3H_8O$; [71-23-8]

**ORIGINAL MEASUREMENTS:**

Gjaldbaek, J. C.; Niemann, H.

*Acta Chem. Scand.* 1958, *12*, 1015-23.

**VARIABLES:**

T/K = 298.06

Total P/kPa = 101.325

**PREPARED BY:**

J. Chr. Gjaldbaek

**EXPERIMENTAL VALUES:**

| T/K | Mol Fraction $X_1 \times 10^3$ | Bunsen Coefficient $\alpha$ | Ostwald Coefficient L |
|---|---|---|---|
| 298.06 | 0.406 | 0.1209 | 0.1319 |
| 298.06 | 0.406 | 0.1208 | 0.1318 |

The mole fraction and Ostwald solubility values were calculated by the compiler.

AUXILIARY INFORMATION

**METHOD/APPARATUS/PROCEDURE:**

A calibrated all-glass combined manometer and bulb was enclosed in an air thermostat and shaken until equilibrium. Mercury was used for calibration and as the confining liquid. The solvents were degassed in the apparatus. Details are in references 1 and 2.

The absorbed volume of gas was calculated from the initial and final amounts, both saturated with solvent vapor. The amount of solvent was determined by the weight of displaced mercury.

The saturation of the liquid with the gas was carried out close to atmospheric pressure. The solubility values were reported for one atmosphere gas pressure assuming Henry's law is obeyed.

**SOURCE AND PURITY OF MATERIALS:**

(1) Nitrogen. Dansk Ilt-og Brint Fabrik. Contained less than 0.13 percent oxygen.

(2) 1-Propanol. Dried and fractionated, boiling point 97.1 - 97.4 °C, refractive index at 20 °C, $n_D$ = 1.3856.

**ESTIMATED ERROR:**

$\delta T/K = 0.05$

$\delta X_1/X_1 = 0.015$

**REFERENCES:**

1. Lannung, A. *J. Am. Chem. Soc.* 1930, *52*, 68.

2. Gjaldbaek, J. C. *Acta Chem. Scand.* 1952, *6*, 623.

COMPONENTS:

(1) Nitrogen; $N_2$; [7727-37-9]

(2) 1-Propanol; $C_3H_8O$; [71-23-8]

ORIGINAL MEASUREMENTS:

Boyer, F. L.; Bircher, L. J.

*J. Phys. Chem.* 1960, *64*, 1330-1.

VARIABLES:

T/K = 298-308

P/kPa = 101.325

PREPARED BY:

M. E. Derrick

EXPERIMENTAL VALUES:

| T/K | Mol Fraction $X_1 \times 10^3$ | Bunsen Coefficient $\alpha$ | Ostwald Coefficient L |
|---|---|---|---|
| 298.15 | 0.409 | 0.122 | 0.133 ± 0.001 |
| 308.15 | 0.437 | 0.129 | 0.146 ± 0.002 |

The Bunsen coefficients were calculated by the compiler.

The observed mole fraction solubilities were taken from Boyer's dissertation (2).

The Dissertation gives the equations $\log X_1 = -3.594 + 0.449 \log C$ at 298.15 K and $\log X_1 = -3.576 + 0.452 \log C$ at 308.15 K where C is the number of carbon atoms in the alcohol. The mole fraction solubility given in the original paper was calculated from the 298.15 K equation.

Smoothed Data: $\Delta G^\circ/J\ mol^{-1} = -RT \ln X_1 = 5016.1 + 48.040\ T$

The equation is tentative.

AUXILIARY INFORMATION

METHOD/APPARATUS/PROCEDURE:

Commercial Van Slyke blood gas apparatus (E. H. Sargent Co.) was modified by the authors for their use.

The total pressure of gas and solvent vapor in the solution chamber was adjusted to a pressure of one atm. The pressure was maintained at one atm during the solution process. The saturated solution was transferred to a bulb below the lower stopcock of the extraction vessel and sealed off. The gas and solvent vapor were then brought to volume over mercury. See (1) for details of the extraction procedure.

SOURCE AND PURITY OF MATERIALS:

(1) Nitrogen. Air Reduction Co. 99.7 per cent nitrogen.

(2) 1-Propanol. Source not given. Treated by standard methods to remove aldehydes and ketones, then dried and distilled.

ESTIMATED ERROR:

$\delta T/K = 0.01$

$\delta L/cm^3 = 0.002$

REFERENCES:

1. Peters, J. P.; Van Slyke, D. D. "Quantitative Clinical Chemistry" Williams and Wilkins Co., Baltimore, MD, 1932, Volume II.
2. Boyer, F. L. Ph.D. Dissertation Vanderbilt U., 1959, Nashville, TN

**COMPONENTS:**

(1) Nitrogen; $N_2$; [7727-37-9]

(2) 1- Propanol; $C_3H_8O$; [71-23-8]

**ORIGINAL MEASUREMENTS:**

Katayama, T.; Nitta, T.

*J. Chem. Eng. Data.* 1976, *21*, 194-6

**VARIABLES:**

T/K = 213-298

**PREPARED BY:**

C.L. Young

**EXPERIMENTAL VALUES:**

| T/K | Ostwald coefficient, L | Henry's Constant /atm | Mole fraction* of nitrogen in liquid, $x_{N_2}$ |
|---|---|---|---|
| 298.15 | 0.131 | 2490 | 0.000402 |
| 273.15 | 0.121 | 2520 | 0.000397 |
| 253.15 | 0.115 | 2500 | 0.000400 |
| 233.15 | 0.111 | 2450 | 0.000408 |
| 213.15 | 0.107 | 2360 | 0.000424 |

* at partial pressure of 1 atmosphere

**AUXILIARY INFORMATION**

**METHOD /APPARATUS/PROCEDURE:**

Volumetric apparatus with equilibrium cell of approximately 0.08 litres. Magnetic stirrer. Solvent carefully degassed. Equilibrium established in 1-1.5 hours. Details in ref. (1).

**SOURCE AND PURITY OF MATERIALS:**

(1) Osaka Oxygen Co. sample, purity 99.9995 mole per cent.

(2) Nakarai Chemicals sample, distilled, purity 99.9 mole per cent.

**ESTIMATED ERROR:**

$\delta T/K = \pm 0.12$ at T = 213.15 K, ±0.05 at other temperatures; $\delta x_{N_2} = \pm 1\%$

**REFERENCES:**

1. Nitta, T.; Tatsuishi. A.; Katayama, T. *J.Chem.Eng. Jpn.* 1973, *6*, 475.

| COMPONENTS: | ORIGINAL MEASUREMENTS: |
|---|---|
| (1) Nitrogen; $N_2$; [7727-37-9]<br>(2) 1-Propanol; $C_3H_8O$; [71-2-8] | Makranczy, J.; Rusz, L.; Balog-Megyery, K.<br>*Hung. J. Ind. Chem.* 1979, *7*, 41-6. |
| VARIABLES: | PREPARED BY: |
| T/K = 298 | C.L. Young. |

EXPERIMENTAL VALUES:

| T/K | $P^+$/kPa | Ostwald coefficient | Mole fraction of nitrogen*, $x_{N_2}$ |
|---|---|---|---|
| 298.15 | 101.3 | 0.141 | 0.000433 |

* calculated by compiler

\+ partial pressure of nitrogen

AUXILIARY INFORMATION

| METHOD/APPARATUS/PROCEDURE: | SOURCE AND PURITY OF MATERIALS: |
|---|---|
| Volumetric method. The apparatus of Bodor, Bor, Mohai, and Sipos (1) was used. | No details given. |
| | ESTIMATED ERROR: |
| | $\delta x_{N_2} = \pm 3\%$ |
| | REFERENCES: |
| | 1. Bodor, E.; Bor, Gy.; Mohai, B.; Sipos, G. *Veszpremi Vegyip. Egy. Kozl.* 1957, *1*, 55. *Chem. Abstr.* 1961, *55*, 3175h. |

**COMPONENTS:**

(1) Nitrogen; $N_2$; [7727-37-9]

(2) 2-Propanol; $C_3H_8O$; [67-63-0]

**ORIGINAL MEASUREMENTS:**

Kretschmer, C. B.; Nowakowska, J.; Wiebe, R.

*Ind. Eng. Chem.* 1946, *38*, 506-9.

**VARIABLES:**

T/K = 273.15 - 323.15
P/kPa = 101.325

**PREPARED BY:**

P. L. Long
H. L. Clever

**EXPERIMENTAL VALUES:**

| T/K | Mol Fraction $x_1 \times 10^3$ | Bunsen Coefficient $\alpha$ | Ostwald Coefficient L |
|---|---|---|---|
| 273.15 | 0.4523 | 0.1353 | 0.1353 |
| 298.15 | 0.4607 | 0.1342 | 0.1465 |
| 323.15 | 0.4821 | 0.1367 | 0.1617 |

The Ostwald coefficients were measured at a total pressure of 101.325 kPa (1 atm). The compiler calculated the Bunsen coefficient and mole fraction values for an oxygen partial pressure of 101.325 kPa (1 atm) assuming the Ostwald coefficient was independent of pressure.

Smoothed Data: $\ln x_1 = -7.2995 - 1.1105/(T/100K)$

Standard error about the regression line = $6.13 \times 10^{-6}$

| T/K | Mol Fraction $x_1 \times 10^3$ |
|---|---|
| 273.15 | 0.450 |
| 283.15 | 0.457 |
| 293.15 | 0.463 |
| 298.15 | 0.466 |
| 303.15 | 0.469 |
| 313.15 | 0.474 |
| 323.15 | 0.479 |

AUXILIARY INFORMATION

**METHOD/APPARATUS/PROCEDURE:**

The apparatus was similar to that of Horiuti (1), it consisted of a gas buret, manometer, and absorption pipet equipped with a magnetic stirrer. All volumes were calibrated by mercury displacement.

The solvent was partly degassed by boiling under reflux. About 40 $cm^3$ of solvent was admitted into the absorption pipet, and pumped on to complete the degassing. Dry gas, measured in the buret, was admitted, final equilibrium was established after 1 to 3 hours stirring. During the solution process the total pressure was kept at one atm by additions of dry gas.

The solubility value was corrected for the increase of solvent volume with gas absorption, and for the nitrogen impurity in the oxygen.

**SOURCE AND PURITY OF MATERIALS:**

(1) Nitrogen. Commercial, but source not given. Passed through alkaline pyrogallol to remove $O_2$, passed through Drierite to dry.

(2) 2-Propanol. Source not given. Distilled, dried with metallic Al, b.p.(760 mmHg)/°C 82.24, density, $\rho_4^{25}/g\ cm^{-3}$ 0.78081.

**ESTIMATED ERROR:**

Gas buret $\delta V/cm^3 = 0.005$
Average deviation from the mean 0.3%, maximum deviation 0.9% (authors).

**REFERENCES:**

1. Horiuti, J.
*J. Sci. Papers Inst. Phys. Chem. Res. (Tokyo)* 1931, *17*, 125.

COMPONENTS:

(1) Nitrogen; $N_2$; [7727-37-9]

(2) 2-Propanol; $C_3H_8O$; [67-63-0]

ORIGINAL MEASUREMENTS:

Katayama, T.; Nitta, T.
*J. Chem. Eng. Data.* 1976, *21*, 194-6.

VARIABLES:

T/K = 213-298

PREPARED BY:

C.L. Young.

EXPERIMENTAL VALUES:

| T/K | Ostwald coefficient, L | Henry's Constant /atm | Mole fraction* of nitrogen in liquid, $x_{N_2}$ |
|---|---|---|---|
| 298.15 | 0.147 | 2170 | 0.000461 |
| 273.15 | 0.135 | 2210 | 0.000452 |
| 253.15 | 0.128 | 2210 | 0.000452 |
| 233.15 | 0.121 | 2190 | 0.000457 |
| 213.15 | 0.118 | 2090 | 0.000478 |

* at a partial pressure of 1 atmosphere

AUXILIARY INFORMATION

METHOD /APPARATUS/PROCEDURE:

Volumetric apparatus with equilibrium cell of approximately 0.08 litres. Magnetic stirrer. Solvent carefully degassed. Equilibrium established in 1-1.5 hours. Details in ref. (1).

SOURCE AND PURITY OF MATERIALS:

(1) Osaka Oxygen Co. sample, purity 99.9995 mole per cent.

(2) Nakarai Chemicals sample, purity 99.9 mole per cent.

ESTIMATED ERROR:

$\delta T/K = \pm 0.12$ at $T = 213.15K$, $\pm 0.05$ at other temperatures; $\delta x_{N_2} = \pm 1\%$

REFERENCES:

1. Nitta, T.; Tatsuishi, A.; Katayama, T. *J. Chem. Eng. Jpn.* 1973, *6*, 475.

| COMPONENTS: | ORIGINAL MEASUREMENTS: |
|---|---|
| (1) Nitrogen; $N_2$; [7727-37-9]<br>(2) 1-Butanol; $C_4H_{10}O$; [71-36-3] | Kretschmer, C. B.; Nowakowska, J.; Wiebe, R.<br>*Ind. Eng. Chem.* 1946, *38*, 506-9. |

| VARIABLES: | PREPARED BY: |
|---|---|
| T/K = 273.15 - 323.15<br>P/kPa = 101.325 | P. L. Long<br>H. L. Clever |

EXPERIMENTAL VALUES:

| T/K | Mol Fraction $x_1 \times 10^3$ | Bunsen Coefficient $\alpha$ | Ostwald Coefficient L |
|---|---|---|---|
| 273.15 | 0.4312 | 0.1075 | 0.1075 |
| 298.15 | 0.4605 | 0.1122 | 0.1225 |
| 323.15 | 0.4821 | 0.1148 | 0.1358 |

The Ostwald coefficients were measured at a total pressure of 101.325 kPa (1 atm). The compiler calculated the Bunsen coefficient and mole fraction values for an oxygen partial pressure of 101.325 kPa (1 atm) assuming the Ostwald coefficient was independent of pressure.

Smoothed Data: $\ln x_1 = -7.0243 - 1.9750/(T/100K)$

Standard error about the regression line = $1.98 \times 10^{-6}$

| T/K | Mol Fraction $x_1 \times 10^3$ |
|---|---|
| 273.15 | 0.4319 |
| 283.15 | 0.4431 |
| 293.15 | 0.4537 |
| 298.15 | 0.4589 |
| 303.15 | 0.4639 |
| 313.15 | 0.4737 |
| 323.15 | 0.4830 |

AUXILIARY INFORMATION

METHOD/APPARATUS/PROCEDURE:

The apparatus was similar to that of Horiuti (1), it consisted of a gas buret, manometer, and absorption pipet equipped with a magnetic stirrer. All volumes were calibrated by mercury displacement.

The solvent was partly degassed by boiling under reflux. About 40 $cm^3$ of solvent was admitted into the absorption pipet, and pumped on to complete the degassing. Dry gas, measured in the buret, was admitted, final equilibrium was established after 1 to 3 hours stirring. During the solution process the total pressure was kept at one atm by additions of dry gas.

The solubility value was corrected for the increase of solvent volume with gas absorption, and for the nitrogen impurity in the oxygen.

SOURCE AND PURITY OF MATERIALS:

(1) Nitrogen. Commercial, but source not given. Passed through alkaline pyrogallol to remove $O_2$, passed through Drierite to dry.

(2) 1-Butanol. Source not given. Distilled, dried with metallic Al, b.p. (760 mmHg)/°C 117.67, density, $\rho_4^{25}/g\ cm^{-3}$ 0.80573.

ESTIMATED ERROR:

Gas buret $\delta V/cm^3 = 0.005$
Average deviation from the mean 0.3%, maximum deviation 0.9% (authors).

REFERENCES:

1. Horiuti, J.
*J. Sci Papers Inst. Phys. Chem. Res. (Tokyo)* 1931, *17*, 125.

| COMPONENTS: | ORIGINAL MEASUREMENTS: |
|---|---|
| (1) Nitrogen; $N_2$; [7727-37-9]<br>(2) 1-Butanol; $C_4H_{10}O$; [71-36-3] | Boyer, F. L.; Bircher, L. J.<br>*J. Phys. Chem.* 1960, *64*, 1330-1. |
| VARIABLES:<br>T/K = 298-308<br>P/kPa = 101.325 | PREPARED BY:<br>M. E. Derrick |

EXPERIMENTAL VALUES:

| T/K | Mol Fraction $X_1 \times 10^3$ | Bunsen Coefficient $\alpha$ | Ostwald Coefficient L |
|---|---|---|---|
| 298.15 | 0.460 | 0.112 | 0.122 ± 0.001 |
| 308.15 | 0.473 | 0.114 | 0.129 ± 0.002 |

The Bunsen coefficients were calculated by the compiler.

The observed mole fraction solubilities were taken from Boyer's dissertation (2).

The Dissertation gives the equations log $X_1$ = -3.594 + 0.449 log C at 298.15 K and log $X_1$ = -3.576 + 0.452 log C at 308.15 K where C is the number of carbon atoms in the alcohol. The mole fraction solubility given in the original paper was calculated from the 298.15 K equation.

Smoothed Data: $\Delta G^{o}/J\ mol^{-1} = -RT \ln X_1 = 2085.8 + 56.891\ T$

The equation is tentative.

AUXILIARY INFORMATION

METHOD/APPARATUS/PROCEDURE:

Commercial Van Slyke blood gas apparatus (E. H. Sargent Co.) was modified by the authors for their use.

The total pressure of gas and solvent vapor in the solution chamber was adjusted to a pressure of one atm. The pressure was maintained at one atm during the solution process. The saturated solution was transferred to a bulb below the lower stopcock of the extraction vessel and sealed off. The gas and solvent vapor were then brought to volume over mercury. See (1) for details of the extraction procedure.

SOURCE AND PURITY OF MATERIALS:

(1) Nitrogen. Air Reduction Co. 99.7 per cent nitrogen.

(2) 1-Butanol. Source not given. Treated by standard methods to remove aldehydes and ketones, then dried and distilled.

ESTIMATED ERROR:

$\delta T/K = 0.01$

$\delta L/cm^3 = 0.001$

REFERENCES:

1. Peters, J. P.; Van Slyke, D. D. "Quantitative Clinical Chemistry" Williams and Wilkins Co., Baltimore, MD, 1932, Volume II.
2. Boyer, F. L. Ph.D. Dissertation Vanderbilt U., 1959, Nashville, TN

| COMPONENTS: | ORIGINAL MEASUREMENTS: |
|---|---|
| (1) Nitrogen; $N_2$; [7727-37-9]<br><br>(2) 1-Butanol; $C_4H_{10}O$; [71-36-3] | Katayama, T.; Nitta, T.<br><br>*J. Chem. Eng. Data.* 1976, *21*, 194-6. |
| VARIABLES:<br>T/K = 213-298 | PREPARED BY:<br>C.L. Young. |

EXPERIMENTAL VALUES:

| T/K | Ostwald coefficient, L | Henry's Constant /atm | Mole fraction* of nitrogen in liquid, $x_{N_2}$ |
|---|---|---|---|
| 298.15 | 0.123 | 2170 | 0.000461 |
| 273.15 | 0.114 | 2190 | 0.000457 |
| 253.15 | 0.107 | 2200 | 0.000455 |
| 233.15 | 0.101 | 2180 | 0.000459 |
| 213.15 | 0.0954 | 2150 | 0.000465 |

* at a partial pressure of 1 atmosphere

AUXILIARY INFORMATION

METHOD /APPARATUS/PROCEDURE:

Volumetric apparatus with equilibrium cell of approximately 0.08 litres. Magnetic stirrer. Solvent carefully degassed. Equilibrium established in 1-1.5 hours. Details in ref. (1).

SOURCE AND PURITY OF MATERIALS:

(1) Osaka Oxygen Co. sample, purity 99.9995 mole per cent.

(2) Nakarai Chemicals sample, distilled purity 99.9 mole per cent.

ESTIMATED ERROR:

$\delta T/K = \pm 0.12$ at T = 213.15K, $\pm 0.05$ at other temperatures; $\delta x_{N_2} = \pm 1\%$

REFERENCES:

1. Nitta, T.; Tatsuishi. A.; Katayama, T. *J. Chem. Eng. Jpn.* 1973, *6*, 475.

**COMPONENTS:**

(1) Nitrogen; $N_2$; [7727-37-9]

(2) 1-Butanol; $C_4H_{10}O$; [71-36-3]

**ORIGINAL MEASUREMENTS:**

Makranczy, J.; Rusz, L.; Balog-Megyery, K.

*Hung. J. Ind. Chem.* 1979, *7*, 41-6.

**VARIABLES:**

T/K = 298

**PREPARED BY:**

C.L. Young

**EXPERIMENTAL VALUES:**

| T/K | $P^+$/kPa | Ostwald coefficient | Mole fraction of nitrogen*, $x_{N_2}$ |
|---|---|---|---|
| 298.15 | 101.3 | 0.120 | 0.000451 |

* calculated by compiler

\+ partial pressure of nitrogen.

AUXILIARY INFORMATION

**METHOD/APPARATUS/PROCEDURE:**

Volumetric method. The apparatus of Bodor, Bor, Mohai, and Sipos (1) was used.

**SOURCE AND PURITY OF MATERIALS:**

No details given.

**ESTIMATED ERROR:**

$\delta x_{N_2} = \pm 3\%$

**REFERENCES:**

1. Bodor, E.; Bor, Gy.; Mohai, B.; Sipos, G. *Veszpremi Vegyip. Egy. Kozl.* 1957, *1*, 55, *Chem. Abstr.* 1961, *55*, 3175h

**COMPONENTS:**

(1) Nitrogen; $N_2$; [7727-37-9]

(2) 2-Methyl-1-propanol or Isobutanol; $C_4H_{10}O$; [78-83-1]

**ORIGINAL MEASUREMENTS:**

Battino, R.; Evans, F. D.; Danforth, W. F.; Wilhelm, E.

*J. Chem. Thermodyn.* 1971, *3*, 743-51.

**VARIABLES:**

T/K = 274.02 - 327.96
P/kPa = 101.325

**PREPARED BY:**

H. L. Clever

**EXPERIMENTAL VALUES:**

| T/K | Mol Fraction $X_1 \times 10^3$ | Bunsen Coefficient $\alpha$ | Ostwald Coefficient L |
|---|---|---|---|
| 274.02 | 0.477 | 0.118 | 0.118 |
| 282.78 | 0.490 | 0.119 | 0.124 |
| 295.83 | 0.495 | 0.120 | 0.130 |
| 298.41 | 0.482 | 0.116 | 0.127 |
| 312.77 | 0.484 | 0.115 | 0.132 |
| 327.96 | 0.475 | 0.112 | 0.134 |

The Bunsen coefficients were calculated by the compiler.
The solubility values were adjusted to a nitrogen partial pressure of 101.325 kPa (1 atm) by Henry's law.

Smoothed Data: $\Delta G^o/J\ mol^{-1} = -RT \ln X_1 = -216.5 + 64.195\ T$

Std. Dev. $\Delta G^o$ = 36.1, Coef. Corr. = 0.9996

$\Delta H^o/J\ mol^{-1} = -216.5$, $\Delta S^o/J\ K^{-1}\ mol^{-1} = -64.195$

| T/K | Mol Fraction $X_1 \times 10^3$ | $\Delta G^o/J\ mol^{-1}$ |
|---|---|---|
| 278.15 | 0.487 | 17,639 |
| 288.15 | 0.485 | 18,281 |
| 298.15 | 0.484 | 18,923 |
| 308.15 | 0.483 | 19,565 |
| 318.15 | 0.481 | 20,207 |
| 328.15 | 0.480 | 20,849 |

**AUXILIARY INFORMATION**

**METHOD/APPARATUS/PROCEDURE:**

The solubility apparatus is based on the design of Morrison and Billett (1) and the version used is described by Battino, Evans, and Danforth (2). The degassing apparatus is that described by Battino, Banzhof, Bogan, and Wilhelm (3).

Degassing. Up to 500 $cm^3$ of solvent is placed in a flask of such size that the liquid is about 4 cm deep. The liquid is rapidly stirred, and vacuum is intermittently applied through a liquid $N_2$ trap until the permanent gas residual pressure drops to 5 microns.

Solubility Determination. The degassed solvent is passed in a thin film down a glass helical tube containing solute gas plus the solvent vapor at a total pressure of one atm. The volume of gas absorbed is found by difference between the initial and final volumes in the buret system. The solvent is collected in a tared flask and weighed.

**SOURCE AND PURITY OF MATERIALS:**

(1) Nitrogen. Matheson Co., Inc. 99.999 per cent.

(2) 2-methyl-1-propanol. Fisher Co. Certified grade (99 mole per cent).

**ESTIMATED ERROR:**

$\delta T/K = 0.03$
$\delta P/mmHg = 0.5$
$\delta X_1/X_1 = 0.01$

**REFERENCES:**


1. Morrison, T. J.; Billett, F. *J. Chem. Soc.* 1948, 2033.
2. Battino, R.; Evans, F. D.; Danforth, W. F. *J. Am. Oil Chem. Soc.* 1968, *45*, 830.
3. Battino, R.; Banzhof, M.; Bogan, M.; Wilhelm, E. *Anal. Chem.* 1971, *43*, 806.

| COMPONENTS: | ORIGINAL MEASUREMENTS: |
|---|---|
| (1) Nitrogen; $N_2$; [7727-37-9]<br>(2) 1-Pentanol; $C_5H_{12}O$; [71-41-0] | Just, G.<br>*Z. Phys. Chem.* 1901, *37*, 342-67. |
| **VARIABLES:** | **PREPARED BY:** |
| T/K = 293-298<br>P/kPa = 101.325 | M. E. Derrick<br>H. L. Clever |

EXPERIMENTAL VALUES:

| T/K | Mol Fraction $x_1 \times 10^4$ | Bunsen Coefficient $\alpha$ | Ostwald Coefficient L |
|---|---|---|---|
| 293.15 | 5.43 | 0.1126 | 0.1208 |
| 298.15 | 5.44 | 0.1122 | 0.1225 |

The author measured the Ostwald coefficient at a pressure of about 746 mmHg. The compiler assumed the Ostwald coefficient to be independent of pressure, and calculated the mole fraction and Bunsen coefficient values at 101.325 kPa (1 atm) partial pressure of the gas.

AUXILIARY INFORMATION

METHOD/APPARATUS/PROCEDURE:

An Ostwald apparatus as modified by Timofejew (1), and Steiner (2) is used. The apparatus consists of a gas buret, an absorption flask, and a mercury manometer. The system is thermostated with a water jacket.

The gas is introduced into the degassed liquid. The gas volume absorbed is determined by the gas buret. The solvent volume is determined at the end of the experiment by pouring the solvent into a graduated flask.

SOURCE AND PURITY OF MATERIALS:

(1) Nitrogen. Prepared by the reaction of sodium nitrite and ammonium nitrate in aqueous solution in the presence of calcium dichromate.

(2) 1-Pentanol. No information.

ESTIMATED ERROR:

$\delta L/L = 0.03$ (compiler)

REFERENCES:

1. Timofejew, W.
*Z. Physik. Chem.* 1890, *6*, 141.

2. Steiner,
*Ann. Phys. (Leipzig)* 1894, *52*, 275.

| COMPONENTS: | ORIGINAL MEASUREMENTS: |
|---|---|
| (1) Nitrogen; $N_2$; [7727-37-9]<br>(2) 1-Pentanol; $C_5H_{12}O$; [71-41-0] | Gjaldbaek, J. C.; Niemann, H.<br>*Acta Chem. Scand.* 1958, *12*, 1015-23. |

| VARIABLES: | PREPARED BY: |
|---|---|
| T/K = 298.06 - 298.15<br>Total P/kPa = 101.325 | J. Chr. Gjaldbaek |

EXPERIMENTAL VALUES:

| T/K | Mol Fraction $X_1 \times 10^3$ | Bunsen Coefficient $\alpha$ | Ostwald Coefficient L |
|---|---|---|---|
| 298.06 | 0.491 | 0.1021 | 0.1114 |
| 298.15 | 0.495 | 0.1013 | 0.1106 |

The mole fraction and Ostwald solubility values were calculated by the compiler.

AUXILIARY INFORMATION

METHOD/APPARATUS/PROCEDURE:

A calibrated all-glass combined manometer and bulb was enclosed in an air thermostat and shaken until equilibrium. Mercury was used for calibration and as the confining liquid. The solvents were degassed in the apparatus. Details are in references 1 and 2.

The absorbed volume of gas was calculated from the initial and final amounts, both saturated with solvent vapor. The amount of solvent was determined by the weight of displaced mercury.

The saturation of the liquid with the gas was carried out close to atmospheric pressure. The solubility values were reported for one atmosphere gas pressure assuming Henry's law is obeyed.

SOURCE AND PURITY OF MATERIALS:

(1) Nitrogen. Dansk Ilt-og Brint Fabrik. Contained less than 0.13 percent oxygen.

(2) 1-Pentanol. Fractionated at about 115 mmHg. Normal boiling point 137.83 - 137.90°C. Refractive index at 20°C, $n_D$ = 1.412.

ESTIMATED ERROR:

$\delta T/K = 0.05$
$\delta X_1/X_1 = 0.015$

REFERENCES:

1. Lannung, A. *J. Am. Chem. Soc.* 1930, *52*, 68.

2. Gjaldbaek, J. C. *Acta Chem. Scand.* 1952, *6*, 623.

**COMPONENTS:**

(1) Nitrogen; $N_2$; [7727-37-9]

(2) 1-Pentanol; $C_5H_{12}O$; [71-41-0]

**ORIGINAL MEASUREMENTS:**

Boyer, F. L.; Bircher, L. J.

*J. Phys. Chem.* 1960, *64*, 1330-1.

**VARIABLES:**

T/K = 298-308

P/kPa = 101.325

**PREPARED BY:**

M. E. Derrick

**EXPERIMENTAL VALUES:**

| T/K | Mol Fraction $X_1 \times 10^3$ | Bunsen Coefficient $\alpha$ | Ostwald Coefficient L |
|---|---|---|---|
| 298.15 | 0.516 | 0.106 | 0.116 ± 0.002 |
| 308.15 | 0.568 | 0.115 | 0.130 ± 0.004 |

The Bunsen coefficients were calculated by the compiler.

The observed mole fraction solubilities were taken from Boyer's dissertation (2).

The Dissertation gives the equations $\log X_1 = -3.594 + 0.449 \log C$ at 298.15 K and $\log X_1 = -3.576 + 0.452 \log C$ at 308.15 K where C is the number of carbon atoms in the alcohol. The mole fraction solubility given in the original paper was calculated from the 298.15 K equation.

Smoothed Data: $\Delta G^\circ/J\ mol^{-1} = -RT \ln X_1 = 7312.1 + 38.408T$

The equation is tentative.

**AUXILIARY INFORMATION**

**METHOD/APPARATUS/PROCEDURE:**

Commercial Van Slyke blood gas apparatus (E. H. Sargent Co.) was modified by the authors for their use.

The total pressure of gas and solvent vapor in the solution chamber was adjusted to a pressure of one atm. The pressure was maintained at one atm during the solution process. The saturated solution was transferred to a bulb below the lower stopcock of the extraction vessel and sealed off. The gas and solvent vapor were then brought to volume over mercury. See (1) for details of the extraction procedure.

**SOURCE AND PURITY OF MATERIALS:**

(1) Nitrogen. Air Reduction Co. 99.7 per cent nitrogen.

(2) 1-Pentanol. Source not given. Treated by standard methods to remove aldehydes and ketones, then dried and distilled.

**ESTIMATED ERROR:**

$\delta T/K$: 0.01

$\delta L/cm^3$: 0.004

**REFERENCES:**

1. Peters, J. P.; Van Slyke, D. D. "Quantitative Clinical Chemistry" Williams and Wilkins Co., Baltimore, MD, 1932, Volume II.
2. Boyer, F. L. Ph.D. Dissertation Vanderbilt U., 1959, Nashville, TN

| COMPONENTS: | ORIGINAL MEASUREMENTS: |
|---|---|
| (1) Nitrogen; $N_2$; [7727-37-9]<br>(2) 1-Pentanol; $C_5H_{12}O$; [71-41-0] or 1-Hexanol; $C_6H_{14}O$; [111-27-3] | Makranczy, J.; Rusz, L.; Balog-Megyery, K.<br>*Hung. J. Ind. Chem.* 1979, *7*, 41-6. |
| VARIABLES: | PREPARED BY: |
| $T/K = 298$ | C.L. Young |

EXPERIMENTAL VALUES:

| T/K | $P^+$/kPa | Ostwald coefficient | Mole fraction of nitrogen*, $x_{N_2}$ |
|---|---|---|---|
| | | 1-Pentanol | |
| 298.15 | 101.3 | 0.123 | 0.000546 |
| | | 1-Hexanol | |
| 298.15 | 101.3 | 0.117 | 0.000599 |

* calculated by compiler

\+ partial pressure of nitrogen

AUXILIARY INFORMATION

| METHOD/APPARATUS/PROCEDURE: | SOURCE AND PURITY OF MATERIALS: |
|---|---|
| Volumetric method. The apparatus of Bodor, Bor, Mohai, and Sipos (1) was used. | No details given. |
| | ESTIMATED ERROR: |
| | $\delta x_{N_2} = \pm 3\%$ |
| | REFERENCES: |
| | 1. Bodor, E.; Bor, Gy.; Mohai, B,; Sipos, G. *Veszpremi Vegyip. Egy. Kozl.* 1957, *1*, 55. *Chem. Abstr.* 1961, *55*, 3175h |

**COMPONENTS:**

(1) Nitrogen; $N_2$; [7727-37-9]

(2) 1-Hexanol; $C_6H_{14}O$; [111-27-3]

**ORIGINAL MEASUREMENTS:**

Boyer, F. L.; Bircher, L. J.

*J. Phys. Chem.* 1960, *64*, 1330-1.

**VARIABLES:**

T/K = 298.15
P/kPa = 101.325

**PREPARED BY:**

M. E. Derrick

**EXPERIMENTAL VALUES:**

| T/K | Mol Fraction $X_1 \times 10^3$ | Bunsen Coefficient $\alpha$ | Ostwald Coefficient L |
|---|---|---|---|
| 298.15 | 0.584 | 0.104 | 0.114 ± 0.003 |

The Bunsen coefficient was calculated by the compiler.

The observed mole fraction solubility was taken from Boyer's dissertation (2).

The Dissertation gives the equations log $X_1$ = -3.594 + 0.449 log C at 298.15 K where C is the number of carbon atoms in the alcohol. The mole fraction solubility given in the original paper was calculated from the 298.15 K equation.

**AUXILIARY INFORMATION**

**METHOD/APPARATUS/PROCEDURE:**

Commercial Van Slyke blood gas apparatus (E. H. Sargent Co.) was modified by the authors for their use.

The total pressure of gas and solvent vapor in the solution chamber was adjusted to a pressure of one atm. The pressure was maintained at one atm during the solution process. The saturated solution was transferred to a bulb below the lower stopcock of the extraction vessel and sealed off. The gas and solvent vapor were then brought to volume over mercury. See (1) for details of the extraction procedure.

**SOURCE AND PURITY OF MATERIALS:**

(1) Nitrogen. Air Reduction Co. 99.7 per cent nitrogen.

(2) 1-Hexanol. Source not given. Treated by standard methods to remove aldehydes and ketones, then dried and distilled.

**ESTIMATED ERROR:**

$\delta T/K = 0.01$
$\delta L/cm^3 = 0.003$

**REFERENCES:**

1. Peters, J. P.; Van Slyke, D. D. "Quantitative Clinical Chemistry" Williams and Wilkins Co., Baltimore, MD, 1932, Volume II.
2. Boyer, F. L. Ph.D. Dissertation Vanderbilt U., 1959, Nashville, TN

| COMPONENTS: | ORIGINAL MEASUREMENTS: |
|---|---|
| (1) Nitrogen; $N_2$; [7727-37-9]<br>(2) Cyclohexanol; $C_6H_{12}O$; [108-93-0] | Gjaldbaek, J. C.; Niemann, H.<br>*Acta Chem. Scand.* 1958, *12*, 1015-23. |
| VARIABLES:<br>T/K = 298.15<br>Total P/kPa = 101.325 | PREPARED BY:<br>J. Chr. Gjaldbaek |

EXPERIMENTAL VALUES:

| T/K | Mol Fraction $X_1 \times 10^3$ | Bunsen Coefficient $\alpha$ | Ostwald Coefficient L |
|---|---|---|---|
| 298.15 | 0.266 | 0.0562 | 0.0613 |

The mole fraction and Ostwald solubility values were calculated by the compiler.

AUXILIARY INFORMATION

METHOD/APPARATUS/PROCEDURE:

A calibrated all-glass combined manometer and bulb was enclosed in an air thermostat and shaken until equilibrium. Mercury was used for calibration and as the confining liquid. The solvents were degassed in the apparatus. Details are in references 1 and 2.

The absorbed volume of gas was calculated from the initial and final amounts, both saturated with solvent vapor. The amount of solvent was determined by the weight of displaced mercury.

The saturation of the liquid with the gas was carried out close to atmospheric pressure. The solubility values were reported for one atmosphere gas pressure assuming Henry's law is obeyed.

SOURCE AND PURITY OF MATERIALS:

(1) Nitrogen. Dansk Ilt-og Brint Fabrik. Contained less than 0.13 percent oxygen.

(2) Cyclohexanol. Fractionated by distillation at low pressure and fractionated by freezing.

Completely solid at 24°C. Estimated water content was 0.05 percent.

ESTIMATED ERROR:

$\delta T/K = 0.05$

$\delta X_1/X_1 = 0.015$

REFERENCES:

1. Lannung, A. *J. Am. Chem. Soc.* 1930, *52*, 68.

2. Gjaldbaek, J. C. *Acta Chem. Scand.* 1952, *6*, 623.

**COMPONENTS:**

(1) Nitrogen; $N_2$; [7727-37-9]

(2) 1-Heptanol; $C_7H_{16}O$; [111-70-6]

**ORIGINAL MEASUREMENTS:**

Boyer, F. L.; Bircher, L. J.

*J. Phys. Chem.* 1960, *64*, 1330-1.

**VARIABLES:**

T/K = 298.15

P/kPa = 101.325

**PREPARED BY:**

M. E. Derrick

**EXPERIMENTAL VALUES:**

| | $X_1 \times 10^3$ | Coefficient $\alpha$ | Coefficient L |
|---|---|---|---|
| 298.15 | 0.610 | 0.0962 | 0.105 ± 0.002 |

The Bunsen coefficient was calculated by the compiler.

The observed mole fraction solubility was taken from Boyer's dissertation (2).

The Dissertation gives the equations $\log X_1 = -3.594 + 0.449 \log C$ at 298.15 K where C is the number of carbon atoms in the alcohol. The mole fraction solubility given in the original paper was calculated from the 298.15 K equation.

AUXILIARY INFORMATION

**METHOD/APPARATUS/PROCEDURE:**

Commercial Van Slyke blood gas apparatus (E. H. Sargent Co.) was modified by the authors for their use.

The total pressure of gas and solvent vapor in the solution chamber was adjusted to a pressure cf one atm. The pressure was maintained at one atm during the solution process. The saturated solution was transferred to a bulb below the lower stopcock of the extraction vessel and sealed off. The gas and solvent vapor were then brought to volume over mercury. See (1) for details of the extraction procedure.

**SOURCE AND PURITY OF MATERIALS:**

(1) Nitrogen. Air Reduction Co. 99.7 per cent nitrogen.

(2) 1-Heptanol. Source not given. Treated by standard methods to remove aldehydes and ketones, then dried and distilled.

**ESTIMATED ERROR:**

$\delta T/K = 0.01$

$\delta L/cm^3 = 0.002$

**REFERENCES:**

1. Peters, J. P.; Van Slyke, D. D. "Quantitative Clinical Chemistry' Williams and Wilkins Co., Baltimore, MD, 1932, Volume II.

2. Boyer, F. L. Ph.D. Dissertation Vanderbilt U., 1959, Nashville,TN

**COMPONENTS:**

(1) Nitrogen; $N_2$; [7727-37-9]

(2) Aliphatic alcohols; $C_7H_{16}O$ and $C_8H_{18}O$

**ORIGINAL MEASUREMENTS:**

Ijams, C. C.

Ph.D. thesis, 1941
Vanderbilt University
Nashville, TN

**VARIABLES:**

T/K = 298.15
P/kPa = 101.325

**PREPARED BY:**

H. L. Clever

**EXPERIMENTAL VALUES:**

| T/K | Mol Fraction $x_1 \times 10^3$ | Bunsen Coefficient $\alpha$ | Ostwald Coefficient L |
|---|---|---|---|
| 1-Heptanol; $C_7H_{16}O$; [111-70-6] | | | |
| 298.15 | 0.609 | 0.0963 | 0.1051 |
| 1-Octanol; $C_8H_{18}O$; [111-87-5] | | | |
| 298.15 | 0.657 | 0.0930 | 0.1015 |

The mole fraction and Ostwald coefficient values were calculated by the compiler.

AUXILIARY INFORMATION

**METHOD/APPARATUS/PROCEDURE:**

A Van Slyke-Neill Manometric Apparatus manufactured by the Eimer and Amend Co. was used.

The procedure of Van Slyke (1) for pure liquids was modified so that small solvent sample (2 $cm^3$) could be used with almost 100 per cent recovery of the solvent.

An improved temperature control system was used.

**SOURCE AND PURITY OF MATERIALS:**

(1) Nitrogen. Air Reduction Co. Purity 99.5 per cent.

(2) Aliphatic alcohols. Purified. Vapor pressure and boiling point also reported in the thesis.

**ESTIMATED ERROR:**

$\delta T/K = 0.05$

**REFERENCES:**

1. Van Slyke, D. D. *J. Biol. Chem.* 1939, *130*, 545.

| COMPONENTS: | ORIGINAL MEASUREMENTS: |
|---|---|
| (1) Nitrogen; $N_2$; [7727-37-9]<br>(2) 1-Heptanol; $C_7H_{16}O$; [111-70-6]<br>or<br>1-Octanol; $C_8H_{18}O$; [111-87-5] | Makranczy, J.; Rusz, L.; Balog-Megyery, K.<br>*Hung. J. Ind. Chem.* 1979, *7*, 41-6. |
| **VARIABLES:** | **PREPARED BY:** |
| T/K = 298 | C.L. Young |

EXPERIMENTAL VALUES:

| T/K | $P^+$/kPa | Ostwald coefficient | Mole fraction of nitrogen*, $x_{N_2}$ |
|---|---|---|---|
| | | 1-Heptanol | |
| 298.15 | 101.3 | 0.114 | 0.000660 |
| | | 1-Octanol | |
| 298.15 | 101.3 | 0.108 | 0.000700 |

* calculated by compiler

\+ partial pressure of nitrogen

AUXILIARY INFORMATION

METHOD/APPARATUS/PROCEDURE:

Volumetric method. The apparatus of Bodor, Bor, Mohai, and Sipos (1) was used.

SOURCE AND PURITY OF MATERIALS:

No details given.

ESTIMATED ERROR:

$\delta x_{N_2} = \pm 3\%$

REFERENCES:

1. Bodor, E.; Bor, Gy.; Mohai, B.; Sipos, G. *Veszpremi Vegyip. Egy. Kozl.* 1957, *1*, 55, *Chem. Abstr.* 1961, *55*, 3175h

**COMPONENTS:**

(1) Nitrogen; $N_2$; [7727-37-9]

(2) 1-Octanol; $C_8H_{18}O$; [111-87-5]

**ORIGINAL MEASUREMENTS:**

Boyer, F. L.; Bircher, L. J.

*J. Phys. Chem.* 1960, *64*, 1330-1.

**VARIABLES:**

T/K = 298-308

P/kPa = 101.325

**PREPARED BY:**

M. E. Derrick

**EXPERIMENTAL VALUES:**

| T/K | Mol Fraction $X_1 \times 10^3$ | Bunsen Coefficient $\alpha$ | Ostwald Coefficient $L$ |
|---|---|---|---|
| 298.15 | 0.680 | 0.0934 | 0.102 $\pm$ 0.002 |
| 308.15 | 0.673 | 0.0940 | 0.106 $\pm$ 0.001 |

The Bunsen coefficients were calculated by the compiler.

The observed mole fraction solubilities were taken from Boyer's dissertation (2).

The Dissertation gives the equations log $X_1$ = -3.594 + 0.449 log C at 298.15 K and log $X_1$ = 3.576 + 0.452 log C at 308.15 K where C is the number of carbon atoms in the alcohol. The mole fraction solubility given in the original paper was calculated from the 298.15 K equation.

Smoothed Data: $\Delta G^o/J\ mol^{-1} = -RT \ln X_1 = -823.32 + 63.400\ T$

The equation is tentative.

**AUXILIARY INFORMATION**

**METHOD/APPARATUS/PROCEDURE:**

Commercial Van Slyke blood gas apparatus (E. H. Sargent Co.) was modified by the authors for their use.

The total pressure of gas and solvent vapor in the solution chamber was adjusted to a pressure of one atm. The pressure was maintained at one atm during the solution process. The saturated solution was transferred to a bulb below the lower stopcock of the extraction vessel and sealed off. The gas and solvent vapor were then brought to volume over mercury. See (1) for details of the extraction procedure.

**SOURCE AND PURITY OF MATERIALS:**

(1) Nitrogen. Air Reduction Co. 99.7 per cent nitrogen.

(2) 1-Octanol. Source not given. Treated by standard methods to remove aldehydes and ketones, then dried and distilled.

**ESTIMATED ERROR:**

$\delta T/K = 0.01$

$\delta L/cm^3 = 0.002$

**REFERENCES:**


1. Peters, J. P.; Van Slyke, D. D. "Quantitative Clinical Chemistry" Williams and Wilkins Co., Baltimore, MD, 1932, Volume II.

2. Boyer, F. L. Ph.D. Dissertation Vanderbilt U., 1959, Nashville, TN

| COMPONENTS: | ORIGINAL MEASUREMENTS: |
|---|---|
| (1) Nitrogen; $N_2$; [7727-37-9]<br>(2) 1-Octanol; $C_8H_{18}O$; [111-87-5] | Wilcock, R. J.; Battino, R.; Danforth, W. F.; Wilhelm, E.<br>*J. Chem. Thermodyn.* 1978, *10*, 817-22. |
| **VARIABLES:** | **PREPARED BY:** |
| T/K = 283.32 - 313.60<br>P/kPa = 101.325 | H. L. Clever |

EXPERIMENTAL VALUES:

| T/K | Mol Fraction $X_1 \times 10^3$ | Bunsen Coefficient $\alpha$ | Ostwald Coefficient L |
|---|---|---|---|
| 283.32 | 0.5727 | 0.08208 | 0.08514 |
| 298.15 | 0.6230 | 0.08822 | 0.09629 |
| 313.60 | 0.6353 | 0.08876 | 0.1019 |

The Bunsen coefficients were calculated by the compiler.

The solubility values were adjusted to a nitrogen partial pressure of 101.325 kPa (1 atm) by Henry's law.

Smoothed Data: $\Delta G^o/J\ mol^{-1} = -RT \ln X_1 = 2495.1 + 53.169\ T$

Std. Dev. $\Delta G^o = 44.0$, Coef. Corr. = 0.9985

$\Delta H^o/J\ mol^{-1} = 2495.1$, $\Delta S^o/J\ K^{-1}\ mol^{-1} = -53.169$

| T/K | Mol Fraction $X_1 \times 10^3$ | $\Delta G^o/J\ mol^{-1}$ |
|---|---|---|
| 278.15 | 0.568 | 17,284 |
| 288.15 | 0.589 | 17,816 |
| 298.15 | 0.610 | 18,348 |
| 308.15 | 0.631 | 18,879 |
| 318.15 | 0.650 | 19,411 |

AUXILIARY INFORMATION

METHOD/APPARATUS/PROCEDURE:

The solubility apparatus is based on the design of Morrison and Billett (1) and the version used is described by Battino, Evans, and Danforth (2). The degassing apparatus is that described by Battino, Banzhof, Bogan, and Wilhelm (3).

Degassing. Up to 500 $cm^3$ of solvent is placed in a flask of such size that the liquid is about 4 cm deep. The liquid is rapidly stirred, and vacuum is intermittently applied through a liquid $N_2$ trap until the permanent gas residual pressure drops to 5 microns.

Solubility Determination. The degassed solvent is passed in a thin film down a glass helical tube containing solute gas plus the solvent vapor at a total pressure of one atm. The volume of gas absorbed is found by difference between the initial and final volumes in the buret system. The solvent is collected in a tared flask and weighed.

SOURCE AND PURITY OF MATERIALS:

(1) Nitrogen. The Matheson Co., Inc. 99.999 per cent.

(2) 1-octanol. Eastman Organic Chemicals. Distilled, density $\rho_{298.15} = 0.8247\ g\ cm^{-3}$

ESTIMATED ERROR:

$\delta T/K = 0.03$
$\delta P/mmHg = 0.5$
$\delta X_1/X_1 = 0.01$

REFERENCES:

1. Morrison, T. J.; Billett, F. *J. Chem. Soc.* 1948, 2033.

2. Battino, R.; Evans, F. D.; Danforth, W. F. *J. Am. Oil Chem. Soc.* 1968, *45*, 830.

3. Battino, R.; Banzhof, M.; Bogan, M.; Wilhelm, E. *Anal. Chem.* 1971, *43*, 806.

**COMPONENTS:**

(1) Nitrogen; $N_2$; [7727-37-9]

(2) 1-Nonanol; $C_9H_{20}O$; [143-08-8]

or

1-Decanol; $C_{10}H_{22}O$; [112-08-8]

**ORIGINAL MEASUREMENTS:**

Makranczy, J.; Rusz, L.; Balog-Megyery, K.

*Hung. J. Ind. Chem.* 1979, *7*, 41-6.

**VARIABLES:**

T/K - 298

**PREPARED BY:**

C.L. Young

**EXPERIMENTAL VALUES:**

| T/K | $P^+$/kPa | Ostwald coefficient | Mole fraction of nitrogen*, $x_{N_2}$ |
|---|---|---|---|
| | | 1-Nonanol | |
| 298.15 | 101.3 | 0.105 | 0.000750 |
| | | 1-Decanol | |
| 298.15 | 101.3 | 0.103 | 0.000805 |

* Calculated by compiler

\+ Partial pressure of nitrogen

**AUXILIARY INFORMATION**

**METHOD/APPARATUS/PROCEDURE:**

Volumetric method. The apparatus of Bodor, Bor, Mohai, and Sipos (1) was used.

**SOURCE AND PURITY OF MATERIALS:**

No details given.

**ESTIMATED ERROR:**

$\delta x_{N_2} = \pm 3\%$.

**REFERENCES:**

1. Bodor, E.; Bor, Gy.; Mohai, B.; Sipos. G. *Veszpremi Vegyip. Egy.Kozl* 1957, *1*,55, *Chem. Abstr.* 1961, *55*, 3175h

| COMPONENTS: | ORIGINAL MEASUREMENTS: |
|---|---|
| (1) Nitrogen; $N_2$; [7727-37-9]<br>(2) 1-Decanol; $C_{10}H_{22}O$; [112-30-1] | Wilcock, R. J.; Battino, R. Danforth, W. F.; Wilhelm, E.<br>*J. Chem. Thermodyn.* 1978, *10*, 817-22. |
| VARIABLES:<br>T/K = 282.57 - 313.42<br>P/kPa = 101.325 | PREPARED BY:<br>A. L. Cramer |

EXPERIMENTAL VALUES:

| T/K | Mol Fraction $X_1 \times 10^3$ | Bunsen Coefficient $\alpha$ | Ostwald Coefficient L |
|---|---|---|---|
| 282.57 | 0.6403 | 0.07590 | 0.07852 |
| 298.08 | 0.6681 | 0.07822 | 0.08536 |
| 313.42 | 0.6859 | 0.07927 | 0.09096 |

The Bunsen coefficients were calculated by the compiler.

The solubility values were adjusted to a nitrogen partial pressure of 101.325 kPa (1 atm) by Henry's law.

Smoothed Data: $\Delta G^o/J\ mol^{-1} = -RT \ln X_1 = 1637.5 + 55.327\ T$

Std. Dev. $\Delta G^o$ = 8.8, Coef. Corr. = 0.9999

$\Delta H^o/J\ mol^{-1} = 1637.5$, $\Delta S^o/J\ K^{-1}\ mol^{-1} = -55.327$

| T/K | Mol Fraction $X_1 \times 10^3$ | $\Delta G^o/J\ mol^{-1}$ |
|---|---|---|
| 278.15 | 0.635 | 17,027 |
| 288.15 | 0.650 | 17,580 |
| 298.15 | 0.666 | 18,133 |
| 308.15 | 0.680 | 18,687 |
| 318.15 | 0.694 | 19,240 |

AUXILIARY INFORMATION

METHOD/APPARATUS/PROCEDURE:

The solubility apparatus is based on the design of Morrison and Billett (1) and the version used is described by Battino, Evans, and Danforth (2). The degassing apparatus is that described by Battino, Banzhof, Bogan, and Wilhelm (3).

Degassing. Up to 500 $cm^3$ of solvent is placed in a flask of such size that the liquid is about 4 cm deep. The liquid is rapidly stirred, and vacuum is intermittently applied through a liquid $N_2$ trap until the permanent gas residual pressure drops to 5 microns.

Solubility Determination. The degassed solvent is passed in a thin film down a glass helical tube containing solute gas plus the solvent vapor at a total pressure of one atm. The volume of gas absorbed is found by difference between the initial and final volumes in the buret system. The solvent is collected in a tared flask and weighed.

SOURCE AND PURITY OF MATERIALS:

(1) Nitrogen. Matheson Co., Inc. 99.999 per cent.

(2) 1-Decanol. Eastman Organic Chemicals. Distilled, density $\rho_{298.15} = 0.8206\ g\ cm^{-3}$.

ESTIMATED ERROR:

$\delta T/K = 0.03$
$\delta P/mmHg = 0.5$
$\delta X_1/X_1 = 0.01$

REFERENCES:


1. Morrison, T. J.; Billett, F. *J. Chem. Soc.* 1948, 2033.
2. Battino, R.; Evans, F. D.; Danforth, W. F. *J. Am. Oil Chem. Soc.* 1968, *45*, 830.
3. Battino, R.; Banzhof, M.; Bogan, M.; Wilhelm, E. *Anal. Chem.* 1971, *43*, 806.

**COMPONENTS:**

(1) Nitrogen; $N_2$; [7727-37-9]

(2) 1-Undecanol; $C_{11}H_{24}O$; [112-42-5]
or
1-Dodecanol; $C_{12}H_{26}O$; [112-53-8]

**ORIGINAL MEASUREMENTS:**

Makranczy, J.; Rusz, L.; Balog-Megyery, K.

*Hung. J. Ind. Chem.* 1979, *7*, 41-6.

**VARIABLES:**

T/K = 298

**PREPARED BY:**

C.L. Young

**EXPERIMENTAL VALUES:**

| T/K | $P^+$/kPa | Ostwald coefficient | Mole fraction of nitrogen*, $x_{N_2}$ |
|---|---|---|---|
| | | 1-Undecanol | |
| 298.15 | 101.3 | 0.100 | 0.000849 |
| | | 1-Dodecanol | |
| 298.15 | 101.3 | 0.098 | 0.000898 |

* calculated by compiler

\+ partial pressure of nitrogen

**AUXILIARY INFORMATION**

**METHOD/APPARATUS/PROCEDURE:**

Volumetric method. The apparatus of Bodor, Bor, Mohai, and Sipos (1) was used.

**SOURCE AND PURITY OF MATERIALS:**

No details given.

**ESTIMATED ERROR:**

$\delta x_{N_2}$ = ±3%.

**REFERENCES:**

1. Bodor, E.; Bor, Gy.; Mohai, B.; Sipos, G.
*Veszpremi Vegyip. Egy. Kozl.* 1957, *1*, 55,
Chem. Abstr. 1961, *55*, 3175h

| COMPONENTS: | EVALUATOR: |
|---|---|
| (1) Nitrogen; $N_2$; [7727-37-9]<br>(2) 2-Propanone(acetone); $C_3H_6O$; [67-64-1] | Rubin Battino<br>Department of Chemistry<br>Wright State University<br>Dayton, Ohio 45435<br>U.S.A.<br><br>December 1981 |

CRITICAL EVALUATION:

The nitrogen/acetone system has been studied by three groups (1,2,3). However, the two data points obtained by Just were so low that they are to be not considered. Smoothing the other points in the temperature range 195 to 314 K gave the following least squares equation:

$$\ln x_1 = -11.3529 + 3.77545/(T/100K) + 2.34933 \ln (T/100K)$$

The standard deviation of the fit in the mole fraction was $3.2 \times 10^{-6}$ or about 0.4% at the mid-range of temperature. Smoothed values at 10K intervals are given below. The $\Delta\bar{C}p°$ value was constant at 20 J $K^{-1}$ $mol^{-1}$. Entropy and enthalpy changes are also given in the table.

| T/K | $10^4 x_1$ | $\Delta\bar{H}°$/J $mol^{-1}$ | $\Delta\bar{S}°$/ J $K^{-1}$ $mol^{-1}$ |
|---|---|---|---|
| 193.15 | 3.891 | 634 | -62.0 |
| 203.15 | 3.979 | 829 | -61.0 |
| 213.15 | 4.083 | 1024 | -60.1 |
| 223.15 | 4.200 | 1220 | -59.2 |
| 233.15 | 4.330 | 1415 | -58.3 |
| 243.15 | 4.471 | 1610 | -57.5 |
| 253.15 | 4.623 | 1806 | -56.7 |
| 263.15 | 4.784 | 2001 | -56.0 |
| 273.15 | 4.955 | 2196 | -55.2 |
| 283.15 | 5.135 | 2392 | -54.5 |
| 293.15 | 5.324 | 2587 | -53.9 |
| 298.15 | 5.421 | 2684 | -53.5 |
| 303.15 | 5.520 | 2782 | -53.2 |
| 313.15 | 5.726 | 2977 | -52.6 |
| 323.15 | 5.939 | 3173 | -51.9 |

References

1. Just, G. *Z. Phys. Chem.* 1901, *37*, 342-67.

2. Horiuti, J. *Sci. Pap. Inst. Phys. Chem. Res. (Jpn)* 1931/32, *17*, 125-256.

3. Kretschmer, C. B.; Nowakowska, J.; Wiebe, R. *Ind. Eng. Chem.* 1946, *38*, 506-9.

COMPONENTS:

(1) Nitrogen; $N_2$; [7727-37-9]

(2) 2-Propanone or acetone; $C_3H_6O$; [67-64-1]

ORIGINAL MEASUREMENTS:

Just, G.

*Z. Phys. Chem.* 1901, *37*, 342-67.

VARIABLES:

T/K = 293-298
P/kPa = 101.325

PREPARED BY:

M. E. Derrick
H. L. Clever

EXPERIMENTAL VALUES:

| T/K | Mol Fraction $x_1 \times 10^4$ | Bunsen Coefficient $\alpha$ | Ostwald Coefficient L |
|---|---|---|---|
| 293.15 | 4.22 | 0.1289 | 0.1383 |
| 298.15 | 4.41 | 0.1338 | 0.1460 |

The author measured the Ostwald coefficient at a pressure of about 746 mmHg. The compiler assumed the Ostwald coefficient to be independent of pressure, and calculated the mole fraction and Bunsen coefficient values at 101.325 kPa (1 atm) partial pressure of the gas.

AUXILIARY INFORMATION

METHOD/APPARATUS/PROCEDURE:

An Ostwald apparatus as modified by Timofejew (1), and Steiner (2) is used. The apparatus consists of a gas buret, an absorption flask, and a mercury manometer. The system is thermostated with a water jacket.

The gas is introduced into the degassed liquid. The gas volume absorbed is determined by the gas buret. The solvent volume is determined at the end of the experiment by pouring the solvent into a graduated flask.

SOURCE AND PURITY OF MATERIALS:

(1) Nitrogen. Prepared by the reaction of sodium nitrite and ammonium nitrate in aqueous solution in the presence of calcium dichromate.

(2) 2-Propanone. No information.

ESTIMATED ERROR:

$\delta L/L = 0.03$ (compiler)

REFERENCES:


1. Timofejew, W. *Z. Physik. Chem.* 1890, *6*, 141.

2. Steiner, *Ann. Phys. (Leipzig)* 1894, *52*, 275.

| COMPONENTS: | ORIGINAL MEASUREMENTS: |
|---|---|
| (1) Nitrogen; $N_2$; [7727-37-9]<br>(2) 2-Propanone or Acetone; $C_3H_6O$; [67-64-1] | Horiuti, J.<br>*Sci. Pap. Inst. Phys. Chem. Res. (Jpn)* 1931/32, *17*, 125-256. |
| VARIABLES: | PREPARED BY: |
| T/K = 195.05 - 314.25<br>Total P/kPa = 101.325 | M. E. Derrick<br>H. L. Clever |

EXPERIMENTAL VALUES:

| T/K | Mol Fraction $X_1 \times 10^3$ | Bunsen Coefficient $\alpha$ | Ostwald Coefficient L |
|---|---|---|---|
| 195.05 | 0.3907 | 0.1354 | 0.0967 |
| 212.85 | 0.4090 | 0.1387 | 0.1081 |
| 232.40 | 0.4300 | 0.1423 | 0.1211 |
| 252.95 | 0.4610 | 0.1486 | 0.1376 |
| 273.15 | 0.4948 | 0.1553 | 0.1553 |
| 293.15 | 0.5328 | 0.1628 | 0.1747 |
| 314.25 | 0.5699 | 0.1691 | 0.1946 |

Smoothed Data: $\ln X_1 = -11.2433 + 3.6459/(T/100) + 2.2850 \ln (T/100)$

Standard error about the regression line = $1.90 \times 10^{-6}$

| T/K | Mol Fraction $X_1 \times 10^3$ |
|---|---|
| 193.15 | 0.3892 |
| 203.15 | 0.3980 |
| 213.15 | 0.4083 |
| 223.15 | 0.4200 |
| 233.15 | 0.4327 |
| 243.15 | 0.4467 |
| 253.15 | 0.4616 |

| T/K | Mol Fraction $X_1 \times 10^3$ |
|---|---|
| 263.15 | 0.4775 |
| 273.15 | 0.4943 |
| 283.15 | 0.5119 |
| 293.15 | 0.5303 |
| 303.15 | 0.5495 |
| 313.15 | 0.5695 |

The Ostwald coefficient was measured at a total pressure of one atm. The Bunsen coefficient and the mole fraction solubility values were calculated by the compiler for a gas pressure of one atm assuming the gas to behave ideally and the Ostwald coefficient to be independent of pressure.

AUXILIARY INFORMATION

METHOD/APPARATUS/PROCEDURE:

The apparatus consists of a gas buret, a solvent reservoir, and an absorption pipet. The volume of the pipet was determined at various miniscus heights by weighing a quantity of water. The meniscus height was read with a cathetometer.

The dry gas introduced into the degassed solvent. The gas and solvent are mixed with a magnetic stirrer until saturation. Care was taken to prevent solvent vapor from mixing with the solute gas in the gas buret. The volume of gas was determined from a gas buret reading, the volume of the solvent was determined from the meniscus height in the absorption pipet.

SOURCE AND PURITY OF MATERIALS:

(1) Nitrogen. Prepared in the laboratory by heating sodium azide (Kahlbaum, recrystalized and dried). Gas dried over phosphorus pentoxide.

(2) 2-Propanone. Nippon Pure Chem. Co. or Merck. Extra pure grade, recrystallized with sodium sulfite. Stored over calcium chloride then fractionated. Normal boiling point 56.09°C.

ESTIMATED ERROR:

$\delta T/K = 0.05$

If the gas and solution are assumed to be ideal, the fitted equation gives the following thermodynamic values.

| T/K | $\Delta G^o$/kJ mol$^{-1}$ | $\Delta H^o$/J mol$^{-1}$ | $\Delta S^o$/J K$^{-1}$ mol$^{-1}$ | $\Delta C_p^o$/J K$^{-1}$ mol$^{-1}$ |
|---|---|---|---|---|
| 273.15 | 17.288 | 2158 | -55.39 | 19.00 |
| 298.15 | 18.652 | 2633 | -53.73 | 19.00 |

**COMPONENTS:**

(1) Nitrogen; $N_2$; [7727-37-9]

(2) 2-Propanone or acetone; $C_3H_6O$; [67-64-1]

**ORIGINAL MEASUREMENTS:**

Kretschmer, C. B.; Nowakowska, J.; Wiebe, R.

*Ind. Eng. Chem.* 1946, *38*, 506-09.

**VARIABLES:**

T/K = 248.15 - 298.15
P/kPa = 101.325

**PREPARED BY:**

P. L. Long
H. L. Clever

**EXPERIMENTAL VALUES:**

| T/K | Mol Fraction $x_1 \times 10^3$ | Bunsen Coefficient $\alpha$ | Ostwald Coefficient L |
|---|---|---|---|
| 248.15 | 0.4550 | 0.1475 | 0.1340 |
| 273.15 | 0.4954 | 0.1564 | 0.1554 |
| 298.15 | 0.5486 | 0.1664 | 0.1816 |

The Ostwald coefficients were measured at a total pressure of 101.325 kPa (1 atm). The compiler calculated the Bunsen coefficient and mole fraction values for an oxygen partial pressure of 101.325 kPa (1 atm) assuming the Ostwald coefficient was independent of pressure.

Smoothed Data: $\ln x_1 = -6.5910 - 2.7528/(T/100K)$

Standard error about the regression line = $6.98 \times 10^{-6}$

| T/K | Mol Fraction $x_1 \times 10^3$ |
|---|---|
| 253.15 | 0.463 |
| 263.15 | 0.482 |
| 273.15 | 0.501 |
| 283.15 | 0.519 |
| 293.15 | 0.537 |
| 298.15 | 0.545 |

AUXILIARY INFORMATION

**METHOD/APPARATUS/PROCEDURE:**

The apparatus was similar to that of Horiuti (1), it consisted of a gas buret, manometer, and absorption pipet equipped with a magnetic stirrer. All volumes were calibrated by mercury displacement.

The solvent was partly degassed by boiling under reflux. About 40 $cm^3$ of solvent was admitted into the absorption pipet, and pumped on to complete the degassing. Dry gas, measured in the buret, was admitted, final equilibrium was established after 1 to 3 hours stirring. During the solution process the total pressure was kept at one atm by additions of dry gas.

The solubility value was corrected for the increase of solvent volume with gas absorption, and for the nitrogen impurity in the oxygen.

**SOURCE AND PURITY OF MATERIALS:**

(1) Nitrogen. Commercial, but source not given. Passed through alkaline pyrogallol to remove $O_2$, passed through Drierite to dry.

(2) 2-Propanone. Source not given. Treated with $Ag_2O$, dried over $CaSO_4$, distilled, b.p. (760 mmHg)/°C 56.10 - 56.14, density, $\rho_4^{25}$/g $cm^{-3}$ 0.78490.

**ESTIMATED ERROR:**

Gas buret $\delta V/cm^3$ = 0.005
Average deviation from the mean 0.3%, maximum deviation 0.9% (authors).

**REFERENCES:**

1. Horiuti, J. *J. Sci Papers Inst. Phys. Chem. Res. (Tokyo)* 1931, *17*, 125.

**COMPONENTS:**

(1) Nitrogen; $N_2$; [7727-37-9]

(2) Cyclohexanone; $C_6H_{10}O$; [108-94-1]

**ORIGINAL MEASUREMENTS:**

Guerry, D. Jr.

Ph.D. thesis, 1944
Vanderbilt University
Nashville, TN

**VARIABLES:**

T/K = 293-298
P/kPa = 101.325

**PREPARED BY:**

H. L. Clever

**EXPERIMENTAL VALUES:**

| T/K | Mol Fraction $x_1 \times 10^4$ | Bunsen Coefficient $\alpha$ | Ostwald Coefficient L |
|---|---|---|---|
| 293.15 | 3.72 | 0.0804 | 0.0863 |
| 298.15 | 3.77 | 0.0812 | 0.0886 |

The Ostwald coefficients were calculated by the compiler.

AUXILIARY INFORMATION

**METHOD/APPARATUS/PROCEDURE:**

A Van Slyke-Neill Manometric Apparatus manufactured by the Eimer and Amend Co. was used.

The procedure of Van Slyke (1) for pure liquids was modified (2) so that small solvent samples (2 $cm^3$) could be used with almost complete recovery of the sample.

An improved temperature control system was used.

**SOURCE AND PURITY OF MATERIALS:**

(1) Nitrogen. Ohio Chemical Co. Oil pumped, 99.8 per cent pure.

(2) Cyclohexanone. Eastman Kodak Co. Purified, distilled, b.p. (754.5 mmHg) t/°C 155.19. Refractive index, density, and vapor pressure data are in the thesis.

**ESTIMATED ERROR:**

$\delta T/K = 0.05$

**REFERENCES:**

1. Van Slyke, D. D. *J. Biol. Chem.* 1939, *130*, 545.

2. Ijams, C. C. Ph.D. thesis, 1941 Vanderbilt University

**COMPONENTS:**

(1) Nitrogen; $N_2$; [7727-37-9]

(2) Acetic Acid; $C_2H_4O_2$; [64-19-7]

**ORIGINAL MEASUREMENTS:**

Just, G.

*Z. Phys. Chem.* 1901, *37*, 342-67.

**VARIABLES:**

T/K = 293-298
P/kPa = 101.325

**PREPARED BY:**

M. E. Derrick
H. L. Clever

**EXPERIMENTAL VALUES:**

| T/K | Mol Fraction $x_1 \times 10^4$ | Bunsen Coefficient $\alpha$ | Ostwald Coefficient L |
|---|---|---|---|
| 293.15 | 2.79 | 0.1092 | 0.1172 |
| 298.15 | 2.80 | 0.1090 | 0.1190 |

The author measured the Ostwald coefficient at a pressure of about 746 mmHg. The compiler assumed the Ostwald coefficient to be independent of pressure, and calculated the mole fraction and Bunsen coefficient values at 101.325 kPa (1 atm) partial pressure of the gas.

AUXILIARY INFORMATION

**METHOD/APPARATUS/PROCEDURE:**

An Ostwald apparatus as modified by Timofejew (1), and Steiner (2) is used. The apparatus consists of a gas buret, an absorption flask, and a mercury manometer. The system is thermostated with a water jacket.

The gas is introduced into the degassed liquid. The gas volume absorbed is determined by the gas buret. The solvent volume is determined at the end of the experiment by pouring the solvent into a graduated flask.

**SOURCE AND PURITY OF MATERIALS:**

(1) Nitrogen. Prepared by the reaction of sodium nitrite and ammonium nitrate in aqueous solution in the presence of calcium dichromate.

(2) Acetic Acid. No information.

**ESTIMATED ERROR:**

$\delta L/L = 0.03$ (compiler)

**REFERENCES:**


1. Timofejew, W.
*Z. Physik. Chem.* 1890, *6*, 141.

2. Steiner,
*Ann. Phys. (Leipzig)* 1894, *52*, 275.

| COMPONENTS: | ORIGINAL MEASUREMENTS: |
|---|---|
| (1) Nitrogen; $N_2$; [7727-37-9]<br>(2) 2-Methylpropanoic acid; $C_4H_8O_2$; [79-31-2] | Drucker, K.; Moles, E.<br>*Z. Physik. Chem.* 1910, *75*, 405-36. |
| VARIABLES:<br>T/K = 298<br>P/kPa = 35-111 | PREPARED BY:<br>C. L. Young |

EXPERIMENTAL VALUES:

| T/K | *p*/mmHg | *p*/kPa | Ostwald coefficient, *L* |
|---|---|---|---|
| 298.20 | 262.6 | 35.01 | 0.1609* |
| | 388.3 | 51.77 | 0.1640 |
| | 566.1 | 75.47 | 0.1647 |
| | 662.4 | 88.31 | 0.1656 |
| | 783.5 | 104.5 | 0.1656 |
| | 832.2 | 111.0 | 0.1666 |

* probably in error.

AUXILIARY INFORMATION

| METHOD/APPARATUS/PROCEDURE: | SOURCE AND PURITY OF MATERIALS: |
|---|---|
| Volumetric method using an Ostwald type gas buret and pipet. Solvent degassed before being admitted to pipet. | (1) From cylinder, passed through concentrated alkali, alkaline pyrogallol, concentrated sulfuric acid and over hot copper.<br>(2) From Kahlbaum, distilled four times. |
| | ESTIMATED ERROR:<br>$\delta L/L$ = ±1-3% for dil. soln.;<br>±5-7% for conc. soln. |
| | REFERENCES: |

COMPONENTS:

(1) Nitrogen; $N_2$; [7727-37-9]

(2) Acetic acid, methyl ester or Methyl acetate; $C_3H_6O_2$; [79-20-9]

ORIGINAL MEASUREMENTS:

Horiuti, J.

*Sci. Pap. Inst. Phys. Chem. Res. (Jpn)* 1931/32, *17*, 125-256.

VARIABLES:

T/K = 194.45 - 313.25

Total P/kPa = 101.325

PREPARED BY:

M. E. Derrick
H. L. Clever

EXPERIMENTAL VALUES:

| T/K | Mol Fraction $X_1 \times 10^3$ | Bunsen Coefficient $\alpha$ | Ostwald Coefficient L |
|---|---|---|---|
| 194.45 | 0.3948 | 0.1264 | 0.0900 |
| 213.05 | 0.4225 | 0.1323 | 0.1032 |
| 232.55 | 0.4573 | 0.1398 | 0.1190 |
| 252.85 | 0.4905 | 0.1462 | 0.1353 |
| 273.15 | 0.5342 | 0.1551 | 0.1551 |
| 293.15 | 0.5761 | 0.1629 | 0.1748 |
| 313.25 | 0.6201 | 0.1706 | 0.1957 |

Smoothed Data: $\ln X_1 = -10.3409 + 2.3823/(T/100) + 1.9228 \ln (T/100)$

Standard error about the regression line = $1.62 \times 10^{-6}$

| T/K | Mol Fraction $X_1 \times 10^3$ |
|---|---|
| 193.15 | 0.3930 |
| 203.15 | 0.4075 |
| 213.15 | 0.4231 |
| 223.15 | 0.4395 |
| 233.15 | 0.4567 |
| 243.15 | 0.4748 |
| 253.15 | 0.4935 |

| T/K | Mol Fraction $X_1 \times 10^3$ |
|---|---|
| 263.15 | 0.5130 |
| 273.15 | 0.5332 |
| 283.15 | 0.5540 |
| 293.15 | 0.5755 |
| 303.15 | 0.5976 |
| 313.15 | 0.6203 |

The Ostwald coefficient was measured at a total pressure of one atm. The Bunsen coefficient and the mole fraction solubility values were calculated by the compiler for a gas pressure of one atm assuming the gas to behave ideally and the Ostwald coefficient to be independent of pressure.

AUXILIARY INFORMATION

METHOD/APPARATUS/PROCEDURE:

The apparatus consists of a gas buret, a solvent reservoir, and an absorption pipet. The volume of the pipet was determined at various miniscus heights by weighing a quantity of water. The meniscus height was read with a cathetometer.

The dry gas introduced into the degassed solvent. The gas and solvent are mixed with a magnetic stirrer until saturation. Care taken to prevent solvent vapor from mixing with the solute gas in the gas buret. The volume of gas was determined from a gas buret reading, the volume of the solvent was determined from the meniscus height in the absorption pipet.

SOURCE AND PURITY OF MATERIALS:

(1) Nitrogen. Prepared in the laboratory by heating sodium azide (Kahlbaum, recrystalized and dried). Gas dried over phosphorus pentoxide.

(2) Acetic acid, methyl ester. Merck. Extra pure grade treated with phosphorous pentoxide several times and distilled several times. Normal boiling point, 57.12°C.

ESTIMATED ERROR:

$\delta T/K = 0.05$

If the gas and solution are assumed to be ideal the fitted equation gives the following thermodynamic values.

| T/K | $\Delta G^o$/kJ mol$^{-1}$ | $\Delta H^o$/J mol$^{-1}$ | $\Delta S^o$/J K$^{-1}$ mol$^{-1}$ | $\Delta C_p^o$/J K$^{-1}$ mol$^{-1}$ |
|---|---|---|---|---|
| 273.15 | 17.116 | 2386 | -53.93 | 15.99 |
| 298.15 | 18.446 | 2785 | -52.53 | 15.99 |

COMPONENTS:

(1) Nitrogen; $N_2$; [7727-37-9]

(2) Acetic acid ethyl ester or ethyl acetate; $C_4H_8O_2$; [141-78-6]

ORIGINAL MEASUREMENTS:

Just, G.

*Z. Phys. Chem.* 1901, *37*, 342-67.

VARIABLES:

T/K = 293-298

P/kPa = 101.325

PREPARED BY:

M. E. Derrick

H. L. Clever

EXPERIMENTAL VALUES:

| T/K | Mol Fraction $x_1 \times 10^4$ | Bunsen Coefficient $\alpha$ | Ostwald Coefficient L |
|---|---|---|---|
| 293.15 | 6.83 | 0.1564 | 0.1678 |
| 298.15 | 6.95 | 0.1582 | 0.1727 |

The author measured the Ostwald coefficient at a pressure of about 746 mmHg. The compiler assumed the Ostwald coefficient to be independent of pressure, and calculated the mole fraction and Bunsen coefficient values at 101.325 kPa (1 atm) partial pressure of the gas.

AUXILIARY INFORMATION

METHOD/APPARATUS/PROCEDURE:

An Ostwald apparatus as modified by Timofejew (1), and Steiner (2) is used. The apparatus consists of a gas buret, an absorption flask, and a mercury manometer. The system is thermostated with a water jacket.

The gas is introduced into the degassed liquid. The gas volume absorbed is determined by the gas buret. The solvent volume is determined at the end of the experiment by pouring the solvent into a graduated flask.

SOURCE AND PURITY OF MATERIALS:

(1) Nitrogen. Prepared by the reaction of sodium nitrite and ammonium nitrate in aqueous solution in the presence of calcium dichromate.

(2) Acetic acid ethyl ester. No information given.

ESTIMATED ERROR:

$\delta L/L = 0.03$ (compiler)

REFERENCES:

1. Timofejew, W. *Z. Physik. Chem.* 1890, *6*, 141.

2. Steiner, *Ann. Phys. (Leipzig)* 1894, *52*, 275.

**COMPONENTS:**

(1) Nitrogen; $N_2$; [7727-37-9]

(2) Acetic acid 2-methylpropyl ester or isobutyl acetate; $C_6H_{11}O_2$; [110-19-0]

**ORIGINAL MEASUREMENTS:**

Just, G.

*Z. Phys. Chem.* 1901, *37*, 342-67.

**VARIABLES:**

T/K = 293-298
P/kPa = 101.325

**PREPARED BY:**

M. E. Derrick
H. L. Clever

**EXPERIMENTAL VALUES:**

| T/K | Mol Fraction $x_1 \times 10^4$ | Bunsen Coefficient $\alpha$ | Ostwald Coefficient L |
|---|---|---|---|
| 293.15 | 9.39 | 0.1585 | 0.1701 |
| 298.15 | 9.47 | 0.1589 | 0.1734 |

The author measured the Ostwald coefficient at a pressure of about 746 mmHg. The compiler assumed the Ostwald coefficient to be independent of pressure, and calculated the mole fraction and Bunsen coefficient values at 101.325 kPa (1 atm) partial pressure of the gas.

**AUXILIARY INFORMATION**

**METHOD/APPARATUS/PROCEDURE:**

An Ostwald apparatus as modified by Timofejew (1), and Steiner (2) is used. The apparatus consists of a gas buret, an absorption flask, and a mercury manometer. The system is thermostated with a water jacket.

The gas is introduced into the degassed liquid. The gas volume absorbed is determined by the gas buret. The solvent volume is determined at the end of the experiment by pouring the solvent into a graduated flask.

**SOURCE AND PURITY OF MATERIALS:**

(1) Nitrogen. Prepared by the reaction of sodium nitrite and ammonium nitrate in aqueous solution in the presence of calcium dichromate.

(2) Acetic acid 2-methylpropyl ester. No information.

**ESTIMATED ERROR:**

$\delta L/L = 0.03$ (compiler)

**REFERENCES:**


1. Timofejew, W. *Z. Physik. Chem.* 1890, *6*, 141.

2. Steiner, *Ann. Phys. (Leipzig)* 1894, *52*, 275.

**COMPONENTS:**

(1) Nitrogen; $N_2$; [7727-37-9]

(2) Acetic acid pentyl ester or amyl acetate; $C_7H_{14}O_2$; [628-63-7]

**ORIGINAL MEASUREMENTS:**

Just, G.

*Z. Phys. Chem.* 1901, *37*, 342-67.

**VARIABLES:**

T/K = 293-298
P/kPa = 101.325

**PREPARED BY:**

M. E. Derrick
H. L. Clever

**EXPERIMENTAL VALUES:**

| T/K | Mol Fraction $x_1 \times 10^4$ | Bunsen Coefficient $\alpha$ | Ostwald Coefficient L |
|---|---|---|---|
| 293.15 | 9.35 | 0.1409 | 0.1512 |
| 298.15 | 9.43 | 0.1413 | 0.1542 |

The author measured the Ostwald coefficient at a pressure of about 746 mmHg. The compiler assumed the Ostwald coefficient to be independent of pressure, and calculated the mole fraction and Bunsen coefficient values at 101.325 kPa (1 atm) partial pressure of the gas.

AUXILIARY INFORMATION

**METHOD/APPARATUS/PROCEDURE:**

An Ostwald apparatus as modified by Timofejew (1), and Steiner (2) is used. The apparatus consists of a gas buret, an absorption flask, and a mercury manometer. The system is thermostated with a water jacket.

The gas is introduced into the degassed liquid. The gas volume absorbed is determined by the gas buret. The solvent volume is determined at the end of the experiment by pouring the solvent into a graduated flask.

**SOURCE AND PURITY OF MATERIALS:**

(1) Nitrogen. Prepared by the reaction of sodium nitrite and ammonium nitrate in aqueous solution in the presence of calcium dichromate.

(2) Acetic acid pentyl ester. No information given.

**ESTIMATED ERROR:**

$\delta L/L = 0.03$ (compiler)

**REFERENCES:**

1. Timofejew, W. *Z. Physik. Chem.* 1890, *6*, 141.

2. Steiner, *Ann. Phys. (Leipzig)* 1894, *52*, 275.

**COMPONENTS:**

(1) Nitrogen; $N_2$; [7727-37-9]

(2) 1,1'-Oxybisethane or diethyl ether; $C_4H_{10}O$; [60-29-7]

**ORIGINAL MEASUREMENTS:**

Christoff, A.

*Z. Phys. Chem.* 1912, *79*, 456-60.

**VARIABLES:**

T/K = 273.15
P/kPa = Atmospheric

**PREPARED BY:**

M. E. Derrick
H. L. Clever

**EXPERIMENTAL VALUES:**

| T/K | Mol Fraction $x_1 \times 10^3$ | Bunsen Coefficient $\alpha$ | Ostwald Coefficient L |
|---|---|---|---|
| 273.15 | 1.16 | 0.258 | 0.258 |

The author measured the Ostwald coefficient. The total pressure was corrected for the vapor pressure of the solvent at the temperature of the measurement. The compiler calculated the mole fraction and Bunsen coefficient values.

AUXILIARY INFORMATION

**METHOD/APPARATUS/PROCEDURE:**

The apparatus is a modified Ostwald apparatus described by Just (1), and modified by Skirrow (2). The apparatus consists of a thermostated gas buret and absorption flask.

The modification involves the use of vapor free gas in the gas buret. A correction is made for the vapor pressure of the liquid. A steel capillary tube with a stopcock, which prevents the gas and the liquid vapor mixing in the buret, is used to connect the absorption flask and the buret.

**SOURCE AND PURITY OF MATERIALS:**

(1) Nitrogen. Prepared by decomposing sodium nitrite. Passed over glowing copper to remove nitric oxide.

(2) Diethyl ether. Merck (Darmstadt). Stated to be pure and anhydrous.

**ESTIMATED ERROR:**

$\delta L/L = 0.05$

**REFERENCES:**


1. Just, G. *Z. Phys. Chem.* 1901, *37*, 342.

2. Skirrow, F. W. *Z. Phys. Chem.* 1902, *41*, 139.

COMPONENTS:

(1) Nitrogen; $N_2$; [7727-37-9]

(2) 1,1-Oxybisethane or Diethylether; $C_4H_{10}O$; [60-29-7]

ORIGINAL MEASUREMENTS:

Horiuti, J.

*Sci. Pap. Inst. Phys. Chem. Res. (Jpn)* 1931/32, *17*, 125-256.

VARIABLES:

T/K = 195.45 - 293.15

Total P/kPa = 101.325

PREPARED BY:

M. E. Derrick
H. L. Clever

EXPERIMENTAL VALUES:

| T/K | Mol Fraction $X_1 \times 10^3$ | Bunsen Coefficient $\alpha$ | Ostwald Coefficient L |
|---|---|---|---|
| 195.45 | 1.157 | 0.2872 | 0.2055 |
| 212.55 | 1.136 | 0.2755 | 0.2144 |
| 232.05 | 1.139 | 0.2691 | 0.2286 |
| 252.65 | 1.155 | 0.2651 | 0.2452 |
| 273.15 | 1.199 | 0.2672 | 0.2672 |
| 293.15 | 0.9113 | 0.2674 | 0.2870 |

Smoothed Data: $\ln X_1 = 4.39833 - 13.8281/(T/100) - 6.14581 \ln (T/100)$

Standard error about the regression line = $9.53 \times 10^{-5}$

| T/K | Mol Fraction $X_1 \times 10^3$ |
|---|---|
| 193.15 | 1.11 |
| 203.15 | 1.15 |
| 213.15 | 1.18 |
| 223.15 | 1.19 |
| 233.15 | 1.19 |
| 243.15 | 1.17 |

| T/K | Mol Fraction $X_1 \times 10^3$ |
|---|---|
| 253.15 | 1.14 |
| 263.15 | 1.11 |
| 273.15 | 1.07 |
| 283.15 | 1.03 |
| 293.15 | 0.98 |

The Ostwald coefficient was measured at a total pressure of one atm. The Bunsen coefficient and the mole fraction solubility values were calculated by the compiler for a gas pressure of one atm assuming the gas to behave ideally and the Ostwald coefficient to be independent of pressure.

AUXILIARY INFORMATION

METHOD/APPARATUS/PROCEDURE:

The apparatus consists of a gas buret, a solvent reservoir, and an absorption pipet. The volume of the pipet was determined at various miniscus heights by weighing a quantity of water. The meniscus height was read with a cathetometer.

The dry gas introduced into the degassed solvent. The gas and solvent are mixed with a magnetic stirrer until saturation. Care taken to prevent solvent vapor from mixing with the solute gas in the gas buret. The volume of gas was determined from a gas buret reading, the volume of the solvent was determined from the meniscus height in the absorption pipet.

SOURCE AND PURITY OF MATERIALS:

(1) Nitrogen. Prepared in the laboratory by heating sodium azide (Kahlbaum, recrystalized and dried) Gas dried over phosphorous pentoxide.

(2) 1,1-Oxybisethane. Merck's "for analysis". Stored over sodium amalgam and distilled. Constant boiling within 0.01°C.

ESTIMATED ERROR:

$\delta T/K = 0.05$

If the gas and solution are assumed to be ideal the fitted equation gives the following thermodynamic values.

| T/K | $\Delta G^o$/kJ $mol^{-1}$ | $\Delta H^o$/J $mol^{-1}$ | $\Delta S^o$/J $K^{-1}$ $mol^{-1}$ | $\Delta C_p^o$/J $K^{-1}$ $mol^{-1}$ |
|---|---|---|---|---|
| 273.15 | 15.533 | -2460 | -65.88 | -51.1 |
| 293.15 | 16.887 | -3482 | -69.49 | -51.1 |

**COMPONENTS:**

(1) Nitrogen; $N_2$; [7727-37-9]

(2) Cyclic ethers: $C_4H_8O$, $C_4H_8O_2$, $C_5H_8O$, and $C_5H_{10}O$

**ORIGINAL MEASUREMENTS:**

Guerry, D. Jr.

Ph.D. thesis, 1944
Vanderbilt University
Nashville, TN

**VARIABLES:**

T/K = 293-298
P/kPa = 101.325

**PREPARED BY:**

H. L. Clever

**EXPERIMENTAL VALUES:**

| T/K | Mol Fraction $x_1 \times 10^4$ | Bunsen Coefficient $\alpha$ | Ostwald Coefficient L |
|---|---|---|---|
| | Tetrahydrofuran; $C_4H_8O$; [109-99-9] | | |
| 293.15 | 5.07 | 0.140 | 0.150 |
| 298.15 | 5.21 | 0.143 | 0.156 |
| | 1,4-Dioxane; $C_4H_8O_2$; [123-91-1] | | |
| 293.15 | 2.29 | 0.0827 | 0.0888 |
| 298.15 | 2.37 | 0.0844 | 0.0921 |
| | 2,3-Dihydropyran; $C_5H_8O$; [25512-65-6] | | |
| 293.15 | 4.90 | 0.121 | 0.130 |
| 298.15 | 4.93 | 0.121 | 0.132 |
| | Tetrahydro-2H-pyran; $C_5H_{10}O$; [142-68-7] | | |
| 293.15 | 5.77 | 0.133 | 0.143 |
| 298.15 | 5.93 | 0.136 | 0.148 |

The Ostwald coefficients were calculated by the compiler.

**AUXILIARY INFORMATION**

**METHOD/APPARATUS/PROCEDURE:**

The apparatus was a modified Van Slyke-Neill Manometric Apparatus manufactured by the Eimer and Amend Co.

The procedure of Van Slyke (1) for pure liquids was modified (2) so that small solvent samples (2 $cm^3$) could be used with almost 100 per cent recovery of the sample.

An improved temperature control system was used.

**SOURCE AND PURITY OF MATERIALS:**

Tetrahydrofuran. Eastman Kodak Co. B.p.(752.7 mmHg) t/°C 65.50 - 65.54.

1,4-Dioxane. Eastman Kodak Co. B.p.(743.7 mmHg) t/°C 100.81-100.82.

Dihydro-2H-pyran. Prepared from tetrahydrofurfuryl alcohol. B.p. (743.6 mmHg) t/°C 84.81 - 84.89.

Tetrahydro-2H-pyran. Prepared by catalytic reduction of dihydro-2H-pyran. B.p.(750.6 mmHg) t/°C 87.51-87.52. All b.p. are corrected.

**SOURCE AND PURITY OF MATERIALS:**

(1) Nitrogen. Ohio Chemical Co. Oil pumped, 99.8 per cent pure.

(2) Cyclic ethers. The ethers were fractionally distilled from over Na in a nitrogen atmosphere. In addition to the solubility data the thesis contains measured values of refractive index, density, vapor pressure and b.p.

**ESTIMATED ERROR:**

$\delta T/K = 0.05$

**REFERENCES:**

1. Van Slyke, D. D. *J. Biol. Chem.* 1939, *130*, 545.

2. Ijams, C. C. Ph.D. thesis, 1941 Vanderbilt University

**COMPONENTS:**

(1) Nitrogen; $N_2$; [7727-37-9]

(2) 1,1'-Oxybispropane or dipropyl ether; $C_6H_{14}O$; [111-43-3]

**ORIGINAL MEASUREMENTS:**

Guerry, D. Jr.

Ph.D. thesis, 1944
Vanderbilt University
Nashville, TN

**VARIABLES:**

T/K = 293-298
P/kPa = 101.325

**PREPARED BY:**

H. L. Clever

**EXPERIMENTAL VALUES:**

| T/K | Mol Fraction $x_1 \times 10^4$ | Bunsen Coefficient $\alpha$ | Ostwald Coefficient L |
|---|---|---|---|
| 293.15 | 12.0 | 0.198 | 0.212 |
| 298.15 | 12.2 | 0.199 | 0.217 |

The Ostwald coefficients were calculated by the compiler.

**AUXILIARY INFORMATION**

**METHOD/APPARATUS/PROCEDURE:**

A Van Slyke-Neill Manometric Apparatus manufactured by the Eimer and Amend Co. was used.

The procedure of Van Slyke (1) for pure liquids was modified (2) so that small solvent samples (2 $cm^3$) could be used with almost complete recovery of the sample.

An improved temperature control system was used.

**SOURCE AND PURITY OF MATERIALS:**

(1) Nitrogen. Ohio Chemical Co. Oil pumped, 99.8 per cent pure.

(2) Dipropyl ether. Eastman Kodak Co. Refluxed four hours over Na, then distilled from Na in a $N_2$ atm. B.p. (746.2 mmHg) t/°C 89.03 - 89.28 (corr.). Refractive index, density, and vapor pressure data are in the thesis.

**ESTIMATED ERROR:**

$\delta T/K = 0.05$

**REFERENCES:**

1. Van Slyke, D. D. *J. Biol. Chem.* 1939, *130*, 545.

2. Ijams, C. C. Ph.D. thesis, 1941 Vanderbilt University

COMPONENTS:

(1) Nitrogen; $N_2$; [7727-37-9]

(2) 1,2-Epoxyethane (ethylene oxide); $C_2H_4O$; [75-21-8]

ORIGINAL MEASUREMENTS:

Olson, J. D.

*J. Chem. Eng. Data* 1977, *22*, 326-9.

VARIABLES:

T/K = 273-323
P/MPa = 0.7-2.2

PREPARED BY:

R. Battino

EXPERIMENTAL VALUES:

| t/°C | $T^a$/K | $H^b$/(atm/mole fraction) | P/atm (Range) | No. of Points |
|---|---|---|---|---|
| 0 | 273.2 | 2800 | 7-20 | 6 |
| 25 | 298.2 | 2180 | 9-22 | 11 |
| 50 | 323.2 | 1820 | 18-21 | 3 |

[a] Calculated by compiler.

[b] Henry's law constant calculated over pressure range and number of experimental data points indicated. H defined as

$H = \lim_{x \to 0} (f_1/x_1)$ where f is the fugacity and x the mole fraction.

AUXILIARY INFORMATION

METHOD/APPARATUS/PROCEDURE:

The nitrogen and ethylene oxide were weighed into a bomb of known volume. Equilibrium pressures were determined at each temperature for several loadings.

SOURCE AND PURITY OF MATERIALS:

(1) Linde high purity - 99.99%.

(2) UCC commercial grade. No volatile impurities above 100 ppm level.

ESTIMATED ERROR:

$\delta T/K = \pm 0.1$
$\delta H/H = \pm 0.01$, compiler's estimate

REFERENCES:

**COMPONENTS:**

(1) Nitrogen; $N_2$; [7727-37-9]

(2) Trichloromethane or chloroform; $CHCl_3$; [67-66-3]

**ORIGINAL MEASUREMENTS:**

Just, G.

*Z. Physik. Chem.* 1901, *37*, 342-67.

**VARIABLES:**

T/K = 293-298
P/kPa = 101.325

**PREPARED BY:**

M. E. Derrick
H. L. Clever

**EXPERIMENTAL VALUES:**

| T/K | Mol Fraction $x_1 \times 10^4$ | Bunsen Coefficient $\alpha$ | Ostwald Coefficient L |
|---|---|---|---|
| 293.15 | 4.27 | 0.1195 | 0.1282 |
| 298.15 | 4.45 | 0.1235 | 0.1348 |

The author measured the Ostwald coefficient at a pressure of about 746 mmHg. The compiler assumed the Ostwald coefficient to be independent of pressure, and calculated the mole fraction and Bunsen coefficient values at 101.325 kPa (1 atm) partial pressure of the gas.

**AUXILIARY INFORMATION**

**METHOD/APPARATUS/PROCEDURE:**

An Ostwald apparatus as modified by Timofejew (1), and Steiner (2) is used. The apparatus consists of a gas buret, an absorption flask, and a mercury manometer. The system is thermostated with a water jacket.

The gas is introduced into the degassed liquid. The gas volume absorbed is determined by the gas buret. The solvent volume is determined at the end of the experiment by pouring the solvent into a graduated flask.

**SOURCE AND PURITY OF MATERIALS:**

(1) Nitrogen. Prepared by the reaction of sodium nitrite and ammonium nitrate in aqueous solution in the presence of calcium dichromate.

(2) Trichloromethane. No information.

**ESTIMATED ERROR:**

$\delta L/L = 0.03$ (compiler)

**REFERENCES:**

1. Timofejew, W. *Z. Physik. Chem.* 1890, *6*, 141.

2. Steiner, *Ann. Phys. (Leipzig)* 1894, *52*, 275.

**COMPONENTS:**

(1) Nitrogen; $N_2$; [7727-37-9]

(2) Trichloromethane or chloroform; $CHCl_3$; [67-66-3]

**ORIGINAL MEASUREMENTS:**

Körösy, F.

*Trans. Faraday Soc.* 1937, *33*, 416-25.

**VARIABLES:**

T/K = 295.15
P/kPa = 101.325

**PREPARED BY:**

H. L. Clever

**EXPERIMENTAL VALUES:**

| T/K | Mol Fraction $x_1 \times 10^4$ | Bunsen Coefficient $\alpha$ | Ostwald Coefficient L |
|---|---|---|---|
| 295.15 | 4.25 | 0.118 | 0.128 |

The mole fraction and Bunsen coefficient values were calculated by the compiler.

**AUXILIARY INFORMATION**

**METHOD/APPARATUS/PROCEDURE:**

The apparatus and method of Winkler (1) were used. However, the apparatus was usually not thermostated, and degassing was by evacuation and shaking the solvent, not by evacuating and boiling the solvent as was done by Winkler.

**SOURCE AND PURITY OF MATERIALS:**

No information

**ESTIMATED ERROR:**

$\delta L/L = 0.05$

**REFERENCES:**

1. Winkler, L. W. *Ber.* 1891, *24*, 89.

| COMPONENTS: | ORIGINAL MEASUREMENTS: |
|---|---|
| (1) Nitrogen; $N_2$; [7727-37-9]<br>(2) Methane, chlorotrifluoro; $CClF_3$; [75-72-9]<br>(2') Tetrafluoromethane; $CF_4$; [75-73-0] | Leites, I.L.; Adlivankina, M.A.<br>*Khim. Prom.* 1966, *42*, 848-50. |
| VARIABLES:<br>T/K = 93-123<br>P/kPa = 101.325 | PREPARED BY:<br>V. Katovic |

EXPERIMENTAL VALUES:

| t/°C | T[b]/K | $\alpha^a$/m$^3$(STP) m$^{-3}$ atm$^{-1}$ (Freon-13) $CClF_3$ | (Freon - 14) $CF_4$ |
|---|---|---|---|
| -150 | 123 | 0.293 | 0.066 |
| -160 | 113 | 0.346 | 0.138 |
| -170 | 103 | 0.415 | 0.231 |
| -180 | 93 | 0.428[c] | 0.346 |

[a] Bunsen coefficient.

[b] Calculated by compiler.

[c] Obtained by extrapolation.

Figure: Solubility of nitrogen in $CF_4$ at -180°C

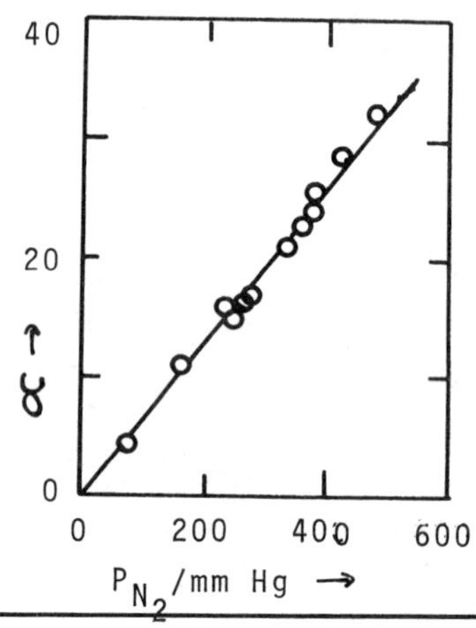


AUXILIARY INFORMATION

| METHOD/APPARATUS/PROCEDURE: | SOURCE AND PURITY OF MATERIALS: |
|---|---|
| Solubility was determined by a static method. The system was thermostated in a specially constructed cryostat. | (1) Purified by removing $O_2$ and $H_2O$<br>(2) 0.2% $N_2$<br>(2') 1.5% $N_2$ |
| | ESTIMATED ERROR:<br>$\delta T/K = \pm 0.1$<br>$\delta\alpha/\alpha = \pm 0.02$, compiler's estimate |
| | REFERENCES: |

**COMPONENTS:**

(1) Nitrogen; $N_2$; [7727-37-9]

(2) Dichlorodifluoromethane or Freon-12; $CCl_2F_2$; [75-71-8]

**ORIGINAL MEASUREMENTS:**

Steinberg, M.; Manowitz, B.; Pruzansky, J.

*US AEC BNL*-542 (T-140).
*Chem. Abstr.* 1959, *53*, 21242g

**VARIABLES:**

T/K = 199.65 - 276.15

**PREPARED BY:**

H. L. Clever

**EXPERIMENTAL VALUES:**

| Temperature $t/^{0}C$ | $T/K$ | Solubility Coefficient[1] | Henry's Constant[2] $H$/atm | Mol Fraction[3] $10^3 x_1$ |
|---|---|---|---|---|
| -73.5 | 199.65 | 0.65 | - | 2.05 |
| -70 | 203.15 | - | 510 | 1.96 |
| -30 | 243.15 | 0.44 | - | 1.51 |
| -12.3 | 260.85 | - | 600 | 1.67 |
| 0.0 | 273.15 | - | 630 | 1.59 |
| + 3.0 | 276.15 | 0.47 | - | 1.73 |

[1] The solubility coefficient is defined $cm^3(288.15K)\ cm^{-3}atm^{-1}$. The gas volume is reduced to 288.15 K and 1 atm, the solvent volume is for the temperature of the measurement.

[2] Henry's constant, $H/atm = (p_1/atm)/\ x_1$.

[3] The mole fraction values were calculated for a gas partial pressure of 101.325 kPa (1 atm) by the compiler. Density data were taken from Bichowsky, F. R.; Gilkey, W. K. *Ind. Eng. Chem.* 1931, *23*, 366 - 7.

**AUXILIARY INFORMATION**

**METHOD/APPARATUS/PROCEDURE:**

Dynamic tracer technique (1).

**SOURCE AND PURITY OF MATERIALS:**

(1) Nitrogen.

(2) Dichlorodifluoromethane.

No information.

**ESTIMATED ERROR:**

$\delta x_1/x_1 = \pm\ 0.05 - 0.10$ (compiler)

**REFERENCES:**

1. Steinberg, M.; Manowitz, B. *Ind. Eng. Chem.* 1959, *51*, 47.

COMPONENTS:

(1) Nitrogen; $N_2$; [7727-37-9]

(2) Tetrachloromethane; $CCl_4$; [56-23-5]

ORIGINAL MEASUREMENTS:

Horiuti, J.

*Sci. Pap. Inst. Phys. Chem. Res. (Jpn)* 1931/32, *17*, 125-256.

VARIABLES:

T/K = 253.45 - 333.25

Total P/kPa = 101.325

PREPARED BY:

M. E. Derrick
H. L. Clever

EXPERIMENTAL VALUES:

| T/K | $X_1 \times 10^3$ | Coefficient $\alpha$ | Coefficient L |
|---|---|---|---|
| 253.45 | 0.5558 | 0.1354 | 0.1256 |
| 273.15 | 0.5894 | 0.1403 | 0.1403 |
| 293.15 | 0.6304 | 0.1465 | 0.1572 |
| 313.15 | 0.6744 | 0.1529 | 0.1754 |
| 333.25 | 0.7208 | 0.1601 | 0.1953 |

Smoothed Data: $\log X_1 = -11.30302 + 4.07891/(T/100) + 2.3634 \ln (T/100)$

Standard error about the regression line = $9.19 \times 10^{-7}$

| T/K | Mol Fraction $X_1 \times 10^3$ |
|---|---|
| 253.15 | 0.5550 |
| 263.15 | 0.5721 |
| 273.15 | 0.5903 |
| 283.15 | 0.6096 |
| 293.15 | 0.6300 |

| T/K | Mol Fraction $X_1 \times 10^3$ |
|---|---|
| 303.15 | 0.6514 |
| 313.15 | 0.6738 |
| 323.15 | 0.6971 |
| 333.15 | 0.7213 |

If the gas and solution are assumed to be ideal the fitted equation gives the following thermodynamic values.

| T/K | $\Delta G^o$/kJ mol$^{-1}$ | $\Delta H^o$/J mol$^{-1}$ | $\Delta S^o$/J K$^{-1}$ mol$^{-1}$ | $\Delta C_p^o$/J K$^{-1}$ mol$^{-1}$ |
|---|---|---|---|---|
| 273.15 | 16.885 | 1976 | -54.58 | 19.65 |
| 298.15 | 18.228 | 2467 | -52.86 | 19.65 |

AUXILIARY INFORMATION

METHOD/APPARATUS/PROCEDURE:

The apparatus consists of a gas buret, a solvent reservoir, and an absorption pipet. The volume of the pipet was determined at various miniscus heights by weighing a quantity of water. The miniscus height was read with a cathetometer.

The dry gas was introduced into the degassed solvent. The gas and solvent were mixed with a magnetic stirrer until saturation. Care was taken to prevent solvent vapor from mixing with the solute gas in the gas buret. The volume of gas was determined from a gas buret reading, the volume of the solvent was determined from the miniscus height in the absorption pipet.

The Ostwald coefficient was measured at a total pressure of one atm. The Bunsen coefficient and the mole fraction solubility values were calculated by the compiler for a gas pressure of one atm assuming the gas to behave ideally and the Ostwald coefficient to be independent of pressure.

SOURCE AND PURITY OF MATERIALS:

(1) Nitrogen. Prepared in the laboratory by heating sodium azide (Kahlbaum, recrystalized and dried). Gas dried over phosphorus pentoxide.

(2) Tetrachloromethane. Kahlbaum. Dried, and distilled. Normal boiling point 76.74°C.

ESTIMATED ERROR:

$\delta T/K = 0.05$

**COMPONENTS:**

(1) Nitrogen; $N_2$; [7727-37-9]

(2) 2,2,2 - Trichloro,-1,1-ethanediol (chloral hydrate); $C_2H_3ClO_2$; [302-17-0]

(3) Water; $H_2O$; [7732-18-5]

**ORIGINAL MEASUREMENTS:**

Hammel, A. von

*Z. Physik. Chem.* 1915, *90*, 121-5.

**VARIABLES:**

T/K = 288
Concentration

**PREPARED BY:**

R. Battino

**EXPERIMENTAL VALUES:**

| wt. % (2) | $10^2$ L[a] | $10^2\alpha$[b] | wt. % (2) | $10^2$ L[a] | $10^2\alpha$[b] |
|---|---|---|---|---|---|
| | | 15°C (288 K) | | | |
| 0 | 1.725 | | 37.6 | 1.23 | 1.30 |
| 0 | 1.675 | 1.796[c] | 48.9 | 1.15 | 1.21 |
| 0 | 1.706 | | 49.3 | 1.18 | 1.24 |
| 6.9 | 1.64 | 1.73 | 61.3 | 1.14 | 1.20 |
| 14.0 | 1.54 | 1.62 | 70.9 | 1.31 | 1.38 |
| 15.0 | 1.52 | 1.60 | 71.2 | 1.30 | 1.37 |
| 23.6 | 1.34 | 1.41 | 78.3 | 1.52 | 1.60 |
| 26.1 | 1.41 | 1.49 | 79.1 | 1.56 | 1.65 |

[a] Ostwald coefficient.

[b] Bunsen coefficient.

[c] Average value. About 6% lower than best modern values

AUXILIARY INFORMATION

**METHOD/APPARATUS/PROCEDURE:**

The Ostwald apparatus as modified by Geffcken (1) was used.

**SOURCE AND PURITY OF MATERIALS:**

No details given.

**ESTIMATED ERROR:**

$\delta L/L = \pm 0.03$, compiler's estimate

**REFERENCES:**

1. Geffcken, G. *Z. Physik. Chem.* 1904, *49*, 257-302.

**COMPONENTS:**

(1) Nitrogen; $N_2$; [7727-37-9]

(2) 1,2-Dichloro-1,1,2,2-tetrafluoroethane or Freon-114; $C_2Cl_2F_4$; [76-14-2]

**ORIGINAL MEASUREMENTS:**

Williams, V. D.

*J. Chem. Eng. Data* 1959, *4*, 92-3.

**VARIABLES:**

T/K = 293 - 333

$N_2$ P/kPa = 70 - 900

**PREPARED BY:**

H. L. Clever

**EXPERIMENTAL VALUES:**

| Temperature t/°F | T/K | Henry's Constant $k/(\text{psia})^{-1} = x_1/p_1/\text{psia}$ |
|---|---|---|
| 68 | 293 | $5.5 \times 10^{-4}$ |
| 104 | 313 | $3.7 \times 10^{-4}$ |
| 140 | 333 | $2.0 \times 10^{-4}$ |

The nine to ten solubility measurements made at each temperature are shown in a graph. Only the Henry's constants were tabulated.

AUXILIARY INFORMATION

**METHOD/APPARATUS/PROCEDURE:**

The method and apparatus were similar to those of Parmlee (1).

A 450 g sample of Freon-114 was placed in a 1000 $cm^3$ bomb equiped with a three-way diaphram valve. The Freon-114 was frozen and pumped on to degas. Its vapor pressure was checked at 293 K as a measure of its purity. Nitrogen gas was added to the bomb. The bomb was rocked for ½ hour after pressure equilibrium was reached.

Both the gas phase and the liquid phase were sampled. They were analyzed by a mass spectroscope.

**SOURCE AND PURITY OF MATERIALS:**

No information

**ESTIMATED ERROR:**

$\delta k/k = 0.05$

**REFERENCES:**

1. Parmelee, H. M. *Refrig. Eng.* 1951, *59*, 573.

**COMPONENTS:**

(1) Nitrogen; $N_2$; [7727-37-9]

(2) 1,1,2-trichloro-1,2,2-trifluoroethane; $C_2Cl_3F_3$ or ($CCl_2F \cdot CClF_2$) [76-13-1]

**ORIGINAL MEASUREMENTS:**

Hiraoka, H.; Hildebrand, J. H.

*J. Phys. Chem.* 1964, *68*, 213-4.

**VARIABLES:**

T/K = 277.15 - 308.15
P/kPa = 101.325

**PREPARED BY:**

M. E. Derrick
A. L. Cramer

**EXPERIMENTAL VALUES:**

| Temperature t/°C | T/K | Mol Fraction $x_1 \times 10^3$ | Bunsen Coefficient $\alpha$ | Ostwald Coefficient L |
|---|---|---|---|---|
| 4.00 | 277.15 | 19.33 | 0.373 | 0.378 |
| 15.00 | 288.15 | 19.46 | 0.370 | 0.390 |
| 24.98 | 298.13 | 19.31 | 0.362 | 0.395 |
| 35.00 | 308.15 | 19.88 | 0.367 | 0.414 |

Smoothed Data: $\ln x_1 = -6.017 - 0.6519/(T/100)$

| T/K | Mol Fraction $x_1 \times 10^4$ |
|---|---|
| 278.15 | 19.27 |
| 283.15 | 19.35 |
| 288.15 | 19.43 |
| 293.15 | 19.50 |
| 298.15 | 19.58 |
| 303.15 | 19.65 |
| 308.15 | 19.72 |

**AUXILIARY INFORMATION**

**METHOD/APPARATUS/PROCEDURE:**

The degassed liquid was saturated with nitrogen gas having a pressure near 1 atm. The volume of gas absorbed by the liquid was determined by P-V measurements in a gas buret (1).

A thermostated gas buret was used for solution measurements. A thermostated pipet containing a magnetic stirrer was used as a solvent reservoir. The liquid volume was determined by the difference in volume between the pipet and the mercury. The connection between the buret and pipet prevents gas and solvent vapor from mixing.

**SOURCE AND PURITY OF MATERIALS:**

(1) Nitrogen. General Dynamic Corp., passed through cold trap.

(2) 1,1,2-Trichloro-1,2,2-trifluoroethane. Union Carbide Co. Distilled, purity checked by ultraviolet absorbance.

**ESTIMATED ERROR:**

$\delta T/K = 0.02$
$\delta x_1/x_1 = 0.003$

**REFERENCES:**

1. Kobatake, Y.; Hildebrand, J. H. *J. Phys. Chem.* 1961, *65*, 331.

| COMPONENTS: | ORIGINAL MEASUREMENTS: |
|---|---|
| (1) Nitrogen; $N_2$; [7727-37-9]<br>(2) Chlorobenzene; $C_6H_5Cl$; [108-90-7] | Horiuti, J.<br>*Sci. Pap. Inst. Phys. Chem. Res. (Jpn)* 1931/32, *17*, 125-256. |
| VARIABLES: | PREPARED BY: |
| T/K = 233.45 - 353.45<br>Total P/kPa = 101.325 | M. E. Derrick<br>H. L. Clever |

EXPERIMENTAL VALUES:

| T/K | Mol Fraction $X_1 \times 10^3$ | Bunsen Coefficient $\alpha$ | Ostwald Coefficient L |
|---|---|---|---|
| 233.45 | 0.349 | 0.0813 | 0.0695 |
| 253.45 | 0.366 | 0.0838 | 0.0778 |
| 273.15 | 0.392 | 0.0881 | 0.0881 |
| 293.15 | 0.420 | 0.0926 | 0.0994 |
| 313.25 | 0.450 | 0.0973 | 0.1116 |
| 333.20 | 0.487 | 0.1032 | 0.1259 |
| 353.45 | 0.521 | 0.1081 | 0.1399 |

Smoothed Data: $\ln X_1 = -12.41926 + 4.99558/(T/100) + 2.73200 \ln (T/100)$

Standard error about the regression line = $1.83 \times 10^{-6}$

| T/K | Mol Fraction $X_1 \times 10^3$ |
|---|---|
| 233.15 | 0.348 |
| 243.15 | 0.357 |
| 253.15 | 0.368 |
| 263.15 | 0.379 |
| 273.15 | 0.392 |
| 283.15 | 0.405 |
| 293.15 | 0.419 |

| T/K | Mol Fraction $X_1 \times 10^3$ |
|---|---|
| 303.15 | 0.434 |
| 313.15 | 0.450 |
| 323.15 | 0.467 |
| 333.15 | 0.485 |
| 343.15 | 0.503 |
| 353.15 | 0.522 |

The Ostwald coefficient was measured at a total pressure of one atm. The Bunsen coefficient and the mole fraction solubility values were calculated by the compiler for a gas pressure of one atm assuming the gas to behave ideally and the Ostwald coefficient to be independent of pressure.

AUXILIARY INFORMATION

METHOD/APPARATUS/PROCEDURE:

The apparatus consists of a gas buret, a solvent reservoir, and an absorption pipet. The volume of the pipet was determined at various miniscus heights by weighing a quantity of water. The miniscus height was read with a cathetometer.

The dry gas introduced into the degassed solvent. The gas and solvent are mixed with a magnetic stirrer until saturation. Care taken to prevent solvent vapor from mixing with the solute gas in the gas buret. The volume of gas was determined from a gas buret reading, the volume of the solvent was determined from the miniscus height in the absorption pipet.

SOURCE AND PURITY OF MATERIALS:

(1) Nitrogen. Prepared in the laboratory by heating sodium azide (Kahlbaum, recrystalized and dried). Gas dried over phosphorous pentoxide.

(2) Chlorobenzene. Kahlbaum. Dried, and distilled. Normal boiling point 131.96°C.

ESTIMATED ERROR:

$\delta T/K = 0.05$

If the gas and solution are assumed to be ideal the fitted equation gives the following thermodynamic values.

| T/K | $\Delta G^o$/kJ mol$^{-1}$ | $\Delta H^o$/J mol$^{-1}$ | $\Delta S^0$/J K$^{-1}$ mol$^{-1}$ | $\Delta C_p^o$/J K$^{-1}$ mol$^{-1}$ |
|---|---|---|---|---|
| 273.15 | 17.817 | 2051 | -57.72 | 22.71 |
| 298.15 | 19.234 | 2619 | -55.73 | 22.71 |

**COMPONENTS:**

(1) Nitrogen; $N_2$; [7727-37-9]

(2) 1-Chlorohexane; $C_6H_{13}Cl$; [544-10-5]

**ORIGINAL MEASUREMENTS:**

Guerry, D. Jr.

Ph.D. thesis, 1944
Vanderbilt University
Nashville, TN

**VARIABLES:**

T/K = 293-298
P/kPa = 101.325

**PREPARED BY:**

H. L. Clever

**EXPERIMENTAL VALUES:**

| T/K | Mol Fraction $x_1 \times 10^4$ | Bunsen Coefficient $\alpha$ | Ostwald Coefficient L |
|---|---|---|---|
| 293.15 | 8.02 | 0.131 | 0.141 |
| 298.15 | 8.19 | 0.133 | 0.145 |

The Ostwald coefficients were calculated by the compiler.

AUXILIARY INFORMATION

**METHOD/APPARATUS/PROCEDURE:**

A Van Slyke-Neill Manometric Apparatus manufactured by the Eimer and Amend Co. was used.

The procedure of Van Slyke (1) for pure liquids was modified (2) so that small solvent samples (2 $cm^3$) could be used with almost complete recovery of the sample.

An improved temperature control system was used.

**SOURCE AND PURITY OF MATERIALS:**

(1) Nitrogen. Ohio Chemical Co. Oil pumped, 99.8 per cent pure.

(2) 1-Chlorohexane. Eastman Kodak Co. Purified, distilled from $P_2O_5$ in a $N_2$ atm. B.p. (746.6 mmHg) t/°C 134.66 (corr.). Refractive index, density, and vapor pressure data are in the thesis.

**ESTIMATED ERROR:**

$\delta T/K = 0.05$

**REFERENCES:**

1. Van Slyke, D. D. *J. Biol. Chem.* 1939, *130*, 545.

2. Ijams, C. C. Ph.D. thesis, 1941 Vanderbilt University

| COMPONENTS: | ORIGINAL MEASUREMENTS: |
|---|---|
| (1) Nitrogen; $N_2$; [7727-37-9]<br>(2) Hexafluorobenzene; $C_6F_6$; [392-56-3] | Evans, F.D.; Battino, R.<br>*J. Chem. Thermodyn.* 1971, *3*, 753-60. |

| VARIABLES: | PREPARED BY: |
|---|---|
| T/K = 282.94 - 297.85<br>P/kPa = 101.325 | H. L. Clever |

EXPERIMENTAL VALUES:

| T/K | $X_1 \times 10^3$ | Coefficient $\alpha$ | Coefficient L |
|---|---|---|---|
| 282.94 | 1.799 | 0.356 | 0.369 |
| 282.94 | 1.794 | 0.343 | 0.368 |
| 297.56 | 1.808 | 0.351 | 0.382 |
| 297.85 | 1.783 | 0.346 | 0.377 |

The Bunsen coefficients were calculated by the compiler.

The solubility values were adjusted to an oxygen partial pressure of 101.325 kPa (1 atm) by Henry's law.

Smoothed Data: $\Delta G^o/J\ mol^{-1} = -RT \ln X_1 = -34.1 + 52.682\ T$

Std. Dev. $\Delta G^o$ = 14.3, Coef. Corr. = 0.9996

$\Delta H^o/J\ mol^{-1} = -34.1$, $\Delta S^o/J\ K^{-1}\ mol^{-1} = -52.682$

| T/K | Mol Fraction $X_1 \times 10^3$ | $\Delta G^o/J\ mol^{-1}$ |
|---|---|---|
| 278.15 | 1.80 | 14,621 |
| 283.15 | 1.80 | 14,884 |
| 288.15 | 1.79 | 15,146 |
| 293.15 | 1.79 | 15,410 |
| 298.15 | 1.79 | 15,673 |

AUXILIARY INFORMATION

METHOD/APPARATUS/PROCEDURE:

The solubility apparatus is based on the design of Morrison and Billett (1) and the version used is described by Battino, Evans, and Danforth (2). The degassing apparatus is that described by Battino, Banzhof, Bogan, and Wilhelm (3).

Degassing: Up to 500 $cm^3$ of solvent is placed in a flask of such size that the liquid is about 4 cm deep. The liquid is rapidly stirred, and vacuum is intermittently applied through a liquid $N_2$ trap until the permanent gas residual pressure drops to 5 microns.

Solubility Determination: The degassed solvent is passed in a thin film down a glass helical tube containing solute gas plus the solvent vapor at a total pressure of one atm. The volume of gas absorbed is found by difference between the initial and final volumes in the buret system. The solvent is collected in a tared flask and weighed.

SOURCE AND PURITY OF MATERIALS:

(1) Nitrogen. Air Products and Chemicals Co. 99.6 per cent.

(2) Hexafluorobenzene. Imperiel Smelting Co., Avnomouth, U.K. GC purity 99.7%, density, $\rho_{298.15} = 1.60596$ g $cm^{-3}$. Purification described *Anal. Chem.* 1968, *40*, 224.

ESTIMATED ERROR:

$\delta T/K = 0.03$
$\delta P/mmHg = 0.5$
$\delta X_1/X_1 = 0.005$

REFERENCES:

1. Morrison, T. J.; Billett, F. *J. Chem. Soc.* 1948, 2033.
2. Battino, R.; Evans, F. D.; Danforth, W. F. *J. Am. Oil Chem. Soc.* 1968, *45*, 830.
3. Battino, R.; Banzhof, M.; Bogan, M.; Wilhelm, E. *Anal. Chem.* 1971, *43*, 806.

**COMPONENTS:**

(1) Nitrogen; $N_2$; [7727-37-9]

(2) 1-Bromoheptane; $C_7H_{15}Br$; [629-04-9]

**ORIGINAL MEASUREMENTS:**

Ijams, C. C.

Ph.D. thesis, 1941
Vanderbilt University
Nashville, TN

**VARIABLES:**

T/K = 298.15
P/kPa = 101.325

**PREPARED BY:**

**EXPERIMENTAL VALUES:**

| T/K | Mol Fraction $x_1 \times 10^3$ | Bunsen Coefficient $\alpha$ | Ostwald Coefficient L |
|---|---|---|---|
| 298.15 | 0.675 | 0.0964 | 0.1052 |

The mole fraction and Ostwald coefficient values were calculated by the compiler.

AUXILIARY INFORMATION

**METHOD/APPARATUS/PROCEDURE:**

A Van Slyke-Neill Manometric Apparatus manufactured by the Eimer and Amend Co. was used.

The procedure of Van Slyke (1) for pure liquids was modified so that small solvent sample (2 $cm^3$) could be used with almost 100 per cent recovery of the solvent.

An improved temperature control system was used.

**SOURCE AND PURITY OF MATERIALS:**

(1) Nitrogen. Air Reduction Co. Purity 99.5 per cent.

(2) 1-Bromoheptane. Purified. Vapor pressure and boiling point also reported in the thesis.

**ESTIMATED ERROR:**

$\delta T/K = 0.05$

**REFERENCES:**

1. Van Slyke, D. D. *J. Biol. Chem.* 1939, *130*, 545.

COMPONENTS:

(1) Nitrogen; $N_2$; [7727-37-9]

(2) Undecafluoro (trifluoromethyl)-cyclohexane or perfluoromethyl cyclohexane; $C_7F_{14}$; [355-02-2]

ORIGINAL MEASUREMENTS:

Gjaldbaek, J. C.; Hildebrand, J. H.

*J. Am. Chem. Soc.* 1949, *71*, 3147-50.

VARIABLES:

T/K = 298.05 - 298.15
Total P/kPa = 101.325

PREPARED BY:

J. Chr. Gjaldbaek

EXPERIMENTAL VALUES:

| | $X_1 \times 10^3$ | Coefficient $\alpha$ | Coefficient L |
|---|---|---|---|
| 298.05 | 3.25 | 0.371 | 0.405 |
| 298.15 | 3.36 | 0.384 | 0.419 |

One value from the paper was rejected.

The mole fraction and Ostwald solubility values were calculated by the compiler.

AUXILIARY INFORMATION

METHOD/APPARATUS/PROCEDURE:

A calibrated all-glass combined manometer and bulb was enclosed in an air thermostat and shaken until equilibrium. Mercury was used for calibration and as the confining liquid. The solvents were degassed in the apparatus. Details are in references 1 and 2.

The absorbed volume of gas was calculated from the initial and final amounts, both saturated with solvent vapor. The amount of solvent was determined by the weight of displaced mercury.

The saturation of the liquid with the gas was carried out close to atmospheric pressure. The solubility values were reported for one atmosphere gas pressure assuming Henry's law is obeyed.

SOURCE AND PURITY OF MATERIALS:

(1) Nitrogen. Stuart Oxygen Co. 99.996 percent $N_2$.

(2) Undecafluoro(trifluoromethyl)-cyclohexane. E. I. duPont Co. Distilled, boiling point 76.5 - 76.6 at 760 mmHg.

ESTIMATED ERROR:

$\delta T/K = 0.05$
$\delta X_1/X_1 = 0.015$

REFERENCES:

1. Lannung, A. *J. Am. Chem. Soc.* 1930, *52*, 68.

2. Gjaldbaek, J. C. *Acta Chem. Scand.* 1952, *6*, 623.

**COMPONENTS:**

(1) Nitrogen; $N_2$; [7727-37-9]

(2) Hexadecafluoroheptane or Perfluoroheptane; $C_7F_{16}$; [335-57-9]

**ORIGINAL MEASUREMENTS:**

Gjaldbaek, J. C.; Hildebrand, J. H.

*J. Am. Chem. Soc.* 1949, *71*, 3147-50.

**VARIABLES:**

T/K = 275.13 - 321.75
Total P/kPa = 101.325

**PREPARED BY:**

J. Chr. Gjaldbaek

**EXPERIMENTAL VALUES:**

| T/K | Mol Fraction $X_1 \times 10^3$ | Bunsen Coefficient $\alpha$ | Ostwald Coefficient $L$ |
|---|---|---|---|
| 275.13 | 3.98 | 0.406 | 0.409 |
| 275.28 | 4.01 | 0.409 | 0.412 |
| 275.60 | 3.96 | 0.404 | 0.408 |
| 277.25 | 3.99 | 0.406 | 0.412 |
| 298.06 | 3.90 | 0.384 | 0.419 |
| 298.12 | 3.93 | 0.387 | 0.422 |
| 320.06 | 3.82 | 0.362 | 0.424 |
| 321.75 | 3.82 | 0.362 | 0.426 |

The mole fraction and Ostwald solubility values were calculated by the compiler.

Smoothed Data: $\Delta G^\circ/J\ mol^{-1} = -RT \ln X_1 = -679.41 + 48.394T$

Std. dev. $\Delta G^\circ$ = 11.0, Coef. Corr. = 0.9998

$\Delta H^\circ/J\ mol^{-1}$ - -679.41, $\Delta S^\circ/J\ K^{-1}\ mol^{-1}$ = -48.394

| T/K | Mol Fraction $X_1 \times 10^3$ | $\Delta G^\circ/J\ mol^{-1}$ |
|---|---|---|
| 273.15 | 4.00 | 12,539 |
| 283.15 | 3.96 | 13,023 |
| 293.15 | 3.92 | 13,507 |
| 298.15 | 3.90 | 13,749 |
| 303.15 | 3.88 | 13,991 |
| 313.15 | 3.85 | 14,475 |
| 323.15 | 3.82 | 14,959 |

## AUXILIARY INFORMATION

**METHOD/APPARATUS/PROCEDURE:**

A calibrated all-glass combined manometer and bulb was enclosed in an air thermostat and shaken until equilibrium. Mercury was used for calibration and as the confining liquid. The solvents were degassed in the apparatus. Details are in references 1 and 2.

The absorped volume of gas was calculated from the initial and final amounts, both saturated with solvent vapor. The amount of solvent was determined by the weight of displaced mercury.

The saturation of the liquid with the gas was carried out close to atmospheric pressure. The solubility values were reported for one atmosphere gas pressure assuming Henry's law is obeyed.

**SOURCE AND PURITY OF MATERIALS:**

(1) Nitrogen. Stuart Oxygen Co. 99.996 percent $N_2$.

(2) Hexadecafluoroheptane. E. I. duPont Co. Distilled, boiling point 82.3 - 82.4°C, at 753.7 mmHg.

**ESTIMATED ERROR:**

$\delta T/K = 0.05$
$\delta X_1/X_1 = 0.015$

**REFERENCES:**

1. Lannung, A. *J. Am. Chem. Soc.* 1930, *52*, 68.

2. Gjaldbaek, J. C. *Acta Chem. Scand.* 1952, *6*, 623.

**COMPONENTS:**

(1) Nitrogen; $N_2$; [7727-37-9]

(2) Decafluorobis (trifluoromethyl) cyclohexane or perfluoro dimethylcyclohexane; $C_8F_{16}$; Isomer not specified.

**ORIGINAL MEASUREMENTS:**

Gjaldbaek, J. C.; Hildebrand, J. H.

*J. Am. Chem. Soc.* 1949, *71*, 3147-50.

**VARIABLES:**

T/K = 298.10 - 320.25
Total P/kPa = 101.325

**PREPARED BY:**

J. Chr. Gjaldbaek

**EXPERIMENTAL VALUES:**

| T/K | Mol Fraction $X_1 \times 10^3$ | Bunsen Coefficient $\alpha$ | Ostwald Coefficient L |
|---|---|---|---|
| 298.15 | 3.19 | 0.329 | 0.359 |
| 311.12 | 3.20 | 0.330 | 0.376 |
| 313.67 | 3.16 | 0.319 | 0.366 |
| 320.25 | 3.19 | 0.319 | 0.374 |

The mole fraction and Ostwald solubility values were calculated by the compiler.

Smoothed Data: $\Delta G^o/J\ mol^{-1} = -RT \ln X_1 = -79.059 + 48.056\ T$

Std. dev. $\Delta G^o$ = 2.3, Coef. Corr. = 0.9999

$\Delta H^o/J\ mol^{-1} = -79.06$, $\Delta S^o/J\ K^{-1}\ mol^{-1} = -48.06$

| T/K | Mol Fraction $X_1 \times 10^3$ | $\Delta G^o/J\ mol^{-1}$ |
|---|---|---|
| 298.15 | 3.19 | 14,249 |
| 308.15 | $3.18_5$ | 14,730 |
| 318.15 | 3.18 | 15,210 |

**AUXILIARY INFORMATION**

**METHOD/APPARATUS/PROCEDURE:**

A calibrated all-glass combined manometer and bulb was enclosed in an air thermostat and shaken until equilibrium. Mercury was used for calibration and as the confining liquid. The solvents were degassed in the apparatus. Details are in references 1 and 2.

The absorbed volume of gas was calculated from the initial and final amounts, both saturated with solvent vapor. The amount of solvent was determined by the weight of displaced mercury.

The saturation of the liquid with the gas was carried out close to atmospheric pressure. The solubility values were reported for one atmosphere gas pressure assuming Henry's law is obeyed.

**SOURCE AND PURITY OF MATERIALS:**

(1) Nitrogen. Dansk Itl-og Brint Fabrik. Contained less than 0.13 percent oxygen.

(2) Decafluorobis (trifluoromethyl)-cyclohexane. E. I. duPont Co. Distilled, boiling point 102.4 - 102.6 at 760.0 mmHg.

**ESTIMATED ERROR:**

$\delta T/K = 0.05$
$\delta X_1/X_1 = 0.015$

**REFERENCES:**

1. Lannung, A. *J. Am. Chem. Soc.* 1930, *52*, 68.

2. Gjaldbaek, J. C. *Acta Chem. Scand.* 1952, *6*, 623.

**COMPONENTS:**

(1) Nitrogen; $N_2$; [7727-37-9]

(2) FC-80, which is mostly a mixture of isomers of perfluorobutyltetrahydrofuran. For example, 2,2,3,3,4,4,5-heptafluorotetrahydro-5(nonafluorobutyl)furan; $C_8F_{16}O$; [335-36-4]

**ORIGINAL MEASUREMENTS:**

Sargent, J. W.; Seffl, R. J.

*Fed. Proc.* 1970, *29*, 1699 - 1703.

**VARIABLES:**

T/K = 298-310

Total P/kPa = 101.325

**PREPARED BY:**

A. L. Cramer

H. L. Clever

**EXPERIMENTAL VALUES:**

| Temperature t/°C | T/K | Nitrogen Solubility g $O_2$ 1000 $g^{-1}$ | cm$^3$ $O_2$ 100 cm$^{-3}$ |
|---|---|---|---|
| 25 | 298.15 | 0.216 | 33.4 |
| 37 | 310.15 | 0.217 | 34.0 |

The authors solubility, cm$^3$ $N_2$ 100 cm$^{-3}$, appears to be the Ostwald coefficient times 100.

AUXILIARY INFORMATION

**METHOD/APPARATUS/PROCEDURE:**

The equilibrium apparatus is a 50 cm$^3$ three neck flask equipped with a water condenser, a thermometer, and a gas dispersing tube extending below the fluid surface level. Between 25 and 35 cm$^3$ of solvent are placed in the flask. Gas is bubbled through the solvent for 2 - 3 h. An 80 microliter sample is slowly taken into a gas tight syringe. The sample is injected into a gas chromatograph. The results of three sampling and analyses are averaged.

The chromatograph is an F and M Model 720 equipped with a thermal conductivity detector at 320 °C, and a 6' by ¼" stainless steel column packed with 13X molecular sieve at room temperature. The apparatus is calibrated with pure gas samples.

Details of the procedure were furnished by R. D. Danielson of the 3 M Co.

**SOURCE AND PURITY OF MATERIALS:**

(1) Nitrogen. Source not given. Commercial cylinder.

(2) FC-80. Minnesota Mining and Manufacturing Co. All H in an organic compound are replaced by F by Simons electrochemical process (1)

**ESTIMATED ERROR:**

$\delta T/K = 0.1$

$\delta c/c = 0.01$ (reproducibility)

$= 0.05$ (absolute)

**REFERENCES:**

1. Simons, J. H., Editor *Fluorine Chemistry* Academic Press, New York, 1950.

COMPONENTS:

(1) Nitrogen; $N_2$; [7727-37-9]

(2) Heptafluorotetrahydro(nonafluorobutyl)-furan or Perfluorobutyl perfluorotetrahydrofuran or FC-80; $C_8F_{16}O$; [40464-54-8]

ORIGINAL MEASUREMENTS:

Tham, M. K.; Walker, R. D. Jr.; Modell, J. H.

*J. Chem. Eng. Data* 1973, *18*, 385-6.

VARIABLES:

T/K = 298.15 - 323.15
Total P/kPa = 101.325

PREPARED BY:

T. D. Kittredge
H. L. Clever

EXPERIMENTAL VALUES:

| T/K | Mol Fraction $X_1 \times 10^3$ | Bunsen Coefficient $\alpha$ | Ostwald Coefficient L |
|---|---|---|---|
| 298.15 | 4.30 | 0.411 | 0.449 |
| 303.15 | 4.18 | 0.396 | 0.439 |
| 310.15 | 4.03 | 0.378 | 0.429 |
| 313.15 | 4.00 | 0.373 | 0.428 |
| 323.15 | 3.80 | 0.349 | 0.413 |

The Bunsen and Ostwald coefficients were calculated by the compiler.

Smoothed Data: $\Delta G^o/J\ mol^{-1} = -RT\ ln\ X_1 = -3910.8 + 58.430\ T$

Std. dev. $\Delta G^o$ = 6.9, Coef. corr. = 0.9998

$\Delta H^o/J\ mol^{-1} = -3910.8$, $\Delta S^o/J\ K^{-1}\ mol^{-1} = -58.430$

| T/K | Mol Fraction $X_1 \times 10^3$ | $\Delta G^o/J\ mol^{-1}$ |
|---|---|---|
| 298.15 | 4.30 | 13,510 |
| 303.15 | 4.19 | 13,802 |
| 308.15 | 4.08 | 14,094 |
| 313.15 | 3.98 | 14,386 |
| 318.15 | 3.89 | 14,679 |
| 323.15 | 3.80 | 14,971 |

AUXILIARY INFORMATION

METHOD/APPARATUS/PROCEDURE:

The apparatus and procedure described by Shoor, Walker and Gubbins (1) were used. The solvent was placed in a saturator immersed in a constant temperature bath. The gas was dispersed through the liquid by a fritted glass disk at a rate of about 40 $cm^3$ $m^{-1}$. After an initial equilibration period of 30 minutes, samples were withdrawn with a Hamilton microliter syringe at 15 minute intervals and analyzed chromatographically (Perkin-Elmer 900 gas chromatograph with thermal conductivity detector). The solubility at one atmosphere partial pressure of the gas was the mean value of three consecutive samples which agreed within the limits of experimental error.

SOURCE AND PURITY OF MATERIALS:

(1) Nitrogen. Source not given. Minimum purity 99.6 per cent.

(2) FC-80. Minnesota Mining and Manufacturing Co. A mixture of several isomers. The authors also measured the density and viscosity of the solvent at several temperatures. At 298.15 K the density is 1.7657 g $cm^{-3}$ and the viscosity is 1.3953 centipoise.

ESTIMATED ERROR:

$\delta T/K = 0.05$
$\delta X_1/X_1 = 0.005$ (compiler)

REFERENCES:

1. Shoor, S. K.; Walker, R. D. Jr.; Gubbins, K. E. *J. Phys. Chem.* 1969, *73*, 312.

**COMPONENTS:**

(1) Nitrogen; $N_2$; [7727-37-9]

(2) 1,1,1,2,2,3,3,4,4,5,5,6,6-Trideca-fluoro-6-[1,2,2,2-tetrafluoro-1-(trifluoromethyl)ethoxy]-hexane or Perfluoro-1-isopropoxy hexane or Caroxin-F; $C_9F_{20}O$; [37340-18-4] or [41719-16-8]

**ORIGINAL MEASUREMENTS:**

Tham, M. K.; Walker, R. D. Jr.; Modell, J. H.

*J. Chem. Eng. Data* 1973, *18*, 385-6.

**VARIABLES:**

T/K = 298.15 - 323.15
Total P/kPa = 101.325

**PREPARED BY:**

T. D. Kittredge
H. L. Clever

**EXPERIMENTAL VALUES:**

| T/K | Mol Fraction $X_1 \times 10^3$ | Bunsen Coefficient $\alpha$ | Ostwald Coefficient L |
|---|---|---|---|
| 298.15 | 5.10 | 0.398 | 0.434 |
| 303.15 | 5.05 | 0.391 | 0.434 |
| 310.15 | 4.97 | 0.381 | 0.433 |
| 313.15 | 4.92 | 0.375 | 0.430 |
| 323.15 | 4.80 | 0.361 | 0.427 |

The Bunsen and Ostwald coefficients were calculated by the compiler.

Smoothed Data: $\Delta G^o/J\ mol^{-1} = -RT\ ln\ X_1 = -1976.7 + 50.498\ T$

Std. dev. $\Delta G^o$ = 5.0, Coef. corr. = 0.9998

$\Delta H^o/J\ mol^{-1} = -1976.7$, $\Delta S^o/J\ K^{-1}\ mol^{-1} = -50.498$

| T/K | Mol Fraction $X_1 \times 10^3$ | $\Delta G^o/J\ mol^{-1}$ |
|---|---|---|
| 298.15 | 5.71 | 13,079 |
| 303.15 | 5.04 | 13,332 |
| 308.15 | 4.98 | 13,584 |
| 313.15 | 4.92 | 13,837 |
| 318.15 | 4.86 | 14,089 |
| 323.15 | 4.81 | 14,341 |

AUXILIARY INFORMATION

**METHOD/APPARATUS/PROCEDURE:**

The apparatus and procedure described by Shoor, Walker and Gubbins (1) were used. The solvent was placed in a saturator immersed in a constant temperature bath. The gas was dispersed through the liquid by a fritted glass disk at a rate of about 40 $cm^3\ m^{-1}$. After an initial equilibration period of 30 minutes, samples were withdrawn with a Hamilton microliter syringe at 15 minute intervals and analyzed chromatographically (Perkin-Elmer 900 gas chromatograph with thermal conductivity detector). The solubility at one atmosphere partial pressure of the gas was the mean value of three consecutive samples which agreed within the limits of experimental error.

**SOURCE AND PURITY OF MATERIALS:**

(1) Nitrogen. Source not given. Minimum purity 99.6 per cent.

(2) Caroxin-F. Allied Chemical Corp. Purity > 99.9 per cent as supplied. The authors also measured the density and viscosity of the solvent at several temperatures. At 298.15 K the density is 1.7449 g $cm^{-3}$ and the viscosity is 1.5405 centipoise.

**ESTIMATED ERROR:**

$\delta T/K = 0.05$
$\delta X_1/X_1 = 0.005$ (Compiler)

**REFERENCES:**

1. Shoor, S. K.; Walker, R. D. Jr.; Gubbins, K. E. *J. Phys. Chem.* 1969, *73*, 312.

**COMPONENTS:**

(1) Nitrogen; $N_2$; [7727-37-9]

(2) 1,1,2,2,3,3,4,4-Octafluoro-1,4-bis(1,2,2,2-tetrafluoro-1-(trifluoromethyl)ethoxy) butane or Perfluoro-1,4-diisopropoxy butane or Caroxin-D; $C_{10}F_{22}O_2$; [23228-90-2]

**ORIGINAL MEASUREMENTS:**

Tham, M. K.; Walker, R. D. Jr.; Modell, J. H.

*J. Chem. Eng. Data* 1973, *18*, 385-6.

**VARIABLES:**

T/K = 298.15 - 323.15
Total P/kPa = 101.325

**PREPARED BY:**

T. D. Kittredge
H. L. Clever

**EXPERIMENTAL VALUES:**

| T/K | Mol Fraction $X_1 \times 10^3$ | Bunsen Coefficient $\alpha$ | Ostwald Coefficient $L$ |
|---|---|---|---|
| 298.15 | 5.00 | 0.345 | 0.377 |
| 303.15 | 4.94 | 0.338 | 0.375 |
| 310.15 | 4.89 | 0.331 | 0.376 |
| 313.15 | 4.83 | 0.326 | 0.374 |
| 323.15 | 4.75 | 0.316 | 0.374 |

The Bunsen and Ostwald coefficients were calculated by the compiler.

Smoothed Data: $\Delta G^o/J\ mol^{-1} = -RT \ln X_1 = -1663.9 + 49.631\ T$

Std. dev. $\Delta G^o = 5.93$, Coef. corr. = 1.000

$\Delta H^o/J\ mol^{-1} = -1663.9$, $\Delta S^o/J\ K^{-1}\ mol^{-1} = -49.631$

| T/K | Mol Fraction $X_1 \times 10^3$ | $\Delta G^o/J\ mol^{-1}$ |
|---|---|---|
| 298.15 | 5.00 | 13,134 |
| 303.15 | 4.95 | 13,382 |
| 308.15 | 4.89 | 13,630 |
| 313.15 | 4.84 | 13,878 |
| 318.15 | 4.79 | 14,126 |
| 323.15 | 4.75 | 14,374 |

AUXILIARY INFORMATION

**METHOD/APPARATUS/PROCEDURE:**

The apparatus and procedure described by Shoor, Walker and Gubbins (1) were used. The solvent was placed in a saturator immersed in a constant temperature bath. The gas was dispersed through the liquid by a fritted glass disk at a rate of about 40 $cm^3\ m^{-1}$. After an initial equilibration period of 30 minutes, samples were withdrawn with a Hamilton microliter syringe at 15 minute intervals and analyzed chromatographically (Perkin-Elmer 900 gas chromatograph with thermal conductivity detector). The solubility at one atmosphere partial pressure of the gas was the mean value of three consecutive samples which agreed within the limits of experimental error.

**SOURCE AND PURITY OF MATERIALS:**

(1) Nitrogen. Source not given. Minimum purity 99.6 per cent.

(2) Caroxin-D. Allied Chemical Corp. Purity > 99.9 per cent as supplied. The authors also measured the density and viscosity of the solvent. At 298.15 K the density is 1.7465 g $cm^{-3}$ and the viscosity is 2.0579 centipoise.

**ESTIMATED ERROR:**

$\delta T/K = 0.05$
$\delta X_1/X_1 = 0.005$ (Compiler)

**REFERENCES:**

1. Shoor, S. K.; Walker, R. D.; Gubbins, K. E. *J. Phys. Chem.* 1969, *73*, 312.

**COMPONENTS:**

(1) Nitrogen; $N_2$; [7727-37-9]

(2) L-1822, which is a mixture of mostly 10-carbon fluorocarbons, including cyclic, open chain, and branched molecules.

**ORIGINAL MEASUREMENTS:**

Sargent, J. W.; Seffl, R. J.

*Fed. Proc.* 1970, *29*, 1699-1703.

**VARIABLES:**

T/K = 298.15, 310.15
Total P/kPa = 101.325

**PREPARED BY:**

A. L. Cramer
H. L. Clever

**EXPERIMENTAL VALUES:**

| Temperature | | Nitrogen Solubility | |
|---|---|---|---|
| t/°C | T/K | g $O_2$ 1000 $g^{-1}$ | $cm^3$ $O_2$ 100 $cm^{-3}$ |
| 25 | 298.15 | 0.184 | 30.6 |
| 37 | 310.15 | 0.162 | 27.8 |

The authors solubility, $cm^3$ $N_2$ 100 $cm^{-3}$, appears to be the Ostwald coefficient times 100.

AUXILIARY INFORMATION

**METHOD/APPARATUS/PROCEDURE:**

The equilibrium apparatus is a 50 $cm^3$ three neck flask equipped with a water condenser, a thermometer, and a gas dispersing tube extending below the fluid surface level. Between 25 and 35 $cm^3$ of solvent are placed in the flask. Gas is bubbled through the solvent for 2 - 3 h. An 80 micro-liter sample is slowly taken into a gas tight syringe. The sample is injected into a gas chromatograph. The results of three sampling and analyses are averaged.

The chromatograph is an F and M Model 720 equipped with a thermal conductivity detector at 320 °C, and a 6' by ¼" stainless steel column packed with 13X molecular sieve at room temperature. The apparatus is calibrated with pure gas samples.

Details of the procedure were furnished by R. D. Danielson of the 3 M Co.

**SOURCE AND PURITY OF MATERIALS:**

(1) Nitrogen. Source not given. Commercial cylinder.

(2) L-1822. Minnesota Mining and Manufacturing Co. All H in an organic compound are replaced by F by Simons electrochemical process (1)

**ESTIMATED ERROR:**

$\delta T/K = 0.1$
$\delta c/c = 0.01$ (reproducibility)
$= 0.05$ (absolute)

**REFERENCES:**

1. Simons, J. H., Editor *Fluorine Chemistry* Academic Press, New York, 1950.

**COMPONENTS:**

(1) Nitrogen; $N_2$; [7727-37-9]

(2) Various solvents, see below

**ORIGINAL MEASUREMENTS:**

Gorbachev, V. M.; Tret'yakov, G. V.

*Zavodsk. Lab.* 1966, *32*, 796-8.

**VARIABLES:**

T/K = 293

**PREPARED BY:**

V. Katovic

**EXPERIMENTAL VALUES:**

| Substance | $C_6H_6$ | $CCl_4$ | $CH_3SiCl_3$ | $GeCl_4$ | $SiCl_4$ |
|---|---|---|---|---|---|
| | | $L^a$ at | 20°C | (293 K) | |
| Standard $C_6H_6$ | 0.1162[b] | 0.155 | 0.190 | 0.208 | 0.224 |
| Standard $CCl_4$ | 0.114 | 0.1572[b] | 0.184 | 0.208 | 0.225 |

[a] Ostwald coefficient.

[b] Literature values (1).

[c] The Ostwald coefficients for nitrogen in each solvent (columns) were determined from ratios against a standard of either benzene or carbon tetrachloride

[d] Benzene; $C_6H_6$; [71-43-2]
Carbon tetrachloride; $CCl_4$; [56-23-5]
Methyl trichloro silane; $CH_3SiCl_3$; [75-79-6]
Germanium tetrachloride; $GeCl_4$; [10038-98-9]
Silicon tetrachloride; $SiCl_4$; [10026-04-7]

AUXILIARY INFORMATION

**METHOD/APPARATUS/PROCEDURE:**

The solubility of $N_2$ in the investigated liquids was determined by gas chromatography. Chromatographs of equal volumes of a standard liquid ($C_6H_6$ or $CCl_4$) saturated with $N_2$ and the investigated saturated liquid were obtained. Solubilities were calculated from a ratio of nitrogen peak height and liquid peak area of the standard and investigated liquids and the known solubility of $N_2$ in the standard liquid.

**SOURCE AND PURITY OF MATERIALS:**

No details given.

**ESTIMATED ERROR:**

$\delta L/L = \pm 0.03$, compiler's estimate

**REFERENCES:**

(1) Horiuti, J.
*Sci. Papers. Inst. Phys. Chem. Res. (Tokyo)* 1931, *17*, 125.

| COMPONENTS: | EVALUATOR: |
|---|---|
| (1) Nitrogen; $N_2$; [7727-37-9]<br>(2) Solvents containing sulfur:<br>Carbon disulfide; $CS_2$; [75-15-0]<br>Sulfinylbismethane or dimethyl sulfoxide; $C_2H_6OS$; [67-68-5] | H. Lawrence Clever<br>Department of Chemistry<br>Emory University<br>Atlanta, GA 30322 USA<br><br>1981, December |

CRITICAL EVALUATION:

The solubility of nitrogen in carbon disulfide is reported in four papers. Just's (1) values at 293.15 and 298.15 K are 34 to 38 per cent lower than the modern values and should be rejected. The other three papers originated from the laboratory of J. H. Hildebrand over a 23 year period. Gjaldbaek and Hildebrand (2) reported solubility values at 298.04 and 298.11 K in 1949, Kobatake and Hildebrand (3) reported four values between 279.37 and 304.35 K in 1961, and Powell (4) reported values over the 273.15 to 303.15 K temperature interval in 1971. Powell reported a solubility value only at 298.15 K and gave the slope of a plot of log $x_1$ *vs.* log $T$.

At 298.15 K the mole fraction solubility values at 101.325 kPa (1 atm) nitrogen partial pressure from the three papers (2, 3, 4) agree within 0.7 percent. The smoothed data of Kobatake and Hildebrand (3) and Powell (4) agree within 0.5 per cent at the temperatures between 283.15 and 303.15 K. The recommended solubility values are based on the four solubility values of Kobatake and Hildebrand.

The equation for use between 278.15 and 303.15 K is

$$\log x_1 = -6.4402 - 5.8881/(T/100\ \text{K})$$

with a standard error about the regression line of $2.03 \times 10^{-7}$.

The temperature independent thermodynamic changes in enthalpy and entropy for the transfer of one mole of nitrogen from the gas phase at a partial pressure of 101.325 kPa to the infinitely dilute solution are

$$\Delta H_1^\circ/\text{kJ mol}^{-1} = 4.895 \text{ and } \Delta S_1^\circ/\text{J K}^{-1}\text{mol}^{-1} = -53.5.$$

The smoothed solubility values and the Gibbs energy of solution are given in Table 1.

Table 1. Solubility of nitrogen in carbon disulfide of a nitrogen partial pressure of 101.325 kPa. Recommended values of the mole fraction solubility and partial molar Gibbs energy of solution as a function of temperature.

| $T$/K | Mol Fraction $10^4 x_2$ | $\Delta G_1^\circ$/kJ mol$^{-1}$ |
|---|---|---|
| 278.15 | 1.922 | 19.789 |
| 283.15 | 1.995 | 20.057 |
| 288.15 | 2.068 | 20.325 |
| 293.15 | 2.142 | 20.593 |
| 298.15 | 2.215 | 20.860 |
| 303.15 | 2.288 | 21.128 |

Only one solubility value is reported for nitrogen + sulfinylbismethane (dimethyl sulfoxide). The mole fraction solubility of $8.33 \times 10^{-5}$ at 298.15 K and 101.325 kPa nitrogen pressure reported by Dymond (5) is classed as tentative.


References:

1. Just, G. *Z. Phys. Chem.* 1901, *37*, 342.
2. Gjaldbaek, J. C.; Hildebrand, J. H. *J. Am. Chem. Soc.* 1949, *71*, 3147.
3. Kobatake, Y.; Hildebrand, J. H. *J. Phys. Chem.* 1961, *65*, 331.
4. Powell, R. J. *J. Chem. Eng. Data* 1972, *17*, 302.
5. Dymond, J. H. *J. Phys. Chem.* 1967, *71*, 1829.

| COMPONENTS: | ORIGINAL MEASUREMENTS: |
|---|---|
| (1) Nitrogen; $N_2$; [7727-37-9]<br>(2) Carbon disulfide; $CS_2$; [75-15-0] | Just, G.<br>*Z. Phys. Chem.* 1901, *37*, 342-67. |
| VARIABLES: | PREPARED BY: |
| T/K = 293-298<br>P/kPa = 101.325 | M. E. Derrick<br>H. L. Clever |

EXPERIMENTAL VALUES:

| T/K | Mol Fraction $x_1 \times 10^4$ | Bunsen Coefficient $\alpha$ | Ostwald Coefficient L |
|---|---|---|---|
| 293.15 | 1.327 | 0.04929 | 0.05290 |
| 298.15 | 1.453 | 0.05369 | 0.05860 |

The author measured the Ostwald coefficient at a pressure of about 746 mmHg. The compiler assumed the Ostwald coefficient to be independent of pressure, and calculated the mole fraction and Bunsen coefficient values at 101.325 kPa (1 atm) partial pressure of the gas.

AUXILIARY INFORMATION

METHOD/APPARATUS/PROCEDURE:

An Ostwald apparatus as modified by Timofejew (1), and Steiner (2) is used. The apparatus consists of a gas buret, an absorption flask, and a mercury manometer. The system is thermostated with a water jacket.

The gas is introduced into the degassed liquid. The gas volume absorbed is determined by the gas buret. The solvent volume is determined at the end of the experiment by pouring the solvent into a graduated flask.

SOURCE AND PURITY OF MATERIALS:

(1) Nitrogen. Prepared by the reaction of sodium nitrite and ammonium nitrate in aqueous solution in the presence of calcium dichromate.

(2) Carbon disulfide. No information.

ESTIMATED ERROR:

$\delta L/L = 0.03$ (compiler)

REFERENCES:

1. Timofejew, W.
*Z. Physik. Chem.* 1890, *6*, 141.

2. Steiner,
*Ann. Phys. (Leipzig)* 1894, *52*, 275.

COMPONENTS:

(1) Nitrogen; $N_2$; [7727-37-9]

(2) Carbon disulfide; $CS_2$; [75-15-0]

ORIGINAL MEASUREMENTS:

Gjaldbaek, J. C.; Hildebrand, J. H.

*J. Am. Chem. Soc.* 1949, *71*, 3147-50.

VARIABLES:

T/K = 298.04 - 298.11

Total P/kPa = 101.325

PREPARED BY:

J. Chr. Gjaldbaek

EXPERIMENTAL VALUES:

| T/K | Mol Fraction $X_1 \times 10^3$ | Bunsen Coefficient $\alpha$ | Ostwald Coefficient L |
|---|---|---|---|
| 298.04 | 0.223 | 0.0822 | 0.0897 |
| 298.11 | 0.223 | 0.0823 | 0.0898 |

The mole fraction and Ostwald solubility values were calculated by the compiler.

AUXILIARY INFORMATION

METHOD/APPARATUS/PROCEDURE:

A calibrated all-glass combined manometer and bulb was enclosed in an air thermostat and shaken until equilibrium. Mercury was used for calibration and as the confining liquid. The solvents were degassed in the apparatus. Details are in references 1 and 2.

The absorbed volume of gas was calculated from the initial and final amounts, both saturated with solvent vapor. The amount of solvent was determined by the weight of displaced mercury.

The saturation of the liquid with the gas was carried out close to atmospheric pressure. The solubility values were reported for one atmosphere gas pressure assuming Henry's law is obeyed.

SOURCE AND PURITY OF MATERIALS:

(1) Nitrogen. Stuart Oxygen Co. 99.996 percent $N_2$.

(2) Carbon disulfide. Mallinckrodt Chemical Co. Analytical reagent grade, boiling point 45.80 - 45.85 at 750.1 mmHg.

ESTIMATED ERROR:

$\delta T/K = 0.05$

$\delta X_1/X_1 = 0.015$

REFERENCES:


1. Lannung, A. *J. Am. Chem. Soc.* 1930, *52*, 68.

2. Gjaldbaek, J. C. *Acta Chem. Scand.* 1952, *6*, 623.

**COMPONENTS:**

(1) Nitrogen; $N_2$; [7727-37-9]

(2) Carbon Disulfide; $CS_2$; [75-15-0]

**ORIGINAL MEASUREMENTS:**

Kobatake, Y.; Hildebrand, J. H.

*J. Phys. Chem.* 1961, *65*, 331-5.

**VARIABLES:**

T/K = 279.37 - 304.35
P/kPa = 101.325

**PREPARED BY:**

M. E. Derrick
H. L. Clever

**EXPERIMENTAL VALUES:**

| T/K | Mol Fraction $X_1 \times 10^3$ | Bunsen Coefficient $\alpha$ | Ostwald Coefficient L |
|---|---|---|---|
| 279.37 | 0.1941 | 0.0733 | 0.0750 |
| 288.56 | 0.2072 | 0.0775 | 0.0819 |
| 298.15 | 0.2215 | 0.0819 | 0.0894 |
| 304.35 | 0.2307 | 0.0847 | 0.0944 |

The Bunsen and Ostwald coefficients were calculated by the compiler.

Smoothed Data: $\Delta G^o/J\ mol^{-1} = -RT \ln X_1 = 4890.5 + 53.563\ T$

Std. Dev. $\Delta G^o = 1.9$, Coef. Corr. = 1.000

$\Delta H^o/J\ mol^{-1} = 4890.5$, $\Delta S^o/J\ K^{-1}\ mol^{-1} = -53.563$

| T/K | Mol Fraction $X_1 \times 10^3$ | $\Delta G^o/J\ mol^{-1}$ |
|---|---|---|
| 283.15 | 0.1995 | 20,057 |
| 288.15 | 0.2068 | 20,325 |
| 293.15 | 0.2142 | 20,593 |
| 298.15 | 0.2215 | 20,860 |
| 303.15 | 0.2288 | 21,128 |

**AUXILIARY INFORMATION**

**METHOD/APPARATUS/PROCEDURE:**

The apparatus consists of a gas-measuring buret, an absorption pipet, and reservoir for solvent. The buret is thermostated at 25°C, the pipet at any temperature from 5 to 30°C.

The solvent is degassed by freezing with liquid nitrogen, evacuating, then boiling with a heat lamp. The solvent never comes into contact with stop-cock grease.

The amount of gas is determined by P-V measurements in the gas buret before and after admission of the gas into the pipet. No solvent vapor is allowed into the buret system. The solvent is magnetically stirred for up to 24 hours until equilibrium is reached.

Ideal gas behavior and Henry's law are assumed to calculate the mole fraction solubility at a gas partial pressure of one atm.

**SOURCE AND PURITY OF MATERIALS:**

(1) Nitrogen. Linde Oxygen Co. Standard grade, 99.9 % nitrogen.

(2) $CS_2$. Analytical Reagent shaken with Hg, $HgCl_2$, distilled, Mallinckrodt Chemical Works.

**ESTIMATED ERROR:**

$\delta X_1/X_1 = 0.003$

**REFERENCES:**

**COMPONENTS:**

(1) Nitrogen; $N_2$; [7727-37-9]

(2) Carbon Disulfide; $CS_2$; [75-15-0]

**ORIGINAL MEASUREMENTS:**

Powell, R. J.

*J. Chem. Eng. Data* 1972, *17*, 302-4.

**VARIABLES:**

T/K = 273.15 - 318.15

P/kPa = 101.325

**PREPARED BY:**

P. L. Long

**EXPERIMENTAL VALUES:**

| T/K | Mol Fraction $x_1 \times 10^4$ | Bunsen Coefficient $\alpha$ | Ostwald Coefficient L | $R\frac{\Delta \log X_1}{\Delta \log T} = N$ |
|---|---|---|---|---|
| 298.15 | 2.22 | 0.0821 | 0.0896 | 4.00 |

The author states that solubility measurements were made between 273.15 and 318.15 K, but only the solubility at 298.15 was given in the paper. The slope $R(\Delta \log X_1/\Delta \log T)$ was given. The smoothed data below were calculated by the compiler from the slope in the form:

$$\log x_1 = \log(2.22 \times 10^{-4}) + (4.00/R)\log(T/298.15)$$

with R = 1.9872 cal $K^{-1}$ $mol^{-1}$.

| T/K | Mol Fraction $x_1 \times 10^4$ |
|---|---|
| 273.15 | 1.86 |
| 278.15 | 1.93 |
| 283.15 | 2.00 |
| 288.15 | 2.07 |
| 293.15 | 2.15 |
| 298.15 | 2.22 |
| 303.15 | 2.30 |

The Bunsen and Ostwald coefficients were calculated by the compiler.

AUXILIARY INFORMATION

**METHOD/APPARATUS/PROCEDURE:**

Dymond and Hildebrand (1) apparatus which uses an all glass pumping system to spray slugs of degassed solvent into the gas. The amount of gas dissolved is calculated from the initial and final gas pressures. The solvent is degassed by freezing and pumping followed by boiling under reduced pressure.

**SOURCE AND PURITY OF MATERIALS:**

(1) Nitrogen. Source not given. Manufacturers research grade, dried over $CaCl_2$ before use.

(2) Carbon disulfide. No source given. Spectrochemical grade.

**ESTIMATED ERROR:**

$\delta N$/cal $K^{-1}$ $mol^{-1}$ = 0.1

$\delta x_1/x_1$ = 0.002

**REFERENCES:**

1. Dymond, J. H.; Hildebrand, J. H. *Ind. Eng. Chem. Fundam.* 1967, *6*, 130.

**COMPONENTS:**

(1) Nitrogen; $N_2$; [7727-37-9]

(2) Sulfinylbismethane or Dimethyl Sulfoxide; $C_2H_6OS$ ($CH_3SOCH_3$); [67-68-5]

**ORIGINAL MEASUREMENTS:**

Dymond, J. H.

*J. Phys. Chem.* 1967, *71*, 1829-31.

**VARIABLES:**

T/K = 298.15
P/kPa = 101.325

**PREPARED BY:**

M. E. Derrick
H. L. Clever

**EXPERIMENTAL VALUES:**

| T/K | Mol Fraction $x_1 \times 10^4$ | Bunsen Coefficient $\alpha$ | Ostwald Coefficient L |
|---|---|---|---|
| 298.15 | 0.833 | 0.0262 | 0.0286 |

The Bunsen and Ostwald coefficients were calculated by the compiler.

**AUXILIARY INFORMATION**

**METHOD/APPARATUS/PROCEDURE:**

The liquid is saturated with the gas at a gas partial pressure of 1 atm.

The apparatus is that described by Dymond and Hildebrand (1). The apparatus uses an all-glass pumping system to spray slugs of degassed solvent into the gas. The amount of gas dissolved is calculated from the initial and final gas pressure.

**SOURCE AND PURITY OF MATERIALS:**

(1) Nitrogen. General Dynamics Corp. Dried.

(2) Dimethyl Sulfoxide. Matheson, Coleman, and Bell Co. Spectro-quality reagent, dried, and a fraction frozen out. Melting pt.: 18.37°C.

**ESTIMATED ERROR:**

**REFERENCES:**

1. Dymond, J.; Hildebrand, J. H. *Ind. Eng. Chem. Fundam.* 1967, *6*, 130.

COMPONENTS:

(1) Nitrogen; $N_2$; [7727-37-9]

(2) Sulfurhexafluoride; $SF_6$; [2551-62-4]

ORIGINAL MEASUREMENTS:

Miller, H. C.; Verdelli, L.S.; Gall, J. F.

*Ind. Eng. Chem.* 1951, *43*, 1126-9.

VARIABLES:

T/K = 300

PREPARED BY:

R. Battino

EXPERIMENTAL VALUES:

| $P^b$/psig | $\alpha^a$/cm$^3$(STP) cm$^{-3}$ atm$^{-1}$ | % by wt.[c] | % by vol.[c] |
|---|---|---|---|
| 356 | | 0 | 0 |
| 365 | 2. | 0.2 | 0.60 |
| 383 | 2.3 | 0.41 | 2.1 |
| 400 | 2.1 | 0.597 | 3.04 |
| 420 | 1.9 | 0.807 | 4.07 |

[a] Bunsen coefficient at 27°C (300K). Pressures up to 420 psig were used.

[b] Total pressure.

[c] Of nitrogen in liquid sulfur hexafluoride at 27°C (300 K).

AUXILIARY INFORMATION

METHOD/APPARATUS/PROCEDURE:

A container with liquid $SF_6$ at 27°C was pressurized with various amounts of nitrogen which dissolved. This was heated up to 50°C where both substances were in the gas phase. By pressure and volume measurements the Bunsen coefficient could be calculated. The paper also reports on the physical properties of $SF_6$.

SOURCE AND PURITY OF MATERIALS:

(1) No details given.

(2) "Purified." Orthobaric liquid density of 1.30 g/ml at 26°C. $t_c$/°C = 45.5; $\rho_c$/g cm$^{-3}$ = 0.727; $P_c$/atm = 36.8.

ESTIMATED ERROR:

$\delta\alpha/\alpha = \pm 0.10$, compiler's estimate.

REFERENCES:

**COMPONENTS:**

(1) Nitrogen; $N_2$; [7727-37-9]

(2) Nitromethane; $CH_3NO_2$; [75-52-5]

**ORIGINAL MEASUREMENTS:**

Friedman, H. L.

*J. Am. Chem. Soc.* 1954, *76*, 3294-7.

**VARIABLES:**

T/K = 298.00
P/kPa = 98.33

**PREPARED BY:**

P. L. Long
H. L. Clever

**EXPERIMENTAL VALUES:**

| T/K | Mol Fraction $x_1 \times 10^4$ | Bunsen Coefficient $\alpha$ | Ostwald Coefficient L |
|---|---|---|---|
| 298.00 | 2.01 | 0.083 | 0.091 |

The author measured the Ostwald coefficient at about 700 mmHg nitrogen pressure. The Bunsen coefficient and the mole fraction solubility at 101.325 kPa (1 atm) nitrogen pressure were calculated by the compiler with the assumptions that the gas is ideal and that Henry's law is obeyed.

**AUXILIARY INFORMATION**

**METHOD/APPARATUS/PROCEDURE:**

The method was essentially that employed by Euken and Herzberg (1). Modifications included a magnetic stirring device instead of shaking the saturation vessel, and balancing the gas pressure against a column of mercury with electrical contacts instead of balancing the gas pressure against the atmosphere.

The solvent was degassed by vacuum. The procedure, repeated 5 to 10 times, was to alternate 5 to 15 s evacuation and rapid stirring to produce cavitation.

To measure the solubility, the gas, presaturated with solvent vapor, was brought into contact with about 80 $cm^3$ of solvent in the absorption vessel. Initial conditions were established by a time extrapolation. Solubility equilibrium was approached from both under- and super-saturation by varying the rate.

**SOURCE AND PURITY OF MATERIALS:**

(1) Nitrogen. Source not given. The gas was dried over $CaH_2$ and CaO.

(2) Nitromethane. Source not given. Distilled, dried by filtering at 253 K.

**ESTIMATED ERROR:**

$\delta T/K = 0.05$
$\delta P/mmHg = 0.3$
$\delta L/L = 0.03$

**REFERENCES:**

1. Eucken, A.; Herzberg, G. *Z. Phys. Chem.* 1950, *195*, 1.

| COMPONENTS: | ORIGINAL MEASUREMENTS: |
|---|---|
| (1) Nitrogen; $N_2$; [7727-37-9]<br>(2) N-Methylacetamide; $C_3H_7NO$; [79-16-3] | Wood, R. H.; DeLaney, D. E.<br>*J. Phys. Chem.* 1968, *72*, 4651-4. |

| VARIABLES: | PREPARED BY: |
|---|---|
| T/K = 308.15 - 343.15<br>P/kPa = 101.325 | P. L. Long<br>H. L. Clever |

EXPERIMENTAL VALUES:

The authors obtained the equation

$$\ln X_1 = -245.7/(T/K) - 7.514$$

by a linear regression of their experimental data. The equation can be rearranged to

$$\Delta G^\circ = - RT \ln X_1 = 245.7R + 7.514\ RT$$

The experimental data were not included in the paper. They are available in a thesis (1). The smoothed mole fraction argon solubilities at 101.325 kPa $N_2$ gas and five degree intervals between 308.15 and 343.15 K were given in the paper. The Bunsen and Ostwald coefficients were calculated by the compiler.

Smoothed Data:

| T/K | Mol Fraction[a] $x_1 \times 10^4$ | Bunsen Coefficient $\alpha$ | Ostwald Coefficient L |
|---|---|---|---|
| 308.15 | 2.461 | 0.7139 | 0.8054 |
| 313.15 | 2.492 | 0.7198 | 0.8252 |
| 318.15 | 2.523 | 0.7256 | 0.8451 |
| 323.15 | 2.554 | 0.7313 | 0.8652 |
| 328.15 | 2.583 | 0.7364 | 0.8847 |
| 333.15 | 2.613 | 0.7417 | 0.9046 |
| 338.15 | 2.641 | 0.7463 | 0.9239 |
| 343.15 | 2.669 | 0.7509 | 0.9434 |

[a] These values were given by the authors and are slightly different than those calculated from their equation given above. These values are too large by about 1.5 ppt, which is less than the experimental error.

AUXILIARY INFORMATION

METHOD/APPARATUS/PROCEDURE:

A gas buret was connected to a solvent buret through a three-way capillary stopcock. A measured volume of gas was transferred to a known volume of solvent; when equilibrium was reached the total pressure and volume of the system was measured (1). The apparatus and procedure were checked by measuring the solubility of Ar in $H_2O$ at 298.15 K. The Bunsen coefficient of 0.03105 checked well with the literature (2).

SOURCE AND PURITY OF MATERIALS:

(1) Nitrogen. Source not given. Purity 99.99 per cent.

(2) N-Methylacetamide. Source not given. Recrystallized three times in a dry box. Typically had a water content of 0.04 mol per cent after a solubility run.

ESTIMATED ERROR:

Duplicate runs checked to within 0.5 per cent.

REFERENCES:

1. DeLaney, D. E. M. S. Thesis, University of Delaware, 1968.

2. Ben-Naim, A.; Baer, S. *Trans. Faraday Soc.* 1963, *59*, 2735; *ibid.* 1964, *60*, 1736.

COMPONENTS:

(1) Nitrogen; $N_2$; [7727-37-9]

(2) Nitric acid, propyl ester or 1-propyl nitrate; $C_3H_7NO_3$; [627-13-4]

ORIGINAL MEASUREMENTS:

Amster, A. B.; Levy, J. B.

*J. Am. Rocket Soc.* 1959, *29*, 870-1.

VARIABLES:

T/K = "Room temperature"
P/kPa = 28.0-85.3

PREPARED BY:

M. E. Derrick
H. L. Clever

EXPERIMENTAL VALUES:

| Pressure P/mmHg[1] | P/kPa | Solubility[1] mol $g^{-1}$ x $10^6$ | Mol Fraction $x_1$ x $10^4$ | Bunsen Coefficient α | Ostwald Coefficient L |
|---|---|---|---|---|---|
| 210 | 28.0 | 1.8 | 1.9 | 0.16 | 0.17 |
| 210 | 28.0 | 1.8 | 1.9 | 0.16 | 0.17 |
| 220 | 29.3 | 1.8 | 1.9 | 0.15 | 0.16 |
| 220 | 29.3 | 1.4 | 1.5 | 0.12 | 0.13 |
| 420 | 56.0 | 3.0 | 3.2 | 0.13 | 0.14 |
| 420 | 56.0 | 3.1 | 3.3 | 0.13 | 0.14 |
| 440 | 58.7 | 3.0 | 3.2 | 0.12 | 0.13 |
| 440 | 58.7 | 2.6 | 2.7 | 0.10 | 0.11 |
| 620 | 82.7 | 4.8 | 5.0 | 0.14 | 0.15 |
| 620 | 82.7 | 4.4 | 4.6 | 0.13 | 0.14 |
| 620 | 82.7 | 4.5 | 4.7 | .13 | 0.14 |
| 640 | 85.3 | 4.4 | 4.6 | 0.12 | 0.13 |

[1] The compiler read values of pressure and solubility from a graph in the author's paper. The pressure is total pressure minus the equilibrium vapor pressure of pure nitric acid propyl ester.

The compiler calculated the mole fraction, Bunsen coefficient, and Ostwald coefficient values assuming a temperature of 293.15 K. Density values for the solvent were available at the temperature.

AUXILIARY INFORMATION

METHOD/APPARATUS/PROCEDURE:

The apparatus consists of an absorption flask connected by a flexible spiral to a mercury filled capillary (1).

A weighed quantity of liquid is placed in the flask. The gas is added , and shaken with the degassed liquid. The volume of the gas is taken as the flask volume minus the solvent volume. The solubility is calculated from

$$\text{Solubility/mol g}^{-1} = V\Delta P/WRT$$

where V is the gas volume,
ΔP is the pressure change on gas dissolution,
W is the weight of solvent,
T is the temperature, and
R the gas constant.

SOURCE AND PURITY OF MATERIALS:

(1) Nitrogen. Technical grade, water pumped, 99.5 per cent. Passed through a drying tube.

(2) Nitric acid, propyl ester. Source and purity not given.

ESTIMATED ERROR:

$\delta P/\text{mmHg} = 5$
$\delta T/K = 2$
$\delta\alpha\ \alpha = 0.15$

REFERENCES:

1. Levy, J. B.; Taft, R. W. jr. Aaron, D.; Hammett, L. P. *J. Am. Chem. Soc.* 1953, *75*, 3955.

**COMPONENTS:**

(1) Nitrogen; $N_2$; [7727-37-9]

(2) Cyclic amines; $C_4H_9N$, $C_5H_5N$, and $C_5H_{10}N$

**ORIGINAL MEASUREMENTS:**

Guerry, D. Jr.

Ph.D. thesis, 1944
Vanderbilt University
Nashville, TN

**VARIABLES:**

T/K = 293-298
P/kPa = 101.325

**PREPARED BY:**

H. L. Clever

**EXPERIMENTAL VALUES:**

| T/K | Mol Fraction $x_1 \times 10^4$ | Bunsen Coefficient $\alpha$ | Ostwald Coefficient L |
|---|---|---|---|
| Pyrrolidine; $C_4H_9N$; [123-75-1] | | | |
| 293.15 | 3.54 | 0.0954 | 0.1024 |
| 298.15 | 3.68 | 0.0990 | 0.1081 |
| Pyridine; $C_5H_5N$; [110-86-1] | | | |
| 293.15 | 2.41 | 0.0673 | 0.0722 |
| 298.15 | 2.50 | 0.0692 | 0.0755 |
| Piperidine; $C_5H_{11}N$; [110-89-4] | | | |
| 293.15 | 4.13 | 0.0939 | 0.1008 |
| 298.15 | 4.22 | 0.0953 | 0.1040 |

The Ostwald coefficients were calculated by the compiler.

AUXILIARY INFORMATION

**METHOD/APPARATUS/PROCEDURE:**

The apparatus was a modified Van Slyke-Neill Manometric Apparatus manufactured by the Eimer and Amend Co.

The procedure of Van Slyke (1) for pure liquids was modified (2) so that small solvent samples (2 $cm^3$) could be used with almost complete recovery of the sample.

An improved temperature control system was used.

SOURCE AND PURITY OF MATERIALS:

(1) Nitrogen. Ohio Chemical Co. Oil pumped, 99.8 per cent pure.

(2) Cyclic amines. The pyridine and pyrrolidine were distilled from BaO under a $N_2$ atmosphere. The piperidine was distilled from KOH under a $N_2$ atmosphere. Experimental data on refractive index, density and vapor pressure are in the thesis.

**SOURCE AND PURITY OF MATERIALS:**

Pyrrolidine. Pyrrole was prepared and catalytically reduced to pyrrolidine. B.p. (750 mmHg) t/°C 88.12 - 88.26 (corr.).

Pyridine. Mallincrodt Chemical Co. Purified and distilled. B.p. (743.9 mmHg) t/°C 114.96 - 115.06 (corr.).

Piperidine. Part was a commercial sample (Eastman Kodak Co.), part prepared by reduction of pyridine. B.p. (752.4 mmHg) t/°C 106.00 - 106.17.

**ESTIMATED ERROR:**

$\delta T/K = 0.05$

**REFERENCES:**


1. Van Slyke, D. D. *J. Biol. Chem.* 1939, *130*, 545.
2. Ijams, C. C. Ph.D. thesis, 1941 Vanderbilt University

**COMPONENTS:**

(1) Nitrogen; $N_2$; [7727-37-9]

(2) Nitrobenzene; $C_6H_5NO_2$; [98-95-3]

**ORIGINAL MEASUREMENTS:**

Just, G.

*Z. Phys. Chem.* 1901, *37*, 342-67.

**VARIABLES:**

T/K = 293-298
P/kPa = 101.325

**PREPARED BY:**

M. E. Derrick
H. L. Clever

**EXPERIMENTAL VALUES:**

| T/K | Mol Fraction $x_1 \times 10^4$ | Bunsen Coefficient $\alpha$ | Ostwald Coefficient L |
|---|---|---|---|
| 293.15 | 2.60 | 0.05667 | 0.06082 |
| 298.15 | 2.64 | 0.05731 | 0.06255 |

The author measured the Ostwald coefficient at a pressure of about 746 mmHg. The compiler assumed the Ostwald coefficient to be independent of pressure, and calculated the mole fraction and Bunsen coefficient values at 101.325 kPa (1 atm) partial pressure of the gas.

## AUXILIARY INFORMATION

**METHOD/APPARATUS/PROCEDURE:**

An Ostwald apparatus as modified by Timofejew (1), and Steiner (2) is used. The apparatus consists of a gas buret, an absorption flask, and a mercury manometer. The system is thermostated with a water jacket.

The gas is introduced into the degassed liquid. The gas volume absorbed is determined by the gas buret. The solvent volume is determined at the end of the experiment by pouring the solvent into a graduated flask.

**SOURCE AND PURITY OF MATERIALS:**

(1) Nitrogen. Prepared by the reaction of sodium nitrite and ammonium nitrate in aqueous solution in the presence of calcium dichromate.

(2) Nitrobenzene. No information.

**ESTIMATED ERROR:**

$\delta L/L = 0.03$ (compiler)

**REFERENCES:**

1. Timofejew, W. *Z. Physik. Chem.* 1890, *6*, 141.

2. Steiner, *Ann. Phys. (Leipzig)* 1894, *52*, 275.

**COMPONENTS:**

(1) Nitrogen; $N_2$; [7727-37-9]

(2) Benzenamine; $C_6H_7N$; [62-53-3]

Nitrobenzene; $C_6H_5NO_2$; [98-95-3]

**ORIGINAL MEASUREMENTS:**

Metschl, J.

*J. Phys. Chem.* 1924, *28*, 417-37.

**VARIABLES:**

T/K = 298.15
p/kPa = 101.325

**PREPARED BY:**

M. E. Derrick
H. L. Clever

**EXPERIMENTAL VALUES:**

| T/K | Mol Fraction $x_1 \times 10^4$ | Bunsen Coefficient $\alpha$ | Ostwald Coefficient L |
|---|---|---|---|
| Benzenamine or aniline; $C_6H_7N$; [62-53-3] | | | |
| 298.15 | 1.3 | 0.033 | 0.036 |
| Nitrobenzene; $C_6H_5NO_2$; [98-95-3] | | | |
| 298.15 | 2.6 | 0.056 | 0.061 |

The compiler calculated the mole fraction and Ostwald coefficient values.

**AUXILIARY INFORMATION**

**METHOD/APPARATUS/PROCEDURE:**

The degassed solvent is saturated with nitrogen at 6 to 10 different pressures between 1.5 and 4.5 atm.

The pressure is reduced to one atm, and the "supersaturated gas" is shaken out of the solvent and measured.

The slope of a plot of the volume of gas reduced to standard conditions shaken out of 100 $cm^3$ of solvent *vs.* the pressure of saturation is nearly identical to the solubility-pressure graph (Henry's law). The slope of the curve is used to obtain the Bunsen coefficient. From the method of calculation the value represents the volume of gas shaken out of the solvent when the pressure is reduced from 2 to 1 atm.

**SOURCE AND PURITY OF MATERIALS:**

(1) Nitrogen. Prepared by the reaction of sodium nitrite and ammonium nitrate in the presence of calcium dichromate in aqueous solution.

(2) Aniline. Allowed to stand over KOH, then distilled.

Nitrobenzene. Chemically pure. Allowed to stand over calcium chloride, then distilled.

**ESTIMATED ERROR:**

**REFERENCES:**

| COMPONENTS: | ORIGINAL MEASUREMENTS: |
|---|---|
| (1) Nitrogen; $N_2$; [7727-37-9]<br>(2) Benzenamine or aniline; $C_6H_7N$; [62-53-3] | Just, G.<br>*Z. Phys. Chem.* 1901, *37*, 342-67. |
| VARIABLES:<br>T/K = 292-298<br>P/kPa = 101.325 | PREPARED BY:<br>M. E. Derrick<br>H. L. Clever |

EXPERIMENTAL VALUES:

| T/K | Mol Fraction $x_1 \times 10^4$ | Bunsen Coefficient $\alpha$ | Ostwald Coefficient L |
|---|---|---|---|
| 293.15 | 1.13 | 0.02788 | 0.02992 |
| 298.15 | 1.15 | 0.02816 | 0.03074 |

The author measured the Ostwald coefficient at a pressure of about 746 mmHg. The compiler assumed the Ostwald coefficient to be independent of pressure, and calculated the mole fraction and Bunsen coefficient values at 101.325 kPa (1 atm) partial pressure of the gas.

AUXILIARY INFORMATION

METHOD/APPARATUS/PROCEDURE:

An Ostwald apparatus as modified by Timofejew (1), and Steiner (2) is used. The apparatus consists of a gas buret, an absorption flask, and a mercury manometer. The system is thermostated with a water jacket.

The gas is introduced into the degassed liquid. The gas volume absorbed is determined by the gas buret. The solvent volume is determined at the end of the experiment by pouring the solvent into a graduated flask.

SOURCE AND PURITY OF MATERIALS:

(1) Nitrogen. Prepared by the reaction of sodium nitrite and ammonium nitrate in aqueous solution in the presence of calcium dichromate.

(2) Benzenamine. No information given.

ESTIMATED ERROR:

$\delta L/L = 0.03$ (compiler)

REFERENCES:

1. Timofejew, W. *Z. Physik. Chem.* 1890, *6*, 141.

2. Steiner, *Ann. Phys. (Leipzig)* 1894, *52*, 275.

**COMPONENTS:**

(1) Nitrogen; $N_2$; [7727-37-9]

(2) 1,1,2,2,3,3,4,4,4-Nonafluoro-N,N-bis(nonafluorobutyl)-1-butanamine; $(C_4F_9)_3N$; [311-89-7]

**ORIGINAL MEASUREMENTS:**

Kobatake, Y.; Hildebrand, J. H.

*J. Phys. Chem.* 1961, *65*, 331-5.

**VARIABLES:**

T/K = 283.83 - 303.60
P/kPa = 101.325

**PREPARED BY:**

M. E. Derrick
H. L. Clever

**EXPERIMENTAL VALUES:**

| T/K | Mol Fraction $X_1 \times 10^3$ | Bunsen Coefficient $\alpha$ | Ostwald Coefficient L |
|---|---|---|---|
| 283.83 | 3.537 | 0.230 | 0.239 |
| 287.18 | 3.524 | 0.227 | 0.239 |
| 293.56 | 3.502 | 0.223 | 0.240 |
| 298.15 | (3.490) | 0.220 | 0.240 |
| 299.15 | 3.488 | 0.219 | 0.240 |
| 303.60 | 3.474 | 0.216 | 0.240 |

The Bunsen and Ostwald coefficients were calculated by the compiler.

Smoothed Data: $\Delta G^o/J\ mol^{-1} = -RT \ln X_1 = -638.71 + 49.183\ T$

Std. Dev. $\Delta G^o = 0.84$, Coef. Corr. = 0.9992

$\Delta H^o/J\ mol^{-1} = -638.71$, $\Delta S^o/J\ K^{-1}\ mol^{-1} = -49.183$

| T/K | Mol Fraction $X_1 \times 10^3$ | $\Delta G^o/J\ mol^{-1}$ |
|---|---|---|
| 283.15 | 3.538 | 13,288 |
| 288.15 | 3.521 | 13,534 |
| 293.15 | 3.505 | 13,779 |
| 298.15 | 3.490 | 14,025 |
| 303.15 | 3.475 | 14,271 |

The value in ( ) is assumed to be the authors smoothed value.

AUXILIARY INFORMATION

**METHOD/APPARATUS/PROCEDURE:**

The apparatus consists of a gas-measuring buret, an absorption pipet, and reservoir for solvent. The buret is thermostated at 25°C, the pipet at any temperature from 5 to 30°C.

The solvent is degassed by freezing with liquid nitrogen, evacuating, then boiling with a heat lamp. The solvent never comes into contact with stop-cock grease.

The amount of gas is determined by P-V measurements in the gas buret before and after admission of the gas into the pipet. No solvent vapor is allowed into the buret system. The solvent is magnetically stirred for up to 24 hours until equilibrium is reached.

Ideal gas behavior and Henry's law are assumed to calculate the mole fraction solubility at a gas partial pressure of one atm.

**SOURCE AND PURITY OF MATERIALS:**

(1) Nitrogen. Linde Oxygen Co. Standard grade, 99.9 % nitrogen.

(2) $(C_4F_9)_3N$. Minnesota Mining and Manufacturing Co. dried, fractionated, b.p. 178.5 - 179.0°C.

**ESTIMATED ERROR:**

$\delta X_1/X_1 = 0.003$

**REFERENCES:**

**COMPONENTS:**

(1) Nitrogen; $N_2$; [7727-37-9]

(2) FC-47, which is largely perfluorotributylamine or 1,1,2,2,3,3,4,4,4-nonafluoro-N,N-bis (nonafluorobutyl)-1-butanamine; $C_{12}F_{27}N$; [311-89-7]

**ORIGINAL MEASUREMENTS:**

Sargent, J. W.; Seffl, R. J.

*Fed. Proc.* 1970, *29*, 1699 - 1703.

**VARIABLES:**

T/K = 298-310
Total P/kPa = 101.325

**PREPARED BY:**

A. L. Cramer
H. L. Clever

**EXPERIMENTAL VALUES:**

| Temperature t/°C | T/K | Nitrogen Solubility g $O_2$ 1000 $g^{-1}$ | $cm^3$ $O_2$ 100 $cm^{-3}$ |
|---|---|---|---|
| 25 | 298.15 | 0.174 | 28.4 |
| 37 | 310.15 | 0.171 | 28.6 |

The authors solubility, $cm^3$ $N_2$ 100 $cm^{-3}$, appears to be the Ostwald coefficient times 100.

**AUXILIARY INFORMATION**

**METHOD/APPARATUS/PROCEDURE:**

The equilibrium apparatus is a 50 $cm^3$ three neck flask equipped with a water condenser, a thermometer, and a gas dispersing tube extending below the fluid surface level. Between 25 and 35 $cm^3$ of solvent are placed in the flask. Gas is bubbled through the solvent for 2 - 3 h. An 80 microliter sample is slowly taken into a gas tight syringe. The sample is injected into a gas chromatograph. The results of three sampling and analyses are averaged.

The chromatograph is an F and M Model 720 equipped with a thermal conductivity detector at 320 °C, and a 6' by ¼" stainless steel column packed with 13X molecular sieve at room temperature. The apparatus is calibrated with pure gas samples.

Details of the procedure were furnished by R. D. Danielson of the 3 M Co.

**SOURCE AND PURITY OF MATERIALS:**

(1) Nitrogen. Source not given. Commercial cylinder.

(2) FC-47. Minnesota Mining and Manufacturing Co. All H in an organic compound are replaced by F by Simons electrochemical process (1)

**ESTIMATED ERROR:**

$\delta T/K = 0.1$
$\delta c/c = 0.01$ (reproducibility)
$= 0.05$ (absolute)

**REFERENCES:**

1. Simons, J. H., Editor *Fluorine Chemistry* Academic Press, New York, 1950.

| COMPONENTS: | ORIGINAL MEASUREMENTS: |
|---|---|
| (1) Nitrogen; $N_2$; [7727-37-9]<br>(2) 1,1,2,2,3,3,4,4,4-nonafluoro-N,N-bis(nonafluorobutyl)-1-butanamine (Perfluorotributylamine); $C_{12}F_{27}N$; [311-89-7] | Powell, R. J.<br>*J. Chem. Eng. Data* 1972, *17*, 302-4. |
| VARIABLES: | PREPARED BY: |
| T/K = 288.15 - 318.15<br>$N_2$ P/kPa = 101.325 | P. L. Long |

EXPERIMENTAL VALUES:

| T/K | Mol Fraction $x_1 \times 10^4$ | Bunsen Coefficient $\alpha$ | Ostwald Coefficient L | $R\frac{\Delta \log X_1}{\Delta \log T} = N$ |
|---|---|---|---|---|
| 298.15 | 42.36 | 0.267 | 0.292 | -0.35 |

The author states that solubility measurements were made between 288.15 and 318.15 K, but only the solubility at 298.15 was given in the paper. The slope $R(\Delta \log X_1/\Delta \log T)$ was given. The smoothed data below were calculated by the compiler from the slope in the form:

$$\log x_1 = \log(42.36 \times 10^{-4}) - (0.35/R)\log(T/298.15)$$

with R = 1.9872 cal $K^{-1}$ $mol^{-1}$.

| T/K | Mol Fraction $x_1 \times 10^4$ |
|---|---|
| 288.15 | 42.62 |
| 293.15 | 42.49 |
| 298.15 | 42.36 |
| 303.15 | 42.23 |
| 308.15 | 42.11 |
| 313.15 | 41.99 |
| 318.15 | 41.88 |

The Bunsen and Ostwald coefficients were calculated by the compiler.

AUXILIARY INFORMATION

METHOD/APPARATUS/PROCEDURE:

Dymond and Hildebrand (1) apparatus which uses an all glass pumping system to spray slugs of degassed solvent into the gas. The amount of gas dissolved is calculated from the initial and final gas pressures. The solvent is degassed by freezing and pumping followed by boiling under reduced pressure.

SOURCE AND PURITY OF MATERIALS:

(1) Nitrogen. Source not given. Manufacturers research grade, dried over $CaCl_2$ before use.

(2) Perfluorotributylamine. Minnesota Mining & Mfg. Co., Column distilled, used portion with b.p. = 447.85 - 448.64 K, & single peak GC.

ESTIMATED ERROR:

$\delta N$/cal $K^{-1}$ $mol^{-1}$ = 0.1

$\delta x_1/x_1$ = 0.002

REFERENCES:

1. Dymond, J. H.; Hildebrand, J. H. *Ind. Eng. Chem. Fundam.* 1967, *6*, 130.

**COMPONENTS:**

(1) Nitrogen; $N_2$; [7727-37-9]

(2) Hydrazine; $NH_2NH_2$; [302-01-2]

**ORIGINAL MEASUREMENTS:**

Chang, E. T.; Gokcen, N. A.; Poston, T. M.

*J. Phys. Chem.* 1968, *72*, 638 -42.

**VARIABLES:**

T/K = 278.15 - 308.15
P/kPa =106.31 - 209.92

**PREPARED BY:**

P. L. Long
H. L. Clever

**EXPERIMENTAL VALUES:**

| T/K | Nitrogen Pressure $P_1$/atm | Henry's Constant[1] $10^5 K/atm^{-1}$ | Mol Fraction $x_1 \times 10^5$ | Bunsen Coefficient $\alpha$ | Ostwald Coefficient L |
|---|---|---|---|---|---|
| 278.15 | 1.1453 | 0.49 | 0.56 | | |
| | 2.0836 | 0.58 | 1.21 | | |
| | 1.0 | 0.54 av. | 0.54 | 0.0038 | 0.0039 |
| 293.15 | 1.0492 | 0.66 | 0.69 | | |
| | 1.1120 | 0.64 | 0.71 | | |
| | 1.9719 | 0.74 | 1.45 | | |
| | 2.0619 | 0.73 | 1.51 | | |
| | 1.0 | 0.69 | 0.69 | 0.0049 | 0.0039 |
| 308.18 | 1.0842 | 0.78 | 0.84 | | |
| | 2.1112 | 0.85 | 1.80 | | |
| | 1.0 | 0.82 av. | 0.82 | 0.0056 | 0.0064 |

[1]Henry's constant, $K/atm^{-1} = x_1/(P_1/atm)$.

The solubility values at 101.325 kPa (1 atm) were calculated by the compiler from the average Henry's constant.

| T/K | Mol Fraction $x_1 \times 10^5$ | Gibbs Energy $\Delta G°/kJ\ mol^{-1}$ |
|---|---|---|
| 278.15 | 0.54 | 28.06 |
| 288.15 | 0.62 | 28.71 |
| 298.15 | 0.72 | 29.36 |
| 308.15 | 0.82 | 30.00 |

Smoothed Data: Mole fraction solubility of nitrogen in hydrazine at one atm $N_2$ pressure.

$$\ln x_1 = -1206.46/(T/K) - 7.7956$$

$$\Delta H°/kJ\ mol^{-1} = 10.0;\ \Delta S°/JK^{-1}\ mol^{-1} = -64.8$$

**AUXILIARY INFORMATION**

**METHOD/APPARATUS/PROCEDURE:**

The solvent was degassed under vacuum in the previously weighed apparatus. Apparatus and degassed solvent weighed. Gas was introduced into the apparatus at a known P and T. The liquid was stirred, and the pressure was observed until there was no further change. Hydrazine did not appear to decompose with time as did the substituted hydrazines.

The apparatus was of all glass construction. It consisted of three calibrated volumes for the measurement of the gas, a container for the solvent, which was stirred with a glass enclosed magnet, and a manometer for measuring the pressure with a microslide cathetometer. The solvent container had a capacity for 100 g of solvent with a 5 ml gas space above the liquid surface. The apparatus sections were calibrated to $\pm$ 0.0002-3 $cm^3$ (1). There are no additional solubility data in the author's other discussions (2,3) of the system.

**SOURCE AND PURITY OF MATERIALS:**

(1) Nitrogen. No information given.

(2) Hydrazine. No information on source. It was freshly distilled before use. The density was measured, and fitted to the equation:

$$\rho/g\ ml^{-1} = 1.02492 - 0.000865t/C.$$

**ESTIMATED ERROR:**

$\delta T/K = 0.03$
$\delta P/mmHg = 0.01$
$\delta X_1/X_1 = 0.05$

**REFERENCES:**


1. Chang, E. T.; Gokcen, N. A. *J. Phys. Chem.* 1966, *70*, 2394.
2. Chang, E. T.; Gokcen, N. A.; Poston, T. M. *J. Spacecr. Rockets* 1969, *6*, 1177.
3. Gokcen, N. A.; Chang, E. T. *J. Chem. Ed.* 1977, *54*, 368.

**COMPONENTS:**

(1) Nitrogen; $N_2$; [7727-37-9]

(2) Methylhydrazine; $NHCH_3NH_2$; [60-34-4]

**ORIGINAL MEASUREMENTS:**

Chang, E. T.; Gokcen, N. A. Poston, T. M.

*J. Phys. Chem.* 1968, *72*, 638 - 42.

**VARIABLES:**

T/K = 253.24 - 298.14
P/kPa = 99.16 - 211.32

**PREPARED BY:**

H. L. Clever
P. L. Long

**EXPERIMENTAL VALUES:**

| T/K | Nitrogen Pressure $P_1$/atm | Henry's Constant[1] $10^5$K/atm$^{-1}$ | Mol Fraction $x_1 \times 10^5$ | Bunsen Coefficient α | Ostwald Coefficient L |
|---|---|---|---|---|---|
| 253.24 | 0.9786 | 6.62 | 6.48 | | |
| | 2.0042 | 6.68 | 13.39 | | |
| | 1.0 | 6.65 av. | 6.65 | 0.0296 | 0.0274 |
| 273.15 | 1.1052 | 7.55 | 8.34 | | |
| | 1.9863 | 7.56 | 15.02 | | |
| | 1.0 | 7.56 av. | 7.56 | 0.0329 | 0.0329 |
| 298.14 | 1.0577 | 9.25 | 9.78 | | |
| | 2.0856 | 9.33 | 19.46 | | |
| | 1.0 | 9.29 | 9.29 | 0.0392 | 0.0428 |

[1] Henry's constant, K/atm$^{-1}$ = $x_1/(P_1/\text{atm})$. The solubility values at one atm nitrogen pressure were calculated by the compiler from the value of the average Henry's constant.

Smoothed Data: Mole fraction of nitrogen in methyl hydrazine at one atm $N_2$ pressure.

$$\ln x_1 = -558.58/(T/K) - 7.422$$

$\Delta H°$/kJ mol$^{-1}$ = 4.64;

$\Delta S°$/J K$^{-1}$ mol$^{-1}$ = -61.7

| T/K | Mol Fraction $x_1 \times 10^5$ | $\Delta G°$/kJ mol$^{-1}$ |
|---|---|---|
| 258.15 | 6.87 | 20.57 |
| 268.15 | 7.45 | 21.19 |
| 278.15 | 8.03 | 21.81 |
| 288.15 | 8.61 | 22.43 |
| 298.15 | 9.18 | 23.04 |

AUXILIARY INFORMATION

**METHOD/APPARATUS/PROCEDURE:**

The solvent was degassed under vacuum in the previously weighed apparatus. Apparatus and degassed solvent were weighed. Gas was introduced into the apparatus at a known P and T, the liquid stirred, and the pressure observed until there was no further change. Equilibrium was established within 10 m and the P was followed for 40 m. Substituted hydrazines appear to decompose with time. For decomposing solvents the P was followed for up to 2 h, and the solubility value was corrected for the gaseous decomp. prod.

The apparatus was of all Pyrex glass construction. It consisted of three calibrated volumes for the measurement of the gas, a container for the solvent, which was stirred with a glass enclosed magnet, and a manometer with a microslide cathetometer for measuring the pressure. The solvent container had a capacity for 100 g of solvent with a 5 ml gas space above the liquid surface. The apparatus sections were calibrated to ± 0.0002 - 3 $cm^3$ (1). See comment on page 232 about additional references.

**SOURCE AND PURITY OF MATERIALS:**

(1) Nitrogen. No information given.

(2) Methylhydrazine. Source not given. Distilled prior to use.

Denstiy, ρ/g ml$^{-1}$ = 0.89338 - 0.000943t/°C.

**ESTIMATED ERROR:**

$\delta T/K = 0.03$
$\delta P/mmHg = 0.01$
$\delta x_1/x_1 = 0.05$

**REFERENCES:**

1. Chang, E. T.; Gokcen, N. A. *J. Phys. Chem.* 1966, *70*, 2394.

**COMPONENTS:**

(1) Nitrogen; $N_2$; [7727-37-9]

(2) 1,1-Dimethylhydrazine; $N(CH_3)_2NH_2$; [57-14-7]

**ORIGINAL MEASUREMENTS:**

Chang, E. T.; Gokcen, N. A. Poston, T. M.

*J. Phys. Chem.* 1968, *72*, 638 - 42.

**VARIABLES:**

T/K = 253.05 - 293.16
P/kPa = 101.38 - 209.92

**PREPARED BY:**

H. L. Clever
P. L. Long

**EXPERIMENTAL VALUES:**

| T/K | Nitrogen Pressure $P_1$/atm | Henry's Constant[1] $10^5$K/atm$^{-1}$ | Mol Fraction $x_1 \times 10^5$ | Bunsen Coefficient α | Ostwald Coefficient L |
|---|---|---|---|---|---|
| 253.05 | 1.2330 | 29.71 | 36.64 | | |
| | 1.9703 | 29.66 | 58.44 | | |
| | 1.0 | 29.7 av. | 29.7 | 0.0919 | 0.0851 |
| 273.15 | 1.0005 | 32.57 | 32.59 | | |
| | 1.3281 | 32.74 | 43.48 | | |
| | 2.0717 | 32.86 | 68.08 | | |
| | 1.0 | 32.7 av. | 32.7 | 0.0990 | 0.0990 |
| 293.16 | 1.1587 | 36.85 | 42.70 | | |
| | 1.9251 | 37.04 | 71.30 | | |
| | 1.0 | 36.9 av. | 36.9 | 0.108 | 0.116 |

[1]Henry's constant, K/atm$^{-1}$ = $X_1/P_1$/atm). The solubility values at 1.0 atm nitrogen pressure were calculated by the compiler from the average value of Henry's constant.

Smoothed Data: Mole fraction solubility of nitrogen at one atm nitrogen pressure.

$\ln x_1 = -404.66/(T/K) - 6.5297$

$\Delta H°/kJ\ mol^{-1} = 3.36$;

$\Delta S°/J\ K^{-1}\ mol^{-1} = -54.3$

| T/K | Mol Fraction $x_1 \times 10^5$ | $\Delta G°/J\ mol^{-1}$ |
|---|---|---|
| 258.15 | 30.4 | 17.38 |
| 268.15 | 32.2 | 17.92 |
| 278.15 | 34.1 | 18.47 |
| 288.15 | 35.8 | 19.01 |
| 298.15 | 37.6 | 19.55 |

AUXILIARY INFORMATION

**METHOD/APPARATUS/PROCEDURE:**

The solvent was degassed under vacuum in the previously weighed apparatus. Apparatus and degassed solvent were weighed. Gas was introduced into the apparatus at a known P and T, the liquid stirred, and the pressure observed until there was no further change. Equilibrium was established within 10 m and the P was followed for 40 m. Substituted hydrazines appear to decompose with time. For decomposing solvents the P was followed for up to 2 h, and the solubility value was corrected for the gaseous decomp. prod. The apparatus was of all Pyrex glass construction. It consisted of three calibrated volumes for the measurement of the gas, a container for the solvent, which was stirred with a glass enclosed magnet, and a manometer with a microslide cathetometer for measuring the pressure. The solvent container had a capacity for 100 g of solvent with a 5 ml gas space above the liquid surface. The apparatus sections were calibrated to ± 0.0002 - 3 $cm^3$ (1). See comment on page 232 about additional references.

**SOURCE AND PURITY OF MATERIALS:**

(1) Nitrogen. No information given.

(2) 1,1-Dimethylhydrazine. Source not given. Distilled just prior to use.

Density, $\rho$/g ml$^{-1}$ = 0.80980 - 0.001030 t/°C

**ESTIMATED ERROR:**

$\delta T/K = 0.03$
$\delta P/mmHg = 0.01$
$\delta X_1/X_1 = 0.05$

**REFERENCES:**

1. Chang, E. T.; Gokcen, N. A. *J. Phys. Chem.* 1966, *70*, 2394.

**COMPONENTS:**

(1) Nitrogen; $N_2$; [7727-37-9]

(2) Hydrazine; $NH_2NH_2$; [302-01-2]

(3) 1,1-Dimethylhydrazine; $N(CH_3)_2NH_2$; [57-14-7]

**ORIGINAL MEASUREMENTS:**

Chang, E. T.; Gokcen, N. A.

*J. Phys. Chem.* 1968, *72*, 2556 - 62.

**VARIABLES:**

T/K = 273.15 - 303.15
P/kPa = 50.66 - 253.31

**PREPARED BY:**

P. L. Long
H. L. Clever

**EXPERIMENTAL VALUES:**

| Mol Fraction 1,1-Dimethyl-hydrazine/$x_3$ | Gibbs Energy $\Delta G°$/cal mol$^{-1}$ = -RT ln K/atm$^{-1}$ | Henry's Constant[1] K/atm$^{-1}$ = $X_1/(P_1$/atm) $10^5$K/atm$^{-1}$ at 288.15K |
|---|---|---|
| 0.0 | 2360 + 15.60 T/K | 0.623 |
| 0.1 | 1830 + 15.59 T/K | 1.60 |
| 0.2 | 1450 + 15.60 T/K | 3.10 |
| 0.3 | 1200 + 15.46 T/K | 5.14 |
| 0.4 | 1150 + 14.82 T/K | 7.74 |
| 0.5 | 1140 + 14.18 T/K | 10.87 |
| 0.6 | 1100 + 13.72 T/K | 14.7 |
| 0.7 | 1050 + 13.38 T/K | 19.0 |
| 0.8 | 980 + 13.16 T/K | 24.0 |
| 0.9 | 900 + 13.03 T/K | 29.5 |
| 1.0 | 800 + 13.00 T/K | 35.7 |

[1] The Henry's constant is numerically equal to the mole fraction solubility at 101.325 kPa (1 atm).

The Gibbs energy equation was fitted to data taken in the 273.15 - 303.15 K temperature range.

The Henry's constant is based on data measured over the 0.5 - 2.5 atm pressure range. The values in the Table above are the Henry's constant at 288.15 K. Values at other temperatures can be calculated from the Gibbs energy equation.

AUXILIARY INFORMATION

**METHOD/APPARATUS/PROCEDURE:**

The solvent was degassed under vacuum in the previously weighed apparatus. Apparatus and degassed solvent were weighed. Gas was introduced into the apparatus at a known P and T, the liquid stirred, and the pressure observed until there was no further change. Equilibrium was established within 10 m and the P was followed for 40 m. Substituted hydrazines appear to decompose with time. For decomposing solvents the P was followed for up to 2 h, and the solubility value was corrected for the gaseous decomp. product.

The apparatus was of all Pyrex glass construction. It consisted of three calibrated volumes for the measurement of the gas, a container for the solvent, which was stirred with a glass enclosed magnet, and a manometer with a microslide cathetometer for measuring the pressure. The solvent container had a capacity for 100 g of solvent with a 5 ml gas space above the liquid surface.

**SOURCE AND PURITY OF MATERIALS:**

The source and purity of the materials were not given. The density and refractive index of the solvent components and several of their mixtures are given. The solvents were freshly distilled before use.

**ESTIMATED ERROR:**

$\delta T/K = 0.03$
$\delta P/mmHg = 0.01$
$\delta X_1/X_1 = 0.05$

**REFERENCES:**

1. Chang, E. T.; Gokcen, N. A. *J. Phys. Chem.* 1966, *70*, 2394.

The apparatus sections were calibrated to $\pm$ 0.0002 - 3 $cm^3$ (1).

**COMPONENTS:**

(1) Nitrogen; $N_2$; [7727-37-9]

(2) 1,2-Dimethylhydrazine; $NHCH_3NHCH_3$; [540-73-8]

**ORIGINAL MEASUREMENTS:**

Chang, E. T.; Gokcen, N. A. Poston, T. M.

*J. Phys. Chem.* 1968, *72*, 638 - 42.

**VARIABLES:**

T/K = 273.15 - 298.15

P/kPa = 101.325

**PREPARED BY:**

P. L. Long

H. L. Clever

**EXPERIMENTAL VALUES:**

The authors made no experimental measurements on this system. The authors give an estimated Gibbs energy equation for the solution of nitrogen in 1,2-dimethylhydrazine. They used logical assumptions to find a linear relationship between the Gibbs energy of solution of nitrogen in hydrazine, methylhydrazine, and 1,1-dimethylhydrazine as a function of $1/r^{12}$, where r is the distance of approach of solvent and solute molecules. The value of r in each solvent was determined from a simple cell model. The linear relationship was extrapolated to obtain the estimated value of the Gibbs energy of solution of nitrogen in 1,2-dimethylhydrazine. The estimated equation is

$$\Delta G°/\text{cal mol}^{-1} = -(R/\text{cal K}^{-1}\text{ mol}^{-1})(T/K)\ \ln\ (K/\text{atm}^{-1}) = 1300 + 8.51\ (T/K)$$

where K is the Henry's constant defined as $K/\text{atm}^{-1} = X_1/(P/\text{atm})$. The Henry's constants tabulated below were calculated from the equation by the compiler. They are numerically equal to the mole fraction solubility at one atm $N_2$ pressure.

| T/K | Henry's Constant $10^5 K/\text{atm}^{-1}$ |
|---|---|
| 273.15 | 126 |
| 278.15 | 131 |
| 283.15 | 137 |
| 288.15 | 143 |
| 293.15 | 148 |
| 298.15 | 154 |

AUXILIARY INFORMATION

**METHOD/APPARATUS/PROCEDURE:**

Estimated data, see above.

**SOURCE AND PURITY OF MATERIALS:**

**ESTIMATED ERROR:**

**REFERENCES:**

**COMPONENTS:**

(1) Nitrogen; $N_2$; [7727-37-9]

(2) Octamethylcyclotetrasiloxane; $C_8H_{24}O_4Si_4$; [556-67-2]

**ORIGINAL MEASUREMENTS:**

Wilcock, R. J.; McHale, J. L.; Battino, R.; Wilhelm, E.

*Fluid Phase Equil.* 1978, *2*, 225-30.

**VARIABLES:**

T/K = 292-313

**PREPARED BY:**

R. Battino

**EXPERIMENTAL VALUES:**

| T/K | $10^3 x_1$ [a] | L [b] |
|---|---|---|
| 292.15 | 2.624 | 0.2035 |
| 292.15 | 2.537 | 0.1967 |
| 292.15 | 2.541 | 0.1970 |
| 298.15 | 2.499 | 0.1963 |
| 298.30 | 2.588 | 0.2034 |
| 298.55 | 2.448 | 0.1925 |
| 313.11 | 2.579 | 0.2090 |

[a] Mole fraction solubility at 101.325 kPa partial pressure of gas.

[b] Ostwald coefficient.

[c] At 298.15 K smoothed values are: $x_1 = 2.551 \times 10^{-3}$; $\Delta H_1^\circ = 368$ J mol$^{-1}$, $\Delta S_1^\circ = -48.5$ J K$^{-1}$ mol$^{-1}$.

**AUXILIARY INFORMATION**

**METHOD/APPARATUS/PROCEDURE:**

The solubility apparatus is based on the design of Morrison and Billett (1) and the version used is described by Battino, et al. (3). Degassing: Up to 1000 cm$^3$ of solvent is placed in a flask of such size that the liquid is about 4 cm deep. The liquid is stirred rapidly and vacuum applied intermittently through a liquid nitrogen trap until the permanent gas residual pressure drops to 5 microns. Solubility determination: The degassed solvent is passed in a thin film down a glass spiral tube containing the solute gas plus the solvent vapor at a total pressure of ca. 1 atm. The volume of gas absorbed is found by difference between the initial and final volumes in the buret system. The solvent is collected in a tared flask and weighed.

**SOURCE AND PURITY OF MATERIALS:**

(1) Matheson, 99.999%

(2) General Electric Co.; density is 0.9500 g cm$^{-3}$ for distilled solvent.

**ESTIMATED ERROR:**

$\delta T/K = \pm 0.02$

$\delta x_1/x_1 = \pm 0.01$

**REFERENCES:**

1. Morrison, T.J.; Billett, F. *J. Chem. Soc.* 1948, 2033.

2. Battino, R.; Evans, F.D.; Danforth, W.F. *J. Am. Oil. Chem. Soc.* 1968, *45*, 830.

3. Battino, R.; Banzhof, M.; Bogan, M; Wilhelm, E. *Anal. Chem.* 1971, *43*, 806.

| COMPONENTS: | ORIGINAL MEASUREMENTS: |
|---|---|
| (1) Nitrogen; $N_2$; [7727-37-9]<br>(2) Animal and vegetable oils | Vibrans, F. C.<br>*Oil and Soap* 1935, *12*, 14 -5. |
| VARIABLES: T/K = 296 - 318 | PREPARED BY: P. L. Long<br>H. L. Clever |

EXPERIMENTAL VALUES:

| T/K | Solubility, $cm^3$ (273.15 K, 1 atm) $H_2$ 100 $cm^{-3}$ Experiment | | | | | Av. |
|---|---|---|---|---|---|---|
| Corn oil | | | | | | |
| 296.15-299.15 | 6.38 | 6.32 | 6.23 | | | 6.31 |
| 318.15 | 6.38 | 6.26 | 6.26 | 6.35 | 6.40 | 6.33 |
| Cottonseed oil | | | | | | |
| 296.15-299.15 | 6.06 | 6.20 | | | | 6.13 |
| 318.15 | 6.06 | 6.22 | | | | 6.14 |
| Hydrogenated cottonseed oil | | | | | | |
| 318.15 | 6.60 | 6.62 | 6.76 | 6.84 | | 6.70 |
| Steam rendered lard | | | | | | |
| 318.15 | 6.23 | 6.22 | | | | 6.22 |

AUXILIARY INFORMATION

METHOD/APPARATUS/PROCEDURE:

Van Slyke manometric method (1).

SOURCE AND PURITY OF MATERIALS:

(1) Nitrogen. From commercial cylinders. Source or purity not stated.

(2) Corn oil. Commercial sample bought on market. Representative of its class.

Cottonseed oil. Commercial sample bought on market. Representative of its class.

Lard. From a mixture of killing and cutting fats. Steam rendered.

ESTIMATED ERROR:

$\delta T/K$ = 3 at room temperature
2 at 318 K
$\delta c/c$ = 0.04 (author)

REFERENCES:

1. Van Slyke, D. D.; *et al.* *J. Biol. Chem.* 1924, *61*, 523 and 575.

**COMPONENTS:**

(1) Nitrogen; $N_2$; [7727-37-9]

(2) Animal and vegetable oils

**ORIGINAL MEASUREMENTS:**

Schaffer, P. S.; Haller, H. S.

*Oil and Soap* 1943, *20*, 161 - 2.

**VARIABLES:**

T/K = 313-333
P/kPa = 101.325

**PREPARED BY:**

P. L. Long
H. L. Clever

**EXPERIMENTAL VALUES:**

| T/K | $cm^3$ (273.15 K, 1 atm) $N_2$ 100 $cm^{-3}$ |
|---|---|
| Cottonseed oil | |
| 313.15 | 6.2 |
| Butter oil | |
| 313.15 | 8.9 |
| 333.15 | 7.9 |
| Lard | |
| 313.15 | 6.6 |

The author's solubility unit is the Bunsen coefficient x 100.

**AUXILIARY INFORMATION**

**METHOD/APPARATUS/PROCEDURE:**

The apparatus consists of a reaction flask connected to a water jacketed gas buret, a mercury leveling bulb, and a differential manometer. The reaction vessel is attached to a shaking mechanism in an air bath.

A 50 $cm^3$ sample of oil is placed in the falsk. Heat and vacuum are applied to remove dissolved gases. The sample is cooled to the temperature of the measurement. Nitrogen gas is admitted, and a zero reading is taken. The sample is shaken until the sample takes up no more gas at a pressure of 1 atm. Three determinations were made on each sample.

**SOURCE AND PURITY OF MATERIALS:**

(1) Nitrogen. From commercial cylinder, used as received.

(2) Cottonseed oil. Good grade commercial product.

Butter oil. Obtained from butter prepared in the lab from fresh cream

Lard. Good grade commercial product.

**ESTIMATED ERROR:**

$\delta T/K = 0.5$
$\delta c/c = 0.003$ (authors, error among three determinations)

**REFERENCES:**

| COMPONENTS: | ORIGINAL MEASUREMENTS: |
|---|---|
| (1) Nitrogen; $N_2$; [7727-37-9]<br>(2) Soybean Oil | Tomoto, N.; Kusano, K.<br>*Yukagaku* 1967, *16*, 108-13.<br>*Chem. Abstr.* 1953, *66*, 1061412. |
| VARIABLES: | PREPARED BY: |
| T/K = 303-343 | R. Battino |

EXPERIMENTAL VALUES:

| t/°C | $T^a$/K | $10^2\alpha^b$/cm (STP) $cm^{-3}$ $atm^{-1}$ |
|---|---|---|
| 30 | 303 | 8.60 |
| 50 | 323 | 6.85 |
| 70 | 343 | 5.22 |

[a] Calculated by compiler.

[b] Bunsen coefficient.

AUXILIARY INFORMATION

METHOD/APPARATUS/PROCEDURE:

Described in the paper.

SOURCE AND PURITY OF MATERIALS:

(1) No details given.

(2) Saponification value: 192.0; acid value: 0.17; iodine value: 128.5; unsaponifiable: 0.33%; $d_{30}$: 0.9132 g $cm^{-3}$; M.W.: 877 g $mol^{-1}$; molar volume: 960.4 $cm^3$ $mol^{-1}$.

ESTIMATED ERROR:

$\delta\alpha/\alpha$ = ±0.02, compiler's estimate.

REFERENCES:

| COMPONENTS: | EVALUATOR: |
|---|---|
| (1) Nitrogen; $N_2$; [7727-37-9]<br>(2) Olive Oil | Rubin Battino<br>Department of Chemistry<br>Wright State University<br>Dayton, OH 45435<br>U.S.A. |

CRITICAL EVALUATION:

Four groups (1-4) have studied the nitrogen/olive oil system and with some disparity between their results. We have rejected the single value determined by Ikels (3) as being too low. The remaining data were smoothed in two ways by least squares analysis to yield:

$$\ln x_1 = -5.6757 - 0.58383/(T/100K) \qquad (1)$$

$$L = 1.9886 \times 10^{-3} + 2.3052 \times 10^{-4} (T/K) \qquad (2)$$

where $x_1$ is the mole fraction solubility at 101 kPa partial pressure of gas and L is the Ostwald coefficient. The fit for both equations was to 3.1% in the solubility (one standard deviation divided by the solubility at the mid-point between 285 and 328 K). Smoothed values for both solubility units at 5K intervals are presented in the table. These are tentative values.

| T/K | $10^3 x_1$ | $10^2 L$ | T/K | $10^3 x_1$ | $10^2 L$ |
|---|---|---|---|---|---|
| 283.15 | 2.79 | 6.73 | 308.15 | 2.84 | 7.30 |
| 288.15 | 2.80 | 6.84 | 313.15 | 2.85 | 7.42 |
| 293.15 | 2.81 | 6.96 | 318.15 | 2.85 | 7.53 |
| 298.15 | 2.82 | 7.07 | 323.15 | 2.86 | 7.65 |
| 303.15 | 2.83 | 7.19 | 328.15 | 2.87 | 7.76 |

References


1. Battino, R.; Evans, F. D.; Danforth, W. F. *J. Am. Oil Chem. Soc.* 1968, *45*, 830-3.
2. Power, G. G.; Stegall, H. *J. Appl. Physiology* 1970, *29*, 145-9.
3. Ikels, K. G. *DDC Report No. SAM-TDR-64-1* 1964.
4. Davidson, D.; Eggleton, P.; Foggie, P. *Quart. J. Exptl. Physiol.* 1952, *37*, 91-105.

**COMPONENTS:**

(1) Nitrogen; $N_2$; [7727-37-9]

(2) Olive Oil, Tetralin, Oleic acid, Ethyl palmitate (liquid), Lard

**ORIGINAL MEASUREMENTS:**

Davidson, D.; Eggleton, P.; Foggie, P.

*Quart. J. Exptl. Physiol.* 1952, 37, 91-105.

**VARIABLES:**

T/K = 298-323

**PREPARED BY:**

R. Battino

**EXPERIMENTAL VALUES:**

| t/°C | $T^a$/K | $L^b$ |
|---|---|---|
| | Olive Oil | |
| 25.2 | 298.4 | 0.0686 |
| | Tetralin | |
| 25.2 | 298.4 | 0.0638 |
| | Oleic Acid | |
| 25.2 | 298.4 | 0.085 |
| | Ethyl Palmitate (liquid) | |
| 25.2 | 298.4 | 0.109 |
| | Lard (liquid) | |
| 50 | 323 | 0.07 |

[a] Calculated by compiler.

[b] Ostwald coefficient.

[c] Tetralin: 1,2,3,4-Tetrahydronaphthalene; $C_{10}H_{12}$; [119-64-2]

[d] Oleic Acid: (Z)-9-Octadecenoic acid; $C_{18}H_{34}O_2$; [112-80-1]

[e] Ethyl Palmitate: Hexadecanoic acid, ethyl ester; $C_{18}H_{36}O_2$; [628-97-7]

**AUXILIARY INFORMATION**

**METHOD/APPARATUS/PROCEDURE:**

Combined with diffusion measurements. Measurement based knowing gas and liquid phase volumes and the initial (before any gas dissolves) pressure and the final equilibrium pressure.

**SOURCE AND PURITY OF MATERIALS:**

No details given.

**ESTIMATED ERROR:**

$\delta L/L$ = ±0.08, compiler's estimate.
$\delta T/K$ = ±0.01, but reported to 0.1°C or 1°C.

**REFERENCES:**

**COMPONENTS:**

(1) Nitrogen; $N_2$; [7727-37-9]

(2) Olive oil, fats

(3) Water; $H_2O$; [7732-18-5]

**ORIGINAL MEASUREMENTS:**

Ikels, K. G.

*DDC, Report No. SAM-TDR-64-1* 1964.

**VARIABLES:**

T/K = 311

**PREPARED BY:**

R. Battino

**EXPERIMENTAL VALUES:**

| Substance | t/°C | $T^a$/K | $10^2\alpha^a$ |
|---|---|---|---|
| Water | 37.6 | 310.8 | 1.206 |
| Olive Oil | 37.4 | 310.6 | 6.092 |
| Human fat (pooled) | 37.4 | 310.6 | 6.170 |
| Human fat (individual) | 37.4 | 310.6 | 6.109 |
| Dog fat | 37.4 | 310.6 | 6.092 |

[a] Calculated by compiler.

[b] Bunsen coefficient in units of $cm^3$ (STP) $cm^{-3}atm^{-1}$. These values are the means of multiple measurements.

**AUXILIARY INFORMATION**

**METHOD/APPARATUS/PROCEDURE:**

Combination of van Slyke and gas chromatography.

**SOURCE AND PURITY OF MATERIALS:**

(1) No details given.

(2) Human adipose tissue from the abdominal region within 3-4 h of post-mortem. Density of pooled sample was 0.906 g $cm^{-3}$. Density of olive oil was 0.907 g $cm^{-3}$ at 310K.

(3) Distilled.

**ESTIMATED ERROR:**

$\delta\alpha/\alpha$ = ±0.01, compiler's estimate.

**REFERENCES:**

COMPONENTS:

(1) Nitrogen; $N_2$; [7727-37-9]

(2) Olive Oil

ORIGINAL MEASUREMENTS:

Battino, R.; Evans, F. D.; Danforth, W. F.

*J. Am. Oil Chem. Soc.* 1968, *45*, 830-3.

VARIABLES:

T/K = 297.93 - 327.71
P/kPa = 101.325

PREPARED BY:

H. L. Clever

EXPERIMENTAL VALUES:

| T/K | Mol Fraction $X_1 \times 10^3$ | Bunsen Coefficient $\alpha$ | Ostwald Coefficient L |
|---|---|---|---|
| 297.93 | 2.94 | $0.0674_8$ | $0.0736_0$ |
| 298.04 | 3.00 | $0.0688_0$ | $0.0750_7$ |
| 307.95 | 2.88 | $0.0657_7$ | $0.0741_5$ |
| 318.20 | 2.84 | $0.0644_5$ | $0.0750_5$ |
| 327.71 | 2.82 | $0.0636_1$ | $0.0763_2$ |

Smoothed Data: $\Delta G^o/J\ mol^{-1} = -RT \ln X_1 = -1460.7 + 53.314\ T$

Std. Dev. $\Delta G^o$ = 23, Coef. Corr. = 0.9992

$\Delta H^o/J\ mol^{-1} = -1460.7$, $\Delta S^o/J\ K^{-1}\ mol^{-1} = -53.314$

| T/K | Mol Fraction $X_1 \times 10^3$ | $\Delta G\ /J\ mol^{-1}$ |
|---|---|---|
| 298.15 | 2.96 | 14,435 |
| 308.15 | 2.90 | 14,968 |
| 318.15 | 2.85 | 15,501 |
| 328.15 | 2.80 | 16,034 |

AUXILIARY INFORMATION

METHOD/APPARATUS/PROCEDURE:

The apparatus is based on the design by Morrison and Billett (1) and the version used is a modification of the apparatus of Clever, Battino, Saylor and Gross (2).

Degassing. The solvent is sprayed into an evacuated chamber of an all glass apparatus; it is stirred and heated until the pressure drops to the vapor pressure of the liquid.

Solubility Determination. The degassed liquid passes in a thin film down a glass spiral tube at a total pressure of one atm of solute gas plus solvent vapor. The gas absorbed is measured in the attached buret system, and the solvent is collected in a tared flask and weighed.

SOURCE AND PURITY OF MATERIALS:

(1) Nitrogen. Matheson Co., Inc. min. vol. % purity 99.999.

(2) Olive oil. A. U.S.P., Fisher Scientific Company., 0.58% free fatty acid.
B. Nutritional Biochemicals Corp., 0.30% free fatty acid. The density was measured and fitted to the equation $\rho/g\ cm^{-3} = 0.9152 - 0.000468t/C$. The average mol wt is 884 ± 45 (3).

ESTIMATED ERROR:

$\delta T/K = 0.03$
$\delta P/mmHg = 0.5$
$\delta X_1/X_1 = 0.005$

REFERENCES:

1. Morrison, T. J.; Billett, F. *J. Chem. Soc.* 1948, 2033.
2. Clever, H. L.; Battino, R.; Saylor, J. H.; Gross, P. M. *J. Phys. Chem.* 1957, *61*, 1078.
3. Johnson, L. F.; Shoolery, J. N. *Anal. Chem.* 1962, *34*, 1136.

| COMPONENTS: | ORIGINAL MEASUREMENTS: |
|---|---|
| (1) Nitrogen; $N_2$; [7727-37-9]<br>(2) Olive oil. | Power, G.G.; Stegall, H.<br>*J. Appl. Physiology* 1970, *29*, 145-9. |
| VARIABLES:<br>T/K = 285-310 | PREPARED BY:<br>C.L. Young. |

EXPERIMENTAL VALUES:

| T/K | Bunsen coefficient, $\alpha$ | S.D.* | No. of measurements |
|---|---|---|---|
| 310.15 | 0.0643 | 0.0005 | 5 |
| 298.15 | 0.0633 | 0.0005 | 5 |
| 285.15 | 0.0622 | 0.0002 | 5 |

* Standard deviation.

AUXILIARY INFORMATION

METHOD/APPARATUS/PROCEDURE:

Liquid saturated with gas in a stirred cell. 5.0 $cm^3$ samples of liquid removed in a calibrated syringe and amount of gas extracted by two extraction in a van Slyke apparatus determined. Details in source and ref. (1).

SOURCE AND PURITY OF MATERIALS:

(1) Matheson Co. sample, purity better than 99.7 mole per cent.

(2) No details given.

ESTIMATED ERROR:

$\delta T/K = \pm 0.1$

REFERENCES:

1. Power, G.G.
*J. Appl. Physiology.* 1968, *24*, 468.

COMPONENTS:

(1) Nitrogen; $N_2$; [7727-37-9]

(2) Human blood, plasma

ORIGINAL MEASUREMENTS:

Farhi, L.E.; Edwards, A.W.T.; Homma, T.

*J. Appl. Physiol.* 1963, *18*, 97-106.

VARIABLES:

T/K = 310

PREPARED BY:

R. Battino

EXPERIMENTAL VALUES:

| Subject | t/°C | $T^a$/K | $10^2\alpha^b$ | Hematocrit |
|---|---|---|---|---|
| | | Human blood | | |
| TV | 37.25 | 310.40 | 1.285 | 43.7 |
| TV | 37.35 | 310.50 | 1.260 | |
| TV | 37.45 | 310.60 | 1.299 | 43.2 |
| LF | 37.35 | 310.50 | 1.260 | 43.7 |
| LF | 37.35 | 310.50 | 1.260 | |
| LF | 37.20 | 310.35 | 1.261 | 46.5 |
| AE | 37.35 | 310.50 | 1.248 | 42.7 |
| AE | 37.25 | 310.40 | 1.250 | 41.8 |
| AE | 37.20 | 310.35 | 1.238 | 38.5 |
| AE | 37.25 | 310.40 | 1.242 | |
| AE | 37.30 | 310.45 | 1.272 | |
| AE | 37.40 | 310.55 | 1.248 | |
| DH | 37.40 | 310.55 | 1.276 | 47.5 |
| WR | 37.10 | 310.25 | 1.291 | 46.5 |
| SF | 37.10 | 310.25 | 1.282 | 46.5 |
| ML | 37.35 | 310.50 | 1.292 | 44.2 |
| HR | 37.20 | 310.35 | 1.268 | 48.9 |
| RM | 37.40 | 310.55 | 1.270 | 43.2 |
| | | Human Plasma | | |
| TV | 37.25 | 310.40 | 1.215 | |
| AE | 37.25 | 310.40 | 1.195 | |

continued on following page

AUXILIARY INFORMATION

METHOD/APPARATUS/PROCEDURE:

Used a combination of the van Slyke method and gas chromatography.

SOURCE AND PURITY OF MATERIALS:

(1) From air.

(2) From nine normal healthy men aged 22-48 years.

ESTIMATED ERROR:

$\delta\alpha/\alpha = \pm 0.02$, compiler's estimate
$\delta T/K = \pm 0.05$

REFERENCES:

| COMPONENTS: | EVALUATOR: |
|---|---|
| (1) Nitrogen; $N_2$; [7727-37-9] (2) Human blood, plasma | Farhi, L.E.; Edwards, A.W.T.; Homma, T. *J. Appl. Physiol.* <u>1963</u>, *18*, 97-106. |

EXPERIMENTAL VALUES:

continued

[a] Calculated by compiler.

[b] Bunsen coefficient in units of $cm^3$ (STP) $cm^{-3}$ $atm^{-1}$.

[c] The average value of the Bunsen coefficient at 37.30°C (310.45K) was 0.01277 at a hematocrit of 45.2.

COMPONENTS:

(1) Nitrogen; $N_2$; [7727-37-9]

(2) Human blood

(3) Water; $H_2O$; [7732-18-5]

ORIGINAL MEASUREMENTS:

Christoforides, C.; Hedley-Whyte, J.

*Fed. Proc.* 1970, *29*, A330.

VARIABLES:

T/K = 298-310

PREPARED BY:

R. Battino

EXPERIMENTAL VALUES:

$\alpha^a$/cm³ (STP) cm⁻³ atm⁻¹

| | 25°C (298K) | 37°C (310K) |
|---|---|---|
| Human blood | 0.0145 | 0.0139 |
| Water | 0.0147 | 0.0128 |

[a] Bunsen coefficient.

AUXILIARY INFORMATION

METHOD/APPARATUS/PROCEDURE:

Measured by "standard Van Slyke manometry."

SOURCE AND PURITY OF MATERIALS:

(1) No details given.

(2) C.C. Hb 15.4g/100ml

(3) No details given.

ESTIMATED ERROR:

$\delta\alpha/\alpha$ = ±0.01, compiler's estimate.

REFERENCES:

| COMPONENTS: | ORIGINAL MEASUREMENTS: |
|---|---|
| (1) Nitrogen; $N_2$; [7727-37-9]<br>(2) Hemoglobin | Conant, J. B.; Scott, N. D.<br>*J. Biol. Chem.* 1926, *68*, 107-21 |
| **VARIABLES:** | **PREPARED BY:** |
| T/K ≅ 296 | R. Battino |

EXPERIMENTAL VALUES:

| Material Used | | $c_{Hb}$/ (g/100 $cm^3$) | $10^3\alpha^a$ | $cm^3$ $N_2$ absorbed /g Hb |
|---|---|---|---|---|
| Solution 4. | Horse Hb. Undiluted. | 22.5 | 4.7 | 0.021 |
| Solution 4. | Diluted to 2/3 concentration. | 15.0 | 3.1 | 0.020 |
| Solution 4. | Diluted to 1/3 concentration. | 7.5 | 1.7 | 0.023 |
| Solution 12. | Undiluted | 15.9 | 4.1 | 0.026 |
| Solution 12. | Diluted 1/2. | 7.95 | 2.1 | 0.026 |
| Solution 5. | Laked with toluene. | 12.23 | 3.4 | 0.028 |
| Solution 6. | After shaking with toluene. | 8.0 | 2.3 | 0.029 |
| Solution 6. | Same as above, laked with water. | 8.0 | 1.1 | 0.013 |
| Solution 7. | Cow Hb in water. | 13.6 | 2.7 | 0.020 |
| Solution 8. | Same as above, shaken with toluene. | 16.1 | 3.9 | 0.021 |
| Solution 7. | M/10 with NaCl. | 12.8 | 3.1 | 0.024 |
| Solution 7. | M/10 with phosphate buffer.pH=6.12. | 12.8 | 2.3 | 0.018 |
| Solution 7. | With phosphate buffer. pH=6.98. | 12.8 | 2.4 | 0.019 |
| Solution 7. | M/10 with borate buffer. pH=8.41. | 12.8 | 3.6 | 0.028 |
| Solution 7. | With borate buffer. pH=9.08. | 12.8 | 2.9 | 0.023 |
| Solution 9. | Purified horse Hb, isoelectric. | 4.5 | 1.5 | 0.034 |
| Solution 10. | Same as above, phosphate buffer. pH=6.12. | 8.0 | 3.8 | 0.047 |
| Solution 11. | Same as above, borate buffer. pH=8.41 | 8.5 | 3.7 | 0.044 |

continued on following page

AUXILIARY INFORMATION

METHOD/APPARATUS/PROCEDURE:

Van Slyke method used.

SOURCE AND PURITY OF MATERIALS:

(1) From air.

(2) Horse hemoglobin except for solutions 7 and 8 which were cow hemoglobin. Details in paper.

ESTIMATED ERROR:

$\delta\alpha/\alpha$ = ±0.08, compiler's estimate.

REFERENCES:

| COMPONENTS: | ORIGINAL MEASUREMENTS: |
|---|---|
| (1) Nitrogen; $N_2$; [7727-37-9] | Conant, J. B.; Scott, N. D. |
| (2) Hemoglobin | *J. Biol. Chem.* 1926, *68*, 107-21. |

continued

[a] Bunsen coefficient in $cm^3$ (STP) $cm^{-3}$ $atm^{-1}$.
[b] $cm^3$ $N_2$ at STP.
[c] At "room temperature." About 23°C (296 K).

| $P_{N_2}$/mmHg | $P^a_{N_2}$/kPa | Solubility[b] | $P_{N_2}$/mmHg | $P^a_{N_2}$/kPa | Solubility[b] |
|---|---|---|---|---|---|
| Solution 5 | | | Solution 9 | | |
| 585 | 78.0 | 0.0279 | 585 | 78.0 | 0.0359 |
| 66.6 | 8.88 | 0.0127 | 585 | 78.0 | 0.0325 |
| 66.6 | 8.88 | 0.0151 | 62.6 | 8.35 | 0.0271 |
| 9.6 | 1.28 | 0.0088 | 62.6 | 8.35 | 0.0271 |
| 9.6 | 1.28 | 0.0113 | 12.3 | 1.64 | 0.0237 |
| | | | 12.3 | 1.64 | 0.0257 |

[a] Calculated by compiler.
[b] $cm^3$ $N_2$ at STP dissolved per gram of hemoglobin.

**COMPONENTS:**

(1) Nitrogen; $N_2$; [7727-37-9]

(2) Ox blood and serum

**ORIGINAL MEASUREMENTS:**

Findlay, A.; Creighton, H.J.M.

*Biochem. J.* 1910, *5*, 294-305.

**VARIABLES:**

T/K = 298
P/kPa = 101-209

**PREPARED BY:**

C.L. Young

**EXPERIMENTAL VALUES:**

| T/K | Density of soln. /g cm$^{-3}$ | $p_{N_2}$/mmHg | $p_{O_2}$/MPa | Solubility, $S^+$ |
|---|---|---|---|---|
| | | Deaerated Ox Blood | | |
| 298.15 | 1.066 | 757 | 0.101 | 0.0041 |
| | | 979 | 0.131 | 0.0080 |
| | | 1089 | 0.145 | 0.0108 |
| | | 1241 | 0.165 | 0.0135 |
| | | 1367 | 0.182 | 0.0143 |
| | | 1564 | 0.209 | 0.0165 |
| | | Deaerated Ox Serum | | |
| 298.15 | 1.028 | 734 | 0.0979 | 0.0008 |
| | | 891 | 0.119 | 0.0014 |
| | | 1056 | 0.141 | 0.0020 |
| | | 1250 | 0.167 | 0.0034 |
| | | 1368 | 0.182 | 0.0047 |
| | | 1529 | 0.204 | 0.0065 |

+ Solubility, $S$, given as $\frac{\text{Conc. of gas in the liquid phase}}{\text{Conc. of gas in the gas phase}}$

**AUXILIARY INFORMATION**

**METHOD/APPARATUS/PROCEDURE:**

Gas buret and absorption pipet similar to that of Geffcken (1) except that the manometer tube was longer to give the higher pressures.

**SOURCE AND PURITY OF MATERIALS:**

(1) Obtained by heating a mixture of sodium nitrite and ammonium chloride.

(2) Obtained from slaughter house. Deaerated blood and serum kept under a reduced pressure of ~2 kPa for one hour.

**ESTIMATED ERROR:**

$\delta T/K = \pm 0.1$; $\delta S/S = \pm 2\%$ (estimated by compiler).

**REFERENCES:**

1. Geffcken, G. *Z. Phys. Chem.* 1904, *49*, 257.

COMPONENTS:

(1) Nitrogen; $N_2$; [7727-37-9]

(2) Ox erythrocytes and horse hemoglobin; see below

ORIGINAL MEASUREMENTS:

Van Slyke, D. D.; Dillon, R. T.; Margaria, R.

*J. Biol. Chem.* 1934, *105*, 571-96.

VARIABLES:

T/K = 298
Concentration

PREPARED BY:

R. Battino

EXPERIMENTAL VALUES:

| Suspension or Solution[a] | Gas Mixture[b] | $P^c_{N_2}$/mmHg | $P^d_{N_2}$/kPa | $10^2\alpha^e$ | $10^2\alpha^f_{Hb}$ |
|---|---|---|---|---|---|
| 1. Laked cells | Air | 581 | 77.5 | 1.527 | 1.96 |
| 2. Suspension | | 575 | 76.7 | 1.523 | 1.89 |
| 5. Suspension | | 576 | 76.8 | 1.527 | 1.93 |
| 6. Hb crystals | | 576 | 76.8 | 1.507 | 1.64 |
| 6. Hb crystals | | 578 | 77.1 | 1.515 | 1.67 |
| 1. Suspension | Air + tank $O_2$ | 119.4 | 15.9 | 1.54 | 2.1 |
| 4. Laked cells | | 99.0 | 13.2 | 1.63 | 1.9 |
| 4. Suspension | | 60.3 | 8.04 | 1.54 | 2.1 |
| 5. Suspension | | 39.4 | 5.25 | 1.56 | 2.3 |
| 1. Suspension[g] | | 34.6 | 4.61 | 1.71 | 3.3 |
| 4. Suspension | | 22.7 | 3.03 | 1.57 | 2.3 |
| 4. Laked cells | Air + pure $O_2$ | 93.5 | 12.5 | 1.48 | 1.7 |
| 3. Suspension | | 39.2 | 5.23 | 1.57 | 2.3 |
| 3. Suspension | | 38.0 | 5.07 | 1.58 | 2.4 |
| 1. Suspension[g] | | 14.5 | 1.93 | 1.21 | |
| 5. Suspension | Air + pure $CO_2$ | 95.6 | 12.8 | 1.49 | 1.6 |
| 5. Suspension | | 63.5 | 8.47 | 1.51 | 1.8 |
| 5. Suspension | | 60.7 | 8.09 | 1.33 | 0.6 |
| 2. Suspension | | 31.2 | 4.16 | 1.46 | 1.7 |
| 5. Suspension | | 25.5 | 3.40 | 1.52 | 2.1 |

continued on following page

AUXILIARY INFORMATION

METHOD/APPARATUS/PROCEDURE:

The solutions were saturated with gas using the rotating double tonometer method of Austin, et al. (1). 2-2.5 h were allowed for saturation. The amount of dissolved nitrogen was determined using the method of Van Slyke and Neil (2).

SOURCE AND PURITY OF MATERIALS:

(1) From air or from diluted mixtures with air (see footnotes).

(2) From ox blood or in one case horse hemoglobin.

ESTIMATED ERROR:

$\delta T/K = \pm 0.04$
$\delta\alpha/\alpha = \pm 0.01$, compiler's estimate (declines with decreasing partial pressure of nitrogen)

REFERENCES:

(1) Austin, J.H.; Cullen, G.E.; Hastings, A.B.; McLean, F.C.; Peters, J.D.; Van Slyke, D.D. *J. Biol. Chem.* 1922, *54*, 134.
(2) Van Slyke, D.D.; Neil, J.M. *J. Biol. Chem.* 1924, *61*, 523.

| COMPONENTS: | ORIGINAL MEASUREMENTS: |
|---|---|
| (1) Nitrogen; $N_2$; [7727-37-9]<br>(2) Ox erythrocytes and horse hemoglobin; see below | Van Slyke, D. D.; Dillon, R. T.; Margaria, R.<br>*J. Biol. Chem.* 1934, *105*, 571-96. |

EXPERIMENTAL VALUES:

continued

[a] See following table for composition of these suspensions or solutions keyed to the indexing number.

[b] Nitrogen from air or air diluted with tank oxygen (99.6% from Linde Air Products Company), pure oxygen (made from heating a mixture of $KClO_3$ and $MnO_2$ and 99 + % pure), or pure carbon dioxide (99.8 + %) made in a Kipp generator.

[c] Partial pressure of nitrogen.

[d] Calculated by compiler.

[e] Bunsen coefficient, $cm^3$ (STP) $cm^{-3}$ $atm^{-1}$. Precision poorer as nitrogen partial pressure is decreased.

[f] $cm^3$ (STP) nitrogen taken up by one gram of hemoglobin.

[g] Authors state that these measurements are of poorer precision.

Composition of Solutions and Cell Suspensions

| Cell Suspension Number | Blood Sample Number | Hb[a] Concentration | $O_2$ Capacity[b] | Salt NaCl | Concentration[c] $NaHCO_3$ | $H_2O$ Content[d] |
|---|---|---|---|---|---|---|
| 1 | 1 | 15.5 | 20.7 | 177 | | 0.870 |
| 2 | 1 | 15.1 | 20.2 | 97 | 79 | 0.881 |
| 3 | 2 | 15.3 | 20.5 | 176 | | 0.874 |
| 4 | 3 | 15.5 | 20.5 | 177 | | 0.870 |
| 5 | 3 | 15.5 | 20.6 | 109 | 69 | 0.874 |
| 6[e] | 4 | 14.5 | | 56 | 47 | 0.885 |

[a] Calculated by the approximate formula, Hb concentration in gram per 100 $cm^3$ = $O_2$ capacity x 0.746.

[b] Oxygen capacity in volume per cent.

[c] Concentrations are in millimoles per kilogram of water.

[d] Concentration in grams of water per $cm^3$ of solution.

[e] Solution of crystallized horse hemoglobin. All other suspensions are of ox erythrocytes.

**COMPONENTS:**

(1) Nitrogen; $N_2$; [7727-37-9]

(2) Blood from eel, cod, trout, and man

**ORIGINAL MEASUREMENTS:**

Steen, J. B.

*Acta Physiol. Scand.* 1963, *58*, 124-37.

**VARIABLES:**

T/K = 280-293
P/kPa = 101.325

**PREPARED BY:**

A. L. Cramer
H. L. Clever

**EXPERIMENTAL VALUES:**

| T/K | Blood Source | pH Range | Number of Analyses | Volume per cent Nitrogen[1] Range | Average |
|---|---|---|---|---|---|
| 279.65 | Eel | 6.20 - 7.60 | 19 | 1.57 - 1.73 | 1.65 |
| | | 7.60 - 8.50 | 19 | 1.71 - 1.85 | 1.81 |
| 293.15 | | 6.60 - 8.50 | 10 | 1.46 - 1.55 | 1.49 |
| 279.65 | Cod | 6.40 - 8.50 | 10 | 1.62 - 1.68 | 1.66 |
| 279.65 | Trout | 6.40 - 8.50 | 12 | 1.64 - 1.75 | 1.74 |
| 279.65 | Man | 6.40 - 8.60 | 10 | 2.00 - 2.11 | 2.04 |

[1] At atm nitrogen

AUXILIARY INFORMATION

**METHOD/APPARATUS/PROCEDURE:**

Blood equilibrated with pure nitrogen or admixed with $CO_2$ and/or $O_2$. The pH was adjusted by adding 0.17 M lactic acid or 0.17 M NaOH.

The gas was presaturated with water vapor by bubbling through an isotonic NaCl solution.

The content of nitrogen was estimated by the method of Edwards, Scholander, and Roughton (1,2)

At a pH of 7.5 - 7.7 there is a sudden significant decrease in the solubility of both argon and nitrogen in eel blood (fig 7 in paper).

**SOURCE AND PURITY OF MATERIALS:**

(1) Nitrogen. Purified.

(2) Blood. Eel blood from female eels kept in running fresh water before heart puncture. Eel, *Anguilla vulgaris*. Cod blood from codfish, *Gadus callarias*, kept in aerated seawater at 12 °C before heart puncture. Trout blood from rainbow trout, *Salmo irrideus*, kept in running tap water at 10 °C before heart puncture. Human blood was supplied by the author.

**ESTIMATED ERROR:**

$\delta T/K = 1$

**REFERENCES:**

1. Roughton, F. J. W.; Scholander, P.F. *J. Biol. Chem.* 1943, *148*, 541, 573.

2. Edwards, G. A.; Scholander, P. F.; Roughton, F. J. W. *J. Biol. Chem.* 1943, *148*, 565.

| COMPONENTS: | ORIGINAL MEASUREMENTS: |
|---|---|
| (1) Nitrogen; $N_2$; [7727-37-9]<br>(2) Rabbit brain and blood and saline solution. | Ohta, Y.; Ar, A.; Farhi, L.E.<br>*J. Appl. Physiology*, 1979, *46*, 1169-70. |
| VARIABLES: | PREPARED BY: |
| T/K = 310 | C.L. Young |

EXPERIMENTAL VALUES:

| T/K | Bunsen coefficient, α | No. of animals |
|---|---|---|
| | "Saline" | |
| 310.15 | 0.0121 ± 0.0001 | 5 |
| | Blood | |
| 310.15 | 0.0124 ± 0.0001 | 5 |
| | Brain | |
| 310.15 | 0.0140 ± 0.0003 | 5 |

The partial pressure of nitrogen was not given but was considerable less than one atmosphere.

AUXILIARY INFORMATION

METHOD/APPARATUS/PROCEDURE:

Saline, rabbit blood and brain were saturated by passing humidified gas through three vessels in series. Brain was prepared by manually squeezing out blood from the brain of a freshly killed rabbit. Volume of brain determined by saline displacement. The tissue was homogenized and diluted with an equal volume of 5% low foam detergent. Blood sample was heparinized. Samples of each of the three solutions were analysed by GC using helium carrier gas, a molecular sieve column and a thermal conductivity detector.

SOURCE AND PURITY OF MATERIALS:

See under method.

ESTIMATED ERROR:

$\delta T/K = \pm 0.1$

REFERENCES:

| COMPONENTS: | ORIGINAL MEASUREMENTS: |
|---|---|
| (1) Nitrogen; $N_2$; [7727-37-9]<br>(2) Protein Solutions, Plasma<br>(3) Water; $H_2O$; [7732-18-5] | Stoddard, J. L.<br>*J. Biol. Chem.* 1926-27, *71*, 629-92. |

| VARIABLES: | PREPARED BY: |
|---|---|
| T/K = 293 | R. Battino |

EXPERIMENTAL VALUES:

| Substance | Concentration/ (g/100 cm$^3$) | $10^2\alpha^a$/ cm$^3$(STP) cm$^{-3}$atm$^{-1}$ |
|---|---|---|
| Hemoglobin | 9.284 | 1.613 |
| Plasma protein | 5.972 | 1.516 |
| | 5.772 | 1.5185 |
| | 6.56 | 1.514 |
| | 5.115 | 1.54 |
| | 6.308 | 1.526 |
| | 5.795 | 1.502 |
| | 5.70 | 1.441 |
| | 6.401 | 1.522 |
| | 5.580 | 1.525 |
| Water | | 1.561 |

[a] Bunsen coefficient. All solubilities reported are averages of multiple measurements.

[b] All measurements at 20°C (298.2K).

[c] The average for all measurements in plasma protein solutions was an $\alpha$ value of 0.01516.

AUXILIARY INFORMATION

METHOD/APPARATUS/PROCEDURE:

A manometric/volumetric apparatus was used which the paper describes exhaustively.

SOURCE AND PURITY OF MATERIALS:

(1) No details given.

(2) From blood from a healthy horse.

(3) Distilled.

ESTIMATED ERROR:

$\delta\alpha/\alpha$ = ±0.01, compiler's estimate.

REFERENCES:

**COMPONENTS:**

(1) Nitrogen; $N_2$; [7727-37-9]

(2) Plasma, cell-plasma suspensions, see below.

**ORIGINAL MEASUREMENTS:**

Van Slyke, D. D.; Dillon, R. T.; Margaria, R.

*J. Biol. Chem.* 1934, *105*, 571-96.

**VARIABLES:**

T/K = 311

Concentration

**PREPARED BY:**

R. Battino

**EXPERIMENTAL VALUES:**

All Measurements at 38°C (311.95 K)

| Liquid Phase | Ox Blood | $P_{N_2}$[a] /mmHg | $P_{N_2}$[b] /kPa | $10^2\alpha$[c]/$cm^3$(STP) $cm^{-3}$ $atm^{-1}$ |
|---|---|---|---|---|
| Plasma[d] | A | 553.0 | 73.73 | 1.141 |
| | B | 554.7 | 73.95 | 1.163 |
| | C | 559.2 | 74.55 | 1.164 |
| Plasma-Cell | C | 558.2 | 74.42 | 1.216 |
| Mixture[e] | A | 534.0 | 71.19 | 1.224 |
| | C | 556.7 | 74.22 | 1.265 |
| | A | 550.7 | 73.42 | 1.299 |
| | B | 552.6 | 73.67 | 1.339 |
| | C | 556.1 | 74.14 | 1.366 |
| | A | 552.0 | 73.59 | 1.391 |
| | C | 556.0 | 74.13 | 1.395 |
| Cells (extrapolated) | | | | 1.46 |

[a] Partial pressure of nitrogen.

[b] Calculated by compiler.

[c] Bunsen coefficient.

[d] Ox plasma.

[e] Ox erythrocytes.

AUXILIARY INFORMATION

**METHOD/APPARATUS/PROCEDURE:**

The solutions were saturated with gas using the rotating double tonometer method of Austin, et al. (1). 2-2.5 h were allowed for saturation. The amount of dissolved nitrogen was determined using the method of Van Slyke and Neil (2).

**SOURCE AND PURITY OF MATERIALS:**

(1) From air.

(2) From ox blood - see paper for details on preparation.

**ESTIMATED ERROR:**

$\delta T/K = \pm 0.04$

$\delta\alpha/\alpha = \pm 0.01$, compiler's estimate

**REFERENCES:**

1. Austin, J.H.; Cullen, G.E.; Hastings, A.B.; McLean, F.C.; Peters, J.D.; Van Slyke, D.D. *J. Biol. Chem.* 1922, *54*, 134.

2. Van Slyke, D.D.; Neil, J.M. *J. Biol. Chem.* 1924, *61*, 523.

COMPONENTS:

(1) Nitrogen; $N_2$; [7727-37-9]

(2) Leguminous hemoglobin

ORIGINAL MEASUREMENTS:

Ewing, G. J.; Ionescu, L. V.

*J. Phys. Chem.* 1972, *76*, 591-6.

VARIABLES:

T/K = 278-298

P/MPa = 0-0.5

PREPARED BY:

R. Battino.

EXPERIMENTAL VALUES:

| t/°C | $T^a$/K | $H^b$/(atm/mol) |
|---|---|---|
| 5 | 278 | 1070 ± 70 |
| 15 | 288 | 1410 ± 30 |
| 25 | 298 | 1600 ± 80 |

[a] Calculated by compiler.

[b] Henry's law constants obtained over a 0-5 atm partial pressure of nitrogen range.

AUXILIARY INFORMATION

METHOD/APPARATUS/PROCEDURE:

Solubilities determined as described in reference 2. Henry's law constants were calculated from the quantity of nitrogen dissolved in each experimental run.

SOURCE AND PURITY OF MATERIALS:

(1) No details given.

(2) From soybean root following similar procedure to reference 1. Average molecular weight of 15,600 g $mol^{-1}$.

ESTIMATED ERROR:

REFERENCES:

1. Ellfolk, N. Acta Chem. Scand. 1960, 14, 609.

2. Maestas, S.; Ewing, G. J. J. Phys. Chem. 1970, 74, 2341.

| COMPONENTS: | ORIGINAL MEASUREMENTS: |
|---|---|
| (1) Nitrogen; $N_2$; [7727-37-9]<br>(2) Serum albumin<br>(3) Water; $H_2O$; [7732-18-5] | Shkol'nikova, R. I.<br>*Uch. Zap. Leningr. Gos. Univ., Ser. Khim. Nauk.* 1959, Nr. 18, 64 - 86.<br>*Chem. Abstr.* 1961, *55*, 25443b. |
| VARIABLES: | PREPARED BY: |
| T/K = 283.15 - 313.15<br>P/kPa = 101.325<br>serum albumin/wt % = 0.575 - 1.99 | A. L. Cramer<br>H. L. Clever |

EXPERIMENTAL VALUES:

| T/K | serum albumin/wt % 0.575 | | 1.15 | |
|---|---|---|---|---|
| | Bunsen Coefficient | Ostwald Coefficient | Bunsen Coefficient | Ostwald Coefficient |
| 283.15 | 14.2 | 14.7 | 13.1 | 13.6 |
| 288.15 | 13.5 | 14.2 | 12.4 | 13.1 |
| 293.15 | 11.7 | 12.6 | 10.8 | 11.6 |
| 298.15 | 11.4 | 12.4 | 10.2 | 11.1 |
| 303.15 | 10.9 | 12.1 | 10.1 | 11.0 |
| 308.15 | 10.5 | 11.8 | 9.9 | 10.8 |
| 313.15 | 10.2 | 11.7 | 9.8 | 10.2 |

| T/K | serum albumin/wt % 1.68 | | 1.99 | |
|---|---|---|---|---|
| | Bunsen Coefficient | Ostwald Coefficient | Bunsen Coefficient | Ostwald Coefficient |
| 283.15 | 12.4 | 12.8 | 12.1 | 12.5 |
| 288.15 | 11.9 | 12.5 | 11.2 | 11.8 |
| 293.15 | 9.4 | 10.1 | 9.1 | 9.7 |
| 298.15 | 9.3 | 10.1 | 8.8 | 9.6 |
| 303.15 | 9.2 | 10.2 | 8.5 | 9.4 |
| 308.15 | 9.1 | 10.2 | 8.1 | 9.1 |
| 313.15 | 9.0 | 10.1 | 8.0 | 9.1 |

The enthalpy of solution of nitrogen was calculated from the temperature coefficient of the Bunsen coefficient. It is 3060, 1250, 890, 300, and 410 cal $mol^{-1}$ in water, and 0.575, 1.15, 1.68, and 1.99 wt % serum albumin respectively.

AUXILIARY INFORMATION

| METHOD/APPARATUS/PROCEDURE: | SOURCE AND PURITY OF MATERIALS: |
|---|---|
| The apparatus and procedure of Lannung were modified (1). | (1) Nitrogen. Source not given. Stated to be near 100 percent.<br>(2) Serum albumin. No information.<br>(3) Water. No information. |
| | ESTIMATED ERROR: |
| | REFERENCES:<br>1. Lannung, A. *J. Am. Chem. Soc.* 1930, *52*, 68. |

**COMPONENTS:**

(1) Nitrogen; $N_2$; [7727-37-9]

(2) Phosphate buffer and human red cell ghost in phosphate buffer.

**ORIGINAL MEASUREMENTS:**

Power, G.G.; Stegall, H.

*J. Appl. Physiology*, 1970, *29*, 145-9.

**VARIABLES:**

T/K = 310

**PREPARED BY:**

C.L. Young

**EXPERIMENTAL VALUES:**

| T/K | Bunsen coefficient, α | S.D.* | No. of measurements |
|---|---|---|---|
| Phosphate Buffer | | | |
| 310.15 | 0.01247 | 0.00006 | 4 |
| Ghosts suspension in phosphate buffer | | | |
| 310.15 | 0.01302 | 0.00003 | 5 |

* Standard deviation.

$$\alpha^\circ\text{ghost} = (\alpha \text{ ghosts suspension} - \alpha \text{ buffer}) / \text{g ghosts cm}^{-3}$$
$$= 0.096 \pm 0.006$$

AUXILIARY INFORMATION

**METHOD/APPARATUS/PROCEDURE:**

5 to 12 $cm^{-3}$ samples placed in a stirrer cell and gas, saturated with water vapor passed through liquid for 30-60 mins. Samples of saturated liquid withdrawn and transferred to Van Slyke apparatus. Dissolved gas removed under reduced pressure. Red cell ghosts prepared by centrifugation of human blood and lysing the cells using phosphate buffer of pH 7.4. Ghost separated by high speed centrifugation. Washed with phosphate buffer containing 0.01 M sodium nitrite.

**SOURCE AND PURITY OF MATERIALS:**

(1) Matheson sample, purity better than 99.7 mole per cent.

(2) See method.

**ESTIMATED ERROR:**

$\delta T/K = \pm 0.1$

**REFERENCES:**

**COMPONENTS:**

(1) Nitrogen; $N_2$; [7727-37-9]

(2) Human urine

**ORIGINAL MEASUREMENTS:**

Farhi, L. E.; Edwards, A. W. T.; Homma, T.

*J. Appl. Physiol.* 1963, *18*, 97-106.

**VARIABLES:**

T/K = 310

**PREPARED BY:**

R. Battino

**EXPERIMENTAL VALUES:**

| Subject | t/°C | $T^a$/K | Specific gravity | $10^2\alpha^b$ |
|---|---|---|---|---|
| DH | 37.40 | 310.55 | 1.004 | 1.221 |
| WR | 37.15 | 310.30 | 1.006 | 1.208 |
| SF | 37.15 | 310.30 | 1.011 | 1.185 |
| TV | 37.45 | 310.60 | 1.020 | 1.120 |
| ML | 37.35 | 310.50 | 1.022 | 1.118 |
| HR | 37.25 | 310.40 | 1.022 | 1.107 |
| RM | 37.40 | 310.55 | 1.022 | 1.101 |
| LF | 37.25 | 310.40 | 1.033 | 1.039 |

[a] Calculated by compiler.

[b] Bunsen coefficient in units of $cm^3$ (STP) $cm^{-3}$ $atm^{-1}$.

**AUXILIARY INFORMATION**

**METHOD/APPARATUS/PROCEDURE:**

Used the Van Slyke method combined with gas chromatography.

**SOURCE AND PURITY OF MATERIALS:**

(1) From air.

(2) From eight normal healthy men aged 22-48 years.

**ESTIMATED ERROR:**

$\delta\alpha/\alpha = \pm 0.02$, compiler's estimate.

**REFERENCES:**

**COMPONENTS:**

(1) Nitrogen; $N_2$; [7727-37-9]

(2) Benzene; $C_6H_6$; [71-43-2]

(3) Cholest-5-en-3β-ol or Cholesterol; $C_{27}H_{46}O$; [57-88-5]

**ORIGINAL MEASUREMENTS:**

Byrne, J. E.; Battino, R.; Danforth, W. F.

*J. Chem. Thermodyn.* 1974, *6*, 245-50.

**VARIABLES:**

T/K = 310.65
Total P/kPa = 101.325

**PREPARED BY:**

H. L. Clever

**EXPERIMENTAL VALUES:**

| T/K | Mass Fraction Cholesterol $W_3$ | Ostwald Coefficient L | "Salting Out" Parameter $k = (1/W_3) \log (L_o/L_w)$ |
|---|---|---|---|
| 310.64 | 0.0 | 0.1323 | |
| 310.69 | 0.0 | 0.1328 | |
| 310.62 | 0.0528 | 0.129 | |
| 310.63 | 0.0528 | 0.130 | 0.19 ± 0.17 |
| 310.57 | 0.1069 | 0.119 | |
| 310.65 | 0.1069 | 0.117 | 0.49 ± 0.09 |
| 310.65 | 1.0 | 0.061* | |

*Extrapolated nitrogen solubility in hypothetical liquid cholesterol.

$L_o$ and $L_w$ are the nitrogen Ostwald coefficients in benzene and in the benzene + cholesterol solution, respectively.

AUXILIARY INFORMATION

**METHOD/APPARATUS/PROCEDURE:**

The solubility apparatus is based on the design of Morrison and Billett (1) and the version used is described by Battino, Evans and Danforth (2). The degassing apparatus is that described by Battino, Banzhof, Bogan, and Wilhelm (3).

Degassing: Up to 500 $cm^3$ of solvent is placed in a flask of such size that the liquid is about 4 cm deep. The liquid is rapidly stirred, and vacuum is intermittently applied through a liquid nitrogen trap until the permanent gas residual pressure drops to 5 microns.

Solubility Determination: The degassed solvent is passed in a thin film down a glass helical tube containing solute gas plus the solvent vapor at a total pressure of one atm. The volume of gas absorbed is found by difference between the initial and final volumes in the buret system. The solvent is collected in a tared flask and weighed.

**SOURCE AND PURITY OF MATERIALS:**

(1) Nitrogen. Matheson Co., Inc. 99.995 per cent.

(2) Benzene. J. T. Baker Analyzed Reagent Grade. 99.9 mol per cent, thiophene free.

(3) Cholesterol. Source not given. Recrystalized from benzene.

**ESTIMATED ERROR:**
$\delta T/K = 0.01$
$\delta P/mmHg = 0.5$
$\delta L_o/L_o = 0.01$
$\delta L_w/L_w = 0.02$

**REFERENCES:**


1. Morrison, T. J.; Billett, F. *J. Chem. Soc.* 1948, 2033.
2. Battino, R.; Evans, F. D.; Danforth, W. F. *J. Am. Oil Chem. Soc.* 1968, *45*, 830.
3. Battino, R.; Banzhof, M.; Bogan, M.; Wilhelm, E. *Anal. Chem.* 1971, *43*, 806.

| COMPONENTS: | ORIGINAL MEASUREMENTS: |
|---|---|
| (1) Nitrogen; $N_2$; [7727-37-9]<br>(2) Benzene; $C_6H_6$; [71-43-2]<br>(3) Lecithin | Byrne, J. E.; Battino, R.; Danforth, W. F.<br>*J. Chem. Thermodyn.* 1974, *6*, 245-50. |
| VARIABLES:<br>T/K = 310.65<br>Total P/kPa = 101.325 | PREPARED BY:<br>H. L. Clever |

EXPERIMENTAL VALUES:

| T/K | Mass Fraction Lecithin $W_3$ | Ostwald Coefficient L | "Salting Out" Parameter $k = (1/W_3) \log (L_o/L_w)$ |
|---|---|---|---|
| 310.64 | 0.0 | 0.1324 | |
| 310.69 | 0.0 | 0.1328 | |
| 310.64 | 0.1993 | 0.115 | |
| 310.65 | 0.1993 | 0.116 | 0.30 ± 0.04 |
| 310.65 | 1.0 | 0.066* | |

*Extrapolated nitrogen solubility in hypothetical liquid lecithin.

$L_o$ and $L_w$ are the nitrogen Ostwald coefficients in benzene and in the benzene + lecithin solution, respectively.

AUXILIARY INFORMATION

METHOD/APPARATUS/PROCEDURE:

The solubility apparatus is based on the design of Morrison and Billett (1) and the version used is described by Battino, Evans and Danforth (2). The degassing apparatus is that described by Battino, Banzhof, Bogan, and Wilhelm (3).

Degassing: Up to 500 $cm^3$ of solvent is placed in a flask of such size that the liquid is about 4 cm deep. The liquid is rapidly stirred, and vacuum is intermittently applied through a liquid nitrogen trap until the permanent gas residual pressure drops to 5 microns.

Solubility Determination: The degassed solvent is passed in a thin film down a glass helical tube containing solute gas plus the solvent vapor at a total pressure of one atm. The volume of gas absorbed is found by difference between the initial and final volumes in the buret system. The solvent is collected in a tared flask and weighed.

SOURCE AND PURITY OF MATERIALS:

(1) Nitrogen. Matheson Co., Inc. 99.995 per cent.

(2) Benzene. J. T. Baker Analyzed Reagent Grade. 99.9 mol per cent, thiophene free.

(3) Lecithin. Nutritional Biochemicals Corp. Vegetable source, about 95 per cent, used as received.

ESTIMATED ERROR:

$\delta T/K = 0.01$
$\delta P/mmHg = 0.5$
$\delta L_o/L_o = 0.01$
$\delta L_w/L_w = 0.02$

REFERENCES:


1. Morrison, T. J.; Billett, F. *J. Chem. Soc.* 1948, 2033.
2. Battino, R.; Evans, F. D.; Danforth, W. F. *J. Am. Oil Chem. Soc.* 1968, *45*, 830.
3. Battino, R.; Banzhof, M.; Bogan, M.; Wilhelm, E. *Anal. Chem.* 1971, *43*, 806.

**COMPONENTS:**

(1) Nitrogen; $N_2$; [7727-37-9]

(2) Benzene; $C_6H_6$; [71-43-2]

(3) Cephalin

**ORIGINAL MEASUREMENTS:**

Byrne, J. E.; Battino, R.; Danforth, W. F.

*J. Chem. Thermodyn.* 1974, *6*, 245-50.

**VARIABLES:**

T/K = 310.65
Total P/kPa = 101.325

**PREPARED BY:**

H. L. Clever

**EXPERIMENTAL VALUES:**

| T/K | Mass Fraction Cephalin $W_3$ | Ostwald Coefficient L | "Salting Out" Parameter $k = (1/W_3) \log (L_o/L_w)$ |
|---|---|---|---|
| 310.64 | 0.0 | 0.1323 | |
| 310.69 | 0.0 | 0.1328 | |
| 310.60 | 0.2003 | 0.118 | |
| 310.63 | 0.2003 | 0.112 | 0.31 ± 0.05 |
| 310.65 | 1.0 | 0.065* | |

*Extrapolated nitrogen solubility in hypothetical liquid cephalin.

$L_o$ and $L_w$ are the nitrogen Ostwald coefficients in benzene and in the benzene + cephalin solution, respectively.

AUXILIARY INFORMATION

**METHOD/APPARATUS/PROCEDURE:**

The solubility apparatus is based on the design of Morrison and Billett (1) and the version used is described by Battino, Evans and Danforth (2). The degassing apparatus is that described by Battino, Banzhof, Bogan, and Wilhelm (3).

Degassing: Up to 500 $cm^3$ of solvent is placed in a flask of such size that the liquid is about 4 cm deep. The liquid is rapidly stirred, and vacuum is intermittently applied through a liquid nitrogen trap until the permanent gas residual pressure drops to 5 microns.

Solubility Determination: The degassed solvent is passed in a thin film down a glass helical tube containing solute gas plus the solvent vapor at a total pressure of one atm. The volume of gas absorbed is found by difference between the initial and final volumes in the buret system. The solvent is collected in a tared flask and weighed.

**SOURCE AND PURITY OF MATERIALS:**

(1) Nitrogen. Matheson Co., Inc. 99.995 per cent.

(2) Benzene. J. T. Baker Analyzed Reagent Grade. 99.9 mol per cent, thiophene free.

(3) Cephalin. Nutritional Biochemicals Corp. Homostatic phospatide obtained from bovine brain tissue, used as received.

**ESTIMATED ERROR:**

$\delta T/K = 0.01$
$\delta P/mmHg = 0.5$
$\delta L_o/L_o = 0.01$
$\delta L_w/L_w = 0.02$

**REFERENCES:**


1. Morrison, T. J.; Billett, F. *J. Chem. Soc.* 1948, 2033.
2. Battino, R.; Evans, F. D.; Danforth, W. F. *J. Am. Oil Chem. Soc.* 1968, *45*, 830.
3. Battino, R.; Banzhof, M.; Bogan, M.; Wilhelm, E. *Anal. Chem.* 1971, *43*, 806.

**COMPONENTS:**

(1) Nitrogen; $N_2$; [7727-37-9]

(2) Lipoidal suspension.

(3) Water; $H_2O$; [7732-18-5]

**ORIGINAL MEASUREMENTS:**

Grollman, A.

*J. Biol. Chem.* 1929, *82*, 317-25.

**VARIABLES:**

T/K = 311

**PREPARED BY:**

R. Battino

**EXPERIMENTAL VALUES:**

37.5°C (310.7K)

| Substance | $10^2\alpha$[a] |
|---|---|
| Water | 1.28 |
| Lipoidal Suspension[b] | 1.33 |

[a] Bunsen coefficient in units of $cm^3$(STP) $cm^{-3}$ $atm^{-1}$.

[b] Made up of 0.3 g blood lipid in 100 $cm^3$ water.

**AUXILIARY INFORMATION**

**METHOD/APPARATUS/PROCEDURE:**

The Van Slyke method was used.

**SOURCE AND PURITY OF MATERIALS:**

(1) No details given.

(2) Extracted from rabbit or dog blood with petroleum ether and ethyl ether.

(3) No details given.

**ESTIMATED ERROR:**

$\delta\alpha/\alpha$ = ±0.02, compiler's estimate.

**REFERENCES:**

| COMPONENTS: | ORIGINAL MEASUREMENTS: |
|---|---|
| (1) Nitrogen; $N_2$; [7727-37-9]<br>(2) Urea; $CH_4N_2O$; [57-13-6]<br>(3) Water; $H_2O$; [7732-18-5] | Braun, L.<br>*Z. Phys. Chem.* 1900, *33*, 721-41. |
| VARIABLES: | PREPARED BY: |
| T/K = 278-298<br>Concentration | C.L. Young |

EXPERIMENTAL VALUES:

| T/K | Wt. of urea per 100g soln./g | Bunsen coefficient, $\alpha$ |
|---|---|---|
| 278.2 | 15.650 | 0.01749 |
| | 14.365 | 0.01836 |
| | 11.900 | 0.01793 |
| | 11.664 | 0.01782 |
| | 10.240 | 0.01835 |
| | 9.418 | 0.01897 |
| | 6.897 | 0.01979 |
| | 6.409 | 0.01987 |
| | 5.194 | 0.01982 |
| | 5.145 | 0.01969 |
| | 2.460 | 0.02003 |
| | 2.287 | 0.01992 |
| | 0.000 | 0.02173 |
| 283.2 | 15.650 | 0.01621 |
| | 14.365 | 0.01657 |
| | 11.900 | 0.01671 |
| | 11.664 | 0.01633 |
| | 10.240 | 0.01732 |
| | 9.418 | 0.01758 |
| | 6.897 | 0.01827 |
| | 6.409 | 0.01832 |
| | 5.194 | 0.01890 |
| | 5.145 | 0.01819 |
| | 2.460 | 0.01889 |
| | 2.287 | 0.01841 |
| | 0.0 | 0.02003 |

AUXILIARY INFORMATION

METHOD/APPARATUS/PROCEDURE:

Ostwald method, using gas buret and pipet. Measurement of volume of gas before and after absorption. Vapor pressure of water was allowed for assuming Raoult's law. Solution was degassed. Concentration of salt solution estimated by titration with silver nitrate solution.

SOURCE AND PURITY OF MATERIALS:

No details given.

ESTIMATED ERROR:

$\delta\alpha/\alpha = \pm 0.03$

REFERENCES:

| COMPONENTS: | ORIGINAL MEASUREMENTS: |
|---|---|
| (1) Nitrogen; $N_2$; [7727-37-9] | Braun, L. |
| (2) Urea; $CH_4N_2O$; [57-13-6] | *Z. Phys. Chem.* <u>1900</u>, *33*, 721-41. |
| (3) Water; $H_2O$; [7732-18-5] | |

EXPERIMENTAL VALUES:

| T/K | Wt. of urea per 100g soln./g | Bunsen coefficient, $\alpha$ |
|---|---|---|
| 288.2 | 15.650 | 0.01503 |
| | 14.365 | 0.01514 |
| | 11.900 | 0.01486 |
| | 11.664 | 0.01504 |
| | 10.240 | 0.01601 |
| | 9.418 | 0.01580 |
| | 6.897 | 0.01649 |
| | 6.409 | 0.01666 |
| | 5.194 | 0.01678 |
| | 5.145 | 0.01651 |
| | 2.460 | 0.01715 |
| | 2.287 | 0.01708 |
| | 0.0 | 0.01789 |
| 293.2 | 15.650 | 0.01396 |
| | 14.365 | 0.01406 |
| | 11.900 | 0.01385 |
| | 11.664 | 0.01395 |
| | 10.240 | 0.01465 |
| | 9.418 | 0.01455 |
| | 6.897 | 0.01510 |
| | 6.409 | 0.01507 |
| | 5.194 | 0.01523 |
| | 5.145 | 0.01509 |
| | 2.460 | 0.01576 |
| | 2.287 | 0.01552 |
| | 0.0 | 0.01621 |
| 298.2 | 15.650 | 0.01300 |
| | 14.365 | 0.01334 |
| | 11.900 | 0.01296 |
| | 11.664 | 0.01315 |
| | 10.240 | 0.01316 |
| | 9.418 | 0.01328 |
| | 6.897 | 0.01369 |
| | 6.409 | 0.01373 |
| | 5.194 | 0.01376 |
| | 5.145 | 0.01350 |
| | 2.460 | 0.01396 |
| | 2.287 | 0.01390 |
| | 0.0 | 0.01432 |

| COMPONENTS: | ORIGINAL MEASUREMENTS: |
|---|---|
| (1) Nitrogen; $N_2$; [7727-37-9]<br>(2) Urea; $CH_4N_2O$ or $(NH_2)_2CO$; [57-13-6]<br>(3) Water; $H_2O$; [7732-18-5] | Hüfner, G.<br>*Z. Physik. Chem.* 1907, *57*, 611-24. |
| **VARIABLES:** | **PREPARED BY:** |
| T/K = 293.25, 293.40<br>$N_2$ P/kPa = 101.325 | H. L. Clever |

EXPERIMENTAL VALUES:

| Temperature t/°C | T/K | $C_{CH_4N_2O}$/mol dm$^{-3}$ | Gas Pressure/ mmHg | Absorbed Gas Volume[1] | Bunsen Coefficient $\alpha \times 10^2$ |
|---|---|---|---|---|---|
| 20.18 | 293.33 | 0 | --- | -- | 1.565[2] |
| 20.25 | 293.40 | 1.0 | 664.4 | 5.34 | 1.490 |
| 20.10 | 293.25 | | 683.6 | 5.40 | 1.465 |
| 20.18 Av. | | | | | 1.477 Av. |

[1]Volume of gas reduced to 273.15 K and 760 mmHg absorbed in 490.94 cm$^3$ solution at the stated pressure.

[2]Average of four runs, see Hüfner's $N_2$ + $H_2O$ data sheet.

AUXILIARY INFORMATION

**METHOD/APPARATUS/PROCEDURE:**

The apparatus, described in reference (1), was the Bunsen absorption type. It consisted of an absorption flask, which holds 409.94 cm$^3$ of solvent, and gas burets.

The final partial pressure of the gas ranged between 648.6 and 720.7 mmHg. The author reported both the volume of gas absorbed at the final pressure in 409.94 cm$^3$ of solvent reduced to 273.15 K and 760 mmHg, and the Bunsen coefficient, which was calculated assuming Henry's law.

Urea name is harnstoff in the paper.

**SOURCE AND PURITY OF MATERIALS:**

(1) Nitrogen. Source not given. Argon free.

(2) Urea. Source not given.

(3) Water. Pure.

**ESTIMATED ERROR:**

$\delta T/K = 0.01$

**REFERENCES:**

1. Hüfner, G. *Archiv. F. Anatomie und Physiologie, Physiolog. Abtly.* 1894, *5*, 191.

**COMPONENTS:**

(1) Nitrogen; $N_2$; [7727-37-9]

(2) Glycine; $C_2H_5NO_2$ or $CH_2(NH_2)COOH$; [56-40-6]

(3) Water; $H_2O$; [7732-18-5]

**ORIGINAL MEASUREMENTS:**

Hüfner, G.

*Z. Physik. Chem.* 1907, *57*, 611-24.

**VARIABLES:**

T/K = 293.25 - 293.43

$N_2$ P/kPa = 101.325

**PREPARED BY:**

H. L. Clever

**EXPERIMENTAL VALUES:**

| Temperature t/°C | T/K | $C_{C_2H_5NO_2}$/mol $dm^{-3}$ | Gas Pressure mmHg | Absorbed Gas Volume[1] | Bunsen Coefficient $\alpha \times 10^2$ |
|---|---|---|---|---|---|
| 20.18 | 293.33 | 0 | --- | -- | 1.565[2] |
| 20.28 | 293.43 | 1.0 | 679.9 | 4.29 | 1.170 |
| 20.1 | 293.25 | | 679.3 | 4.51 | 1.231 |
| 20.1 | 293.25 | | 691.2 | 4.61 | 1.236 |
| 20.16 Av. | | | | | 1.212 Av. |

[1] Volume of gas reduced to 273.15 K and 760 mmHg absorbed in 490.94 $cm^3$ solution at the stated pressure.

[2] Average of four runs, see Hüfner's $N_2$ + $H_2O$ data sheet.

AUXILIARY INFORMATION

**METHOD/APPARATUS/PROCEDURE:**

The apparatus, described in reference (1), was the Bunsen absorption type. It consisted of an absorption flask, which holds 409.94 $cm^3$ of solvent, gas burets.

The final partial pressure of the gas ranged between 648.6 and 720.7 mmHg. The author reported both the volume of gas absorbed at the final pressure in 409.94 $cm^3$ of solvent reduced to 273.15 K and 760 mmHg, and the Bunsen coefficient, which was calculated assuming Henry's law.

D-Glycine name glykoholl (amino-essigsäure) in the paper.

**SOURCE AND PURITY OF MATERIALS:**

(1) Nitrogen. Source not given. Argon free.

(2) Glycine. Merck.

(3) Water. Pure.

**ESTIMATED ERROR:**

$\delta T/K = 0.01$

**REFERENCES:**

1. Hüfner, G. *Archiv. F. Anatomie und Physiologie, Physiolog. Abtly.* 1894, *5*, 191.

| COMPONENTS: | ORIGINAL MEASUREMENTS: |
|---|---|
| (1) Nitrogen; $N_2$; [7727-37-9]<br>(2) D-Alanine; $C_3H_7NO_2$ or $CH_3CH(NH_2)COOH$; [338-69-2]<br>(3) Water; $H_2O$; [7732-18-5] | Hüfner, G.<br>*Z. Physik. Chem.* 1907, *57*, 611-24. |
| VARIABLES:<br>T/K = 293.33<br>$N_2$ P/kPa = 101.325 | PREPARED BY:<br>H. L. Clever |

EXPERIMENTAL VALUES:

| Temperature t/°C | T/K | $C_{C_3H_7NO_2}$/mol $dm^{-3}$ | Gas Pressure/ mmHg | Absorbed Gas Volume[1] | Bunsen Coefficient $\alpha \times 10^2$ |
|---|---|---|---|---|---|
| 20.18 | 293.33 | 0 | --- | -- | 1.565[2] |
| 20.18 | 293.33 | 1.0 | 674.5 | 4.38 | 1.204 |
| 20.2 | 293.35 | | 684.1 | 4.51 | 1.222 |
| 20.19 Av. | | | | | 1.213 Av. |

[1]Volume of gas reduced to 273.15 K and 760 mmHg absorbed in 490.94 $cm^3$ solution at the stated pressure.

[2]Average of four runs, see Hüfner's $N_2$ + $H_2O$ data sheet.

AUXILIARY INFORMATION

METHOD/APPARATUS/PROCEDURE:

The apparatus, described in reference (1), was the Bunsen absorption type. It consisted of an absorption flask, which holds 409.94 $cm^3$ of solvent, and gas burets.

The final partial pressure of the gas ranged between 648.6 and 720.7 mmHg. The author reported both the volume of gas absorbed at the final pressure in 409.94 $cm^3$ of solvent reduced to 273.15 K and 760 mmHg, and the Bunsen coefficient, which was calculated assuming Henry's law.

SOURCE AND PURITY OF MATERIALS:

(1) Nitrogen. Source not given. Argon free.

(2) D-Alanine. Merck.

(3) Water. Pure.

ESTIMATED ERROR:

$\delta T/K = 0.01$

REFERENCES:

1. Hüfner, G. *Archiv. F. Anatomie und Physiologie, Physiolog. Abtly.* 1894, *5*, 191.

| COMPONENTS: | ORIGINAL MEASUREMENTS: |
|---|---|
| (1) Nitrogen; $N_2$; [7727-37-9]<br>(2) (R*,R*)-1,2,3,4-Butanetetrol; $C_4H_{10}O_4$; [7493-90-5]<br>(3) Water; $H_2O$; [7732-18-5] | Hüfner, G.<br>*Z. Physik. Chem.* 1907, *57*, 611-24. |

| VARIABLES: | PREPARED BY: |
|---|---|
| T/K = 293.25 - 293.50<br>$N_2$ P/kPa = 101.325 | H. L. Clever |

EXPERIMENTAL VALUES:

| Temperature t/°C | Temperature T/K | $C_{C_4H_{10}O_4}$/mol $dm^{-3}$ | Gas Pressure/ mmHg | Absorbed Gas Volume[1] | Bunsen Coefficient $\alpha \times 10^2$ |
|---|---|---|---|---|---|
| 20.18 | 293.33 | 0 | --- | -- | 1.565[2] |
| 20.1 | 293.25 | 1.0 | 672.3 | 4.79 | 1.321 |
| 20.3 | 293.45 | | 707.0 | 4.91 | 1.287 |
| 20.35 | 293.50 | | 672.4 | 4.92 | 1.356 |
| 20.25 Av. | | | | | 1.321 Av. |

[1] Volume of gas reduced to 273.15 K and 760 mmHg absorbed in 490.94 $cm^3$ solution at the stated pressure.

[2] Average of four runs, see Hüfner's $N_2$ + $H_2O$ data sheet.

AUXILIARY INFORMATION

METHOD/APPARATUS/PROCEDURE:

The apparatus, described in reference (1), was the Bunsen absorption type. It consisted of an absorption flask, which holds 409.94 $cm^3$ of solvent, and gas burets.

The final partial pressure of the gas ranged between 648.6 and 720.7 mmHg. The author reported both the volume of gas absorbed at the final pressure in 409.94 $cm^3$ of solvent reduced to 273.15 K and 760 mmHg, and the Bunsen coefficient, which was calculated assuming Henry's law.

(R, R)-1,2,3,4-Butanetetrol name erythrit (erythritol) in the paper.

SOURCE AND PURITY OF MATERIALS:

(1) Nitrogen. Source not given. Argon free.

(2) Erythritol. Merck.

(3) Water. Pure.

ESTIMATED ERROR:

$\delta T/K = 0.01$

REFERENCES:

1. Hüfner, G. *Archiv. F. Anatomie und Physiologie, Physiolog. Abtly.* 1894, *5*, 191.

COMPONENTS:

(1) Nitrogen; $N_2$; [7727-37-9]

(2) L-Arabinose, $C_5H_{10}O_5$; [5328-37-0]

(3) Water; $H_2O$; [7732-18-5]

ORIGINAL MEASUREMENTS:

Hüfner, G.

*Z. Physik. Chem.* 1907, *57*, 611-24.

VARIABLES:

T/K = 293

$N_2$ P/kPa = 101.325

PREPARED BY:

H. L. Clever

EXPERIMENTAL VALUES:

| Temperature t/°C | T/K | $C_{C_5H_{10}O_5}$/mol dm$^{-3}$ | Gas Pressure/ mmHg | Absorbed Gas Volume[1] | Bunsen Coefficient $\alpha \times 10^2$ |
|---|---|---|---|---|---|
| 20.18 | 293.33 | 0 | --- | -- | 1.565[2] |
| 20.15 | 293.30 | 1.0 | 680.40 | 4.47 | 1.218 |
| 20.28 | 293.43 | | 676.78 | 4.34 | 1.189 |
| 20.21 Av. | | | | | 1.203 Av. |

[1] Volume of gas reduced to 273.15 K and 760 mmHg absorbed in 490.94 cm$^3$ solution at the stated pressure.

[2] Average of four runs, see Hüfner's $N_2$ + $H_2O$ data sheet.

AUXILIARY INFORMATION

METHOD/APPARATUS/PROCEDURE:

The apparatus, described in reference (1), was the Bunsen absorption type. It consisted of an absorption flask, which holds 409.94 cm$^3$ of solvent, and gas burets.

The final partial pressure of the gas ranged between 648.6 and 720.7 mmHg. The author reported both the volume of gas absorbed at the final pressure in 409.94 cm$^3$ of solvent reduced to 273.15 K and 760 mmHg, and the Bunsen coefficient, which was calculated assuming Henry's law.

SOURCE AND PURITY OF MATERIALS:

(1) Nitrogen. Source not given. Argon free.

(2) L-Arabinose. Merck.

(3) Water. Pure.

ESTIMATED ERROR:

$\delta T/K = 0.01$

REFERENCES:

1. Hüfner, G. *Archiv. F. Anatomie und Physiologie, Physiolog. Abtly.* 1894, *5*, 191.

| COMPONENTS: | ORIGINAL MEASUREMENTS: |
|---|---|
| (1) Nitrogen; $N_2$; [7727-37-9]<br>(2) D-Fructose; $C_6H_{12}O_6$; [57-48-7]<br>(3) Water; $H_2O$; [7732-18-5] | Hüfner, G.<br>*Z. Physik. Chem.* 1907, *57*, 611- 24. |
| VARIABLES: | PREPARED BY: |
| T/K = 293.40<br>$N_2$ P/kPa = 101.325 | H. L. Clever |

EXPERIMENTAL VALUES:

| Temperature t/°C | T/K | $C_{C_6H_{12}O_6}$/mol dm$^{-3}$ | Gas Pressure/ mmHg | Absorbed Gas Volume[1] | Bunsen Coefficient $\alpha \times 10^2$ |
|---|---|---|---|---|---|
| 20.18 | 293.33 | 0 | --- | -- | 1.565[2] |
| 20.25 | 293.40 | 1.0 | 648.6 | 4.27 | 1.221 |

[1] Volume of gas reduced to 273.15 K and 760 mmHg absorbed in 490.94 $cm^3$ solution at the stated pressure.

[2] Average of four runs, see Hüfner's $N_2$ + $H_2O$ data sheet.

AUXILIARY INFORMATION

METHOD/APPARATUS/PROCEDURE:

The apparatus, described in reference (1), was the Bunsen absorption type. It consisted of an absorption flask, which holds 409.94 $cm^3$ of solvent, and gas burets.

The final partial pressure of the gas ranged between 648.6 and 720.7 mmHg. The author reported both the volume of gas absorbed at the final pressure in 409.94 $cm^3$ of solvent reduced to 273.15 K and 760 mmHg, and the Bunsen coefficient, which was calculated assuming Henry's law.

D-Fructose name Fruchtzucker or levulose in the paper.

SOURCE AND PURITY OF MATERIALS:

(1) Nitrogen. Source not given. Argon free.

(2) D-Fructose. Merck.

(3) Water. Pure

ESTIMATED ERROR:

$\delta T/K = 0.01$

REFERENCES:

1. Hüfner, G. *Archiv. F. Anatomie und Physiologie, Physiolog. Abtly.* 1894, *5*, 191.

**COMPONENTS:**

(1) Nitrogen; $N_2$; [7727-37-9]

(2) D-Glucose; $C_6H_{12}O_6$; [50-99-7]

(3) Water; $H_2O$; [7732-18-5]

**ORIGINAL MEASUREMENTS:**

Hüfner, G.

*Z. Physik. Chem.* 1907, *57*, 611-24.

**VARIABLES:**

T/K = 293.20 - 293.57

$N_2$ P/kPa = 101.325

**PREPARED BY:**

H. L. Clever

**EXPERIMENTAL VALUES:**

| Temperature t/°C | T/K | $C_{C_6H_{12}O_6}$/mol dm$^{-3}$ | Gas Pressure/ mmHg | Absorbed Gas Volume[1] | Bunsen Coefficient $\alpha \times 10^2$ |
|---|---|---|---|---|---|
| 20.18 | 293.33 | 0 | ---- | -- | 1.565[2] |
| 20.25 | 293.40 | 0.25 | 693.9 | 5.56 | 1.485 |
| 20.10 | 293.25 | | 697.0 | 5.57 | 1.482 |
| 20.25 | 293.40 | | 680.2 | 5.40 | 1.472 |
| 20.2 Av. | | | | | 1.480 Av. |
| 20.05 | 293.20 | 0.50 | 681.3 | 5.05 | 1.374 |
| 20.2 | 293.35 | | 675.8 | 4.98 | 1.366 |
| 20.3 | 293.45 | | 704.5 | 5.17 | 1.361 |
| 20.3 | 293.45 | | 700.2 | 5.36 | 1.419 |
| 20.21 Av. | | | | | 1.380 Av. |
| 20.42 | 293.57 | 1.00 | 672.1 | 4.51 | 1.244 |
| 20.10 | 293.25 | | 688.3 | 4.45 | 1.199 |
| 20.20 | 293.35 | | 720.7 | 4.67 | 1.201 |
| 20.15 | 293.30 | | 704.3 | 4.74 | 1.248 |
| 20.05 | 293.20 | | 688.2 | 4.38 | 1.180 |
| 20.18 Av. | | | | | 1.215 Av. |

[1] Volume of gas reduced to 273.15 K and 760 mmHg absorbed in 490.94 $cm^3$ solution at the stated pressure.

[2] Average of four runs, see Hüfner's $N_2$ + $H_2O$ data sheet.

**AUXILIARY INFORMATION**

**METHOD/APPARATUS/PROCEDURE:**

The apparatus, described in reference (1), was the Bunsen absorption type. It consisted of an absorption flask, which holds 409.94 $cm^3$ of solvent, and gas burets.

The final partial pressure of the gas ranged between 648.6 and 720.7 mmHg. The author reported both the volume of gas absorbed at the final pressure in 409.94 $cm^3$ of solvent reduced to 273.15 K and 760 mmHg, and the Bunsen coefficient, which was calculated assuming Henry's law.

D-glucose name tranbenzucker (grape sugar) in the paper.

**SOURCE AND PURITY OF MATERIALS:**

(1) Nitrogen. Source not given. Argon free.

(2) D-Glucose. Source not given.

(3) Water. Pure.

**ESTIMATED ERROR:**

$\delta T/K = 0.01$

**REFERENCES:**

1. Hüfner, G. *Archiv. F. Anatomie und Physiologie, Physiolog. Abtly.* 1894, *5*, 191.

| COMPONENTS: | ORIGINAL MEASUREMENTS: |
|---|---|
| (1) Nitrogen; $N_2$; [7727-37-9]<br>(2) Gelatin<br>(3) Water; $H_2O$; [7732-18-5] | Shkol'nikova, R. I.<br>*Uch. Zap. Leningr. Gos. Univ., Ser. Khim. Nauk.* 1959, Nr. 18, 64 - 86.<br>*Chem. Abstr.* 1961, *55*, 25443b. |

| VARIABLES: | PREPARED BY: |
|---|---|
| T/K = 283.15 - 313.15<br>P/kPa = 101.325<br>Gelatin/wt % = 1 - 10 | A. L. Cramer<br>H. L. Clever |

EXPERIMENTAL VALUES:

| T/K | Gelatin/wt % 1 | | Gelatin/wt % 5 | |
|---|---|---|---|---|
| | Bunsen Coefficient | Ostwald Coefficient | Bunsen Coefficient | Ostwald Coefficient |
| 283.15 | 15.6 | 16.2 | 14.3 | 14.8 |
| 288.15 | 14.2 | 14.9 | 13.6 | 14.1 |
| 293.15 | 13.0 | 14.0 | 12.5 | 13.5 |
| 298.15 | 12.6 | 13.8 | 11.4 | 12.4 |
| 303.15 | 11.5 | 12.8 | 10.4 | 11.5 |
| 308.15 | 10.7 | 12.1 | 9.8 | 11.1 |
| 313.15 | 10.1 | 11.6 | 9.2 | 10.6 |

| T/K | Gelatin/wt % 10 | |
|---|---|---|
| | Bunsen Coefficient | Ostwald Coefficient |
| 283.15 | 13.8 | 14.3 |
| 288.15 | 12.2 | 13.0 |
| 293.15 | 10.8 | 11.6 |
| 298.15 | 10.2 | 11.1 |
| 303.15 | 9.8 | 10.9 |
| 308.15 | 8.3 | 9.4 |
| 313.15 | 7.2 | 8.2 |

The enthalpy of solution of nitrogen was calculated from the temperature coefficient of the Bunsen coefficient. It is 3060, 2540, 2590, and 3820 cal $mol^{-1}$ in water, and 1, 5, and 10 wt % gelatin respectively.

AUXILIARY INFORMATION

METHOD/APPARATUS/PROCEDURE:

The apparatus and procedure of Lannung were modified (1).

SOURCE AND PURITY OF MATERIALS:

(1) Nitrogen. Source not given. Stated to be near 100 percent.

(2) Gelatin. No information.

(3) Water. No information.

ESTIMATED ERROR:

REFERENCES:

1. Lannung, A.
*J. Am. Chem. Soc.* 1930, *52*, 68.

| COMPONENTS: | ORIGINAL MEASUREMENTS: |
|---|---|
| (1) Nitrogen; $N_2$; [7727-37-9]<br>(2) Sucrose; $C_{12}H_{22}O_{11}$; [57-50-1]<br>(3) Water; $H_2O$; [7732-18-5] | Müller, C.<br>*Z. Physik. Chem.* 1912, *81*, 483-503. |
| VARIABLES: | PREPARED BY: |
| T/K = 290-291<br>P/kPa = 101.325 | R. Battino |

EXPERIMENTAL VALUES:

| $T^a$/K | Weight Percentage Sucrose | Ostwald Coeff.[b] $10^2L$ | Bunsen Coeff.[c] $10^2\alpha$ |
|---|---|---|---|
| 289.95 | 11.38 | 1.520 | 1.432 |
| 290.05 | 20.00 | 1.309 | 1.233 |
| 290.15 | 29.93 | 1.089 | 1.025 |
| 290.95 | 30.12 | 1.100 | 1.033 |
| 291.15 | 47.89 | 0.791 | 0.742 |
| 290.85 | 48.57 | 0.701 | 0.658 |

[a] Temperature reported to 0.1°C.

[b] Ostwald coefficient calculated by compiler.

[c] Bunsen coefficient.

AUXILIARY INFORMATION

METHOD/APPARATUS/PROCEDURE:

Water is degassed by pumping under vacuum. The absorption vessel is about 600 $cm^3$ and is separately thermostatted from the buret gas measuring system. Both are connected by a copper capillary tube. Gas uptake is read on the gas burets. The absorption flask volume and all liquid volumes were calibrated using mercury. The original paper contains details of operation and a drawing of the apparatus. Solubilities were also measured in pure water, glycerol and chloral hydrate.

SOURCE AND PURITY OF MATERIALS:

(1) Nitrogen - from heating potassium nitrite and ammonium nitrite. Washed with potassium hydroxide solution, ferrous sulfate solution and sulfuric acid; and finally passed through a glowing copper tube.
(2) Sucrose - chemically pure from Merck in Darmstadt.
(3) Water - no comment by author.

ESTIMATED ERROR:

$\delta\alpha/\alpha = \pm 0.01$, estimate by compiler.

REFERENCES:

| COMPONENTS: | ORIGINAL MEASUREMENTS: |
|---|---|
| (1) Nitrogen; $N_2$; [7727-37-9]<br>(2) Sulfur dioxide; $O_2S$; [7446-09-5] | Dornte, R.W.; Ferguson, C.V.<br>*Ind. Eng. Chem.* 1939, *31*, 112-3. |
| VARIABLES: | PREPARED BY: |
| T/K = 213-253 | R. Battino |

EXPERIMENTAL VALUES:

| T/°C | T/K[a] | K[b] Static | K[c] Circ. |
|---|---|---|---|
| -60 | 213.2 | 2.86 | 2.24 |
| -60 | 213.2 | | 1.97 |
| -60 | 213.2 | | 2.23 |
| -50 | 223.2 | 6.35 | 4.50 |
| -50 | 223.2 | | 4.32 |
| -40 | 233.2 | 14.0 | 7.55 |
| -40 | 233.2 | 10.6 | 8.00 |
| -40 | 233.2 | | 8.09 |
| -30 | 243.2 | 27.2 | |
| -20 | 253.2 | 56.6 | 26.7 |

[a] Calculated by compiler.

[b] Kuenan coefficient, $cm^3$ (STP) of gas dissolved in 1 g of solvent at 101.325kPa partial pressure of gas static method.

[c] Kuenan coefficient, circulation method. Authors give $\log_{10}K = 8.729-8180/(4.58\ T/K)$ as a smoothing equation to ±10%.

AUXILIARY INFORMATION

METHOD/APPARATUS/PROCEDURE:

The apparatus consisted of a gas buret, a differential manometer, and two 10 $cm^3$ bulbs. The bulbs were in a dewar flask whose temperature was controlled manually by adding dry ice. Measurements could be taken while the gas was circulating or static. Gas uptake was read on the buret. Corrections were applied for the lines being at different temperatures.

SOURCE AND PURITY OF MATERIALS:

No details given.

ESTIMATED ERROR:

$\delta K/K = \pm 0.1$
$\delta T/K = \pm 0.5$

REFERENCES:

**COMPONENTS:**

(1) Nitrogen; $N_2$; [7727-37-9]

(2) Nitrous oxide; $N_2O$; [10024-97-2]

**ORIGINAL MEASUREMENTS:**

Steinberg, M.; Manowitz, B.; Pruzansky, J.

*US AEC BNL*-542 (T-140).
*Chem. Abstr.* 1959, *53*, 21242g

**VARIABLES:**

$T/K$ = 186.15 - 243.65

**PREPARED BY:**

H. L. Clever

**EXPERIMENTAL VALUES:**

| Temperature $t/^{0}C$ | Temperature $T/K$ | Solubility Coefficient[1] | Henry's Constant[2] $H$/atm | Mol Fraction[3] $10^3x_1$ |
|---|---|---|---|---|
| -87 | 186.15 | 0.93 | - | 1.41 |
| -80 | 193.15 | 0.87 | - | 1.33 |
| -72 | 201.15 | - | 670 | 1.49 |
| -56 | 217.15 | 1.17 | - | 1.82 |
| -45 | 228.15 | 0.85 | - | 1.34 |
| -35.5 | 237.65 | 1.03 | - | 1.65 |
| -33 | 240.15 | - | 560 | 1.79 |
| -29.5 | 243.65 | - | 560 | 1.79 |
| +32 | 305.15 | 0.99[4] | - | 1.75 |

[1] The solubility coefficient units are $cm^3(288.15\ K)cm^{-3}atm^{-1}$. The gas volume is reduced to 288.15 K and 1 atm and the solvent volume is for the temperature of the measurement.

[2] Henry's constant, $H/\text{atm} = (p_1/\text{atm})/x_1$. The values are apparently from the data smoothed by the authors.

[3] The mole fraction values were calculated for a gas partial pressure of 101.325 kPa (1 atm) by the compiler. The density values were estimated from a single value given in the *International Critical Tables*, Vol. III.

[4] Solubiluty coefficient value calculated by the authors from literature value based on the average residual nitrogen in liquid nitrous oxide cylinders. No reference given.

**AUXILIARY INFORMATION**

**METHOD/APPARATUS/PROCEDURE:**

See earlier paper for some details of the procedure(1). The measurements were made using a modified McDaniel method (2), a static method.

**SOURCE AND PURITY OF MATERIALS:**

(1) Nitrogen.

(2) Nitrous oxide.

No information.

**ESTIMATED ERROR:**

$\delta x_1/x_1 = \pm\ 0.05 - 0.10$ (compiler)

**REFERENCES:**

1. Steinberg, M.; Manowitz, B. *Ind. Eng. Chem.* 1959, *51*, 47.

2. Furman, N.H. "Scott's Standard Methods of Chemical Analysis," vol. II, 5th ed., Van Nostrand, New York, 1939.

| COMPONENTS: | ORIGINAL MEASUREMENTS: |
|---|---|
| (1) Nitrogen; $N_2$; [7727-37-9]<br><br>(2) Nitrogen Oxide; $N_2O_4$; [10544-72-6] | Chang, E. T.; Gokcen, N. A.<br><br>*J. Phys. Chem.* 1966, *70*, 2394-9.<br><br>NASA *Accession No. N66-14076, Rept. No.* ATN=64-(9228)-4, 22p (1964). *From Sci. Tech. Aerospace Rept.* 4(4), 640(1964). |
| VARIABLES:<br>T/K = 262.02 - 303.16<br>$N_2$ P/kPa = 35.50 - 197.58 | PREPARED BY:<br><br>P. L. Long<br>H. L. Clever |

EXPERIMENTAL VALUES:

Smoothed Data: Mole fraction solubility at 1 atm $N_2$ pressure.

$$\ln x_1 = -360.31/(T/K) - 6.106$$

$$\Delta H°/kJ\ mol^{-1} = 3.00,\ \Delta S°/J\ K^{-1}\ mol^{-1} = -50.8$$

| T/K | Mol Fraction $10^4 x_1$ | Gibbs Energy $\Delta G°/kJ\ mol^{-1}$ |
|---|---|---|
| 263.15 | 5.67 | 16.36 |
| 268.15 | 5.81 | 16.61 |
| 273.15 | 5.96 | 16.86 |
| 278.15 | 6.10 | 17.12 |
| 283.15 | 6.24 | 17.37 |
| 288.15 | 6.38 | 17.63 |
| 293.15 | 6.52 | 17.88 |
| 298.15 | 6.66 | 18.13 |
| 303.15 | 6.79 | 18.39 |

The mole fraction solubility of one atm $N_2$ pressure is numerically the value of Henry's constant at that temperature.

The authors discuss the system further, but present no additional solubility data (1,2).

continued on following page

AUXILIARY INFORMATION

METHOD/APPARATUS/PROCEDURE:

The solvent was degassed under vacuum in the previously weighed apparatus. Apparatus and degassed solvent were weighed. Gas was introduced into the apparatus at a known P and T, the liquid stirred, and the pressure observed until there was no further change. Equilibrium was established within 10 m and the P was followed for 40 m.

The apparatus was of all glass construction. It consisted of three calibrated volumes for the measurement of the gas, a container for the solvent, which was stirred with a glass enclosed magnet, and a manometer for measuring the pressure with a microslide cathetometer. The solvent container had a capacity for 100 g of solvent with a 5 ml gas space above the liquid surface. The apparatus sections were calibrated to ± 0.0002-3 $cm^3$.

SOURCE AND PURITY OF MATERIALS:

(1) Nitrogen. No source given.

(2) Nitrogen Oxide. Source not given. Research grade. 99.5% min. purity. Density,

$$\rho/g\ ml^{-1} = 1.4916 - 0.00226\ t/°C.$$

ESTIMATED ERROR:

$\delta T/K = 0.03$
$\delta P/mmHg = 0.01$
$\delta X_1/X_1 = 0.05$

REFERENCES:


1. Chang, E. T.; Gokcen, N. A.; Poston, T. M. *J. Spacecr. Rockets* 1969, *6*, 1177.
2. Gokcen, N. A.; Chang, E. T. *J. Chem. Ed.* 1977, *54*, 368.

| COMPONENTS: | ORIGINAL MEASUREMENTS: |
|---|---|
| (1) Nitrogen; $N_2$; [7727-37-9] | Chang, E. T.; Gokcen, N. A. |
| (2) Nitrogen oxide; $N_2O_4$; [10544-72-6] | *J. Phys. Chem.* 1966, *70*, 2394-9. |
| | *NASA Accession No. N66-14076, Rept. No. ATN-64-(9228)-4*, 22p (1964). *From Sci. Tech. Aerospace Rept.* *4(4)*, 640(1964). |

EXPERIMENTAL VALUES: continued

| T/K | Nitrogen Pressure $p_1$/atm | Henry's Constant[1] $10^4K$/atm$^{-1}$ | Mol Fraction $10^4x_1$ | Bunsen Coefficient $\alpha$/cm$^3$ (STP) cm$^{-3}$ atm$^{-1}$ | Ostwald Coefficient $L$/cm$^3$ cm$^{-3}$ |
|---|---|---|---|---|---|
| 262.02 | 0.5916 | 5.54 | 3.28 | | |
| | 0.6186 | 5.56 | 3.44 | | |
| | 1.2964 | 5.69 | 7.38 | | |
| | 1.3192 | 5.69 | 7.51 | | |
| | 1.9440 | 5.61 | 10.91 | | |
| | 1.9500 | 5.68 | 11.08 | | |
| | 1.0 | | 5.62 | 0.208 | 0.199 |
| 273.15 | 0.3504 | 5.96 | 2.09 | | |
| | 0.3811 | 6.09 | 2.32 | | |
| | 0.4716[2] | 6.00 | 2.83 | | |
| | 0.6830 | 5.94 | 4.06 | | |
| | 0.8972 | 6.01 | 5.39 | | |
| | 0.9389[2] | 6.00 | 5.63 | | |
| | 1.4283 | 6.02 | 8.60 | | |
| | 1.5043 | 5.95 | 8.95 | | |
| | 1.7951 | 6.00 | 10.77 | | |
| | 1.0 | | 5.99 | 0.218 | 0.218 |
| 298.15 | 0.3801 | 6.50 | 2.47 | | |
| | 0.5909 | 6.52 | 3.85 | | |
| | 0.7222 | 6.67 | 4.82 | | |
| | 0.7393 | 6.60 | 4.88 | | |
| | 0.8457 | 6.55 | 5.54 | | |
| | 1.1125 | 6.66 | 7.41 | | |
| | 1.3032 | 6.59 | 8.59 | | |
| | 1.0 | | 6.58 | 0.230 | 0.251 |
| 303.16 | 0.5604[2] | 6.80 | 3.81 | | |
| | 0.8380[2] | 6.85 | 5.74 | | |
| | 1.0 | | 6.85 | 0.238 | 0.268 |

[1] Henry's constant, $K/\text{atm}^{-1} = X_1/(P_1/\text{atm})$

[2] Data from a new apparatus having a greater degree of accuracy, but similar to the original apparatus.

The values at 101.325 kPa (1.0 atm) partial pressure nitrogen were calculated by the compiler from the average value of Henry's constant.

COMPONENTS:

(1) Nitrogen; $N_2$; [7727-37-9]

(2) Oils.

ORIGINAL MEASUREMENTS:

Baldwin, R. R.; Daniel, S. G.;

*J. Appl. Chem.*, 1952, *2* (Apr.), 161-5.

*J. Inst. Petrol.*, London, 1953, *39*, 105-24.

VARIABLES:

T/K = 273.15 - 373.15

P/kPa = 101.325

PREPARED BY:

P.L. Long
H.L. Clever

EXPERIMENTAL VALUES:

| T/K | Bunsen Coefficient $\alpha$ | Ostwald Coefficient L |
|---|---|---|
| Oil A1, viscosity 615 centistokes at $100^{o}F$, mean mol wt 670. | | |
| 293.15 | 0.0662 | 0.0711 |
| Oil A2, viscosity 268 centistokes at $100^{o}F$, mean mol wt 610. | | |
| 273.15 | 0.0632 | 0.0632 |
| 293.15 | 0.0664 | 0.0713 |
| 333.15 | 0.0738 | 0.0901 |
| 373.15 | 0.0739 | 0.100 |
| Oil A3, viscosity 181 centistokes at $100^{o}F$, mean mol wt 570. | | |
| 293.15 | 0.0717 | 0.0770 |
| Oil A4, viscosity 80.3 centistokes at $100^{o}F$, mean mol wt 530. | | |
| 293.15 | 0.0748 | 0.0803 |
| Oil A5, viscosity 34.9 centistokes at $100^{o}F$, mean mol wt 400. | | |
| 273.15 | 0.0773 | 0.0773 |
| 293.15 | 0.0762 | 0.0818 |
| 333.15 | 0.0785 | 0.0957 |
| 373.15 | 0.0818 | 0.1113 |
| Oil B1, viscosity 260 centistokes at $100^{o}F$, mean mol wt 630. | | |
| 293.15 | 0.0644 | 0.0691 |

METHOD/APPARATUS/PROCEDURE:

The method consists of two operations: Saturation of the liquid with the gas under consideration; and determination of the amount of dissolved gas in the liquid saturated with the gas. All gases were passed through a liquid air trap to remove water and $CO_2$ vapors before passage through the liquid. All gases were passed through a long coil, thermostated, before the saturation process.

Oil or similar liquid is deaerated by passing it through a continuously evacuated vessel. Gas is then passed through the liquid to saturation. Liquid is then degassed, and transferred to a storage burette, disconnected from the degassing apparatus, thermostated, and saturated with the desired gas. Burette is now connected to another degassing apparatus, portions of the oil degassed and gas collected.

SOURCE AND PURITY OF MATERIALS:

(1) Nitrogen. From commercial cylinders, and gas analysis on a Bone and Wheeler apparatus shows it to be at least 99.5% pure; no source given.

(2) Oils. The oils conform to DTD-472B specifications. Composition and physical properties of the oils are given in the appendix of the 1953 paper. The molecular weights were determined by freezing point depression in naphthalene to within $\pm 5$ percent.

REFERENCES:

COMPONENTS:

(1) Nitrogen; $N_2$; [7727-37-9]

(2) Mineral oil

ORIGINAL MEASUREMENTS:

Rodman, C. J.; Maude, A. H.

*Trans. Am. Electrochem. Soc.* 1925, *47*, 71 - 92.

VARIABLES:

$T$/K = 298.15, 353.15
$p_1$/kPa = 101.3 (760 mmHg)

PREPARED BY:

H. L. Clever

EXPERIMENTAL VALUES:

| Temperature $t/^{0}C$ | $T/K$ | Bunsen Coefficient $\alpha/cm^3(STP)cm^{-3}atm^{-1}$ | Ostwald Coefficient $L/cm^3 cm^{-3}$ | Solubility $g\ kg^{-1}$ |
|---|---|---|---|---|
| 25 | 298.15 | 0.0848 | 0.0925 | 0.127 |
| 80 | 353.15 | 0.0916 | 0.1185 | 0.144 |

These values appear in the International Critical Tables, McGraw-Hill Book Co., New York and London, Vol. III, pp. 261 - 270 where they are credited to an industrial report edited by A. H. Maude.

AUXILIARY INFORMATION

METHOD/APPARATUS/PROCEDURE:

The apparatus consists of an 180 $cm^3$ absorption bottle connected to a 100 $cm^3$ gas buret. The absorption bottle sits in a thermostat, which is attached to a shaking machine.

A weighed sample of oil is introduced into the absorption vessel. The sample is degassed by vacuum taking care to avoid excessive foaming. The gas is brought into the system. An initial buret reading taken, and the shaker is started and reading taken every 5 minutes until 2 or 3 constant readings are obtained.

SOURCE AND PURITY OF MATERIALS:

1. Nitrogen. No information.

2. Mineral oil. A Pennsylvania base oil, 96 per cent saturated hydrocarbons, and distilling between 300 and 400 $^{0}C$. Density at 25 $^{0}C$ = 0.840 and at 80 $^{0}C$ = 0.800 g $cm^{-3}$. As a commercial product the oil is known as "Wemco A".

ESTIMATED ERROR:

REFERENCES:

| COMPONENTS: | ORIGINAL MEASUREMENTS: |
|---|---|
| (1) Nitrogen; $N_2$; [7727-37-9]<br>(2) Mineral oil, white | Kubie, L. S.<br>*J. Biol. Chem.* 1927, *72*, 548-8. |
| VARIABLES:<br>T/K = 295.15 - 295.65<br>P/kPa = 101.325 | PREPARED BY:<br>H. L. Clever |

EXPERIMENTAL VALUES:

| T/K | Bunsen Coefficient $\alpha/cm^3(STP)\ cm^{-3}atm^{-1}$ |
|---|---|
| 295.15-295.65 | 0.071 ± 0.002 |

AUXILIARY INFORMATION

METHOD/APPARATUS/PROCEDURE:

The appartaus was the constant volume Van Slyke and Neill type (1).

Saturation was attained by bubbling the gas through the oil for several hours.

A 1.990 ± 0.007 $cm^3$ volume of saturated solution was analyzed. The Van Slyke correction factor for unextracted and redissolved gas was 1.04.

SOURCE AND PURITY OF MATERIALS:

(1) Oxygen. Commercial cylinder.

(2) Mineral oil. "Amalie" brand, L. Sonneborn Sons, Inc., NY. A medicinal white oil which meets US Pharmacopeia standards. Prepared by fractional distillation from a naphthalene base, unsaturated hydrocarbons removed. Specific gravity 0.890 - 0.895 at 15 $^0$C, viscosity 285 at 100 $^0$F.

ESTIMATED ERROR:

$\delta T/K = 1$

REFERENCES:

1. Van Slyke, D. D.; Neill, J. M. *J. Biol. Chem.* 1924, *56*, 523.

COMPONENTS:

(1) Nitrogen; $N_2$; [7727-37-9]

(2) Paraffin oil

ORIGINAL MEASUREMENTS:

Nasini, A. G.; Corinaldi, G.

*Soc. Ital. Prog. Sci. Atti Rium.* 1932, *20*, 264-6.

VARIABLES:

T/K = 305.15
Total P/kPa = 101.325

PREPARED BY:

A. L. Cramer
H. L. Clever

EXPERIMENTAL VALUES:

| T/K | Bunsen Coefficient $\alpha$ | Ostwald Coefficient L |
|---|---|---|
| 305.15 | .0736 | .0822 |

The Ostwald coefficient was calculated by the compiler.

AUXILIARY INFORMATION

METHOD/APPARATUS/PROCEDURE:

The apparatus and procedure were based on those of Lannung (1).

SOURCE AND PURITY OF MATERIALS:

(1) Nitrogen. Laboratory preparation from ammonium nitrate followed by "conventional purification".

(2) Paraffin oil. Density, $\rho_{288.15} = 0.87$ g cm$^{-3}$.

ESTIMATED ERROR:

$\delta T/K = 0.1$

REFERENCES:

1. Lannung, A. *J. Am. Chem. Soc.* 1930, *52*, 68.

**COMPONENTS:**

(1) Nitrogen; $N_2$; [7727-37-9]

(2) Paraffin Oil

**ORIGINAL MEASUREMENTS:**

Luther, H; Rottger, H.

*Elektrotech. Z.* 1957, *A78*, 462-4.

**VARIABLES:**

T/K = 318
P/kPa = 27-103

**PREPARED BY:**

R. Battino

**EXPERIMENTAL VALUES:**

| t/°C | $T^a$/K | P/mm Hg | S/$cm^3$ $g^{-1}$ |
|---|---|---|---|
| 45 | 318 | 200 | 0.023 |
| 45 | 318 | 300 | 0.034 |
| 45 | 318 | 500 | 0.054 |
| 45 | 318 | 700 | 0.076 |
| 45 | 318 | 770 | 0.084 |

[a] Calculated by compiler.

**AUXILIARY INFORMATION**

**METHOD/APPARATUS/PROCEDURE:**

Solubility determined by an appartus described in the paper by a combined volume and pressure measurement.

**SOURCE AND PURITY OF MATERIALS:**

(1) No details given.

(2) Molecular weight of 357 g $mol^{-1}$, $d_{20}$=0.8420 g $cm^{-3}$, $n_D^{20}$= 1.4597.

**ESTIMATED ERROR:**

$\delta S/S$ = ±0.05, compiler's estimate.

**REFERENCES:**

| COMPONENTS: | ORIGINAL MEASUREMENTS: |
|---|---|
| (1) Nitrogen; $N_2$; [7727-37-9]<br>(2) Paraffin oil and a 50:50 mixture of 1-decanol and 1-dodecanol | Luther, H.; Hiemenz, W.<br>*Chem. Ingr.-Tech.* 1957, *29*, 530-5. |
| **VARIABLES:** | **PREPARED BY:** |
| T/K = 293-355 | R. Battino |

EXPERIMENTAL VALUES:

| t/°C | $T^a$/K | $10^4 x_1$ [b] | H [c] | $10^6 S$ [d] | $10^2 L$ [e] |
|---|---|---|---|---|---|
| | | Paraffin Oil | | | |
| 20 | 293 | 16.4 | 608 | | |
| 20.0 | 293.2 | | | 1.89 | 8.33 |
| 41.0 | 314.2 | | | 1.87 | 8.82 |
| 61.5 | 334.7 | | | 1.83 | 9.20 |
| 82.0 | 355.2 | | | 1.80 | 9.57 |
| | | Alcohol Mixture [f] | | | |
| 20 | 293 | 6.2 | 1621 | | |
| 20.0 | 293.2 | | | 1.69 | 7.54 |
| 41.0 | 314.2 | | | 1.93 | 9.28 |
| 61.5 | 334.7 | | | 2.07 | 10.5 |
| 82.0 | 355.2 | | | 2.24 | 12.0 |

[a] Calculated by compiler.

[b] Mole fraction solubility at 101.325kPa partial pressure of gas.

[c] Henry's law constant, $P_1/\text{atm} = (H/\text{atm})x_1$.

[d] Solubility in mol $cm^{-3}$ at $P_{N2}$=400 torr.

[e] Ostwald coefficient.

[f] 50 mol % 1-decanol + 50 mol % 1-dodecanol.

AUXILIARY INFORMATION

| METHOD/APPARATUS/PROCEDURE: | SOURCE AND PURITY OF MATERIALS: |
|---|---|
| Solubilities determined by an apparatus described in the paper by a combined pressure and volume measurement. | (1) No details given.<br>(2) Paraffin oil - density of 0.8795 g $cm^{-3}$ and molecular weight of 405 g $mol^{-1}$.<br>(3) Alcohol mixture is 50 mol % 1-decanol + 50 mol % 1-dodecanol. |
| | **ESTIMATED ERROR:**<br>$\delta H/H = \pm 0.05$, compiler's estimate. |
| | **REFERENCES:** |

| COMPONENTS: | ORIGINAL MEASUREMENTS: |
|---|---|
| (1) Nitrogen; $N_2$; [7727-37-9]<br>(2) Paraffin wax | Ridenour, W. P.; Weatherford, W. D.; Capell, R. G.<br>*Ind. Eng. Chem.* 1954, *46*, 2376-81. |
| VARIABLES:<br>T/K = 349.55<br>P/kPa = 35.26 - 103.19 | PREPARED BY:<br>A. L. Cramer<br>H. L. Clever |

EXPERIMENTAL VALUES:

| T/K | Pressure/ mmHg | Mol Fraction $x_1 \times 10^3$ | Bunsen Coefficient $\alpha$ | Solubility Coefficient $cm^3$ $N_2$ (STP) $g^{-1}$ wax |
|---|---|---|---|---|
| 349.55 | 264.5 | 0.506 | 0.071 | 0.033 |
| | 442.7 | 0.906 | 0.076 | 0.058 |
| | 590.5 | 1.27 | 0.080 | 0.081 |
| | 774.0 | 1.66 | 0.080 | 0.106 |
| | 760.0 | (1.66) | 0.080 | 0.104 |

The mole fraction solubility at a nitrogen partial pressure of 760 mmHg was calculated by the compiler.

AUXILIARY INFORMATION

METHOD/APPARATUS/PROCEDURE:

The apparatus was similar to the equilibrium adsorption apparatus described by Brunauer, Emmett, and Teller (1) for the measurement of the surface area of a solid catalyst.

A weighed amount of wax was placed in the apparatus. The gas and solvent were equilibrated for 20 to 60 minutes. The gas volume in the buret system was calculated from the ideal gas law.

The results of the absorption measurements were checked by a desorption measurement. The two measurements agreed very well.

SOURCE AND PURITY OF MATERIALS:

(1) Nitrogen. Air Reduction Co. 99.96% $N_2$, 0.04 % $O_2$.

(2) Paraffin wax. Described as 122°F English melting point wax. Molecular weight 350, actual melt point 123.2°F (323.8 K), density 0.7716 g $cm^{-3}$ at 293.3 K and 0.7662 g $cm^{-3}$ at 298.0 K.

ESTIMATED ERROR:

$\delta T/K = 2.$

$\delta P/mmHg = 0.2$

$\delta\alpha/cm^3 = 0.004$ (low pressure)
$-0.001$ (high pres.)

REFERENCES:

1. Brunauer, S.; Emmett, P. H.; Teller, E.
*J. Am. Chem. Soc.* 1938, *60*, 309.

**COMPONENTS:**

(1) Nitrogen; $N_2$; [7727-37-9]

(2) Petrowax A

**ORIGINAL MEASUREMENTS:**

Ridenour, W. P.; Weatherford, W. D.; Capell, R. G.

*Ind. Eng. Chem.* 1954, *46*, 2376-81.

**VARIABLES:**

T/K = 354.75
P/kPa = 32.88 - 105.26

**PREPARED BY:**

A. L. Cramer
H. L. Clever

**EXPERIMENTAL VALUES:**

| T/K | Pressure/ mmHg | Mol Fraction $X_1 \times 10^3$ | Bunsen Coefficient $\alpha$ | Solubility Coefficient $cm^3$ $N_2$ (STP) $g^{-1}$ wax |
|---|---|---|---|---|
| 354.75 | 246.4 | 0.54 | 0.040 | 0.016 |
| | 419.3 | 1.27 | 0.056 | 0.038 |
| | 654.4 | 2.11 | 0.058 | 0.063 |
| | 789.5 | 2.57 | 0.060 | 0.077 |

AUXILIARY INFORMATION

**METHOD/APPARATUS/PROCEDURE:**

The apparatus was similar to the equilibrium adsorption apparatus described by Brunauer, Emmett, and Teller (1) for the measurement of the surface area of a solid catalyst.

A weighed amount of wax was placed in the apparatus. The gas and solvent were equilibrated for 20 to 60 minutes. The gas volume in the buret system was calculated from the ideal gas law.

The results of the absorption measurements were checked by a desorption measurement. The two measurements agreed very well.

**SOURCE AND PURITY OF MATERIALS:**

(1) Nitrogen. Air Reduction Co. 99.96% $N_2$, 0.04% $O_2$.

(2) Petrowax A. Molecular weight 750, melting point 159.6°F (352.8 K) density 0.8039 g $cm^{-3}$ at 355.6 K and 0.7662 g $cm^{-3}$ at 370.2 K.

**ESTIMATED ERROR:**

$\delta T/K = 2.$
$\delta P/mmHg = 0.2$
$\delta \alpha/cm^3 = 0.004$ (low pressure)
-0.001 (high pres.)

**REFERENCES:**

1. Brunauer, S.; Emmett, P. H.; Teller, E.
*J. Am. Chem. Soc.* 1938, *60*, 309.

COMPONENTS:

(1) Nitrogen; $N_2$; [7727-37-9]

(2) Hydrocarbon fuels

ORIGINAL MEASUREMENTS:

Baldwin, R. R.; Daniel, S. G.

*J. Appl. Chem.* 1952, *2*, 161-5.

*J. Inst. Petrol*, London, 1953, *39*, 105-24.

VARIABLES:

T/K = 273.15 - 293.15
P/kPa = 101.325

PREPARED BY:

P. L. Long
H. L. Clever

EXPERIMENTAL VALUES:

| T/K | Bunsen Coefficient $\alpha$ | Ostwald Coefficient L |
|---|---|---|
| 100 octane fuel D | | |
| 293.15 | 0.204 | 0.219 |
| Kerosene | | |
| 293.15 | 0.1185 | 0.127 |
| 273.15 | 0.116 | 0.116 |

AUXILIARY INFORMATION

METHOD/APPARATUS/PROCEDURE:

1. Degassing. The liquid was deaerated by passage through a continuously evacuated vessel. The liquid was saturated with gas, then degassed again.

2. Saturation of liquid with gas. The gas was passed through a liquid air trap to remove water and carbon dioxide, then through a long coil thermostated at the temperature of the solubility measurement, and then bubbled through the liquid up to five hours.

3. Determination of the amount of gas liberated under vacuum from a known volume of the saturated liquid. The saturated liquid is placed in a vessel attached to a calibrated buret. The whole apparatus was initially evacuated. The gas released from the liquid was transferred to the calibrated buret by a Töpler pump. Five operations and transfers recover all but a trace of the gas. The last trace (2-5%) removed by boiling the liquid.

SOURCE AND PURITY OF MATERIALS:

(1) Nitrogen. Commercial cylinder, source not given. 99.5 per cent pure by gas analysis on Bone and Wheeler apparatus.

(2) 100 Octane Fuel. Molecular weight estimated to be 100. Kerosene. Mean molecular weight measured, obtained 165.

ESTIMATED ERROR:

$\delta P/P = 0.01$
$\delta \alpha/\alpha = 0.01$

REFERENCES:

COMPONENTS:

(1) Nitrogen; $N_2$; [7727-37-9]

(2) Hydrocarbon fuels and oil

ORIGINAL MEASUREMENTS:

Logvinyuk, V. P.; Makarenkov, V. V.; Malyshev, V. V.; Panchenkov, G. M.

*Khim. Tekhnol. Topl. Masel* 1970, *15*, (No. 5), 27-9.
*Chem. Technol. Fuels Oils (Eng.trans)* 1970, *15*, 353-5.

VARIABLES:

T/K= 253.15 - 365.15
$N_2$ P/kPa= 101.325

PREPARED BY:

S. A. Johnson
H. L. Clever

EXPERIMENTAL VALUES:

| T/K | Bunsen Coefficient α | Ostwald Coefficient L |
|---|---|---|
| Hydrocarbon fuel, T-1, density $\rho_4^{20}$ = 0.816 | | |
| 253.15 | 0.122 | 0.113 |
| 273.15 | 0.122 | 0.122 |
| 293.15 | 0.122 | 0.131 |
| 323.15 | 0.099 | 0.117 |
| Hydrocarbon fuel, TS-1, density $\rho_4^{20}$ = 0.775 | | |
| 293.15 | 0.129 | 0.138 |
| Hydrocarbon fuel, T-6, density $\rho_4^{20}$ = 0.84 | | |
| 253.15 | 0.099 | 0.092 |
| 273.15 | 0.099 | 0.099 |
| 293.15 | 0.114 | 0.122 |
| 323.15 | 0.114 | 0.135 |
| 365.15 | 0.099 | 0.132 |
| Hydrogenated fuel, $\rho_4^{20}$ = 0.832 | | |
| 293.15 | 0.114 | 0.122 |
| Oil, MK-8, density $\rho_4^{20}$ = o.855 | | |
| 293.15 | 0.076 | 0.082 |

The compiler calculated the Ostwald coefficients.

AUXILIARY INFORMATION

METHOD/APPARATUS/PROCEDURE:

The apparatus and procedure were described in reference (1) which was not available to the compiler.

**Mashinostroenie*, see *Izv. Vyssh. Uchebn. Zaved., Mashinostr.*

SOURCE AND PURITY OF MATERIALS:

(1) Nitrogen. No information given.

(2) Fuels and oil. Solvent densities in text above.

ESTIMATED ERROR:

$\delta\alpha/\alpha$ = 0.06 (authors)

REFERENCES:

1. Gogitidize, L. D.; Logvinyuk, V. P.; Makarenkov, V.V.; Panchenkov, G.M.; Malyshev, V.V.; Yakovlevskii, V.V. *"Method of Evaluating the Operating Properties of Jet Fuels and Lubricating Materials" (Russ.)*, *Mashinostroenie** 1966.

| COMPONENTS: | ORIGINAL MEASUREMENTS: |
|---|---|
| (1) Nitrogen; $N_2$; [7727-37-9]<br>(2) Petroleum | Gniewosz, S.; Walfisz, A.<br>*Z. Phys. Chem.* 1887, *1*, 70-2. |
| VARIABLES:<br>T/K = 283-293<br>Pressure ="Atmospheric" | PREPARED BY:<br>M. E. Derrick<br>H. L. Clever |

EXPERIMENTAL VALUES:

| T/K | Bunsen Coefficient $\alpha$ | Ostwald Coefficient L |
|---|---|---|
| 283.15 | 0.135 | 0.140 |
| 293.15 | 0.117 | 0.126 |

The Ostwald coefficients were calculated by the compiler.

AUXILIARY INFORMATION

METHOD/APPARATUS/PROCEDURE:

The apparatus consisted of an absorption flask connected to a gas buret by a flexible lead capillary. The system was thermostated in a large water bath.

The volume of nitrogen absorbed in a known volume of degassed petroleum was directly measured on the gas buret.

SOURCE AND PURITY OF MATERIALS:

(1) Nitrogen. No information.

(2) Petroleum. Russian petroleum. Cleaned by boiling in a large copper flask.

ESTIMATED ERROR:

$\delta\alpha/\alpha = 0.05$ (compiler)

REFERENCES:

**COMPONENTS:**

(1) Nitrogen; $N_2$; [7727-37-9]

(2) Amsco 123-15

**ORIGINAL MEASUREMENTS:**

Steinberg, M.; Manowitz, B.

*Ind. Eng. Chem.* 1959, *51*, 47 - 51.

**VARIABLES:**

T/K = 218.15 - 373.15
P/kPa = 101.325

**PREPARED BY:**

H. L. Clever

**EXPERIMENTAL VALUES:**

| T/K | Absorption Coefficient $\beta$ |
|---|---|
| 218.15 | 0.125 |
| 297.15 | 0.126 |
| 373.15 | 0.136 |

The authors define the absorption coefficient as the volume of gas, corrected to 288.15 K and 101.325 kPa, absorbed under a total system pressure of 101.325 kPa per unit volume (at 288.15 K) of solvent.

**AUXILIARY INFORMATION**

**METHOD/APPARATUS/PROCEDURE:**

Van Slyke Method.

**SOURCE AND PURITY OF MATERIALS:**

(1) Nitrogen. Matheson Co.

(2) Amsco 123-15. American Mineral Spirits Co. No. 140. Paraffin 59.6 wt %, naphthene 27.3 wt %, and aromatics 13.2 wt %.

**ESTIMATED ERROR:**

$\delta\beta/\beta$ = 0.05 - 0.10 (authors)

**REFERENCES:**

COMPONENTS:

(1) Nitrogen; $N_2$; [7727-37-9]

(2) Crude oils from U.S.S.R.

ORIGINAL MEASUREMENTS:

Safranova, T. P.; Zhuze, T. P.

*Khim. i. Tekhnol. Top. liv. Masel* 1958, 41-6.

*Chem. Abstr.* 1958, *52*, 8518e.

VARIABLES:

T/K = 373

P/MPa = 5.1 - 30.4

PREPARED BY:

V. Katovic

EXPERIMENTAL VALUES:

| Crude Oil | $\alpha$/cm$^3$(STP)cm$^{-3}$ atm$^{-1}$ at t/°C = 100(373 K) P/atm 50 | 100 | 200 | 300 |
|---|---|---|---|---|
| (1) "Surakhanskaya otbornaya" | 0.101 | 0.091 | 0.078 | 0.080 |
| (2) "Nebitdagskaya" | 0.077 | 0.075 | 0.066 | 0.064 |
| (3) "Romashinskaya" | 0.118 | 0.100 | 0.088 | 0.087 |
| P/MPa (compiler) | 5.07 | 10.1 | 20.3 | 30.4 |

| Crude Oil | Density $d_4^{20}$ | Viscosity /cs | Fraction <200°C % | Fraction 200-300°C % |
|---|---|---|---|---|
| (1) | 0.8494 | 5.14 | 18 | 23.5 |
| (2) | 0.8713 | 8.70 | 20 | 20 |
| (3) | 0.8530 | 6.54 | 25 | 19.5 |

AUXILIARY INFORMATION

METHOD/APPARATUS/PROCEDURE:

The apparatus consists of a thermostated steel vessel (450 cm$^3$) equipped with a magnetic stirrer, heaters, temperature controls, auxilliary pumps and analytical equipment.

The amounts of liquid and gas were determined volumetrically. The original paper contains details and a drawing of apperatus.

SOURCE AND PURITY OF MATERIALS:

(1) Stated to contain 3.09% of air

(2) See details above.

ESTIMATED ERROR:

REFERENCES:

# NITROGEN SOLUBILITIES ABOVE 200 kPa

| COMPONENTS: | EVALUATOR: |
|---|---|
| (1) Nitrogen; $N_2$; [7727-37-9]<br><br>(2) Water; $H_2O$; [7732-18-5] | Colin L. Young,<br>School of Chemistry,<br>University of Melbourne,<br>Parkville, Victoria 3052,<br>Australia.<br><br>June 1981 |

CRITICAL EVALUATION:

The most extensive studies of the solubility of nitrogen in water are those of Wiebe *et al.* (1), (2), Krase and coworkers (3), (4) and of Smith and coworkers (5), (6), (7). The data of Smith and coworkers (5), (6), (7) agree within about 2% with those of Wiebe *et al.* (2). However, there are larger discrepancies between the data of Krase and coworkers (3), (4) and Wiebe *et al.* (2). There is also some uncertainty with regard to the data of Suciu (8), McKee (9) and Suciu, Zoss and Sibbitt (10) which are classified as doubtful. The relatively low pressure data of Pray *et al.* (11) were incorporated with additional new data to give the data reported in references (8) and (10). The data of Smith and coworkers (5), (6), (7) and Wiebe *et al.* (1), (2) are classified as recommended for the temperature range 298.15 K to 398.15 K. Above this temperature range up to 513 K, the data of Saddington and Krase (4) are the most reliable and are classified as tentative. The data of Pray *et al.* (11) give an indication of the solubility at higher temperatures but must be regarded with some caution.

The data of Bassett and Dode (12) cover pressures up to 450 MPa but are rejected since it appears that the water became contaminated with oil from the high pressure compressor. These workers' data disagree with the data of Wiebe *et al.* (2) at 100 MPa. Bassett and Dode found that the solubility of nitrogen in water at 18°C initially increased with increasing pressure and then decreased with further increases in pressure. This finding is supported by work by Tsiklis and Maslennikova (13) who found similar behaviour at 330°C. These latter workers (13) found the system showed the phenomenon of gas-gas immiscibility of the second kind.

The data of Frolich *et al.* (14) are not of high accuracy and are presented in graphical form. Consequently these data are rejected for the present system.

References


1. Wiebe, R.; Gaddy, V. L.; Heins, C. *Ind. Eng. Chem.* 1932, *24*, 927.
2. Wiebe, R.; Gaddy, V. L.; Heins, C. *Ind. Eng. Chem.* 1933, *55*, 947.
3. Goodman, J. B.; Krase, N. W. *Ind. Eng. Chem.* 1931, *23*, 401.
4. Saddington, A. W.; Krase, N. W. *J. Am. Chem. Soc.* 1934, *56*, 353.
5. Smith, N. O.; Kelemen, S.; Nagy, B. *Geochim. Cosmochim. Acta* 1962, *26*, 921.
6. O'Sullivan, T. D.; Smith, N. O.; Nagy, B. *Geochim. Cosmochim. Acta* 1966, *30*, 617.
7. O'Sullivan, T. D.; Smith, N. O. *J. Phys. Chem.* 1970, *74*, 1460.
8. Suciu, S. *Ph.D. Thesis*, 1951, Purdue University.
9. McKee, O. L. Jr. *Ph.D. Thesis*, 1953, Purdue University.
10. Suciu, S. N.; Zoss, L. M.; Sibbitt, W. L. *Am. Soc. Mech. Eng.* Paper No. 53-A-64. Presented at the Annual Meeting, New York, 1953.
11. Pray, H. A.; Schweichert, C. E.; Minnich, B. H. *Ind. Eng. Chem.* 1952, *44*, 1147.
12. Bassett, J.; Dode, M. *Compt. Rend.* 1936, *203*, 775.
13. Tsiklis, D. S.; Maslennikova, V. Ya. *Doklady Akad. Nauk. SSSR* 1965, *161*, 645.
14. Frolich, P. K.; Trauch, E. J.; Hogan, J. J.; Peer, A. A. *Ind. Eng. Chem.* 1931, *23*, 548.

| COMPONENTS: | ORIGINAL MEASUREMENTS: |
|---|---|
| (1) Nitrogen; $N_2$; [7727-37-9]<br><br>(2) Water; $H_2O$; [7732-18-5] | Goodman, J. B. and Krase, N. W., *Ind. Eng. Chem.*, 1931, *23*, 401-4/ |
| VARIABLES:<br>T/K = 273-442<br>P/MPa = 10.1-30.4 | PREPARED BY:<br><br>C. L. Young |

EXPERIMENTAL VALUES:

| T/K | *P*/bar | $10^3$ Mole fraction of nitrogen in water, $10^3 x_{N_2}$ | T/K | *P*/bar | $10^3$ Mole fraction of nitrogen in water, $10^3 x_{N_2}$ |
|---|---|---|---|---|---|
| 273.15 | 101.3 | 1.04 | 353.15 | 202.6 | 1.62 |
| | 126.7 | 1.26 | | 304.0 | 2.04 |
| | 202.6 | 2.27 | 373.15 | 101.3 | 0.681 |
| | 304.0 | 2.57 | | 126.7 | 0.835 |
| 298.15 | 101.3 | 0.764 | | 202.6 | 1.61 |
| | 126.7 | 1.03 | | 304.0 | 2.08 |
| | 202.6 | 1.97 | 417.15 | 101.3 | 0.732 |
| | 304.0 | 2.32 | | 126.7 | 0.928 |
| 323.15 | 101.3 | 0.716 | | 207.6 | 0.91 |
| | 126.7 | 0.885 | | 304.0 | 2.47 |
| | 102.6 | 1.78 | 442.15 | 101.3 | 0.771 |
| | 304.0 | 2.13 | | 126.7 | 1.09 |
| 353.15 | 101.3 | 0.667 | | 202.6 | 2.35 |
| | 126.7 | 0.821 | | 304.0 | 2.73 |

AUXILIARY INFORMATION

METHOD/APPARATUS/PROCEDURE:

Magnetically stirred equilibrium cell of about ∿ 0.02 ℓ. Pressure measured with dead weight gauge. Copper block thermostat used at temperatures above 273.15K. Cell charged with gas and liquid; after equilibrium established liquid sample removed from bottom of cell and amount of dissolved gas estimated volumetrically. Details in source.

SOURCE AND PURITY OF MATERIALS:

(1) Commercial sample oxygen; carbon dioxide and water vapor removed before use.

(2) Distilled.

ESTIMATED ERROR:

$\delta T/K = \pm 0.2$; $\delta P/\text{bar} = \pm 0.5$;

$\delta x_{N_2} = \pm 1\text{-}2\%$ (estimated by compiler).

REFERENCES:

**COMPONENTS:**

(1) Nitrogen; $N_2$; [ 7727-37-9 ]

(2) Water; $H_2O$; [ 7732-18-5 ]

**ORIGINAL MEASUREMENTS:**

Wiebe, R.; Gaddy, V. L. ; Heins, C.
*Ind. Eng. Chem.* 1932, *24*, 927.

**VARIABLES:**

T/K = 298

P/MPa=2.53-101.3

**PREPARED BY:**

C. L. Young

**EXPERIMENTAL VALUES:**

| T/K | P/bar | $10^3$ × Mole fraction of nitrogen in liquid, $10^3 x_{N_2}$ |
|---|---|---|
| 298.15 | 25.33 | 0.249 |
| | 50.66 | 0.267 |
| | 101.33 | 0.903 |
| | 202.65 | 1.616 |
| | 303.97 | 2.180 |
| | 506.62 | 3.160 |
| | 810.60 | 4.361 |
| | 1013.25 | 5.076 |

**AUXILIARY INFORMATION**

**METHOD/APPARATUS/PROCEDURE:**

One pass flow method with two vessel adsorption train. Second vessel used as source of sample for analysis. Pressure measured with dead weight gauges. Measurements taken both for a high pressure and low pressure approach to equilibrium. Details in source and ref. 1.

**SOURCE AND PURITY OF MATERIALS:**

(1) Best commercial sample purified by passage over finely divided copper. Major impurity argon and traces of oxygen.

(2) No details given.

**ESTIMATED ERROR:**

$\delta T/K = \pm 0.05$; $\delta P/\text{bar} = \pm 0.5$;

$\delta x_{N_2} = \pm 1\%$.

**REFERENCES:**

1. Wiebe, R. ; Gaddy, V. L. ; Heins C. *Ind. Eng. Chem.* 1931, *23*, 401.

COMPONENTS:

(1) Nitrogen; $N_2$; [7727-37-9]

(2) Water; $H_2O$; [7732-18-5]

ORIGINAL MEASUREMENTS:

Wiebe, R., Gaddy, V. L.; Heins, C. *J. Am. Chem. Soc.* 1933, *55*, 947.

VARIABLES:

T/K = 298-373

P/MPa = 2.5-101.3

PREPARED BY:

C. L. Young

EXPERIMENTAL VALUES:

| T/K | $P$/bar | $10^2$ Mole fraction of nitrogen in liquid, $10^2x_{N_2}$ | T/K | $P$/bar | $10^2$ Mole fraction of nitrogen in liquid, $10^2x_{N_2}$ |
|---|---|---|---|---|---|
| 298.15 | 25.33 | 0.0249 | 348.15 | 25.33 | 0.0181 |
| | 50.66 | 0.0481 | | 50.66 | 0.0353 |
| | 101.33 | 0.0902 | | 101.33 | 0.0676 |
| | 202.65 | 0.1610 | | 202.65 | 0.1236 |
| | 303.97 | 0.2183 | | 303.97 | 0.1721 |
| | 506.62 | 0.3163 | | 506.62 | 0.2554 |
| | 810.60 | 0.4364 | | 810.60 | 0.3604 |
| | 1013.25 | 0.5083 | | 1013.25 | 0.4222 |
| 323.15 | 25.33 | 0.0195 | 373.15 | 25.33 | 0.0190 |
| | 50.66 | 0.0381 | | 50.66 | 0.0369 |
| | 101.33 | 0.0722 | | 101.33 | 0.0704 |
| | 202.65 | 0.1306 | | 202.65 | 0.1300 |
| | 303.97 | 0.1807 | | 303.97 | 0.1816 |
| | 506.62 | 0.2651 | | 506.62 | 0.2707 |
| | 810.60 | 0.3717 | | 810.60 | 0.3819 |
| | 1013.25 | 0.4356 | | 1013.25 | 0.4450 |

AUXILIARY INFORMATION

METHOD/APPARATUS/PROCEDURE:

One pass flow method with two vessel adsorption train. Second vessel used as source of sample for analysis. Pressure measured with dead weight gauges. Measurements taken both for a high pressure and low pressure approach to equilibrium. Details in source and ref. 1.

SOURCE AND PURITY OF MATERIALS:

(1) Purity 99.9 mole percent; major impurity argon, trace of oxygen.

(2) No details given.

ESTIMATED ERROR:

$\delta T/K = \pm 0.05$; $\delta P/\text{bar} = \pm 0.5$;

$\delta x_{N_2} = \pm 1\%$.

REFERENCES:

1. Wiebe, R. Gaddy, V. L.; Heins, C. *Ind. Eng. Chem.* 1931, *23*, 401.

| COMPONENTS: | ORIGINAL MEASUREMENTS: |
|---|---|
| (1) Nitrogen; $N_2$; [7727-37-9]<br>(2) Water; $H_2O$; [7732-18-5] | Saddington, A. W.; Krase, N. W.<br>*J. Am. Chem. Soc.* 1934, *56*, 353-61. |
| VARIABLES:<br>T/K = 338-503<br>P/MPa = 10.1-30.4 | PREPARED BY:<br>C. L. Young |

EXPERIMENTAL VALUES:

| T/K | *P*/bar | Mole fraction of nitrogen in liquid, $x_{N_2}$ |
|---|---|---|
| 338.15 | 101.3 | 0.000700 |
| 353.15 | 101.3 | 0.000698 |
| 398.15 | 101.3 | 0.000855 |
| 453.15 | 101.3 | 0.001173 |
| 483.15 | 101.3 | 0.001297 |
| 513.15 | 101.3 | 0.001446 |
| 323.15 | 202.7 | 0.001290 |
| 353.15 | 202.7 | 0.001248 |
| 373.15 | 202.7 | 0.001302 |
| 423.15 | 202.7 | 0.001550 |
| 473.15 | 202.7 | 0.002343 |
| 513.15 | 202.7 | 0.003119 |
| 323.15 | 304.0 | 0.001834 |
| 343.15 | 304.0 | 0.001730 |
| 378.15 | 304.0 | 0.001853 |
| 408.15 | 304.0 | 0.002230 |
| 438.15 | 304.0 | 0.002783 |
| 503.15 | 304.0 | 0.004313 |

AUXILIARY INFORMATION

METHOD/APPARATUS/PROCEDURE:

Single pass flow system with five pipet adsorption train. Last pipet used as source of sample for analysis. Temperature measured with three chromel-alumel thermocouples. Pressure measured with dead weight piston gauge. Gas bubbled through water for at least three hours. Liquid phase sample allowed to expand and volume of gas measured. Details in source.

SOURCE AND PURITY OF MATERIALS:

(1) Commercial sample passed through charcoal adsorption trap.

(2) No details given.

ESTIMATED ERROR:

$\delta T/K = \pm 1.0$; $\delta P/bar = \pm 0.5$;

$\delta x_{N_2} = \pm 0.002$ (estimated by compiler).

REFERENCES:

COMPONENTS:

(1) Nitrogen; $N_2$; [7727-37-9]

(2) Water; $H_2O$; [7732-18-5]

ORIGINAL MEASUREMENTS:

Suciu, S.
Ph. D. thesis, 1951
Purdue University

VARIABLES:

T/K = 397.0 - 535.9
Total P/kPa = 10,440

PREPARED BY:

H. L. Clever

EXPERIMENTAL VALUES:

| Temperature t/°F | Temperature T/K | Pressure psig[1] | Nitrogen Solubility $S/cm^3$ (STP) $g^{-1}$ | Vapor Composition $c_{H_2O}/mol\ dm^{-3}$ |
|---|---|---|---|---|
| 255 | 397.0 | 1500 | 1.05 | 0.16 |
| 303 | 423.7 | | 1.35 | 0.30 |
| 392 | 473.15 | | 2.18 | 0.98 |
| 505 | 535.9 | | 2.50 | 2.67 |

[1] Total pressure in pounds per square inch gage. Equivalent to 103 atm or 10,440 kPa.

The values above were read from large scale graphs (Fig 9 and 10) in the thesis by the compiler. The curves were drawn using Suciu's own data plus the values from Saddington, A. W.; Krase, N. W. *J. Am. Chem. Soc.* 1934, *56*, 353, and from Pray, H. A. H.; Schweickert, C. E.; Minnich, B. H. Battelle Memorial Institute, BMI-T-25, 1950

The work was carried out in the Department of Mechanical Engineering under the direction of W. L. Sibbitt.

AUXILIARY INFORMATION

METHOD/APPARATUS/PROCEDURE:

The absorption apparatus was a stainless steel high pressure bomb. It consisted of three chambers arranged in line axially, and connected by valve openings for control. There was a vapor chamber, a two phase chamber, and a liquid chamber. The entire bomb with valves was kept at the temperature of the measurement. The system was inside a 1½ inch aluminum cover.

The bomb was charged with water and the gas. There was a flexible lead to the high pressure gas supply to the pressure could be maintained at a constant value. The apparatus was rocked through a 60 degree angle at 30 oscillations per minute until equilibrium was established.

The vapor phase was analysed by ex- oanding the compressed gas and vapor mixture from the vapor chamber through a drying train and a wet test meter.

SOURCE AND PURITY OF MATERIALS:

(1) Nitrogen. Purchased in standard steel gas bottles usually pressurized at 2100 psig.

(2) Water. No information given.

ESTIMATED ERROR:

The solubility values were reproducible to 1 % with a maximum deviation of 3 %.

METHOD (continued)

The liquid phase was analysed by slowly transferring the contents of the liquid chamber to a buret system at atmospheric pressure and room temperature. The gas volume was measured, the water absorbed and weighed.

**COMPONENTS:**

(1) Nitrogen; $N_2$; [7727-37-9]

(2) Water; $H_2O$; [7732-18-5]

**ORIGINAL MEASUREMENTS:**

Pray, H. A.; Schweichert, C. E.; Minnich, B. H.

*Ind. Eng. Chem.* 1952, *44*, 1146-51.

**VARIABLES:**

T/K = 533-589
P/MPa = 1.03-3.45

**PREPARED BY:**

C. L. Young

**EXPERIMENTAL VALUES:**

| T/K | $P$/bar | $10^3$ Mole fraction of nitrogen in water, $10^3 x_{N_2}$ |
|---|---|---|
| 533.1 | 10.34 | 0.31 |
| | 34.47 | 0.89 |
| 588.7 | 10.34 | 0.39 |
| | 20.68 | 1.11 |
| | 34.47 | 1.66 |

AUXILIARY INFORMATION

**METHOD/APPARATUS/PROCEDURE:**

Rocking equilibrium cell of 3 ℓ capacity; pressure measured with dead weight gauge and temperature measured using chromel-alumel thermocouple. Cell contents equilibrated and liquid sample removed. The amount of dissolved gas estimated volumetrically.

**SOURCE AND PURITY OF MATERIALS:**

No details given.

**ESTIMATED ERROR:**

$\delta T/K = \pm 1$; $\delta P/\text{bar} = \pm 1$;

$\delta x_{N_2} = \pm 1\text{-}5\%$.

**REFERENCES:**

COMPONENTS:

(1) Nitrogen; $N_2$; [7727-37-9]

(2) Water; $H_2O$; [7732-18-5]

ORIGINAL MEASUREMENTS:

McKee, O. L. Jr.
Ph. D. thesis, 1953
Purdue University

VARIABLES: T/K = 273-295

Total P/kPa = 3450 - 20,680

PREPARED BY:

H. L. Clever

EXPERIMENTAL VALUES:

| Total Pressure P/lb in$^{-2}$ | Temperature t/°F | T/K | Nitrogen Solubility S/cm$^3$(STP) g$^{-1}$ | Temperature t/°F | T/K | Nitrogen Solubility S/cm$^3$(STP) g$^{-1}$ |
|---|---|---|---|---|---|---|
| 500 | 32 | 273.15 | 0.71 | 72 | 295.4 | 0.53 |
| | | | 0.76 | | | 0.76 |
| 1000 | 32 | 273.15 | 1.41 | 71 | 294.8 | 0.99 |
| | | | 1.43 | | | 1.38 |
| 1500 | 32 | 273.15 | 2.00 | 71 | 294.8 | 1.43 |
| | | | 2.05 | | | 1.37 |
| 2000 | 32 | 273.15 | 2.47 | 71 | 294.8 | 1.77 |
| | | | 2.53 | | | 1.76 |
| 3000 | 32 | 273.15 | 3.51 | 70 | 294.3 | 2.42 |
| | | | 3.53 | | | 2.37 |
| | | | | 71 | 294.8 | 2.44 |

The work was carried out in the Department of Mechanical Engineering under the direction of W. L. Sibbitt.

AUXILIARY INFORMATION

METHOD/APPARATUS/PROCEDURE:

The apparatus of Zoss (1) modified to improve the temperature control at 32 °F, and to improve the sampling of the liquid and vapor phases.

Total Pressure:

| P/lb in$^{-2}$ | P/atm | P/kPa |
|---|---|---|
| 500 | 34.0 | 3450 |
| 1000 | 68.0 | 6895 |
| 1500 | 102.1 | 10340 |
| 2000 | 136.1 | 13790 |
| 3000 | 204.1 | 20680 |

SOURCE AND PURITY OF MATERIALS:

(1) Nitrogen. No information given.

(2) Water. Distilled. Boiled three hours prior to use.

ESTIMATED ERROR: Calculated by author:

| Solubility | Max Error |
|---|---|
| 0.8 | 16.3 % |
| 1.5 | 3.5 |
| 3.5 | 2.0 |

REFERENCES:

1. Zoss, L. M.
Ph. D. thesis, 1952
Purdue University

| COMPONENTS: | ORIGINAL MEASUREMENTS: |
|---|---|
| (1) Nitrogen; $N_2$; [7727-37-9]<br>(2) Water; $H_2O$; [7732-18-5] | Eichelberger, W. C.<br>*Ind. Eng. Chem.* 1955, *47*, 2223-8. |
| VARIABLES:<br>T/K = 338<br>P/MPa = 7.1-20.3 | PREPARED BY:<br>R. Battino |

EXPERIMENTAL VALUES:

| $P_{N_2}$/psig | $P_{N_2}$[a]/MPa | Solubility[b] |
|---|---|---|
| 1010 | 7.063 | 0.68 |
| 1370 | 9.545 | 0.91 |
| 1475 | 10.27 | 0.97 |
| 2230 | 15.47 | 1.39 |
| 2810 | 19.47 | 1.69 |
| 2930 | 20.30 | 1.76 |

[a] Calculated by compiler. Partial pressure of nitrogen (absolute).

[b] Solubility in units of ml $N_2$ (STP) per ml water at 65°C (338.2 K) Gas volumes not corrected for non-ideality of gas at pressure of measurement.

AUXILIARY INFORMATION

METHOD/APPARATUS/PROCEDURE:

A pressure vessel is charged up to pressure with gas and liquid and rocked for several hours. After equilibration a sample is removed into a gas buret where the liquid and gas volumes are measured at atmospheric pressure.

SOURCE AND PURITY OF MATERIALS:

(1) Tank.

(2) Distilled.

ESTIMATED ERROR:

$\delta T/K = \pm 0.1$
$\delta$Solubility/Solubility = ±0.03, compiler's estimate.

REFERENCES:

**COMPONENTS:**

(1) Nitrogen; $N_2$; [7727-37-9]

(2) Water; $H_2O$; [7732-18-5]

**ORIGINAL MEASUREMENTS:**

Smith, N. O.; Kelemen, S.; Nagy, B.

*Geochim. Cosmochim. Acta* 1962, *26*, 921-6.

**VARIABLES:**

T/K = 303
P/MPa = 1.1-5.9

**PREPARED BY:**

C. L. Young

**EXPERIMENTAL VALUES:**

| T/K | $P$/bar | $10^3$ Mole fraction of nitrogen in liquid, $10^3 x_{N_2}$ |
|---|---|---|
| 303.15 | 11.0 | 0.110 |
| | 31.0 | 0.335 |
| | 40.7 | 0.428 |
| | 50.5 | 0.552 |
| | 59.0 | 0.637 |

AUXILIARY INFORMATION

**METHOD/APPARATUS/PROCEDURE:**

Rocking equilibrium cell. Pressure measured with a Bourdon pressure gauge. Cell charged with boiled water; gas admitted to known pressure. Cell contents allowed to equilibrate. Final pressure measured and used to calculate amount of gas dissolved. Details in ref. 1.

**SOURCE AND PURITY OF MATERIALS:**

(1) Purity 99.6 mole per cent.

(2) Degassed.

**ESTIMATED ERROR:**

$\delta T/K = \pm 1.0$; $\delta P/\text{bar} = \pm 0.3$;

$\delta x_{N_2} = \pm 0.02 \times 10^{-4}$ (estimated by compiler).

**REFERENCES:**

1. Duffy, J. R.; Smith, N. O.; Nagy, B. *Geochim. Cosmochim. Acta* 1961, *24*, 23-31.

COMPONENTS:

(1) Nitrogen; $N_2$; [7727-37-9]

(2) Water; $H_2O$; [7732-18-5]

ORIGINAL MEASUREMENTS:

O'Sullivan, T. D.; Smith, N. O.

*Geochim. Cosmochim. Acta* 1966, *30*, 617-9.

VARIABLES:

T/K = 325
P/MPa = 10.1-60.8

PREPARED BY:

C. L. Young

EXPERIMENTAL VALUES:

| T/K | $P$/bar | $10^3$ Mole fraction of nitrogen in liquid, $10^3 x_{N_2}$ |
|---|---|---|
| 324.65 | 101.3 | 0.810 |
| | 202.6 | 1.47 |
| | 304.0 | 2.05 |
| | 405.3 | 2.54 |
| | 506.6 | 3.00 |
| | 607.9 | 3.44 |

AUXILIARY INFORMATION

METHOD /APPARATUS/PROCEDURE:

Large steel autoclave (∿ 4.5 ℓ) cell. Pressure measured with Bourdon gauge. Temperature measured with iron-constantan thermocouple. Cell charged with liquid, compressed gas added. After equilibrium attained, samples removed and analysed using volumetric techniques. Details in source.

SOURCE AND PURITY OF MATERIALS:

(1) Purity 99.90 mole per cent.

(2) Distilled and de-ionised.

ESTIMATED ERROR:

$\delta T/K = \pm 0.05$; $\delta P/\text{bar} = \pm 1$;

$\delta x_{N_2} = \pm 10^{-5}$ (estimated by compiler).

REFERENCES:

COMPONENTS:

(1) Nitrogen; $N_2$; [7727-37-9]

(2) Water; $H_2O$; [7732-18-5 ]

ORIGINAL MEASUREMENTS:

O'Sullivan, T. D.; Smith, N. O.

*J. Phys. Chem.* 1970, *74*, 1460-6.

(Correction in *J. Phys. Chem.* 1970, *74*, 4612.)

VARIABLES:

T/K = 324-398
P/MPa = 10.1-61.6

PREPARED BY:

C. L. Young

EXPERIMENTAL VALUES:

| T/K | $P$/bar | $10^3$ Mole fraction of nitrogen in liquid, $10^3 x_{N_2}$ |
|---|---|---|
| 324.65 | 101.3 | 0.799 |
| | 202.6 | 1.454 |
| | 304.0 | 2.017 |
| | 405.3 | 2.49 |
| | 506.6 | 2.92 |
| | 607.9 | 3.35 |
| 375.65 | 102.3 | 0.777 |
| | 203.7 | 1.447 |
| | 306.0 | 2.005 |
| | 408.3 | 2.52 |
| | 509.7 | 2.98 |
| | 612.0 | 3.37 |
| 398.15 | 104.4 | 0.808 |
| | 206.7 | 1.492 |
| | 309.0 | 2.047 |
| | 410.4 | 2.57 |
| | 513.7 | 3.06 |
| | 616.1 | 3.51 |

AUXILIARY INFORMATION

METHOD/APPARATUS/PROCEDURE:

Large steel autoclave (∿ 4.51 ℓ) cell. Pressure measured with Bourdon gauge. Temperature measured with iron-constantan thermocouple. Cell charged with liquid, compressed gas added. After equilibrium obtained, samples removed and analysed using volumetric techniques. Details in ref. 1.

SOURCE AND PURITY OF MATERIALS:

(1) Matheson Co. sample, purity 99.996 mole per cent.

(2) Distilled and de-ionised, air removed.

ESTIMATED ERROR:

$\delta T/K = \pm 0.5$; $\delta P/bar = \pm 0.05\%$;

$\delta x_{N_2} = \pm 0.4\%$.

REFERENCES:

1. O'Sullivan, T. D.; Smith, N.O. *Geochem. Cosmochim. Acta* 1966, *30*, 617-9.

COMPONENTS:

(1) Nitrogen; $N_2$; [7727-37-9]

(2) Magnesium Sulfate; $MgSO_4$; [7487-88-9]

(3) Water; $H_2O$; [7732-18-5]

ORIGINAL MEASUREMENTS:

Smith, N. O.; Kelemen, S.; Nagy, B.

*Geochim. Cosmochim. Acta* 1962, *26*, 921-6.

VARIABLES:

T/K = 303
P/MPa = 1.5-7.0
Concentration

PREPARED BY:

C. L. Young

EXPERIMENTAL VALUES:

| T/K | Conc./$10^3$ mol $m^{-3}$ | $P$/bar | $10^4$ Mole fraction of nitrogen in liquid, $10^4\ x_{N_2}$ |
|---|---|---|---|
| 303.15 | 1.25 | 15.17 | 0.65 |
| | | 26.54 | 1.34 |
| | | 40.89 | 2.23 |
| | | 48.95 | 3.11 |
| | | 63.09 | 3.51 |
| | 2.49 | 16.89 | 0.41 |
| | | 45.51 | 1.25 |
| | | 59.78 | 1.70 |
| | | 69.84 | 2.28 |

AUXILIARY INFORMATION

METHOD /APPARATUS/PROCEDURE:

Rocking equilibrium cell. Pressure measured with a Bourdon pressure gauge. Cell charged with salt solution; gas admitted to known pressure. Cell contents allowed to equilibrate. Final pressure measured and used to calculate amount of gas dissolved. Details in ref. 1.

SOURCE AND PURITY OF MATERIALS:

(1) Purity 99.6 mole per cent.

(2) Fisher Scientific Certified Chemical.

(3) Degassed.

ESTIMATED ERROR:

$\delta T/K = \pm 1.0$; $\delta P/\text{bar} \simeq \pm 0.3$; $\delta x_{N_2} = \pm 0.2 \times 10^{-4}$ (estimated by compiler).

REFERENCES:

1. Duffy, J. R.; Smith, N. O.; Nagy, B. *Geochim. Cosmochim. Acta* 1961, *24*, 23-31.

COMPONENTS:

(1) Nitrogen; $N_2$; [7727-37-9]

(2) Calcium Chloride; $CaCl_2$; [10043-53-4]

(3) Water; $H_2O$; [7732-18-5]

ORIGINAL MEASUREMENTS:

Smith, N. O.; Kelemen, S.; Nagy, B. *Geochim. Cosmochim. Acta* 1962, *26*, 921-6.

VARIABLES:

T/K = 303
P/MPa = 1.2-7.3
Concentration

PREPARED BY:

C. L. Young

EXPERIMENTAL VALUES:

| T/K | Conc./$10^3$ mol $m^{-3}$ | $P$/bar | $10^4$ Mole fraction of nitrogen in liquid phase, $10^4$ $x_{N_2}$ |
|---|---|---|---|
| 303.15 | 0.50 | 12.4 | 1.03 |
| | | 24.6 | 1.82 |
| | | 34.5 | 2.86 |
| | | 45.8 | 3.84 |
| | | 54.1 | 4.73 |
| | | 60.5 | 5.68 |
| | 1.75 | 14.8 | 0.56 |
| | | 26.9 | 1.46 |
| | | 38.3 | 2.13 |
| | | 48.3 | 2.44 |
| | | 55.8 | 3.10 |
| | | 64.7 | 3.62 |
| | 3.50 | 16.2 | 0.38 |
| | | 29.9 | 0.93 |
| | | 44.3 | 1.37 |
| | | 58.8 | 1.78 |
| | | 66.4 | 2.38 |
| | | 76.0 | 2.64 |
| | 5.60 | 26.2 | 0.32 |
| | | 34.1 | 0.52 |
| | | 54.0 | 0.90 |
| | | 72.6 | 1.83 |

AUXILIARY INFORMATION

METHOD/APPARATUS/PROCEDURE:

Rocking equilibrium cell. Pressure measured with a Bourdon pressure gauge. Cell charged with salt solution; gas admitted to known pressure. Cell contents allowed to equilibrate. Final pressure measured and used to calculate amount of gas dissolved. Details in ref. 1.

SOURCE AND PURITY OF MATERIALS:

(1) Purity 99.6 mole per cent.

(2) Fisher Scientific Certified Chemical.

(3) Degassed

ESTIMATED ERROR:

$\delta T/K = \pm 1.0$; $\delta P/\text{bar} \simeq \pm 0.3$; $\delta x_{N_2} = 0.2 \times 10^{-4}$ (estimated by compiler).

REFERENCES:

1. Duffy, J. R.; Smith, N. O.; Nagy, B. *Geochim. Cosmochim. Acta* 1961, *24*, 23-31.

**COMPONENTS:**

(1) Nitrogen; $N_2$; [7727-37-9]

(2) Sodium Sulfate; $Na_2SO_4$; [7757-82-6]

(3) Water; $H_2O$; [7732-18-5]

**ORIGINAL MEASUREMENTS:**

Smith, N. O.; Kelemen, S.; Nagy, B. *Geochim. Cosmochim. Acta* 1962, *26*, 921-6.

**VARIABLES:**

T/K = 303
P/MPa = 1.3-7.0
Concentration

**PREPARED BY:**

C. L. Young

**EXPERIMENTAL VALUES:**

| T/K | Conc./$10^3$ mol $m^{-3}$ | $P$/bar | $10^4$ Mole fraction of nitrogen in liquid, $10^4\ x_{N_2}$ |
|---|---|---|---|
| 303.15 | 0.50 | 12.6 | 0.79 |
| | | 24.8 | 1.78 |
| | | 42.1 | 3.24 |
| | | 54.5 | 3.88 |
| | 1.25 | 15.5 | 0.58 |
| | | 27.6 | 1.23 |
| | | 47.2 | 2.01 |
| | | 60.3 | 2.99 |
| | 2.10 | 16.5 | 0.38 |
| | | 30.7 | 0.84 |
| | | 45.5 | 1.39 |
| | | 57.4 | 2.13 |
| | | 69.8 | 3.18 |

**AUXILIARY INFORMATION**

**METHOD/APPARATUS/PROCEDURE:**

Rocking equilibrium cell. Pressure measured with a Bourdon pressure gauge. Cell charged with salt solution; gas admitted to known pressure. Cell contents allowed to equilibrate. Final pressure measured and used to calculate amount of gas dissolved. Details in ref. 1.

**SOURCE AND PURITY OF MATERIALS:**

(1) Purity 99.6 mole per cent.

(2) Fisher Scientific N.F. grade chemical.

(3) Degassed.

**ESTIMATED ERROR:**

$\delta T/K = \pm 1.0$; $\delta P/\text{bar} \simeq \pm 0.3$; $\delta x_{N_2} = \pm 0.2 \times 10^{-4}$ (estimated by compiler).

**REFERENCES:**


1. Duffy, J. R.; Smith, N. O.; Nagy, B. *Geochim. Cosmochim. Acta* 1961, *24*, 23-31.

**COMPONENTS:**

(1) Nitrogen; $N_2$; [7727-37-9]

(2) Sodium Chloride; NaCl; [7647-14-5]

(3) Water; $H_2O$; [7732-18-5]

**ORIGINAL MEASUREMENTS:**

Smith, N. O.; Kelemen, S.; Nagy, B.

*Geochim. Cosmochim. Acta* 1962, *26*, 921-6.

**VARIABLES:**

T/K = 303

P/MPa = 1.2-7.1

Concentration

**PREPARED BY:**

C. L. Young

**EXPERIMENTAL VALUES:**

| T/K | Conc./$10^3$ mol $m^{-3}$ | $P$/bar | $10^4$ Mole fraction of nitrogen in liquid phase, $10^4 x_{N_2}$ |
|---|---|---|---|
| 303.15 | 1.00 | 12.4 | 0.94 |
| | | 23.8 | 1.74 |
| | | 35.2 | 2.56 |
| | | 45.9 | 3.70 |
| | | 55.0 | 4.68 |
| | | 65.2 | 5.50 |
| 303.15 | 2.70 | 14.1 | 0.57 |
| | | 27.2 | 1.29 |
| | | 38.3 | 2.02 |
| | | 51.7 | 2.81 |
| | | 61.4 | 3.46 |
| | | 72.6 | 4.14 |
| 303.15 | 5.47 | 16.9 | 0.32 |
| | | 31.5 | 0.88 |
| | | 44.1 | 1.39 |
| | | 56.5 | 1.86 |
| | | 71.2 | 2.72 |

AUXILIARY INFORMATION

**METHOD/APPARATUS/PROCEDURE:**

Rocking equilibrium cell. Pressure measured with a Bourdon pressure gauge. Cell charged with salt solution; gas admitted to known pressure. Cell contents allowed to equilibrate. Final pressure measured and used to calculate amount of gas dissolved. Details in ref. 1.

**SOURCE AND PURITY OF MATERIALS:**

(1) Purity 99.6 mole per cent.

(2) Fisher Scientific Certified Chemical.

(3) Degassed.

**ESTIMATED ERROR:**

$\delta T/K = \pm 1.0$; $\delta P/\text{bar} \simeq \pm 0.3$; $\delta x_{N_2} = \pm 0.2 \times 10^{-4}$ (estimated by compiler).

**REFERENCES:**

1. Duffy, J. R.; Smith, N. O.; Nagy, B. *Geochim. Cosmochim. Acta* 1961, *24*, 23-31.

| COMPONENTS: | ORIGINAL MEASUREMENTS: |
|---|---|
| (1) Nitrogen; $N_2$; [7727-37-9]<br>(2) Sodium Chloride; NaCl; [7647-14-5]<br>(3) Water; $H_2O$; [7732-18-5] | O'Sullivan, T. D.; Smith, N. O.<br>*Geochim. Cosmochim. Acta* 1966, *30*, 617-9. |
| VARIABLES:<br>T/K - 325-376<br>P/MPa = 10.1-61.2<br>Concentration | PREPARED BY:<br>C. L. Young |

EXPERIMENTAL VALUES:

| T/K | Conc./$10^3$ mol $m^{-3}$ | $P$/bar | $10^4$ Mole fraction of nitrogen in liquid, $10^4\ x_{N_2}$ |
|---|---|---|---|
| 324.65 | 1.0 | 101.3 | 6.00 |
| | | 202.6 | 10.9 |
| | | 304.0 | 15.2 |
| | | 405.3 | 19.0 |
| | | 506.6 | 22.6 |
| | | 607.9 | 25.9 |
| 375.65 | 1.0 | 102.3 | 6.10 |
| | | 203.7 | 11.3 |
| | | 306.0 | 15.6 |
| | | 408.3 | 19.5 |
| | | 509.7 | 23.0 |
| | | 612.0 | 26.6 |

AUXILIARY INFORMATION

METHOD /APPARATUS/PROCEDURE:

Large steel autoclave (∿ 4.5 ℓ) cell. Pressure measured with Bourdon gauge. Temperature measured with iron-constantan thermocouple. Cell charged with liquid, compressed gas added. After equilibrium attained, samples removed and analysed using volumetric techniques. Details in source.

SOURCE AND PURITY OF MATERIALS:

(1) Purity 99.90 mole per cent.

(2) Baker analyzed reagent grade, dried at 388 K.

(3) Distilled and de-ionized.

ESTIMATED ERROR:

$\delta T/K = \pm 0.05$; $\delta P/\text{bar} = \pm 1$; $\delta x_{N_2} = \pm 10^{-5}$ (estimated by compiler).

REFERENCES:

**COMPONENTS:**

(1) Nitrogen; $N_2$; [7727-37-9]

(2) Sodium Chloride; NaCl; [7647-14-5]

(3) Water; $H_2O$; [7732-18-5]

**ORIGINAL MEASUREMENTS:**

O'Sullivan, T. D.; Smith, N. O.
*J. Phys. Chem.* 1970, *74*, 1460-6.
(Corrn.-*J. Phys. Chem.* 1970, *74*, 4612)

**VARIABLES:**

T/K = 325-398
P/MPa = 10.1-61.2
Concentration

**PREPARED BY:**

C. L. Young

**EXPERIMENTAL VALUES:**

| T/K | Conc./$10^3$ mol $m^{-3}$ | $P$/bar | $10^3$ Mole fraction of nitrogen in liquid, $10^3 x_{N_2}$ |
|---|---|---|---|
| 324.65 | 1.000 | 101.3 | 0.593 |
| | | 202.6 | 1.076 |
| | | 304.0 | 1.497 |
| | | 405.3 | 1.860 |
| | | 506.6 | 2.216 |
| | | 607.9 | 2.53 |
| 375.65 | | 101.3 | 0.603 |
| | | 203.7 | 1.113 |
| | | 306.0 | 1.538 |
| | | 408.3 | 1.920 |
| | | 509.7 | 2.252 |
| | | 612.0 | 2.60 |
| 398.15 | | 104.4 | 0.632 |
| | | 206.7 | 1.102 |
| | | 309.0 | 1.533 |
| | | 410.4 | 1.883 |
| | | 513.7 | 2.231 |
| | | 616.1 | 2.55 |
| 324.65 | 4.00 | 202.6 | 0.500 |
| | | 304.0 | 0.700 |
| | | 405.3 | 0.878 |
| | | 506.6 | 1.034 |
| | | 607.9 | 1.179 |

(cont.)

**AUXILIARY INFORMATION**

**METHOD/APPARATUS/PROCEDURE:**

Large steel autoclave (∿ 4.5 ℓ) cell. Pressure measured with Bourdon gauge. Temperature measured with iron-constantan thermocouple. Cell charged with liquid, compressed gas added. After equilibrium attained samples removed and analysed using volumetric techniques. Details in ref. 1.

**SOURCE AND PURITY OF MATERIALS:**

(1) Matheson Co. sample; purity 99.996 mole per cent.

(2) Baker analyzed reagent dried at 388 K.

(3) Distilled and de-ionized air removed.

**ESTIMATED ERROR:**

$\delta T/K = \pm 0.5$; $\delta P/\text{bar} = \pm 0.05$;
$\delta x_{N_2} = \pm 0.4\%$

**REFERENCES:**

1. O'Sullivan, T. D.; Smith, N. O. *Geochim. Cosmochim. Acta* 1966, *30*, 617-9.

COMPONENTS:

(1) Nitrogen; $N_2$; [7727-37-9]
(2) Sodium Chloride, NaCl; [7647-14-5]
(3) Water; $H_2O$; [7732-18-5]

ORIGINAL MEASUREMENTS:

O'Sullivan, T. D.; Smith, N. O.
*J. Phys. Chem.* 1970, *74*, 1460 - 6.
(*Corrn.-J. Phys. Chem.* 1970, *74*, 4612)

EXPERIMENTAL VALUES:

continued

| T/K | Conc./$10^3$ mol $m^{-3}$ | $P$/bar | $10^4$ Mole fraction of nitrogen in liquid, $10^4 x_{N_2}$ |
|---|---|---|---|
| 375.65 | 4.000 | 203.7 | 0.523 |
| | | 306.0 | 0.731 |
| | | 408.3 | 0.899 |
| | | 509.7 | 1.047 |
| | | 612.0 | 1.205 |
| 398.15 | | 206.7 | 0.567 |
| | | 309.0 | 0.740 |
| | | 410.4 | 0.921 |
| | | 513.7 | 1.041 |
| | | 616.1 | 1.227 |

**COMPONENTS:**

(1) Nitrogen; $N_2$; [7727-37-9]

(2) Detergent

(3) Water; $H_2O$; [7732-18-5]

**ORIGINAL MEASUREMENTS:**

Enns, T.; Scholander, P. F.; Bradstreet, E. D.

*J. Phys. Chem.* 1965, *69*, 389-91.

**VARIABLES:**

T/K = 298
P/MPa = 0-10

**PREPARED BY:**

R. Battino

**EXPERIMENTAL VALUES:**

| t/°C | $T^a$/K | $P^b$ = 0 (0 MPa) | $P^b$ = 34 atm (3.45MPa) | $P^b$ = 68 atm (6.89MPa) | $P^b$ = 102 atm (10.34MPa) |
|---|---|---|---|---|---|
| | | | Water | | |
| 25 | 298.15 | 733 | 773 | 802 | 843 |
| 25 | 298.15 | 705 | 744 | 777 | 811 |
| 25 | 298.15 | 732 | 769 | 806.5 | 844 |
| | | | Detergent | | |
| 25 | 298.15 | 712 | 742 | 774 | 811 |

[a] Calculated by compiler.

[b] Hydrostatic pressure.

[c] The values in the table are the nitrogen equilibrium pressures in units of mm Hg. The table shows the effect of hydrostatic pressure on the nitrogen equilibrium pressure.

**AUXILIARY INFORMATION**

**METHOD/APPARATUS/PROCEDURE:**

Water was gas-extracted before the experiment. Nitrogen at the desired initial pressure was dissolved in it at controlled temperature. The equilibrium vessel was shaken for 0.5 to 1 h during the nitrogen uptake. The solution was then transferred to a syringe which had teflon tubing attached to it. Pressure was then measured by a null point measurement of the pressure developed inside the teflon tube. Details and a diagram are given in the original paper.

**SOURCE AND PURITY OF MATERIALS:**

(1) No details given.

(2) Surface tension equal to 30% that of water.

(3) No details given.

**ESTIMATED ERROR:**

**REFERENCES:**

**COMPONENTS:**

(1) Nitrogen; $N_2$; [7727-37-9]

(2) Oxygen; $O_2$; [7782-44-7]

(3) Cyclohexane; $C_6H_{12}$; [110-82-7]

**ORIGINAL MEASUREMENTS:**

Khodeeva, S. M.

*Tr. Naucho-Issledov. Proekt. Inst. Azotn. Prod. Prom. Org. Sin.* 1971, *12*, 30-39.

**VARIABLES:**

T/K = 373-433

P/MPa = 4-11

**PREPARED BY:**

C. L. Young

**EXPERIMENTAL VALUES:**

| T/K | *P*/atm | *P*/MPa | α cm³/g⁻¹ | Total Mole fraction of gases in liquid | Volume fraction of $O_2$ in gas phase | Volume fraction of $O_2$ in dissolved gas |
|---|---|---|---|---|---|---|
| 373.15 | 41.3 | 4.18 | 12.1 | 0.043 | 0.082 | 0.118 |
| | 44.6 | 4.52 | 11.9 | 0.043 | 0.082 | - |
| | 47.5 | 4.81 | 13.3 | 0.048 | 0.082 | - |
| | 54.2 | 5.49 | 14.8 | 0.053 | 0.082 | 0.088 |
| | 55.0 | 5.57 | 15.3 | 0.054 | 0.082 | - |
| | 58.1 | 5.89 | 15.9 | 0.057 | 0.082 | 0.092 |
| | 59.8 | 6.06 | 16.8 | 0.060 | 0.082 | 0.113 |
| | 65.9 | 6.68 | 18.9 | 0.066 | 0.082 | 0.111 |
| | 67.3 | 6.82 | 19.4 | 0.068 | 0.082 | - |
| | 72.4 | 7.34 | 20.3 | 0.071 | 0.082 | 0.111 |
| | 73.5 | 7.45 | 21.4 | 0.075 | 0.082 | 0.125 |
| | 78.7 | 7.97 | 22.5 | 0.078 | 0.082 | 0.135 |
| | 82.8 | 8.39 | 24.2 | 0.083 | 0.082 | 0.113 |
| | 91.3 | 9.25 | 27.0 | 0.092 | 0.082 | 0.127 |
| 403.15 | 42.1 | 4.27 | 11.8 | 0.042 | 0.082 | - |
| | 46.8 | 4.74 | 13.5 | 0.048 | 0.082 | 0.118 |
| | 51.8 | 5.25 | 15.0 | 0.053 | 0.082 | 0.113 |
| | 57.1 | 5.79 | 16.2 | 0.058 | 0.082 | 0.116 |
| | 61.3 | 6.21 | 17.9 | 0.063 | 0.082 | 0.121 |
| | 65.6 | 6.65 | 19.3 | 0.068 | 0.082 | 0.127 |
| | 89.0 | 9.02 | 27.3 | 0.093 | 0.082 | 0.130 |
| | 96.8 | 9.81 | 30.7 | 0.103 | 0.082 | 0.115 |
| | 104.0 | 10.54 | 33.2 | 0.111 | 0.082 | 0.123 |

continued on following page

**AUXILIARY INFORMATION**

**METHOD/APPARATUS/PROCEDURE:**

Volumetric method in which saturated liquid sample was stripped of gas under reduced pressure. Amount of liquid in sample determined gravimetrically. Amount of dissolved gas determined by measuring the volume at known temperature and pressure. Some details in source and ref. (1).

**SOURCE AND PURITY OF MATERIALS:**

(1) Specially purified sample, less than 0.005 volume per cent impurity.

(2) No details given.

(3) 0.5% of α-naphthylamine added.

**ESTIMATED ERROR:**

$\delta T/K = \pm 0.5$; $\delta P/atm = \pm 0.2$; $\delta\alpha/\alpha = \pm 1\%$.

**REFERENCES:**


1. Khodeeva, S. M.; Dymova, R. P. *Tr. Naucho-Issledov. Proekt. Inst. Azotn. Prom. Prod. Org. Sin.* 1971, *12*, 39.

| COMPONENTS: | ORIGINAL MEASUREMENTS: |
|---|---|
| (1) Nitrogen; $N_2$; [7727-37-9]<br>(2) Oxygen; $O_2$; [7782-44-7]<br>(3) Cyclohexane; $C_6H_{12}$; [110-82-7] | Khodeeva, S. M.<br>*Tr. Naucho-Issledov. Proekt. Inst. Azotn. Prod. Prom. Org. Sin.* 1971, *12*, 30-39. |

EXPERIMENTAL VALUES:

continued

| T/K | *P*/atm | *P*/MPa | $\alpha$ $cm^3/g^{-1}$ | Total Mole fraction of gases in liquid | Volume fraction of $O_2$ in gas phase | Volume fraction of $O_2$ in dissolved gas |
|---|---|---|---|---|---|---|
| 433.15 | 38.7 | 3.92 | 12.2 | 0.044 | 0.082 | 0.122 |
| | 45.5 | 4.61 | 14.2 | 0.050 | 0.082 | - |
| | 49.4 | 5.01 | 16.0 | 0.056 | 0.082 | 0.117 |
| | 51.8 | 5.25 | 17.1 | 0.060 | 0.082 | 0.119 |
| | 60.7 | 6.15 | 20.4 | 0.071 | 0.082 | 0.122 |
| | 68.0 | 6.89 | 22.5 | 0.078 | 0.082 | 0.121 |
| | 76.3 | 7.73 | 25.5 | 0.088 | 0.082 | 0.120 |
| | 82.3 | 8.34 | 28.2 | 0.096 | 0.082 | 0.123 |
| | 87.6 | 8.88 | 30.3 | 0.102 | 0.082 | 0.125 |
| 373.15 | 45.5 | 4.61 | 12.6 | 0.045 | 0.200 | - |
| | 43.5 | 4.41 | 12.4 | 0.045 | 0.200 | - |
| | 46.5 | 4.71 | 13.5 | 0.048 | 0.200 | - |
| | 54.3 | 5.50 | 16.2 | 0.057 | 0.200 | 0.259 |
| | 69.2 | 7.01 | 21.2 | 0.074 | 0.200 | 0.248 |
| | 74.6 | 7.56 | 23.2 | 0.080 | 0.200 | 0.265 |
| | 80.0 | 8.11 | 25.3 | 0.087 | 0.200 | 0.264 |
| | 87.1 | 8.83 | 27.3 | 0.093 | 0.200 | 0.256 |
| | 94.6 | 9.59 | 29.1 | 0.099 | 0.200 | 0.266 |
| 403.15 | 44.6 | 4.52 | 13.0 | 0.046 | 0.200 | 0.266 |
| | 48.4 | 4.90 | 14.7 | 0.052 | 0.200 | 0.269 |
| | 54.7 | 5.54 | 16.6 | 0.059 | 0.200 | 0.257 |
| | 61.5 | 6.23 | 18.9 | 0.067 | 0.200 | 0.254 |
| | 63.4 | 6.42 | 20.2 | 0.071 | 0.200 | 0.273 |
| | 66.8 | 6.77 | 21.2 | 0.074 | 0.200 | - |
| | 73.1 | 7.41 | 23.3 | 0.081 | 0.200 | 0.252 |
| | 79.7 | 8.08 | 25.3 | 0.087 | 0.200 | 0.249 |
| | 86.9 | 8.81 | 28.1 | 0.096 | 0.200 | 0.251 |
| | 93.9 | 9.51 | 30.9 | 0.105 | 0.200 | 0.249 |
| 433.15 | 50.3 | 5.10 | 16.4 | 0.058 | 0.200 | 0.246 |
| | 55.9 | 5.66 | 18.8 | 0.066 | 0.200 | - |
| | 62.3 | 6.31 | 20.9 | 0.073 | 0.200 | 0.248 |
| | 68.2 | 6.91 | 22.9 | 0.079 | 0.200 | 0.236 |
| | 73.3 | 7.43 | 25.0 | 0.086 | 0.200 | 0.255 |
| | 82.8 | 8.39 | 27.5 | 0.093 | 0.200 | - |
| | 88.6 | 8.98 | 30.6 | 0.103 | 0.200 | 0.245 |
| | 95.8 | 9.71 | 33.5 | 0.112 | 0.200 | 0.248 |
| | 102.3 | 10.37 | 35.3 | 0.117 | 0.200 | 0.246 |

COMPONENTS:

(1) Nitrogen; $N_2$; [7727-37-9]

(2) Methylcyclohexane; $C_7H_{14}$; [108-87-2]

(3) Heptane; $C_7H_{16}$; [142-82-5]

ORIGINAL MEASUREMENTS:

Brunner, G; Peter, S.; Wenzel, H.

*Chem. Eng. J.* 1974, 7, 99.

VARIABLES:

T/K = 453-492
P/MPa = 17-33
Concentration

PREPARED BY:

C. L. Young

EXPERIMENTAL VALUES:

| T/K | P/bar | Mole fraction in liquid phase $x_{N_2}$ | $x_{C_7H_{14}}$ | $x_{C_7H_{16}}$ | Mole fraction in gas phase $y_{N_2}$ | $y_{C_7H_{14}}$ | $y_{C_7H_{16}}$ |
|---|---|---|---|---|---|---|---|
| 453.15 | 235 | 0.371 | 0.160 | 0.469 | 0.882 | 0.028 | 0.090 |
| | | 0.336 | 0.329 | 0.335 | 0.885 | 0.054 | 0.061 |
| | | 0.307 | 0.503 | 0.190 | 0.901 | 0.070 | 0.029 |
| | 275 | 0.454 | 0.138 | 0.408 | 0.869 | 0.031 | 0.100 |
| | | 0.404 | 0.296 | 0.300 | 0.869 | 0.062 | 0.069 |
| | | 0.363 | 0.466 | 0.171 | 0.896 | 0.073 | 0.031 |
| | 329 | 0.747 | 0.063 | 0.190 | 0.757 | 0.061 | 0.182 |
| | | 0.539 | 0.230 | 0.231 | 0.862 | 0.066 | 0.072 |
| | | 0.469 | 0.391 | 0.140 | 0.881 | 0.086 | 0.033 |
| 472.15 | 168 | 0.306 | 0.177 | 0.517 | 0.829 | 0.041 | 0.130 |
| | | 0.278 | 0.359 | 0.363 | 0.844 | 0.073 | 0.083 |
| | | 0.249 | 0.553 | 0.198 | 0.863 | 0.098 | 0.039 |
| | 201 | 0.374 | 0.159 | 0.467 | 0.824 | 0.042 | 0.134 |
| | | 0.336 | 0.334 | 0.330 | 0.846 | 0.073 | 0.081 |
| | | 0.297 | 0.514 | 0.189 | 0.867 | 0.095 | 0.038 |
| | 246 | 0.558 | 0.112 | 0.330 | 0.817 | 0.044 | 0.139 |
| | | 0.441 | 0.282 | 0.277 | 0.839 | 0.077 | 0.083 |
| | | 0.380 | 0.452 | 0.168 | 0.860 | 0.099 | 0.041 |

continued on following page

AUXILIARY INFORMATION

METHOD/APPARATUS/PROCEDURE:

Static cell of about 1000 $cm^3$ capacity fitted with magnetic stirrer. Pressure measured with Bourdon gauge. Temperature measured with nickel-chromium alloy nickel thermocouple. Each phase sampled, heptane and methylcyclohexane frozen out; total amount estimated gravimetrically and composition of frozen portion determined by gas chromatography; nitrogen estimated volumetrically.

SOURCE AND PURITY OF MATERIALS:

(1) Linde sample; purity 99.98 mole per cent.

(2) Fluka sample; purity 99.4 mole per cent.

(3) Fluka sample; purity 99.7 mole per cent.

ESTIMATED ERROR:

$\delta T/K = \pm 0.5$; $\delta P/bar = \pm 0.5$; $\delta x_{N_2}$, $\delta x_{C_7H_{14}}$, $\delta x_{C_7H_{16}}$, $\delta y_{N_2}$, $\delta y_{C_7H_{14}}$, $\delta y_{C_7H_{16}} = \pm 0.006$.

REFERENCES:

COMPONENTS:

(1) Nitrogen; $N_2$; [7727-37-9]

(2) Methylcyclohexane; $C_7H_{14}$; [108-87-2]

(3) Heptane; $C_7H_{16}$; [142-82-5]

ORIGINAL MEASUREMENTS:

Brunner, G.; Peter. S.; Wenzel, H.

*Chem. Eng. J.* 1974, *7*, 99.

EXPERIMENTAL VALUES:

continued

| T/K | P/bar | Mole fraction in liquid phase $x_{N_2}$ | $x_{C_7H_{14}}$ | $x_{C_7H_{16}}$ | Mole fraction in gas phase $y_{N_2}$ | $y_{C_7H_{14}}$ | $y_{C_7H_{16}}$ |
|---|---|---|---|---|---|---|---|
| 492.15 | 129 | 0.279 | 0.183 | 0.538 | 0.747 | 0.060 | 0.193 |
| | | 0.245 | 0.376 | 0.379 | 0.760 | 0.113 | 0.127 |
| | | 0.221 | 0.575 | 0.204 | 0.771 | 0.164 | 0.065 |
| | 158 | 0.384 | 0.156 | 0.460 | 0.743 | 0.061 | 0.196 |
| | | 0.320 | 0.339 | 0.341 | 0.775 | 0.107 | 0.118 |
| | | 0.281 | 0.531 | 0.188 | 0.797 | 0.145 | 0.058 |
| | 198 | 0.559 | 0.111 | 0.330 | 0.569 | 0.108 | 0.323 |
| | | 0.484 | 0.255 | 0.261 | 0.749 | 0.121 | 0.130 |
| | | 0.385 | 0.452 | 0.163 | 0.799 | 0.145 | 0.056 |

**COMPONENTS:**

(1) Nitrogen; $N_2$; [7727-37-9]
(2) Cyclohexanone; $C_6H_{10}O$; [108-94-1]
(3) Cyclohexanol; $C_6H_{12}O$; [108-93-0]
(4) Acetic acid; $C_2H_4O_2$; [64-19-7]

**ORIGINAL MEASUREMENTS:**

Khodeeva, S.M.; Dymova, R.P.

*Tr. Nauch-Issled. Proekt. Inst. Azotn. Prom. Prod. Org. Sin.* 1971, *12*, 18-30.

**VARIABLES:**

T/K = 373-448
P/MPa = 3-12
Concentration

**PREPARED BY:**

C.L. Young

**EXPERIMENTAL VALUES:**

| T/K | Wt. fractions before absorption. $C_6H_{12}O$ | $C_6H_{10}O$ | $CH_3COOH$ | P/atm | P/MPa | $\alpha^+$/cm g$^{-3}$ | Mole fraction of nitrogen § $x_{N_2}$ |
|---|---|---|---|---|---|---|---|
| 373.15 | 0.10 | 0.225 | 0.675 | 35.4 | 3.59 | 4.7 | 0.014 |
| | | | | 50.9 | 5.16 | 6.4 | 0.020 |
| | | | | 51.8 | 5.25 | 6.8 | 0.021 |
| | | | | 55.9 | 5.66 | 7.2 | 0.022 |
| | | | | 67.1 | 6.80 | 8.6 | 0.026 |
| | | | | 84.7 | 8.58 | 10.8 | 0.033 |
| | | | | 88.1 | 8.93 | 11.1 | 0.034 |
| | | | | 91.9 | 9.31 | 12.1 | 0.037 |
| 423.15 | | | | 32.9 | 3.33 | 5.1 | 0.016 |
| | | | | 37.8 | 3.83 | 5.8 | 0.018 |
| | | | | 41.2 | 4.17 | 6.3 | 0.019 |
| | | | | 53.3 | 5.40 | 7.9 | 0.024 |
| | | | | 58.2 | 5.90 | 8.9 | 0.027 |
| | | | | 61.5 | 6.23 | 9.3 | 0.029 |
| | | | | 107.7 | 10.91 | 15.9 | 0.049 |
| | | | | 113.5 | 11.50 | 16.7 | 0.051 |
| | | | | 117.2 | 11.88 | 18.9 | 0.058 |
| 373.15 | 0.20 | 0.52 | 0.28 | 31.0 | 3.14 | 3.8 | 0.014 |
| | | | | 36.0 | 3.65 | 4.3 | 0.016 |
| | | | | 38.2 | 3.87 | 4.6 | 0.017 |
| | | | | 41.3 | 4.18 | 4.7 | 0.018 |
| | | | | 42.0 | 4.26 | 5.1 | 0.019 |
| | | | | 45.3 | 4.59 | 5.5 | 0.025 |
| | | | | 54.7 | 5.54 | 6.4 | 0.024 |

AUXILIARY INFORMATION

**METHOD/APPARATUS/PROCEDURE:**

Volumetric method in which saturated liquid sample was stripped of gas under reduced pressure. Amount of liquid in sample determined by weighing and composition determined by gas chromatography. Amount of gas dissolved determined by measuring the volume at a known temperature and pressure. Some details in source and ref. (1).

**SOURCE AND PURITY OF MATERIALS:**

(1) Specially purified sample, less than 0.005 volume per cent impurity

(2) Distilled no impurity detected by GC.

(3) Twice distilled dried over freshly fused sodium sulfate.

(4) Chemically pure.

**ESTIMATED ERROR:**

$\delta T/K = \pm 0.5$; $\delta P/atm = \pm 0.2$;
$\delta\alpha/\alpha = \pm 0.01$.

**REFERENCES:**

1. Khodeeva, S.M.; Dymova, R.P.
*Tr. Naucho. Issledov. Proekt. Inst. Azotn. Prom. Prod. Org. Sin.* 1971, *12*, 39.

COMPONENTS:

(1) Nitrogen; $N_2$; [7727-37-9]
(2) Cyclohexanone; $C_6H_{10}O$; [108-94-1]
(3) Cyclohexanol; $C_6H_{12}O$; [108-93-0]
(4) Acetic acid; $C_2H_4O_2$; [64-19-7]

ORIGINAL MEASUREMENTS:

Khodeeva, S.M.; Dymova, R.P.

*Tr. Nauch.-Issled. Proekt. Inst. Azotn. Prom. Prod. Org. Sin.* 1971, 12, 18-30.

EXPERIMENTAL VALUES:

| T/K | Wt.fractions before absorption | | | $P$/atm | $P$/MPa | $\alpha^+$/cm g$^{-3}$ | Mole fraction of nitrogen § $x_{N_2}$ |
|---|---|---|---|---|---|---|---|
| | $C_6H_{12}O$ | $C_6H_{10}O$ | $CH_3COOH$ | | | | |
| 373.15 | 0.20 | 0.52 | 0.28 | 58.6 | 5.94 | 7.3 | 0.027 |
| | | | | 65.5 | 6.64 | 7.7 | 0.029 |
| | | | | 68.4 | 6.93 | 8.5 | 0.032 |
| | | | | 70.7 | 7.16 | 8.6 | 0.032 |
| 423.15 | | | | 31.6 | 3.20 | 4.7 | 0.018 |
| | | | | 34.6 | 3.51 | 5.3 | 0.020 |
| | | | | 52.2 | 5.29 | 7.2 | 0.027 |
| | | | | 54.5 | 5.52 | 7.5 | 0.028 |
| | | | | 66.6 | 6.75 | 9.5 | 0.035 |
| | | | | 70.4 | 7.13 | 9.9 | 0.037 |
| | | | | 78.3 | 7.93 | 10.7 | 0.040 |
| 448.15 | | | | 36.5 | 3.70 | 5.7 | 0.021 |
| | | | | 38.8 | 3.93 | 6.1 | 0.023 |
| | | | | 51.7 | 5.24 | 7.9 | 0.029 |
| | | | | 53.6 | 5.43 | 8.3 | 0.031 |
| 373.15 | 0.10 | 0.585 | 0.315 | 30.1 | 3.05 | 3.8 | 0.014 |
| | | | | 45.0 | 4.56 | 5.8 | 0.021 |
| | | | | 48.3 | 4.89 | 6.0 | 0.022 |
| | | | | 58.5 | 5.93 | 7.2 | 0.026 |
| | | | | 64.3 | 6.52 | 8.1 | 0.030 |
| | | | | 67.2 | 6.81 | 8.2 | 0.030 |
| | | | | 72.2 | 7.32 | 9.0 | 0.033 |
| | | | | 74.3 | 7.53 | 9.0 | 0.033 |
| | | | | 80.0 | 8.11 | 10.1 | 0.037 |
| 423.15 | | | | 26.9 | 2.73 | 3.8 | 0.014 |
| | | | | 29.7 | 3.01 | 4.4 | 0.016 |
| | | | | 41.2 | 4.17 | 5.8 | 0.021 |
| | | | | 44.6 | 4.52 | 6.6 | 0.024 |
| | | | | 53.5 | 5.42 | 7.6 | 0.028 |
| | | | | 59.0 | 5.98 | 8.4 | 0.031 |
| | | | | 62.5 | 6.33 | 9.6 | 0.035 |
| | | | | 66.7 | 6.76 | 9.9 | 0.036 |
| | | | | 80.2 | 8.13 | 12.2 | 0.045 |
| | | | | 84.5 | 8.56 | 12.5 | 0.046 |
| 448.15 | | | | 31.9 | 3.23 | 5.3 | 0.019 |
| | | | | 40.9 | 4.14 | 7.3 | 0.027 |
| | | | | 47.8 | 4.84 | 8.1 | 0.030 |
| | | | | 50.0 | 5.07 | 8.5 | 0.031 |
| | | | | 55.7 | 5.64 | 9.6 | 0.035 |
| | | | | 59.5 | 6.03 | 10.1 | 0.037 |
| | | | | 66.7 | 6.76 | 11.2 | 0.041 |
| | | | | 70.3 | 7.12 | 11.0 | 0.040 |
| | | | | 80.0 | 8.11 | 13.5 | 0.049 |
| | | | | 84.5 | 8.56 | 14.5 | 0.053 |

+$\alpha$ = Volume of gas, measured at 273.15 K and 101.3 kPa pressure dissolved by 1g of solvent.

§ calculated by compiler.

COMPONENTS:

(1) Nitrogen; $N_2$; [7727-37-9]

(2) Acetic Acid; $C_2H_4O_2$; [64-19-7]

(3) Cyclohexanone; $C_6H_{10}O$; [108-94-1]

ORIGINAL MEASUREMENTS:

Khodeeva, S.M.

*Tr. Naucho-Issledov.Proekt.Inst. Azotn. Prom. Prod. Org. Sin.* 1971, *12*, 18-30.

VARIABLES:

T/K = 348-473
P/MPa = 2-9
Concentration

PREPARED BY:

C.L. Young

EXPERIMENTAL VALUES:

| T/K | Wt.per cent of cyclohexanone before absorption | $P$/atm | $P$/MPa | $\alpha^+$ /cm$^3$g$^{-1}$ | Mole fraction of nitrogen § $x_{N_2}$ |
|---|---|---|---|---|---|
| 348.15 | 25 | 31.0 | 3.14 | 3.5 | 0.010 |
| | | 35.3 | 3.58 | 3.9 | 0.012 |
| | | 38.0 | 3.85 | 4.4 | 0.013 |
| | | 43.6 | 4.42 | 4.8 | 0.014 |
| | | 55.4 | 5.61 | 6.7 | 0.020 |
| | | 59.5 | 6.03 | 6.3 | 0.019 |
| | | 67.3 | 6.82 | 7.8 | 0.023 |
| | | 69.4 | 7.03 | 8.1 | 0.024 |
| | | 73.6 | 7.46 | 8.3 | 0.025 |
| 373.15 | 25 | 37.7 | 3.82 | 4.3 | 0.013 |
| | | 40.6 | 4.11 | 4.6 | 0.014 |
| | | 43.4 | 4.40 | 5.1 | 0.015 |
| | | 47.0 | 4.76 | 5.4 | 0.016 |
| | | 49.1 | 4.98 | 5.7 | 0.017 |
| | | 51.5 | 5.22 | 6.6 | 0.020 |
| | | 53.0 | 5.37 | 6.4 | 0.019 |
| | | 63.3 | 6.41 | 7.5 | 0.022 |
| | | 66.9 | 6.78 | 8.4 | 0.025 |
| | | 76.8 | 7.78 | 9.6 | 0.029 |
| | | 82.2 | 8.33 | 10.2 | 0.030 |
| 423.15 | | 20.4 | 2.07 | 3.3 | 0.010 |
| | | 27.8 | 2.82 | 4.2 | 0.012 |
| | | 30.4 | 3.08 | 4.9 | 0.015 |
| | | 33.2 | 3.36 | 5.2 | 0.015 |
| | | 37.5 | 3.80 | 5.6 | 0.017 |
| | | 37.8 | 3.83 | 5.9 | 0.018 |

AUXILIARY INFORMATION

METHOD/APPARATUS/PROCEDURE:

Volumetric method in which saturated liquid sample was stripped of gas under reduced pressure. Amount of liquid in sample determined by weighing and composition determined by gas chromatography. Amount of dissolved gas determined by measuring the volume at a known temperature and pressure. Some details in source and ref. (1).

SOURCE AND PURITY OF MATERIALS:

(1) Specially purified sample, less than 0.005 volume per cent impurity.

(2) Chemically "pure"

(3) Distilled,no impurity detected by GC.

ESTIMATED ERROR:

$\delta T/K = \pm 0.5$; $\delta P/atm = \pm 0.2$; $\delta\alpha/\alpha = \pm 0.01$.

REFERENCES:


1. Khodeeva, S.M.; Dymova, R.P. *Tr. Naucho-Issledov. Proekt. Azotn.Prom.Prod.Org.Sin.* 1971, *12*, 39.

| COMPONENTS: | ORIGINAL MEASUREMENTS: |
|---|---|
| (1) Nitrogen; $N_2$; [7727-37-9] | Khodeeva, S.M. |
| (2) Acetic Acid; $C_2H_4O_2$; [64-19-7] | *Tr.Naucho-Issledov.Proekt.Inst. Azotn. Prom. Prod. Org. Sin.* 1971, *12*, 18-30. |
| (3) Cyclohexanone; $C_6H_{10}O$; [108-94-1] | |

EXPERIMENTAL VALUES:

| T/K | Wt.per cent of cyclohexanone before absorption | $P$/atm | $P$/MPa | $\alpha^+$ /$cm^3 g^{-1}$ | Mole fraction of nitrogen § $x_{N_2}$ |
|---|---|---|---|---|---|
| 423.15 | | 41.2 | 4.17 | 6.5 | 0.019 |
| | | 42.6 | 4.32 | 6.5 | 0.019 |
| | | 44.4 | 4.50 | 6.1 | 0.018 |
| | | 48.4 | 4.90 | 6.5 | 0.019 |
| | | 51.8 | 5.25 | 7.2 | 0.021 |
| | | 56.6 | 5.73 | 7.7 | 0.023 |
| | | 60.2 | 6.10 | 8.7 | 0.026 |
| | | 65.5 | 6.64 | 9.2 | 0.027 |
| | | 67.7 | 6.86 | 9.9 | 0.029 |
| | | 75.1 | 7.61 | 11.2 | 0.033 |
| | | 79.2 | 8.02 | 12.0 | 0.036 |
| 473.15 | | 29.5 | 2.99 | 4.6 | 0.014 |
| | | 31.2 | 3.16 | 5.2 | 0.015 |
| | | 52.2 | 5.29 | 7.9 | 0.023 |
| | | 55.1 | 5.58 | 8.6 | 0.026 |
| | | 58.3 | 5.91 | 9.5 | 0.028 |
| | | 69.9 | 7.08 | 12.0 | 0.036 |
| | | 74.6 | 7.56 | 12.9 | 0.038 |
| | | 79.0 | 8.00 | 14.1 | 0.042 |
| | | 86.7 | 8.78 | 14.9 | 0.044 |
| 373.15 | 64 | 28.8 | 2.92 | 3.7 | 0.013 |
| | | 40.7 | 4.12 | 5.1 | 0.018 |
| | | 44.2 | 4.48 | 5.3 | 0.019 |
| | | 48.0 | 4.86 | 6.0 | 0.021 |
| | | 56.3 | 5.70 | 7.4 | 0.026 |
| | | 63.4 | 6.42 | 8.0 | 0.029 |
| | | 67.4 | 6.83 | 8.6 | 0.031 |
| | | 76.6 | 7.76 | 9.8 | 0.035 |
| | | 81.5 | 8.26 | 10.4 | 0.037 |
| 423.15 | | 24.3 | 2.46 | 3.6 | 0.013 |
| | | 40.9 | 4.14 | 5.4 | 0.019 |
| | | 43.2 | 4.38 | 6.0 | 0.021 |
| | | 51.9 | 5.26 | 7.4 | 0.026 |
| | | 71.0 | 7.19 | 9.7 | 0.035 |
| | | 74.1 | 7.51 | 10.1 | 0.036 |
| | | 76.4 | 7.74 | 10.6 | 0.038 |

$^+\alpha$ = Volume of gas, measured at 273.15 K and 101.3 kPa pressure dissolved by 1g of solvent.

§ calculated by compiler.

| COMPONENTS: | EVALUATOR: |
|---|---|
| (1) Nitrogen; $N_2$; [7727-37-9]<br>(2) Methane; $CH_4$; [74-82-8] | Colin L. Young,<br>School of Chemistry,<br>University of Melbourne,<br>Parkville, Victoria 3052,<br>Australia.<br>June 1981 |

CRITICAL EVALUATION:

This system has been extensively studied but since the critical temperatures of the two components are relatively similar, much of the data fall more naturally into the category of vapor-liquid equilibrium data rather than gas-liquid solubility data.

The data given in the three earliest investigations (1), (2), (3), were of limited accuracy and are rejected. The data of Vellinger and Pons (4) are also rejected because the results were limited to 190 K and were only presented in graphical form. The data of Brandt and Stroud (5) are rejected since they are of fairly low accuracy by recent standards and furthermore the data were only included in the original paper for comparison purposes and are not extensive. Sprow and Prausnitz's data (6) were limited to 97.6 K and will not be considered further.

The data of Miller *et al.* (7) and of Parrish and Hiza (8) are recommended for the temperature range 90 K to 120 K. The data of Cheung and Wang (9) cover a similar temperature and are classified as tentative since the pressure range is limited and the accuracy less than that of Miller *et al.* (7) and of Parrish and Hiza (8). The data of Stryjek *et al.* (10) and Kidnay *et al.* (11) are classified as recommended for the temperature range 120 K to 183 K. The data of Bloomer and Parent (12), (13) cover the temperature range of approximately 95 to 185 K but are less accurate than the recommended values but still of reasonable accuracy. The investigation was made using a dew point-buttle point apparatus and the results are not the most suitable for solubility evaluation. These data are therefore classified as tentative. The high pressure results of Cines *et al.* (14) show considerable scatter and differ somewhat from those of Bloomer and Parent (12), (13) and Stryjek *et al.* (10), hence are classified as doubtful. The data of Chang and Lu (15) are in fair agreement with the results of Stryjek *et al.* (10) and Bloomer and Parent (12), (13) and the results at -150°F and -240°F (122 K and 172 K) are classified as tentative.

The data of Fastovsky and Petrovsky (16) are rejected. Their work is mainly concerned with vapor-liquid equilibrium where the mole fraction of the more volatile component (nitrogen) is greater than 0.2. The limited comparison which can be made with the data of Stryjek *et al.* (10) shows that the two sets of data are consistent but the data in reference (16) are less precise.

Similarly the data of Skripka *et al.* (17) are rejected since they are limited in scope consisting mainly of data for which the mole fraction of nitrogen in the liquid phase is greater than 0.2.

References


1. McTaggart, H. A.; Edwards, E. *Trans. Roy. Soc., Canada III* 1919, *13*, 57.

2. Rysakov, M. W.; Teodarovich, V. P.; Kozyreva, O. V. *Bull. GIVD* 1934 *(2)*, 12.

3. Torochesnikova, N. S.; Levius, L. A. *Zh. Khim. Prom.* 1939, *16*, 19.

4. Vellinger, E.; Pons, E. *Comp. Rend.* 1943, *217*, 689.


continued on following page

| COMPONENTS: | EVALUATOR: |
| --- | --- |
| (1) Nitrogen; $N_2$; [7727-37-9]<br><br>(2) Methane; $CH_4$; [74-82-8] | Colin L. Young,<br>School of Chemistry,<br>University of Melbourne,<br>Parkville, Victoria 3052,<br>Australia.<br><br>June 1981 |

continued

References (cont.)


5. Brandt, L. W.; Stroud, L. *Ind. Eng. Chem.* 1958, *50*, 849.

6. Sprow, F. B.; Prausnitz, J. M. *Am. Inst. Chem. Engng. J.* 1966, *12*, 780.

7. Miller, R. C.; Kidnay, A. J.; Hiza, M. J. *Am. Inst. Chem. Engng. J.* 1973, *19*, 145.

8. Parrish, W. R.; Hiza, M. J. *Adv. Cryog. Eng.* 1973, *19*, 300.

9. Cheung, H.; Wang, D. I. J. *Ind. Eng. Chem. Fundam.* 1964, *3*, 355.

10. Stryjek, R.; Chappelear, P. S.; Kobayashi, R. *J. Chem. Eng. Data* 1974, *19*, 334.

11. Kidnay, A. J.; Miller, R. C.; Parrish, W. R.; Hiza, M. J. *Cryogenics* 1975, *15*, 531.

12. Bloomer, O. T.; Parent, J. D. *Inst. Gas Technol., Res. Bull. No. 17*, 1962.

13. Bloomer, O. T.; Parent, J. D. *Chem. Eng. Prog. Symp. Ser.* 1953, *49* (6), 11.

14. Cines, M. R.; Roach, J. T.; Hogan, R. J.; Roland, C. H. *Chem. Eng. Prog. Symp. Ser.* 1953, *49* (6), 1.

15. Chang, S. D.; Lu, B. C.-Y. *Chem. Eng. Prog. Symp. Ser.* 1967, *63* (81), 18.

16. Fastovsky, V. G.; Petrovsky, Yu. V. *J. Phys. Chem. (USSR)* 1957, *31*, 2317.

17. Skripka, V. G.; Nikitina, I. E.; Zhadanovich, L. A.; Sirotin, A. G.; Ben'yaminovich, O. A. *Gazov. Prom.* 1970, *15* (12), 35.

| COMPONENTS: | ORIGINAL MEASUREMENTS: |
|---|---|
| (1) Nitrogen; $N_2$; [7727-37-9]<br>(2) Methane; $CH_4$; [74-82-8] | Bloomer, O.T.; Parent, J.D. D.<br>*Chem. Eng. Progr. Symp. Ser.* 1953, No. 6, *49*, 11. |
| VARIABLES:<br>T/K = 91-174<br>P/MPa = 0.1-51 | PREPARED BY:<br>C. L. Young |

EXPERIMENTAL VALUES:

| T/K | *P*/bar | Mole fraction of nitrogen in liquid, $x_{N_2}$ | T/K | *P*/bar | Mole fraction of nitrogen in liquid, $x_{N_2}$ |
|---|---|---|---|---|---|
| 145.0 | 9.149 | 0.0140 | 163.48 | 23.96 | 0.0611 |
| 161.1 | 18.00 | 0.0140 | 166.37 | 26.52 | 0.0611 |
| 172.8 | 27.23 | 0.0140 | 166.37 | 26.34 | 0.0611 |
| 184.6 | 40.84 | 0.0140 | 171.76 | 31.15 | 0.0611 |
| 184.6 | 39.89 | 0.0140 | 176.9 | 36.93 | 0.0611 |
| 187.1 | 43.31 | 0.0140 | 177.1 | 36.93 | 0.0611 |
| 188.1 | 44.55 | 0.0140 | 180.40 | 40.47 | 0.0611 |
| 189.1 | 45.73 | 0.0140 | 183.15 | 43.82 | 0.0611 |
| 189.7 | 46.49 | 0.0140 | 186.32 | 47.24 | 0.0611 |
| 189.8 | 46.42 | 0.0140 | 187.01 | 47.82 | 0.0611 |
| 140.0 | 8.274 | 0.0295 | 187.09 | 47.79 | 0.0611 |
| 155.6 | 15.69 | 0.0295 | 122.83 | 5.87 | 0.1002 |
| 166.1 | 23.20 | 0.0295 | 141.37 | 12.87 | 0.1002 |
| 174.0 | 30.04 | 0.0295 | 155.17 | 20.96 | 0.1002 |
| 181.8 | 38.37 | 0.0295 | 163.57 | 27.53 | 0.1002 |
| 184.6 | 41.83 | 0.0295 | 170.39 | 33.74 | 0.1002 |
| 187.1 | 44.76 | 0.0295 | 172.22 | 35.97 | 0.1002 |
| 187.6 | 45.46 | 0.0295 | 174.81 | 38.56 | 0.1002 |
| 187.9 | 45.68 | 0.0295 | 177.96 | 42.09 | 0.1002 |
| 188.6 | 46.59 | 0.0295 | 181.36 | 45.62 | 0.1002 |
| 188.9 | 47.06 | 0.0295 | 183.15 | 47.22 | 0.1002 |
| 132.98 | 7.391 | 0.0611 | 184.26 | 48.04 | 0.1002 |
| 153.40 | 16.95 | 0.0611 | | | |

(cont.)

AUXILIARY INFORMATION

METHOD/APPARATUS/PROCEDURE:

Bubble point-dew point apparatus with glass equilibrium cell. Temperature measured with copper-constantan thermocouple. Pressure measured using dead weight piston balance. Gaseous mixture compressed through dew point and the bubble point by adding mercury to fixed volume. Dew and bubble points observed visually and also determined by discontinuities in pressure-volume data. Dew point data and other details in source.

SOURCE AND PURITY OF MATERIALS:

(1) Purity 99.99+ mole per cent, dried.

(2) Pure sample mole per cent fractionated final purity 99.97 mole per cent.

ESTIMATED ERROR:

$\delta T/K = \pm 0.05$; $\delta P/\text{bar} = \pm 0.07$; $\delta x_{N_2} = \pm 0.0005$.

REFERENCES:

| COMPONENTS: | ORIGINAL MEASUREMENTS: |
|---|---|
| (1) Nitrogen; $N_2$; [7727-37-9]<br><br>(2) Methane; $CH_4$; [74-82-8] | Bloomer, O.T.; Parent, J.D.<br>*Chem. Eng. Progr. Symp. Ser.* 1953, No. 6, *49*, 11. |

EXPERIMENTAL VALUES:

| T/K | *P*/bar | Mole fraction of nitrogen in liquid, $x_{N_2}$ | T/K | *P*/bar | Mole fraction of nitrogen in liquid, $x_{N_2}$ |
|---|---|---|---|---|---|
| 109.82 | 2.942 | 0.1000 | 145.08 | 43.76 | 0.697 |
| 103.43 | 2.015 | 0.1000 | 148.71 | 44.92 | 0.697 |
| 97.82 | 1.407 | 0.100 | 110.24 | 10.80 | 0.697 |
| 109.54 | 5.87 | 0.2879 | 136.99 | 34.83 | 0.697 |
| 121.54 | 10.43 | 0.2879 | 141.73 | 40.38 | 0.697 |
| 133.64 | 17.03 | 0.2879 | 117.48 | 15.64 | 0.697 |
| 143.68 | 23.86 | 0.2879 | 110.24 | 10.79 | 0.697 |
| 152.04 | 31.47 | 0.2879 | 104.68 | 7.86 | 0.697 |
| 159.65 | 38.33 | 0.2879 | 100.83 | 6.11 | 0.697 |
| 165.43 | 43.86 | 0.2879 | 131.42 | 28.50 | 0.697 |
| 168.49 | 47.04 | 0.2879 | 124.72 | 21.75 | 0.697 |
| 170.93 | 49.14 | 0.2879 | 91.21 | 2.972 | 0.6972 |
| 173.14 | 50.28 | 0.2879 | 137.27 | 40.02 | 0.8422 |
| 173.71 | 50.65 | 0.2879 | 136.71 | 39.67 | 0.8422 |
| 98.71 | 3.070 | 0.2883 | 137.83 | 40.18 | 0.8422 |
| 91.48 | 1.821 | 0.2883 | 135.43 | 38.76 | 0.8422 |
| 139.93 | 30.62 | 0.5088 | 132.81 | 35.60 | 0.8422 |
| 158.79 | 48.85 | 0.5088 | 129.58 | 31.38 | 0.8422 |
| 161.48 | 49.13 | 0.5088 | 125.56 | 26.70 | 0.8422 |
| 131.82 | 23.01 | 0.5088 | 120.26 | 20.90 | 0.8422 |
| 152.59 | 43.97 | 0.5088 | 113.86 | 15.20 | 0.8422 |
| 156.36 | 47.55 | 0.5088 | 107.82 | 10.86 | 0.8422 |
| 121.93 | 15.43 | 0.5088 | 101.64 | 7.41 | 0.8422 |
| 110.17 | 8.729 | 0.5088 | 96.89 | 5.31 | 0.8422 |
| 104.63 | 6.371 | 0.5088 | 129.47 | 35.71 | 0.9515 |
| 160.37 | 49.02 | 0.5088 | 129.74 | 35.85 | 0.9515 |
| 161.80 | 48.88 | 0.5088 | 128.63 | 34.75 | 0.9515 |
| 159.82 | 48.92 | 0.5088 | 126.40 | 31.65 | 0.9515 |
| 125.81 | 38.38 | 0.5088 | 132.33 | 27.15 | 0.9515 |
| 91.48 | 2.522 | 0.5095 | 116.92 | 20.15 | 0.9515 |
| 147.04 | 44.68 | 0.697 | 108.57 | 12.69 | 0.9515 |
| 148.15 | 45.06 | 0.697 | 100.79 | 7.76 | 0.9515 |
| 149.09 | 44.88 | 0.697 | 93.96 | 4.71 | 0.9515 |

| COMPONENTS: | ORIGINAL MEASUREMENTS: |
|---|---|
| (1) Nitrogen; $N_2$; [7727-37-9]<br>(2) Methane; $CH_4$; [74-82-8] | Cines, M. R.; Roach, J. T.; Hogan, J. J.; Roland, C. H.<br>*Chem. Eng. Prog. Symp. Ser.* 1953, *49*, 1-10. |
| VARIABLES:<br>T/K = 135-208<br>P/MPa = 0.1-4 | PREPARED BY:<br>C. L. Young |

EXPERIMENTAL VALUES:

| T/K | *P*/psia | *P*/MPa | Mole fraction of nitrogen in liquid, $x_{N_2}$ | in vapor, $y_{N_2}$ |
|---|---|---|---|---|
| 135.4 | 20 | 0.138 | 0.079 | 0.748 |
| | 25 | 0.172 | 0.111 | 0.791 |
| | 30 | 0.207 | 0.144 | 0.822 |
| | 35 | 0.241 | 0.180 | 0.848 |
| | 40 | 0.276 | 0.221 | 0.868 |
| | 50 | 0.345 | 0.312 | 0.899 |
| | 60 | 0.414 | 0.427 | 0.922 |
| | 70 | 0.483 | 0.560 | 0.940 |
| | 80 | 0.552 | 0.682 | 0.954 |
| | 90 | 0.621 | 0.793 | 0.968 |
| | 100 | 0.689 | 0.901 | 0.985 |
| 146.5 | 25 | 0.172 | 0.032 | 0.427 |
| | 30 | 0.207 | 0.047 | 0.524 |
| | 35 | 0.241 | 0.063 | 0.598 |
| | 40 | 0.276 | 0.080 | 0.649 |
| | 50 | 0.345 | 0.115 | 0.721 |
| | 60 | 0.414 | 0.152 | 0.768 |
| | 70 | 0.483 | 0.190 | 0.801 |
| | 80 | 0.552 | 0.234 | 0.829 |
| | 90 | 0.621 | 0.283 | 0.851 |
| | 100 | 0.689 | 0.338 | 0.870 |
| | 125 | 0.862 | 0.486 | 0.907 |
| | 150 | 1.03 | 0.638 | 0.933 |
| | 175 | 1.21 | 0.780 | 0.955 |

(cont.)

AUXILIARY INFORMATION

METHOD/APPARATUS/PROCEDURE:

Vapor recirculating apparatus. Temperature measured with thermocouple and pressure with Bourdon gauge. Liquid and gaseous samples taken and analysed by gas density balance. Details in source.

SOURCE AND PURITY OF MATERIALS:

(1) Mass spectrometric analysis revealed only 0.03 mole per cent oxygen.

(2) Purified natural gas, 99.67 mole per cent methane, 0.24 mole per cent nitrogen and 0.09 mole per cent ethane.

ESTIMATED ERROR:

$\delta T/K = \pm 0.2$; $\delta P/MPa = \pm 0.007$;
$\delta x_{N_2}$, $\delta y_{N_2}$ = ±0.009 or less.

REFERENCES:

| COMPONENTS: | ORIGINAL MEASUREMENTS: |
|---|---|
| (1) Nitrogen; $N_2$; [7727-37-9] | Cines, M. R.; Roach, J. T.; Hogan, J. J.; Roland, C. H. |
| (2) Methane; $CH_4$; [74-82-8] | *Chem. Eng. Prog. Symp. Ser.* 1953, *49*, 1-10. |

EXPERIMENTAL VALUES:

| T/K | $P$/psia | $P$/MPa | Mole fraction of nitrogen in liquid $x_{N_2}$ | in vapor, $y_{N_2}$ |
|---|---|---|---|---|
| 146.5 | 200 | 1.38 | 0.902 | 0.977 |
| 157.6 | 50 | 0.345 | 0.036 | 0.324 |
| | 60 | 0.414 | 0.056 | 0.445 |
| | 70 | 0.483 | 0.077 | 0.529 |
| | 80 | 0.552 | 0.099 | 0.586 |
| | 90 | 0.621 | 0.121 | 0.630 |
| | 100 | 0.689 | 0.144 | 0.668 |
| | 125 | 0.862 | 0.203 | 0.737 |
| | 150 | 1.03 | 0.272 | 0.782 |
| | 175 | 1.21 | 0.348 | 0.818 |
| | 200 | 1.38 | 0.432 | 0.846 |
| | 250 | 1.72 | 0.600 | 0.894 |
| | 300 | 2.07 | 0.763 | 0.934 |
| | 350 | 2.41 | 0.893 | 0.969 |
| | 400 | 2.76 | 0.989 | 0.997 |
| 161.6 | 50 | 0.345 | 0.014 | 0.163 |
| | 60 | 0.414 | 0.031 | 0.295 |
| | 70 | 0.483 | 0.049 | 0.389 |
| | 80 | 0.552 | 0.066 | 0.461 |
| | 90 | 0.621 | 0.085 | 0.519 |
| | 100 | 0.689 | 0.104 | 0.566 |
| | 125 | 0.862 | 0.153 | 0.657 |
| | 150 | 1.03 | 0.207 | 0.717 |
| | 175 | 1.21 | 0.266 | 0.760 |
| | 200 | 1.38 | 0.334 | 0.793 |
| | 250 | 1.72 | 0.480 | 0.846 |
| | 300 | 2.07 | 0.618 | 0.888 |
| | 350 | 2.41 | 0.744 | 0.922 |
| | 400 | 2.76 | 0.854 | 0.952 |
| | 450 | 3.10 | 0.944 | 0.980 |
| 168.7 | 70 | 0.483 | 0.008 | 0.079 |
| | 80 | 0.552 | 0.021 | 0.178 |
| | 90 | 0.621 | 0.036 | 0.260 |
| | 100 | 0.689 | 0.050 | 0.330 |
| | 125 | 0.862 | 0.086 | 0.460 |
| | 150 | 1.03 | 0.125 | 0.550 |
| | 175 | 1.21 | 0.164 | 0.618 |
| | 200 | 1.38 | 0.205 | 0.668 |
| | 250 | 1.72 | 0.301 | 0.739 |
| | 300 | 2.07 | 0.406 | 0.789 |
| | 350 | 2.41 | 0.513 | 0.829 |
| | 400 | 2.76 | 0.621 | 0.860 |
| | 450 | 3.10 | 0.725 | 0.888 |
| | 500 | 3.45 | 0.822 | 0.912 |
| 179.8 | 125 | 0.862 | 0.009 | 0.070 |
| | 150 | 1.03 | 0.036 | 0.210 |
| | 175 | 1.21 | 0.063 | 0.317 |
| | 200 | 1.38 | 0.089 | 0.395 |
| | 250 | 1.72 | 0.145 | 0.512 |
| | 300 | 2.07 | 0.206 | 0.592 |
| | 350 | 2.41 | 0.277 | 0.651 |
| | 400 | 2.76 | 0.352 | 0.692 |
| | 450 | 3.10 | 0.431 | 0.726 |
| | 500 | 3.45 | 0.509 | 0.752 |
| | 550 | 3.79 | 0.588 | 0.770 |
| | 600 | 4.14 | 0.668 | 0.783 |

(cont.)

| COMPONENTS: | ORIGINAL MEASUREMENTS: |
|---|---|
| (1) Nitrogen; $N_2$; [7727-37-9] <br> (2) Methane; $CH_4$; [74-82-8] | Cines, M. R.; Roach, J. T.; Hogan, J. J.; Roland, C. H. *Chem. Eng. Prog. Symp. Ser.* 1953, *49*, 1-10. |

EXPERIMENTAL VALUES:

| T/K | *P*/psia | *P*/MPa | Mole fraction of nitrogen in liquid $x_{N_2}$ | in vapor, $y_{N_2}$ |
|---|---|---|---|---|
| 190.0 | 200 | 1.38 | 0.006 | 0.040 |
| | 250 | 1.72 | 0.046 | 0.213 |
| | 300 | 2.07 | 0.089 | 0.334 |
| | 350 | 2.41 | 0.136 | 0.416 |
| | 400 | 2.76 | 0.185 | 0.479 |
| | 450 | 3.10 | 0.238 | 0.526 |
| | 500 | 3.45 | 0.295 | 0.565 |
| | 550 | 3.79 | 0.354 | 0.597 |
| | 600 | 4.14 | 0.415 | 0.620 |
| | 650 | 4.48 | 0.479 | 0.634 |
| 207.6 | 400 | 2.76 | 0.018 | 0.079 |
| | 450 | 3.10 | 0.049 | 0.162 |
| | 500 | 3.45 | 0.082 | 0.226 |
| | 550 | 3.79 | 0.118 | 0.278 |
| | 600 | 4.14 | 0.161 | 0.318 |
| | 650 | 4.48 | 0.215 | 0.351 |

COMPONENTS:

(1) Nitrogen; $N_2$; [7727-37-9]

(2) Methane; $CH_4$; [74-82-8]

ORIGINAL MEASUREMENTS:

Cheung, H.; Wang, D. I. J.

*Ind. Eng. Chem. Fund.* 1964, *3*, 355.

VARIABLES:

T/K = 92-124

P/kPa = 21-567

PREPARED BY:

C. L. Young

EXPERIMENTAL VALUES:

| T/K | *P*/bar | Mole fraction of nitrogen in liquid, $x_{N_2}$ | in vapor, $y_{N_2}$ |
|---|---|---|---|
| 91.7 | 0.212 | 0.009 | 0.295 |
| 91.6 | 0.292 | 0.019 | 0.476 |
| | 0.445 | 0.037 | 0.644 |
| | 1.147 | 0.147 | 0.861 |
| 97.2 | 0.361 | 0.009 | 0.230 |
| 97.1 | 0.478 | 0.018 | 0.405 |
| | 0.679 | 0.036 | 0.570 |
| | 1.680 | 0.146 | 0.827 |
| 104.9 | 0.703 | 0.009 | 0.211 |
| 105.1 | 0.875 | 0.018 | 0.335 |
| 105.1 | 1.200 | 0.036 | 0.512 |
| 105.2 | 2.748 | 0.145 | 0.790 |
| 114.4 | 1.519 | 0.009 | 0.144 |
| 114.6 | 1.759 | 0.018 | 0.247 |
| 114.7 | 2.198 | 0.036 | 0.394 |
| | 4.639 | 0.144 | 0.709 |
| 119.2 | 5.642 | 0.144 | 0.663 |
| 123.9 | 2.824 | 0.009 | 0.098 |
| 124.1 | 2.842 | 0.017 | 0.162 |
| | 3.786 | 0.035 | 0.304 |

AUXILIARY INFORMATION

METHOD/APPARATUS/PROCEDURE:

Static equilibrium cell fitted with magnetic stirrer. Pressure measured on mercury manometer. Temperature measured by thermocouple. Details in source reference. Known quantity solvent added to evacuated cell. Metered quantity of solvent added. Solubility determined from equilibrium pressure and mass balance. Details in source reference.

SOURCE AND PURITY OF MATERIALS:

Not given.

ESTIMATED ERROR:

$\delta T/K = \pm 0.2$; $\delta P/bar = \pm 0.0015$;

$\delta x_{N_2} = \delta y_{N_2} = \pm 2\%$ (estimated by compiler).

REFERENCES:

COMPONENTS:

(1) Nitrogen; $N_2$; [7727-37-9]

(2) Methane; $CH_4$; [74-82-8]

ORIGINAL MEASUREMENTS:

Chang, S. D.; Lu, B. C.-Y.
*Chem. Eng. Prog. Symp. Ser.* 1967, *63* (81), 18-27.

VARIABLES:

T/K = 122-171
P/MPa = 0.4-4.5

PREPARED BY:

C. L. Young

EXPERIMENTAL VALUES:

| T/K | $P$/bar | Mole fraction of nitrogen in liquid, $x_{N_2}$ | in vapor, $y_{N_2}$ |
|---|---|---|---|
| 171.4 | 26.59 | 0.01720 | 0.06380 |
| | 26.90 | 0.01770 | 0.04630 |
| | 28.11 | 0.02510 | 0.07550 |
| | 28.76 | 0.03110 | 0.08010 |
| | 31.76 | 0.06380 | 0.1577 |
| | 31.83 | 0.05360 | 0.1496 |
| | 33.07 | 0.07310 | 0.1836 |
| | 35.49 | 0.1017 | 0.2217 |
| | 36.97 | 0.1081 | 0.2407 |
| | 36.97 | 0.1182 | 0.2477 |
| | 37.80 | 0.1164 | 0.2535 |
| | 39.90 | 0.1445 | 0.2861 |
| | 41.18 | 0.1634 | 0.2896 |
| | 43.07 | 0.1803 | 0.3105 |
| | 43.21 | 0.1949 | 0.3103 |
| | 44.69 | 0.1908 | 0.3146 |
| 122.0 | 3.50 | 0.03371 | 0.3599 |
| | 6.96 | 0.1503 | 0.7070 |

AUXILIARY INFORMATION

METHOD/APPARATUS/PROCEDURE:

Recirculating vapor flow apparatus with cell consisting of 100 ml Jerguson gauge with stainless steel body. Temperature measured using copper-constantan thermocouples, pressure measured using Bourdon gauges. Magnetic circulating pump. Details of apparatus in source ref. Cell charged vapor recirculated for 2 or more hours. Samples of vapor and liquid removed at constant pressure and analysed using gas chromatography. Details in source.

SOURCE AND PURITY OF MATERIALS:

(1) Matheson research grade, purity 99.999 mole per cent.

(2) Matheson research grade, purity 99.99 mole per cent.

ESTIMATED ERROR:

$\delta T$/K = ±0.1; $\delta P$/bar = ±0.5%;
$\delta x_{N_2} = \delta y_{N_2} \leq 10\%$ (estimated by compiler)

REFERENCES:

**COMPONENTS:**

(1) Nitrogen; $N_2$; [7727-37-9];

(2) Methane; $CH_4$; [74-82-8]

**ORIGINAL MEASUREMENTS:**

Miller, R. C.; Kidnay, A. J.; Hiza, M. J.

*Am. Inst. Chem. Engrs. J.* 1973, *19*, 145.

**VARIABLES:**

T/K = 112
P/MPa = 0.2-1.3

**PREPARED BY:**

C. L. Young

**EXPERIMENTAL VALUES:**

| T/K | $P$/bar | Mole fraction of nitrogen in liquid, $x_{N_2}$ | in vapor, $y_{N_2}$ |
|---|---|---|---|
| 112.00 | 1.862 | - | 0.4478 |
| | 1.838 | 0.0338 | - |
| | 2.192 | - | 0.5311 |
| | 2.181 | 0.0545 | - |
| | 4.027 | - | 0.7476 |
| | 4.013 | 0.1336 | - |
| | 4.335 | - | 0.7660 |
| | 4.285 | 0.1505 | - |
| | 5.559 | - | 0.8190 |
| | 5.485 | 0.2170 | - |
| | 6.990 | - | 0.8626 |
| | 6.985 | 0.3082 | - |
| | 9.229 | - | 0.9041 |
| | 9.120 | 0.4701 | - |
| | 9.460 | - | 0.9086 |
| | 9.490 | 0.5172 | - |
| | 10.852 | - | 0.9287 |
| | 10.847 | 0.6103 | - |
| | 11.970 | - | 0.9433 |
| | 11.960 | 0.7068 | - |
| | 12.960 | - | 0.9552 |
| | 12.957 | 0.7811 | - |

**AUXILIARY INFORMATION**

**METHOD/APPARATUS/PROCEDURE:**

Recirculating vapor-flow apparatus. Temperature measured with platinum resistance thermometer. Pressure measured with Bourdon gauge. Gas and liquid samples analysed by gas chromatography. Details in source and refs. 1 and 2.

**SOURCE AND PURITY OF MATERIALS:**

No details given.

**ESTIMATED ERROR:**

$\delta T/K = \pm 0.05$; $\delta P/\text{bar} = \pm 0.02$;

$\delta x_{N_2}$, $\delta y_{N_2}$ = ±1%.

**REFERENCES:**

1. Kidnay, A.J.; Miller, R.C.; Hiza, M.J. *Ind. Eng. Chem. Fund.* 1971, *10*, 459.

2. Miller, R.C.; Kidnay, A.J.; Hiza, M.J. *J. Chem. Thermodynamics* 1972, *4*, 807.

**COMPONENTS:**

(1) Nitrogen; $N_2$; [7727-37-9]

(2) Methane; $CH_4$; [74-82-8 ]

**ORIGINAL MEASUREMENTS:**

Parrish, W. R.; Hiza, M. J.
*Adv. Cryog. Eng.* 1973, *19*, 300.

**VARIABLES:**

T/K = 95-120
P/MPa = 0.2-1.0

**PREPARED BY:**

C. L. Young

**EXPERIMENTAL VALUES:**

| T/K | $P$/bar | Mole fraction of nitrogen in liquid, $x_{N_2}$ | in gas, $y_{N_2}$ |
|---|---|---|---|
| 95.00 | 2.254 | 0.2679 | 0.9216 |
| | 2.737 | 0.3562 | 0.9374 |
| 100.00 | 1.988 | 0.1329 | 0.8336 |
| | 2.960 | 0.2397 | 0.8907 |
| 105.00 | 3.649 | 0.2101 | 0.8533 |
| | 4.512 | 0.2869 | 0.8855 |
| 110.00 | 4.810 | 0.2093 | 0.8237 |
| | 5.916 | 0.2856 | 0.8615 |
| 115.00 | 5.882 | 0.1891 | 0.7782 |
| | 7.725 | 0.2868 | 0.8363 |
| 120.00 | 4.984 | 0.0977 | 0.6122 |
| | 7.549 | 0.1938 | 0.7458 |
| | 9.982 | 0.2990 | 0.8125 |

**AUXILIARY INFORMATION**

**METHOD/APPARATUS/PROCEDURE:**

Recirculating vapor flow apparatus; details given in source and refs. 1-4. Temperature measured using platinum resistance thermometer. Pressure measured with Bourdon gauge. Mixture equilibrated and samples analysed using gas chromatography with thermal conductivity detector.

**SOURCE AND PURITY OF MATERIALS:**

No details given.

**ESTIMATED ERROR:**

$\delta T/K = \pm 0.02$; $\delta P/\text{bar} = \pm 0.02$ (up to 20 bar); $= \pm 0.05$ (above 20 bar); $\delta x_{N_2} = \delta y_{N_2} = \pm 1\%$ or $\pm 0.001$ whichever is greater.

**REFERENCES:**


1. Hiza, M.J.; Duncan, A.G. *Rev. Sci. Inst.* 1969, *40*, 513.
2. Duncan, A.G.; Hiza, M.J. *Adv. Cryog. Eng.* 1970, *15*, 42.
3. Kidnay, A.J.; Miller, R.C.; Hiza, M.J., *Ind. Eng. Chem. Fund.* 1971, *10*, 459.
4. Miller, R.C.; Kidnay, A.J.; Hiza, M.J. *J. Chem. Thermodynamics* 1972, *4*, 807.

| COMPONENTS: | ORIGINAL MEASUREMENTS: |
|---|---|
| (1) Nitrogen; $N_2$; [7727-37-9]<br>(2) Methane; $CH_4$; [74-82-8] | Stryjek, R.; Chappelear, P. S.; Kobayashi, R.<br>*J. Chem. Eng. Data* 1974, *19*, 334. |
| VARIABLES:<br>T/K = 114-183<br>P/MPa = 0.5-5.0 | PREPARED BY:<br>C. L. Young |

EXPERIMENTAL VALUES:

| T/K | P/bar | Mole fraction of nitrogen in liquid, $x_{N_2}$ | Mole fraction of nitrogen in vapor, $y_{N_2}$ | T/K | P/bar | Mole fraction of nitrogen in liquid, $x_{N_2}$ | Mole fraction of nitrogen in vapor, $y_{N_2}$ |
|---|---|---|---|---|---|---|---|
| 113.71 | 3.44 | 0.089407 | 0.6491 | 149.82 | 34.13 | 0.3778 | 0.6480 |
| | 5.17 | 0.1671 | 0.7660 | | 40.95 | 0.5034 | 0.6867 |
| | 6.89 | 0.2609 | 0.8290 | 160.93 | 17.24 | 0.009750 | 0.04452 |
| | 10.31 | 0.4984 | 0.8990 | | 19.06 | 0.02968 | 0.1221 |
| 122.04 | 2.79 | 0.016538 | 0.19633 | | 20.75 | 0.04895 | 0.1835 |
| | 3.44 | 0.03514 | 0.3466 | | 24.68 | 0.09572 | 0.2986 |
| | 5.17 | 0.086024 | 0.5588 | | 27.20 | 0.1257 | 0.3471 |
| | 6.88 | 0.1411 | 0.6655 | | 31.37 | 0.1774 | 0.4130 |
| | 13.79 | 0.4283 | 0.8411 | | 36.27 | 0.2457 | 0.4671 |
| 127.59 | 3.53 | 0.009745 | 0.1144 | | 41.23 | 0.3128 | 0.5131 |
| | 5.17 | 0.04714 | 0.3786 | | 44.61 | 0.3683 | 0.5284 |
| | 6.88 | 0.09116 | 0.5318 | | 48.33 | 0.4382 | 0.5296 |
| | 13.79 | 0.3040 | 0.7662 | | 48.88 | 0.4514 | 0.5181 |
| 138.44 | 6.91 | 0.01852 | 0.1499 | | 48.95 | 0.4527 | 0.5178 |
| | 8.20 | 0.04106 | 0.2724 | 172.04 | 25.55 | 0.005951 | 0.01938 |
| | 10.34 | 0.08104 | 0.4183 | | 26.89 | 0.01689 | 0.05251 |
| | 13.79 | 0.1490 | 0.5568 | | 27.65 | 0.02446 | 0.07397 |
| | 20.68 | 0.3037 | 0.6983 | | 31.23 | 0.05764 | 0.1532 |
| | 27.03 | 0.4593 | 0.7677 | | 34.47 | 0.09245 | 0.2157 |
| 149.82 | 12.34 | 0.02838 | 0.1529 | | 37.92 | 0.1287 | 0.2646 |
| | 13.96 | 0.05140 | 0.2404 | | 41.23 | 0.1657 | 0.3009 |
| | 17.20 | 0.09849 | 0.3683 | | 48.47 | 0.2599 | 0.3445 |
| | 20.62 | 0.1503 | 0.4579 | | 50.33 | 0.3156 | 0.3156 |
| | 20.68 | - | 0.4615 | 177.59 | 31.03 | 0.004056 | 0.01200 |
| | 27.34 | 0.2579 | 0.5779 | | | | (cont.) |

AUXILIARY INFORMATION

METHOD/APPARATUS/PROCEDURE:

Vapor recycled through equilibrium cell by magnetic pump. Temperature measured with platinum resistance thermometer using Mueller bridge. Pressure measured with Heise gauge. After equilibrium established, vapor phase analysed then recycle pump stopped and liquid phase sampled. Samples of both phases analysed with a thermal conductivity detector. Details in source and refs. 1 and 2. Additional vapor-liquid equilibrium data in source.

SOURCE AND PURITY OF MATERIALS:

(1) Matheson sample; purity 99.99 mole per cent.

(2) Purity specified as 99.99 mole per cent.

ESTIMATED ERROR:

$\delta T/K = \pm 0.02$; $\delta P/\text{bar} < \pm 0.07$; $\delta x_{N_2}$, $\delta y_{N_2} = \pm 0.2\text{-}1.0\%$.

REFERENCES:


1. Stryjek, R.; *Low Temperature Vapor-Liquid Equilibria of the Nitrogen-Methane, Nitrogen-Ethane and Nitrogen-Methane-Ethane Systems*, Rice University, Houston, Tex., 1972.
2. Wichterle, I.; Kobayashi, R. *J. Chem. Eng. Data* 1972, *17*, 4.

COMPONENTS:

(1) Nitrogen; $N_2$; [7727-37-9]

(2) Methane; $CH_4$; [74-82-8]

ORIGINAL MEASUREMENTS:

Stryjek, R.; Chappelear, P. S.; Kobayashi, R.

*J. Chem. Eng. Data* 1974, *19*, 334.

EXPERIMENTAL VALUES: continued

| T/K | P/bar | Mole fraction of nitrogen in liquid, $x_{N_2}$ | Mole fraction of nitrogen in vapor, $y_{N_2}$ | T/K | P/bar | Mole fraction of nitrogen in liquid, $x_{N_2}$ | Mole fraction of nitrogen in vapor, $y_{N_2}$ |
|---|---|---|---|---|---|---|---|
| 177.59 | 34.51 | 0.03257 | 0.07937 | 177.59 | 49.78 | 0.2272 | 0.2272 |
| | 36.82 | 0.05547 | 0.1266 | 183.15 | 39.71 | 0.02544 | 0.04931 |
| | 39.58 | 0.08284 | 0.1693 | | 41.23 | 0.04160 | 0.07742 |
| | 41.99 | 0.1097 | 0.2023 | | 42.95 | 0.05676 | 0.09635 |
| | 44.82 | 0.1391 | 0.2286 | | 45.09 | 0.07848 | 0.1251 |
| | 46.61 | 0.1613 | 0.2459 | | 46.47 | 0.09106 | 0.1357 |
| | 48.26 | 0.1816 | 0.2495 | | 47.50 | 0.1060 | 0.1425 |
| | 48.88 | 0.1929 | 0.2488 | | 48.26 | 0.1165 | 0.1449 |
| | 49.64 | 0.2026 | 0.2458 | | 48.47 | 0.1175 | 0.1425 |

COMPONENTS:

(1) Nitrogen; $N_2$; [7727-37-9]

(2) Methane; $CH_4$; [74-82-8]

ORIGINAL MEASUREMENTS:

Kidnay, A. J.; Miller, R. C.; Parrish, W. R.; Hiza, M. J.

*Cryogenics* 1975, *15*, 531.

VARIABLES:

T/K = 112-180
P/MPa = 0.2-4.9

PREPARED BY:

C. L. Young

EXPERIMENTAL VALUES:

| T/K | P/bar | Mole fraction of nitrogen in liquid, $x_{N_2}$ | Mole fraction of nitrogen in vapor, $y_{N_2}$ | T/K | P/bar | Mole fraction of nitrogen in liquid, $x_{N_2}$ | Mole fraction of nitrogen in vapor, $y_{N_2}$ |
|---|---|---|---|---|---|---|---|
| 112.00 | 1.965 | 0.0376 | 0.4720 | 150.00 | 14.228 | 0.0518 | 0.2479 |
| | 4.725 | 0.1504 | 0.7654 | | 18.088 | 0.1082 | 0.3938 |
| 120.00 | 5.234 | 0.1048 | 0.6327 | | 22.55 | 0.1788 | 0.5026 |
| | 10.058 | 0.3012 | 0.8182 | | 27.21 | 0.2510 | 0.5742 |
| 130.00 | 4.155 | 0.0097 | 0.1098 | 160.00 | 16.813 | 0.0095 | 0.0438 |
| | 4.552 | 0.0186 | 0.1862 | | 17.968 | 0.0223 | 0.0979 |
| | 5.025 | 0.0292 | 0.2640 | | 18.988 | 0.0334 | 0.1394 |
| | 6.594 | 0.0641 | 0.4339 | | 19.913 | 0.0448 | 0.1742 |
| | 7.797 | 0.0926 | 0.5201 | | 21.94 | 0.0684 | 0.2406 |
| | 10.203 | 0.1547 | 0.6316 | | 22.10 | 0.0714 | 0.2458 |
| | 12.579 | 0.2188 | 0.7018 | | 26.19 | 0.1205 | 0.3442 |
| | 15.038 | 0.2941 | 0.7508 | | 30.38 | 0.1756 | 0.4184 |
| 140.00 | 7.221 | 0.0132 | 0.1051 | | 33.95 | 0.2243 | 0.4657 |
| | 8.298 | 0.0316 | 0.2166 | | 38.46 | 0.2820 | 0.5051 |
| | 9.258 | 0.0485 | 0.2952 | 170.00 | 24.47 | 0.0082 | 0.0291 |
| | 10.108 | 0.0637 | 0.3498 | | 25.39 | 0.0189 | 0.0652 |
| | 14.056 | 0.1372 | 0.5228 | | 26.20 | 0.0276 | 0.0889 |
| | 18.047 | 0.2165 | 0.6214 | | 27.47 | 0.0399 | 0.1219 |
| | 21.94 | 0.3008 | 0.6850 | | 28.43 | 0.0495 | 0.1458 |
| 150.00 | 10.673 | 0.0038 | 0.0224 | | 30.36 | 0.0888 | 0.1875 |
| | 11.213 | 0.0106 | 0.0675 | | 34.43 | 0.1115 | 0.2593 |
| | 12.908 | 0.0333 | 0.1776 | | 38.58 | 0.1562 | 0.3135 |

(cont.)

AUXILIARY INFORMATION

METHOD/APPARATUS/PROCEDURE:

Recirculating vapor flow apparatus. Details given in source and refs. 1-4. Temperature measured using platinum resistance thermometer. Pressure measured with Bourdon gauge. Mixture equilibrated and samples analysed using gas chromatography with thermal conductivity detector.

SOURCE AND PURITY OF MATERIALS:

No details given.

ESTIMATED ERROR:

$\delta T/K = \pm 0.02$; $\delta P/\text{bar} = \pm 0.02$ (up to 20 bar); $= \pm 0.05$ (above 20 bar); $\delta x_{N_2} = \delta y_{N_2} = \pm 1\%$ or $\pm 0.001$ whichever is greater.

REFERENCES:


1. Hiza, M.J; Duncan, A.G. *Rev. Sci. Inst.* 1969, *40*, 513.
2. Duncan, A.G.; Hiza, M.J. *Adv. Cryog. Eng.* 1970, *15*, 42.
3. Kidnay, A.J.; Miller, R.C.; Hiza, M.J. *Ind. Eng. Chem. Fund.* 1971, *10*, 459.
4. Miller, R.C.; Kidnay, A.J.; Hiza, M.J. *J. Chem. Thermodynamics* 1972, *4*, 807.

COMPONENTS:

(1) Nitrogen; $N_2$; [7727-37-9]

(2) Methane; $CH_4$; [74-82-8]

ORIGINAL MEASUREMENTS:

Kidnay, A. J.; Miller. R. C.; Parrish, W. R.; Hiza, M. J. *Cryogenics* 1975, *15*, 531.

EXPERIMENTAL VALUES:

| T/K | $P$/bar | Mole fraction of nitrogen in liquid, $x_{N_2}$ | Mole fraction of nitrogen in vapor, $y_{N_2}$ | T/K | $P$/bar | Mole fraction of nitrogen in liquid, $x_{N_2}$ | Mole fraction of nitrogen in vapor, $y_{N_2}$ |
|---|---|---|---|---|---|---|---|
| 170.00 | 42.52 | 0.2001 | 0.3519 | 180.00 | 37.87 | 0.0434 | 0.0936 |
| | 46.31 | 0.2466 | 0.3743 | | 39.45 | 0.0588 | 0.1192 |
| | 48.33 | 0.2729 | 0.3784 | | 40.24 | 0.0632 | 0.1260 |
| | 48.95 | 0.2915 | 0.3798 | | 42.72 | 0.0863 | 0.1572 |
| | 49.39 | 0.3066 | 0.3801 | | 44.75 | 0.1083 | 0.1786 |
| 180.00 | 34.11 | 0.0089 | 0.0230 | | 46.56 | 0.1254 | 0.1919 |
| | 35.40 | 0.0202 | 0.0494 | | 48.70 | 0.1567 | 0.1992 |
| | 36.72 | 0.0327 | 0.0738 | | | | |

| COMPONENTS: | ORIGINAL MEASUREMENTS: |
|---|---|
| (1) Nitrogen; $N_2$; [7727-37-9]<br>(2) Methane; $CH_4$; [74-82-8]<br>(3) Ethane; $C_2H_6$; [74-84-0] | Cosway, H. F.; Katz, D. L.<br>*Am. Inst. Chem. Engnrs. J.* 1959, *5*, 46-50. |
| VARIABLES:<br>T/K = 144-200<br>P/MPa = 3.5-6.9<br>Composition | PREPARED BY:<br>C. L. Young |

EXPERIMENTAL VALUES:

| Mole fractions in liquid | | | Mole fractions in gas | | |
|---|---|---|---|---|---|
| $x_{N_2}$ | $x_{CH_4}$ | $x_{C_2H_6}$ | $y_{N_2}$ | $y_{CH_4}$ | $y_{C_2H_6}$ |
| | T/K = 199.8 | | $P$/psia = 500 | $P$/MPa = 3.45 | |
| 0.0956 | 0.0000 | 0.9044 | 0.8922 | 0.0000 | 0.1078 |
| 0.0715 | 0.1989 | 0.7296 | 0.6029 | 0.2979 | 0.0992 † |
| 0.0000 | 0.6600 | 0.3400 | 0.0000 | 0.9300 | 0.0700 † |
| | T/K = 199.8 | | $P$/psia = 1000 | $P$/MPa = 6.89 | |
| 0.2129 | 0.0000 | 0.7871 | 0.9055 | 0.0000 | 0.0945 |
| 0.2064 | 0.1564 | 0.6372 | 0.7460 | 0.1642 | 0.0898 |
| 0.1752 | 0.5612 | 0.2636 | 0.3435 | 0.5608 | 0.0957 |
| | T/K = 144.3 | | $P$/psia = 1000 | $P$/MPa = 6.89 | |
| 0.1826 | 0.0000 | 0.8174 | 0.9923 | 0.0000 | 0.00766 |
| 0.2403 | 0.2525 | 0.5072 | 0.8810 | 0.1131 | 0.00593 |
| 0.3506 | 0.4499 | 0.1995 | 0.7975 | 0.1979 | 0.00463 * |
| 0.4900 | 0.5100 | 0.0000 | 0.7300 | 0.2700 | 0.00 § |
| 0.5090 | 0.4910 | 0.0000 | 0.7520 | 0.2480 | 0.00 |

† Data quoted in original but taken from Bloomer, O. T.; Gami, D. C.; Parent, J. D. *Inst. Gas. Tech. Res. Bull.* 1953, *22*, 1.

(cont.)

AUXILIARY INFORMATION

METHOD/APPARATUS/PROCEDURE:

Recirculating vapor flow system based on that described in ref. (1). Details of present apparatus in source and ref. (2). Temperature measured with thermocouple. Pressure measured with Bourdon gauge. High pressure magnetic circulating pump employed.
Cell charged under pressure. Samples of each phase expanded to room temperature and pressure; analysed by mass spectrometry.

SOURCE AND PURITY OF MATERIALS:

(1) No details given.

(2 and 3) Phillips Petroleum Co. samples.

ESTIMATED ERROR:

$\delta T/K = \pm 0.2$; $\delta P/MPa = \pm 0.03$; $\delta x$, $\delta y = \pm 2\%$ (estimated by compiler).

REFERENCES:

1. Dodge, B. F.; Dunbar, A. K. *J. Am. Chem. Soc.* 1927, *49*, 591.

2. Aroyan, H. J.; Katz, D. L. *Ind. Eng. Chem.* 1951, *43*, 185.

| COMPONENTS: | ORIGINAL MEASUREMENTS: |
|---|---|
| (1) Nitrogen; $N_2$; [7727-37-9] | Cosway, H. F.; Katz, D. L. |
| (2) Methane; $CH_4$; [74-82-8] | *Am. Inst. Chem. Engnrs. J.* 1959, 5, 46-50. |
| (3) Ethane; $C_2H_6$; [74-84-0] | |

EXPERIMENTAL VALUES:

* Data quoted in original but taken from
Bloomer, O. T.; Parent, J. D. *Chem. Eng. Progr. Symposium Ser.* 1953, No. 6, *49*, 11.

§ Data quoted in original but taken from
Cines, M. R.; Roach, J. T.; Hogan, R. J.; Roland, C. H. *Chem. Eng. Progr. Symposium Ser.* 1953, No. 6, *49*, 1

COMPONENTS:

(1) Nitrogen; $N_2$; [7727-37-9]

(2) Methane; $CH_4$; [74-82-8]

(3) Ethane; $C_2H_6$; [74-84-0]

ORIGINAL MEASUREMENTS:

Chang, S.D.; Lu B.C.Y.

*Chem. Eng. Prog. Symp. Ser.* 1967, 62, (81) 18-27.

VARIABLES:

T/K = 171
P/MPa = 0.2-2.8
Composition

PREPARED BY:

C.L. Young

EXPERIMENTAL VALUES:

T/K = 171.4

| Pressure /MPa | Composition, mole fraction Vapor $y_{N_2}$ | $y_{CH_4}$ | $y_{C_2H_6}$ | Liquid $x_{N_2}$ | $x_{CH_4}$ | $x_{C_2H_6}$ |
|---|---|---|---|---|---|---|
| 1.377 | 0.5628 | 0.3956 | 0.04158 | 0.04082 | 0.2201 | 0.7391 |
| 1.370 | 0.6222 | 0.3368 | 0.04109 | 0.02853 | 0.1686 | 0.8029 |
| 1.373 | 0.1200 | 0.8497 | 0.03357 | 0.009581 | 0.4957 | 0.4948 |
| 1.381 | 0.6241 | 0.3331 | 0.04288 | 0.04912 | 0.1873 | 0.7636 |
| 1.369 | 0.4788 | 0.4838 | 0.03748 | 0.03694 | 0.2751 | 0.6900 |
| 1.356 | 0.2000 | 0.7661 | 0.03385 | 0.01901 | 0.3888 | 0.5922 |
| 1.380 | 0.0742 | 0.8937 | 0.03213 | 0.00542 | 0.5258 | 0.4687 |
| 2.066 | 0.7065 | 0.2634 | 0.03008 | 0.03361 | 0.1648 | 0.8016 |
| 2.066 | 0.7417 | 0.2261 | 0.03213 | 0.04383 | 0.1446 | 0.8116 |
| 2.059 | 0.4193 | 0.5564 | 0.02436 | 0.04599 | 0.4182 | 0.5358 |
| 2.070 | 0.7443 | 0.2220 | 0.03371 | 0.04360 | 0.1482 | 0.8082 |
| 2.058 | 0.6457 | 0.3208 | 0.03350 | 0.03811 | 0.2259 | 0.7359 |
| 2.073 | 0.3843 | 0.5915 | 0.02421 | 0.04622 | 0.4832 | 0.4706 |
| 2.087 | 0.1884 | 0.7954 | 0.01625 | 0.03092 | 0.6674 | 0.3016 |
| 2.077 | 0.1364 | 0.8498 | 0.01387 | 0.02511 | 0.7213 | 0.2535 |
| 2.756 | 0.7693 | 0.1989 | 0.03180 | 0.05924 | 0.1564 | 0.7843 |
| 2.756 | 0.5599 | 0.4148 | 0.02533 | - | - | - |
| 2.756 | 0.8030 | 0.1669 | 0.03011 | 0.06841 | 0.1503 | 0.7813 |
| 2.770 | 0.7391 | 0.2330 | 0.02786 | 0.06195 | 0.2071 | 0.7309 |
| 2.766 | 0.5279 | 0.4504 | 0.02167 | 0.06417 | 0.4455 | 0.4903 |
| 2.763 | 0.3804 | 0.6025 | 0.01714 | 0.07453 | 0.6324 | 0.2931 |
| 2.763 | 0.3329 | 0.6511 | 0.01602 | 0.07475 | 0.6805 | 0.2448 |
| 0.3168 | 0.6249 | 0.3712 | 0.001856 | 0.03356 | 0.5000 | 0.4663 |
| 0.7894 | 0.8409 | 0.1574 | 0.001740 | 0.09687 | 0.3638 | 0.5393 |
| 1.308 | 0.9026 | 0.09737 | 0.000696 | 0.1646 | 0.3556 | 0.4800 |

cont.

AUXILIARY INFORMATION

METHOD/APPARATUS/PROCEDURE:

Recirculating vapor flow apparatus with cell consisting of 100 ml Jerguson gauge with stainless steel body. Temperature measured using copper-constantan thermocouples. Pressure measured using Bourdon gauges Magnetic circulating pump. Cell charged, vapor recirculated for 2 or more hours. Samples of vapor and liquid removed at constant pressure and analysed using gas chromatography. Details in source.

SOURCE AND PURITY OF MATERIALS:

(1) Matheson research grade, purity 99.999 mole per cent.

(2) Matheson research grade, purity 99.99 mole per cent.

(3) Matheson research grade, purity 99.9 mole per cent.

ESTIMATED ERROR:

$\delta T/K = \pm 0.1$; $\delta P/MPa = \pm 0.5\%$
$\delta x_{N_2} \simeq \delta y_{N_2} < 10\%$ (estimated by compiler).

REFERENCES:

| COMPONENTS: | ORIGINAL MEASUREMENTS: |
|---|---|
| (1) Nitrogen; $N_2$; [7727-37-9] | Chang, S.D.; Lu, B.C.Y. |
| (2) Methane; $CH_4$; [74-82-8] | *Chem. Eng. Prog. Symp. Ser.* 1967, *62*, (81) 18-27. |
| (3) Ethane; $C_2H_6$; [74-84-0] | |

EXPERIMENTAL VALUES: T/K = 171.4

| Pressure /MPa | Vapor $y_{N_2}$ | Vapor $y_{CH_4}$ | Vapor $y_{C_2H_6}$ | Liquid $x_{N_2}$ | Liquid $x_{CH_4}$ | Liquid $x_{C_2H_6}$ |
|---|---|---|---|---|---|---|
| | Composition, mole fraction | | | | | |
| 1.797 | 0.9313 | 0.06777 | 0.00095 | 0.3188 | 0.2966 | 0.3846 |
| 2.228 | 0.9597 | 0.04133 | 0.00112 | 0.5940 | 0.1779 | 0.2281 |
| 2.364 | 0.9729 | 0.02598 | 0.00115 | 0.7605 | 0.1078 | 0.1317 |
| 2.628 | 0.9873 | 0.01144 | 0.001276 | 0.9358 | 0.03627 | 0.02793 |
| 0.1924 | 0.04142 | 0.9537 | 0.004872 | 0.07798 | 0.3348 | 0.5871 |
| 0.5695 | 0.8446 | 0.1521 | 0.00335 | 0.09513 | 0.3405 | 0.5643 |
| 1.011 | 0.7817 | 0.2169 | 0.001316 | 0.08766 | 0.2419 | 0.6705 |
| 1.591 | 0.9242 | 0.07415 | 0.001652 | 0.1659 | 0.2413 | 0.5928 |
| 1.991 | 0.9529 | 0.04571 | 0.001380 | 0.2772 | 0.1863 | 0.5365 |
| 2.124 | 0.9582 | 0.03966 | 0.002098 | BL0.4219 | 0.1458 | 0.4323 |
| | | | | *TL0.5322 | 0.1529 | 0.3149 |
| 2.255 | 0.9700 | 0.02891 | 0.001710 | 0.5450 | 0.1644 | 0.2906 |
| 2.293 | 0.9722 | 0.02661 | 0.001161 | BL0.4715 | 0.1513 | 0.3772 |
| | | | | 0.6628 | 0.1418 | 0.1954 |
| 2.406 | 0.9807 | 0.01880 | 0.0005356 | 0.6996 | 0.07831 | 0.2221 |
| 2.453 | 0.9837 | 0.01510 | 0.001157 | 0.7509 | 0.06091 | 0.1882 |
| 0.2296 | 0.7110 | 0.2867 | 0.002279 | 0.02279 | 0.2783 | 0.6989 |
| 0.2151 | 0.7914 | 0.2086 | 0.003105 | 0.03949 | 0.2750 | 0.6855 |
| 0.4881 | 0.9582 | 0.04112 | 0.0006961 | 0.1095 | 0.2642 | 0.6263 |
| 1.018 | 0.9571 | 0.04216 | 0.0006863 | 0.1026 | 0.1672 | 0.7302 |
| 1.349 | 0.9673 | 0.03206 | 0.0006209 | 0.1507 | 0.1629 | 0.6864 |
| 1.718 | 0.9715 | 0.02675 | 0.0006855 | 0.2048 | 0.1469 | 0.6482 |
| 2.180 | 0.9780 | 0.02107 | 0.0008995 | 0.3158 | 0.1260 | 0.5582 |
| 2.424 | 0.9809 | 0.01785 | 0.001251 | BL0.3658 | 0.1025 | 0.5317 |
| | | | | *TL0.7047 | 0.07117 | 0.2241 |
| 2.522 | 0.9901 | 0.008430 | 0.001499 | 0.8971 | 0.03577 | 0.06715 |
| 0.1744 | 0.09370 | 0.9045 | 0.001847 | | | |
| 0.3403 | 0.5256 | 0.4731 | 0.001238 | 0.03182 | 0.6916 | 0.2766 |
| 0.6836 | 0.7535 | 0.2456 | 0.0009046 | 0.09313 | 0.6348 | 0.2680 |
| 1.400 | 0.8888 | 0.1106 | 0.0006336 | 0.3072 | 0.4874 | 0.2054 |
| 2.296 | 0.9625 | 0.03672 | 0.0007767 | 0.8206 | 0.1318 | 0.04758 |
| 2.337 | 0.9641 | 0.03515 | 0.0007576 | 0.8396 | 0.1171 | 0.04330 |
| 2.728 | 0.9872 | 0.01229 | 0.0005563 | 0.9605 | 0.02907 | 0.01039 |
| 2.803 | 0.9920 | 0.007547 | 0.0004530 | 0.8523 | 0.1429 | 0.004819 |
| 2.470 | 0.9853 | 0.01352 | 0.001279 | BL0.3500 | 0.08683 | 0.5632 |
| | | | | TL0.8496 | 0.06764 | 0.08278 |
| 2.348 | 0.9770 | 0.02171 | 0.001226 | BL0.4389 | 0.1464 | 0.4147 |
| | | | | TL0.7208 | 0.1228 | 0.1564 |
| 2.337 | 0.9742 | 0.02312 | 0.001278 | BL0.4731 | 0.1439 | 0.3830 |
| | | | | TL0.7217 | 0.1233 | 0.1551 |

BL Bottom Liquid Layer
TL Top Liquid Layer
* Taken before allow 1 hour for layers to separate

COMPONENTS:

(1) Nitrogen; $N_2$; [7727-37-9]

(2) Methane; $CH_4$; [74-82-8]

(3) Ethane; $C_2H_6$; [74-84-0]

ORIGINAL MEASUREMENTS:

Gupta, M.K.; Gardner, G.C.;

*J. Chem. Eng. Data* 1980, *25*, 313-8.

VARIABLES:

T/K = 260-280
P/MPa = 5-9
Composition

PREPARED BY:

C.L. Young

EXPERIMENTAL VALUES:

| Total pressure p/atm | p/MPa | $x_{N_2}$ | $x_{CH_4}$ | $x_{C_2H_6}$ | $y_{N_2}$ | $y_{CH_4}$ | $y_{C_2H_6}$ |
|---|---|---|---|---|---|---|---|
| | | | Mole fractions | | | | |
| | | | T/K = 260.00 | | | | |
| 50.05 | 5.071 | 0.0139 | 0.2665 | 0.7196 | 0.0533 | 0.5049 | 0.4418 |
| 50.01 | 5.067 | 0.0296 | 0.2205 | 0.7499 | 0.1200 | 0.4293 | 0.4507 |
| 49.90 | 5.056 | 0.0452 | 0.1734 | 0.7814 | 0.1940 | 0.3464 | 0.4596 |
| 50.02 | 5.068 | 0.0592 | 0.1300 | 0.8107 | 0.2667 | 0.2652 | 0.4681 |
| 49.92 | 5.058 | 0.0723 | 0.0890 | 0.8387 | 0.3396 | 0.1852 | 0.4752 |
| 50.07 | 5.073 | 0.0847 | 0.0486 | 0.8667 | 0.4147 | 0.1032 | 0.4821 |
| 65.13 | 6.599 | 0.1172 | 0.1114 | 0.7714 | 0.3811 | 0.1839 | 0.4350 |
| 65.00 | 6.586 | 0.1198 | 0.1021 | 0.7781 | 0.3945 | 0.1692 | 0.4362 |
| 65.09 | 6.595 | 0.1230 | 0.0929 | 0.7841 | 0.4091 | 0.1546 | 0.4363 |
| 64.96 | 6.582 | 0.0767 | 0.2400 | 0.6832 | 0.2091 | 0.3681 | 0.4227 |
| 65.08 | 6.594 | 0.0886 | 0.2047 | 0.7068 | 0.2541 | 0.3203 | 0.4256 |
| 65.16 | 6.602 | 0.0260 | 0.3897 | 0.5843 | 0.0537 | 0.5316 | 0.4147 |
| 65.02 | 6.588 | - | - | - | 0.1262 | 0.4574 | 0.4164 |
| 65.00 | 6.586 | 0.1327 | 0.0570 | 0.8103 | 0.4624 | 0.0969 | 0.4407 |
| 65.00 | 6.586 | 0.1351 | 0.0499 | 0.8150 | 0.4744 | 0.0848 | 0.4408 |
| 64.98 | 6.584 | 0.0463 | 0.3297 | 0.6240 | 0.1081 | 0.4749 | 0.4170 |
| 65.00 | 6.586 | 0.1494 | 0.0000 | 0.8506 | 0.5541 | 0.0000 | 0.4459 |
| 65.00 | 6.586 | 0.0000 | 0.4585 | 0.5415 | 0.0000 | 0.5833 | 0.4167 |

cont.

AUXILIARY INFORMATION

METHOD/APPARATUS/PROCEDURE:

Recirculating vapor flow apparatus with diaphragm pump. Temperature measured with platinum resistance thermometer and pressure with Bourdon gauges. Cell stirred with double propeller stirrer. Vapor and liquid samples analysed by gas chromatography using a thermal conductivity detector. Details in ref. (1).

SOURCE AND PURITY OF MATERIALS:

(1) Purity at least 99.7 mole per cent.

(2) Purity at least 99 mole per cent.

No extraneous peaks were found when samples analysed by gas chromatography.

ESTIMATED ERROR:

$\delta T/K = \pm 0.02$; $\delta p/Mpa = \pm 0.03$ up to 3.4 MPa, $\pm$ 0.1 above 3.4 MPa; $\delta x_{N_2}$, $\delta y_{N_2}$ = $\pm$ 0.002

REFERENCES:

1. Somait, F.; Kidnay, A.J. *J. Chem. Eng. Data* 1978, *23*, 301.

| COMPONENTS: | ORIGINAL MEASUREMENTS: |
|---|---|
| (1) Nitrogen; $N_2$; [7727-37-9]<br>(2) Methane; $CH_4$; [74-82-8]<br>(3) Ethane; $C_2H_6$; [74-84-0] | Gupta, M.K.; Gardner, G.C.;<br>*J. Chem. Eng. Data*, 1980, *25*, 313-8. |

| Total pressure p/atm | p/MPa | $x_{N_2}$ | $x_{CH_4}$ | $x_{C_2H_6}$ | $y_{N_2}$ | $y_{CH_4}$ | $y_{C_2H_6}$ |
|---|---|---|---|---|---|---|---|
| | | | T/K = 260.00 | | | | |
| 74.80 | 7.579 | 0.1734 | 0.0464 | 0.7802 | 0.4963 | 0.0703 | 0.4336 |
| 75.18 | 7.618 | 0.1750 | 0.0453 | 0.7797 | 0.4984 | 0.0684 | 0.4332 |
| 75.08 | 7.607 | 0.1645 | 0.0821 | 0.7533 | 0.4463 | 0.1221 | 0.4315 |
| 75.13 | 7.613 | 0.1663 | 0.0795 | 0.7542 | 0.4507 | 0.1176 | 0.4317 |
| 75.06 | 7.605 | 0.1532 | 0.1219 | 0.7249 | 0.3925 | 0.1773 | 0.4302 |
| 75.13 | 7.613 | 0.1565 | 0.1146 | 0.7290 | 0.4031 | 0.1666 | 0.4303 |
| 75.13 | 7.613 | 0.1415 | 0.1676 | 0.6910 | 0.3350 | 0.2357 | 0.4293 |
| 75.00 | 7.599 | 0.1472 | 0.1469 | 0.7058 | 0.3611 | 0.2094 | 0.4295 |
| 74.90 | 7.589 | 0.1349 | 0.1874 | 0.6676 | 0.3103 | 0.2606 | 0.4290 |
| 75.02 | 7.601 | 0.1274 | 0.2162 | 0.6564 | 0.2765 | 0.2940 | 0.4295 |
| 75.08 | 7.607 | 0.1141 | 0.2629 | 0.6229 | 0.2245 | 0.3442 | 0.4313 |
| 75.09 | 7.608 | 0.1245 | 0.2270 | 0.6484 | 0.2636 | 0.3051 | 0.4312 |
| 75.20 | 7.620 | 0.1040 | 0.3046 | 0.5914 | 0.1824 | 0.3811 | 0.4365 |
| 75.10 | 7.610 | 0.1145 | 0.2643 | 0.6213 | 0.2237 | 0.3453 | 0.4309 |
| 75.17 | 7.617 | 0.0926 | 0.3548 | 0.5526 | 0.1366 | 0.4143 | 0.4490 |
| 75.07 | 7.606 | 0.0987 | 0.3239 | 0.5775 | 0.1641 | 0.3966 | 0.4392 |
| 74.90 | 7.589 | 0.0878 | 0.3848 | 0.5274 | 0.1121 | 0.4245 | 0.4635 |
| 75.07 | 7.606 | 0.0919 | 0.3579 | 0.5502 | 0.1340 | 0.4159 | 0.4501 |
| | | | T/K = 270.00 | | | | |
| 35.00 | 3.546 | 0.0341 | 0.0160 | 0.9500 | 0.2297 | 0.0470 | 0.7232 |
| 35.01 | 3.547 | 0.0285 | 0.0322 | 0.9393 | 0.1892 | 0.0936 | 0.7172 |
| 35.01 | 3.547 | 0.0223 | 0.0505 | 0.9272 | 0.1446 | 0.1450 | 0.7104 |
| 35.01 | 3.547 | 0.0196 | 0.0576 | 0.9228 | 0.1270 | 0.1652 | 0.7078 |
| 35.00 | 3.546 | 0.0116 | 0.0814 | 0.9069 | 0.0736 | 0.2284 | 0.6979 |
| 35.00 | 3.546 | 0.0077 | 0.0921 | 0.9002 | 0.0482 | 0.2572 | 0.6946 |
| 35.00 | 3.546 | - | - | - | 0.0797 | 0.2201 | 0.7001 |
| 50.00 | 5.066 | 0.0783 | 0.0231 | 0.8986 | 0.3546 | 0.0503 | 0.5951 |
| 50.00 | 5.066 | 0.0749 | 0.0347 | 0.8904 | 0.3335 | 0.0740 | 0.5925 |
| 50.00 | 5.066 | 0.0602 | 0.0766 | 0.8631 | 0.2564 | 0.1602 | 0.5834 |
| 50.00 | 5.066 | 0.0463 | 0.1173 | 0.8364 | 0.1862 | 0.2392 | 0.5746 |
| 50.00 | 5.066 | 0.0331 | 0.1535 | 0.8133 | 0.1283 | 0.3060 | 0.5657 |
| 50.01 | 5.067 | 0.0297 | 0.1619 | 0.8084 | 0.1139 | 0.3222 | 0.5638 |
| 50.00 | 5.066 | 0.0019 | 0.2361 | 0.7619 | 0.0066 | 0.4466 | 0.5468 |
| 50.00 | 5.066 | 0.0096 | 0.2089 | 0.7815 | 0.0362 | 0.4114 | 0.5524 |
| 50.01 | 5.067 | 0.0177 | 0.1948 | 0.7875 | 0.0664 | 0.3775 | 0.5561 |
| 50.00 | 5.066 | - | - | - | 0.0627 | 0.3807 | 0.5565 |
| 49.99 | 5.065 | 0.0827 | 0.0000 | 0.9173 | 0.3955 | 0.0000 | 0.6045 |
| 64.95 | 6.581 | 0.1213 | 0.0539 | 0.8247 | 0.3712 | 0.0906 | 0.5382 |
| 64.85 | 6.570 | 0.1205 | 0.0548 | 0.8247 | 0.3701 | 0.0923 | 0.5376 |
| 65.15 | 6.601 | 0.0937 | 0.1381 | 0.7682 | 0.2526 | 0.2192 | 0.5281 |
| 64.92 | 6.578 | 0.0935 | 0.1361 | 0.7704 | 0.2546 | 0.2167 | 0.5287 |
| 65.00 | 6.586 | 0.0948 | 0.1333 | 0.7719 | 0.2589 | 0.2124 | 0.5286 |
| 64.98 | 6.584 | 0.0622 | 0.2245 | 0.7133 | 0.1449 | 0.3348 | 0.5203 |
| 64.95 | 6.581 | 0.0610 | 0.2285 | 0.7106 | 0.1412 | 0.3375 | 0.5212 |
| 65.05 | 6.591 | 0.0646 | 0.2212 | 0.7142 | 0.1510 | 0.3284 | 0.5207 |
| 65.10 | 6.596 | 0.0343 | 0.3047 | 0.6610 | 0.0667 | 0.4152 | 0.5181 |
| 64.92 | 6.578 | 0.0313 | 0.3104 | 0.6583 | 0.0610 | 0.4213 | 0.5178 |
| 65.06 | 6.592 | - | - | - | 0.0582 | 0.4238 | 0.5180 |
| 65.00 | 6.586 | 0.0760 | 0.1714 | 0.7526 | 0.2026 | 0.2723 | 0.5250 |
| 65.00 | 6.586 | 0.0846 | 0.1633 | 0.7521 | 0.2194 | 0.2543 | 0.5263 |
| 65.00 | 6.586 | 0.1008 | 0.1061 | 0.7931 | 0.2921 | 0.1755 | 0.5324 |
| 75.00 | 7.599 | 0.1525 | 0.0716 | 0.7759 | 0.3550 | 0.1135 | 0.5306 |
| 74.97 | 7.596 | 0.1673 | 0.0317 | 0.8010 | 0.4200 | 0.0476 | 0.5323 |
| 74.94 | 7.593 | 0.1747 | 0.0080 | 0.8173 | 0.4540 | 0.0124 | 0.5339 |
| 74.90 | 7.589 | 0.1117 | 0.2604 | 0.6279 | 0.1444 | 0.2917 | 0.5639 |

| COMPONENTS: | ORIGINAL MEASUREMENTS: |
|---|---|
| (1) Nitrogen; $N_2$; [7727-37-9]<br>(2) Methane; $CH_4$; [74-82-8]<br>(3) Ethane; $C_2H_6$; [74-84-0] | Gupta, M.K.; Gardner, G.C.<br>*J. Chem. Eng. Data*, 1980, *25*, 313-8. |

| Total pressure p/atm | p/MPa | Mole fractions $x_{N_2}$ | $x_{CH_4}$ | $x_{C_2H_6}$ | $y_{N_2}$ | $y_{CH_4}$ | $y_{C_2H_6}$ |
|---|---|---|---|---|---|---|---|
| | | | T/K = 270.00 | | | | |
| 75.00 | 7.599 | 0.1130 | 0.2519 | 0.6351 | 0.1545 | 0.2891 | 0.5564 |
| 75.10 | 7.610 | 0.1130 | 0.2467 | 0.6403 | 0.1586 | 0.2861 | 0.5553 |
| 74.92 | 7.591 | 0.1213 | 0.2023 | 0.6764 | 0.2065 | 0.2556 | 0.5378 |
| 75.08 | 7.607 | 0.1205 | 0.2050 | 0.6745 | 0.2030 | 0.2581 | 0.5389 |
| 74.95 | 7.594 | 0.1331 | 0.1559 | 0.7110 | 0.2590 | 0.2089 | 0.5321 |
| 75.00 | 7.599 | 0.1483 | 0.0993 | 0.7524 | 0.3282 | 0.1410 | 0.5308 |
| 75.10 | 7.610 | - | - | - | 0.2987 | 0.1705 | 0.5307 |
| 84.90 | 8.602 | 0.2202 | 0.0396 | 0.7403 | 0.4034 | 0.0506 | 0.5460 |
| 85.10 | 8.623 | - | - | - | 0.4232 | 0.0320 | 0.5449 |
| 84.97 | 8.610 | 0.2229 | 0.0242 | 0.7530 | 0.4242 | 0.0316 | 0.5443 |
| 84.90 | 8.602 | - | - | - | 0.3534 | 0.0922 | 0.5543 |
| 84.97 | 8.610 | 0.2152 | 0.0918 | 0.6930 | 0.3288 | 0.1093 | 0.5619 |
| 85.10 | 8.623 | 0.2177 | 0.1083 | 0.6739 | 0.3024 | 0.1241 | 0.5734 |
| 85.02 | 8.615 | 0.2178 | 0.1106 | 0.6716 | 0.2991 | 0.1260 | 0.5749 |
| 85.17 | 8.630 | 0.2394 | 0.1287 | 0.6319 | 0.2602 | 0.1382 | 0.6016 |
| 84.90 | 8.602 | - | - | - | 0.3628 | 0.0848 | 0.5524 |
| 84.90 | 8.602 | 0.2153 | 0.0675 | 0.7172 | 0.3643 | 0.0837 | 0.5520 |
| | | | T/K = 280.00 | | | | |
| 50.00 | 5.066 | 0.0709 | 0.0000 | 0.9291 | 0.2835 | 0.0000 | 0.7165 |
| 50.00 | 5.066 | 0.0000 | 0.1820 | 0.8180 | 0.0000 | 0.3352 | 0.6648 |
| 50.10 | 5.076 | 0.0609 | 0.0290 | 0.9101 | 0.2336 | 0.0600 | 0.7064 |
| 50.15 | 5.081 | 0.0467 | 0.0666 | 0.8866 | 0.1706 | 0.1340 | 0.6954 |
| 49.90 | 5.056 | 0.0296 | 0.1077 | 0.8627 | 0.1030 | 0.2113 | 0.6857 |
| 50.12 | 5.078 | 0.0139 | 0.1489 | 0.8372 | 0.0455 | 0.2814 | 0.6730 |
| 65.08 | 6.594 | 0.0900 | 0.1051 | 0.8048 | 0.2021 | 0.1575 | 0.6404 |
| 65.10 | 6.596 | 0.1037 | 0.0668 | 0.8295 | 0.2511 | 0.1041 | 0.6448 |
| 65.05 | 6.591 | 0.1173 | 0.0286 | 0.8541 | 0.3046 | 0.0460 | 0.6494 |
| 65.00 | 6.586 | 0.0723 | 0.1540 | 0.7737 | 0.1451 | 0.2183 | 0.6366 |
| 64.90 | 6.576 | 0.0565 | 0.1944 | 0.7491 | 0.1018 | 0.2620 | 0.6362 |
| 65.00 | 6.586 | 0.0483 | 0.2211 | 0.7306 | 0.0787 | 0.2830 | 0.6383 |
| 65.00 | 6.586 | 0.0530 | 0.2078 | 0.7392 | 0.0909 | 0.2725 | 0.6366 |
| 65.05 | 6.591 | 0.0438 | 0.2369 | 0.7193 | 0.0667 | 0.2925 | 0.6408 |
| 65.10 | 6.596 | 0.0471 | 0.2271 | 0.7258 | 0.0746 | 0.2861 | 0.6392 |
| 65.10 | 6.596 | 0.0454 | 0.2290 | 0.7256 | 0.0717 | 0.2892 | 0.6392 |
| 65.15 | 6.601 | 0.0425 | 0.2407 | 0.7168 | 0.0633 | 0.2947 | 0.6420 |
| 65.15 | 6.601 | 0.0423 | 0.2437 | 0.7139 | 0.0623 | 0.2957 | 0.6420 |
| 65.10 | 6.596 | 0.0400 | 0.2488 | 0.7112 | 0.0574 | 0.2985 | 0.6441 |
| 64.88 | 6.574 | 0.0360 | 0.2616 | 0.7024 | 0.0480 | 0.3021 | 0.6499 |
| 64.92 | 6.578 | 0.0394 | 0.2477 | 0.7129 | 0.0575 | 0.2989 | 0.6436 |
| 64.90 | 6.576 | 0.0349 | 0.2699 | 0.6951 | 0.0430 | 0.2996 | 0.6574 |
| 65.00 | 6.586 | 0.1277 | 0.0000 | 0.8723 | 0.3463 | 0.0000 | 0.6537 |
| 75.03 | 7.602 | 0.1663 | 0.0297 | 0.8040 | 0.3115 | 0.0400 | 0.6485 |
| 74.98 | 7.597 | 0.1573 | 0.0691 | 0.7736 | 0.2579 | 0.0873 | 0.6548 |
| 75.00 | 7.599 | 0.1534 | 0.1054 | 0.7412 | 0.2091 | 0.1222 | 0.6687 |
| 75.05 | 7.604 | 0.1566 | 0.1126 | 0.7307 | 0.1980 | 0.1260 | 0.6760 |
| 74.89 | 7.588 | 0.1596 | 0.1182 | 0.7222 | 0.1849 | 0.1273 | 0.6878 |
| 74.95 | 7.594 | 0.1580 | 0.1139 | 0.7281 | 0.1941 | 0.1255 | 0.6803 |
| 75.00 | 7.599 | 0.1743 | 0.0000 | 0.8257 | 0.3521 | 0.0000 | 0.6479 |

**COMPONENTS:**

(1) Nitrogen; $N_2$; [7727-37-9]

(2) Methane; $CH_4$; [74-82-8]

(3) Propane; $C_3H_8$; [74-98-6]

**ORIGINAL MEASUREMENTS:**

Poon, D. P. L.; Lu, B. C.-Y.

*Adv. Cryogen. Engng.* 1973, *19*, 292-9.

**VARIABLES:**

T/K = 114-122
P/MPa = 0.1-2.3
Composition

**PREPARED BY:**

C. L. Young

**EXPERIMENTAL VALUES:**

| T/K | $P$/psia | $P$/MPa | Mole fractions in liquid $x_{N_2}$ | $x_{CH_4}$ | Mole fractions in vapor $y_{N_2}$ | $y_{CH_4}$ |
|---|---|---|---|---|---|---|
| 114.1 | 70.9 | 0.4888 | 0.0245 | 0.0161 | 0.9886 | 0.0112 |
| | 127.2 | 0.8770 | 0.0446 | 0.0159 | 0.9901 | 0.0098 |
| | 193.4 | 1.333 | 0.0637 | 0.0136 | 0.9932 | 0.0067 |
| | 82.8 | 0.5709 | 0.0388 | 0.2624 | 0.8986 | 0.1013 |
| | 124.5 | 0.8584 | 0.0620 | 0.2606 | 0.9314 | 0.0685 |
| | 165.0 | 1.138 | 0.1252 | 0.2602 | 0.9496 | 0.0504 |
| | 65.8 | 0.4537 | 0.0562 | 0.6126 | 0.7851 | 0.2149 |
| | 101.4 | 0.6991 | 0.0989 | 0.5779 | 0.8644 | 0.1356 |
| | 124.1 | 0.8556 | 0.1797 | 0.5525 | 0.9074 | 0.0926 |
| | 16.6 | 0.1145 | 0.0000 | 0.9320 | 0.0000 | 0.9999 |
| | 43.4 | 0.2992 | 0.0503 | 0.8352 | 0.6241 | 0.3759 |
| | 41.8 | 0.2882 | 0.0525 | 0.8657 | 0.6057 | 0.3943 |
| | 77.6 | 0.5350 | 0.1270 | 0.7679 | 0.7902 | 0.2098 |
| | 131.7 | 0.9080 | 0.2999 | 0.6180 | 0.8866 | 0.1134 |
| | 158.3 | 1.091 | 0.4184 | 0.5117 | 0.9145 | 0.0855 |
| | 173.0 | 1.193 | 0.5072 | 0.4373 | 0.9283 | 0.0717 |
| | 185.3 | 1.278 | 0.6106 | 0.3566 | 0.9353 | 0.0647 |
| | 220.8 | 1.522 | 0.8059 | 0.1800 | 0.9593 | 0.0407 |
| | 233.7 | 1.611 | 0.8644 | 0.1248 | 0.9699 | 0.0301 |
| | 253.7 | 1.749 | 0.9415 | 0.0547 | 0.9811 | 0.0189 |
| | 260.6 | 1.797 | 0.9750 | 0.0248 | 0.9922 | 0.0078 |

(cont.)

**AUXILIARY INFORMATION**

**METHOD/APPARATUS/PROCEDURE:**

Recirculating vapor flow apparatus with cell consisting of 100 ml Jerguson gauge with stainless steel body. Temperature measured using copper-constantan thermocouples. Pressure measured using Bourdon gauges. Magnetic circulating pump. Cell charged, vapor recirculated for 2 or more hours and samples of vapor and liquid removed at constant pressure and analysed using gas chromatography. Details in source.

**SOURCE AND PURITY OF MATERIALS:**

(1) Matheson research grade sample, purity 99.997 mole per cent.

(2 and 3) Phillips Petroleum Co. sample, research grade, purity 99.99 mole per cent.

**ESTIMATED ERROR:**

$\delta T/K = \pm 0.05$; $\delta P/MPa = \pm 0.005$; $\delta x$, $\delta y = \pm 1\%$.

**REFERENCES:**

| COMPONENTS: | ORIGINAL MEASUREMENTS: |
|---|---|
| (1) Nitrogen; $N_2$; [7727-37-9] | Poon, D. P. L.; Lu, B. C.-Y. |
| (2) Methane; $CH_4$; [74-82-8] | *Adv. Cryogen. Engng.* 1973, *19*, 292-9. |
| (3) Propane; $C_3H_8$; [74-98-6] | |

EXPERIMENTAL VALUES:

| T/K | $P$/psia | $P$/MPa | Mole fractions in liquid $x_{N_2}$ | $x_{CH_4}$ | Mole fractions in vapor $y_{N_2}$ | $y_{CH_4}$ |
|---|---|---|---|---|---|---|
| 118.3 | 50.8 | 0.3503 | 0.0559 | 0.9126 | 0.5299 | 0.4701 |
| | 89.6 | 0.6178 | 0.1501 | 0.8217 | 0.7491 | 0.2509 |
| | 127.1 | 0.8763 | 0.2509 | 0.7187 | 0.8279 | 0.1721 |
| | 155.9 | 1.075 | 0.3559 | 0.6287 | 0.8633 | 0.1367 |
| | 175.8 | 1.212 | 0.4403 | 0.5481 | 0.8896 | 0.1104 |
| | 191.5 | 1.320 | 0.4954 | 0.4979 | 0.8938 | 0.1062 |
| | 214.4 | 1.478 | 0.5946 | 0.4007 | 0.9156 | 0.0844 |
| | 48.8 | 0.3365 | 0.0207 | 0.4689 | 0.6498 | 0.3501 |
| | 78.8 | 0.5433 | 0.0399 | 0.4499 | 0.7845 | 0.2155 |
| | 142.0 | 0.9791 | 0.0820 | 0.4169 | 0.8815 | 0.1184 |
| | 178.6 | 1.231 | 0.1221 | 0.3954 | 0.9022 | 0.0977 |
| | 19.8 | 0.1365 | 0.0000 | 0.7492 | - | - |
| | 106.8 | 0.7364 | 0.0970 | 0.6262 | 0.8147 | 0.1853 |
| | 150.5 | 1.038 | 0.1683 | 0.5672 | 0.8791 | 0.1209 |
| | 169.7 | 1.170 | 0.2258 | 0.5226 | 0.8969 | 0.1031 |
| | 193.8 | 1.336 | 0.2843 | 0.4869 | 0.9122 | 0.0878 |
| | 200.7 | 1.384 | 0.3033 | 0.4792 | 0.9044 | 0.0956 |
| | 11.8 | 0.0814 | 0.0000 | 0.2901 | - | - |
| | 59.0 | 0.4068 | 0.0227 | 0.2713 | 0.8067 | 0.1931 |
| | 102.2 | 0.7046 | 0.0470 | 0.2713 | 0.8836 | 0.1154 |
| | 123.7 | 0.8529 | 0.0601 | 0.2631 | 0.9098 | 0.0892 |
| | 147.0 | 1.014 | 0.0743 | 0.2531 | 0.9190 | 0.0802 |
| | 185.8 | 1.281 | 0.0967 | 0.2383 | 0.9324 | 0.0672 |
| | 24.6 | 0.1696 | 0.0071 | 0.0628 | 0.8782 | 0.1217 |
| | 64.3 | 0.4433 | 0.0199 | 0.0603 | 0.9505 | 0.0494 |
| | 105.3 | 0.7260 | 0.0324 | 0.0578 | 0.9671 | 0.0328 |
| | 140.8 | 0.9708 | 0.0443 | 0.0558 | 0.9795 | 0.0205 |
| | 245.9 | 1.695 | 0.0858 | 0.0491 | 0.9820 | 0.0179 |
| 122.2 | 3.5 | 0.0241 | 0.0000 | 0.0543 | - | - |
| | 30.6 | 0.2110 | 0.0080 | 0.0501 | 0.8948 | 0.1048 |
| | 64.8 | 0.4468 | 0.0190 | 0.0458 | 0.9564 | 0.0434 |
| | 142.8 | 0.9846 | 0.0505 | 0.0435 | 0.9794 | 0.0205 |
| | 15.3 | 0.1055 | 0.0000 | 0.2825 | - | - |
| | 67.3 | 0.4640 | 0.0203 | 0.2628 | 0.7711 | 0.2288 |
| | 108.6 | 0.7488 | 0.0360 | 0.2462 | 0.8591 | 0.1409 |
| | 145.2 | 1.001 | 0.0500 | 0.2388 | 0.8951 | 0.1048 |
| | 186.8 | 1.288 | 0.0767 | 0.2327 | 0.9207 | 0.0793 |
| | 23.2 | 0.1600 | 0.0000 | 0.5158 | - | - |
| | 48.3 | 0.3330 | 0.0156 | 0.4876 | 0.5374 | 0.4625 |
| | 88.8 | 0.6123 | 0.0378 | 0.4576 | 0.7492 | 0.2508 |
| | 173.8 | 1.198 | 0.0824 | 0.4073 | 0.8680 | 0.1319 |
| | 240.3 | 1.657 | 0.1443 | 0.3789 | 0.8875 | 0.1125 |
| | 26.8 | 0.1848 | 0.0000 | 0.7003 | - | - |
| | 51.8 | 0.3571 | 0.0215 | 0.6813 | 0.4699 | 0.5301 |
| | 79.4 | 0.5474 | 0.0450 | 0.6528 | 0.6528 | 0.3472 |
| | 199.2 | 1.373 | 0.1699 | 0.5419 | 0.8696 | 0.1304 |
| | 244.6 | 1.686 | 0.2678 | 0.4635 | 0.9054 | 0.0946 |
| | 30.1 | 0.2075 | 0.0000 | 0.9042 | - | - |
| | 83.8 | 0.5778 | 0.0846 | 0.8225 | 0.6484 | 0.3517 |
| | 137.3 | 0.9467 | 0.1853 | 0.7294 | 0.7908 | 0.2092 |
| | 185.1 | 1.276 | 0.3532 | 0.6102 | 0.8561 | 0.1439 |
| | 255.1 | 1.759 | 0.5875 | 0.4019 | 0.9023 | 0.0967 |
| | 327.4 | 2.257 | 0.8468 | 0.1510 | 0.9497 | 0.0503 |

COMPONENTS:

(1) Nitrogen; $N_2$; [7727-37-9]

(2) Methane; $CH_4$; [74-82-8]

(3) Butane; $C_4H_{10}$; [106-97-8]

ORIGINAL MEASUREMENTS:

Roberts, L. R.; McKetta, J. J.

*J. Chem. Eng. Data* 1963, *8*, 161-3.

VARIABLES:

T/K = 311-411
P/MPa = 3.4-20.7
Composition

PREPARED BY:

C. L. Young

EXPERIMENTAL VALUES:

| T/K | P/MPa | Mole fractions in liquid $x_{C_4H_{10}}$ | $x_{CH_4}$ | $x_{N_2}$ | Mole fractions in vapor $y_{C_4H_{10}}$ | $y_{CH_4}$ | $y_{N_2}$ |
|---|---|---|---|---|---|---|---|
| 310.9 | 3.447 | 0.895 | 0.074 | 0.031 | 0.150 | 0.434 | 0.416 |
| | | 0.856 | 0.131 | 0.013 | 0.149 | 0.691 | 0.160 |
| | | 0.918 | 0.040 | 0.042 | 0.154 | 0.187 | 0.659 |
| | 6.895 | 0.774 | 0.164 | 0.062 | 0.112 | 0.480 | 0.408 |
| | | 0.809 | 0.113 | 0.078 | 0.110 | 0.327 | 0.563 |
| | | 0.746 | 0.214 | 0.044 | 0.115 | 0.609 | 0.276 |
| | | 0.689 | 0.300 | 0.011 | 0.121 | 0.808 | 0.071 |
| | 10.34 | 0.731 | 0.165 | 0.104 | 0.109 | 0.344 | 0.547 |
| | | 0.661 | 0.242 | 0.097 | 0.114 | 0.475 | 0.411 |
| | | 0.567 | 0.384 | 0.049 | 0.128 | 0.708 | 0.164 |
| | | 0.746 | 0.115 | 0.139 | 0.096 | 0.246 | 0.658 |
| | 13.79 | 0.519 | 0.348 | 0.133 | 0.151 | 0.523 | 0.326 |
| | | 0.507 | 0.373 | 0.120 | 0.141 | 0.564 | 0.295 |
| | | 0.451 | 0.456 | 0.093 | 0.152 | 0.638 | 0.210 |
| | | 0.430 | 0.497 | 0.073 | 0.168 | 0.698 | 0.134 |
| | | 0.462 | 0.434 | 0.104 | 0.143 | 0.624 | 0.233 |
| | | -- | -- | -- | 0.133 | 0.540 | 0.327 |
| | | 0.663 | 0.137 | 0.200 | 0.113 | 0.233 | 0.654 |
| | | 0.540 | 0.149 | 0.201 | 0.110 | 0.249 | 0.641 |
| | | -- | -- | -- | 0.117 | 0.280 | 0.603 |
| | 17.24 | 0.546 | 0.196 | 0.258 | 0.129 | 0.269 | 0.602 |
| | | 0.456 | 0.302 | 0.242 | 0.146 | 0.399 | 0.455 |
| | | -- | -- | -- | 0.161 | 0.455 | 0.384 |

(cont.)

AUXILIARY INFORMATION

METHOD/APPARATUS/PROCEDURE:

Recirculating vapor flow apparatus with windowed high pressure cell. After equilibrium established fluids allowed to stand for 1 hour. Samples withdrawn and analysed either by mass spectrometry or gas chromatography.

SOURCE AND PURITY OF MATERIALS:

(1) Minimum purity 99.9 mole per cent.

(2) Pure grade minimum purity 99 mole per cent.

(3) Instrument grade minimum purity 99.5 mole per cent.

ESTIMATED ERROR:

$\delta T/K = \pm 0.06$; $\delta P/MPa = \pm 0.03$;
$\delta x_{N_2}$, $\delta x_{CH_4}$, $\delta x_{C_4H_{10}}$, $\delta y_{N_2}$, $\delta y_{CH_4}$, $\delta y_{C_4H_{10}} = \pm 0.002$.

REFERENCES:

| COMPONENTS: | ORIGINAL MEASUREMENTS: |
|---|---|
| (1) Nitrogen; $N_2$; [7727-37-9] | Roberts, L. R.; McKetta, J. J. |
| (2) Methane; $CH_4$; [74-82-8] | *J. Chem. Eng. Data* 1963, *8*, 161-3. |
| (3) Butane; $C_4H_{10}$; [106-97-8] | |

EXPERIMENTAL VALUES:

| T/K | P/MPa | Mole fractions in liquid $x_{C_4H_{10}}$ | $x_{CH_4}$ | $x_{N_2}$ | Mole fractions in vapor $y_{C_4H_{10}}$ | $y_{CH_4}$ | $y_{N_2}$ |
|---|---|---|---|---|---|---|---|
| 310.9 | 17.24 | 0.400 | 0.379 | 0.221 | 0.175 | 0.472 | 0.353 |
| | | 0.404 | 0.371 | 0.225 | 0.163 | 0.464 | 0.373 |
| | | 0.429 | 0.338 | 0.233 | 0.155 | 0.429 | 0.416 |
| | 20.68 | 0.538 | 0.092 | 0.370 | 0.134 | 0.119 | 0.747 |
| | | 0.508 | 0.136 | 0.356 | 0.158 | 0.167 | 0.675 |
| | | 0.450 | 0.195 | 0.355 | 0.167 | 0.234 | 0.599 |
| 344.3 | 3.447 | 0.907 | 0.067 | 0.026 | 0.338 | 0.334 | 0.328 |
| | 6.895 | 0.807 | 0.137 | 0.056 | 0.239 | 0.409 | 0.352 |
| | | 0.869 | 0.039 | 0.092 | 0.228 | 0.132 | 0.640 |
| | | 0.883 | 0.022 | 0.095 | 0.230 | 0.073 | 0.697 |
| | | 0.884 | 0.013 | 0.103 | 0.235 | 0.043 | 0.722 |
| | 10.34 | 0.750 | 0.117 | 0.133 | 0.218 | 0.249 | 0.533 |
| | | 0.733 | 0.143 | 0.124 | 0.235 | 0.295 | 0.470 |
| | | 0.699 | 0.192 | 0.109 | 0.228 | 0.386 | 0.386 |
| | | 0.661 | 0.257 | 0.082 | 0.242 | 0.490 | 0.268 |
| | | 0.639 | 0.301 | 0.060 | 0.252 | 0.570 | 0.178 |
| | | 0.642 | 0.296 | 0.062 | 0.245 | 0.561 | 0.194 |
| | 13.79 | 0.554 | 0.323 | 0.123 | 0.281 | 0.462 | 0.257 |
| | | 0.560 | 0.307 | 0.133 | 0.285 | 0.440 | 0.275 |
| | | 0.578 | 0.291 | 0.131 | 0.281 | 0.432 | 0.287 |
| | | 0.535 | 0.356 | 0.109 | 0.291 | 0.487 | 0.222 |
| | | 0.713 | 0.070 | 0.217 | 0.217 | 0.122 | 0.661 |
| | | 0.650 | 0.158 | 0.192 | 0.237 | 0.267 | 0.496 |
| | | 0.600 | 0.245 | 0.155 | 0.267 | 0.376 | 0.357 |
| | | 0.570 | 0.290 | 0.140 | 0.292 | 0.415 | 0.293 |
| | 17.34 | 0.507 | 0.203 | 0.290 | 0.347 | 0.237 | 0.416 |
| | | 0.630 | 0.067 | 0.303 | 0.265 | 0.095 | 0.640 |
| | | 0.543 | 0.166 | 0.291 | 0.313 | 0.213 | 0.474 |
| | | 0.531 | 0.189 | 0.280 | 0.318 | 0.232 | 0.450 |
| | | 0.952 | 0.0 | 0.048 | 0.604 | 0.0 | 0.396 |
| 377.6 | 3.447 | 0.934 | 0.051 | 0.015 | 0.588 | 0.273 | 0.139 |
| | | 0.918 | 0.078 | 0.004 | 0.602 | 0.356 | 0.042 |
| | | 0.942 | 0.029 | 0.029 | 0.601 | 0.143 | 0.256 |
| | 6.895 | 0.758 | 0.236 | 0.006 | 0.449 | 0.516 | 0.035 |
| | | 0.810 | 0.144 | 0.046 | 0.439 | 0.358 | 0.203 |
| | | 0.834 | 0.098 | 0.068 | 0.444 | 0.284 | 0.272 |
| | | 0.810 | 0.129 | 0.061 | 0.440 | 0.310 | 0.250 |
| | | 0.833 | 0.088 | 0.079 | 0.428 | 0.232 | 0.340 |
| | 10.34 | 0.730 | 0.089 | 0.181 | 0.415 | 0.162 | 0.423 |
| | | 0.739 | 0.073 | 0.188 | 0.395 | 0.129 | 0.475 |
| | | 0.720 | 0.122 | 0.158 | 0.417 | 0.218 | 0.365 |
| | | -- | -- | -- | 0.407 | 0.200 | 0.393 |
| | | 0.639 | 0.254 | 0.107 | 0.466 | 0.353 | 0.181 |
| | | -- | -- | -- | 0.401 | 0.218 | 0.381 |
| | | 0.673 | 0.188 | 0.139 | 0.430 | 0.280 | 0.290 |
| | | 0.628 | 0.265 | 0.107 | 0.470 | 0.354 | 0.176 |
| | | 0.640 | 0.246 | 0.114 | 0.455 | 0.357 | 0.188 |
| 410.9 | 3.447 | 0.989 | 0.002 | 0.009 | 0.943 | 0.014 | 0.043 |
| | | 0.987 | 0.004 | 0.009 | 0.940 | 0.016 | 0.044 |
| | 6.895 | 0.883 | 0.101 | 0.016 | 0.770 | 0.192 | 0.038 |
| | | 0.904 | 0.046 | 0.050 | 0.769 | 0.106 | 0.125 |
| | 10.34 | 0.849 | 0.052 | 0.099 | 0.754 | 0.080 | 0.166 |
| | | 0.849 | 0.041 | 0.110 | 0.746 | 0.071 | 0.183 |

COMPONENTS:

(1) Nitrogen; $N_2$; [7727-37-9]

(2) Ethene; $C_2H_4$; [74-85-1]

ORIGINAL MEASUREMENTS:

Cannon, W. A.; Robson, J.H.; English, W.D.

Report DAC-60510-F2. Contract NAS7-548 (1968). (Tech. Report Douglas Missile & Space Systems Division, Astropower Laboratory.)

VARIABLES:

T/K = 144-228
P/MPa = 2.1-4.8

PREPARED BY:

R. Battino

EXPERIMENTAL VALUES:

| $P_{N_2}$/psia | $P_{N_2}^a$/MPa | $x_1^b$ -200.2°F (144.2 K) | -150.1°F (172.0 K) | -100.1°F (199.6 K) | -50.2°F (227.5 K) |
|---|---|---|---|---|---|
| 300 | 2.07 | 0.0914 | 0.0660 | 0.0481 | 0.0263 |
| 300 | 2.07 | 0.0900 | 0.0643 | 0.0455 | 0.0264 |
| 300 | 2.07 | 0.0907 | 0.0648 | 0.0455 | 0.0258 |
| 300 | 2.07 | 0.0907 | 0.0655 | 0.0475 | 0.0258 |
| 700 | 4.83 | 0.241 | 0.161 | 0.137 | 0.108 |
| 700 | 4.83 | 0.251 | 0.166 | 0.131 | 0.109 |
| 700 | 4.83 | 0.246 | 0.160 | 0.132 | 0.111 |
| 700 | 4.83 | 0.246 | 0.168 | 0.133 | 0.114 |

[a] Calculated by compiler.

[b] Mole fraction solubility at indicated partial pressure.

AUXILIARY INFORMATION

METHOD/APPARATUS/PROCEDURE:

About 200 $cm^3$ of liquid propellant was condensed into a pressure vessel of about 1 liter capacity. Nitrogen was added to the proper pressure and the apparatus agitated for about one hour. A sample of the liquid phase was withdrawn and flashed to a gas at a lower temperature. The gas sample was then analyzed by gas chromatography.

SOURCE AND PURITY OF MATERIALS:

(1) Air Products and Chemicals, high purity grade, 99.9%.

(2) J. T. Baker Chemical Company, C. P. Grade, 99.0%.

ESTIMATED ERROR:

$\delta T/K = \pm 0.2$
$\delta P/P = \pm 0.02$
$\delta x_1/x_1 = \pm 0.02$

REFERENCES:

COMPONENTS:

(1) Nitrogen; $N_2$; [7727-37-9]

(2) Ethene; $C_2H_4$; [74-85-1]

ORIGINAL MEASUREMENTS:

Grauso, L.; Fredenslund, A.; Mollerup, J.

*Fluid Phase Equilibria* 1977, *1*, 13.

VARIABLES:

T/K = 200-260
P/MPa = 0.5-11

PREPARED BY:

C. L. Young

EXPERIMENTAL VALUES:

| T/K | $P$/bar | Mole fraction of nitrogen in liquid, $x_{N_2}$ | in vapor, $y_{N_2}$ |
|---|---|---|---|
| 200.0 | 4.543 | 0.0000 | 0.0000 |
| | 15.41 | 0.0307 | 0.6504 |
| | 28.94 | 0.0714 | 0.7762 |
| | 39.57 | 0.1058 | 0.8121 |
| | 60.33 | 0.1811 | 0.8294 |
| | 81.83 | 0.2742 | 0.8158 |
| | 102.64 | 0.3940 | 0.7654 |
| | 110.38 | 0.4735 | 0.7133 |
| 260.0 | 30.01 | 0.0000 | 0.0000 |
| | 36.80 | 0.0199 | 0.1144 |
| | 43.66 | 0.0413 | 0.1949 |
| | 52.41 | 0.0716 | 0.2595 |
| | 62.12 | 0.1080 | 0.3007 |
| | 70.92 | 0.1498 | 0.3139 |
| | 76.85 | 0.1939 | 0.2932 |

AUXILIARY INFORMATION

METHOD /APPARATUS/PROCEDURE:

Recirculating vapor flow apparatus. Temperature measured with platinum resistance thermometer. Pressure measured with dead weight piston balance. Coexisting samples analysed by gas chromatography. Details in source and ref. 1.

SOURCE AND PURITY OF MATERIALS:

(1) Purity 99.998 mole per cent.

(2) Purity 99.95 mole per cent.

ESTIMATED ERROR:

$\delta T/K = \pm 0.01$; $\delta P/\text{bar} = \pm 0.02\%$;

$\delta x_{N_2}$, $\delta y_{N_2} \leq \pm 0.003$.

REFERENCES:

1. Fredenslund, A.; Mollerup, J.; Christiansen, L. J. *Cryogenics* 1973, *13*, 405.

COMPONENTS:

(1) Nitrogen; $N_2$; [7727-37-9]

(2) Ethene; $C_2H_4$; [74-85-1]

ORIGINAL MEASUREMENTS:

Gasem, K. A. M.; Hiza, M. J.; Kidnay, A. J.

*Fluid Phase Equilib.* 1981, *6*, 181-9.

VARIABLES:

T/K = 120-200
P/MPa = 0.001-9.3

PREPARED BY:

C. L. Young

EXPERIMENTAL VALUES:

| T/K | P/MPa | Mole fraction of nitrogen in liquid*, $x_{N_2}$ | in gas, $y_{N_2}$ | T/K | P/MPa | Mole fraction of nitrogen in liquid*, $x_{N_2}$ | in gas, $y_{N_2}$ |
|---|---|---|---|---|---|---|---|
| 120.00 | 0.0012 | 0.0000 | 0.0000 | 140.00 | 0.0118 | 0.0000 | 0.0000 |
| | 0.644 | 0.0451 | 1.0000 | | 0.606 | -- | 0.9787 |
| | 1.035 | 0.0719 | 1.0000 | | 1.047 | 0.0469 | 0.9878 |
| | 1.503 | 0.1083 | 1.0000 | | 1.600 | 0.0736 | 0.9917 |
| | 2.086 | 0.1666 | 1.0000 | | 2.193 | 0.1038 | 0.9931 |
| | 2.285 ** | 0.2024 | 1.0000 | | 2.875 | 0.1424 | 0.9925 |
| 125.28 | 0.0027 | 0.0000 | 0.0000 | | 3.599 | 0.1947 | 0.9902 |
| | 0.496 | 0.0291 | 0.9970 | | 4.299 | 0.2373 | 0.9855 |
| | 1.006 | 0.0601 | 0.9994 | | 4.788 | 0.2792 | -- |
| | 1.507 | 0.0953 | 1.0000 | | 4.903 | -- | 0.9731 |
| | 1.818 | 0.1129 | 1.0000 | | 5.026 ** | 0.3019 | 0.9646 |
| | 2.150 | 0.1414 | 1.0000 | | 5.045 | -- | 0.9714 |
| | 2.635 | 0.1954 | 0.9994 | | 5.127 | 0.9316 | 0.9737 |
| | 2.880 ** | 0.2282 | 0.9981 | | 5.260 | -- | 0.9682 |
| | 2.910 | -- | 0.9996 | | 5.360 | 0.9481 | 0.9690 |
| | 3.031 | 0.9188 | - | | 5.410 | 0.9658 | 0.9717 |
| 134.00 | 0.0069 | 0.0000 | 0.0000 | 144.92 | 0.00182 | 0.0000 | 0.0000 |
| | 0.671 | 0.0338 | 0.9906 | | 1.602 | 0.0716 | 0.9864 |
| | 1.666 | 0.0851 | 0.9965 | | 3.097 | 0.1502 | 0.9882 |
| | 2.646 | 0.1517 | 0.9966 | | 4.536 | 0.2357 | 0.9826 |
| | 3.692 | 0.2302 | 0.9918 | | 5.110 | 0.2776 | 0.9776 |
| | 4.033 | 0.2516 | 0.9880 | | 5.593 | 0.3070 | 0.9658 |
| | 4.110 ** | 0.2710 | 0.9900 | | | | |

continued on following page

AUXILIARY INFORMATION

METHOD/APPARATUS/PROCEDURE:

Recirculating vapor flow apparatus. Temperature measured using a platinum resistance thermometer. Pressure measured using Bourdon gauge. Mixture equilibrated and samples analyzed using gas chromatography with a thermal conductivity detector. Details in source and refs. (1), (2) and (3).

SOURCE AND PURITY OF MATERIALS:

No details given.

ESTIMATED ERROR:

$\delta T/K = \pm 0.02$; $\delta P/MPa = \pm 0.005$ up to 5 MPa; $\pm 0.02$ above 5 MPa; $\delta x_{N_2}$, $\delta y_{N_2} = \pm 0.002$.

REFERENCES:


1. Kidnay, A.J.; Miller, R.C.; Parrish, W.R.; Hiza, M.J. *Cryogenics* 1975, *15*, 531.
2. Hiza, M.J.; Duncan, A.G. *Rev. Sci. Inst.* 1969, *40*, 513.
3. Duncan, A.G.; Hiza, M.J. *Adv. Cryog. Engng.* 1970, *15*, 42.

| COMPONENTS: | ORIGINAL MEASUREMENTS: |
|---|---|
| (1) Nitrogen; $N_2$; [7727-37-9]<br>(2) Ethene; $C_2H_4$; [74-85-1] | Gasem, K. A. M.; Hiza, M. J.; Kidnay, A. J.<br>*Fluid Phase Equilib.* 1981, *6*, 181-9. |

EXPERIMENTAL VALUES:

| T/K | *P*/MPa | Mole fraction of nitrogen in liquid*, $x_{N_2}$ | in gas, $y_{N_2}$ | T/K | *P*/MPa | Mole fraction of nitrogen in liquid*, $x_{N_2}$ | in gas, $y_{N_2}$ |
|---|---|---|---|---|---|---|---|
| 144.92 | 5.928 | 0.3502 | 0.9508 | 160.00 | 2.795 | 0.1062 | 0.9660 |
| | 5.978 | 0.3326 | 0.9508 | | 3.497 | 0.1350 | 0.9676 |
| | 6.008 | 0.6123 | 0.9570 | | 4.080 | 0.1615 | 0.9670 |
| | 6.353 | 0.7838 | 0.9574 | | 4.830 | 0.2023 | 0.9631 |
| | 6.360 | 0.6376 | - | | 6.558 | 0.2923 | 0.9473 |
| | 6.444 | 0.8004 | 0.9507 | | 8.200 | 0.4315 | 0.9079 |
| 160.00 | 0.0563 | 0.0000 | 0.0000 | | 9.342 | 0.6538 | 0.8631 |
| | 0.430 | 0.0167 | 0.8587 | 200.00 | 0.455 | 0.0000 | 0.0000 |
| | 0.580 | 0.0219 | 0.8927 | | 0.837 | 0.0132 | 0.4166 |
| | 0.720 | 0.0247 | 0.9113 | | 1.227 | -- | 0.5800 |
| | 0.817 | 0.0285 | 0.9210 | | 1.320 | 0.0236 | 0.6035 |
| | 0.897 | 0.0310 | 0.9367 | | 1.388 | -- | 0.6186 |
| | 1.120 | 0.0398 | 0.9391 | | 1.576 | 0.0351 | -- |
| | 1.152 | 0.0406 | 0.9402 | | 1.760 | 0.0394 | 0.6860 |
| | 1.470 | 0.0525 | 0.9511 | | 2.637 | 0.0648 | 0.7642 |
| | 1.809 | 0.0663 | 0.9567 | | 3.797 | 0.1021 | 0.8093 |
| | 2.150 | 0.0797 | 0.9618 | | 4.730 | 0.1332 | 0.8248 |

* Where liquid phase separation occurs, the liquid compositions are for the bottom liquid layer.

** Pressure at which phase separation occurs.

| COMPONENTS: | EVALUATOR: |
|---|---|
| (1) Nitrogen; $N_2$; [7727-37-9]<br>(2) Ethane; $C_2H_6$; [74-84-0] | Colin L. Young,<br>School of Chemistry,<br>University of Melbourne,<br>Parkville, Victoria 3052,<br>AUSTRALIA.<br>June 1981. |

CRITICAL EVALUATION:

This system has been investigated by seven different groups. The experimental data of Stryjek *et al.* (1) for the temperature range 140-195 K, of Gupta *et al.* (2) for the temperature range 260-280 K and of Grauso *et al.* (3) for the temperature range 200-290 K are all thought to be reliable and show good agreement when interpolated or extrapolated. The older data of Eakin *et al.* (4) and (5) are more or less consistent with the data of the above three groups but show considerably more scatter. The data of Chang and Lu (6) are only reported at two temperatures and their vapor-phase concentrations of ethane are almost certainly too large being two or three times those obtained by Eakin *et al.* (4), (5); therefore, the data are classified as doubtful. However the liquid phase compositions are in fair agreement with values interpolated or extrapolated from the data of Stryjek *et al.* (1). The data of Yu *et al.* (7) on this system are not considered here as they are mainly concerned with liquid-liquid-gas equilibrium exhibited by this system at low temperatures (below 135 K). The limited data of Cannon *et al.* (8) are classified as tentative, the data being in fair agreement with the data of Grauso *et al.* (3) although considerably less accurate.


References

1. Stryjek, R.; Chappelear, P. S.; Kobayashi, R. *J. Chem. Eng. Data* 1974, *19*, 340.

2. Gupta, M. K.; Gardner, G. C.; Hegarty, M. J.; Kidnay, A. J. *J. Chem. Eng. Data* 1980, *25*, 313.

3. Grauso, L.; Fredenslund, A.; Mollerup, J. *Fluid Phase Equilibria* 1977, *1*, 13.

4. Eakin, B. E.; Ellington, R. T.; Gami, D. C. *Inst. Gas Technol., Res. Bull. No. 26*, 1955.

5. Ellington, R. T.; Eakin, B. E.; Parent, J. D.; Bloomer, O. T. *Thermodynamic Properties of Gases, Liquids and Solids*, Amer. Soc. Mech. Engnrs., New York, 1959, p.180.

6. Chang, S. D.; Lu, B. C. Y. *Chem. Eng. Prog. Symp. Ser.* 1967, *63* (81), 18.

7. Yu, P.; Elshayal, I. M.; Lu, B. C. Y. *Can. J. Chem. Eng.* 1969, *47*, 495.

8. Cannon, W. A.; Robson, J. H.; English, W. D. *Report DAC-60510-F2*, 1968, (Tech. Report, Douglas Missile and Space Systems Division, Astropower Laboratory.)

COMPONENTS:

(1) Nitrogen; $N_2$; [7727-37-9]

(2) Ethane; $C_2H_6$; [74-84-0]

ORIGINAL MEASUREMENTS:

Ellington, R.T.; Eakin, B.E.; Parent, J.D.; Gami, D.C.; Bloomer, O.T.

*Thermodynamic Properties of Gases, Liquids and Solids,* Amer. Soc. Mech. Engineers: New York, 1959, p.180.

VARIABLES:

T/K = 101-298
P/MPa - 0.7-5.5

PREPARED BY:

C.L. Young,

EXPERIMENTAL VALUES:

| T/K | $P$/bar | Mole fraction of nitrogen in liquid, $x_{N_2}$ | T/K | $P$/bar | Mole fraction of nitrogen in liquid, $x_{N_2}$ |
|---|---|---|---|---|---|
| 159.9 | 10.34 | 0.0498 | 301.0 | 55.16 | 0.0498 |
| 170.5 | 13.79 | 0.0498 | 100.8 | 6.89 | 0.1501 |
| 193.5 | 17.24 | 0.0498 | 109.3 | 10.34 | 0.1501 |
| 213.4 | 20.68 | 0.0498 | 116.5 | 13.79 | 0.1501 |
| 228.7 | 24.13 | 0.0498 | 123.5 | 17.24 | 0.1501 |
| 240.7 | 27.56 | 0.0498 | 130.3 | 20.68 | 0.1501 |
| 251.0 | 31.03 | 0.0498 | 137.0 | 24.13 | 0.1501 |
| 260.0 | 34.47 | 0.0498 | 143.8 | 27.56 | 0.1501 |
| 267.7 | 37.92 | 0.0498 | 151.2 | 31.03 | 0.1501 |
| 274.5 | 41.37 | 0.0498 | 158.8 | 34.47 | 0.1501 |
| 280.8 | 44.82 | 0.0498 | 166.8 | 37.92 | 0.1501 |
| 286.9 | 48.26 | 0.0498 | 175.7 | 41.37 | 0.1501 |
| 292.7 | 51.71 | 0.0498 | 186.3 | 44.82 | 0.1501 |
| 297.7 | 55.16 | 0.0498 | 198.3 | 48.26 | 0.1501 |

AUXILIARY INFORMATION

METHOD/APPARATUS/PROCEDURE:

Dew point-bubble point apparatus with glass equilibrium cell. Temperature measured with copper-constantan thermocouple. Pressure measured using dead weight tester.

A variety of experimental procedures were adopted. Details in source and ref. 1. Bubble point taken as re-appearance of small vapor bubble on cooling at constant pressure.

SOURCE AND PURITY OF MATERIALS:

(1) Stated purity 99.99 mole per cent.

(2) Phillips Petroleum Research grade sample; purity 99.88 mole per cent.

ESTIMATED ERROR:

$\delta T/K = 0.03$; $\delta P/bar = 0.1$; $\delta x = 0.5\%$

(estimated by compiler).

REFERENCES:

1. Eakin, B.E.;Ellington, R.T.; Gami, D.C. *Inst. Gas Technol., Res. Bull No. 26* 1955 .

COMPONENTS:

(1) Nitrogen; $N_2$; [7727-37-9]

(2) Ethane; $C_2H_6$; [74-84-0]

ORIGINAL MEASUREMENTS:

Chang, S. D.; Lu, B. C. Y.
*Chem. Eng. Prog. Symp. Ser.*, 1967, 63(81), 18-27.

VARIABLES:

T/K = 122-171
P/MPa = 0.4-3.4

PREPARED BY:

C. L. Young

EXPERIMENTAL VALUES:

| T/K | P/bar | Mole fraction of nitrogen in liquid, $x_{N_2}$ | in vapor, $y_{N_2}$ |
|---|---|---|---|
| 171.4 | 3.87 | 0.02495 | 0.5564 |
| | 7.22 | 0.03795 | 0.8140 |
| | 10.29 | 0.05210 | 0.8665 |
| | 14.11 | 0.05823 | 0.9081 |
| | 21.01 | 0.06228 | 0.9300 |
| | 24.52 | 0.07307 | - |
| | 27.83 | 0.1213 | 0.9254 |
| | 34.11 | 0.08764 | 0.9184 |
| 122.0 | 3.44 | 0.04956 | 0.9958 |
| | 6.82 | 0.05040 | 0.9972 |
| | 10.11 | 0.07889 | 0.9980 |
| | 13.74 | 0.1317 | 0.9985 |
| | 20.60 | 0.1529 | 0.9982 |

AUXILIARY INFORMATION

METHOD/APPARATUS/PROCEDURE:

Recirculating vapor flow apparatus with cell consisting of 100 ml Jerguson gauge with stainless steel body. Temperature measured using copper-constantan thermocouples. Pressure measured using Bourdon gauges. Magnetic circulating pump. Cell charged vapor recirculated for 2 or more hours. Samples of vapor and liquid removed at constant pressure and analyzed using gas chromatography. Details in source.

SOURCE AND PURITY OF MATERIALS:

(1) Matheson research grade, purity 99.999 mole per cent.

(2) Matheson research grade, purity 99.9 mole per cent.

ESTIMATED ERROR:

$\delta T/K = \pm 0.1$; $\delta P/bar = \pm 0.5\%$;

$\delta x_{N_2} \simeq \delta y_{N_2} \leq 10\%$ (estimated by compiler).

REFERENCES:

| COMPONENTS: | ORIGINAL MEASUREMENTS: |
|---|---|
| (1) Nitrogen; $N_2$; [7727-37-9]<br>(2) Ethane; $C_2H_6$; [74-84-0] | Cannon, W. A.; Robson, J. H.; English, W.D.<br>Report DAC-60510-F2. Contract NAS7-548 (1968). (Tech. Report Douglas Missile & Space Systems Division, Astropower Laboratory.) |
| VARIABLES:<br>T/K = 144-228<br>P/MPa = 2.1-4.8 | PREPARED BY:<br>R. Battino |

EXPERIMENTAL VALUES:

| $P_{N_2}$/psia | $P_{N_2}^a$/MPa | $x_1^b$ -199.8°F (144.4 K) | -150.4°F (171.8 K) | -99.6°F (200.0 K) | -50.1°F (227.5 K) |
|---|---|---|---|---|---|
| 300 | 2.07 | 0.106 | 0.0714 | 0.0571 | 0.0478 |
| 300 | 2.07 | 0.106 | 0.0747 | 0.0571 | 0.0450 |
| 300 | 2.07 | 0.105 | 0.0737 | 0.0543 | 0.0487 |
| 300 | 2.07 | 0.107 | 0.0760 | 0.0558 | 0.0467 |
| 700 | 4.83 | 0.278 | 0.179 | 0.144 | 0.130 |
| 700 | 4.83 | 0.268 | 0.173 | 0.137 | 0.127 |
| 700 | 4.83 | 0.274 | 0.180 | 0.149 | 0.119 |
| 700 | 4.83 | 0.264 | 0.175 | 0.141 | 0.116 |

[a] Calculated by compiler.

[b] Mole fraction solubility at indicated partial pressure.

AUXILIARY INFORMATION

| METHOD/APPARATUS/PROCEDURE: | SOURCE AND PURITY OF MATERIALS: |
|---|---|
| About 200 $cm^3$ of liquid propellant was condensed into a pressure vessel of about 1 liter capacity. Nitrogen was added to the proper pressure and the apparatus agitated for about one hour. A sample of the liquid phase was withdrawn and flashed to a gas at a lower temperature. The gas sample was then analyzed by gas chromatography. | (1) Air Products and Chemicals, high purity grade, 99.9%.<br>(2) J.T. Baker, C.P. grade, 99.0%. |
| | ESTIMATED ERROR:<br>$\delta T/K = \pm 0.2$<br>$\delta P/P = \pm 0.02$<br>$\delta x_1/x_1 = \pm 0.02$ |
| | REFERENCES: |

| COMPONENTS: | ORIGINAL MEASUREMENTS: |
|---|---|
| (1) Nitrogen; $N_2$; [7727-37-9]<br>(2) Ethane; $C_2H_6$; [74-84-0] | Stryjek, R.; Chappelear, P. S.; Kobayashi, R.<br>*J. Chem. Eng. Data* 1974, *19*, 340. |
| VARIABLES:<br>T/K = 139-194<br>P/MPa - 0.4-13.5 | PREPARED BY:<br>C. L. Young |

EXPERIMENTAL VALUES:

| T/K | $P$/bar | Mole fraction of nitrogen in liquid, $x_{N_2}$ | in vapor, $y_{N_2}$ | T/K | $P$/bar | Mole fraction of nitrogen in liquid, $x_{N_2}$ | in vapor, $y_{N_2}$ |
|---|---|---|---|---|---|---|---|
| 138.71 | 3.53 | 0.01882 | 0.98645 | 149.82 | 119.14[a] | 0.7083 | 0.7083 |
| | 6.89 | 0.03870 | 0.99164 | | 88.87[b] | 0.465 | 0.9001 |
| | 10.34 | 0.05790 | 0.99338 | | 119.14[a,b] | 0.732 | 0.732 |
| | 13.79 | 0.0776 | 0.99409 | 172.04 | 3.62 | 0.01229 | 0.8449 |
| | 20.68 | 0.1192 | 0.99444 | | 6.94 | 0.02320 | 0.9132 |
| | 27.56 | 0.1637 | 0.99420 | | 13.72 | 0.04863 | 0.94746 |
| | 34.47 | 0.2120 | 0.99308 | | 13.79 | 0.04845 | 0.94778 |
| | 69.09 | 0.3740 | 0.96216 | | 27.56 | 0.09903 | 0.96368 |
| | 68.95 | 0.3897 | 0.95515 | | 41.78 | 0.1552 | 0.96635 |
| 149.82 | 3.45 | 0.01563 | 0.96043 | | 55.30 | 0.2090 | 0.96215 |
| | 6.93 | 0.03178 | 0.97867 | | 68.81 | 0.2689 | 0.95600 |
| | 10.34 | 0.04796 | 0.98411 | | 90.39 | 0.3755 | 0.92771 |
| | 13.81 | 0.06400 | 0.98752 | | 103.42 | 0.4327 | 0.9012 |
| | 20.68 | 0.09698 | 0.98937 | | 120.66 | 0.5626 | 0.8138 |
| | 27.56 | 0.1343 | 0.98985 | | 124.93 | 0.6341 | 0.7704 |
| | 34.47 | 0.1685 | 0.98834 | | 125.35[a] | 0.7080 | 0.7080 |
| | 41.44 | 0.2058 | 0.98656 | 194.26 | 3.52 | 0.005715 | 0.4862 |
| | 55.50 | 0.2839 | 0.98097 | | 6.92 | 0.01612 | 0.7146 |
| | 69.09 | 0.3531 | 0.95918 | | 13.79 | 0.03722 | 0.8360 |
| | 88.87 | 0.4336 | 0.9001 | | | | |

(cont.)

[a] Critical point; [b] Smoothed value.

AUXILIARY INFORMATION

METHOD /APPARATUS/PROCEDURE:

Vapor recycled through equilibrium cell by magnetic pump. Temperature measured by platinum resistance thermometer using a Mueller bridge. Pressure measured with Heise gauge. Samples of both phases analyzed with a thermal conductivity detector (1-3). After equilibrium established vapor phase analyzed then recycle pump stopped and liquid phase sampled (1-3).

SOURCE AND PURITY OF MATERIALS:

(1) Matheson sample; purity 99.99 mole per cent.

(2) Specified purity 99.99 mole per cent.

ESTIMATED ERROR:

$\delta T/K \sim 0.02$; $\delta P/\text{bar} < 1$; $\delta x_{N_2} \sim \delta y_{N_2} = 0.2\text{-}1.0\%$.

REFERENCES:


1. Stryjek, R. Monograph. Rice University, 1972.
2. Wichterle, I.; Kobayashi, R. *J. Chem. Eng. Data* 1972, *17*, 4.
3. Stryjek, R.; Chappelear, P. S.; Kobayashi, R. *J. Chem. Eng. Data* 1974 *19*, 334.

| COMPONENTS: | ORIGINAL MEASUREMENTS: |
|---|---|
| (1) Nitrogen; $N_2$; [7727-37-9] | Stryjek, R.; Chappelear, P. S.; Kobayashi, R. |
| (2) Ethane; $C_2H_6$; [74-84-0] | *J. Chem. Eng. Data* 1974, *19*, 340. |

EXPERIMENTAL VALUES:

| T/K | $P$/bar | Mole fraction of nitrogen in liquid, $x_{N_2}$ | Mole fraction of nitrogen in vapor, $y_{N_2}$ | T/K | $P$/bar | Mole fraction of nitrogen in liquid, $x_{N_2}$ | Mole fraction of nitrogen in vapor, $y_{N_2}$ |
|---|---|---|---|---|---|---|---|
| 194.26 | 27.56 | 0.08081 | 0.8966 | 194.26 | 120.24 | 0.4534 | 0.8355 |
| | 41.51 | 0.1273 | 0.9126 | | 121.35 | 0.4577 | 0.8304 |
| | 55.71 | 0.1756 | 0.9154 | | 130.86 | 0.5510 | 0.7714 |
| | 69.09 | 0.2225 | 0.9126 | | 134.17 | 0.6472 | 0.7144 |
| | 89.77 | 0.3024 | 0.8977 | | 134.65[a] | 0.6746 | 0.6746 |
| | 97.56 | 0.3338 | 0.8885 | | | | |

[a] Critical point; [b] Smoothed value.

**COMPONENTS:**

(1) Nitrogen; $N_2$; [7727-37-9]

(2) Ethane; $C_2H_6$; [74-84-0]

**ORIGINAL MEASUREMENTS:**

Grauso, L.; Fredenslund, A.; Mollerup, J.

*Fluid Phase Equilibria* 1977, *1*, 13.

**VARIABLES:**

T/K = 200-230
P/MPa = 0.2-13

**PREPARED BY:**

C. L. Young

**EXPERIMENTAL VALUES:**

| T/K | $P$/bar | Mole fraction of nitrogen in liquid, $x_{N_2}$ | Mole fraction of nitrogen in vapor, $y_{N_2}$ |
|---|---|---|---|
| 200.0 | 2.182 | 0.0000 | 0.0000 |
| | 6.66 | 0.0137 | 0.6389 |
| | 10.43 | 0.0264 | 0.7533 |
| | 21.17 | 0.0647 | 0.8568 |
| | 41.54 | 0.1360 | 0.9029 |
| | 60.99 | 0.2063 | 0.9064 |
| | 80.48 | 0.2764 | 0.8981 |
| | 102.42 | 0.3751 | 0.8716 |
| | 124.79 | 0.5112 | 0.8116 |
| | 131.95 | 0.5738 | 0.7620 |
| 230.0 | 7.005 | 0.0000 | 0.0000 |
| | 10.54 | 0.0094 | 0.2927 |
| | 15.70 | 0.0240 | 0.4889 |
| | 21.56 | 0.0399 | 0.6029 |
| | 34.52 | 0.0776 | 0.7129 |
| | 54.05 | 0.1372 | 0.7660 |
| | 80.72 | 0.2265 | 0.7796 |
| | 101.77 | 0.3120 | 0.7566 |
| 230.0 | 119.58 | 0.4110 | 0.7047 |
| | 124.76 | 0.4543 | 0.6650 |
| 260.0 | 17.12 | 0.0000 | 0.0000 |
| | 25.72 | 0.0235 | 0.2562 |
| | 31.15 | 0.0393 | 0.3478 |
| | 44.83 | 0.0796 | 0.4753 |
| | 56.46 | 0.1160 | 0.5287 |
| | 71.06 | 0.1647 | 0.5576 |
| | 88.49 | 0.2373 | 0.5558 |
| | 100.55 | 0.3077 | 0.5209 |
| 290.0 | 35.13 | 0.0000 | 0.0000 |
| | 42.68 | 0.0242 | 0.0970 |
| | 49.54 | 0.0464 | 0.1586 |
| | 55.49 | 0.0709 | 0.1940 |
| | 61.30 | 0.0936 | 0.2163 |
| | 68.32 | 0.1322 | 0.2196 |
| | 70.18 | 0.1594 | 0.2115 |

**AUXILIARY INFORMATION**

**METHOD /APPARATUS/PROCEDURE:**

Recirculating vapor flow apparatus. Temperature measured with platinum resistance thermometer. Pressure measured with dead weight piston balance. Coexisting samples analysed by gas chromatography. Details in source and ref. 1.

**SOURCE AND PURITY OF MATERIALS:**

(1) Purity 99.998 mole per cent.

(2) Purity 99.95 mole per cent.

**ESTIMATED ERROR:**

$\delta T/K = \pm 0.001$; $\delta P/\text{bar} = \pm 0.02\%$;
$\delta x_{N_2} = \delta y_{N_2} \leq \pm 0.003$.

**REFERENCES:**

1. Fredenslund, A.; Mollerup, J.; Christiansen, L. J. *Cryogenics* 1973, *13*, 405.

COMPONENTS:

(1) Nitrogen; $N_2$; [7727-37-9]

(2) Ethane; $C_2H_6$; [74-84-0]

ORIGINAL MEASUREMENTS:

Gupta, M.K.; Gardner, G.C.; Hegarty, M. J.; Kidnay, A. J.

*J. Chem. Eng. Data* 1980, *25*, 313-8.

VARIABLES:

T/K = 260-280
P/MPa = 2-10

PREPARED BY:

C.L. Young

EXPERIMENTAL VALUES:

| T/K | Total pressure p/atm | p/MPa | Mole fraction in nitrogen in liquid $x_{N_2}$ | in gas $y_{N_2}$ |
|---|---|---|---|---|
| 260.00 | 16.80 | 1.702 | 0.0000 | 0.0000 |
| | 18.10 | 1.834 | 0.0036 | 0.0523 |
| | 20.59 | 2.086 | 0.0106 | 0.1375 |
| | 22.55 | 2.285 | 0.0161 | - |
| | 25.63 | 2.597 | 0.0254 | - |
| | 35.36 | 3.583 | 0.0541 | 0.4065 |
| | 44.88 | 4.547 | 0.0828 | 0.4823 |
| | 65.05 | 6.591 | 0.1494 | 0.5541 |
| | 75.20 | 7.620 | 0.1873 | 0.5653 |
| | 84.27 | 8.539 | 0.2258 | 0.5632 |
| | 93.80 | 9.504 | 0.2714 | 0.5473 |
| 270.00 | 21.68 | 2.197 | 0.0000 | 0.0000 |
| | 23.50 | 2.381 | 0.0052 | 0.0523 |
| | 24.35 | 2.467 | 0.0077 | 0.0770 |
| | 25.13 | 2.546 | 0.0098 | 0.0942 |
| | 27.65 | 2.802 | 0.0157 | 0.1489 |
| | 28.63 | 2.901 | 0.0193 | 0.1698 |
| | 30.00 | 3.040 | 0.0233 | 0.1889 |
| | 30.40 | 3.080 | 0.0251 | - |
| | 32.52 | 3.295 | 0.0302 | 0.2378 |
| | 32.62 | 3.305 | 0.0302 | 0.2262 |
| | 34.52 | 3.498 | 0.0359 | 0.2636 |
| | 34.70 | 3.516 | 0.0382 | 0.2646 |
| | 35.26 | 3.573 | 0.0390 | 0.2668 |
| | 37.31 | 3.780 | 0.0435 | 0.2938 |

AUXILIARY INFORMATION

METHOD/APPARATUS/PROCEDURE:

Recirculating vapor flow apparatus with diaphragm pump. Temperature measured with platinum resistance thermometer and pressure with Bourdon gauges. Cell stirred with double propeller stirrer. Vapor and liquid samples analysed by gas chromatography using a thermal conductivity detector. Details in ref. (1).

SOURCE AND PURITY OF MATERIALS:

(1) Purity at least 99.7 mole per cent.

(2) Purity at least 99 mole per cent

No extraneous peaks were found when samples analysed by gas chromatography.

ESTIMATED ERROR:

$\delta T/K = \pm 0.02$; $\delta p/MPa = \pm 0.03$ up to 3.4 MPa, $\pm$ 0.1 above 3.4 MPa; $\delta x_{N_2}$, $\delta y_{N_2} = \pm 0.002$

REFERENCES:

1. Somait, F.; Kidnay, A.J. *J. Chem. Eng. Data* 1978, *23*, 301.

| COMPONENTS: | ORIGINAL MEASUREMENTS: |
|---|---|
| (1) Nitrogen; $N_2$; [7727-37-9]<br>(2) Ethane; $C_2H_6$; [74-84-0] | Gupta, M.K.; Gardner, G.C.; Hegarty, M. J.; Kidnay, A. J.<br>*J. Chem. Eng. Data* 1980, *25*, 313-8. |

| T/K | Total pressure p/atm | Total pressure p/MPa | Mole fraction in nitrogen in liquid $x_{N_2}$ | Mole fraction in nitrogen in gas $y_{N_2}$ |
|---|---|---|---|---|
| 270.00 | 40.39 | 4.093 | 0.0543 | 0.3216 |
| | 40.53 | 4.107 | 0.0563 | 0.3321 |
| | 41.19 | 4.174 | 0.0554 | 0.3339 |
| | 45.50 | 4.610 | 0.0696 | 0.3618 |
| | 47.42 | 4.805 | 0.0687 | 0.3811 |
| | 50.31 | 5.098 | 0.0848 | 0.3921 |
| | 53.47 | 5.418 | 0.0951 | 0.4156 |
| | 55.20 | 5.593 | 0.0989 | 0.4127 |
| | 57.70 | 5.846 | 0.1076 | 0.4290 |
| | 60.18 | 6.098 | 0.1165 | 0.4291 |
| | 65.62 | 6.649 | - | 0.4506 |
| | 70.43 | 7.136 | 0.1555 | 0.4545 |
| | 80.50 | 8.157 | 0.2030 | 0.4631 |
| | 80.59 | 8.166 | - | 0.4631 |
| | 81.42 | 8.250 | 0.2079 | 0.4611 |
| | 84.12 | 8.523 | 0.2263 | - |
| | 85.45 | 8.658 | 0.2306 | - |
| | 89.81 | 9.100 | 0.2572 | 0.4652 |
| | 90.71 | 9.191 | 0.2636 | - |
| | 93.25 | 9.449 | 0.2849 | - |
| | 93.50 | 9.474 | - | 0.4594 |
| | 93.70 | 9.494 | 0.2931 | - |
| | 96.42 | 9.770 | - | 0.4446 |
| | 96.45 | 9.773 | - | 0.4475 |
| | 96.67 | 9.795 | 0.3361 | - |
| | 97.72 | 9.901 | 0.3742 | - |
| 280.00 | 27.60 | 2.797 | 0.0000 | 0.0000 |
| | 29.50 | 2.989 | 0.0061 | 0.0437 |
| | 30.72 | 3.113 | 0.0097 | 0.0684 |
| | 31.80 | 3.222 | 0.0128 | 0.0866 |
| | 36.23 | 3.671 | 0.0265 | 0.1558 |
| | 40.32 | 4.085 | 0.0394 | 0.2047 |
| | 45.35 | 4.595 | 0.0556 | 0.2511 |
| | 51.69 | 5.237 | 0.0767 | 0.2943 |
| | 55.89 | 5.663 | 0.0921 | 0.3167 |
| | 60.20 | 6.100 | 0.1078 | 0.3331 |
| | 65.30 | 6.617 | 0.1284 | 0.3471 |
| | 71.68 | 7.263 | 0.1570 | 0.3542 |
| | 76.90 | 7.792 | 0.1852 | 0.3484 |
| | 80.25 | 8.131 | 0.2088 | 0.3360 |
| | 80.90 | 8.197 | 0.2162 | 0.3339 |
| | 82.25 | 8.334 | 0.2311 | 0.3177 |
| | 83.55 | 8.466 | single phase | |

| COMPONENTS: | ORIGINAL MEASUREMENTS: |
|---|---|
| (1) Nitrogen; $N_2$; [7727-37-9]<br>(2) Propene; $C_3H_6$; [115-07-1] | Blagoi, Yu. P.; Orobinskii, N. A.<br>*Zhur. Fiz. Khim.* 1965, *39*, 2022-4. |

| VARIABLES: | PREPARED BY: |
|---|---|
| T/K = 79-91<br>P/KPa = 31-258 | C. L. Young |

EXPERIMENTAL VALUES:

| T/K | $P/10^5$Pa | Mole fraction of nitrogen in liquid, $x_{N_2}$ | T/K | $P/10^5$Pa | Mole fraction of nitrogen in liquid, $x_{N_2}$ |
|---|---|---|---|---|---|
| 90.5 | 0.4523 | 0.004 | 85.95 | 2.502 | 0.030 |
| 88.3 | 0.4123 | 0.004 | 85.75 | 2.474 | 0.030 |
| 85.55 | 0.3664 | 0.004 | 83.95 | 2.084 | 0.030 |
| 82.00 | 0.3111 | 0.004 | 83.90 | 2.064 | 0.030 |
| 90.5 | 0.933 | 0.009 | 81.50 | 1.621 | 0.030 |
| 85.95 | 0.779 | 0.009 | 79.4 | 1.272 | 0.030 |
| 82.35 | 0.666 | 0.009 | 85.95 | 2.531 | 0.033 |
| 83.25 | 1.199 | 0.015 | 85.85 | 2.498 | 0.033 |
| 81.30 | 1.092 | 0.015 | 84.10 | 2.116 | 0.033 |
| 78.6 | 0.975 | 0.015 | 82.75 | 1.845 | 0.033 |
| 90.2 | 2.158 | 0.021 | 82.35 | 1.776 | 0.033 |
| 87.5 | 1.896 | 0.021 | 79.40 | 1.282 | 0.033 |
| 84.10 | 1.601 | 0.021 | 79.05 | 1.224 | 0.033 |
| 79.6 | 1.271 | 0.021 | 85.85 | 2.498 | 0.070 |
| 89.55 | 2.451 | 0.024 | 84.20 | 2.134 | 0.070 |
| 86.00 | 2.053 | 0.024 | 82.35 | 1.781 | 0.070 |
| 81.05 | 1.505 | 0.024 | 78.00 | 1.091 | 0.070 |
| 87.10 | 2.531 | 0.027 | 85.95 | 2.531 | 0.342 |
| 85.05 | 2.125 | 0.027 | 83.00 | 1.906 | 0.342 |
| 81.95 | 1.679 | 0.027 | 79.6 | 1.298 | 0.342 |
| 86.25 | 2.578 | 0.030 | | | |

AUXILIARY INFORMATION

METHOD/APPARATUS/PROCEDURE:

Recirculating vapor flow apparatus fitted with magnetic pump. Temperature measured with platinum resistance thermometer; pressure measured with Bourdon gauge. Samples of liquid phase analysed by gas chromatography. Details in ref. (1).

SOURCE AND PURITY OF MATERIALS:

No details given.

ESTIMATED ERROR:

$\delta T/K = \pm 0.1$; $\delta P/kPa = \pm 0.5$; $\delta x_{N_2} = \pm 0.001$.

REFERENCES:

1. Blagoi, Yu. P.; Orobinskii, N. A. *Ukrain. Fiz. Zhur.* 1963, *8*, 1378.

COMPONENTS:

(1) Nitrogen; $N_2$; [7727-37-9]

(2) Propene; $C_3H_6$; [115-07-1]

ORIGINAL MEASUREMENTS:

Grauso, L.; Fredenslund, A.; Mollerup, J.

*Fluid Phase Equilibria* 1977, *1*, 13.

VARIABLES:

T/K = 260-290

P/MPa = 0.4-16.5

PREPARED BY:

C. L. Young

EXPERIMENTAL VALUES:

| T/K | P/bar | Mole fraction of nitrogen in liquid, $x_{N_2}$ | in vapor, $y_{N_2}$ |
|---|---|---|---|
| 260.0 | 3.829 | 0.0000 | 0.0000 |
| | 17.16 | 0.0234 | 0.7382 |
| | 36.70 | 0.0595 | 0.8461 |
| | 56.70 | 0.0966 | 0.8764 |
| | 86.32 | 0.1536 | 0.8869 |
| | 118.95 | 0.2195 | 0.8790 |
| | 149.86 | 0.2874 | 0.8617 |
| | 173.03 | 0.3448 | 0.8393 |
| | 191.11 | 0.3945 | 0.8129 |
| | 214.44 | 0.4882 | 0.7503 |
| 290.0 | 9.370 | 0.0000 | 0.0000 |
| | 17.32 | 0.0144 | 0.3843 |
| | 23.11 | 0.0248 | 0.5068 |
| | 31.97 | 0.0413 | 0.6069 |
| | 42.64 | 0.0604 | 0.6734 |
| | 55.37 | 0.0855 | 0.7122 |
| | 70.44 | 0.1148 | 0.7399 |
| | 86.20 | 0.1473 | 0.7500 |
| | 111.38 | 0.2028 | 0.7505 |
| | 133.14 | 0.2589 | 0.7335 |
| | 150.07 | 0.3171 | 0.7104 |
| | 165.39 | 0.3734 | 0.6677 |

AUXILIARY INFORMATION

METHOD/APPARATUS/PROCEDURE:

Recirculating vapor flow apparatus. Temperature measured with platinum resistance thermometer. Pressure measured with dead weight piston balance. Coexisting samples analysed by gas chromatography. Details in source and ref. 1.

SOURCE AND PURITY OF MATERIALS:

(1) Purity 99.998 mole per cent.

(2) Purity 99.99 mole per cent.

ESTIMATED ERROR:

$\delta T/K = \pm 0.01$; $\delta P/bar = \pm 0.02\%$; $\delta x_{N_2}$, $\delta y_{N_2} \leq 0.003$.

REFERENCES:

1. Fredenslund, A.; Mollerup, J.; Christiansen, L. J. *Cryogenics* 1973, *13*, 405.

| COMPONENTS: | EVALUATOR: |
|---|---|
| (1) Nitrogen; $N_2$; [7727-37-9]<br><br>(2) Propane; $C_3H_8$; [74-98-6] | Colin L. Young,<br>School of Chemistry,<br>University of Melbourne,<br>Parkville, Victoria 3052,<br>AUSTRALIA.<br><br>June 1981. |

CRITICAL EVALUATION:

This system has been investigated by five groups. The study of Roof and Baron (1) was restricted to critical loci and is not considered further. Cheung and Wang (2) made a fairly limited study of the system at temperatures between 91 and 128 K at pressures up to 0.5 MPa. These data are classified as tentative but are not considered further because of their limited nature. The data of Grauso *et al.* (3) are thought to be reliable. The isotherms studied by Grauso *et al.* (3) at 230, 260 and 290 K are in good agreement with values interpolated from the data of Schindler *et al.* (4). However there are considerable differences between the data of Poon and Lu (5) and Schindler *et al.* (4). The isotherms of Poon and Lu at 114.1, 118.3 and 122.2 K are probably more accurate than those of Schindler *et al.* (4) at similar temperatures since it is probable that the sampling technique used by Schindler *et al.* (4) is less reliable for low nitrogen concentrations in the liquid than the technique used by Poon and Lu (5). Therefore the data of Grauso *et al.* (3) and Poon and Lu (5) are classified as tentative whereas the data of Schindler *et al.* (4) are classified as doubtful at low temperatures (below 250 K). The limited data of Cannon et al (6) are classified as tentative, the data being in fair agreement with the data of Grauso et al. (3) but considerably less accurate.

References

1. Roof, J. G.; Baron, J. D.
   *J. Chem. Eng. Data* 1967, *12*, 292.

2. Cheung, H.; Wang, D. I. J.
   *Ind. Eng. Chem. Fund.* 1964, *3*, 355.

3. Grauso, L.; Fredenslund, A.; Mollerup, J.
   *Fluid Phase Equilibria* 1977, *1*, 13.

4. Schindler, D. L.; Swift, G. W.; Kurata, F.
   *Hydrocarbon Process.* 1966, *45*, 205.

5. Poon, D. P. L.; Lu, B. C. Y.
   *Adv. Cryog. Engng.* 1973, *19*, 292.

6. Cannon, W. A.; Robson, J. H.; English, W. D.;
   *Report DAC-60510-F2*, 1968, (Tech Report, Douglas Missile and Space Systems Division, Astropower Laboratory).

**COMPONENTS:**

(1) Nitrogen; $N_2$; [7727-37-9]

(2) Propane; $C_3H_8$; [74-98-6]

**ORIGINAL MEASUREMENTS:**

Cheung, H.; Wang, D. I. J.
*Ind. Eng. Chem. Fund.* 1964, *3*, 355.

**VARIABLES:**

T/K = 92-128

P/MPa = 0.1-0.6

**PREPARED BY:**

C. L. Young

**EXPERIMENTAL VALUES:**

| T/K | $P$/bar | Mole fraction of nitrogen in liquid, $x_{N_2}$ |
|---|---|---|
| 91.9 | 1.195 | 0.0157 |
| 91.9 | 4.046 | 0.0670 |
| 97.4 | 4.969 | 0.0610 |
| 103.7 | 5.838 | 0.0569 |
| 110.2 | 1.846 | 0.0124 |
| 128.2 | 2.387 | 0.0107 |

The mole fraction of nitrogen in the gas phase is essentially unity for these pressures and temperatures (authors).

**AUXILIARY INFORMATION**

**METHOD/APPARATUS/PROCEDURE:**

Static equilibrium cell fitted with magnetic stirrer. Pressure measured on mercury manometer. Temperature measured by thermocouple. Details of apparatus in source reference. Known quantity solvent added to evacuated cell. Metered quantity of solute added. Solubility determined from equilibrium pressure and mass balance. Details in source ref.

**SOURCE AND PURITY OF MATERIALS:**

Not given.

**ESTIMATED ERROR:**

$\delta$T/K = ±0.2; $\delta P$/bar = ±0.0015;

$\delta x_{N_2}$ = ±2% (estimated by compiler).

**REFERENCES:**

**COMPONENTS:**

(1) Nitrogen; $N_2$; [7727-37-9]

(2) Propane; $C_3H_8$; [74-98-6]

**ORIGINAL MEASUREMENTS:**

Schindler, D. L.; Swift, G. W.; Kurata, F.

*Hydrocarbon Process.*, 1966, *45* (11), 205.

**VARIABLES:**

T/K = 103-353

P/MPa = 0.7 - 13.8

**PREPARED BY:**

C. L. Young

**EXPERIMENTAL VALUES:**

| T/K | P/bar | Mole fraction of nitrogen in liquid, $x_{N_2}$ | Mole fraction of nitrogen in vapor, $y_{N_2}$ | T/K | P/bar | Mole fraction of nitrogen in liquid, $x_{N_2}$ | Mole fraction of nitrogen in vapor, $y_{N_2}$ |
|---|---|---|---|---|---|---|---|
| 353.15 | 41.37 | 0.032 | 0.143 | 248.15 | 96.53 | 0.185 | 0.926 |
| | 55.16 | 0.053 | 0.230 | 223.15 | 13.79 | 0.025 | 0.9290 |
| | 68.95 | 0.139 | 0.264 | | 41.37 | 0.077 | 0.9680 |
| 343.15 | 41.37 | 0.040 | 0.225 | | 68.95 | 0.129 | 0.9686 |
| | 55.16 | 0.084 | 0.307 | | 96.53 | 0.181 | 0.9639 |
| | 68.95 | 0.128 | 0.358 | 198.15 | 13.79 | 0.029 | 0.9812 |
| | 82.74 | 0.187 | 0.375 | | 55.16 | 0.085 | 0.9900 |
| 333.15 | 41.37 | 0.049 | 0.350 | | 68.95 | 0.137 | 0.9880 |
| | 68.95 | 0.128 | 0.475 | | 96.53 | 0.186 | 0.9842 |
| 323.15 | 41.37 | 0.055 | 0.422 | 173.15 | 13.79 | 0.032 | 0.9952 |
| | 68.95 | 0.124 | 0.547 | | 41.37 | 0.091 | 0.9965 |
| 298.15 | 13.79 | 0.009 | 0.239 | | 68.95 | 0.142 | 0.9947 |
| | 41.37 | 0.067 | 0.652 | | 96.53 | 0.184 | 0.9907 |
| | 68.95 | 0.127 | 0.728 | 143.15 | 13.79 | 0.046 | 0.9997 |
| | 96.53 | 0.191 | 0.743 | | 41.37 | 0.113 | 0.9993 |
| 273.15 | 13.79 | 0.018 | 0.593 | | 68.95 | 0.152 | 0.9962 |
| | 41.37 | 0.074 | 0.818 | | 96.53 | 0.171 | 0.9904 |
| | 68.95 | 0.130 | 0.854 | | 124.11 | 0.182 | 0.9852 |
| | 96.53 | 0.187 | 0.859 | | 137.90 | 0.188 | 0.9828 |
| 248.15 | 13.79 | 0.024 | 0.815 | 123.15 | 13.79 | 0.062 | 0.9996 |
| | 41.37 | 0.078 | 0.914 | | 27.58 | 0.111 | 0.9997 |
| | 68.95 | 0.132 | 0.927 | | | | |

(cont.)

**AUXILIARY INFORMATION**

**METHOD:/APPARATUS/PROCEDURE:**

Static equilibrium cell with magnetic stirrer. Temperature measured with platinum resistance thermometer; pressure measured with Bourdon gauge. Propane charged into cell, nitrogen added. Samples of both phases analysed by gas chromatography. Details in ref. 1.

**SOURCE AND PURITY OF MATERIALS:**

(1) High purity dry grade, purity 99.995 mole per cent.

(2) Instrument grade, purity 99.96 mole per cent.

**ESTIMATED ERROR:**

$\delta T/K = \pm 0.01$; $\delta P/\text{bar} = \pm 0.15$;

$\delta x_{N_2} = \pm 1.5$ to 4%; $\delta y_{N_2} = \pm 2\%$.

**REFERENCES:**

1. Sinor, J. E.; Schindler, D. L.; Kurata, F. *Am. Inst. Chem. Engnrs. J.* 1966, *12*, 357.

| COMPONENTS: | ORIGINAL MEASUREMENTS: |
|---|---|
| (1) Nitrogen; $N_2$; [7727-37-9] <br> (2) Propane; $C_3H_8$; [74-98-6] | Schindler, D. L.; Swift, G. W.; Kurata, F. <br> *Hydrocarbon Process.*, 1966, *45* (11), 205. |

EXPERIMENTAL VALUES:

| T/K | P/bar | Mole fraction of nitrogen in liquid, $x_{N_2}$ | Mole fraction of nitrogen in vapor, $y_{N_2}$ | T/K | P/bar | Mole fraction of nitrogen in liquid, $x_{N_2}$ | Mole fraction of nitrogen in vapor, $y_{N_2}$ |
|---|---|---|---|---|---|---|---|
| 123.15 | 41.37* | 0.119 | 0.9947 | 103.15 | 13.79* | 0.085 | 0.9953 |
| | 68.95* | 0.128 | 0.9927 | | 41.37* | 0.087 | 0.9943 |
| | 96.53* | 0.136 | 0.9905 | | 68.95* | 0.089 | 0.9933 |
| | 124.11* | 0.141 | 0.9885 | | 96.53* | 0.091 | 0.9923 |
| | 137.90* | 0.143 | 0.9877 | | 124.11* | 0.093 | 0.9913 |
| 103.15 | 6.89 | 0.060 | - | | 137.90* | 0.094 | 0.9908 |

* liquid-liquid insolubility.

| COMPONENTS: | ORIGINAL MEASUREMENTS: |
|---|---|
| (1) Nitrogen; $N_2$; [7727-37-9]<br>(2) Propane; $C_3H_8$; [74-98-6] | Cannon, W. A.; Robson, J. H.; English, W.D.<br>Report DAC-60510-F2. Contract NAS7-548 (1968). (Tech. Report Douglas Missile & Space Systems Division, Astropower Laboratory.) |
| VARIABLES: | PREPARED BY: |
| T/K = 172-256<br>P/MPa = 2.1-4.8 | R. Battino |

EXPERIMENTAL VALUES:

| $P_{N_2}$/psia | $P_{N_2}$[a]/MPa | $x_1$[b] -150.0°F (172.0 K) | -100.0°F (199.8 K) | -50.2°F (227.5 K) | 0.4°F (255.6 K) |
|---|---|---|---|---|---|
| 300 | 2.07 | 0.0490 | 0.0405 | 0.0405 | 0.0355 |
| 300 | 2.07 | 0.0482 | 0.0393 | 0.0403 | 0.0348 |
| 300 | 2.07 | 0.0504 | 0.0406 | 0.0392 | 0.0342 |
| 300 | 2.07 | 0.0510 | 0.0424 | 0.0405 | 0.0330 |
| 700 | 4.83 | 0.112 | 0.107 | 0.104 | 0.0965 |
| 700 | 4.83 | 0.114 | 0.104 | 0.102 | 0.0970 |
| 700 | 4.83 | 0.112 | 0.105 | 0.108 | 0.0977 |
| 700 | 4.83 | 0.110 | 0.106 | 0.109 | 0.0978 |

[a] Calculated by compiler.

[b] Mole fraction solubility at indicated partial pressure.

AUXILIARY INFORMATION

| METHOD/APPARATUS/PROCEDURE: | SOURCE AND PURITY OF MATERIALS: |
|---|---|
| About 200 $cm^3$ of liquid propellant was condensed into a pressure vessel of about 1 liter capacity. Nitrogen was added to the proper pressure and the apparatus agitated for about one hour. A sample of the liquid phase was withdrawn and flashed to a gas at a lower temperature. The gas sample was then analyzed by gas chromatography. | (1) Air Products and Chemicals, high purity grade, 99.9%.<br>(2) J.T. Baker Chemical Company, C.P. grade, 99.9%. |
| | ESTIMATED ERROR: |
| | $\delta T/K = \pm 0.2$<br>$\delta P/P = \pm 0.02$<br>$\delta x_1/x_1 = \pm 0.02$ |
| | REFERENCES: |

COMPONENTS:

(1) Nitrogen; $N_2$; [7727-37-9]

(2) Propane; $C_3H_8$; [74-98-6]

ORIGINAL MEASUREMENTS:

Poon, D. P. L.; Lu, B. C. Y.

*Advan. Cryog. Eng.* 1973, *19*, 292.

VARIABLES:

T/K = 114-122
P/MPa = 0.2-2.8

PREPARED BY:

C. L. Young

EXPERIMENTAL VALUES:

| T/K | $P$/bar | Mole fraction of nitrogen in liquid, $x_{N_2}$ | Mole fraction of nitrogen in vapor, $y_{N_2}$ | T/K | $P$/bar | Mole fraction of nitrogen in liquid, $x_{N_2}$ | Mole fraction of nitrogen in vapor, $y_{N_2}$ |
|---|---|---|---|---|---|---|---|
| 114.1 | 1.50 | 0.0075 | 0.9995 | 118.3 | 14.88 | 0.0621 | 0.9999 |
| | 1.63 | 0.0086 | - | | 19.31 | 0.0760 | 0.9999 |
| | 3.41 | 0.0169 | 0.9996 | | 20.51 | 0.0801 | 0.9999 |
| | 6.58 | 0.0335 | 0.9997 | | 21.53 | 0.0854 | 1.0 |
| | 10.01 | 0.0486 | 0.9998 | 122.2 | 3.03 | 0.0107 | 0.9989 |
| | 13.73 | 0.0640 | 0.9999 | | 3.57 | 0.0133 | 0.9989 |
| | 16.03 | 0.0736 | 0.9999 | | 6.95 | 0.0254 | 0.9994 |
| | 16.52 | 0.0773 | 1.0 | | 10.38 | 0.0380 | 0.9995 |
| | 17.79 | 0.0842 | 1.0 | | 11.45 | 0.0410 | 0.9995 |
| | 18.00 | 0.0845 | 1.0 | | 13.99 | 0.0486 | 0.9996 |
| 118.3 | 1.88 | 0.0079 | - | | 19.31 | 0.0672 | 0.9999 |
| | 4.26 | 0.0175 | 0.9994 | | 23.56 | 0.0812 | 0.9999 |
| | 7.40 | 0.0321 | 0.9994 | | 25.48 | 0.0858 | 0.9999 |
| | 10.02 | 0.0433 | 0.9998 | | 27.58 | 0.0869 | 1.0 |
| | 12.24 | 0.0511 | 0.9998 | | 27.58 | 0.0881 | 1.0 |
| | 12.98 | 0.0557 | 0.9998 | | 27.99 | 0.0888 | 1.0 |

AUXILIARY INFORMATION

METHOD/APPARATUS/PROCEDURE:

Recirculating vapor flow apparatus with cell consisting of 100 ml Jerguson gauge with stainless steel body. Temperature measured using copper-constantan thermocouples. Pressure measured using Bourdon gauges. Magnetic circulating pump. Cell charged, vapor recirculated for 2 or more hours and samples of vapor and liquid removed at constant pressure and analysed using gas chromatography. Details in source.

SOURCE AND PURITY OF MATERIALS:

(1) Matheson research grade sample, purity 99.997 mole per cent.

(2) Phillips Petroleum Co. sample research grade purity 99.99 mole per cent.

ESTIMATED ERROR:

$\delta T/K = \pm 0.05$; $\delta P/\text{bar} = \pm 0.05$;

$\delta x_{N_2} = \pm 1\%$; $\delta y_{N_2} = \pm 0.0001$.

REFERENCES:

COMPONENTS:

(1) Nitrogen; $N_2$; [7727-37-9]

(2) Propane; $C_3H_8$; [74-98-6]

ORIGINAL MEASUREMENTS:

Grauso, L.; Fredenslund, A.; Mollerup, J.

*Fluid Phase Equilibria* 1977, *1*, 13.

VARIABLES:

T/K = 230-290
P/MPa = 0.1-21.9

PREPARED BY:

C. L. Young

EXPERIMENTAL VALUES:

| T/K | P/bar | Mole fraction of nitrogen in liquid, $x_{N_2}$ | Mole fraction of nitrogen in vapor, $y_{N_2}$ | T/K | P/bar | Mole fraction of nitrogen in liquid, $x_{N_2}$ | Mole fraction of nitrogen in vapor, $y_{N_2}$ |
|---|---|---|---|---|---|---|---|
| 230.0 | 0.966 | 0.0000 | 0.0000 | 260.0 | 122.15 | 0.2354 | 0.8939 |
| | 4.08 | - | 0.7498 | | 152.30 | 0.2993 | 0.8756 |
| | 19.66 | 0.0337 | 0.9343 | | 179.83 | 0.3721 | 0.8478 |
| | 36.84 | 0.0745 | 0.9556 | | 219.19 | 0.5280 | 0.7524 |
| | 61.17 | 0.1251 | 0.9608 | 290.0 | 7.692 | 0.0000 | 0.0000 |
| | 83.16 | 0.1676 | 0.9601 | | 20.11 | 0.0238 | 0.5438 |
| | 107.79 | 0.2070 | 0.9545 | | 29.47 | 0.0432 | 0.6583 |
| | 134.90 | 0.2635 | 0.9445 | | 45.71 | 0.0746 | 0.7423 |
| | 169.44 | 0.3244 | 0.9261 | | 56.69 | 0.0968 | 0.7684 |
| | 194.84 | 0.3774 | 0.9071 | | 75.79 | 0.1379 | 0.7904 |
| | 213.55 | 0.4193 | 0.8882 | | 85.30 | 0.1603 | 0.7948 |
| 260.0 | 3.103 | 0.0000 | 0.0000 | | 93.23 | 0.1780 | 0.7966 |
| | 11.89 | 0.0183 | 0.7116 | | 100.10 | 0.1932 | 0.7978 |
| | 21.32 | 0.0376 | 0.8225 | | 111.90 | 0.2216 | 0.7938 |
| | 32.76 | 0.0554 | 0.8623 | | 124.01 | 0.2539 | 0.7874 |
| | 51.23 | 0.0903 | 0.8909 | | 137.13 | 0.2908 | 0.7742 |
| | 76.12 | 0.1389 | 0.9024 | | 150.26 | 0.3310 | 0.7580 |
| | 103.52 | 0.1939 | 0.8997 | | 159.03 | 0.3617 | 0.7410 |

AUXILIARY INFORMATION

METHOD/APPARATUS/PROCEDURE:

Recirculating vapor flow apparatus. Temperature measured with platinum resistance thermometer. Pressure measured with dead weight piston balance. Coexisting samples analysed by gas chromatography. Details in source and ref. 1.

SOURCE AND PURITY OF MATERIALS:

(1) Purity 99.998 mole per cent.

(2) Purity 99.95 mole per cent.

ESTIMATED ERROR:

$\delta T/K = \pm 0.01$; $\delta P/bar = \pm 0.02\%$;

$\delta x_{N_2} \approx \delta y_{N_2} \leq \pm 0.003$.

REFERENCES:

1. Fredenslund, A.; Mollerup, J.; Christiansen, L. J. *Cryogenics* 1973, *13*, 405.

| COMPONENTS: | EVALUATOR: |
|---|---|
| (1) Nitrogen; $N_2$; [7727-37-9]<br>(2) Butane; $C_4H_{10}$; [106-97-8] | Colin L. Young,<br>School of Chemistry,<br>University of Melbourne,<br>Parkville, Victoria 3052,<br>AUSTRALIA.<br>June 1981. |

CRITICAL EVALUATION:

This system has been studied by five groups of workers. The data of Lehigh and McKetta (1) were part of a more extensive study of the ternary system ethane-butane-nitrogen and are in only fair agreement with McKetta's earlier data (2). Since Lehigh and McKetta studied just one isotherm and it is not of the highest accuracy it is not considered further here.
The data of Frolich *et al.* (3) are in marked disagreement with the data of other workers and are only included in view of the fact that they are often quoted and used. The data are classified as doubtful. The data of Skripka *et al.* (4), McKetta and Roberts (2) and Akers *et al.* (5) are more or less consistent and extrapolated values agree fairly well. These three sets of data are classified as tentative.

References


1. Lehigh, W. R.; McKetta, J. J. *J. Chem. Eng. Data* 1966, *11*, 180.

2. Roberts, L. R.; McKetta, J. J. *Am. Inst. Chem. Engnrs. J.* 1961, *7*, 173.

3. Frolich, P. K.; Tauch, E. J.; Hoban, J. J.; Peer, A. A. *Ind. Eng. Chem.* 1931, *23*, 548.

4. Skripka, V. G.; Barsuk, S. D.; Nikitina, I. E.; Ben'yaninovic, O. A. *Gazov. Prom.* 1964, *14*, no. 4, 41.

5. Akers, W. W.; Atwell, L. L.; Robinson, J. A. *Ind. Eng. Chem.* 1954, *46*, 2539.

COMPONENTS:

(1) Nitrogen; $N_2$; [7727-37-9]

(2) Butane; $C_4H_{10}$; [106-97-8]

ORIGINAL MEASUREMENTS:

Frolich, P. K.; Tauch, E. J.; Hogan, J.J.; Peer, A. A.

*Ind. Eng. Chem.* 1931, *23*, 548-50.

VARIABLES:

T/K = 298
P/MPa = 1-18

PREPARED BY:

C. L. Young

EXPERIMENTAL VALUES:

| T/K | $P$/MPa | Solubility*, $S$ | Mole fraction of nitrogen,† $x_{N_2}$ |
|---|---|---|---|
| 298.15 | 1.0 | 1.35 | 0.00577 |
| | 2.0 | 2.7 | 0.0115 |
| | 3.0 | 4.0 | 0.0169 |
| | 4.0 | 5.3 | 0.0223 |
| | 5.0 | 6.65 | 0.0278 |
| | 6.0 | 8.0 | 0.0332 |
| | 7.0 | 9.3 | 0.0384 |
| | 8.0 | 10.65 | 0.0437 |
| | 9.0 | 12.0 | 0.0490 |
| | 10.0 | 13.3 | 0.0540 |
| | 11.0 | 14.65 | 0.0592 |
| | 12.0 | 16.0 | 0.0643 |
| | 13.0 | 17.3 | 0.0692 |
| | 14.0 | 18.65 | 0.0742 |
| | 15.0 | 20.0 | 0.0791 |
| | 16.0 | 21.3 | 0.0838 |
| | 17.0 | 22.65 | 0.0887 |
| | 18.0 | 24.0 | 0.0935 |

* Data taken from graph in original article. Volume of gas measured at 101.325 kPa pressure and 298.15K dissolved by unit volume of liquid measured under the same conditions.

+ calculated by compiler.

AUXILIARY INFORMATION

METHOD APPARATUS/PROCEDURE:

Static equilibrium cell. Liquid saturated with gas and after equilibrium established samples removed and analysed by volumetric method. Allowance was made for vapor pressure of liquid and the solubility of the gas at atmospheric pressure. Details in source.

SOURCE AND PURITY OF MATERIALS:

(1) Highest purity available.

(2) Purity 98 to 99 mole per cent.

ESTIMATED ERROR:

$\delta T/K = \pm 0.1$; $\delta x_{N_2} = \pm 5\%$.

REFERENCES:

**COMPONENTS:**

(1) Nitrogen; $N_2$; [7727-37-9]

(2) n-Butane; $C_4H_{10}$; [106-97-8]

**ORIGINAL MEASUREMENTS:**

Akers, W. W.; Atwell, L. L.; Robinson, J. A.

*Ind. Eng. Chem.* 1954, *46*, 2539.

**VARIABLES:**

T/K = 311-422
P/MPa = 3.6-29.1

**PREPARED BY:**

C. L. Young

**EXPERIMENTAL VALUES:**

| T/K | *P*/bar | Mole fraction of nitrogen in liquid, $x_{N_2}$ | Mole fraction of nitrogen in vapor, $y_{N_2}$ | T/K | *P*/bar | Mole fraction of nitrogen in liquid, $x_{N_2}$ | Mole fraction of nitrogen in vapor, $y_{N_2}$ |
|---|---|---|---|---|---|---|---|
| 310.93 | 35.7 | 0.0394 | 0.8611 | 399.82 | 70.3 | 0.1538 | 0.6739 |
| | 70.0 | 0.0899 | 0.8975 | | 104.5 | 0.2742 | 0.7173 |
| | 138.9 | 0.1955 | 0.8908 | | 138.9 | 0.4678 | 0.6734 |
| | 178.9 | 0.2688 | 0.8879 | 422.04 | 42.6 | - | 0.3762 |
| | 221.8 | 0.3829 | 0.8672 | | 49.6 | - | 0.4729 |
| | 277.2 | 0.5449 | 0.7996 | | 56.5 | - | 0.4973 |
| | 290.9 | - | 0.7239 | | 56.5 | - | 0.4900 |
| 366.48 | 35.7 | 0.0498 | 0.6664 | | 63.4 | 0.1650 | 0.5247 |
| | 70.3 | 0.1282 | 0.7975 | | 70.3 | - | 0.5486 |
| | 104.5 | 0.2171 | 0.8433 | | 70.3 | 0.2167 | 0.5442 |
| | 173.3 | 0.3868 | 0.7912 | | 77.2 | 0.2776 | 0.5485 |
| | 201.0 | - | 0.7509 | | 84.0 | 0.3753 | 0.5429 |
| | 207.9 | 0.5430 | 0.7235 | | 84.0 | - | 0.5361 |
| 399.82 | 35.7 | 0.0116 | 0.5084 | | | | |

**AUXILIARY INFORMATION**

**METHOD/APPARATUS/PROCEDURE:**

Vapor flow apparatus with vapor recirculating via external loop. Equilibrium established with fixed quantity of vapor and liquid. Phases analysed by thermal conductivity. Details in ref. 1.

**SOURCE AND PURITY OF MATERIALS:**

No details given.

**ESTIMATED ERROR:**

$\delta T/K = \pm 0.3$; $\delta P/\text{bar} = \pm 0.3$;

$\delta x_{N_2}$, $\delta y_{N_2} = \pm 0.001$.

**REFERENCES:**

1. Akers, W. W.; Burns, J. F.; Fairchild, W. R. *Ind. Eng. Chem.* 1954, *46*, 2531.

COMPONENTS:

(1) Nitrogen; $N_2$; [7727-37-9]

(2) n-Butane; $C_4H_{10}$; [106-97-8]

ORIGINAL MEASUREMENTS:

Roberts, L. R.; McKetta, J. J.

*Am. Inst. Chem. Engnrs. J.* 1961, *7*, 173.

VARIABLES:

T/K = 311-411

P/MPa = 1.6-23.5

PREPARED BY:

C. L. Young

EXPERIMENTAL VALUES:

| T/K | *P*/bar | Mole fraction of nitrogen in liquid, $x_{N_2}$ | Mole fraction of nitrogen in vapor, $y_{N_2}$ |
|---|---|---|---|
| 310.9 | 16.3 | 0.025 | 0.759 |
| | 64.2 | 0.121 | 0.890 |
| | 124.1 | 0.164 | 0.900 |
| | 171.1 | 0.258 | 0.897 |
| | 207.4 | 0.332 | 0.979 |
| | 234.6 | 0.388 | 0.849 |
| 344.3 | 35.6 | 0.058 | 0.639 |
| | 73.4 | 0.105 | 0.776 |
| | 97.0 | 0.149 | 0.791 |
| | 118.9 | 0.202 | 0.779 |
| | 138.0 | 0.256 | 0.758 |
| | 169.0 | 0.321 | 0.757 |
| 377.6 | 45.6 | 0.066 | 0.479 |
| | 58.4 | 0.112 | 0.548 |
| 377.6 | 83.0 | 0.148 | 0.585 |
| | 94.6 | 0.177 | 0.594 |
| | 128.1 | 0.297 | 0.580 |
| | 138.2 | 0.311 | 0.534 |
| | 138.9 | 0.311 | 0.526 |
| | 143.5 | 0.331 | 0.501 |
| 410.9 | 36.9 | 0.016 | - |
| | 41.1 | 0.034 | 0.144 |
| | 54.7 | 0.073 | 0.222 |
| | 57.7 | 0.078 | 0.231 |
| | 61.2 | 0.085 | 0.240 |
| | 67.6 | 0.113 | 0.243 |
| | 70.7 | - | 0.257 |

AUXILIARY INFORMATION

METHOD/APPARATUS/PROCEDURE:

Cell fitted with sight glass. Vapor recirculated by means of piston pump. Temperature measured with thermocouple and pressure measured with Bourdon gauge. Components charged into cell and vapor recirculated for 4 hours. Samples withdrawn by mercury displacement. Details in source but no details of analysis given.

SOURCE AND PURITY OF MATERIALS:

(1) Research grade, purity better than 99.9 mole per cent.

(2) Research grade, purity over 99.9 mole per cent.

ESTIMATED ERROR:

$\delta T/K = \pm 0.06$; $\delta P/\text{bar} = \pm 0.3$;

$\delta x_{N_2}$, $\delta y_{N_2}$ = ±0.002.

REFERENCES:

**COMPONENTS:**

(1) Nitrogen; $N_2$; [ 7727-37-9 ]

(2) n-Butane; $C_4H_{10}$; [106-97-8 ]

**ORIGINAL MEASUREMENTS:**

Skripka, V. G.; Barsuk, S. D.; Nikitina, I. E; Ben'yaminovic, O. A.

*Gazov. Prom.* 1964, *14*, no. 4, 41.

**VARIABLES:**

T/K = 153-273
P/MPa = 0.5-19.6

**PREPARED BY:**

C. L. Young

**EXPERIMENTAL VALUES:**

| T/K | P/bar | Mole fraction of nitrogen in liquid, $x_{N_2}$ | Mole fraction of nitrogen in vapor, $y_{N_2}$ |
|---|---|---|---|
| 273.15 | 4.9 | 0.0008 | - |
| | 9.8 | 0.0110 | 0.8810 |
| | 39.2 | 0.0616 | 0.9308 |
| | 78.4 | 0.1195 | 0.9524 |
| | 117.7 | 0.1740 | 0.9660 |
| | 156.9 | 0.2286 | 0.9689 |
| | 196.1 | 0.2828 | 0.9689 |
| 253.15 | 4.9 | 0.0050 | 0.8750 |
| | 9.8 | 0.0132 | 0.9163 |
| | 39.2 | 0.0540 | 0.9707 |
| | 78.4 | 0.1072 | 0.9808 |
| | 117.7 | 0.1626 | 0.9792 |
| | 156.9 | 0.2172 | 0.9741 |
| | 196.1 | 0.2710 | 0.9729 |
| 233.5 | 4.9 | 0.0062 | 0.9650 |
| | 9.8 | 0.0145 | 0.9800 |
| | 39.2 | 0.0552 | 0.9905 |
| | 78.4 | 0.1050 | 0.9911 |
| | 117.7 | 0.1556 | 0.9868 |
| | 156.9 | 0.2028 | 0.9825 |
| | 196.1 | 0.2464 | 0.9800 |
| 213.15 | 4.9 | 0.0074 | 0.9903 |
| | 9.8 | 0.0144 | 0.9928 |
| | 39.2 | 0.0576 | 0.9972 |
| | 78.4 | 0.1080 | 0.9982 |
| 213.15 | 117.7 | 0.1508 | 0.9922 |
| | 156.9 | 0.1894 | 0.9864 |
| | 196.1 | 0.2236 | 0.9853 |
| 193.15 | 4.9 | 0.0084 | 0.9960 |
| | 9.8 | 0.0164 | 0.9982 |
| | 39.2 | 0.0590 | 0.9998 |
| | 78.4 | 0.1120 | 0.9990 |
| | 117.7 | 0.1500 | 0.9947 |
| | 156.9 | 0.1824 | 0.9916 |
| | 196.1 | 0.2068 | 0.9905 |
| 173.15 | 4.9 | 0.0100 | 1.000 |
| | 9.8 | 0.0186 | 1.000 |
| | 39.2 | 0.0606 | 1.000 |
| | 78.4 | 0.1072 | 0.9993 |
| | 117.7 | 0.1388 | 0.9953 |
| | 156.9 | 0.1608 | 0.9929 |
| | 196.1 | 0.1774 | 0.9925 |
| 153.15 | 4.9 | 0.0140 | 1.000 |
| | 9.8 | 0.0248 | 1.000 |
| | 39.2 | 0.0718 | 1.000 |
| | 78.4 | 0.1142 | 0.9999 |
| | 117.7 | 0.1370 | 0.9965 |
| | 159.9 | 0.1484 | 0.9945 |
| | 196.1 | 0.1560 | 0.9938 |

**AUXILIARY INFORMATION**

**METHOD/APPARATUS/PROCEDURE:**

Recirculating vapor flow apparatus fitted with stirrer. Temperature measured with platinum resistance thermometer. Liquid and gas phases analysed by gas chromatography. Details in source.

**SOURCE AND PURITY OF MATERIALS:**

(1) Purity 99.9 mole per cent.

(2) Purity 99.5 mole per cent.

**ESTIMATED ERROR:**

$\delta T/K = \pm 0.2$; $\delta P/bar \pm 1\%$; $\delta x_{N_2} = \pm 5\%$
$\delta(1-y_{N_2}) = \pm 5\%$ (estimated by compiler).

**REFERENCES:**

**COMPONENTS:**

(1) Nitrogen; $N_2$; [7727-37-9]

(2) 2-Methylpropane (isobutane); $C_4H_{10}$; [75-28-5]

(3) n-Butane; $C_4H_{10}$; [106-97-8]

**ORIGINAL MEASUREMENTS:**

Ryabtsev, N. I.; Khuchua, K. S.;

*Gazov. Delo* 1970, 25-8

*Chem Abstr.* 1971, *74*, 57804X.

**VARIABLES:**

T/K = 258-358
P/MPa = 0.2-9.8

**PREPARED BY:**

V. Katovic

**EXPERIMENTAL VALUES:**

t/°C = 25 (298 K), $P_o^c$ /kg cm$^{-2}$ = 2.7, n-Butane

| $P^a_{N_2}$ | $C^b$ |
|---|---|
| 4.9 | 3.2 |
| 9.6 | 6.1 |
| 13.3 | 8.2 |
| 20.1 | 12.4 |
| 27.6 | 17.0 |
| 50.5 | 31.8 |
| 66.7 | 44.8 |
| 87.9 | 61.2 |
| 98.9 | 70.5 |

t/°C = 45 (318 K), $P_o^c$ /kg cm$^{-2}$ = 6.2, Isobutane

| $P^a_{N_2}$ | $C^b$ |
|---|---|
| 2. | 1.3 |
| 4.5 | 3.3 |
| 9.5 | 6.8 |
| 19.5 | 14.2 |
| 28.3 | 21.4 |
| 37.1 | 27.9 |
| 54.5 | 43.1 |
| 68.0 | 54.4 |
| 76.0 | 64.1 |
| 95.0 | 85.3 |

t/°C = 25 (298 K), $P_o^c$ /kg cm$^{-2}$ = 3.6, Isobutane

| $P^a_{N_2}$ | $C^b$ |
|---|---|
| 4. | 2.8 |
| 6.7 | 4.5 |
| 8.8 | 6.0 |
| 15.1 | 10.5 |
| 20.0 | 13.6 |
| 29.6 | 20.5 |
| 43.2 | 31.3 |
| 58.8 | 43.8 |
| 77.0 | 58.6 |
| 85.0 | 66.7 |
| 97.0 | 79.0 |

t/°C = 0 (273 K), $P_o^c$/ kg cm$^{-2}$ = 7.0, Isobutane

| $P^a_{N_2}$ | $C^b$ |
|---|---|
| 11.0 | 7.0 |
| 17.4 | 11.3 |
| 31.3 | 20.4 |
| 52.3 | 36.1 |
| 68.1 | 48.8 |
| 86.1 | 65.2 |
| 99.6 | 77.1 |

t/°C = -15 (258 K), $P_o^c$ /kg cm$^{-2}$ = 0.9, Isobutane

| $P^a_{N_2}$ | $C^b$ |
|---|---|
| 5.3 | 3.3 |
| 8.7 | 5.5 |
| 15.0 | 9.2 |
| 21.2 | 13.9 |
| 28.8 | 18.4 |
| 49.9 | 32.5 |
| 65.7 | 43.7 |
| 82.0 | 57.4 |
| 100.2 | 73.4 |

continued on following page

**AUXILIARY INFORMATION**

**METHOD/APPARATUS/PROCEDURE:**

The apparatus and procedure are described in reference 1.

**SOURCE AND PURITY OF MATERIALS:**

(1) Source not given. Minimum pruity 99.997 per cent.

(2) Purity of 95.9-99.9 per cent.

(3) Purity of 95.9-99.9 per cent

**ESTIMATED ERROR:**

$\delta T/K = \pm 0.1$ author's estimate

**REFERENCES:**

1. Ryabtsev, N. I., Khuchua, R. S. *Gazov. Delo* 1968, (10), 26-8; *Chem. Abstr.* 1968, *69*, 69816h.

| COMPONENTS: | ORIGINAL MEASUREMENTS: |
|---|---|
| (1) Nitrogen; $N_2$; [7727-37-9] | Ryabtsev, N. I.; Khuchua, K. S. |
| (2) 2-Methylpropane (isobutane); $C_4H_{10}$; [75-28-5] | *Gazov. Delo* 1970, 25-8; |
| (3) n-Butane; $C_4H_{10}$; [106-97-8] | *Chem. Abstr.* 1971, *74*, 57804X. |

EXPERIMENTAL VALUES:

continued

[a] Partial pressure of nitrogen in kg $cm^{-2}$.

[b] Solubility of nitrogen in $cm^3$, reduced to STP, per one gram of isobutane or n-butane.

[c] Pressure of isobutane or n-butane at given temperature.

[d] The solubility at different pressures for each temperature was calculated according to the following empirical relationship:

$$C^b = A\,P_{N_2}^3 + B\,P_{N_2}^2 + E\,P_{N_2}$$

where $P_{N_2}$ is the nitrogen partial pressure in units of kg $cm^{-2}$ and A, B, and E are constants. The values of A, B, and E are given in the following table.

| System | t/°C | T/K | $10^{-5}A$ | $10^{-5}B$ | $10^{-5}E$ |
|---|---|---|---|---|---|
| Nitrogen/ isobutane | 45 | 318 | 3.31 | -194.2 | 78741.1 |
| | 25 | 298 | 2.5 | -170.0 | 74500.0 |
| | 0 | 273 | 0.04 | -170.7 | 60195.0 |
| | -15 | 258 | 1.5 | - 62.0 | 64530.0 |
| Nitrogen/ n-butane | 25 | 298 | 0.62 | 54.3 | 59950.0 |

| COMPONENTS: | ORIGINAL MEASUREMENTS: |
|---|---|
| (1) Nitrogen; $N_2$; [7727-37-9]<br><br>(2) 2-Methylpropane; $C_4H_{10}$; [75-28-5] | Robinson, D.B.; Kalra, H.;.; Krishnan, T.: Miranda, R. D.<br>*Proc. Annu. Conv., Gas Process. Assoc., Tech. Pap.* 1975, 54, 25. |
| VARIABLES: | PREPARED BY: |
| T/K = 283-339<br>P/MPa = 0.6-21 | C. L. Young |

EXPERIMENTAL VALUES:

| T/K | P/bar | Mole fraction of nitrogen in liquid, $x_{N_2}$ | Mole fraction of nitrogen in vapor, $y_{N_2}$ | T/K | P/bar | Mole fraction of nitrogen in liquid, $x_{N_2}$ | Mole fraction of nitrogen in vapor, $y_{N_2}$ |
|---|---|---|---|---|---|---|---|
| 283.21 | 5.65 | 0.0062 | 0.5880 | 338.65 | 34.82 | 0.0547 | 0.6282 |
| | 9.31 | 0.0122 | 0.7390 | | 52.81 | 0.0941 | 0.7168 |
| | 13.79 | 0.0198 | 0.8117 | | 77.29 | 0.1509 | 0.7586 |
| | 21.37 | 0.0335 | 0.8739 | | 108.32 | 0.2271 | 0.7637 |
| | 32.89 | 0.0556 | 0.9038 | | 138.52 | 0.3125 | 0.7590 |
| | 52.33 | 0.0899 | 0.9293 | | 171.61 | 0.4128 | 0.6897 |
| | 83.63 | 0.1486 | 0.9379 | | 182.78 | 0.4624 | 0.6310 |
| | 124.38 | 0.2302 | 0.9358 | 390.93 | 31.10 | 0.0080 | 0.0512 |
| | 169.89 | 0.2949 | 0.9159 | | 35.58 | 0.0236 | 0.1286 |
| | 207.39 | 0.3686 | 0.8942 | | 43.78 | 0.0540 | 0.2338 |
| 338.65 | 12.55 | 0.0065 | 0.1799 | | 56.12 | 0.1024 | 0.3107 |
| | 15.65 | 0.0111 | 0.3140 | | 71.77 | 0.1623 | 0.3451 |
| | 23.37 | 0.0297 | 0.5045 | | 88.60 | 0.2270 | 0.3520 |

AUXILIARY INFORMATION

METHOD /APPARATUS/PROCEDURE:

Cell fitted with two moveable pistons which enabled cell contents to be circulated in external line. Fitted with optical system which allowed measurement of refractive index. Temperature measured with iron-constantan thermocouple and pressure with strain gauge transducers. Components charged into cell, mixed by piston movement. Samples withdrawn and analysed by gas chromatography. Details in ref. 1.

SOURCE AND PURITY OF MATERIALS:

(1) Linde sample, purity 99.9 mole per cent or better.

(2) Phillips Petroleum Research grade, purity 99.9 mole per cent or better.

ESTIMATED ERROR:

$\delta T/K = \pm 0.05$; $\delta P/bar = \pm 0.2$; $\delta x_{N_2}$, $\delta y_{N_2} = \pm 0.003$.

REFERENCES:

1. Besserer, G. J.; Robinson, D. B. *Can. J. Chem. Eng.* 1971, 49, 651.

| COMPONENTS: | ORIGINAL MEASUREMENTS: |
|---|---|
| (1) Nitrogen; $N_2$; [7727-37-9]<br>(2) Propane, 2-methyl-, (isobutane); $C_4H_{10}$; [75-28-5] | Kalra, H.; Ng, H.-J.; Miranda, R.D.; Robinson, D. B.<br>*J. Chem. Eng. Data.* 1978, *23*, 321-4. |
| VARIABLES:<br>T/K = 255-394<br>P/MPa = 0.2-20.8 | PREPARED BY:<br>C.L. Young |

EXPERIMENTAL VALUES:

| T/K | P/bar | Mole fraction of nitrogen in liquid $x_{N_2}$ | in gas $y_{N_2}$ | T/K | P/bar | Mole fraction of nitrogen in liquid $x_{N_2}$ | in gas $y_{N_2}$ |
|---|---|---|---|---|---|---|---|
| 255.37 | 2.38 | 0.0071* | 0.651 | 310.87 | 66.6 | 0.114 | 0.859 |
| | 4.19 | 0.0066 | 0.802 | | 91.2 | 0.171 | 0.870 |
| | 7.52 | 0.0143 | 0.892 | | 123.4 | 0.237 | 0.869 |
| | 14.2 | 0.0251* | 0.947 | | 153.1 | 0.308 | 0.858 |
| | 24.6 | 0.0445 | 0.963 | | 182.8 | 0.371 | 0.836 |
| | 41.4 | 0.0758 | 0.973 | | 207.7 | 0.436 | 0.803 |
| | 69.1 | 0.119 | 0.968 | 338.71 | 12.5 | 0.0065 | 0.180 |
| | 97.3 | 0.165 | 0.967 | | 15.7 | 0.0111* | 0.314 |
| | 116.3 | 0.197 | 0.964 | | 23.4 | 0.0297* | 0.505 |
| | 138.8 | 0.238 | 0.962 | | 34.8 | 0.0547 | 0.628 |
| | 173.1 | 0.290 | 0.957 | | 36.7 | 0.0634 | 0.644 |
| | 205.0 | 0.334* | 0.943 | | 77.3 | 0.151* | 0.759 |
| 283.21 | 5.69 | 0.0062* | 0.588 | | 104.4 | 0.228 | 0.764 |
| | 9.34 | 0.0122 | 0.739 | | 138.7 | 0.324* | 0.751 |
| | 13.8 | 0.0198 | 0.812 | | 170.3 | 0.458 | 0.676 |
| | 21.4 | 0.0335 | 0.874 | 366.32 | 21.1 | 0.005* | 0.103 |
| | 32.9 | 0.0556 | 0.904 | | 27.2 | 0.026* | 0.243 |
| | 52.3 | 0.0899 | 0.929 | | 45.1 | 0.074* | 0.445 |
| | 83.6 | 0.149 | 0.938 | | 68.1 | 0.138 | 0.542 |
| | 124.4 | 0.230 | 0.936 | | 93.4 | 0.214* | 0.566 |
| | 169.9 | 0.295 | 0.916 | | 118.9 | 0.233* | 0.505 |
| | 207.4 | 0.369* | 0.894 | 394.26 | 39.0 | 0.040 | 0.131 |
| 310.87 | 7.14 | 0.0056* | 0.259 | | 51.0 | 0.088 | 0.215 |
| | 11.7 | 0.0136* | 0.525 | | 60.9 | 0.130 | 0.247 |
| | 21.9 | 0.0332 | 0.717 | | 64.4 | 0.158 | 0.231 |
| | 40.7 | 0.0679 | 0.818 | | 64.9 | 0.166 | 0.230 |

*values did not pass the Christiansen-Fredenslund consistency text, ref.(2).

AUXILLARY INFORMATION

METHOD: /APPARATUS/PROCEDURE:

Glass windowed cell fitted with two moveable pistons which enabled cell contents to be circulated in external line. Fitted with optical system which allowed measurement of refractive index. Temperature measured with iron-constantan thermocouple and pressure with strain gauge transducer. Components charged into cell and mixed by piston movement. Samples withdrawn and analysed by gas chromatography. Details in ref. (1).

SOURCE AND PURITY OF MATERIALS:

(1) Stated minimum purity 99.99 mole per cent.

(2) Phillips Petroleum sample, research grade, purity better than 99.9 mole per cent.

ESTIMATED ERROR:

$\delta T/K = \pm 0.05$; $\delta P/\text{bar} \leq \pm 0.3$;

$\delta x_{N_2}$, $\delta y_{N_2} = \pm 0.005$.

REFERENCES:


1. Besserer, G.J.; Rolinson, D.B.; *Can. J. Chem.* **Eng.** 1971, *49*, 651.
2. Christiansen, L.J.; Fredenslund, A. *Am. Inst. Chem. Eng. J.* 1975, *21*, 49.

COMPONENTS:

(1) Nitrogen; $N_2$; [7727-37-9]

(2) Three butanes, see below

ORIGINAL MEASUREMENTS:

Steinbach, H.G.; Steinbrecher, M.

*Chem. Tech.* (*Berlin*) 1966, *18*, 633.

VARIABLES:

T/K = 273-293
P/MPa = 0.41-1.11

PREPARED BY:

R. Battino

EXPERIMENTAL VALUES:

| Substance | $\alpha^a$ |
|---|---|
| 1,3-Butadiene; $C_4H_6$; [106-99-0] | 0.33 |
| Isobutene; $C_4H_8$; [115-11-7] | 0.31 |
| n-Butane; $C_4H_{10}$; [106-97-8] | 0.29 |

[a] Bunsen coefficient in $cm^3$ (STP) $cm^{-3}$ $atm^{-1}$. Average of values measured over a pressure range of 4 to 11 atm (0.41 to 1.11 MPa) and a temperature range of 0 to 20°C (273 to 293 K). The authors state that within their experimental error of ±5% that the Bunsen coefficient changes negligibly.

AUXILIARY INFORMATION

METHOD/APPARATUS/PROCEDURE:

The nitrogen was added to a steel bomb of 800 $cm^3$ capacity containing the butane. Analysis after equilibration was by gas chromatography at atmospheric pressure.

SOURCE AND PURITY OF MATERIALS:

No details given.

ESTIMATED ERROR:

$\delta\alpha/\alpha = \pm 0.05$

REFERENCES:

| COMPONENTS: | EVALUATOR: |
|---|---|
| (1) Nitrogen; $N_2$; [7727-37-9]<br><br>(2) $C_5$ to $C_{16}$ Alkanes (excluding heptane) | Colin L. Young<br>Department of Physical Chemistry<br>University of Melbourne<br>Parkville, Victoria 3052<br>Australia.<br><br>October 1980 |

CRITICAL EVALUATION:

The solubility of nitrogen in alkanes at low pressure and the high pressure solubility in heptane are treated separately. Data on the solubility of nitrogen in alkanes at high pressure are relatively scarce and it is not possible to recommend any data. The following data are classified as tentative.

| | |
|---|---|
| Pentane + nitrogen | Kalra, Robinson and Besserer (1) |
| Hexane + nitrogen | Poston and McKetta (2) |
| | Baranovich and Smirnova (3) |
| Octane + nitrogen | Baranovich and Smirnova (3) |
| | Graham and Weale (4) |
| Decane + nitrogen | Azarnoosh and McKetta (5) |
| Hexadecane + nitrogen | Sultanov, Skripta and Namoit (6). |

The system hexane + nitrogen was also investigated by Tsiklis and Khodeeva (7) but the study was not extensive and the work is rejected for the present purpose.

References


1. Kalra, H.; Robinson, D. B.; Besserer, G. J. *J. Chem. Eng. Data* 1977, *22*, 215.
2. Poston, R. S.; McKetta, J. J. *J. Chem. Eng. Data* 1966, *11*, 364.
3. Baranovich, Z. N.; Smirnova, A. M. *Zh. Prikl. Khim.* 1972, *45*, 2776.
4. Graham, E. B.; Weale, K. E. *Progr. Internat. Res. Thermodynamic Transport Props. Symp. on Thermophysical Props.* Princeton, *153* (1962).
5. Azarnoosh, A.; McKetta, J. J. *J. Chem. Eng. Data* 1963, *8*, 494.
6. Sultanov, R. G.; Skripta, V. G.; Namoit, A. Yu. *Gazov. Delo.* 1972, *no. 10*, 43.
7. Tsiklis, D. S.; Khodeeva, S. M. *Inzh.-Fiz. Khur. Acad. Nauk. Belorus S.S.R.* 1958, *11*, 62

**COMPONENTS:**

(1) Nitrogen; $N_2$; [7727-37-9]

(2) n-Pentane; $C_5H_{12}$; [109-66-0]

**ORIGINAL MEASUREMENTS:**

Kalra, H.; Robinson, D. B.; Besserer, G. J.

*J. Chem. Eng. Data* 1977, *22*, 215.

**VARIABLES:**

T/K = 277-378
P/MPa = 0.3-20.8

**PREPARED BY:**

C. L. Young

**EXPERIMENTAL VALUES:**

| T/K | P/bar | Mole fraction of nitrogen in liquid, $x_{N_2}$ | in vapor, $y_{N_2}$ | T/K | P/bar | Mole fraction of nitrogen in liquid, $x_{N_2}$ | in vapor, $y_{N_2}$ |
|---|---|---|---|---|---|---|---|
| 277.43 | 16.63 | 0.025 | 0.981 | 344.26 | 4.14 | 0.0022 | 0.323 |
| | 41.44 | 0.055 | 0.990 | | 8.20 | 0.0086 | 0.622 |
| | 69.22 | 0.104 | 0.988 | | 14.20 | 0.019 | 0.770 |
| | 104.0 | 0.149 | 0.988 | | 21.44 | 0.032 | 0.838 |
| | 138.4 | 0.194 | 0.988 | | 30.75 | 0.048 | 0.877 |
| | 172.6 | 0.223 | 0.987 | | 42.20 | 0.066 | 0.901 |
| | 207.3 | 0.251 | 0.985 | | 64.33 | 0.104 | 0.924 |
| 310.70 | 2.50 | 0.0027 | 0.577 | | 92.25 | 0.146 | 0.930 |
| | 4.50 | 0.0055 | 0.757 | | 122.1 | 0.196 | 0.931 |
| | 7.79 | 0.011 | 0.856 | | 161.7 | 0.252 | 0.930 |
| | 13.72 | 0.020 | 0.913 | | 207.8 | 0.330 | 0.922 |
| | 21.37 | 0.031 | 0.940 | 377.59 | 8.14 | 0.0032 | 0.179 |
| | 29.03 | 0.043 | 0.950 | | 9.93 | 0.0079 | 0.320 |
| | 41.30 | 0.061 | 0.960 | | 12.89 | 0.015 | 0.470 |
| | 63.57 | 0.095 | 0.968 | | 19.03 | 0.035 | 0.609 |
| | 84.25 | 0.125 | 0.970 | | 37.37 | 0.071 | 0.774 |
| | 105.2 | 0.155 | 0.970 | | 65.84 | 0.124 | 0.845 |
| | 125.6 | 0.185 | 0.969 | | 105.6 | 0.196 | 0.858 |
| | 153.2 | 0.223 | 0.967 | | 141.1 | 0.266 | 0.862 |
| | 180.9 | 0.258 | 0.964 | | 182.3 | 0.339 | 0.852 |
| | 207.5 | 0.288 | 0.961 | | 207.9 | 0.400 | 0.837 |

**AUXILIARY INFORMATION**

**METHOD/ APPARATUS/PROCEDURE:**

Glass windowed cell fitted with two moveable pistons which enabled cell contents to be circulated in external line. Fitted with optical system which allowed measurement of refractive index. Temperature measured with iron-constantan thermocouple and pressure with strain gauge transducer. Components charged into cell, mixed by piston movement. Samples withdrawn and analysed by gas chromatography. Details in ref. 1.

**SOURCE AND PURITY OF MATERIALS:**

(1) Purity 99.99 mole per cent.

(2) Fischer Scientific pesticide grade, purity 99.9 mole per cent or better.

**ESTIMATED ERROR:**

$\delta T/K = \pm 0.05$; $\delta P/\text{bar} = \pm 0.2$;

$\delta x_{N_2} \simeq \delta y_{N_2} = \pm 0.005$.

**REFERENCES:**

1. Besserer, G. J.; Robinson, D. B. *Can. J. Chem. Eng.* 1971, *49*, 651.

| COMPONENTS: | ORIGINAL MEASUREMENTS: |
|---|---|
| (1) Nitrogen; $N_2$; [7727-37-9]<br>(2) 2-Methylbutane; $C_5H_{12}$; [78-78-4] | Krishnan, T. R.; Kalra, H.; Robinson, D. B.<br>*J. Chem. Eng. Data* 1977, *22*, 282. |
| VARIABLES: | PREPARED BY: |
| T/K = 278-377<br>P/MPa = 0.2-20.8 | C. L. Young |

EXPERIMENTAL VALUES:

| T/K | $P$/bar | Mole fraction of nitrogen in liquid, $x_{N_2}$ | Mole fraction of nitrogen in vapor, $y_{N_2}$ | T/K | $P$/bar | Mole fraction of nitrogen in liquid, $x_{N_2}$ | Mole fraction of nitrogen in vapor, $y_{N_2}$ |
|---|---|---|---|---|---|---|---|
| 377.37 | 9.69 | 0.0037 | 0.163 | 310.82 | 2.24 | 0.0010 | 0.322 |
| | 12.24 | 0.0085 | 0.310 | | 3.28 | 0.0030 | 0.546 |
| | 17.06 | 0.0205 | 0.486 | | 5.21 | 0.0062 | 0.725 |
| | 28.96 | 0.0435 | 0.674 | | 8.24 | 0.0113 | 0.824 |
| | 44.51 | 0.0753 | 0.757 | | 13.76 | 0.0226 | 0.892 |
| | 63.32 | 0.112 | 0.805 | | 24.10 | 0.0373 | 0.931 |
| | 90.94 | 0.170 | 0.827 | | 37.75 | 0.0597 | 0.949 |
| | 115.38 | 0.221 | 0.835 | | 61.12 | 0.0970 | 0.962 |
| | 158.06 | 0.300 | 0.830 | | 96.49 | 0.152 | 0.965 |
| | 207.43 | 0.437 | 0.780 | | 132.00 | 0.192 | 0.965 |
| 344.15 | 5.07 | 0.0027 | 0.255 | | 157.65 | 0.244 | 0.961 |
| | 6.45 | 0.0054 | 0.418 | | 181.57 | 0.276 | 0.958 |
| | 9.20 | 0.0106 | 0.583 | | 207.77 | 0.305 | 0.953 |
| | 12.10 | 0.0153 | 0.675 | 277.59 | 1.83 | 0.0024 | 0.822 |
| | 16.65 | 0.0238 | 0.752 | | 4.55 | 0.0051 | 0.931 |
| | 26.65 | 0.0420 | 0.831 | | 9.89 | 0.0151 | 0.968 |
| | 39.54 | 0.0598 | 0.879 | | 19.17 | 0.0354 | 0.982 |
| | 55.47 | 0.0929 | 0.899 | | 32.13 | 0.0504 | 0.986 |
| | 77.74 | 0.130 | 0.912 | | 55.61 | 0.0866 | 0.991 |
| | 109.25 | 0.185 | 0.918 | | 91.63 | 0.143 | 0.994 |
| | 138.89 | 0.238 | 0.919 | | 130.97 | 0.201 | 0.991 |
| | 174.44 | 0.297 | 0.916 | | 159.85 | 0.240 | 0.985 |
| | 207.78 | 0.350 | 0.905 | | 184.33 | 0.260 | 0.987 |
| | | | | | 205.02 | 0.281 | 0.982 |

AUXILIARY INFORMATION

METHOD/APPARATUS/PROCEDURE:

Cell fitted with two moveable pistons which enabled cell contents to be circulated in external line. Fitted with optical system which allowed measurement of refractive index. Temperature measured with iron-constantan thermocouple and pressure with strain gauge transducer. Components charged into cell, mixed by piston movement. Samples withdrawn and analysed by gas chromatography. Details in ref. 1.

SOURCE AND PURITY OF MATERIALS:

(1) Purity 99.99 mole per cent or better.

(2) Phillips Petroleum sample, purity better than 99.99 mole per cent.

ESTIMATED ERROR:

$\delta T/K = \pm 0.05$; $\delta P/\text{bar} = \pm 0.2$;

$\delta x_{N_2}$, $\delta y_{N_2} = \pm 0.005$.

REFERENCES:

1. Besserer, G. J.; Robinson, D. B. *Can. J. Chem. Eng.* 1971, *49*, 651.

**COMPONENTS:**

(1) Nitrogen; $N_2$; [7727-37-9]

(2) n-Hexane; $C_6H_{14}$; [110-54-3]

**ORIGINAL MEASUREMENTS:**

Poston, R. S.; McKetta, J. J.
*J. Chem. Eng. Data*, 1966, *11*, 364-5.

**VARIABLES:**

T/K = 311-444
P/MPa = 1.7-34.5

**PREPARED BY:**

C. L. Young

**EXPERIMENTAL VALUES:**

| T/K | P/bar | Mole fraction of nitrogen in liquid, $x_{N_2}$ | Mole fraction of nitrogen in vapor, $y_{N_2}$ | T/K | P/bar | Mole fraction of nitrogen in liquid, $x_{N_2}$ | Mole fraction of nitrogen in vapor, $y_{N_2}$ |
|---|---|---|---|---|---|---|---|
| 310.93 | 17.24 | 0.026 | 0.958 | 377.55 | 17.24 | 0.023 | 0.789 |
| | 34.46 | 0.053 | 0.981 | | 34.46 | 0.047 | 0.869 |
| | 68.93 | 0.101 | 0.983 | | 68.93 | 0.095 | 0.921 |
| | 103.39 | 0.137 | 0.989 | | 103.39 | 0.150 | 0.928 |
| | 137.85 | 0.155 | 0.985 | | 137.85 | 0.199 | 0.928 |
| | 172.32 | 0.190 | 0.988 | | 172.32 | 0.242 | 0.927 |
| | 206.78 | 0.218 | 0.986 | | 206.78 | 0.283 | 0.926 |
| | 241.25 | 0.253 | 0.984 | | 241.25 | 0.335 | 0.911 |
| | 275.72 | 0.290 | 0.983 | | 275.72 | 0.380 | 0.899 |
| | 310.18 | 0.330 | 0.975 | | 310.18 | 0.439 | 0.893 |
| | 344.65 | 0.359 | 0.980 | | 344.65 | 0.479 | 0.867 |
| 344.26 | 17.24 | 0.025 | 0.904 | 410.95 | 17.24 | 0.021 | 0.614 |
| | 34.46 | 0.051 | 0.942 | | 34.46 | 0.049 | 0.771 |
| | 68.93 | 0.099 | 0.964 | | 68.93 | 0.105 | 0.835 |
| | 103.39 | 0.144 | 0.973 | | 103.39 | 0.162 | 0.864 |
| | 137.85 | 0.182 | 0.973 | | 137.85 | 0.220 | 0.882 |
| | 172.32 | 0.211 | 0.972 | | 172.32 | 0.282 | 0.884 |
| | 206.78 | 0.244 | 0.964 | | 206.78 | 0.342 | 0.884 |
| | 241.25 | 0.278 | 0.959 | | 241.25 | 0.411 | 0.854 |
| | 275.72 | 0.311 | 0.958 | | 275.72 | 0.507 | 0.816 |
| | 310.18 | 0.330 | 0.960 | | 290.94 | 0.576 | 0.777 |
| | 344.65 | 0.391 | 0.958 | | 290.87 | 0.601 | 0.772 |

(cont.)

**AUXILIARY INFORMATION**

**METHOD/APPARATUS/PROCEDURE:**

Stainless steel glass windowed cell; vapor recycled using high pressure magnetic pump. Pressure measured using Bourdon gauge. Temperature measured using thermocouples. Samples of both phases withdrawn at constant pressure and analysed by gas chromatography. Details in ref. 1.

**SOURCE AND PURITY OF MATERIALS:**

(1) Dry research grade, 99.85 mole per cent.

(2) Research grade, purity 99.9 mole per cent.

**ESTIMATED ERROR:**

$\delta T/K = \pm 0.1$; $\delta P/bar = \pm 0.14$;
$\delta x_{N_2}$, $\delta y_{N_2} = \pm 0.002$.

**REFERENCES:**

1. Roberts, L. R.; Azarnoosh, A.; Wang, R.; McKetta, J. J. *J. Chem. Eng. Data* 1962, *7*, 484.

| COMPONENTS: | ORIGINAL MEASUREMENTS: |
|---|---|
| (1) Nitrogen; $N_2$; [7727-37-9] | Poston, R. S.; McKetta, J. J. |
| (2) n-Hexane; $C_6H_{14}$; [110-54-3] | *J. Chem. Eng. Data* 1966, *11*, 364-5. |

EXPERIMENTAL VALUES:

| T/K | *P*/bar | Mole fraction of nitrogen in liquid, $x_{N_2}$ | Mole fraction of nitrogen in vapor, $y_{N_2}$ | T/K | *P*/bar | Mole fraction of nitrogen in liquid, $x_{N_2}$ | Mole fraction of nitrogen in vapor, $y_{N_2}$ |
|---|---|---|---|---|---|---|---|
| 444.25 | 34.46 | 0.049 | 0.567 | 444.25 | 172.32 | 0.331 | 0.740 |
| | 68.93 | 0.116 | 0.739 | | 186.93 | 0.398 | 0.708 |
| | 103.39 | 0.192 | 0.765 | | 198.65 | 0.485 | 0.693 |
| | 137.85 | 0.258 | 0.751 | | 201.89 | 0.505 | 0.624 |

COMPONENTS:

(1) Nitrogen; $N_2$; [7727-37-9]

(2) n-Hexane; $C_6H_{14}$; [110-54-3]

ORIGINAL MEASUREMENTS:

Baranovich, Z. N.; Smirnova, A. M.
*Zhur. Prikl. Khim.* 1972, *45*, 2776-8.

VARIABLES:

T/K = 233-293
P/MPa = 0.2-0.7

PREPARED BY:

C. L. Young

EXPERIMENTAL VALUES:

| T/K | $P$/bar | Mole fraction of nitrogen in liquid, $x_{N_2}$ |
|---|---|---|
| 233.15 | 2.0 | 0.00184 |
| | 3.0 | 0.00344 |
| | 5.1 | 0.00590 |
| | 7.1 | 0.00849 |
| 253.15 | 2.0 | 0.00175 |
| | 3.0 | 0.00241 |
| | 5.1 | 0.00542 |
| | 7.1 | 0.00755 |
| 263.15 | 3.0 | 0.00316 |
| | 7.1 | 0.00821 |
| 273.15 | 2.0 | 0.00175 |
| | 3.0 | 0.00283 |
| | 5.1 | 0.00462 |
| 283.15 | 2.0 | 0.00224 |
| | 7.1 | 0.00862 |
| 293.15 | 2.0 | 0.00224 |
| | 3.0 | 0.00279 |
| | 5.1 | 0.00583 |
| | 7.1 | 0.00928 |

AUXILIARY INFORMATION

METHOD/APPARATUS/PROCEDURE:

Static equilibrium cell fitted with agitator. Liquid composition determined by stripping and measuring gas volumetrically. Details in ref. 1.

SOURCE AND PURITY OF MATERIALS:

No details given.

ESTIMATED ERROR:

$\delta T/K = \pm 0.1$; $\delta P/\text{bar} = \pm 0.1$;
$\delta x_{N_2} = \pm 0.2 \times 10^{-4}$ (estimated by compiler).

REFERENCES:

1. Baranovich, Z. N.; Bogdanova, L. P.; Smirnova, A. M.
*Zhur. Prikl. Khim.* 1969, *42*, 1393.

| COMPONENTS: | EVALUATOR: |
|---|---|
| (1) Nitrogen; $N_2$; [7727-37-9]<br><br>(2) Heptane; $C_7H_{16}$; [142-82-5] | Colin Young,<br>School of Chemistry,<br>University of Melbourne,<br>Parkville, Victoria 3052,<br>Australia.<br>August 1980 |

CRITICAL EVALUATION:

This system has been investigated at high pressures by four groups. The data of Boomer *et al.* (1) are restricted to a pressure of 10.22 MPa at temperatures between 298 K and 388 K. These data are in broad agreement with the results of other workers but are not considered further because of their limited nature. The data of Akers *et al.* (2) cover the temperature range 305 K to 455 K and are broadly in agreement with the data of Peter and Eicke (3) and Brunner *et al.* (4) when interpolated to 453 K. The data of Peter and Eicke (3) cover the temperature range 376 K to 453 K and that of Brunner, Peter and Wenzel (4) cover the temperature range from 453 K to 497 K. There is rather poor agreement between the two sets of data by Peter and coworkers at 453 K. (Some very small extrapolation is needed as the actual temperatures were 453.95 K and 453.15 K.) The discrepancy is particularly large in the gas phase compositions which are considerably smaller in the nitrogen mole fraction in the 1974 paper than in the 1970 paper.

The three sets of data by Akers *et al.* (2), Peter and Eicke (3) and by Brunner *et al.* (4) are all classified as tentative but the estimated errors quoted are in all three cases too small. The liquid phase compositions are probably not, in general, better than ±5% for all three sets. The data by Figuiere et al. (5) are classed as tentative.

References:

1. Boomer, E. H.; Johnson, C. A.; Piercey, G. A. *Can. J. Research* 1938, *16B*, 396.

2. Akers, W. W.; Kohn, D. M.; Kilgore, C. H. *Ind. Eng. Chem.* 1954, *46*, 2536.

3. Peter, S.; Eicke, H. F. *Ber. Bunsenges. Phys. Chem.* 1970, *74*, 190.

4. Brunner, G.; Peter, S.; Wenzel, H. *Chem. Eng. J.* 1974, *7*, 99.

5. Figuiere, P.; Hom, J.F.; Laugier, S.; Renon, H.; Richon, D.; Szwarc, H. *Am. Inst. Chem. Engnrs. J.* 1980, *26*, 872-5.

**COMPONENTS:**

(1) Nitrogen; $N_2$; [7727-37-9]

(2) n-Heptane; $C_7H_{16}$; [142-82-5]

**ORIGINAL MEASUREMENTS:**

Boomer, E. H.; Johnson, C. A.; Piercey, G. A.

*Can. J. Research*, 1938, *16B*, 396.

**VARIABLES:**

T/K = 298-388
P/MPa = 10.2

**PREPARED BY:**

C. L. Young

**EXPERIMENTAL VALUES:**

| T/K | $P$/bar | Mole fraction of nitrogen in liquid, $x_{N_2}$ | in vapor, $y_{N_2}$ |
|---|---|---|---|
| 298.15 | 102.2 | 0.120 | 0.9896 |
| 328.15 | 102.2 | 0.128 | 0.9824 |
| | | 0.133 | 0.9825 |
| 358.15 | 102.2 | 0.1396 | 0.9756 |
| | | 0.1396 | 0.9746 |
| 388.15 | 102.2 | 0.1498 | 0.9495 |

**AUXILIARY INFORMATION**

**METHOD/APPARATUS/PROCEDURE:**

Rocking autoclave stirred by steel piston falling under gravity. Samples of vapor and liquid trapped in two auxiliary high pressure cells. Equilibrium samples analysed in complicated volumetric and combustion apparatus. Details in refs. 1 and 2.

**SOURCE AND PURITY OF MATERIALS:**

Not specified.

**ESTIMATED ERROR:**

$\delta T/K = \pm 0.1$; $\delta P/\mathrm{bar} \pm 0.2$;

$\delta x_{N_2}$, $\delta y_{N_2} = \pm 1\%$ (estimated by compiler)

**REFERENCES:**


1. Boomer, E. H.; Johnson, C. A.; Argue, G. H. *Can. J. Res. B* 1937, *15*, 367.
2. Boomer, E. H.; Johnson, C. A. *Can. J. Res. B* 1937, *15*, 363.

COMPONENTS:

(1) Nitrogen; $N_2$; [7727-37-9]

(2) n-Heptane; $C_7H_{16}$; [142-82-5]

ORIGINAL MEASUREMENTS:

Akers, W. W.; Kohn, D. M.; Kilgore, C. H.

*Ind. Eng. Chem.*, 1954, *46*, 2536.

VARIABLES:

T/K = 305-455
P/MPa = 7.0-69.1

PREPARED BY:

C. L. Young

EXPERIMENTAL VALUES:

| T/K | $P$/bar | Mole fraction of nitrogen in liquid, $x_{N_2}$ | in vapor, $y_{N_2}$ | T/K | $P$/bar | Mole fraction of nitrogen in liquid, $x_{N_2}$ | in vapor, $y_{N_2}$ |
|---|---|---|---|---|---|---|---|
| 305.37 | 71.02 | 0.080 | 0.99276 | 352.59 | 553.30 | 0.547 | 0.9592 |
| | 105.15 | 0.128 | 0.9900 | | 691.20 | 0.705 | 0.9412 |
| | 122.31 | 0.145 | 0.9961 | 399.82 | 73.08 | 0.116 | 0.9652 |
| | 122.31 | 0.151 | 0.9975 | | 139.41 | 0.213 | 0.9645 |
| | 139.41 | 0.152 | 0.9880 | | 208.36 | 0.301 | 0.9772 |
| | 158.65 | 0.170 | 0.9950 | | 277.65 | 0.387 | 0.9704 |
| | 184.16 | 0.201 | 0.9940 | | 450.57 | 0.634 | 0.9463 |
| | 208.36 | 0.215 | 0.9950 | | 518.83 | 0.721 | 0.828 |
| | 229.53 | 0.244 | 0.9900 | | 518.83 | 0.711 | 0.828 |
| | 243.04 | 0.236 | 0.9930 | | 551.93 | 0.723 | 0.834 |
| | 276.96 | 0.266 | 0.9921 | | 620.87 | 0.753 | - |
| | 346.46 | 0.318 | 0.9893 | 455.37 | 77.98 | 0.128 | 0.9353 |
| | 415.75 | 0.363 | 0.9861 | | 173.75 | 0.321 | 0.9486 |
| | 553.30 | 0.437 | 0.9782 | | 243.04 | 0.442 | 0.9506 |
| | 691.20 | 0.505 | 0.9700 | | 243.04 | 0.428 | 0.9541 |
| 352.59 | 70.33 | 0.096 | 0.9822 | | 277.65 | - | 0.9379 |
| | 145.55 | 0.187 | 0.9882 | | 277.65 | 0.554 | - |
| | 208.36 | 0.247 | 0.9869 | | 309.92 | - | 0.818 |
| | 277.65 | 0.291 | 0.9846 | | 435.96 | 0.752 | - |
| | 415.75 | 0.433 | 0.9757 | | | | |

AUXILIARY INFORMATION

METHOD/APPARATUS/PROCEDURE:

Vapor flow apparatus with vapor recirculating via external loop. Equilibrium established with fixed quantity of vapor and liquid. Phases analysed by thermal conductivity. Details in ref. 1.

SOURCE AND PURITY OF MATERIALS:

(1) Dry, purity greater than 99.99 mole per cent.

(2) Phillips Petroleum Co. research grade, purity 99.58 mole per cent.

ESTIMATED ERROR:

$\delta T/K = \pm 0.3$; $\delta P/\text{bar} = \pm 0.3$;

$\delta x_{N_2}$, $\delta y_{N_2} = \pm 0.001$.

REFERENCES:

1. Akers, W. W.; Burns, J. F.; Fairchild, W. R. *Ind. Eng. Chem.* 1954, *46*, 2531.

**COMPONENTS:**

(1) Nitrogen; $N_2$; [7727-37-9]

(2) n-Heptane; $C_7H_{16}$; [142-82-5]

**ORIGINAL MEASUREMENTS:**

Peter, S.; Eicke, H. F.

*Ber. Bunsenges, Phys. Chem.* 1970, *74*, 190.

**VARIABLES:**

T/K = 376-453
P/MPa = 0.5-61.3

**PREPARED BY:**

C. L. Young

**EXPERIMENTAL VALUES:**

| T/K | P/bar | Mole fraction of nitrogen in liquid, $x_{N_2}$ | in vapor, $y_{N_2}$ | T/K | P/bar | Mole fraction of nitrogen in liquid, $x_{N_2}$ | in vapor, $y_{N_2}$ |
|---|---|---|---|---|---|---|---|
| 376.35 | 24.5 | 0.0362 | 0.951 | 413.15 | 240.3 | 0.372 | 0.961 |
| | 49.0 | 0.0747 | 0.976 | | 343.2 | 0.528 | 0.941 |
| | 147.1 | 0.217 | 0.973 | | 392.3 | 0.639 | 0.861 |
| | 294.2 | 0.392 | 0.975 | | 411.9 | 0.686 | 0.816 |
| | 392.3 | 0.500 | 0.981 | | 416.8 | 0.742 | - |
| | 490.3 | 0.597 | 0.974 | 452.95 | 19.6 | 0.0266 | 0.671 |
| | 588.4 | 0.726 | 0.954 | | 49.0 | 0.0831 | 0.852 |
| | 612.9 | 0.780 | 0.924 | | 88.3 | 0.157 | 0.935 |
| 413.15 | 4.9 | - | 0.870 | | 132.4 | 0.232 | 0.945 |
| | 7.8 | - | 0.910 | | 215.7 | 0.378 | 0.956 |
| | 19.6 | 0.0323 | 0.923 | | 274.6 | 0.502 | 0.925 |
| | 39.2 | 0.0667 | 0.946 | | 304.0 | 0.673 | 0.790 |
| | 85.3 | 0.143 | 0.948 | | 313.8 | 0.742 | 0.812 |
| | 142.2 | 0.234 | 0.955 | | | | |

**AUXILIARY INFORMATION**

**METHOD/APPARATUS/PROCEDURE:**

Details of apparatus in ref. 1. Temperature measured with platinum resistance thermometer. Cell contents stirred with magnetic stirrer. Cell contents equilibrated, sample removed at various levels in cell. Sample analysed by freezing out the hydrocarbon in an acetone/$CO_2$ trap and weighing. Nitrogen estimated volumetrically. Critical data in source reference.

**SOURCE AND PURITY OF MATERIALS:**

(1) Linde sample, purity 99.99 mole per cent.

(2) Phillips Petroleum Co. "Pure" grade, purity > 99 mole per cent.

**ESTIMATED ERROR:**

$\delta T/K = \pm 0.05$; $\delta P/bar = \pm 0.5$;

$\delta x_{N_2} = \pm 0.003$; $\delta y_{N_2} = \pm 0.007$.

**REFERENCES:**

1. Peter, S.; Reinhartz, K. *Z. physik. Chem. NF* 1960, *24*, 103.

| COMPONENTS: | ORIGINAL MEASUREMENTS: |
|---|---|
| (1) Nitrogen; $N_2$; [7727-37-9]<br>(2) n-Heptane; $C_7H_{16}$; [142-82-5] | Brunner, G.; Peter, S.; Wenzel, H.<br>*Chem. Eng. J.* 1974, *7*, 99. |
| VARIABLES:<br>T/K = 453-497<br>P/MPa = 0.5-27.8 | PREPARED BY:<br>C. L. Young |

EXPERIMENTAL VALUES:

| T/K | $P$/bar | Mole fraction of nitrogen in liquid, $x_{N_2}$ | in vapor, $y_{N_2}$ |
|---|---|---|---|
| 453.15 | 53 | 0.101 | 0.827 |
| | 93 | 0.164 | 0.885 |
| | 127 | 0.222 | 0.903 |
| | 138 | 0.244 | 0.910 |
| | 190 | 0.328 | 0.882 |
| | 227 | 0.397 | 0.881 |
| | 278 | 0.543 | 0.851 |
| 472.15 | 44 | 0.075 | 0.735 |
| | 97 | 0.188 | 0.820 |
| | 164 | 0.347 | 0.838 |
| | 168 | 0.396 | 0.836 |
| | 187 | 0.416 | 0.837 |
| | 218 | 0.578 | 0.790 |
| | 234 | 0.586 | 0.771 |
| 497.15 | 48 | 0.087 | 0.598 |
| | 76 | 0.173 | 0.706 |
| | 109 | 0.281 | 0.717 |
| | 120 | 0.322 | 0.730 |
| | 142 | 0.403 | 0.721 |
| | 158 | 0.461 | 0.737 |
| | 183 | 0.588 | 0.602 |

AUXILIARY INFORMATION

METHOD/APPARATUS/PROCEDURE:

Static cell of about 1000 $cm^3$ capacity magnetically stirred. Pressure measured with Bourdon gauge. Temperature measured with nickel-chromium alloy-nickel thermocouple. Each phase sampled, n-heptane frozen and amount estimated, nitrogen estimated volumetrically.

SOURCE AND PURITY OF MATERIALS:

(1) Linde sample, purity 99.98 mole per cent by gas chromatography.

(2) Fluka sample, purity 99.7 mole per cent by gas chromatography.

ESTIMATED ERROR:

$\delta T/K = \pm 0.5$; $\delta P/\text{bar} = \pm 0.5$; $\delta x_{N_2}$, $\delta y_{N_2} = \pm 0.5\%$.

REFERENCES:

**COMPONENTS:**

(1) Nitrogen; $N_2$; [7727-37-9]

(2) Heptane; $C_7H_{16}$; [142-82-5]

**ORIGINAL MEASUREMENTS:**

Figuiere, P.; Hom, J. F.; Laugier, S.; Renon, H.; Richon, D.; Szwarc, H.

*Am. Inst. Chem. Engnrs. J.* 1980, *26*, 872-5.

**VARIABLES:**

T/K = 453-497
P/MPa = 1-29

**PREPARED BY:**

C. L. Young

**EXPERIMENTAL VALUES:**

| T/K | $P$/MPa | Mole fraction of nitrogen in liquid, $x_{N_2}$ | Mole fraction of nitrogen in gas, $y_{N_2}$ | T/K | $P$/MPa | Mole fraction of nitrogen in liquid, $x_{N_2}$ | Mole fraction of nitrogen in gas, $y_{N_2}$ |
|---|---|---|---|---|---|---|---|
| 453.2 | 29.45 | 0.67 | 0.69 | 463.2 | 6.0 | 0.11 | 0.75(5) |
| | 27.65 | 0.56 | 0.79(5) | | 4.3 | 0.07 | 0.69 |
| | 25.05 | 0.49# | 0.84 | | 3.55 | 0.06# | 0.65 |
| | 22.3 | 0.43 | 0.86# | | 3.15 | 0.05(5) | 0.61(5) |
| | 20.1 | 0.39 | 0.87 | | 2.1 | 0.02(7)# | 0.48(5) |
| | 16.1 | 0.305 | 0.88 | | 1.2 | 0.008 | 0.170 |
| | 11.6 | 0.215 | 0.87 | 497.2* | 13.05 | 0.33 | 0.69 |
| | 9.15 | 0.16(8)# | 0.86(5) | | 11.1 | 0.25 | 0.70 |
| | 7.2 | 0.12 | 0.85 | | 9.75 | 0.23 | 0.69 |
| | 5.6 | 0.09(8)# | 0.83(5) | | 8.05 | 0.18 | 0.68# |
| | 3.7 | 0.06(2)# | 0.77 | | 7.8 | 0.17 | 0.67(5) |
| | 2.65 | 0.04# | 0.67 | | 6.05 | 0.12 | 0.61(5) |
| | 2.55 | 0.04# | 0.66 | | 4.35 | 0.08 | 0.56 |
| | 2.1 | 0.03# | 0.61 | | 3.1 | 0.04# | 0.43(5) |
| | 1.75 | 0.02(3)# | 0.54(5) | 497.2** | 2.8 | 0.03 | 0.38 |
| | 1.55 | 0.01(8)# | 0.47(5) | | 3.75 | 0.06(3)# | 0.48 |
| 463.2 | 23.85 | 0.63 | 0.68(5) | | 4.95 | 0.09(4)# | 0.57 |
| | 20.55 | 0.46 | 0.80(5) | | 6.75 | 0.14(2)# | 0.63(5) |
| | 18.65 | 0.40 | 0.80(5) | | 8.6 | 0.19(2)# | 0.69(5) |
| | 16.55 | 0.35 | 0.80 | | 14.05 | 0.36(5) | 0.66(5) |
| | 13.9 | 0.30 | 0.80(5) | | 14.65 | 0.41 | 0.65 |
| | 12.5 | 0.26(2) | 0.80(5) | | 15.05 | 0.44 | 0.62(5) |
| | 9.45 | 0.19 | 0.80(5) | | 15.48 | 0.49 | 0.59 |

* first set of experiments; ** second set of experiments; # smoothed values. Values in parentheses are uncertain.

**AUXILIARY INFORMATION**

**METHOD/APPARATUS/PROCEDURE:**

Static equilibrium cell fitted with sampling valves which removed samples of $10^{-9}$ $m^3$. Samples analysed by gas chromatography. Temperature measured using thermocouples and pressure measured using pressure transducer calibrated against a dead weight gauge.

**SOURCE AND PURITY OF MATERIALS:**

(1) Certified purity 99.9 mole per cent.

(2) Certified purity 99 mole per cent as determined by GC.

**ESTIMATED ERROR:**

$\delta T/K = \pm 0.5$; $\delta P/MPa = \pm 0.04$; $\delta x_{N_2}$, $\delta y_{N_2} = \pm 0.015$.

**REFERENCES:**

**COMPONENTS:**

(1) Nitrogen; $N_2$; [7727-37-9]

(2) n-Octane; $C_8H_{18}$; [111-65-9]

**ORIGINAL MEASUREMENTS:**

Graham, E. B.; Weale, K. E.
*Progr. Internat. Res. Thermodynamic and Transport Properties, Symposium on Thermophysical Properties*
Princeton, 1962, 153.

**VARIABLES:**

T/K = 323-373
P/MPa = 10.1-30.4

**PREPARED BY:**

C. L. Young

**EXPERIMENTAL VALUES:**

| T/K | P/bar | Mole fraction of nitrogen in liquid, $x_{N_2}$ | Mole fraction of nitrogen in gas, $y_{N_2}$ | T/K | P/bar | Mole fraction of nitrogen in liquid, $x_{N_2}$ | Mole fraction of nitrogen in gas, $y_{N_2}$ |
|---|---|---|---|---|---|---|---|
| 323.15 | 101.3 | 0.117 | - | 348.15 | 253.3 | 0.257 | - |
| | 126.7 | 0.143 | - | | 278.6 | 0.277 | - |
| | 152.0 | 0.166 | 0.9975 | | 283.7 | - | 0.9938 |
| | 177.3 | 0.188 | - | | 304.0 | 0.296 | - |
| | 202.7 | 0.209 | - | 373.15 | 50.7 | 0.070 | - |
| | 228.0 | 0.229 | - | | 76.0 | 0.102 | - |
| | 253.3 | 0.249 | - | | 101.3 | 0.132 | - |
| | 278.6 | 0.267 | - | | 121.6 | - | 0.9908 |
| | 283.7 | - | - | | 126.7 | 0.160 | - |
| | 304.0 | 0.285 | 0.9969 | | 152.0 | 0.186 | 0.9903 |
| 348.15 | 101.3 | 0.122 | - | | 177.3 | 0.210 | - |
| | 126.7 | 0.148 | - | | 202.7 | 0.234 | 0.989 |
| | 152.0 | 0.172 | 0.995 | | 228.0 | 0.257 | - |
| | 177.3 | 0.194 | - | | 243.2 | - | 0.9885 |
| | 202.7 | 0.216 | 0.9948 | | 253.3 | 0.278 | - |
| | 228.0 | 0.237 | - | | 283.7 | - | 0.989 |
| | 243.2 | - | 0.9945 | | | | |

**AUXILIARY INFORMATION**

**METHOD /APPARATUS/PROCEDURE:**

Equilibrium vessel made of beryllium copper of 200 $cm^3$ capacity attached to small sampling vessel. Pressure measured with Bourdon gauge, oil thermostat, magnetic stirrer in equilibrium vessel. Solvent introduced into equilibrium vessel, compressed gas admitted and equilibrated. Samples analysed by gravimetric (liquid) and volumetric (gas) techniques. Details in ref. 1.

**SOURCE AND PURITY OF MATERIALS:**

(1) Commercial sample purity 99.95 mole per cent.

(2) Dried and distilled.

**ESTIMATED ERROR:**

$\delta T/K = \pm 0.1$; $\delta P/bar = \pm 0.5$;

$\delta x_{N_2} = \pm 1\%$; $\delta y_{N_2} = \pm 0.1\%$.

**REFERENCES:**

1. Graham, E. B., *Ph.D. Thesis, University of London*, 1958.

COMPONENTS:

(1) Nitrogen; $N_2$; [7727-37-9]

(2) n-Octane; $C_8H_{18}$; [111-65-9]

ORIGINAL MEASUREMENTS:

Baranovich, Z. N.; Smirnova, A. M.
*Zhur. Prikl. Khim.* 1972, *45*, 2776-8.

VARIABLES:

T/K = 233-293
P/MPa = 0.2-0.7

PREPARED BY:

C. L. Young

EXPERIMENTAL VALUES:

| T/K | $P$/bar | Mole fraction of nitrogen in liquid, $x_{N_2}$ |
|---|---|---|
| 233.15 | 2.0 | 0.00109 |
| | 3.0 | 0.00182 |
| | 5.1 | 0.00487 |
| | 7.1 | 0.00720 |
| 253.15 | 2.0 | 0.00131 |
| | 3.0 | 0.00233 |
| | 5.1 | 0.00465 |
| | 7.1 | 0.00749 |
| 263.15 | 2.0 | 0.00109 |
| | 3.0 | 0.00211 |
| | 5.1 | 0.00436 |
| | 7.1 | 0.00728 |
| 273.15 | 2.0 | 0.00116 |
| | 3.0 | 0.00218 |
| | 5.1 | 0.00473 |
| | 7.1 | 0.00785 |
| 293.15 | 2.0 | 0.00160 |
| | 3.0 | 0.00236 |
| | 5.1 | 0.00436 |
| | 7.1 | 0.00762 |

AUXILIARY INFORMATION

METHOD/APPARATUS/PROCEDURE:

Static equilibrium cell fitted with agitator. Liquid composition determined by stripping and measuring gas volumetrically. Details in ref. 1.

SOURCE AND PURITY OF MATERIALS:

No details given.

ESTIMATED ERROR:

$\delta T/K = \pm 0.1$; $\delta P/\text{bar} = \pm 0.1$;
$\delta x_{N_2} = \pm 0.2 \times 10^{-4}$ (estimated by compiler).

REFERENCES:

1. Baranovich, Z. N.; Bogdanovo, L. P. Smirnova, A. M. *Zhur. Prikl. Khim.* 1969, *42*, 1393.

COMPONENTS:

(1) Nitrogen; $N_2$; [7727-37-9]

(2) 2,2,4-Trimethylpentane; $C_8H_{18}$; [540-84-1]

ORIGINAL MEASUREMENTS:

Graham, E. B.; Weale, K. E.

*Progr. Internat. Res. Thermodynamic and Transport Properties, Symposium on Thermophysical Properties* Princeton, 1962, 153.

VARIABLES:

T/K = 323-423
P/MPa = 0.1-30.4

PREPARED BY:

C. L. Young

EXPERIMENTAL VALUES:

| T/K | P/bar | Mole fraction of nitrogen in liquid, $x_{N_2}$ | in gas, $y_{N_2}$ | T/K | P/bar | Mole fraction of nitrogen in liquid, $x_{N_2}$ | in gas, $y_{N_2}$ |
|---|---|---|---|---|---|---|---|
| 323.15 | 1.0 | 0.00159 | - | 348.15 | 177.3 | 0.252 | - |
| | 30.4 | 0.048 | - | | 202.7 | 0.280 | 0.9875 |
| | 50.7 | 0.079 | - | | 228.0 | 0.307 | - |
| | 101.3 | 0.149 | - | | 253.3 | 0.331 | 0.9855 |
| | 121.6 | - | 0.9942 | | 283.7 | - | 0.9844 |
| | 152.0 | 0.213 | 0.9951 | | 304.0 | 0.396 | - |
| | 202.7 | 0.268 | 0.9962 | 423.15 | 50.7 | 0.092 | - |
| | 228.0 | 0.292 | - | | 101.3 | 0.171 | - |
| | 253.3 | 0.314 | 0.9944 | | 152.0 | 0.243 | 0.980 |
| | 278.6 | 0.333 | - | | 177.3 | 0.278 | - |
| | 304.0 | 0.348 | - | | 202.7 | 0.311 | 0.9805 |
| 348.15 | 50.7 | 0.083 | - | | 228.0 | 0.343 | - |
| | 101.3 | 0.155 | - | | 253.3 | 0.374 | 0.980 |
| | 152.0 | 0.221 | 0.9895 | | 304.0 | 0.430 | - |

AUXILIARY INFORMATION

METHOD /APPARATUS/PROCEDURE:

Equilibrium vessel made of beryllium copper of $2 \times 10^5$ $mm^3$ capacity attached to small sampling vessel. Pressure measured with Bourdon gauge; oil thermostat, magnetic stirrer in equilibrium vessel. Solvent introduced into equilibrium vessel, compressed gas admitted and equilibrated. Samples analysed by gravimetric (liquid) and volumetric (gas) techniques. Details in ref. 1.

SOURCE AND PURITY OF MATERIALS:

(1) Purity 99.95 mole per cent.

(2) Distilled and dried.

ESTIMATED ERROR:

$\delta T/K = \pm 0.1$; $\delta P/bar = \pm 0.5$;

$\delta x_{N_2} = \pm 1\%$; $\delta y_{N_2} \simeq \pm 0.1\%$.

REFERENCES:

1. Graham, E. B. *Ph.D. Thesis, University of London*, 1958.

COMPONENTS:

(1) Nitrogen; $N_2$; [7727-37-9]

(2) 2,2,4-Trimethylpentane; $C_8H_{18}$; [540-84-1]

ORIGINAL MEASUREMENTS:

Peter, S.; Ficke, H. F.
*Ber. Bunsenges. Phys. Chem.* 1970, 74, 190.

VARIABLES:

T/K = 376-453
P/MPa = 2.0-61.8

PREPARED BY:

C. L. Young

EXPERIMENTAL VALUES:

| T/K | $P$/bar | Mole fraction of nitrogen in liquid, $x_{N_2}$ | in gas, $y_{N_2}$ |
|---|---|---|---|
| 376.45 | 19.6 | 0.0284 | 0.950 |
| | 49.0 | 0.0723 | 0.970 |
| | 98.1 | 0.143 | 0.976 |
| | 245.2 | 0.335 | 0.976 |
| | 392.3 | 0.485 | 0.980 |
| | 490.3 | 0.578 | 0.965 |
| | 588.4 | 0.704 | 0.932 |
| | 617.8 | 0.748 | 0.870 |
| 413.15 | 19.6 | 0.0301 | 0.866 |
| | 49.0 | 0.0827 | 0.942 |
| | 107.9 | 0.182 | 0.950 |
| | 196.1 | 0.306 | 0.956 |
| | 294.2 | 0.447 | 0.959 |
| | 382.5 | 0.573 | 0.940 |
| | 402.1 | 0.603 | 0.917 |
| | 416.8 | 0.706 | 0.849 |
| 452.95 | 19.6 | 0.0267 | 0.653 |
| | 49.0 | 0.0890 | 0.876 |
| | 98.1 | 0.181 | 0.934 |
| | 196.1 | 0.336 | 0.942 |
| | 274.6 | 0.469 | 0.919 |
| | 304.0 | 0.619 | 0.877 |
| | 313.8 | 0.730 | 0.836 |

AUXILIARY INFORMATION

METHOD /APPARATUS/PROCEDURE:

Temperature measured with platinum resistance thermometer. Cell contents stirred with magnetic stirrer. Cell contents equilibrated. Sample removed at various levels in cell. Sample analysed by freezing out the hydrocarbon in an acetone/carbon dioxide trap and weighing. Nitrogen estimated volumetrically. Details in ref. 1. Critical data in source ref.

SOURCE AND PURITY OF MATERIALS:

(1) Linde sample, purity 99.99 mole per cent.

(2) Shell AG sample.

ESTIMATED ERROR:

$\delta T/K = \pm 0.05$; $\delta P/\text{bar} = \pm 0.5$;
$\delta x_{N_2} = \pm 0.003$; $\delta y_{N_2} = \pm 0.007$.

REFERENCES:

1. Peter, S.; Reinhartz, K. *Z. Phys. Chem. N.F.* 1960, 24, 103.

| COMPONENTS: | ORIGINAL MEASUREMENTS: |
|---|---|
| (1) Nitrogen; $N_2$; [7727-37-9]<br>(2) Decane; $C_{10}H_{22}$; [124-18-5] | Azarnoosh, A.; McKetta, J. J.<br>*J. Chem. Eng. Data* 1963, *8*, 494. |
| VARIABLES:<br>T/K = 311-411<br>P/MPa = 0.3-34.5 | PREPARED BY:<br>C. L. Young |

EXPERIMENTAL VALUES:

| T/K | P/bar | Mole fraction of nitrogen in liquid, $x_{N_2}$ | in vapor, $y_{N_2}$ | T/K | P/bar | Mole fraction of nitrogen in liquid, $x_{N_2}$ | in vapor, $y_{N_2}$ |
|---|---|---|---|---|---|---|---|
| 310.9 | 5.5 | - | 0.99676 | 310.9 | 234.4 | 0.2710 | 0.99870 |
| | 6.9 | - | 0.99776 | | 234.4 | 0.2690 | 0.99950 |
| | 10.3 | - | 0.99756 | | 248.2 | 0.2822 | 0.99930 |
| | 13.8 | - | 0.99940 | | 275.8 | 0.2990 | 0.99850 |
| | 17.2 | - | 0.99964 | | 275.8 | 0.3050 | 0.99960 |
| | 27.58 | 0.0385 | - | | 275.8 | 0.3030 | 0.99870 |
| | 34.47 | 0.0482 | 0.9977 | | 293.0 | 0.3180 | 0.99860 |
| | 46.88 | 0.0665 | 0.9993 | | 324.1 | 0.3380 | - |
| | 51.71 | 0.0710 | 0.99976 | | 344.7 | 0.3550 | 0.99680 |
| | 55.16 | 0.0735 | - | 344.3 | 6.9 | - | 0.9928 |
| | 55.16 | 0.0763 | 0.99972 | | 10.3 | - | 0.9970 |
| | 68.95 | 0.0930 | 0.9997 | | 13.8 | - | 0.9967 |
| | 72.39 | 0.1000 | - | | 17.2 | - | 0.99923 |
| | 93.08 | 0.1242 | 0.99971 | | 28.96 | 0.0422 | - |
| | 96.53 | 0.1268 | 0.99940 | | 34.47 | 0.0513 | 0.9990 |
| | 117.21 | 0.1530 | 0.99940 | | 48.26 | 0.0602 | 0.9993 |
| | 131.00 | 0.1715 | 0.99920 | | 55.16 | 0.0780 | - |
| | 144.79 | 0.1875 | 0.99974 | | 62.05 | 0.0890 | - |
| | 165.4 | 0.2130 | 0.99840 | | 65.50 | 0.0912 | 0.99775 |
| | 165.4 | 0.2110 | 0.99959 | | 82.74 | 0.1130 | 0.99910 |
| | 172.4 | 0.2130 | 0.99920 | | 86.18 | 0.1186 | 0.9970 |
| | 182.7 | 0.2280 | 0.99980 | | 117.21 | 0.1713 | 0.99896 |

(cont.)

AUXILIARY INFORMATION

| METHOD /APPARATUS/PROCEDURE: | SOURCE AND PURITY OF MATERIALS: |
|---|---|
| Stirred, static cell with windows. Pressure measured with Bourdon gauge. Cell charged with components and equilibrated. Samples analysed using gas chromatography. Details in source ref. | (1) Minimum purity 99.9 mole per cent.<br>(2) Phillips Petroleum Co., research grade, purity $\geq$ 99.43 mole per cent. |
| | ESTIMATED ERROR:<br>$\delta T/K = \pm 0.2$; $\delta P/bar = \pm 0.3$;<br>$\delta x_{N_2} = \pm 0.002$; $\delta y_{N_2} = \pm 0.001$ to 0.00001. |
| | REFERENCES: |

| COMPONENTS: | ORIGINAL MEASUREMENTS: |
|---|---|
| (1) Nitrogen; $N_2$; [7727-37-9]<br>(2) Decane; $C_{10}H_{22}$; [124-18-5] | Azarnoosh, A.; McKetta, J. J.<br>*J. Chem. Eng. Data* 1963, *8*, 494. |

EXPERIMENTAL VALUES:

| T/K | *P*/bar | Mole fraction of nitrogen in liquid, $x_{N_2}$ | Mole fraction of nitrogen in vapor, $y_{N_2}$ | T/K | *P*/bar | Mole fraction of nitrogen in liquid, $x_{N_2}$ | Mole fraction of nitrogen in vapor, $y_{N_2}$ |
|---|---|---|---|---|---|---|---|
| 344.3 | 120.66 | 0.1630 | 0.99900 | 377.6 | 172.4 | 0.2350 | 0.99550 |
| | 120.66 | 0.1700 | - | | 186.2 | 0.2410 | 0.99330 |
| | 137.90 | 0.1980 | 0.99871 | | 189.6 | 0.2410 | 0.9958 |
| | 144.79 | 0.1935 | 0.99841 | | 206.8 | 0.260 | 0.9956 |
| | 172.4 | 0.2190 | - | | 227.5 | 0.2760 | - |
| | 172.4 | 0.2280 | 0.99817 | | 234.4 | 0.2850 | 0.9951 |
| | 179.3 | 0.2270 | 0.99819 | | 258.6 | 0.3070 | 0.9950 |
| | 193.1 | 0.2440 | 0.99844 | | 268.9 | 0.3130 | 0.9945 |
| | 206.8 | 0.2590 | 0.99824 | | 275.8 | 0.3230 | 0.9955 |
| | 206.8 | 0.2510 | 0.99845 | | 293.0 | 0.3330 | 0.99312 |
| | 224.0 | 0.2755 | 0.99850 | | 310.3 | 0.3420 | 0.9942 |
| | 234.4 | 0.2780 | - | | 327.5 | 0.3650 | 0.9932 |
| | 241.3 | 0.2910 | 0.99802 | | 341.3 | 0.370 | 0.9923 |
| | 258.5 | 0.3010 | 0.99757 | 410.9 | 2.8 | - | 0.6640 |
| | 282.7 | 0.3270 | 0.99744 | | 4.1 | - | 0.8140 |
| | 293.0 | 0.3339 | 0.99649 | | 5.5 | - | 0.8536 |
| | 296.5 | 0.3350 | 0.99846 | | 6.9 | - | 0.9030 |
| | 310.3 | 0.3500 | 0.99660 | | 10.3 | - | 0.9506 |
| | 324.1 | 0.3590 | 0.99734 | | 17.2 | 0.0329 | 0.9268 |
| | 337.8 | 0.3730 | 0.99660 | | 34.5 | 0.0550 | 0.98990 |
| | 344.7 | 0.3630 | 0.99500 | | 55.2 | 0.0825 | 0.99365 |
| 377.6 | 5.5 | - | 0.9797 | | 62.1 | 0.0966 | - |
| | 6.9 | - | 0.9847 | | 68.9 | 0.1064 | 0.9938 |
| | 13.8 | - | 0.9943 | | 93.1 | 0.1365 | 0.9910 |
| | 20.7 | 0.0350 | 0.9962 | | 103.4 | 0.1520 | 0.99354 |
| | 34.5 | 0.0575 | 0.9965 | | 134.4 | 0.1930 | 0.99302 |
| | 51.7 | 0.0682 | 0.9970 | | 148.2 | 0.2150 | - |
| | 55.2 | 0.0804 | - | | 172.4 | 0.2380 | 0.9932 |
| | 62.1 | 0.0925 | 0.9970 | | 172.4 | 0.2420 | 0.9924 |
| | 68.9 | 0.1070 | 0.9967 | | 203.4 | 0.2670 | 0.99336 |
| | 75.8 | 0.1075 | 0.99617 | | 213.7 | 0.2770 | 0.9940 |
| | 96.5 | 0.1360 | 0.9939 | | 224.1 | 0.2800 | 0.9924 |
| | 103.4 | 0.1520 | 0.99679 | | 241.3 | 0.297 | 0.9920 |
| | 110.3 | 0.1590 | 0.99788 | | 262.0 | 0.3245 | 0.9914 |
| | 131.0 | 0.1860 | - | | 275.8 | 0.3320 | 0.9906 |
| | 137.9 | 0.1990 | 0.99614 | | 293.0 | 0.3510 | 0.9902 |
| | 144.8 | 0.2030 | 0.99612 | | 306.8 | 0.3600 | - |
| | 158.6 | 0.2180 | 0.9963 | | 324.1 | 0.3820 | 0.9883 |
| | 158.6 | 0.2170 | 0.9870 | | 324.1 | 0.3770 | - |
| | 158.6 | 0.2170 | 0.99590 | | 344.7 | 0.3980 | 0.9860 |

COMPONENTS:

(1) Nitrogen; $N_2$; [7727-37-9]

(2) n-Dodecane; $C_{12}H_{26}$; [112-40-3]

ORIGINAL MEASUREMENTS:

Rupprecht, S. D.; Faeth, G.M.

NASA CR 1981, NASA-CR-3422, 111 pp.

*Chem. Abstr.* 1981, 95, 153333x.

VARIABLES:

T/K = 297-373
P/MPa = 1-10

PREPARED BY:

R. Battino

EXPERIMENTAL VALUES:

| T/K | $P^a$/MPa | $x_1$ |
|---|---|---|
| 300.15 | 1.03 | 0.0146 |
| 373.15 | 1.03 | 0.0143 |
| 298.15 | 2.07 | 0.0265 |
| 373.15 | 2.07 | 0.0296 |
| 297.15 | 4.82 | 0.0616 |
| 373.15 | 4.82 | 0.0685 |
| 297.15 | 10.34 | 0.1319 |
| 373.15 | 10.34 | 0.1404 |

[a] Total pressure.

AUXILIARY INFORMATION

METHOD/APPARATUS/PROCEDURE:

A 1000 $cm^3$ chamber is filled with ca. 500 $cm^3$ of liquid. Gas is added to the desired pressure. The chamber is agitated to attain equilibrium. Temperatures are read via thermocouples and pressures via Heisse bourdon tube gages. A liquid sample is removed and allowed to de-pressurize at atmospheric pressure. The solubility is calculated from the measurement of liquid and gas volumes of the de-pressurized sample. Residual dissolved gas is considered to be negligible

SOURCE AND PURITY OF MATERIALS:

No details given.

ESTIMATED ERROR:

$\delta x_1/x_1 = \pm 0.03$, compiler's estimate.

REFERENCES:

| COMPONENTS: | ORIGINAL MEASUREMENTS: |
|---|---|
| (1) Nitrogen; $N_2$; [7727-37-9]<br>(2) Hexadecane; $C_{16}H_{34}$; [544-76-3] | Sultanov, R. G.; Skripka, V. G.; Namiot, A. Yu., *Gazov. Delo.*, 1972, *10*, 43-6.<br>*Chem. Abstr.* 1974, *80*, 38956f |
| VARIABLES: | PREPARED BY: |
| T/K = 323-523<br>P/MPa = 5-49 | C. L. Young |

EXPERIMENTAL VALUES:

| T/K | P/bar | Mole fraction of nitrogen in liquid, $x_{N_2}$ | in vapor, $y_{N_2}$ | T/K | P/bar | Mole fraction of nitrogen in liquid, $x_{N_2}$ | in vapor, $y_{N_2}$ |
|---|---|---|---|---|---|---|---|
| 323.15 | 49 | 0.073 | - | 473.15 | 490 | 0.527 | - |
| | 98 | 0.135 | - | 523.15 | 49 | 0.1135 | 0.985 |
| | 196 | 0.223 | - | | 98 | 0.199 | 0.99 |
| | 294 | 0.282 | - | | 196 | 0.369 | 0.99 |
| | 392 | 0.326 | - | | 294 | 0.4315 | 0.9885 |
| | 490 | 0.360 | - | | 392 | 0.516 | 0.985 |
| 373.15 | 49 | 0.078 | - | | 441 | 0.555 | 0.9845 |
| | 98 | 0.142 | - | | 490 | 0.593 | 0.983 |
| | 196 | 0.239 | - | 573.15 | 49 | 0.127 | 0.9595 |
| | 294 | 0.306 | - | | 98 | 0.228 | 0.973 |
| | 392 | 0.364 | - | | 196 | 0.373 | 0.980 |
| | 490 | 0.413 | - | | 294 | 0.492 | 0.973 |
| 423.15 | 49 | 0.093 | - | | 392 | 0.598 | 0.966 |
| | 98 | 0.158 | - | | 490 | 0.6945 | 0.959 |
| | 196 | 0.253 | - | | 539 | 0.738 | 0.948 |
| | 294 | 0.331 | - | | 588 | 0.780 | 0.924 |
| | 392 | 0.399 | - | 523.15 | 49 | 0.150 | 0.902 |
| | 490 | 0.460 | - | | 98 | 0.2695 | 0.929 |
| 473.15 | 49 | 0.1015 | - | | 196 | 0.44 | 0.935 |
| | 98 | 0.176 | - | | 294 | 0.608 | 0.930 |
| | 196 | 0.287 | - | | 392 | 0.837 | 0.869 |
| | 294 | 0.377 | - | | 393.7 | 0.855 | 0.855 |
| | 392 | 0.455 | - | | | | |

AUXILIARY INFORMATION

| METHOD/APPARATUS/PROCEDURE: | SOURCE AND PURITY OF MATERIALS: |
|---|---|
| Static equilibrium cell fitted with magnetic stirrer, details in ref. 1. Samples of coexisting phases analysed by freezing out n-hexadecane and estimating nitrogen volumetrically. | (1) No details given.<br>(2) "Pure sample"; boiling point 286.3-287.3 °C.<br><br>ESTIMATED ERROR:<br>$\delta T/K = \pm 0.03$; $\delta P/bar = \pm 1$;<br>$\delta x_{N_2} \simeq \delta y_{N_2} = \pm 0.002$ (estimated by compiler).<br><br>REFERENCES:<br>1. Sultanov, R. G.; Skripka, V. G.; Namiot, A. Yu. *Gazov. Prom.* 1971, *16* (4), 6. |

COMPONENTS:

(1) Nitrogen; $N_2$; [7727-37-9]

(2) Cyclohexane; $C_6H_{12}$; [110-82-7]

ORIGINAL MEASUREMENTS:

Khodeeva, S.M.

*Tr. Naucho-Issledov. Proekt. Inst. Azotn. Prom. Prod. Org. Sin.* 1971, *12*, 30-9.

VARIABLES:

T/K = 373-433
P/MPa = 5-10

PREPARED BY:

C.L. Young

EXPERIMENTAL VALUES:

| T/K | $P$/atm | $P$/MPa | $\alpha^+$/cm$^3$g$^{-1}$ | Mole fraction of nitrogen in liquid, $x_{N_2}$ |
|---|---|---|---|---|
| 373.15 | 78.2 | 7.92 | 21.0 | 0.073 |
| | 84.2 | 8.53 | 23.1 | 0.080 |
| | 88.8 | 9.00 | 24.3 | 0.084 |
| | 94.2 | 9.54 | 26.0 | 0.089 |
| | 101.6 | 10.29 | 28.5 | 0.097 |
| 403.15 | 46.9 | 4.75 | 12.5 | 0.045 |
| | 49.8 | 5.05 | 13.9 | 0.050 |
| | 53.0 | 5.37 | 14.8 | 0.053 |
| | 62.2 | 6.30 | 17.5 | 0.064 |
| | 69.2 | 7.01 | 19.8 | 0.069 |
| | 76.7 | 7.77 | 22.4 | 0.078 |
| | 79.2 | 8.02 | 22.8 | 0.079 |
| | 83.2 | 8.43 | 24.7 | 0.085 |
| | 99.7 | 10.10 | 29.5 | 0.100 |
| 433.15 | 64.9 | 6.58 | 20.6 | 0.072 |
| | 69.4 | 7.03 | 21.2 | 0.074 |
| | 73.6 | 7.46 | 24.6 | 0.084 |
| | 80.7 | 8.18 | 26.9 | 0.092 |
| | 88.6 | 8.98 | 28.3 | 0.096 |

\+ $\alpha$ = Volume of gas, measured at 273.15 K and 101.3 kPa pressure dissolved by 1g of solvent.

AUXILIARY INFORMATION

METHOD/APPARATUS/PROCEDURE:

Volumetric method in which saturated liquid sample was stripped of gas under reduced pressure. Amount of liquid in sample determined by weighing. Amount of dissolved gas determined by measuring the volume at a known temperature and pressure. Some details in source and ref. (1).

SOURCE AND PURITY OF MATERIALS:

(1) Specially purified sample, less than 0.005 volume per cent.

(2) No details given.

ESTIMATED ERROR:

$\delta T/K = \pm 0.5$; $\delta P/\text{atm} = \pm 0.2$
$\delta x_{N_2} = \pm 1\%$.

REFERENCES:

1. Khodeeva, S.M.; Dymova, R.P. *Tr. Naucho-Issledov.Proekt.Inst. Azotn.Prom. Prod. Org. Sin.* 1971, *12*, 39.

**COMPONENTS:**

(1) Nitrogen; $N_2$; [7727-37-9]

(2) Cyclohexane; $C_6H_{12}$; [110-82-7]

**ORIGINAL MEASUREMENTS:**

Wild, J.D.; Sridhar, T. Potter, O.E.

*Chem. Eng. J.* 1978, *15*, 209-14.

**VARIABLES:**

T/K = 300-443

P/MPa = 0.1-1.9

**PREPARED BY:**

C. L. Young, R. Battino

**EXPERIMENTAL VALUES:**

| t/°C | T/K | Pressure range (psi gauge) | $10^4$ Henry's Law Constant /atm$^{-1}$ | $10^4$ x Mole fraction $x_{N_2}$ § |
|---|---|---|---|---|
| 35 | 308 | 110-260 | 8.11 | 0.1233 |
| 80.5 | 353.7 | 110-258 | 9.461 | 0.1057 |
| 110 | 383 | 110-255 | 10.7 | 0.0935 |
| 133 | 406 | 110-242 | 11.4 | 0.0877 |
| 150 | 423 | 102-260 | 12.7 | 0.0787 |
| 170 | 443 | 152-268 | 12.9 | 0.0775 |
| 26.5 | 299.7 | 0+ | 7.82 | 0.1279 |
| 35 | 308.0 | 0+ | 8.17 | 0.1224 |
| 45 | 318.0 | 0+ | 8.32 | 0.1202 |
| 59.5 | 332.5 | 0+ | 8.94 | 0.1119 |

\+ Solubilities determined by gas chromatography

§ Calculated mole fraction at 1 atm partial pressure assuming Henry's law obeyed.

AUXILIARY INFORMATION

**METHOD/APPARATUS/PROCEDURE:**

Static equilibrium cell. Liquid saturated with gas and after equilibrium established samples removed and analysed by a volumetric method. Allowance was made for the vapor pressure of the liquid and the solubility of the gas at atmospheric pressure. Details in source and ref. (1). In chromatographic method samples of cyclohexane saturated with nitrogen were analysed on a molecular sieve column using helium carrier gas and a thermal conductivity detector.

**SOURCE AND PURITY OF MATERIALS:**

(1) High purity, oxygen free.
(2) Distilled, boiling point 80.7°C Chromatographic analysis revealed 0.1 mole per cent of hydrocarbon impurity. Refractive index of 1.4201.

**ESTIMATED ERROR:**

$\delta T/K = \pm 0.5$; $\delta x_{N_2} = \pm 1.0\%$

**REFERENCES:**

1. Frolich, P.K.; Tauch, E.J.; Hogan, J.J.; Peer, A.A.

*Ind. Eng. Chem.* 1931, *23*, 548.

| COMPONENTS: | ORIGINAL MEASUREMENTS: |
|---|---|
| (1) Nitrogen; $N_2$; [7727-37-9]<br>(2) Methylcyclohexane; $C_7H_{14}$; [108-87-2] | Peter, S.; Eicke, H. F.<br>*Ber. Bunsenges. Phys. Chem.* 1970, *74*, 190. |
| VARIABLES:<br>T/K = 376-453<br>P/MPa = 2.0-88.2 | PREPARED BY:<br>C. L. Young |

EXPERIMENTAL VALUES:

| T/K | $P$/bar | Mole fraction of nitrogen in liquid, $x_{N_2}$ | in gas, $y_{N_2}$ |
|---|---|---|---|
| 376.45 | 29.4 | 0.0293 | 0.949 |
| | 49.0 | 0.0488 | 0.966 |
| | 294.2 | 0.281 | 0.976 |
| | 490.3 | 0.428 | 0.976 |
| | 686.5 | 0.531 | 0.973 |
| | 882.6 | 0.630 | 0.975 |
| 413.15 | 19.6 | 0.0217 | 0.856 |
| | 49.0 | 0.0589 | 0.935 |
| | 294.2 | 0.331 | 0.959 |
| | 490.3 | 0.511 | 0.963 |
| | 686.5 | 0.648 | 0.967 |
| | 882.6 | 0.759 | 0.965 |
| 452.95 | 29.4 | 0.0340 | 0.762 |
| | 49.0 | 0.0630 | 0.856 |
| | 98.1 | 0.139 | 0.934 |
| | 294.2 | 0.387 | 0.950 |
| | 490.3 | 0.596 | 0.953 |
| | 637.4 | 0.730 | 0.929 |
| | 676.7 | 0.791 | 0.900 |
| | 686.5 | 0.809 | 0.846 |

AUXILIARY INFORMATION

METHOD /APPARATUS/PROCEDURE:

Temperature measured with platinum resistance thermometer. Cell contents stirred with magnetic stirrer. Cell contents equilibrated, sample removed at various levels in cell. Sample analysed by freezing out the hydrocarbon in an acetone/carbon dioxide trap and weighing. Nitrogen estimated volumetrically. Details in ref. 1. Critical data in source ref.

SOURCE AND PURITY OF MATERIALS:

(1) Linde sample, purity 99.99 mole per cent.

(2) Phillips Petroleum pure grade sample, minimum purity 99 mole per cent.

ESTIMATED ERROR:

$\delta T/K = \pm 0.05$; $\delta P/\text{bar} = \pm 0.5$; $\delta x_{N_2} = \pm 0.003$; $\delta y_{N_2} = \pm 0.007$.

REFERENCES:

1. Peter, S; Reinhartz, K. *Z. Phys. Chem., N.F.* 1960, *24*, 103.

| COMPONENTS: | ORIGINAL MEASUREMENTS: |
|---|---|
| (1) Nitrogen; $N_2$; [7727-37-9]<br><br>(2) Methylcyclohexane; $C_7H_{14}$; [108-87-2] | Brunner, G.; Peter, S.; Wenzel, H.<br>*Chem. Eng. J.* 1974, *7*, 99. |
| **VARIABLES:** | **PREPARED BY:** |
| T/K = 453-492<br>P/MPa = 5.2-37.1 | C. L. Young |

EXPERIMENTAL VALUES:

| T/K | $P$/bar | Mole fraction of nitrogen in liquid, $x_{N_2}$ | in vapor, $y_{N_2}$ |
|---|---|---|---|
| 453.15 | 85 | 0.102 | 0.882 |
| | 130 | 0.174 | 0.905 |
| | 141 | 0.178 | 0.911 |
| | 271 | 0.335 | 0.904 |
| | 371 | 0.433 | 0.866 |
| 472.15 | 67 | 0.082 | 0.835 |
| | 84 | 0.120 | 0.862 |
| | 139 | 0.184 | 0.866 |
| | 151 | 0.216 | 0.883 |
| | 212 | 0.285 | 0.874 |
| | 281 | 0.395 | 0.855 |
| | 320 | 0.483 | 0.843 |
| 492.15 | 52 | 0.064 | 0.699 |
| | 102 | 0.146 | 0.791 |
| | 153 | 0.235 | 0.818 |
| | 209 | 0.346 | 0.825 |
| | 221 | 0.383 | 0.818 |
| | 245 | 0.488 | 0.796 |
| | 256 | 0.470 | 0.806 |
| | 277 | 0.650 | 0.698 |
| | 290 | 0.629 | 0.658 |

AUXILIARY INFORMATION

METHOD/APPARATUS/PROCEDURE:

Static cell of about 1000 $cm^3$ capacity, magnetically stirred. Pressure measured with Bourdon gauge. Temperature measured with nickel chromium alloy-nickel thermocouple. Each phase sampled; methylcyclohexane frozen and amount estimated; nitrogen estimated volumetrically.

SOURCE AND PURITY OF MATERIALS:

(1) Linde sample; purity 99.98 mole per cent by gas chromatography.

(2) Fluka sample; purity 99.4 mole per cent by gas chromatography.

ESTIMATED ERROR:

$\delta T/K = \pm 0.5$; $\delta P/\text{bar} = \pm 0.5$;

$\delta x_{N_2}$, $\delta y_{N_2} = \pm 5\%$

REFERENCES:

| COMPONENTS: | ORIGINAL MEASUREMENTS: |
|---|---|
| (1) Nitrogen; $N_2$; [7727-37-9]<br>(2) Benzene; $C_6H_6$; [71-43-2] | Miller. P.; Dodge, B.F.<br>*Ind. Eng. Chem.* 1940, *32*, 434-8. |
| VARIABLES:<br>T/K = 303-423<br>P/MPa = 6.2-30.7 | PREPARED BY:<br>C.L. Young |

EXPERIMENTAL VALUES:

| T/K | *P*/atm | Mole fraction of Nitrogen Liquid, $x_{N_2}$ | Vapor, $y_{N_2}$ |
|---|---|---|---|
| 303.15 | 120.7 | 0.0509 | |
| | 218.1 | 0.08345 | |
| | 297.5 | 0.1051 | |
| 348.15 | 61.3 | 0.0345 | 0.97547 |
| | 109.0 | 0.0598 | 0.98022 |
| | 155.5 | 0.08165 | 0.98048 |
| | 208.0 | 0.1062 | 0.97953 |
| | 252.2 | 0.1265 | 0.97767 |
| | 298.4 | 0.1458 | 0.97547 |
| 373.15 | 61.9 | 0.0396 | 0.9534 |
| | 106.3 | 0.0663 | 0.96265 |
| | 155.5 | 0.0945 | 0.96547 |
| | 198.3 | 0.1165 | 0.96383 |
| | 269.0 | 0.1550 | - |
| | 301.6 | 0.1750 | - |
| 398.15 | 63.6 | 0.0430 | 0.9168 |
| | 103.6 | 0.0711 | 0.9347 |
| | 151.4 | 0.1037 | 0.9406 |
| | 201.5 | 0.1388 | 0.93945 |
| | 201.5 | 0.1379 | 0.9409 |
| | 251.0 | 0.1716 | 0.9391 |
| | 303.1 | 0.2044 | 0.93495 |
| 423.15 | 297.0 | 0.2396 | - |

AUXILIARY INFORMATION

| METHOD/APPARATUS/PROCEDURE: | SOURCE AND PURITY OF MATERIALS: |
|---|---|
| Single pass flow system. Nitrogen bubbled through a series of three pressure cylinders, two being pre-saturators the third being an equilibrium cell which was continually agitated. Gas phase analysed by freezing out benzene and estimating nitrogen volumetrically. Liquid sample analysed using same principle. | (1) Commercial sample.<br>(2) Merck C.P. grade, thiophene-free fractionally crystallized, distilled and dried. |
| | ESTIMATED ERROR:<br>$\delta T/K = \pm 0.1$; $\delta P/atm = \pm 0.2$; $\delta x_{N_2} = \pm 0.0005$; $\delta y_{N_2} = \pm 0.0001$.<br>(estimated by compiler). |
| | REFERENCES: |

COMPONENTS:

(1) Nitrogen, $N_2$; [7727-37-9]

(2) Benzene; $C_6H_6$; [71-43-2]

ORIGINAL MEASUREMENTS:

Krichevskii, I. R.; Efremova, G. D.
*Zhur. Fiz. Khim.* 1948, *22*, 1116.

VARIABLES:

T/K = 288
P/MPa = 2.7-78.5

PREPARED BY:

C. L. Young

EXPERIMENTAL VALUES:

| T/K | $P$/bar | Mole fraction of nitrogen in liquid, $x_{N_2}$ |
|---|---|---|
| 288.15 | 27.2 | 0.0116 |
| | 50.0 | 0.0204 |
| | 99.1 | 0.0395 |
| | 143.9 | 0.0547 |
| | 196.6 | 0.0717 |
| | 260.4 | 0.0886 |
| | 306.0 | 0.100 |
| | 482.3 | 0.139 |
| | 543 | 0.150 |
| | 570 | 0.154 |
| | 785 | 0.191 |

AUXILIARY INFORMATION

METHOD/APPARATUS/PROCEDURE:

Static equilibrium cell fitted with magnetic stirrer. Pressure measured with Bourdon gauge. Sample of liquid analysed by separating the two components, weighing the benzene and estimating the nitrogen volumetrically. Details in source.

SOURCE AND PURITY OF MATERIALS:

(1) Dried.

(2) Dried and purified by fractional crystallisation; final purity 99.65 mole per cent.

ESTIMATED ERROR:

$\delta T/K = \pm 0.1$; $\delta P/\text{bar} = 0.35\%$;
$\delta x_{N_2} = \pm 0.5\%$.

REFERENCES:

| COMPONENTS: | ORIGINAL MEASUREMENTS: |
|---|---|
| (1) Nitrogen; $N_2$; [7727-37-9]<br>(2) Toluene; $C_7H_8$; [108-88-3] | Tsiklis, D.S.; Shenderei, L.I.; El'natanov, A.I.<br>*Khim Prom.* 1963, 348-53.<br>*Khim. Promysl.* 1963, 348-53.<br>*Chem. Abstr.* 1963, 59, 14647F. |
| **VARIABLES:** | **PREPARED BY:** |
| T/K = 473-548<br>P/MPa = 1.5-40.5 | V. Katovic |

EXPERIMENTAL VALUES:

| | | Mole fraction in gas $x_{C_7H_8}$ | | | | | Mole fraction in liquid $x_{C_7H_8}$ | | | | |
|---|---|---|---|---|---|---|---|---|---|---|---|
| P[a]/MPa | P/atm | 205°C (478K) | 215°C (488K) | 225°C (498K) | 250°C (523K) | 275°C (548K) | 205°C (478K) | 215°C (488K) | 225°C (498K) | 250°C (523K) | 275°C (548K) |
| 1.52 | 15 | 0.610 | 0.700 | 0.775 | | | 0.995 | 0.992 | 0.995 | | |
| 2.53 | 25 | 0.430 | 0.500 | 0.575 | 0.705 | 0.905 | 0.985 | 0.980 | 0.985 | 0.990 | 0.995 |
| 4.05 | 40 | 0.305 | 0.350 | 0.410 | 0.500 | 0.675 | 0.965 | 0.965 | 0.970 | 0.965 | 0.970 |
| 6.08 | 60 | 0.225 | 0.260 | 0.310 | 0.360 | 0.495 | 0.945 | 0.945 | 0.945 | 0.940 | 0.940 |
| 8.11 | 80 | 0.165 | 0.200 | 0.215 | 0.275 | 0.330 | 0.920 | 0.925 | 0.920 | 0.910 | 0.940 |
| 10.13 | 100 | 0.300 | 0.160 | 0.155 | 0.220 | 0.225 | 0.900 | 0.900 | 0.895 | 0.880 | 0.908 |
| 15.20 | 150 | 0.075 | 0.105 | 0.125 | 0.170 | 0.260 | 0.845 | 0.895 | 0.840 | 0.800 | 0.775 |
| 20.27 | 200 | 0.070 | 0.085 | 0.105 | 0.155 | 0.275 | 0.790 | 0.785 | 0.765 | 0.735 | 0.695 |
| 25.33 | 250 | 0.065 | 0.090 | 0.115 | 1.175[b] | 0.400 | 0.745 | 0.725 | 0.700 | 0.665 | 0.600 |
| 30.40 | 300 | 0.065 | 0.105 | 0.115 | 0.195 | | 0.695 | 0.675 | 0.645 | 0.590 | |
| 40.53 | 400 | 0.115 | 0.175 | 0.240 | | | 0.580 | 0.560 | 0.500 | | |

[a] Calculated by compiler.

[b] Probably 0.175.

AUXILIARY INFORMATION

| METHOD/APPARATUS/PROCEDURE: | SOURCE AND PURITY OF MATERIALS: |
|---|---|
| Solubility was measured in a thermostatted autoclave. Samples of solution and gas phase were analyzed for nitrogen and toluene (1). | (1) Purified from oxygen by ammoniacal copper solution.<br>(2) Purified by extraction with $H_2SO_4$, Na amalgam and water. Dried with $P_4O_{10}$. |
| | **ESTIMATED ERROR:**<br>$\delta x/x = \pm 0.03$, compiler's estimate. |
| | **REFERENCES:**<br>(1) Tsiklis, D.S.; Kofman, A.M., *Russ. J. Phys. Chem.* 1961, 35, 1120. |

| COMPONENTS: | ORIGINAL MEASUREMENTS: |
|---|---|
| (1) Nitrogen; $N_2$; [7727-37-9]<br>(2) Methanol; $CH_4O$; [67-56-1] | Krichevskii, I.R.; Lebedeva, E.S.<br>*Zhur. Fiz. Khim.* 1947, *21*, 715-7. |
| **VARIABLES:** | **PREPARED BY:** |
| T/K = 273-343<br>P/MPa = 4.9-30.0 | C.L. Young |

EXPERIMENTAL VALUES:

| T/K | $P$/bar | Mole fraction of nitrogen in liquid, $x_{N_2}$ |
|---|---|---|
| 273.15 | 49.0 | 0.0109 |
| | 99.1 | 0.0211 |
| | 147 | 0.0304 |
| | 197 | 0.0391 |
| 298.15 | 49.0 | 0.0116 |
| | 99.1 | 0.0229 |
| | 147 | 0.0321 |
| | 197 | 0.0418 |
| | 245 | 0.0505 |
| 323.15 | 49.0 | 0.0123 |
| | 99.1 | 0.0243 |
| | 147 | 0.0346 |
| | 197 | 0.0452 |
| | 245 | 0.0542 |
| | 295 | 0.0634 |
| 343.15 | 49.0 | 0.0131 |
| | 99.1 | 0.0260 |
| | 147 | 0.0373 |
| | 197 | 0.0489 |
| | 245 | 0.0598 |
| | 284 | 0.0687 |

AUXILIARY INFORMATION

METHOD/APPARATUS/PROCEDURE:

Not clear from original paper.

SOURCE AND PURITY OF MATERIALS:

(1) Contained 0.4 mole per cent oxygen.

(2) Purity better than 99.9 mole per cent.

ESTIMATED ERROR:

$\delta T/K = \pm 0.1$; $\delta P/bar = \pm 0.5\%$; $\delta x_{N_2} = \pm 1\%$. (estimated by compiler).

REFERENCES:

COMPONENTS:

(1) Nitrogen; $N_2$; [7727-37-9]

(2) Ethanol; $C_2H_6O$; [64-17-5]

ORIGINAL MEASUREMENTS:

Frolich, P. K.; Tauch, E. J.; Hogan, J. J.; Peer, A. A.

*Ind. Eng. Chem.* 1931, *23*, 548-50.

VARIABLES:

T/K = 298
P/MPa = 1-17

PREPARED BY:

C. L. Young

EXPERIMENTAL VALUES:

| T/K | $P$/MPa | Solubility*, $S$ | Mole fraction of nitrogen† $x_{N_2}$ |
|---|---|---|---|
| 298.15 | 1.0 | 1.2 | 0.0029 |
| | 2.0 | 2.3 | 0.0055 |
| | 3.0 | 3.3 | 0.0079 |
| | 4.0 | 4.2 | 0.0100 |
| | 5.0 | 5.15 | 0.0122 |
| | 6.0 | 6.05 | 0.0143 |
| | 7.0 | 7.0 | 0.0165 |
| | 8.0 | 7.95 | 0.0187 |
| | 9.0 | 8.9 | 0.0209 |
| | 10.0 | 9.85 | 0.0231 |
| | 11.0 | 10.85 | 0.0254 |
| | 12.0 | 11.85 | 0.0276 |
| | 13.0 | 12.8 | 0.0298 |
| | 14.0 | 13.7 | 0.0318 |
| | 15.0 | 14.5 | 0.0336 |
| | 16.0 | 15.4 | 0.0356 |
| | 17.0 | 16.2 | 0.0374 |

* Data taken from graph in original article. Volume of gas measured at 101.325 kPa pressure and 298.15 K dissolved by unit volume of liquid measured under the same conditions.

† Calculated by compiler.

AUXILIARY INFORMATION

METHOD/APPARATUS/PROCEDURE:

Static equilibrium cell. Liquid saturated with gas and after equilibrium established samples removed and analysed by volumetric method. Allowance was made for vapor pressure of liquid and the solubility of the gas at atmospheric pressure. Details in source.

SOURCE AND PURITY OF MATERIALS:

Stated that the materials were the highest purity available.

ESTIMATED ERROR:

$\delta T/K = \pm 0.1$; $\delta x_{N_2} = \pm 5\%$.

REFERENCES:

COMPONENTS:

(1) Nitrogen; $N_2$; [7727-37-9]

(2) 4-Methyl-1,3-dioxolan-2-one (Propylene carbonate); $C_4H_6O_3$; [108-32-7]

ORIGINAL MEASUREMENTS:

Shakhova, S. F.; Zubchenko, Yu. P.; Kaplan, L. K.

*Khim. Prom.* 1973, *49*, 108.

VARIABLES:

T/K = 283-323
P/MPa = 6.4-12.5

PREPARED BY:

C. L. Young

EXPERIMENTAL VALUES:

| T/K | *P*/bar | Mole fraction of nitrogen in liquid, $x_{N_2}$ | vol $\alpha^*$/vol |
|---|---|---|---|
| 283.15 | 66.67 | 0.00849 | 2.25 |
| | 83.39 | 0.0104 | 2.76 |
| | 90.28 | 0.0111 | 2.95 |
| | 95.14 | 0.0115 | 3.06 |
| | 107.91 | 0.0130 | 3.46 |
| 298.15 | 63.73 | 0.00886 | 2.35 |
| | 75.49 | 0.0101 | 2.68 |
| | 77.92 | 0.0107 | 2.84 |
| | 80.96 | 0.0112 | 2.97 |
| | 88.25 | 0.0117 | 3.10 |
| | 91.70 | 0.0121 | 3.22 |
| | 100.72 | 0.0136 | 3.62 |
| | 112.77 | 0.0154 | 4.11 |
| 323.15 | 63.23 | 0.00987 | 2.62 |
| | 71.64 | 0.0111 | 2.95 |
| | 79.34 | 0.0120 | 3.18 |
| | 91.09 | 0.0138 | 3.69 |
| | 102.84 | 0.0155 | 4.13 |
| | 109.32 | 0.0162 | 4.33 |
| | 115.21 | 0.0168 | 4.50 |
| | 125.04 | 0.0181 | 4.82 |

* quoted in original paper, appears to be volume of gas at T/K = 273.15 and *P* = 1 atmosphere absorbed by unit volume of liquid at room temperature.

AUXILIARY INFORMATION

METHOD/APPARATUS/PROCEDURE:

Mixture stirred by ball in rocking autoclave. Samples of liquid analysed by volumetric method. Details in source.

SOURCE AND PURITY OF MATERIALS:

(1) Purity 99.99 mole per cent.

(2) Distilled; purity 99.9 mole per cent determined by gas chromatography.

ESTIMATED ERROR:

$\delta T/K = \pm 0.1$; $\delta P/\text{bar} = \pm 0.1$; $\delta x_{N_2} = \pm 5\%$ (estimated by compiler).

REFERENCES:

COMPONENTS:

(1) Nitrogen; $N_2$; [7727-37-9]

(2) Cyclohexanone; $C_6H_{10}O$; [108-94-1]

ORIGINAL MEASUREMENTS:

Khodeeva, S. M.; Dymova, R. P.

*Tr. Naucho-Issledov. Proekt. Inst. Azotn. Prom. Org. Sin.* 1971, *12*, 39-46.

VARIABLES:

$T/K = 348-473$
$P/MPa = 4-10$

PREPARED BY:

C. L. Young

EXPERIMENTAL VALUES:

| T/K | P/atm | P/MPa | Mole fraction of nitrogen in liquid, $x_{N_2}$ | in gas, $y_{N_2}$ |
|---|---|---|---|---|
| 348.15 | 39.7 | 4.02 | 0.018 | - |
| | 42.3 | 4.29 | 0.019 | - |
| | 45.0 | 4.56 | 0.021 | - |
| | 47.2 | 4.78 | - | 0.997 |
| | 51.6 | 5.23 | - | 0.998 |
| | 55.3 | 5.60 | 0.025 | - |
| | 59.6 | 6.04 | 0.027 | - |
| | 64.5 | 6.54 | - | 0.993 |
| | 68.5 | 6.94 | - | 0.995 |
| | 72.3 | 7.33 | 0.032 | - |
| | 78.7 | 7.97 | - | 0.991 |
| | 82.4 | 8.35 | - | 0.992 |
| | 90.2 | 9.14 | 0.038 | - |
| | 93.1 | 9.43 | 0.038 | - |
| 373.15 | 24.0 | 2.43 | 0.013 | - |
| | 34.9 | 3.54 | 0.017 | - |
| | 38.7 | 3.92 | - | 0.992 |
| | 42.4 | 4.30 | - | 0.989 |
| | 47.1 | 4.77 | 0.023 | - |
| | 49.4 | 5.01 | 0.023 | - |
| | 53.0 | 5.37 | - | 0.991 |
| | 57.2 | 5.80 | - | 0.989 |

(cont.)

AUXILIARY INFORMATION

METHOD/APPARATUS/PROCEDURE:

Volumetric method in which saturated liquid sample was stripped of gas under reduced pressure. Amount of liquid in sample determined by weighing and amount of gas dissolved by measuring the volume at known temperature and pressure. Some details in source.

SOURCE AND PURITY OF MATERIALS:

(1) Specially purified sample, less than 0.005 volume per cent impurity.

(2) Distilled, no impurity detected by GC.

ESTIMATED ERROR:

$\delta T/K = \pm 0.5$; $\delta P/atm = \pm 0.2$; $\delta x_{N_2} = \pm 1\%$.

REFERENCES:

| COMPONENTS: | ORIGINAL MEASUREMENTS: |
|---|---|
| (1) Nitrogen; $N_2$; [7727-37-9] | Khodeeva, S. M.; Dymova, R. P. |
| (2) Cyclohexanone; $C_6H_{10}O$; [108-94-1] | *Tr. Naucho-Issledov. Proekt. Inst. Azotn. Prom. Org. Sin.* 1971, 12, 39-46. |

EXPERIMENTAL VALUES:

| T/K | $P$/atm | $P$/MPa | Mole fraction of nitrogen in liquid, $x_{N_2}$ | in gas, $y_{N_2}$ |
|---|---|---|---|---|
| 373.15 | 62.3 | 6.31 | 0.030 | - |
| | 64.8 | 6.57 | 0.032 | - |
| | 65.0 | 6.59 | 0.034 | - |
| | 68.3 | 6.92 | - | 0.990 |
| | 71.6 | 7.25 | - | 0.989 |
| | 77.2 | 7.82 | 0.043 | - |
| | 80.2 | 8.13 | 0.041 | - |
| | 83.2 | 8.43 | - | 0.991 |
| | 84.6 | 8.57 | - | 0.988 |
| | 90.4 | 9.16 | - | 0.986 |
| | 92.4 | 9.36 | 0.049 | - |
| | 96.1 | 9.74 | 0.045 | - |
| 423.15 | 46.0 | 4.66 | 0.027 | - |
| | 47.1 | 4.77 | - | 0.980 |
| | 48.8 | 4.94 | 0.028 | - |
| | 53.4 | 5.41 | - | 0.987 |
| | 57.3 | 5.81 | - | 0.978 |
| | 65.5 | 6.64 | 0.039 | - |
| | 68.3 | 6.92 | 0.041 | - |
| | 68.6 | 6.95 | 0.040 | - |
| | 71.0 | 7.19 | 0.041 | - |
| | 73.7 | 7.47 | 0.044 | - |
| | 74.6 | 7.56 | - | 0.977 |
| | 75.1 | 7.61 | - | 0.975 |
| | 76.6 | 7.76 | - | 0.979 |
| | 77.4 | 7.84 | - | 0.982 |
| | 80.1 | 8.12 | - | 0.982 |
| | 81.0 | 8.21 | - | 0.975 |
| | 84.3 | 8.54 | - | 0.979 |
| | 86.5 | 8.76 | 0.051 | - |
| | 92.0 | 9.32 | 0.054 | - |
| 473.15 | 47.4 | 4.80 | 0.036 | - |
| | 49.8 | 5.05 | 0.035 | - |
| | 56.4 | 5.71 | 0.038 | - |
| | 56.9 | 5.77 | 0.038 | - |
| | 61.9 | 6.27 | - | 0.963 |
| | 65.9 | 6.68 | - | 0.958 |
| | 67.9 | 6.88 | - | 0.961 |
| | 71.8 | 7.28 | - | 0.967 |

**COMPONENTS:**

(1) Nitrogen; $N_2$; [7727-37-9]

(2) Pentanedioic acid, -dimethyl ester, (dimethyl glutarate); $C_7H_{12}O_4$; [1119-40-0]

**ORIGINAL MEASUREMENTS:**

Shakhova, S. F.; Zubchenko, Yu. P.; Kaplan, L. K.

*Khim. Prom.* 1973, *49*, 108-10.

**VARIABLES:**

T/K = 313
P/MPa = 5.6-14.2

**PREPARED BY:**

C. L. Young

**EXPERIMENTAL VALUES:**

| T/K | $P$/bar | Mole fraction of nitrogen in liquid, $x_{N_2}$ | vol$^{\alpha *}$/vol |
|---|---|---|---|
| 313.15 | 55.9 | 0.0246 | 3.83 |
| | 86.8 | 0.0355 | 5.60 |
| | 110.3 | 0.0451 | 7.18 |
| | 142.3 | 0.0580 | 9.37 |

* quoted in original paper, appears to be volume of gas at T/K = 273.14 and $P$ = 1 atmosphere absorbed by unit volume of liquid at room temperature.

**AUXILIARY INFORMATION**

**METHOD /APPARATUS/PROCEDURE:**

Mixture stirred by ball in rocking autoclave. Sample of liquid analysed by a volumetric method. Details in source.

**SOURCE AND PURITY OF MATERIALS:**

(1) Purity 99.99 mole per cent.

(2) Analytic grade sample.

**ESTIMATED ERROR:**

$\delta T/K = \pm 0.1$; $\delta P/\text{bar} = \pm 0.1$;

$\delta x_{N_2} = \pm 5\%$ (estimated by compiler).

**REFERENCES:**

COMPONENTS:

(1) Nitrogen; $N_2$; [7727-37-9]

(2) 1,2,3-Propanetriol, triacetate; (Glycerol triacetate); $C_9H_{14}O_6$; [102-76-1]

ORIGINAL MEASUREMENTS:

Shakhova, S. F.; Zubchenko, Yu. P.; Kaplan, L. K.

*Khim. Prom.* 1973, *49*, 108-10.

VARIABLES:

T/K = 288-323
P/MPa = 6.8-14.7

PREPARED BY:

C. L. Young

EXPERIMENTAL VALUES:

| T/K | *P*/bar | Mole fraction of nitrogen in liquid, $x_{N_2}$ | vol$^{\alpha *}$/vol | T/K | *P*/bar | Mole fraction of nitrogen in liquid, $x_{N_2}$ | vol$^{\alpha *}$/vol |
|---|---|---|---|---|---|---|---|
| 288.15 | 80.96 | 0.0286 | 3.49 | 308.15 | 97.58 | 0.0381 | 4.69 |
| | 82.38 | 0.0303 | 3.70 | | 105.88 | 0.0410 | 5.07 |
| | 90.69 | 0.0320 | 3.92 | | 120.68 | 0.0458 | 5.63 |
| | 91.19 | 0.0318 | 3.89 | | 143.68 | 0.0544 | 6.81 |
| | 92.12 | 0.0331 | 4.06 | 323.15 | 79.13 | 0.0391 | 3.91 |
| | 101.53 | 0.0354 | 4.35 | | 102.03 | 0.0523 | 5.23 |
| | 109.53 | 0.0358 | 4.40 | | 107.40 | 0.0551 | 5.51 |
| | 111.05 | 0.0375 | 4.61 | | 107.91 | 0.0549 | 5.49 |
| | 131.92 | 0.0449 | 5.57 | | 108.11 | 0.0617 | 6.17 |
| 308.15 | 67.69 | 0.0273 | 3.33 | | 124.53 | 0.0630 | 6.30 |
| | 82.88 | 0.0346 | 4.24 | | 135.88 | 0.0691 | 6.91 |
| | 94.13 | 0.0369 | 4.54 | | 146.62 | 0.0730 | 7.30 |

* Values quoted in original paper, appears to be the volume of gas at T/K = 273.15 and *P* = 1 atmosphere absorbed by unit volume of liquid at room temperature.

AUXILIARY INFORMATION

METHOD /APPARATUS/PROCEDURE:

Mixture stirred by ball in rocking autoclave. Sample of liquid analysed by volumetric method. Details in source.

SOURCE AND PURITY OF MATERIALS:

(1) Purity 99.99 mole per cent.

(2) Distilled and dried; purity 99.4 mole per cent.

ESTIMATED ERROR:

$\delta T/K = \pm 0.1$; $\delta P/\text{bar} = \pm 0.1$; $\delta x_{N_2} = \pm 5\%$ (estimated by compiler).

REFERENCES:

| COMPONENTS: | ORIGINAL MEASUREMENTS: |
|---|---|
| (1) Nitrogen; $N_2$; [7727-37-9]<br>(2) 2,5,8,11,14-Pentaoxapentadecane, (Tetramethylene glycol dimethyl-ester); $C_{10}H_{22}O_5$; [143-24-8] | Zubchenko, Yu.P.; Shakhova, S.F.<br>*Tr. N.-i.i Proekt.In-ta Azat. Prom-sti i Produktov Organ. Sinteza* 1975, (33), 13-15.<br>*Chem. Abstr.* 1977, *86*, 89113K. |
| VARIABLES:<br>T/K = 313<br>P/MPa = 2.8-14.0 | PREPARED BY:<br>C.L. Young |

EXPERIMENTAL VALUES:

| T/K | *P*/atm | *P*/MPa | α+ vol/vol | Mole fraction of nitrogen in liquid, $x_{N_2}$ |
|---|---|---|---|---|
| 313.15 | 27.15 | 2.79 | 1.73 | 0.0146 |
| | 49.4 | 5.01 | 2.97 | 0.0247 |
| | 79.4 | 8.05 | 4.68 | 0.0384 |
| | 111.3 | 11.27 | 6.80 | 0.0549 |
| | 138.4 | 14.02 | 7.99 | 0.0639 |

\+ quoted in original paper, appears to be volume of gas at T/K = 273.15 and *P* = 1 atmosphere absorbed by unit volume of liquid at room temperature.

AUXILIARY INFORMATION

| METHOD/APPARATUS/PROCEDURE: | SOURCE AND PURITY OF MATERIALS: |
|---|---|
| Mixture stirred by ball in rocking autoclave. Sample of liquid analysed by a volumetric method. Details in ref. (1). | (1) Purity 99.99 mole per cent.<br>(2) No details given. |
| | ESTIMATED ERROR:<br>$\delta T/K = \pm 0.1$; $\delta P/atm = \pm 0.1$; $\delta x_{N_2} = \pm 5\%$.<br>(estimated by compiler). |
| | REFERENCES:<br>1. Shakhova, S.F.; Zubchenko, Yu.P.; Kaplan, L.K. *Khim. Prom.* 1973, *5*, 108. |

COMPONENTS:

(1) Nitrogen; $N_2$; [7727-37-9]

(2) 1,2-Epoxyethane (ethylene oxide); $C_2H_4O$; [75-21-8]

ORIGINAL MEASUREMENTS:

Hess, L. G.; Tilton, V. V.

*Ind. Eng. Chem.* 1950, *42*, 1251-8.

VARIABLES:

T/K = 303-318K
P/MPa = 0.21-0.48

PREPARED BY:

R. Battino

EXPERIMENTAL VALUES:

| t/°C | $T^a$/K | P/psig | $P^a$/MPa | Solubility[b] |
|---|---|---|---|---|
| 30 | 303 | 30 | 0.21 | 0.02 |
| 30 | 303 | 40 | 0.28 | 0.1 |
| 30 | 303 | 50 | 0.34 | 0.19 |
| 45 | 318 | 50 | 0.34 | 0.11 |
| 45 | 318 | 60 | 0.41 | 0.18 |
| 45 | 318 | 70 | 0.48 | 0.25 |

[a] Calculated by compiler.

[b] In weigh per cent of nitrogen in 1,2-epoxyethane at pressure indicated.

AUXILIARY INFORMATION

METHOD/APPARATUS/PROCEDURE:

These are average values of solubility as determined on a laboratory scale and using apparatus and procedures similar to that of Wan and Dodge (1).

SOURCE AND PURITY OF MATERIALS:

No details given.

ESTIMATED ERROR:

Solubility values are crude.

REFERENCES:

1. Wan, S.W.; Dodge, B.F. *Ind. Eng. Chem.* 1940, *32*, 95.

| COMPONENTS: | ORIGINAL MEASUREMENTS: |
|---|---|
| (1) Nitrogen; $N_2$; [7727-37-9]<br>(2) Methane, chlorodifluoro-; $CHClF_2$; [75-45-6] | Maslennikova, V. Ya.; Goryunova, N. P.; Tsiklis, D. S.<br>*Zhur. Fiz. Khim.* 1967, *41*, 735. |
| VARIABLES:<br>T/K = 295-333<br>P/MPa = 2.5-19.5 | PREPARED BY:<br>C. L. Young |

EXPERIMENTAL VALUES:

| T/K | $P$/bar | Mole fraction of nitrogen in liquid, $x_{N_2}$ | in gas, $y_{N_2}$ |
|---|---|---|---|
| 295.15 | 25.3 | 0.035 | 0.59 |
| | 50.7 | 0.085 | 0.74 |
| | 76.0 | 0.135 | 0.79 |
| | 101.3 | 0.185 | 0.80 |
| | 126.7 | 0.24 | 0.79 |
| | 152.0 | 0.315 | 0.763 |
| | 177.3 | 0.41 | 0.705 |
| | 194.5 | 0.57 | 0.57 |
| 313.15 | 25.3 | 0.025 | 0.365 |
| | 50.7 | 0.085 | 0.585 |
| | 76.0 | 0.14 | 0.66 |
| | 101.3 | 0.205 | 0.685 |
| | 126.7 | 0.29 | 0.673 |
| | 152.0 | 0.42 | 0.61 |
| | 162.1 | 0.515 | 0.515 |
| 333.15 | 25.3 | 0.003 | 0.045 |
| | 50.7 | 0.067 | 0.39 |
| | 76.0 | 0.14 | 0.52 |
| | 101.3 | 0.23 | 0.545 |
| | 126.7 | 0.39 | 0.48 |
| | 128.7 | 0.435 | 0.435 |

AUXILIARY INFORMATION

METHOD/APPARATUS/PROCEDURE:

Few details given in source. Ref. 1 given as reference in which apparatus described. It is not clear which of the several apparatus described were used.

SOURCE AND PURITY OF MATERIALS:

(1) Nitrogen containing up to 0.5 mole per cent oxygen.

(2) Technical grade sample.

ESTIMATED ERROR:

$\delta T/K = \pm 0.1$; $\delta P/\text{bar} = \pm 0.05$; $\delta x_{N_2}$, $\delta y_{N_2} = \pm 0.01$ (estimated by compiler).

REFERENCES:

1. Tsiklis, D. S. *Technique of Physico-chemical Experiment at High and Ultra High Pressures*, Isd. Khimiya, Moscow, 1965.

| COMPONENTS: | ORIGINAL MEASUREMENTS: |
|---|---|
| (1) Nitrogen; $N_2$; [7727-37-9]<br>(2) Methane, chlorodifluoro-; $CHClF_2$; [75-45-6] | Nohka, J.; Sarashina, E.; Arai, Y.; Saito, S.<br>*J. Chem. Eng. Japan* 1973, *6*, 10. |

| VARIABLES: | PREPARED BY: |
|---|---|
| T/K = 273-348<br>P/MPa = 2.5-21.5 | C. L. Young |

EXPERIMENTAL VALUES:

| T/K | $P$/bar | Mole fraction of nitrogen in liquid, $x_{N_2}$ | in gas, $y_{N_2}$ | T/K | $P$/bar | Mole fraction of nitrogen in liquid, $x_{N_2}$ | in gas, $y_{N_2}$ |
|---|---|---|---|---|---|---|---|
| 273.15 | 25.3 | 0.0367 | 0.755 | 298.15 | 154.2 | 0.315 | 0.730 |
| | 50.7 | 0.0842 | 0.852 | | 165.9 | 0.347 | 0.704 |
| | 76.0 | 0.129 | 0.873 | | 174.0 | 0.382 | 0.680 |
| | 101.3 | 0.177 | 0.876 | | 179.0 | 0.409 | 0.660 |
| | 126.7 | 0.227 | 0.869 | | 185.9 | 0.463 | 0.618 |
| | 144.3 | 0.265 | 0.859 | 323.15 | 30.4 | 0.0217 | 0.258 |
| | 160.5 | 0.301 | 0.845 | | 50.7 | 0.0638 | 0.472 |
| | 183.1 | 0.350 | 0.825 | | 76.0 | 0.121 | 0.561 |
| | 200.8 | 0.399 | 0.801 | | 101.3 | 0.185 | 0.580 |
| | 215.3 | 0.447 | 0.766 | | 130.7 | 0.291 | 0.540 |
| 298.15 | 25.3 | 0.0268 | 0.498 | | 134.4 | 0.327 | 0.524 |
| | 50.7 | 0.0749 | 0.688 | | 139.3 | 0.359 | 0.507 |
| | 50.7 | 0.0753 | 0.697 | 348.15 | 43.9 | 0.0261 | 0.149 |
| | 76.0 | 0.125 | 0.741 | | 49.7 | 0.0429 | 0.198 |
| | 95.8 | 0.164 | 0.755 | | 64.3 | 0.0820 | 0.278 |
| | 101.3 | 0.178 | 0.763 | | 74.0 | 0.114 | 0.308 |
| | 101.3 | 0.181 | 0.761 | | 82.6 | 0.145 | 0.316 |
| | 126.7 | 0.232 | 0.753 | | 91.2 | 0.198 | 0.281 |
| | 142.4 | 0.279 | 0.748 | | | | |

AUXILIARY INFORMATION

METHOD /APPARATUS/PROCEDURE:

Static cell fitted with magnetic stirrer. Temperature measured with liquid in glass thermometer and pressure measured with Bourdon gauge. After equilibrium established vapor and liquid samples analysed by gas chromatography. Details in refs. 1 and 2.

SOURCE AND PURITY OF MATERIALS:

(1) Purity better than 99.9 mole per cent.

(2) No details given.

ESTIMATED ERROR:

$\delta T/K = \pm 0.1$; $\delta P/bar = \pm 0.1$; $\delta x_{N_2}$, $\delta y_{N_2}$ = ±1% (estimated by compiler).

REFERENCES:

1. Kaminishi, G.; Arai, Y.; Saito, S.; Maeda, S. *J. Chem. Eng. Japan* 1968, *1*, 109.
2. Sarashina, E.; Arai, Y.; Saito, S. *J. Chem. Eng. Japan* 1971, *4*, 377.

COMPONENTS:

(1) Nitrogen; $N_2$; [7727-39-9]

(2) Methane, dichlorodifluoro-; $CCl_2F_2$; [75-71-8]

ORIGINAL MEASUREMENTS:

Maslennikova, V. Ya.; Goryunova, N. P.; Tsiklis, D. S.

*Zhur. Fiz. Khim.* 1967, *41*, 735.

VARIABLES:

T/K = 295-348
P/MPa = 2.5-17.8

PREPARED BY:

C. L. Young

EXPERIMENTAL VALUES:

| T/K | $P$/bar | Mole fraction of nitrogen in liquid, $x_{N_2}$ | in gas, $y_{N_2}$ |
|---|---|---|---|
| 295.15 | 25.3 | 0.07 | 0.665 |
| | 50.7 | 0.105 | 0.82 |
| | 76.0 | 0.165 | 0.855 |
| | 101.3 | 0.223 | 0.863 |
| | 126.7 | 0.29 | 0.86 |
| | 152.0 | 0.37 | 0.83 |
| | 178.3 | 0.61 | 0.61 |
| 323.15 | 25.3 | 0.03 | 0.44 |
| | 50.7 | 0.10 | 0.635 |
| | 76.0 | 0.155 | 0.693 |
| | 101.3 | 0.24 | 0.705 |
| | 126.7 | 0.325 | 0.70 |
| | 152.0 | 0.465 | 0.635 |
| | 157.1 | 0.65 | 0.65 |
| 348.15 | 25.3 | 0.015 | 0.18 |
| | 50.7 | 0.095 | 0.48 |
| | 76.0 | 0.16 | 0.545 |
| | 101.3 | 0.26 | 0.55 |
| | 126.7 | 0.405 | 0.495 |
| | 129.7 | 0.45 | 0.45 |

AUXILIARY INFORMATION

METHOD/APPARATUS/PROCEDURE;

Few details given in source. Ref. 1 given as reference in which apparatus described. It is not clear which of the several apparatus described were used.

SOURCE AND PURITY OF MATERIALS:

(1) Nitrogen containing up to 0.5 mole per cent oxygen.

(2) Technical grade sample.

ESTIMATED ERROR:

$\delta T/K = \pm 0.1$; $\delta P/\text{bar} = \pm 0.05$; $\delta x_{N_2}$, $\delta y_{N_2} = \pm 0.01$ (estimated by compiler).

REFERENCES:

1. Tsiklis, D. S. *Technique of Physico-chemical Experiment at High and Ultra High Pressures*, Isd. Khimiya, Moscow, 1965.

**COMPONENTS:**

(1) Nitrogen; $N_2$; [7727-37-9]

(2) Methane, tetrachloro- (Carbon tetrachloride); $CCl_4$; [56-23-5]

**ORIGINAL MEASUREMENTS:**

Frolich, P. K.; Tauch, E. J.; Hogan, J. J.; Peer, A. A.

*Ind. Eng. Chem.* 1931, *23*, 548-50.

**VARIABLES:**

T/K = 298
P/MPa = 1.0-12.0

**PREPARED BY:**

C. L. Young

**EXPERIMENTAL VALUES:**

| T/K | $P$/MPa | Solubility*, $S$ | Mole fraction of nitrogen in liquid, $x_{N_2}$† |
|---|---|---|---|
| 298.15 | 1.0 | 1.55 | 0.0061 |
| | 2.0 | 3.05 | 0.0120 |
| | 3.0 | 4.6 | 0.0179 |
| | 4.0 | 6.1 | 0.0236 |
| | 5.0 | 7.65 | 0.0295 |
| | 6.0 | 9.2 | 0.0352 |
| | 7.0 | 10.7 | 0.0407 |
| | 8.0 | 12.25 | 0.0464 |
| | 9.0 | 13.75 | 0.0517 |
| | 10.0 | 15.3 | 0.0572 |
| | 11.0 | 16.8 | 0.0625 |
| | 12.0 | 18.35 | 0.0679 |

* Data taken from graph in original article. Volume of gas measured at 101.325 kPa pressure and 298.15 K dissolved by unit volume of liquid measured under the same conditions.

† Calculated by compiler.

**AUXILIARY INFORMATION**

**METHOD/APPARATUS/PROCEDURE:**

Static equilibrium cell. Liquid saturated with gas and after equilibrium established samples removed and analysed by volumetric method. Allowance was made for vapor pressure of liquid and the solubility of the gas at atmospheric pressure. Details in source.

**SOURCE AND PURITY OF MATERIALS:**

Stated that the materials were the highest purity available.

**ESTIMATED ERROR:**

$\delta T/K = \pm 0.1$; $\delta x_{N_2} = \pm 5\%$.

**REFERENCES:**

COMPONENTS:

(1) Nitrogen; $N_2$; [7727-37-9]

(2) Methane, tetrafluoro-; $CF_4$; [75-73-0]

ORIGINAL MEASUREMENTS:

Eckert, C. A.; Prausnitz, J. M. *Am. Inst. Chem. Eng. J.* 1965, *11*, 886.

VARIABLES:

T/K = 70-116
P/kPa = 28-862

PREPARED BY:

C. L. Young

EXPERIMENTAL VALUES:

| T/K | $P$/bar | Mole fraction of nitrogen in liquid, $x_{N_2}$ | in vapor, $y_{N_2}$ |
|---|---|---|---|
| 69.57 | 0.2958 | 0.452 | 0.999+ |
| 69.90 | 0.2782 | 0.268 | 0.999+ |
| 75.81 | 0.6116 | 0.185 | 0.999 |
| 78.06 | 0.3570 | 0.067 | 0.999 |
| 85.28 | 1.7580 | 0.516 | 0.999 |
| 81.51 | 1.1363 | 0.238 | 0.999 |
| 81.83 | 0.6327 | 0.113 | 0.996 |
| 85.28 | 1.7580 | 0.479 | 0.999 |
| 89.96 | 1.1570 | 0.105 | 0.998 |
| 95.33 | 3.717 | 0.329 | 0.997 |
| 107.94 | 8.619 | 0.564 | 0.992 |
| 116.24 | 1.9443 | 0.025 | 0.963 |

AUXILIARY INFORMATION

METHOD /APPARATUS/PROCEDURE:

Recirculating vapor flow apparatus with a heavy-wall glass cell. Temperature measured with thermocouple. Pressure measured with Bourdon gauge. Samples of both phases analysed by gas chromatography.

SOURCE AND PURITY OF MATERIALS:

(1) Matheson sample, purity 99.998 mole per cent.

(2) DuPont sample, purity 99.8 mole per cent.

ESTIMATED ERROR:

$\delta T/K = \pm 0.05$; $\delta P/bar = \pm 0.007$ up to 2 bar; $\pm 0.02$ above 2 bar; $\delta x_{N_2}$, $\delta y_{N_2} = \pm 0.005$ or better.

REFERENCES:

COMPONENTS:

(1) Nitrogen; $N_2$ [7727-37-9]

(2) Chlorotrifluoromethane; $CClF_3$; [75-72-9]

(2') Tetrafluoromethane; $CF_4$; [75-73-0]

ORIGINAL MEASUREMENTS:

Chaikovskii, V.F.; Zakharov, N.D.; Grezin, A.K.; Matyash, Yu. I.

*Kholod. Tekh. Tekhnol* 1976, *22*, 51-4.

*Chem. Abstr.* 1977, *86*, 74947n.

VARIABLES:

T/K = 193.15 - 253.15

P/MPa = 3.0-9.5

PREPARED BY:

V. Katovic

EXPERIMENTAL VALUES:

Mole Fraction of Nitrogen

| P/MPa | In Liquid $x_{N_2}$ | In Gas $y_{N_2}$ | In Liquid $x_{N_2}$ | In Gas $y_{N_2}$ | In Liquid $x_{N_2}$ | In Gas $y_{N_2}$ |
|---|---|---|---|---|---|---|
| Freon-14; $CF_4$; Tetrafluoromethane | | | | | | |
| | T/K = 193.15 | | T/K = 198.15 | | T/K = 203.15 | |
| 3.4 | 0.189 | 0.610 | 0.168 | 0.535 | 0.145 | 0.460 |
| 3.6 | 0.206 | 0.631 | 0.185 | 0.560 | 0.163 | 0.490 |
| 3.8 | 0.224 | 0.641 | 0.202 | 0.577 | 0.180 | 0.513 |
| 4.0 | 0.241 | 0.651 | 0.220 | 0.590 | 0.198 | 0.530 |
| 4.2 | 0.258 | 0.655 | 0.238 | 0.597 | 0.215 | 0.540 |
| 4.4 | 0.275 | 0.657 | 0.255 | 0.601 | 0.235 | 0.546 |
| 4.6 | 0.292 | 0.659 | 0.272 | 0.603 | 0.253 | 0.550 |
| 4.8 | 0.310 | 0.660 | 0.290 | 0.605 | 0.270 | 0.551 |
| 5.0 | 0.327 | 0.660 | 0.307 | 0.605 | 0.287 | 0.551 |
| 5.2 | 0.345 | 0.659 | 0.327 | 0.604 | 0.310 | 0.546 |
| 5.4 | 0.363 | 0.656 | 0.347 | 0.601 | 0.335 | 0.533 |
| 5.6 | 0.387 | 0.652 | 0.370 | 0.594 | 0.365 | 0.511 |
| 5.8 | 0.413 | 0.646 | 0.400 | 0.577 | 0.413 | 0.460 |
| 6.0 | 0.442 | 0.633 | 0.450 | 0.535 | | |
| 6.2 | 0.490 | 0.600 | | | | |

continued on following page

AUXILIARY INFORMATION

METHOD/APPARATUS/PROCEDURE:

Solubilities were determined using an apparatus described in the paper. The composition of the liquid and gas phase was determined by a gas chromatographic method.

SOURCE AND PURITY OF MATERIALS:

(1) 99.9% pure.
(2) 99.5% pure.
(2') 99.5% pure.

ESTIMATED ERROR:

$\delta T/K = 0.1$, estimate by author
$\delta x/x = \pm 0.02$, compiler's estimate

REFERENCES:

| COMPONENTS: | ORIGINAL MEASUREMENTS: |
|---|---|
| (1) Nitrogen; $N_2$; [7727-37-9]<br>(2) Chlorotrifluoromethane; $CClF_3$; [75-72-9]<br>(2') Tetrafluoromethane; $CF_4$; [75-73-0] | Chaikovskii, V.F.; Zakharov, N.D.; Grezin, A.K.; Matyash, Yu. I.<br>*Kholod. Tekh. Tekhnol* 1976, *22*, 51-4.<br>*Chem. Abstr.* 1977, *86*, 74947n. |

EXPERIMENTAL VALUES: continued

| P/MPa | In Liquid $x_{N_2}$ | In Gas $y_{N_2}$ | In Liquid $x_{N_2}$ | In Gas $y_{N_2}$ |
|---|---|---|---|---|
| Freon-13; $CClF_3$; Chlorotrifluoromethane | T/K = 233.15 | | T/K = 258.15 | |
| 3.0 | 0.120 | 0.710 | 0.090 | 0.550 |
| 3.5 | 0.145 | 0.740 | 0.115 | 0.590 |
| 4.0 | 0.170 | 0.770 | 0.140 | 0.625 |
| 4.5 | 0.195 | 0.780 | 0.160 | 0.645 |
| 5.0 | 0.220 | 0.795 | 0.185 | 0.665 |
| 5.5 | 0.240 | 0.805 | 0.210 | 0.675 |
| 6.0 | 0.265 | 0.810 | 0.235 | 0.680 |
| 6.5 | 0.290 | 0.815 | 0.260 | 0.680 |
| 7.0 | 0.315 | 0.815 | 0.285 | 0.680 |
| 7.5 | 0.340 | 0.820 | 0.310 | 0.685 |
| 8.0 | 0.365 | 0.815 | 0.330 | 0.670 |
| 8.5 | 0.385 | 0.810 | 0.355 | 0.660 |
| 9.0 | 0.410 | 0.805 | 0.380 | 0.650 |
| 9.5 | 0.440 | 0.795 | | |

| COMPONENTS: | ORIGINAL MEASUREMENTS: |
|---|---|
| (1) Nitrogen; $N_2$; [7727-37-9]<br><br>(2) Pyrrolidinone, 1-methyl-; (N-methylpyrrolidinone); $C_5H_9NO$; [872-50-4] | Shakhova, S. F.; Zubchenko, Yu. P.; Kaplan, L. K.<br><br>*Khim. Prom.* 1973, *49*, 108-10. |
| VARIABLES: | PREPARED BY: |
| T/K = 283-333<br>P/MPa = 4.3-13.5 | C. L. Young |

EXPERIMENTAL VALUES:

| T/K | *P*/bar | Mole fraction of nitrogen in liquid, $x_{N_2}$ | α* vol/vol | T/K | *P*/bar | Mole fraction of nitrogen in liquid, $x_{N_2}$ | α* vol/vol |
|---|---|---|---|---|---|---|---|
| 283.15 | 52.49 | 0.0085 | 1.99 | 298.15 | 92.21 | 0.0158 | 3.73 |
| | 66.67 | 0.0104 | 2.43 | | 100.51 | 0.0168 | 3.95 |
| | 66.67 | 0.0108 | 2.54 | | 106.49 | 0.0178 | 4.20 |
| | 75.49 | 0.0119 | 2.80 | | 107.40 | 0.0183 | 4.33 |
| | 82.88 | 0.0128 | 3.00 | | 114.80 | 0.0194 | 4.58 |
| | 85.82 | 0.0134 | 3.16 | | 124.53 | 0.0215 | 5.10 |
| | 99.60 | 0.0159 | 3.75 | 333.15 | 52.99 | 0.0112 | 2.63 |
| | 106.90 | 0.0172 | 4.05 | | 73.06 | 0.0147 | 3.47 |
| | 107.40 | 0.0170 | 4.00 | | 81.47 | 0.0163 | 3.83 |
| | 111.86 | 0.0175 | 4.12 | | 92.21 | 0.0180 | 4.26 |
| 298.15 | 42.66 | 0.0074 | 1.72 | | 100.21 | 0.0188 | 4.43 |
| | 52.99 | 0.0095 | 2.23 | | 103.45 | 0.0203 | 4.80 |
| | 62.25 | 0.0115 | 2.69 | | 120.17 | 0.0233 | 5.53 |
| | 81.36 | 0.0141 | 3.32 | | 134.86 | 0.0264 | 6.29 |

* Values quoted in original paper, appears to be volume of gas at T/K = 273.15 and *P* = 1 atmosphere absorbed by unit volume of liquid at room temperature.

AUXILIARY INFORMATION

| METHOD /APPARATUS/PROCEDURE: | SOURCE AND PURITY OF MATERIALS: |
|---|---|
| Mixture stirred by ball in rocking autoclave. Sample of liquid analysed by a volumetric method. Details in source. | (1) Purity 99.99 mole per cent.<br><br>(2) Purity 99.9 mole per cent. |
| | ESTIMATED ERROR:<br>$\delta T/K = \pm 0.1$; $\delta P/bar = \pm 0.1$; $\delta x_{N_2} = \pm 5\%$ (estimated by compiler). |
| | REFERENCES: |

| COMPONENTS: | ORIGINAL MEASUREMENTS: |
|---|---|
| (1) Nitrogen; $N_2$; [7727-37-9]<br>(2) 2-Pyrrolidinone, 1,5-dimethyl- (5-methyl-N-methylpyrrolidinone); $C_6H_{11}NO$; [5075-92-3] | Shakhova, S. F.; Zubchenko, Yu. P.; Kaplan, L. K.<br>*Khim. Prom.* 1973, *49*, 108-10. |
| VARIABLES: | PREPARED BY: |
| T/K = 283-323<br>P/MPa = 7.5-14.2 | C. L. Young |

EXPERIMENTAL VALUES:

| T/K | $P$/bar | Mole fraction of nitrogen in liquid, $x_{N_2}$ | $vol^{\alpha *}/vol$ |
|---|---|---|---|
| 283.15 | 85.32 | 0.0191 | 3.96 |
| | 102.95 | 0.0232 | 4.85 |
| | 118.14 | 0.0262 | 5.48 |
| 298.15 | 74.58 | 0.0189 | 3.93 |
| | 89.26 | 0.0220 | 4.58 |
| | 90.28 | 0.0220 | 4.58 |
| | 103.96 | 0.0256 | 5.35 |
| | 106.90 | 0.0258 | 5.41 |
| | 113.79 | 0.0265 | 5.55 |
| | 123.62 | 0.0290 | 6.10 |
| | 124.53 | 0.0290 | 6.08 |
| 323.15 | 84.81 | 0.0232 | 4.85 |
| | 104.47 | 0.0289 | 6.06 |
| | 124.12 | 0.0325 | 6.81 |
| | 141.75 | 0.0356 | 7.53 |

* quoted in original paper, appears to be volume of gas at T/K = 273.15 and $P$ = 1 atmosphere absorbed by unit volume of liquid at room temperature.

AUXILIARY INFORMATION

METHOD /APPARATUS/PROCEDURE:

Mixture stirred by ball in rocking autoclave. Sample of liquid analysed by a volumetric method. Details in source.

SOURCE AND PURITY OF MATERIALS:

(1) Purity 99.9 mole per cent.

(2) Purity 99.3 mole per cent.

ESTIMATED ERROR:

$\delta T/K = \pm 0.1$; $\delta P/bar = \pm 0.1$; $\delta x_{N_2} = \pm 5\%$ (estimated by compiler).

REFERENCES:

| COMPONENTS: | ORIGINAL MEASUREMENTS: |
|---|---|
| (1) Nitrogen; $N_2$; [7727-37-9]<br>(2) Dog blood. | Hawkins, J.A.; Shilling, C.W.<br>*J. Biol. Chem.* 1936, *113*, 273-8. |

| VARIABLES: | PREPARED BY: |
|---|---|
| T/K = 311<br>P/kPa = 100-600 | C.L. Young |

EXPERIMENTAL VALUES:

$t/^{\circ}C = 38$

| $P^{+}$/mmHg | $P^{+}$/bar | $O_2$ capacity | Water Content /g $cm^{-3}$ | $L$ | $L_0$ /$cm^3 g^{-1}$ |
|---|---|---|---|---|---|
| 715 | 0.953 | 20.14 | 0.8069 | 0.0138 | 0.0171 |
| 1481 | 1.975 | | | 0.0136 | 0.0169 |
| 3071 | 4.094 | | | 0.0136 | 0.0169 |
| 4508 | 6.010 | | | 0.0134 | 0.0166 |
| 701 | 0.935 | 20.06 | 0.8965 | 0.0141 | 0.0157 |
| 1460 | 1.947 | | | 0.0142 | 0.0158 |
| 3006 | 4.008 | | | 0.0135 | 0.0151 |
| 4524 | 6.032 | | | 0.0138 | 0.0154 |
| 712 | 0.949 | 21.55 | 0.7767 | 0.0139 | 0.0179 |
| 1471 | 1.961 | | | 0.0127 | 0.0164 |
| 3006 | 4.008 | | | 0.0137 | 0.0176 |
| 4434 | 5.912 | | | 0.0135 | 0.0174 |
| 698 | 0.931 | Not given | Not given | 0.0148 | |
| 1474 | 1.965 | | | 0.0146 | |
| 4494 | 5.992 | | | 0.0142 | |

$P^{+}$ partial pressure of nitrogen

$L$ Ostwald coefficient

$L_0$ Volume of nitrogen absorbed per gram of water.

AUXILIARY INFORMATION

METHOD/APPARATUS/PROCEDURE:

Blood collected from cannulated femoral artery of dogs anesthetized with nembutal. Experiment carried out in a recompression chamber. Blood saturated with nitrogen and amount estimated using a Van Slyke apparatus (1).

SOURCE AND PURITY OF MATERIALS:

(1) No details given.

(2) Oxalated and used immediately or chilled to $0^{\circ}C$ and kept in stoppered container and used within a day.

ESTIMATED ERROR:

$\delta T/K = \pm 0.005$; $\delta L = \pm 0.05$
(estimated by compiler)

REFERENCES:

1. Van Slyke, D.D.; Neill, J.M. *J. Biol. Chem.* 1924, *61*, 523.

**COMPONENTS:**

(1) Nitrogen; $N_2$; [7727-37-9]

(2) Ox blood.

**ORIGINAL MEASUREMENTS:**

Hawkins, J.A.; Shilling, C.W.

*J. Biol. Chem.* 1936, *113*, 273-8.

**VARIABLES:**

T/K = 311
P/kPa = 76-606

**PREPARED BY:**

C.L. Young

**EXPERIMENTAL VALUES:**

$t/^{0}C = 38$

| $P^{+}$/mmHg | $P^{+}$/bar | $O_2$ capacity | Water Content /g cm$^{-3}$ | $L$ | $L_0$ /cm$^3$g$^{-1}$ |
|---|---|---|---|---|---|
| 567 | 0.756 | 15.70 | 0.8223 | 0.0135 | 0.0164 |
| 1562 | 2.082 | | | 0.0133 | 0.0162 |
| 2913 | 3.884 | | | 0.0126 | 0.0154 |
| 703 | 0.937 | 18.18 | 0.8113 | 0.0140 | 0.0173 |
| 1510 | 2.013 | | | 0.0137 | 0.0169 |
| 2936 | 3.914 | | | 0.0132 | 0.0163 |
| 4215 | 5.620 | | | 0.0133 | 0.0164 |
| 699 | 0.932 | 18.50 | 0.8079 | 0.0134 | 0.0166 |
| 2960 | 3.946 | | | 0.0132 | 0.0163 |
| 4472 | 5.962 | | | 0.0128 | 0.0158 |
| 701 | 0.935 | 20.01 | 0.8799 | 0.0132 | 0.0150 |
| 2967 | 3.956 | | | 0.0132 | 0.0150 |
| 4544 | 6.058 | | | 0.0124 | 0.0141 |
| 698 | 0.931 | 18.86 | 0.8122 | 0.0133 | 0.0164 |
| 4461 | 5.948 | | | 0.0133 | 0.0164 |

$P^{+}$ partial pressure of nitrogen

$L$ Ostwald coefficient

$L_0$ Volume of nitrogen absorbed per gram of water.

## AUXILIARY INFORMATION

**METHOD/APPARATUS/PROCEDURE:**

Blood collected from jugular vein of calves. Experiment carried out in a recompression chamber. Blood saturated with nitrogen and amount estimated using a Van Slyke apparatus (1).

**SOURCE AND PURITY OF MATERIALS:**

(1) No details given.

(2) Oxalated and used immediately or chilled to $0^{0}C$ and kept in stoppered container and used within a day.

**ESTIMATED ERROR:**

$\delta T/K = \pm 0.05; \delta L = \pm 0.001$.
(estimated by compiler)

**REFERENCES:**

1. Van Slyke, D.D.; Neill, J.M. *J. Biol. Chem.* 1924, *61*, 523.

| COMPONENTS: | ORIGINAL MEASUREMENTS: |
|---|---|
| (1) Nitrogen; $N_2$; [7727-37-9]<br>(2) Oxygen; $O_2$; [7782-44-7]<br>(3) Carbon dioxide; $CO_2$; [124-38-9] | Muirbrook, N. K.; Prausnitz, J. M.<br>*Am. Inst. Chem. Engnrs. J.* <u>1965</u>, *11*, 1092-6. |

| VARIABLES: | PREPARED BY: |
|---|---|
| T/K = 273<br>P/MPa = 5-11 | C. L. Young |

EXPERIMENTAL VALUES:

| | | | Mole fractions in liquid | | | in gas | | |
|---|---|---|---|---|---|---|---|---|
| T/K | P/psi | P/MPa | $x_{N_2}$ | $x_{O_2}$ | $x_{CO_2}$ | $y_{N_2}$ | $y_{O_2}$ | $y_{CO_2}$ |
| 273.15 | 755 | 5.21 | 0.006 | 0.031 | 0.963 | 0.048 | 0.174 | 0.778 |
| | 822 | 5.67 | 0.007 | 0.042 | 0.951 | 0.043 | 0.220 | 0.737 |
| | 974 | 6.72 | 0.032 | 0.040 | 0.928 | 0.151 | 0.170 | 0.679 |
| | 952 | 6.56 | 0.031 | 0.036 | 0.933 | 0.158 | 0.154 | 0.688 |
| | 1064 | 7.34 | 0.050 | 0.033 | 0.917 | 0.217 | 0.127 | 0.656 |
| | 1035 | 7.14 | 0.051 | 0.028 | 0.921 | 0.228 | 0.110 | 0.662 |
| | 1174 | 8.09 | 0.047 | 0.062 | 0.891 | 0.166 | 0.207 | 0.627 |
| | 1104 | 7.61 | 0.040 | 0.055 | 0.905 | 0.154 | 0.200 | 0.646 |
| | 1185 | 8.17 | 0.051 | 0.058 | 0.891 | 0.186 | 0.189 | 0.625 |
| | 1340 | 9.24 | 0.080 | 0.055 | 0.865 | 0.242 | 0.151 | 0.607 |
| | 1440 | 9.93 | 0.082 | 0.081 | 0.837 | 0.214 | 0.183 | 0.603 |
| | 1535 | 10.58 | 0.087 | 0.105 | 0.808 | 0.186 | 0.207 | 0.607 |

AUXILIARY INFORMATION

| METHOD/APPARATUS/PROCEDURE: | SOURCE AND PURITY OF MATERIALS: |
|---|---|
| Recirculating flow apparatus in which liquid from bottom of cell is returned to top of cell. The samples of both phases were analysed as vapor. Carbon dioxide removed with "Ascarite" and oxygen removed by reaction with copper at 200 °C. Details in source. | All components were Coleman grade and had specified purity of at least 99.99 mole per cent.<br><br>ESTIMATED ERROR:<br>$\delta T/K = \pm 0.01$; $\delta x$, $\delta y = \pm 1\%$<br>(estimated by compiler).<br><br>REFERENCES: |

| COMPONENTS: | ORIGINAL MEASUREMENTS: |
|---|---|
| (1) Nitrogen; $N_2$; [7727-37-9]<br>(2) Oxygen fluoride; $OF_2$; [7783-41-7] | Cannon, W. A.; Robson, J. H.; English, W. D.<br>Report DAC-60510-F2. Contract NAS7-548 (1968). (Tech. Report Douglas Missile & Space Systems Division, Astropower Laboratory.) |
| VARIABLES:<br>T/K = 145-172<br>P/MPa = 2.1-4.8 | PREPARED BY:<br>R. Battino |

EXPERIMENTAL VALUES:

| $P_{N_2}$/psia | $P^a_{N_2}$/MPa | $x_1$[b] -199.4°F (144.6 K) | -175.1°F (158.1 K) | -149.3°F (172.4 K) |
|---|---|---|---|---|
| 300 | 2.07 | 0.231 | 0.141 | 0.0617 |
| 300 | 2.07 | 0.234 | 0.140 | 0.0645 |
| 300 | 2.07 | 0.228 | 0.143 | 0.0571 |
| 300 | 2.07 | 0.228 | 0.140 | 0.0615 |
| 600 | 4.14 | 0.617[c] | | |
| 600 | 4.14 | 0.616[c] | | |
| 700 | 4.83 | 0.604 | 0.509 | 0.362 |
| 700 | 4.83 | 0.622 | 0.494 | 0.356 |
| 700 | 4.83 | | 0.496 | 0.359 |
| 700 | 4.83 | | 0.499 | 0.356 |

[a] Calculated by compiler.

[b] Mole fraction solubility at indicated partial pressure.

[c] Said to be measurements at 600 psia, but appear to be too high (compiler).

AUXILIARY INFORMATION

METHOD/APPARATUS/PROCEDURE:

About 200 $cm^3$ of liquid propellant was condensed into a pressure vessel of about 1 liter capacity. Nitrogen was added to the proper pressure and the apparatus agitated for about one hour. A sample of the liquid phase was withdrawn and flashed to a gas at a lower temperature. The gas sample was then analyzed by gas chromatography.

SOURCE AND PURITY OF MATERIALS:

(1) Air Products and Chemicals, high purity grade, 99.9%.

(2) Allied Chemical Corporation, 98.5%.

ESTIMATED ERROR:

$\delta T/K = \pm 0.2$
$\delta P/P = \pm 0.02$
$\delta x_1/x_1 = \pm 0.02$

REFERENCES:

| COMPONENTS: | ORIGINAL MEASUREMENTS: |
|---|---|
| (1) Nitrogen; $N_2$; [7727-37-9]<br>(2) Hydrogen sulfide; $H_2S$; [7783-06-4] | Besserer, G. J.; Robinson, D. B. *J. Chem. Eng. Data* 1975, *20*, 157. |
| **VARIABLES:** | **PREPARED BY:** |
| T/K = 256-344<br>P/MPa = 1.2-20.7 | C. L. Young |

EXPERIMENTAL VALUES:

| T/K | P/bar | Mole fraction of nitrogen in liquid, $x_{N_2}$ | Mole fraction of nitrogen in gas, $y_{N_2}$ | T/K | P/bar | Mole fraction of nitrogen in liquid, $x_{N_2}$ | Mole fraction of nitrogen in gas, $y_{N_2}$ |
|---|---|---|---|---|---|---|---|
| 344.26 | 71.2 | 0.0148 | 0.148 | 300.04 | 51.4 | 0.0153 | 0.488 |
| | 86.5 | 0.0290 | 0.242 | | 70.0 | 0.0239 | 0.576 |
| | 103.6 | 0.0448 | 0.281 | | 87.2 | 0.0317 | 0.628 |
| | 120.7 | 0.0636 | 0.321 | | 103.2 | 0.0401 | 0.653 |
| | 137.9 | 0.0849 | 0.342 | | 120.6 | 0.0495 | 0.674 |
| | 155.1 | 0.111 | 0.357 | | 137.5 | 0.0570 | 0.684 |
| | 172.4 | 0.146 | 0.330 | | 155.0 | 0.0663 | 0.691 |
| | 177.2 | 0.161 | 0.331 | | 172.9 | 0.0762 | 0.695 |
| 321.87 | 34.7 | 0.0000 | 0.000 | | 189.7 | 0.0844 | 0.694 |
| | 40.7 | 0.0040 | 0.108 | | 206.8 | 0.0957 | 0.693 |
| | 54.5 | 0.0129 | 0.265 | 277.65 | 11.7 | 0.0000 | 0.000 |
| | 69.4 | 0.0220 | 0.364 | | 17.3 | 0.0035 | 0.271 |
| | 85.5 | 0.0325 | 0.429 | | 34.1 | 0.0102 | 0.582 |
| | 102.5 | 0.0451 | 0.478 | | 52.7 | 0.0186 | 0.698 |
| | 119.8 | 0.0571 | 0.516 | | 68.3 | 0.0241 | 0.743 |
| | 139.8 | 0.0702 | 0.525 | | 86.2 | 0.0314 | 0.769 |
| | 157.0 | 0.0825 | 0.538 | | 103.5 | 0.0377 | 0.791 |
| | 172.4 | 0.0979 | 0.539 | | 121.0 | 0.0416 | 0.798 |
| | 189.6 | 0.113 | 0.543 | | 137.9 | 0.0468 | 0.806 |
| | 206.8 | 0.126 | 0.533 | | 155.1 | 0.0540 | 0.810 |
| 300.04 | 20.7 | 0.0000 | 0.000 | | 172.4 | 0.0586 | 0.806 |
| | 35.2 | 0.0071 | 0.332 | | 189.8 | 0.0642 | 0.811 |

(cont.)

AUXILIARY INFORMATION

METHOD /APPARATUS/PROCEDURE:

Cell fitted with two moveable pistons which enabled cell contents to be circulated in an external line. Fitted with optical system which allowed measurement of refractive index. Temperature measured with iron-constantan thermocouple and pressure with strain gauge transducer. Components charged into cell and mixed by piston movement. Samples withdrawn and analysed by gas chromatography. Details in source and ref. 1.

SOURCE AND PURITY OF MATERIALS:

(1) High purity sample with stated purity of 99.993 mole per cent.

(2) Matheson sample, CP grade purity better than 99.8 mole per cent.

ESTIMATED ERROR:

$\delta T/K = \pm 0.05$; $\delta P/bar = \pm 0.2$;

$\delta x_{N_2}$, $\delta y_{N_2} = \pm 0.003$.

REFERENCES:

1. Besserer, G. J.; Robinson, D. B. *Can. J. Chem. Eng.* 1971, *49*, 651.

| COMPONENTS: | ORIGINAL MEASUREMENTS: |
|---|---|
| (1) Nitrogen; $N_2$; [7727-37-9]<br>(2) Hydrogen sulfide; $H_2S$; [7783-06-4] | Besserer, G. J.; Robinson, D. B.<br>*J. Chem. Eng. Data* 1975, *20*, 157. |

EXPERIMENTAL VALUES:

continued

| T/K | P/bar | Mole fraction of nitrogen in liquid, $x_{N_2}$ | Mole fraction of nitrogen in gas, $y_{N_2}$ | T/K | P/bar | Mole fraction of nitrogen in liquid, $x_{N_2}$ | Mole fraction of nitrogen in gas, $y_{N_2}$ |
|---|---|---|---|---|---|---|---|
| 277.65 | 206.8 | 0.0685 | 0.809 | 256.43 | 121.8 | 0.0312 | 0.887 |
| 256.43 | 17.7 | 0.0049 | 0.625 | | 138.4 | 0.0348 | 0.887 |
| | 34.3 | 0.0084 | 0.776 | | 155.8 | 0.0385 | 0.892 |
| | 53.3 | 0.0145 | 0.839 | | 172.9 | 0.0390 | 0.891 |
| | 69.6 | 0.0176 | 0.866 | | 190.2 | 0.0430 | 0.886 |
| | 86.5 | 0.0230 | 0.873 | | 207.0 | 0.0448 | 0.890 |
| | 104.5 | 0.0268 | 0.881 | | | | |

COMPONENTS:

(1) Nitrogen; $N_2$; [7727-37-9]

(2) Hydrogen sulfide; $H_2S$; [7783-06-4]

ORIGINAL MEASUREMENTS:

Kalra, H.; Krishnan, T. R.; Robinson, D. B.

*J. Chem. Eng. Data* 1976, *21*, 222.

VARIABLES:

T/K = 200-228
P/MPa = 0.1-13.8

PREPARED BY:

C. L. Young

EXPERIMENTAL VALUES:

| T/K | $P$/bar | Mole fraction of nitrogen in liquid, $x_{N_2}$ | in gas, $y_{N_2}$ |
|---|---|---|---|
| 200.15 | 1.41 | 0.0005 | 0.6542 |
| | 5.16 | 0.0011 | 0.9022 |
| | 10.40 | 0.0019 | 0.9507 |
| | 17.11 | 0.0027 | 0.9680 |
| | 27.65 | 0.0040 | 0.9788 |
| | 47.99 | 0.0069 | 0.9858 |
| | 70.12 | 0.0089 | 0.9873 |
| | 97.22 | 0.0111 | 0.9872 |
| | 137.48 | 0.0135 | 0.9843 |
| 228.00 | 3.34 | 0.0004 | 0.3804 |
| | 4.83 | 0.0008 | 0.5658 |
| | 8.17 | 0.0017 | 0.7379 |
| | 17.21 | 0.0033 | 0.8653 |
| | 34.27 | 0.0066 | 0.9285 |
| | 55.30 | 0.0103 | 0.9452 |
| | 83.08 | 0.0153 | 0.9540 |
| | 110.7 | 0.0200 | 0.9555 |
| | 137.5 | 0.0232 | 0.9564 |

AUXILIARY INFORMATION

METHOD/APPARATUS/PROCEDURE:

Cell fitted with two moveable pistons which enabled cell contents to be circulated in external line. Fitted with optical system which allowed measurement of refractive index. Temperature measured with iron-constantan thermocouple and pressure with strain gauge transducer. Components charged into cell, mixed by piston movement. Samples withdrawn and analysed by gas chromatography. Details in ref. 1.

SOURCE AND PURITY OF MATERIALS:

(1) Linde sample, purity 99.99+ mole per cent.

(2) Matheson sample, purity 99.8+ mole per cent.

ESTIMATED ERROR:

$\delta T/K = \pm 0.05$; $\delta P/\text{bar} = \pm 0.2$;
$\delta x_{N_2}$, $\delta y_{N_2} = \pm 0.003$.

REFERENCES:

1. Besserer, G. J.; Robinson, D. B. *Can. J. Chem. Eng.* 1971, *49*, 651.

**COMPONENTS:**

(1) Nitrogen; $N_2$; [7727-37-9]

(2) Chlorine trifluoride; $ClF_3$; [7790-91-2]

**ORIGINAL MEASUREMENTS:**

Cannon, W.A.; Robson, J.H.; English, W.D.

Report DAC-60510-F2. Contract NAS7-548 (1968). (Tech. Report Douglas Missle & Space Systems Division, Astropower Laboratory.)

**VARIABLES:**

T/K = 200-283

P/MPa = 2.1-4.8

**PREPARED BY:**

R. Battino

**EXPERIMENTAL VALUES:**

| $P_{N_2}$/psia | $P_{N_2}^{a}$/MPa | $10^2 x_1^{b}$ -99.1°F (200.3K) | -49.8°F (227.7K) | -0.2°F (255.3K) | 50.3°F (283.3K) |
|---|---|---|---|---|---|
| 300 | 2.07 | 0.729 | 0.837 | 0.867 | 1.04 |
| 300 | 2.07 | 0.790 | 0.867 | 0.917 | 1.07 |
| 300 | 2.07 | 0.760 | 0.994 | 0.950 | 1.07 |
| 300 | 2.07 | 0.802 | 0.961 | 0.850 | 1.10 |
| 700 | 4.83 | 1.93 | 1.81 | 1.98 | 2.34 |
| 700 | 4.83 | 1.88 | 1.87 | 2.13 | 2.40 |
| 700 | 4.83 | 1.87 | 1.81 | 2.04 | 2.49 |
| 700 | 4.83 | 1.94 | 1.83 | 2.14 | 2.48 |

[a] Calculated by compiler.

[b] Mole fraction solubility at indicated partial pressure.

AUXILIARY INFORMATION

**METHOD/APPARATUS/PROCEDURE:**

About 200 $cm^3$ of liquid propellant was condensed into a pressure vessel of about 1 liter capacity. Nitrogen was added to the proper pressure and the apparatus agitated for about one hour. A sample of the liquid phase was withdrawn and flashed to a gas at a lower temperature. The gas sample was then analyzed by gas chromatography.

**SOURCE AND PURITY OF MATERIALS:**

(1) Air Products and Chemicals, high purity grade, 99.9%.

(2) Allied Chemical, MIL-0-81399, 99.0%.

**ESTIMATED ERROR:**

$\delta T/K = \pm 0.2$
$\delta P/P = \pm 0.02$
$\delta x_1/x_1 = \pm 0.02$

**REFERENCES:**

**COMPONENTS:**

(1) Nitrogen; $N_2$; [7727-37-9]

(2) Chlorine pentafluoride; $ClF_5$; [13637-63-3]

**ORIGINAL MEASUREMENTS:**

Sukhoverkhov, V. F.; Garanin, V. F.; Podzolko, L. G.

*Doklady Chem. Akad. Nauk SSR*, 1979, *246*, 1379-83.

*Chm. Abstr.* 1979, *91*, 113159f.

**VARIABLES:**

T/K = 293
P/MPa = 0.5-9

**PREPARED BY:**

R. Battino

**EXPERIMENTAL VALUES:**

| P/MPa | Solubility/volume per cent |
|---|---|
| 0.49 | 1.62 |
| 1.13 | 2.40 |
| 2.26 | 4.41 |
| 3.72 | 5.91 |
| 5.98 | 7.80 |
| 7.95 | 12.40 |
| 8.88 | 11.80 |

[a] At 20°C or 293 K.

[b] Extrapolation to zero pressure gave a Henry's Law constant of 42.8 MPa/mole fraction.

**AUXILIARY INFORMATION**

**METHOD/APPARATUS/PROCEDURE:**

Saturation and equilibration were carried out in a thermostatted conical nickel vessel of 42 $cm^3$ capacity. Equilibration took about 48 h. Analysis was by gas-adsorption chromatography.

**SOURCE AND PURITY OF MATERIALS:**

(1) 99.9 vol. per cent.

(2) 99.9 per cent or better by gas chromatography

**ESTIMATED ERROR:**

$\delta$Solubility/Solubility = ±0.015, authors' estimate.

**REFERENCES:**

**COMPONENTS:**

(1) Nitrogen; $N_2$; [7727-37-9]

(2) Chlorine pentafluoride; $ClF_5$; [13637-63-3]

**ORIGINAL MEASUREMENTS:**

Cannon, W. A.; Robson, J. H.; English, W.D.
Report DAC-60510-F2. Contract NAS7-548 (1968). (Tech. Report Douglas Missile & Sapce Systems Division, Astropower Laboratory.)

**VARIABLES:**

T/K = 200-283
P/MPa = 2.1-4.8

**PREPARED BY:**

R. Battino

**EXPERIMENTAL VALUES:**

| $P_{N_2}$/psia | $P^a_{N_2}$/MPa | $x_1$[b] -100.0°F (199.8 K) | -50.8°F (227.2 K) | 0.0°F (255.4 K) | 50.3°F (283.3 K) |
|---|---|---|---|---|---|
| 300 | 2.07 | 0.0357 | 0.0331 | 0.0333 | 0.0291 |
| 300 | 2.07 | 0.0366 | 0.0326 | 0.0340 | 0.0311 |
| 300 | 2.07 | 0.0340 | 0.0330 | 0.0336 | 0.0294 |
| 300 | 2.07 | 0.0348 | 0.0323 | 0.0333 | 0.0289 |
| 700 | 4.83 | 0.0834 | 0.0745 | 0.0754 | 0.0775 |
| 700 | 4.83 | 0.0825 | 0.0731 | 0.0769 | 0.0752 |
| 700 | 4.83 | 0.0802 | 0.0748 | 0.0758 | 0.0758 |
| 700 | 4.83 | 0.0808 | 0.0747 | 0.0776 | 0.0760 |

[a] Calculated by compiler.

[b] Mole fraction solubility at indicated partial pressure.

AUXILIARY INFORMATION

**METHOD/APPARATUS/PROCEDURE:**

About 200 $cm^3$ of liquid propellant was condensed into a pressure vessel of about 1 liter capacity. Nitrogen was added to the proper pressure and the apparatus agitated for about one hour. A sample of the liquid phase was withdrawn and flashed to a gas at a lower temperature. The gas sample was then analyzed by gas chromatography.

**SOURCE AND PURITY OF MATERIALS:**

(1) Air Products and Chemicals, high purity grade, 99.9%.

(2) Probable purity greater than 99% by weight.

**ESTIMATED ERROR:**

$\delta T/K = \pm 0.2$
$\delta P/P = \pm 0.02$
$\delta x_1/x_1 = \pm 0.02$

**REFERENCES:**

COMPONENTS:

(1) Nitrogen; $N_2$; [7727-37-9]

(2) Perchloryl fluoride; $ClF_6$; [56077-96-4]

ORIGINAL MEASUREMENTS:

Cannon, W. A.; Robson, J. H.; English, W.D.

Report DAC-60510-F2. Contract NAS7-548 (1968). (Tech. Report Douglas Missile & Space Systems Division, Astropower Laboratory.)

VARIABLES:

T/K = 172-256
P/MPa = 2.1-4.8

PREPARED BY:

R. Battino

EXPERIMENTAL VALUES:

| $P_{N_2}$/psia | $P_{N_2}$[a]/MPa | $x_1$[b] -150.0°F (172.0 K) | -100.1°F (199.8 K) | -50.3°F (227.4 K) | 0.0°F (255.4 K) |
|---|---|---|---|---|---|
| 300 | 2.07 | 0.0568 | 0.0495 | 0.0448 | 0.0400 |
| 300 | 2.07 | 0.0543 | 0.0500 | 0.0460 | 0.0382 |
| 300 | 2.07 | 0.0592 | 0.0487 | 0.0443 | 0.0388 |
| 300 | 2.07 | 0.0598 | 0.0504 | 0.0446 | 0.0387 |
| 700 | 4.83 | 0.128 | 0.115 | 0.109 | 0.0960 |
| 700 | 4.83 | 0.130 | 0.114 | 0.107 | 0.0964 |
| 700 | 4.83 | 0.131 | 0.111 | 0.108 | 0.0948 |
| 700 | 4.83 | 0.126 | 0.113 | 0.105 | 0.0942 |

[a] Calculated by compiler.

[b] Mole fraction solubility at indicated partial pressure.

AUXILIARY INFORMATION

METHOD/APPARATUS/PROCEDURE:

About 200 $cm^3$ of liquid propellant was condensed into a pressure vessel of about 1 liter capacity. Nitrogen was added to the proper pressure and the apparatus agitated for about one hour. A sample of the liquid phase was withdrawn and flashed to a gas at a lower temperature. The gas sample was then analyzed by gas chromatography.

SOURCE AND PURITY OF MATERIALS:

(1) Air Products and Chemicals, high purity grade, 99.9%.

(2) Pennsalt Chemical Company, 99.8%.

ESTIMATED ERROR:

$\delta T/K = \pm 0.2$
$\delta P/P = \pm 0.02$
$\delta x_1/x_1 = \pm 0.02$

REFERENCES:

COMPONENTS:

(1) Nitrogen; $N_2$; [7727-37-9]

(2) Sulphur Dioxide; $SO_2$; [7446-09-5]

ORIGINAL MEASUREMENTS:

Dean, M. R.; Walls, W. S.
*Ind. Eng. Chem.* 1947, *39*, 1049.

VARIABLES:
$T$/K = 241-301
$P$/MPa = 1.6-3.6

PREPARED BY:

C. L. Young

EXPERIMENTAL VALUES:

| T/K | $P$/bar | Mole fraction of nitrogen in liquid, $x_{N_2}$ | in gas, $y_{N_2}$ |
|---|---|---|---|
| 301.48 | 35.5 | 0.0143 | 0.843 |
| 301.48 | 35.5 | 0.0134 | 0.849 |
| 209.26 | 15.5 | 0.0051 | 0.693 |
| 253.15 | 17.4 | 0.0055 | 0.958 |
| 241.10 | 35.5 | 0.0078 | 0.989 |
| 241.10 | 18.1 | 0.0033 | 0.972 |

AUXILIARY INFORMATION

METHOD /APPARATUS/PROCEDURE:

Twin steel static cell. Pressure and volume of cell varied by introducing mercury. Pressure measured with Bourdon gauge. Analysis of samples of both gas and liquid phases carried out by Orsat analysis. Details in source.

SOURCE AND PURITY OF MATERIALS:

(1) Purity about 99.7 mole per cent.

(2) Refrigeration grade sample from Virginia Smelting Co. Purity about 99.6 mole per cent.

ESTIMATED ERROR:

$\delta T$/K = ±0.1; $\delta P$/bar = ±0.1; $\delta x_{N_2}$ = ±0.0003; $\delta y_{N_2}$ = ±0.002 (estimated by compiler).

REFERENCES:

| COMPONENTS: | EVALUATOR: |
|---|---|
| (1) Nitrogen; $N_2$; [7727-37-9]<br>(2) Ammonia; $NH_3$; [7664-41-7] | Colin L. Young,<br>School of Chemistry<br>University of Melbourne<br>Parkville, Victoria 3052<br>Australia.<br>November 1981 |

CRITICAL EVALUATION:

This system has been studied in detail by Wiebe and coworkers (1), (2) and by Lindroos and Dodge (3). The data of Wiebe and coworkers cover the pressure range 2.5 to 100 MPa whereas the data of Lindroos and Dodge are for a higher pressure range from 98 to 380 MPa. There is moderate agreement between the two sets of data although the data of Lindroos and Dodge are less precise. The data of Heise (4), which are limited to 298.15 K, are in fair agreement with those of Wiebe and Tremearne (1). All four sets of data are classified as tentative. There are a considerable number of studies of the phase equilibria of this system which are not evaluated in detail here since they do not fall into the category of gas solubility.

Tskilis (5) studied the solubility of ammonia in nitrogen at 75°C and 80°C at pressures between 10 and 450 MPa and made a limited study of the solubility of nitrogen in ammonia at 75°C between 165 and 260 MPa. Krichevskii and coworkers (6), (7) have also made a study of the phase behaviour of this mixture.

Matous, Sobr and Novak (8) studied the quaternary system argon + methane + nitrogen + ammonia. These data are not evaluated here.

Cannon, Robson and English (9) studied the solubility of nitrogen in ammonia at temperatures between 200 K and 283 K at pressures of 2 MPa and 4.8 MPa. Their data are classified as tentative. While the data are broadly consistent with the data of Wiebe and coworkers (1), (2) no detailed comparison is possible because of the lack of any other data over an overlapping range of temperature and pressure.

References:

1. Wiebe, R.; Tremearne, T. H. *J. Amer. Chem. Soc.* 1933, *55*, 975.
2. Wiebe, R.; Gaddy, V. L. *J. Amer. Chem. Soc.* 1937, *59*, 1984.
3. Lindroos, A. E.; Dodge, B. F. *Chem. Eng. Progr. Symp. Ser. No. 3* 1952, *48*, 10.
4. Heise, F. *Ber. Bunsenges Phys. Chem.* 1972, *76*, 938.
5. Tskilis, D. S. *Trudy Gos. Inst. Azot. Prom.* 1951, *3*, 12.
6. Krichevskii, I. R.; Bol'shakov, P. E. *Zh. Fiz. Khim.* 1941, *15*, 184.
7. Krichevskii, I. R.; Efremova, G. D. *Zh. Fiz. Khim.* 1952, *26*, 1117.
8. Matous, J.; Sobr, J.; Novak, J. P. *Coll. Czech. Chem. Comm.* 1970, *35*, 3757.
9. Cannon, W. A.; Robson, J. H.; English, W. D. Report DAC-60510-F2, 1968. (Tech. Report, Douglas Missile and Space Systems Division Astropower Laboratory).

| COMPONENTS: | ORIGINAL MEASUREMENTS: |
|---|---|
| (1) Nitrogen; $N_2$; [7727-37-9]<br>(2) Ammonia; $NH_3$; [7664-41-7] | Wiebe, R.; Tremearne, T. H.<br>*J. Amer. Chem. Soc.* 1933, 55, 975-8. |
| VARIABLES:<br>T/K = 298<br>P/MPa = 3-101 | PREPARED BY:<br>C. L. Young |

EXPERIMENTAL VALUES:

| T/K | $P$/MPa | Solubility, $S^{\dagger}$ | *Mole fraction of nitrogen in liquid, $x_{N_2}$ |
|---|---|---|---|
| 298.15 | 2.5 | 2.22 | 0.00168 |
| | 5.1 | 5.73 | 0.00434 |
| | 10.1 | 12.04 | 0.00907 |
| | 20.3 | 22.48 | 0.01679 |
| | 40.5 | 37.02 | 0.02736 |
| | 60.8 | 45.43 | 0.03337 |
| | 81.1 | 51.10 | 0.03738 |
| | 101.3 | 54.83 | 0.03999 |

* Calculated by compiler.

† Volume of gas reduced to 273.15 K and 1 atmosphere pressure, dissolved by 1 gram of ammonia.

AUXILIARY INFORMATION

| METHOD/APPARATUS/PROCEDURE: | SOURCE AND PURITY OF MATERIALS: |
|---|---|
| Rocking equilibrium cell. Pressure measured with dead weight tester. Liquid samples analysed by stripping out nitrogen at low pressures and low temperatures. | (1) Purity 99.9 mole per cent, impurity argon and trace of oxygen.<br>(2) Water content 0.07 mole per cent. |
| | ESTIMATED ERROR:<br>$\delta T/K = \pm 0.05$; $\delta P/MPa = \pm 0.5\%$; $\delta x_{N_2} = \pm 1\%$. |
| | REFERENCES: |

**COMPONENTS:**

(1) Nitrogen; $N_2$; [7727-37-9]

(2) Ammonia; $NH_3$; [7664-41-7]

**ORIGINAL MEASUREMENTS:**

Wiebe, R.; Gaddy, V.L.

*J. Am. Chem. Soc.* 1937, *59*, 1984-7.

**VARIABLES:**

T/K = 273-373
P/MPa = 5.1-101.3

**PREPARED BY:**

C.L. Young

**EXPERIMENTAL VALUES:**

| T/K | P/bar | $10^2$x Mole fraction of nitrogen in liquid, $10^2x_{N_2}$ |
|---|---|---|
| 273.15 | 101.32 | 0.597 |
| | 202.65 | 1.032 |
| | 405.30 | 1.553 |
| | 607.95 | 1.860 |
| | 810.60 | 2.088 |
| | 1013.25 | 2.206 |
| 323.15 | 50.66 | 0.501 |
| | 101.32 | 1.289 |
| | 202.65 | 2.680 |
| | 405.30 | 4.730 |
| | 607.95 | 6.052 |
| | 810.60 | 6.877 |
| | 1013.25 | 7.362 |
| 348.15 | 101.32 | 1.598 |
| | 202.65 | 4.045 |
| | 405.30 | 8.398 |
| | 607.95 | 11.910 |
| | 810.60 | 14.265 |
| | 1013.25 | 15.518 |
| 363.15 | 354.64 | 11.170 |
| | 506.62 | 19.095 |
| | 557.29 | 24.660 |
| 373.15 | 101.32 | 1.534 |
| | 202.65 | 6.155 |
| | 303.97 | 12.798 |
| | 329.31 | 15.202 |

**AUXILIARY INFORMATION**

**METHOD/APPARATUS/PROCEDURE:**

Rocking equilibrium cell. Pressure measured with dead weight tester. Liquid samples analysed by stripping out nitrogen at low pressure and low temperatures. Details in ref. 1.

**SOURCE AND PURITY OF MATERIALS:**

No details given but thought to be of high purity as in ref. 1.

**ESTIMATED ERROR:**

$\delta T/K = \pm 0.1$; $\delta P/bar = \pm 0.1$;
$\delta x_{N_2} = \pm 0.5\%$. (estimated by compiler)

**REFERENCES:**

1. Wiebe, R.; Tremearne, T.H. *J. Amer. Chem. Soc.* 1933, *55*, 975.

COMPONENTS:

(1) Nitrogen; $N_2$; [7727-37-9]

(2) Ammonia; $NH_3$; [7664-41-7]

ORIGINAL MEASUREMENTS:

Lindroos, A. E.; Dodge, B. F.
*Chem. Eng. Progr. Symp. Ser. No. 3*, 1952, *48*, 10.

VARIABLES:

T/K = 303-383
P/MPa = 76-381

PREPARED BY:

C. L. Young

EXPERIMENTAL VALUES:

| T/K | $P$/bar | Mole fraction of nitrogen* in liquid, $x_{N_2}$ | Mole fraction of nitrogen* in gas, $y_{N_2}$ |
|---|---|---|---|
| 303.15 | 1073 | - | 0.915 |
| | 1062 | 0.044 | - |
| | 1241 | 0.049 | - |
| | 1231 | - | 0.916 |
| | 1845 | 0.062 | - |
| | 1832 | - | 0.954 |
| | 2272 | - | 0.054 |
| | 2257 | 0.941 | - |
| | 2927 | 0.942 | - |
| | 2913 | - | 0.050 |
| | 3060 | 0.885 | - |
| | 3042 | - | 0.051 |
| | 3655 | 0.959 | - |
| | 3628 | - | 0.048 |
| 323.15 | 984 | 0.072 | - |
| | 973 | - | 0.849 |
| | 1161 | 0.071 | - |
| | 1159 | - | 0.858 |
| | 1370 | - | 0.867 |
| | 1362 | 0.074 | - |
| | 1851 | 0.082 | - |
| | 1837 | - | 0.908 |
| | 2276 | 0.919 | - |
| | 2265 | - | 0.077 |
| 323.15 | 2867 | - | 0.060 |
| | 2841 | 0.915 | - |
| | 3301 | - | 0.062 |
| | 3271 | 0.912 | - |
| | 3814 | - | 0.073 |
| | 3807 | 0.938 | - |
| 348.15 | 1098 | 0.147 | - |
| | 1086 | - | 0.689 |
| | 1364 | - | 0.728 |
| | 1359 | 0.158 | - |
| | 1834 | 0.148 | - |
| | 1818 | - | 0.776 |
| | 2094 | - | 0.150 |
| | 2084 | 0.792 | - |
| | 2276 | 0.903 | - |
| | 2264 | - | 0.096 |
| | 2871 | - | 0.116 |
| | 2856 | 0.908 | - |
| | 3256 | - | 0.072 |
| | 3230 | 0.921 | - |
| | 3716 | 0.916 | - |
| | 3704 | - | 0.106 |
| 358.15 | 1121 | 0.259 | - |

(cont.)

AUXILIARY INFORMATION

METHOD/APPARATUS/PROCEDURE:

Details in source. High pressure static cell fitted with stirrer. Pressure measured with manganin resistance gauge.

SOURCE AND PURITY OF MATERIALS:

(1) Purity 99.99 mole per cent.

(2) Purity 99.98 mole per cent.

ESTIMATED ERROR:

$\delta T/K = \pm 0.2$; $\delta P/\text{bar} = \pm 10$; $\delta x_{N_2} = \delta y_{N_2} = \pm 0.5\%$ (estimated by compiler).

REFERENCES:

COMPONENTS:

(1) Nitrogen; $N_2$; [7727-37-9]

(2) Ammonia; $NH_3$; [7664-41-7]

ORIGINAL MEASUREMENTS:

Lindroos, A. E.; Dodge, B. F.

*Chem. Eng. Progr. Symp. Ser. No. 3*, 1952, *48*, 10.

EXPERIMENTAL VALUES:

continued

| T/K | P/bar | Mole fraction of nitrogen* in liquid $x_{N_2}$ | in gas, $y_{N_2}$ | T/K | P/bar | Mole fraction of nitrogen* in liquid, $x_{N_2}$ | in gas, $y_{N_2}$ |
|---|---|---|---|---|---|---|---|
| 358.15 | 1114 | - | 0.522 | 373.15 | 1781 | - | 0.360 |
| | 1353 | - | 0.562 | | 1892 | 0.386 | - |
| | 1345 | 0.244 | - | | 1979 | - | 0.360 |
| | 1369 | - | 0.641 | | 1934 | 0.409 | - |
| | 1364 | 0.276 | - | | 1920 | - | 0.362 |
| | 1834 | 0.236 | - | | 1902 | 0.496 | - |
| | 1823 | - | 0.654 | | 1988 | - | 0.395 |
| | 2099 | 0.739 | - | | 2060 | 0.472 | - |
| | 2084 | - | 0.186 | | 2031 | - | 0.378 |
| | 2101 | - | 0.218 | | 2090 | - | 0.382 |
| | 2087 | 0.674 | - | | 2084 | 0.489 | - |
| | 2295 | - | 0.127 | | 2094 | - | 0.356 |
| | 2274 | 0.801 | - | | 2084 | 0.830 | - |
| | 2313 | - | 0.202 | | 2148 | 0.809 | - |
| | 2298 | 0.609 | - | | 2145 | - | 0.317 |
| | 3329 | 0.921 | - | | 2217 | - | 0.190 |
| | 3312 | - | 0.080 | | 2194 | 0.915 | - |
| | 3806 | 0.925 | - | | 2283 | 0.892 | - |
| | 3780 | - | 0.103 | | 2275 | - | 0.260 |
| 365.15 | 1369 | - | 0.587 | | 2340 | - | 0.149 |
| | 1367 | 0.569 | - | | 2317 | 0.829 | - |
| | 1654 | - | 0.361 | | 2535 | - | 0.162 |
| | 1642 | 0.365 | - | | 2525 | 0.916 | - |
| | 1829 | 0.465 | - | | 2556 | 0.808 | - |
| | 1819 | - | 0.627 | | 2525 | - | 0.132 |
| | 1841 | - | 0.641 | | 2817 | - | 0.145 |
| | 1834 | 0.647 | - | | 2803 | 0.916 | - |
| | 1931 | - | 0.443 | | 3305 | - | 0.120 |
| | 1921 | 0.452 | - | | 3285 | 0.917 | - |
| | 1928 | 0.462 | - | | 3791 | 0.922 | - |
| | 1920 | - | 0.471 | | 3779 | - | 0.115 |
| | 1930 | - | 0.502 | 383.15 | 2053 | 0.429 | - |
| | 1923 | 0.514 | - | | 2042 | - | 0.425 |
| | 1930 | 0.545 | - | | 2505 | 0.404 | - |
| | 1923 | - | 0.537 | | 2494 | - | 0.403 |
| | 1982 | - | 0.541 | | 2851 | 0.404 | - |
| | 1949 | 0.547 | - | | 2840 | - | 0.398 |
| | 2076 | 0.559 | - | | 3059 | 0.568 | - |
| | 2066 | - | 0.548 | | 3045 | - | 0.207 |
| | 2111 | - | 0.312 | | 3311 | - | 0.137 |
| | 2101 | 0.548 | - | | 3289 | 0.915 | - |
| | 2243 | 0.579 | - | | 3498 | - | 0.154 |
| | 2222 | - | 0.282 | | 3482 | 0.916 | - |
| | 2300 | 0.899 | - | | 3801 | 0.918 | - |
| | 2276 | - | 0.176 | | 3782 | - | 0.147 |
| | 3315 | 0.920 | - | 323.15 | 763 | 0.065 | - |
| | 3302 | - | 0.098 | | 758 | 0.068 | - |
| | 3756 | 0.925 | - | | 754 | 0.073 | - |
| | 3701 | - | 0.116 | | 703 | 0.052 | - |
| 373.15 | 1795 | 0.364 | - | | 988 | 0.072 | - |

* This mixture exhibits the barotropic effect therefore the term "liquid" should be interpreted as "more dense phase".

| COMPONENTS: | ORIGINAL MEASUREMENTS: |
|---|---|
| (1) Nitrogen; $N_2$; [7727-37-9]<br><br>(2) Ammonia; $NH_3$; [7664-41-7] | Cannon, W.A.; Robson, J.H.; English, W.D.<br><br>Report DAC-60510-F2. Contract NAS7-548 (1968). (Tech. Report Douglas Missile & Space Systems Division, Astropower Laboratory.) |
| VARIABLES: | PREPARED BY: |
| T/K = 200-283<br>P/MPa = 2.1-4.8 | R. Battino |

EXPERIMENTAL VALUES:

| $P_{N_2}$/psia | $P^a_{N_2}$/MPa | $10^2\ x_1$<br>-100.0°F (199.8 K) | -50.0°F (244.3 K) | 0.0°F (255.4 K) | 50.0°F (283.2 K) |
|---|---|---|---|---|---|
| 300 | 2.07 | 0.026 | 0.070 | 0.104 | 0.136 |
| 300 | 2.07 | 0.028 | 0.078 | 0.094 | 0.130 |
| 300 | 2.07 | 0.034 | 0.064 | 0.102 | 0.131 |
| 300 | 2.07 | 0.031 | 0.060 | 0.100 | 0.122 |
| 700 | 4.83 | 0.065 | 0.130 | 0.236 | 0.353 |
| 700 | 4.83 | 0.071 | 0.142 | 0.222 | 0.342 |
| 700 | 4.83 | 0.072 | 0.140 | 0.220 | 0.334 |
| 700 | 4.83 | 0.071 | 0.133 | 0.222 | 0.352 |

[a] Calculated by compiler.

[b] Mole fraction solubility at indicated partial pressure.

AUXILIARY INFORMATION

| METHOD/APPARATUS/PROCEDURE: | SOURCE AND PURITY OF MATERIALS: |
|---|---|
| About 200 $cm^3$ of liquid propellant was condensed into a pressure vessel of about 1 liter capacity. Nitrogen was added to the proper pressure and the apparatus agitated for about one hour. A sample of the liquid phase was withdrawn and flashed to a gas at a lower temperature. The gas sample was then analyzed by gas chromatography. | (1) Air Products and Chemicals, high purity grade, 99.9%.<br><br>(2) J. T. Baker Chemical Company, 99.9%. |
| | ESTIMATED ERROR: |
| | $\delta T/K = \pm 0.2$<br>$\delta P/P = \pm 0.02$<br>$\delta x_1/x_1 = \pm 0.02$ |
| | REFERENCES: |

**COMPONENTS:**

(1) Argon; Ar; [7440-37-1] + Methane; $CH_4$; [74-82-8] + Nitrogen; $N_2$; [7727-37-9]

(2) Ammonia; $NH_3$; [7664-41-7]

**ORIGINAL MEASUREMENTS:**

Matous, J.; Sobr, J.; Novak, J.P.

*Coll. Czech. Chem. Commun.* 1970, 35, 3757-61.

**VARIABLES:**

T/K = 243-303
P/MPa = 2.8-4.7

**PREPARED BY:**

R. Battino

**EXPERIMENTAL VALUES:**

| t/°C | $T^a$/K | S°/ml $g^{-1}$ | $10^3x_1$ |
|---|---|---|---|
| | 46.5 atm (4.71 MPa) | | |
| -30 | 243.15 | 5.40 | 4.09 |
| 0 | 273.15 | 8.30 | 6.27 |
| 10 | 283.15 | 8.90 | 6.73 |
| 30 | 303.15 | 9.90 | 7.48 |
| | 27.3 atm (2.77 MPa) | | |
| -30 | 243.15 | 3.55 | 2.69 |
| -20 | 253.15 | 4.00 | 3.03 |
| -10 | 263.15 | 4.35 | 3.30 |
| 0 | 273.15 | 4.70 | 3.56 |
| 10 | 283.15 | 4.75 | 3.60 |
| 20 | 293.15 | 4.80 | 3.64 |
| 30 | 303.15 | 4.65 | 3.53 |

[a] Calculated by compiler.

[b] S° is N $cm^3$/g $NH_3$. Presumably N $cm^3$ is at STP.

[c] Mole fraction of gaseous mixture (all components) dissolved at indicated total gas pressure. Composition of the gaseous mixture was 12.9 mol per cent Ar, 26.5 mol per cent $CH_4$ and 60.6 mol per cent $N_2$.

**AUXILIARY INFORMATION**

**METHOD/APPARATUS/PROCEDURE:**

Rocking autoclave apparatus. Details in paper.

**SOURCE AND PURITY OF MATERIALS:**

(1) Supplied by MCHZ - Ostrava: Ar - 99.99%; $N_2$ - 99.99%; $CH_4$ - 95-96% (remainder $N_2$).

(2) Better than 99%. Distilled twice.

**ESTIMATED ERROR:**

$\delta x/x = \pm 0.02$

**REFERENCES:**

| COMPONENTS: | ORIGINAL MEASUREMENTS: |
|---|---|
| (1) Nitrogen; $N_2$; [7727-37-9]<br><br>(2) Ammonia; $NH_3$; [7664-41-7] | Heise, F.<br><br>*Ber. Bunsenges. Phys. Chem.* 1972, 76, 956. |
| VARIABLES: | PREPARED BY: |
| T/K = 298<br>P/MPa = 3.0-20.5 | C. L. Young |

EXPERIMENTAL VALUES:

| T/K | $P$/bar | Mole fraction of nitrogen in liquid, $x_{N_2}$ | in vapor, $y_{N_2}$ |
|---|---|---|---|
| 298.15 | 29.5 | 0.00209 | 0.6283 |
| | 58.4 | 0.00515 | 0.7848 |
| | 85.45 | 0.00800 | 0.8422 |
| | 205.1 | 0.01780 | 0.9074 |

AUXILIARY INFORMATION

METHOD/APPARATUS/PROCEDURE:

Static rocking equilibrium cell. Liquid and gas samples removed after equilibrium established. Samples analysed by freezing out ammonia in liquid nitrogen trap. Details in source and ref. 1.

SOURCE AND PURITY OF MATERIALS:

(1) Messer-Griessheim sample, purity better than 99.95 mole per cent.

(2) Gerling and Holtz sample, purity better than 99.8 mole per cent as determined by gas chromatography and mass spectrometry.

ESTIMATED ERROR:

$\delta T/K = \pm 0.1$; $\delta P/\text{bar} = \pm 0.3$ below 100 bar; $\pm 0.6$ above 100 bar; $\delta x_{N_2}$, $\delta y_{N_2}$ = $\pm 2\%$ (estimated by compiler).

REFERENCES:

1. Heise, F.; *Dissertation*, Göttingen, 1971.

| COMPONENTS: | ORIGINAL MEASUREMENTS: |
|---|---|
| (1) Nitrogen; $N_2$; [7727-37-9]<br>(2) Nitrous oxide; $N_2O$; [10024-97-2] | Zeininger, H.<br>*Chemie-Ing.-Techn.* 1972, *44*, 607-12. |
| VARIABLES:<br>T/K = 213-253<br>P/MPa = 0.6-8.2 | PREPARED BY:<br>C. L. Young |

EXPERIMENTAL VALUES:

| T/K | $P/10^5$Pa | Mole fraction of nitrogen in liquid, $x_{N_2}$ | Mole fraction of nitrogen in gas, $y_{N_2}$ | T/K | $P/10^5$Pa | Mole fraction of nitrogen in liquid, $x_{N_2}$ | Mole fraction of nitrogen in gas, $y_{N_2}$ |
|---|---|---|---|---|---|---|---|
| 213.15 | 5.7 | <0.01 | 0.198 | 233.15 | 59.9 | 0.072 | 0.705 |
| | 10.4 | <0.01 | 0.537 | | 67.6 | 0.096 | 0.739 |
| | 16.3 | 0.008 | 0.700 | | 75.0 | 0.109 | 0.749 |
| | 22.6 | 0.026 | 0.780 | | 81.2 | 0.124 | 0.786 |
| | 29.2 | 0.023 | 0.807 | 253.15 | 23.8 | <0.01 | 0.179 |
| | 35.3 | 0.052 | 0.827 | | 28.0 | <0.01 | 0.269 |
| | 42.6 | 0.058 | 0.828 | | 32.4 | 0.012 | 0.382 |
| | 47.9 | 0.083 | 0.833 | | 36.1 | 0.023 | 0.412 |
| | 56.8 | 0.087 | 0.852 | | 39.2 | 0.023 | 0.464 |
| | 66.0 | 0.106 | 0.857 | | 43.6 | 0.027 | 0.484 |
| | 73.4 | 0.151 | 0.863 | | 46.3 | 0.027 | 0.507 |
| 233.15 | 16.3 | <0.01 | 0.407 | | 51.6 | 0.055 | 0.523 |
| | 19.3 | <0.01 | 0.445 | | 56.9 | 0.053 | 0.524 |
| | 24.1 | <0.01 | 0.532 | | 63.0 | 0.075 | 0.549 |
| | 31.9 | <0.01 | 0.632 | | 69.8 | 0.084 | 0.570 |
| | 41.6 | 0.042 | 0.660 | | 75.2 | 0.093 | 0.594 |
| | 47.5 | 0.058 | 0.689 | | 82.0 | 0.102 | 0.603 |
| | 52.7 | 0.062 | 0.704 | | | | |

AUXILIARY INFORMATION

METHOD/APPARATUS/PROCEDURE:

Static equilibrium cell stirred with a steel ball. Samples of gas and liquid phases removed and analysed by mass spectrometry. Care was taken to avoid large changes in pressure during sampling by taking small samples. Details in source.

SOURCE AND PURITY OF MATERIALS:

Nitrous oxide was "pure" as determined by gas chromatography.

ESTIMATED ERROR:

$\delta T/K = \pm 0.3$; $\delta P/10^5 Pa = \pm 0.2$;
$\delta x_{N_2} = \pm 0.008$; $\delta y_{N_2} = \pm 0.009$.

REFERENCES:

**COMPONENTS:**

(1) Nitrogen; $N_2$; [7727-37-9]

(2) Nitryl Fluoride; $FNO_2$; [10022-50-1]

**ORIGINAL MEASUREMENTS:**

Cannon, W. A.; Robson, J. H.; English, W. D.

Report DAC-60510-F2. Contract NAS7-548 (1968). (Tech. Report Douglas Missile & Space Systems Division, Astropower Laboratory.)

**VARIABLES:**

T/K = 172-256
P/MPa = 2.1-4.8

**PREPARED BY:**

R. Battino

**EXPERIMENTAL VALUES:**

| $P_{N_2}$/psia | $P_{N_2}^{a}$/MPa | $10^2x_1^{b}$ -149.5°F (172.3 K) | -99.1°F (200.3 K) | -49.6°F (227.7 K) | 0.8°F (255.8 K) |
|---|---|---|---|---|---|
| 300 | 2.07 | 3.04 | 1.91 | 1.92 | 1.34 |
| 300 | 2.07 | 2.75 | 2.05 | 1.86 | 1.41 |
| 300 | 2.07 | 2.78 | | 1.92 | |
| 300 | 2.07 | 2.98 | | 1.86 | |
| 700 | 4.83 | 6.58 | 4.95 | 4.26 | 4.20 |
| 700 | 4.83 | 6.66 | 4.69 | 4.33[c] | 4.28 |

[a] Calculated by compiler.

[b] Mole fraction solubility at indicated partial pressure.

[c] Measured at 680 psia (4.69 MPa).

**AUXILIARY INFORMATION**

**METHOD/APPARATUS/PROCEDURE:**

About 200 $cm^3$ of liquid propellant was condensed into a pressure vessel of about 1 liter capacity. Nitrogen was added to the proper pressure and the apparatus agitated for about one hour. A sample of the liquid phase was withdrawn and flashed to a gas at a lower temperature. The gas sample was then analyzed by gas chromatography.

**SOURCE AND PURITY OF MATERIALS:**

(1) Air Products and Chemicals, high purity grade, 99.9%.

(2) Ozark Mahoning Company, 95%.

**ESTIMATED ERROR:**

$\delta T/K = \pm 0.2$
$\delta P/P = \pm 0.02$
$\delta x_1/x_1 = \pm 0.02$

**REFERENCES:**

**COMPONENTS:**

(1) Nitrogen; $N_2$; [7727-37-9]

(2) Nitrogen trifluoride; $NF_3$; [7783-54-2]

**ORIGINAL MEASUREMENTS:**

Cannon, W. A.; Robson, J. H.; English, W. D.

Report DAC-60510-F2. Contract NAS7-548 (1968). (Tech. Report Douglas Missile & Space Systems Division, Astropower Laboratory.)

**VARIABLES:**

T/K = 133-172
P/MPa = 2.1-4.8

**PREPARED BY:**

R. Battino

**EXPERIMENTAL VALUES:**

| $P_{N_2}$/psia | $P_{N_2}$[a] /MPa | $x_1$[b] -220.2°F (133.0 K) | -199.6°F (144.5 K) | -175.1°F (158.1 K) | -150.1°F (172.0 K) |
|---|---|---|---|---|---|
| 300 | 2.07 | 0.441 | 0.264 | 0.183 | 0.134 |
| 300 | 2.07 | 0.446 | 0.264 | 0.192 | 0.133 |
| 300 | 2.07 | 0.432 | 0.268 | 0.183 | 0.135 |
| 300 | 2.07 | 0.434 | 0.264 | 0.183 | 0.137 |
| 700 | 4.83 | | 0.715 | 0.522 | 0.382 |
| 700 | 4.83 | | 0.713 | 0.535 | 0.367 |
| 700 | 4.83 | | 0.731 | 0.531 | 0.394 |
| 700 | 4.83 | | 0.731 | 0.526 | 0.389 |

[a] Calculated by compiler.

[b] Mole fraction solubility at indicated partial pressure.

AUXILIARY INFORMATION

**METHOD/APPARATUS/PROCEDURE:**

About 200 $cm^3$ of liquid propellant was condensed into a pressure vessel of about 1 liter capacity. Nitrogen was added to the proper pressure and the apparatus agitated for about one hour. A sample of the liquid phase was withdrawn and flashed to a gas at a lower temperature. The gas sample was then analyzed by gas chromatography.

**SOURCE AND PURITY OF MATERIALS:**

(1) Air Products and Chemicals, high purity grade, 99.9%.

(2) Air Products and Chemicals, 99.8%.

**ESTIMATED ERROR:**

$\delta T/K = \pm 0.2$
$\delta P/P = \pm 0.02$
$\delta x_1/x_1 = \pm 0.02$

**REFERENCES:**

COMPONENTS:

(1) Nitrogen; $N_2$; [7727-37-9]

(2) Tetrafluorohydrazine; $N_2F_4$; [10036-47-2]

ORIGINAL MEASUREMENTS:

Cannon, W.A.; Robson, J.H.; English, W.D.

Report DAC-60510-F2. Contract NAS7-548 (1968). (Tech. Report Douglas Missile & Space Systems Division, Astropower Laboratory.)

VARIABLES:

T/K = 144-228
P/MPa = 2.1-4.8

PREPARED BY:

R. Battino

EXPERIMENTAL VALUES:

| $P_{N_2}$/psia | $P^a_{N_2}$/MPa | $x_1^b$ -200.2°F (144.2 K) | -150.2°F (171.9 K) | -100.7°F (199.4 K) | -50.1°F (227.5 K) |
|---|---|---|---|---|---|
| 300 | 2.07 | 0.189 | 0.111 | 0.0893 | 0.0611 |
| 300 | 2.07 | 0.190 | 0.112 | 0.0904 | 0.0634 |
| 300 | 2.07 | 0.183 | 0.114 | 0.0899 | 0.0634 |
| 300 | 2.07 | 0.182 | 0.112 | 0.0875 | 0.0619 |
| 700 | 4.83 | 0.421 | 0.240 | 0.206 | 0.167 |
| 700 | 4.83 | 0.421 | 0.240 | 0.209 | 0.167 |
| 700 | 4.83 | 0.425 | 0.257 | 0.206 | 0.166 |
| 700 | 4.83 | 0.425 | 0.253 | 0.215 | 0.165 |

[a] Calculated by compiler.

[b] Mole fraction solubility at indicated partial pressure.

AUXILIARY INFORMATION

METHOD/APPARATUS/PROCEDURE:

About 200 $cm^3$ of liquid propellant was condensed into a pressure vessel of about 1 liter capacity. Nitrogen was added to the proper pressure and the apparatus agitated for about one hour. A sample of the liquid phase was withdrawn and flashed to a gas at a lower temperature. The gas sample was then analyzed by gas chromatography.

SOURCE AND PURITY OF MATERIALS:

(1) Air Products and Chemicals, high purity grade, 99.9%.

(2) Air products and chemicals, 99.5% by weight.

ESTIMATED ERROR:

$\delta T/K = \pm 0.2$
$\delta P/P = \pm 0.02$
$\delta x_1/x_1 = \pm 0.02$

REFERENCES:

| COMPONENTS: | EVALUATOR: |
|---|---|
| (1) Nitrogen; $N_2$; [7727-37-9]<br>(2) Carbon dioxide; $CO_2$; [124-38-9] | Colin L. Young,<br>School of Chemistry,<br>University of Melbourne,<br>Parkville, Victoria 3052,<br>Australia.<br>Novemoer 1981 |

CRITICAL EVALUATION:

This system has been studied in detail by Zenner and Dana (1) at 273, 232 and 218 K. Their measurements are in good agreement with the data of Somait and Kidnay (2) at 270 K and the limited data of Arai *et al.* (3) who studied this mixture at 253, 273 and 288 K. Kaminishi and Toriumi (4) studied this system at a series of temperatures between 233 and 298 K while their data are not of the highest accuracy there is fair agreement both with Zenner and Dana's results (1) at 273 K and with the results of Arai *et al.* (3) at 253 K. The data of Yorizane *et al.* (5) are in agreement with those of Zenner and Dana at 273 K. The data in these five papers (1-5) are in substantial agreement and are classified as tentative. The data of Somait and Kidnay (2) are almost certainly the most reliable of these five sets of data. We do not feel that any particular set of data is sufficiently accurate and over a sufficiently wide range of temperature and pressure to be classified as recommended.

The data of Krichevskii *et al.* (6) cover the temperature range 288 to 303 K. Their data are not in good agreement with the data of Arai *et al.* (3) at 288 K and are not of the highest accuracy. Their data are therefore classified as doubtful.

Tskilis (7) studied the phase behaviour of this system at pressures between 400 and 700 MPa at 273 K. These data are not generally considered to be solubility data and are not evaluated here. The data of Abdulayev (8) and Kulikov (9) are presented in the form of graphs only and are not considered further.

References

1. Zenner, G. H.; Dana, L. I. *Chem. Eng. Progr. Symp. Ser. No. 44* 1963, *59*, 36.
2. Somait, F. A.; Kidnay, A. J. *J. Chem. Eng. Data* 1978, *23*, 301.
3. Arai, Y.; Kaminishi, G.; Saito, S. *J. Chem. Eng. Japan* 1971, *2*, 113.
4. Kaminishi, G.; Toriumi, T. *Kogyo Kagaku Zaashi.* 1966, *69*, 175.
5. Yorizane, M.; Yoshimura, S.; Masuoka, M. *Kagaku Kogaku* 1970, *34*, 953.
6. Krichevskii, I. R.; Khazonova, N. E.; Lesnevskaya, L. S.; Sandalova, L. Yu. *Khim. Prom.* 1962, *38*, 169.
7. Tsiklis, D. S. *Zh. Fiz. Khim.* 1946, *20*, 18.
8. Abdulayev, Ya. A. *Zh. Fiz. Khim.* 1939, *13*, 986.
9. Kulikov, N. E. *Tr. Gork. Politekh. Inst.* 1969, *25*, 18.

| COMPONENTS: | ORIGINAL MEASUREMENTS: |
|---|---|
| (1) Nitrogen; $N_2$; [7727-37-9]<br>(2) Carbon Dioxide; $CO_2$; [124-38-9] | Krichevskii, I. R.; Khazonova, N. E.; Lesnevskaya, L. S.; Sandalova, L. Yu.<br>*Khim. Prom.* 1962, *38*, 169. |
| VARIABLES:<br>T/K = 288-303<br>P/MPa = 6.1-10.2 | PREPARED BY:<br>C. L. Young |

EXPERIMENTAL VALUES: (SMOOTHED)

| T/K | $P$/bar | Mole fraction of nitrogen in liquid, $x_{N_2}$ | in gas, $y_{N_2}$ |
|---|---|---|---|
| 288.15 | 60.8 | 0.025 | 0.090 |
| | 70.9 | 0.050 | 0.158 |
| | 81.1 | 0.074 | 0.195 |
| | 91.2 | 0.102 | 0.210 |
| | 101.3 | 0.161 | 0.205 |
| | 101.5 | 0.187 | 0.187 |
| 293.15 | 60.8 | 0.009 | 0.032 |
| | 70.9 | 0.037 | 0.104 |
| | 81.1 | 0.068 | 0.143 |
| | 91.2 | 0.102 | 0.159 |
| | 98.8 | 0.140 | 0.140 |
| 298.15 | 70.9 | 0.024 | 0.055 |
| | 81.1 | 0.059 | 0.099 |
| | 91.2 | 0.086 | 0.100 |
| | 92.2 | 0.093 | 0.093 |
| 301.15 | 70.9 | 0.006 | 0.012 |
| | 81.1 | 0.043 | 0.063 |
| | 86.1 | 0.060 | 0.067 |
| | 87.1 | 0.064 | 0.064 |
| 303.15 | 76.0 | 0.015 | 0.026 |
| | 81.1 | 0.031 | 0.041 |
| | 84.1 | 0.033 | 0.033 |

AUXILIARY INFORMATION

METHOD /APPARATUS/PROCEDURE:

Static cell fitted with magnetic stirrer. Pressure measured with Bourdon gauge. Details in source.

SOURCE AND PURITY OF MATERIALS:

No details given.

ESTIMATED ERROR:

$\delta T/K = \pm 0.05$; $\delta P/\text{bar} = \pm 0.3$; $\delta x_{N_2} \simeq \delta y_{N_2} = \pm 0.003$.

REFERENCES:

| COMPONENTS: | ORIGINAL MEASUREMENTS: |
|---|---|
| (1) Nitrogen; $N_2$; [7727-37-9]<br>(2) Carbon dioxide; $CO_2$; [124-38-9] | Zenner, G. H.; Dana, L. I.<br>*Chem. Eng. Prog. Symp. Ser.* 1963, 59, No. 44, 36-41. |
| VARIABLES:<br>T/K = 218-273<br>P/MPa = 0.1-1.4 | PREPARED BY:<br>C. L. Young |

EXPERIMENTAL VALUES:

| T/K | P/MPa | Mole fraction of nitrogen in liquid, $x_{N_2}$ | in gas, $y_{N_2}$ | T/K | P/MPa | Mole fraction of nitrogen in liquid, $x_{N_2}$ | in gas, $y_{N_2}$ |
|---|---|---|---|---|---|---|---|
| 273.15 | 6.04 | 0.0521 | 0.2885 | 232.85 | 12.64 | 0.219 | 0.765 |
| | 6.03 | 0.0512 | 0.292 | | 9.68 | 0.155 | 0.756 |
| | 4.83 | 0.0265 | 0.198 | | 7.97 | 0.121 | 0.756 |
| | 6.31 | 0.058 | 0.310 | | 6.66 | 0.096 | 0.738 |
| | 7.99 | 0.0952 | 0.373 | | 13.89 | 0.268 | 0.708 |
| | 9.59 | 0.1408 | 0.396 | | 13.06 | 0.237 | 0.747 |
| | 10.97 | 0.1923 | 0.3845 | 218.15 | 1.28 | 0.009 | 0.525 |
| | 11.66 | 0.228 | 0.354 | | 2.28 | 0.023 | 0.710 |
| | 11.86 | 0.2975 | 0.3045 | | 3.85 | 0.048 | 0.804 |
| 232.85 | 3.55 | 0.0375 | 0.628 | | 5.85 | 0.066 | 0.836 |
| | 1.76 | 0.0208 | 0.372 | | 7.83 | 0.117 | 0.844 |
| | 11.97 | 0.168 | 0.740 | | 9.87 | 0.1522 | 0.847 |
| | 9.88 | 0.141 | - | | 11.77 | 0.1892 | 0.830 |
| | 7.52 | 0.109 | 0.740 | | 13.16 | 0.218 | 0.815 |
| | 10.23 | 0.1565 | 0.746 | | 13.15 | 0.2183 | 0.813 |
| | 11.11 | 0.1857 | 0.753 | | | | |

AUXILIARY INFORMATION

METHOD/APPARATUS/PROCEDURE:

Static bomb. Pressure measured with dead weight tester. Temperature measured with resistance thermometer. Samples of liquid and gas analysed by removing carbon dioxide from sample with potassium hydroxide solution. Nitrogen determined volumetrically.

SOURCE AND PURITY OF MATERIALS:

(1) No details given.

(2) Purified - no other details given.

ESTIMATED ERROR:

$\delta T/K = \pm 0.2$; $\delta P/MPa = \pm 0.05$;

$\delta x_{N_2}$, $\delta y_{N_2}$ = $\pm 0.002$ (estimated by compiler).

REFERENCES:

COMPONENTS:

(1) Nitrogen; $N_2$; [7727-37-9]

(2) Carbon dioxide; $CO_2$; [124-38-9]

ORIGINAL MEASUREMENTS:

Kaminishi, G.; Toriumi, T.

*Kogyo Kagaku Zaashi* 1966, *69*, 175-8.

VARIABLES:

T/K = 233-298
P/MPa = 0.4-1.3

PREPARED BY:

C. L. Young

EXPERIMENTAL VALUES:

| T/K | $P$/MPa | Mole fraction of nitrogen in liquid, $x_{N_2}$ | in gas, $y_{N_2}$ |
|---|---|---|---|
| 298.15 | 7.42 | - | 0.067 |
| 293.15 | 8.22 | - | 0.144 |
| | 8.44 | 0.072 | 0.148 |
| | 8.84 | 0.090 | 0.152 |
| 288.15 | 7.42 | 0.051 | 0.175 |
| | 8.44 | 0.083 | 0.202 |
| 273.15 | 7.42 | 0.076 | 0.353 |
| | 9.63 | 0.133 | 0.395 |
| | 11.01 | 0.188 | 0.377 |
| | 11.38 | 0.215 | 0.366 |
| 253.15 | 3.70 | - | 0.382 |
| | 5.27 | 0.059 | 0.509 |
| | 6.84 | 0.095 | 0.568 |
| | 10.18 | - | 0.596 |
| | 11.73 | 0.210 | 0.582 |
| | 12.74 | 0.246 | 0.557 |
| 233.15 | 5.07 | 0.066 | 0.700 |

AUXILIARY INFORMATION

METHOD/APPARATUS/PROCEDURE:

Static equilibrium cell with agitator. Pressure measured with Bourdon gauge. After equilibrium established sample removed and analysed by volumetric and gravimetric techniques. Carbon dioxide absorbed in potassium hydroxide solution.

SOURCE AND PURITY OF MATERIALS:

Better than 99.5 mole per cent.

ESTIMATED ERROR:

$\delta T/K = \pm 0.05$; $\delta P/MPa = \pm 0.01$;
$\delta x_{N_2}$, $\delta y_{N_2}$ = $\pm 0.003$ (estimated by compiler)

REFERENCES:

COMPONENTS:

(1) Nitrogen; $N_2$; [7727-37-9]

(2) Carbon dioxide; $CO_2$; [124-38-9]

ORIGINAL MEASUREMENTS:

Yorizane, M.; Yoshimura, S.; Masuoka, H.

*Kagaku Kogaku* 1970, *34*, 953-7.

VARIABLES:

T/K = 273
P/MPa = 3.5-11.8

PREPARED BY:

C.L. Young

EXPERIMENTAL VALUES:

| T/K | $P_{total}$/MPa | Mole fraction of nitrogen in liquid, $x_{N_2}$ | in vapor $y_{N_2}$ |
|---|---|---|---|
| 273.15 | 3.49 | 0.0 | 0.0 |
| | 4.05 | 0.015 | 0.0964 |
| | 4.05 | 0.011 | 0.0910 |
| | 5.07 | 0.031 | 0.215 |
| | 6.08 | 0.0561 | 0.290 |
| | 6.08 | 0.0510 | 0.300 |
| | 7.09 | 0.0710 | 0.347 |
| | 8.11 | 0.0967 | 0.370 |
| | 8.11 | 0.101 | 0.375 |
| | 9.12 | 0.126 | 0.392 |
| | 9.63 | 0.145 | 0.390 |
| | 10.13 | 0.158 | 0.397 |
| | 11.15 | 0.200 | 0.390 |
| | 11.65 | 0.251 | 0.356 |
| | 11.75 | 0.272 | 0.338 |
| | 11.82 | 0.297 | 0.297 |

AUXILIARY INFORMATION

METHOD APPARATUS/PROCEDURE:

Vapor recycle equilibrium cell with magnetic pump. Temperature measured with platinum resistance thermometer and pressure measured with Bourdon gauge. Vapor and liquid sampled simultaneously and expanded to 200 $cm^3$ at room temperature, analysed using GC. Details in source.

SOURCE AND PURITY OF MATERIALS:

No details given.

ESTIMATED ERROR:

$\delta T/K = \pm 0.1$; $\delta P/MPa = \pm 0.03$;
$\delta x_{N_2}$, $\delta y_{N_2}$ = ±0.003 - 0.005.
(estimated by compiler).

REFERENCES:

| COMPONENTS: | ORIGINAL MEASUREMENTS: |
|---|---|
| (1) Nitrogen; $N_2$; [7727-37-9]<br><br>(2) Carbon dioxide; $CO_2$; [124-38-9] | Arai, Y.; Kaminishi, G.; Saito, S.<br>*J. Chem. Eng. Japan* 1971, *2*, 113. |
| VARIABLES:<br>T/K = 253-288<br>P/MPa = 7.1-14.3 | PREPARED BY:<br>C. L. Young |

EXPERIMENTAL VALUES:

| T/K | $P$/bar | Mole fraction of nitrogen in liquid, $x_{N_2}$ |
|---|---|---|
| 253.15 | 89.8 | 0.132 |
| | 106.2 | 0.174 |
| | 127.9 | 0.243 |
| | 143.2 | 0.347 |
| 273.15 | 71.1 | 0.071 |
| | 94.5 | 0.128 |
| | 118.0 | 0.254 |
| 288.15 | 77.8 | 0.060 |
| | 90.9 | 0.103 |
| | 96.3 | 0.132 |
| | 97.6 | 0.155 |

Dew point data also given in source.

AUXILIARY INFORMATION

METHOD /APPARATUS/PROCEDURE:

Glass capillary cell fitted with magnetic stirrer. Pressure measured with a dead weight gauge. Temperature measured with mercury in glass thermometer. Details in source. Mixtures of known composition charged into cell. Bubble point determined from plots of volume against pressure. Dew point determined visually. Details in source.

SOURCE AND PURITY OF MATERIALS:

(1) Nippon Sanso Industry Co. sample, purity 99.9 mole per cent.

(2) Showa Tansan Industry Co. sample, purity > 99.9 mole per cent.

ESTIMATED ERROR:

$\delta T/K = \pm 0.01$; $\delta P/bar = \pm 0.1$;

$\delta x_{N_2} \simeq \pm 1\%$ (estimated by compiler).

REFERENCES:

COMPONENTS:

(1) Nitrogen; $N_2$; [7727-37-9]

(2) Carbon dioxide; $CO_2$; [124-38-9]

ORIGINAL MEASUREMENTS:

Somait, F.A. Kidnay, A.J.

*J. Chem. Eng. Data.* 1978, 23, 301-5.

VARIABLES:

T/K = 270
P/MPa = 3.2-12.3

PREPARED BY:

C.L. Young

EXPERIMENTAL VALUES:

| T/K | P/bar | Mole fraction of nitrogen in liquid $x_{N_2}$ | in gas $y_{N_2}$ | T/K | P/bar | Mole fraction of nitrogen in liquid $x_{N_2}$ | in gas, $y_{N_2}$ |
|---|---|---|---|---|---|---|---|
| 270.00 | 31.99 | 0.0000 | 0.0000 | 270.00 | 86.13 | 0.1142 | 0.4139 |
| | 34.25 | 0.0040 | 0.0472 | | 89.82 | 0.1231 | 0.4190 |
| | 36.27 | 0.0078 | 0.0851 | | 92.92 | 0.1319 | 0.4173 |
| | 38.00 | 0.0108 | 0.1140 | | 96.26 | 0.1430 | 0.4210 |
| | 39.52 | 0.0135 | 0.1331 | | 102.04 | 0.1585 | 0.4188 |
| | 41.22 | 0.0168 | 0.1598 | | 107.14 | 0.1769 | 0.4134 |
| | 42.05 | 0.0182 | 0.1683 | | 110.53 | 0.1904 | 0.4094 |
| | 42.81 | 0.0197 | 0.1783 | | 112.52 | 0.1986 | 0.4061 |
| | 45.90 | 0.0263 | 0.2156 | | 115.49 | 0.2142 | 0.3986 |
| | 47.60 | 0.0289 | 0.2323 | | 118.18 | - | 0.3857 |
| | 47.74 | 0.0292 | 0.2325 | | 118.86 | 0.2332 | 0.3800 |
| | 51.52 | 0.0368 | 0.2674 | | 118.86 | 0.2360 | 0.3808 |
| | 56.99 | 0.0476 | 0.3069 | | 119.48 | 0.2486 | 0.3762 |
| | 60.49 | 0.0545 | 0.3280 | | 120.24 | 0.2454 | 0.3720 |
| | 64.44 | 0.0630 | 0.3473 | | 120.73 | 0.2505 | 0.3664 |
| | 70.93 | 0.0778 | 0.3770 | | 121.35 | 0.2556 | 0.3640 |
| | 77.21 | 0.0921 | 0.3961 | | 121.96 | - | 0.3524 |
| | 83.80 | 0.1080 | 0.4126 | | 123.41 | 0.3530 | 0.3530 |

AUXILIARY INFORMATION

METHOD /APPARATUS/PROCEDURE:

Recirculating vapor flow apparatus with diaphragm pump. Temperature measured with platinum resistance thermometer and pressure with Bourdon gauge. Cell stirred with two propeller stirrer. Vapor and liquid samples analysed by gas chromatography using a thermal conductivity detector. Details in source.

SOURCE AND PURITY OF MATERIALS:

No details given.

ESTIMATED ERROR:

$\delta T/K = \pm 0.2$; $\delta P/\text{bar} = \pm 0.015$ up to 100 bar; $\pm 0.1$ above 100 bar; $\delta x_{N_2}$; $\delta y_{N_2} = \pm 0.002$.

REFERENCES:

| COMPONENTS: | ORIGINAL MEASUREMENTS: |
|---|---|
| (1) Nitrogen; $N_2$; [7727-37-9]<br>(2) Carbon dioxide; $CO_2$; [124-38-9]<br>(3) Hydrogen sulfide; $H_2S$; [7783-06-4]<br>(4) Methanol; $CH_4O$; [67-56-1] | Rousseau, R. W.; Matange, J. N.; Ferrell, J. K.<br>*Am. Inst. Chem. Engnrs. J.* 1981, *27*, 605-13. |
| VARIABLES: | PREPARED BY: |
| T/K = 258-273<br>P/MPa = 1-4 | C. L. Young |

EXPERIMENTAL VALUES:

| | | Mole fractions | | | | | | |
|---|---|---|---|---|---|---|---|---|
| | | in liquid | | | | in vapor | | |
| $P$/atm | $P$/MPa | $x_{CH_3OH}$ | $x_{CO_2}$ | $x_{N_2}$ | $x_{H_2S}$ | $x_{CO_2}$ | $x_{N_2}$ | $x_{H_2S}$ |
| | | | T/K = 258.15 | | | | | |
| 9.3 | 0.94 | 0.908 | 0.074 | 0.003 | 0.015 | 0.544 | 0.403 | 0.052 |
| 20.8 | 2.11 | 0.768 | 0.210 | 0.003 | 0.019 | 0.611 | 0.365 | 0.024 |
| 29.6 | 3.00 | 0.646 | 0.331 | 0.005 | 0.021 | 0.589 | 0.400 | 0.010 |
| 40.1 | 4.06 | 0.742 | 0.235 | 0.008 | 0.016 | 0.458 | 0.542 | --- |
| | | | T/K = 273.15 | | | | | |
| 9.2 | 0.93 | 0.900 | 0.085 | 0.002 | 0.013 | 0.685 | 0.284 | 0.031 |
| 21.2 | 2.15 | 0.743 | 0.234 | 0.002 | 0.021 | 0.801 | 0.165 | 0.035 |
| 29.0 | 2.94 | 0.696 | 0.283 | 0.003 | 0.018 | 0.774 | 0.204 | 0.022 |
| 39.8 | 4.03 | 0.687 | 0.285 | 0.007 | 0.020 | 0.585 | 0.376 | 0.039 |

AUXILIARY INFORMATION

| METHOD/APPARATUS/PROCEDURE: | SOURCE AND PURITY OF MATERIALS: |
|---|---|
| Recirculating vapor-liquid equilibrium cell. Pressure in cell measured with Heise gauge calibrated against a dead weight tester. Temperature measured with copper-constantan thermocouple. Composition of gas and liquid samples determined by gas chromatography. | (1) Purity 99.999 mole per cent.<br>(2) Purity 99.99 mole per cent.<br>(3) A 15.1 mole per cent mixture in pure nitrogen.<br>(4) Purity 99.95 mole per cent. |
| | ESTIMATED ERROR:<br>$\delta T/K = \pm 0.1$; $\delta P/MPa = \pm 0.01$;<br>$\delta x$, $\delta y = \pm 0.005$ (estimated by compiler) |
| | REFERENCES: |

COMPONENTS:

(1) Nitrogen; $N_2$; [7727-37-9]

(2) Diborane; $B_2H_6$; [19287-45-7]

ORIGINAL MEASUREMENTS:

Cannon, W.A.; Robson, J.H.; English, W.D.
Report DAC-60510-F2. Contract NAS7-548 (1968). (Tech. Report Douglas Missile & Space Systems Division, Astropower Laboratory.)

VARIABLES:

T/K = 144-227
P/MPa = 2.1-4.8

PREPARED BY:

R. Battino

EXPERIMENTAL VALUES:

| $P_{N_2}$ /psia | $P_{N_2}$ [a]/MPa | $x_1$ [b] -200.0°F (144.3K) | $x_1$ [b] -149.9°F (172.1K) | $x_1$ [b] -99.8°F (199.9K) | $x_1$ [b] -50.4°F (227.4K) |
|---|---|---|---|---|---|
| 300 | 2.07 | 0.114 | 0.0768 | 0.0613 | 0.0423 |
| 300 | 2.07 | 0.115 | 0.0775 | 0.0605 | 0.0406 |
| 300 | 2.07 | 0.114 | 0.0788 | 0.0578 | 0.0396 |
| 300 | 2.07 | 0.116 | 0.0781 | 0.0592 | 0.0395 |
| 700 | 4.83 | 0.284 | 0.195 | 0.166 | 0.140 |
| 700 | 4.83 | 0.284 | 0.196 | 0.168 | 0.140 |
| 700 | 4.83 | 0.294 | | 0.166 | 0.137 |
| 700 | 4.83 | 0.294 | | 0.166 | 0.139 |

[a] Calculated by compiler.

[b] Mole fraction solubility at indicated nitrogen partial pressure.

AUXILIARY INFORMATION

METHOD/APPARATUS/PROCEDURE:

About 200 $cm^3$ of liquid propellant was condensed into a pressure vessel of about 1 liter capacity. Nitrogen was added to the proper pressure and the apparatus agitated for about one hour. A sample of the liquid phase was withdrawn and flashed to a gas at a lower temperature. The gas sample was then analyzed by gas chromatography.

SOURCE AND PURITY OF MATERIALS:

(1) Air Products and Chemicals, high pruity grade, 99.9%.

(2) Callery Chemical Company, 99.0%.

ESTIMATED ERROR:

$\delta T/K = \pm 0.2$
$\delta P/P = \pm 0.02$
$\delta x_1/x_1 = \pm 0.02$

REFERENCES:

**COMPONENTS:**

(1) Nitrogen; $N_2$; [7727-37-9]

(2) Trimethylborane; $C_3H_9B$; [593-90-8]

**ORIGINAL MEASUREMENTS:**

Cannon, W.A.; Robson, J.H.: English, W.D.

Report DAC-60510-F2. Contract NAS7-548 (1968). (Tech. Report Douglas Missile & Space Systems Division, Astropower Laboratory.)

**VARIABLES:**

T/K = 200-283
P/MPa = 2.1-4.8

**PREPARED BY:**

R. Battino

**EXPERIMENTAL VALUES:**

| $P_{N_2}$/psia | $P_{N_2}^a$/MPa | $10^2x_1^b$ -99.8°F (199.9 K) | -45.5°F (230.1 K) | 0.0°F (255.4 K) | 50.0°F (283.2 K) |
|---|---|---|---|---|---|
| 300 | 2.07 | 3.88 | 3.68 | 3.65 | 3.60 |
| 300 | 2.07 | 3.98 | 3.68 | 3.78 | 3.60 |
| 300 | 2.07 | 4.04 | 3.70 | 3.77 | 3.46 |
| 300 | 2.07 | 4.02 | 3.66 | 3.72 | 3.38 |
| 700 | 4.83 | 8.60 | 8.40 | 8.28 | 8.94 |
| 700 | 4.83 | 8.74 | 8.48 | 8.18 | 9.08 |
| 700 | 4.83 | | | | 9.06 |
| 700 | 4.83 | | | | 9.18 |

[a] Calculated by compiler.

[b] Mole fraction solubility at indicated partial pressure.

**AUXILIARY INFORMATION**

**METHOD/APPARATUS/PROCEDURE:**

About 20 $cm^3$ of liquid propellant was condensed into a pressure vessel of about 1 liter capacity. Nitrogen was added to the proper pressure and the apparatus agitated for about one hour. A sample of the liquid phase was withdrawn and flashed to a gas at a lower temperature. The gas sample was then analyzed by gas chromatography.

**SOURCE AND PURITY OF MATERIALS:**

(1) Air Products and Chemicals, high purity grade, 99.9%.

(2) Orgmet Chemical, Hempstead, New Hampshire, 95-98%.

**ESTIMATED ERROR:**

$\delta T/K = \pm 0.2$
$\delta P/P = \pm 0.02$
$\delta x_1/x_1 = \pm 0.02$

**REFERENCES:**

COMPONENTS:

(1) Nitrogen; $N_2$; [7727-37-9]

(2) Methane; $CH_4$; [74-82-8]

(3) Carbon dioxide; $CO_2$; [124-38-9]

ORIGINAL MEASUREMENTS:

Sarashina, E.; Arai, Y.; Saito, S.

*J. Chem. Eng. Japan* 1971, *4*, 377.

VARIABLES:

T/K = 233-273
P/MPa = 6-10

PREPARED BY:

C. L. Young

EXPERIMENTAL VALUES:

| T/K | $P$/atm | $P$/MPa | Mole fraction in liquid $x_{N_2}$ | Mole fraction in liquid $x_{CH_4}$ | Mole fraction in vapor $y_{N_2}$ | Mole fraction in vapor $y_{CH_4}$ |
|---|---|---|---|---|---|---|
| 273.15 | 60 | 6.08 | 0.037 | 0.013 | 0.229 | 0.046 |
| | | | 0.032 | 0.033 | 0.169 | 0.116 |
| | | | 0.023 | 0.052 | 0.124 | 0.170 |
| | | | 0.015 | 0.075 | 0.072 | 0.237 |
| | 80 | 8.11 | 0.079 | 0.026 | 0.306 | 0.064 |
| | | | 0.071 | 0.050 | 0.253 | 0.120 |
| | | | 0.060 | 0.077 | 0.205 | 0.169 |
| | | | 0.057 | 0.091 | 0.182 | 0.196 |
| | | | 0.049 | 0.112 | 0.149 | 0.237 |
| | | | 0.038 | 0.146 | 0.100 | 0.290 |
| | | | 0.030 | 0.165 | 0.075 | 0.315 |
| | | | 0.024 | 0.179 | 0.061 | 0.334 |
| | | | 0.020 | 0.190 | 0.047 | 0.344 |
| | 100 | 10.13 | 0.141 | 0.004 | 0.381 | 0.007 |
| | | | 0.132 | 0.039 | 0.316 | 0.070 |
| | | | 0.126 | 0.082 | 0.247 | 0.123 |
| | | | 0.125 | 0.103 | 0.220 | 0.149 |
| | | | 0.120 | 0.115 | 0.200 | 0.158 |
| 253.15 | 60 | 6.08 | 0.052 | 0.038 | 0.342 | 0.195 |
| | | | 0.032 | 0.110 | 0.174 | 0.374 |
| | | | 0.024 | 0.131 | 0.113 | 0.439 |

(cont.)

AUXILIARY INFORMATION

METHOD/APPARATUS/PROCEDURE:

Static cell fitted with magnetic stirrer. Temperature measured with liquid in glass thermometer and pressure measured with Bourdon gauge. Gas-liquid chromatography was used to analyse the equilibrium compositions.

SOURCE AND PURITY OF MATERIALS:

No details given.

ESTIMATED ERROR:

$\delta T/K = \pm 0.05$; $\delta P/MPa = \pm 0.01$; $\delta x$, $\delta y = \pm 2\%$ (estimated by compiler).

REFERENCES:

1. Kaminishi, G.; Arai, Y.; Saito, S.; Maeda, S. *Chem. Eng. Japan* 1968, *1*, 109.

| COMPONENTS: | ORIGINAL MEASUREMENTS: |
|---|---|
| (1) Nitrogen; $N_2$; [7727-37-9]<br>(2) Methane; $CH_4$; [74-82-8]<br>(3) Carbon dioxide; $CO_2$; [124-38-9] | Sarashina, E.; Arai, Y.; Saito, S.<br>*J. Chem. Eng. Japan* 1971, *4*, 377. |

EXPERIMENTAL VALUES:

| T/K | *P*/atm | *P*/MPa | Mole fraction in liquid $x_{N_2}$ | Mole fraction in liquid $x_{CH_4}$ | Mole fraction in vapor $y_{N_2}$ | Mole fraction in vapor $y_{CH_4}$ |
|---|---|---|---|---|---|---|
| 253.15 | 80 | 8.11 | 0.102 | 0.022 | 0.525 | 0.062 |
| | | | 0.094 | 0.046 | 0.439 | 0.144 |
| | | | 0.086 | 0.072 | 0.372 | 0.210 |
| | | | 0.075 | 0.117 | 0.289 | 0.297 |
| | | | 0.071 | 0.135 | 0.259 | 0.324 |
| | | | 0.065 | 0.153 | 0.230 | 0.348 |
| | | | 0.054 | 0.206 | 0.169 | 0.409 |
| | | | 0.047 | 0.240 | 0.127 | 0.456 |
| | | | 0.035 | 0.282 | 0.092 | 0.494 |
| | | | 0.028 | 0.312 | 0.066 | 0.510 |
| | 100 | 10.13 | 0.136 | 0.097 | 0.393 | 0.182 |
| | | | 0.131 | 0.143 | 0.331 | 0.239 |
| | | | 0.127 | 0.163 | 0.301 | 0.259 |
| | | | 0.123 | 0.178 | 0.277 | 0.280 |
| 233.15 | 60 | 6.08 | 0.063 | 0.044 | 0.561 | 0.168 |
| | | | 0.059 | 0.080 | 0.444 | 0.279 |
| | | | 0.046 | 0.126 | 0.286 | 0.439 |
| | | | 0.034 | 0.174 | 0.205 | 0.521 |
| | | | 0.024 | 0.223 | 0.124 | 0.610 |
| | 80 | 8.11 | 0.104 | 0.023 | 0.682 | 0.060 |
| | | | 0.100 | 0.064 | 0.574 | 0.176 |
| | | | 0.086 | 0.108 | 0.464 | 0.279 |
| | | | 0.086 | 0.197 | 0.332 | 0.406 |
| | | | 0.083 | 0.228 | 0.285 | 0.449 |
| | | | 0.066 | 0.236 | 0.247 | 0.486 |
| | | | 0.064 | 0.318 | 0.185 | 0.541 |
| | | | 0.059 | 0.398 | 0.121 | 0.589 |
| | 100 | 10.13 | 0.144 | 0.127 | 0.486 | 0.244 |
| | | | 0.139 | 0.174 | 0.399 | 0.316 |
| | | | 0.140 | 0.206 | 0.368 | 0.339 |
| | | | 0.148 | 0.224 | 0.361 | 0.349 |
| | | | 0.156 | 0.246 | 0.345 | 0.357 |

**COMPONENTS:**

(1) Nitrogen; $N_2$; [7727-37-9]

(2) Methane; $CH_4$; [74-82-8]

(3) Carbon Dioxide; $CO_2$; [124-38-9]

**ORIGINAL MEASUREMENTS:**

Somait, F.A.; Kidnay, A.J.

*J. Chem. Eng. Data.* 1978, *23*, 301-5.

**VARIABLES:**

T/K = 270
P/MPa = 4.6-8.6

**PREPARED BY:**

C.L. Young

**EXPERIMENTAL VALUES:**

| T/K | P/bar | Mole fractions in liquid $x_{N_2}$ | $x_{CH_4}$ | $x_{CO_2}$ | Mole fractions in gas $y_{N_2}$ | $y_{CH_4}$ | $y_{CO_2}$ |
|---|---|---|---|---|---|---|---|
| 270.00 | 45.60 | 0.0000 | 0.0538 | 0.9462 | 0.0000 | 0.2300 | 0.7700 |
| | | 0.0129 | 0.0271 | 0.9599 | 0.0959 | 0.1251 | 0.7791 |
| | | 0.0226 | 0.0060 | 0.9714 | 0.1850 | 0.0269 | 0.7881 |
| | | 0.0250 | 0.0000 | 0.9750 | 0.2115 | 0.0000 | 0.7885 |
| | 60.79 | 0.0000 | 0.1226 | 0.8774 | 0.0000 | 0.3519 | 0.6481 |
| | | 0.0071 | 0.1071 | 0.8858 | 0.0336 | 0.3145 | 0.6520 |
| | | 0.0093 | 0.1024 | 0.8883 | 0.0441 | 0.3035 | 0.6525 |
| | | - | - | - | 0.1173 | 0.2233 | 0.6594 |
| | | 0.0239 | 0.0690 | 0.9072 | 0.1237 | 0.2178 | 0.6585 |
| | | 0.0297 | 0.0564 | 0.9139 | 0.1563 | 0.1822 | 0.6615 |
| | | 0.0442 | 0.0240 | 0.9318 | 0.2508 | 0.0815 | 0.6677 |
| | | 0.0461 | 0.0192 | 0.9347 | 0.2656 | 0.0655 | 0.6689 |
| | | 0.0560 | 0.0000 | 0.9440 | 0.3290 | 0.0000 | 0.6710 |
| | 86.13 | 0.1142 | 0.0000 | 0.8858 | 0.4139 | 0.0000 | 0.5861 |
| | | 0.0887 | 0.0671 | 0.8442 | 0.2689 | 0.1406 | 0.5905 |
| | | 0.0885 | 0.0688 | 0.8428 | 0.2656 | 0.1433 | 0.5912 |
| | | 0.0667 | 0.1233 | 0.8099 | 0.1747 | 0.2322 | 0.5931 |
| | | 0.0652 | 0.1302 | 0.8046 | 0.1670 | 0.2394 | 0.5936 |
| | | 0.0600 | 0.1442 | 0.7958 | 0.1473 | 0.2592 | 0.5936 |
| | | - | - | - | 0.1172 | 0.2874 | 0.5954 |
| | | 0.0422 | 0.1929 | 0.7649 | 0.0877 | 0.3105 | 0.6018 |
| | | 0.0408 | 0.1959 | 0.7634 | 0.0825 | 0.3161 | 0.6014 |
| | | 0.0301 | 0.2322 | 0.7377 | 0.0535 | 0.3374 | 0.6091 |
| | | 0.0152 | 0.2863 | 0.6985 | 0.0210 | 0.3474 | 0.6316 |
| | | 0.0155 | 0.2864 | 0.6981 | 0.0208 | 0.3479 | 0.6313 |

cont.

**AUXILIARY INFORMATION**

**METHOD:/APPARATUS/PROCEDURE:**

Recirculating vapor flow apparatus with diaphragm pump. Temperature measured with platinum resistance thermometer and pressure with Bowdon gauge. Cell stirred with two propeller stirrer. Vapor and liquid samples analysed by gas chromatography using a thermal conductivity detector. Details in source.

**SOURCE AND PURITY OF MATERIALS:**

No details given

**ESTIMATED ERROR:**

$\delta T/K = \pm 0.2$; $\delta P/bar = \pm 0.015$ up to 100 bar; $\pm 0.1$ above 100 bar; $\delta x$, $\delta y$ = $\pm$ 0.002.

**REFERENCES:**

COMPONENTS

(1) Nitrogen; $N_2$; [7727-37-9]

(2) Methane; $CH_4$; [74-82-8]

(3) Carbon Dioxide; $CO_2$; [124-38-9]

ORIGINAL MEASUREMENTS:

Somait, F.A.; Kidnay, A.J.

*J. Chem. Eng. Data.* 1978, *23*, 301-5.

EXPERIMENTAL VALUES:

| T/K | P/bar | Mole fractions in liquid | | | Mole fractions in gas | | |
|---|---|---|---|---|---|---|---|
| | | $x_{N_2}$ | $x_{CH_4}$ | $x_{CO_2}$ | $y_{N_2}$ | $y_{CH_4}$ | $y_{CO_2}$ |
| 270.00 | 86.13 | 0.0112 | 0.3026 | 0.6862 | - | - | - |
| | | 0.0076 | 0.3280 | 0.6645 | 0.0097 | 0.3323 | 0.6579 |
| | 96.26 | 0.1430 | 0.0000 | 0.8570 | 0.4210 | 0.0000 | 0.5790 |
| | | - | - | - | 0.3237 | 0.0897 | 0.5866 |
| | | 0.1216 | 0.0696 | 0.8088 | 0.2900 | 0.1207 | 0.5893 |
| | | 0.1159 | 0.0818 | 0.8023 | 0.2611 | 0.1452 | 0.5937 |
| | | 0.1092 | 0.0993 | 0.7915 | 0.2326 | 0.1711 | 0.5963 |
| | | 0.1079 | 0.1085 | 0.7836 | 0.2248 | 0.1769 | 0.5983 |
| | | 0.1090 | 0.1098 | 0.7812 | 0.2238 | 0.1775 | 0.5986 |
| | | 0.1045 | 0.1287 | 0.7668 | 0.1988 | 0.1975 | 0.6037 |
| | | 0.1015 | 0.1522 | 0.7463 | 0.1728 | 0.2166 | 0.6107 |
| | | - | - | - | 0.1631 | 0.2205 | 0.6164 |
| | | 0.0971 | 0.1635 | 0.7394 | 0.1597 | 0.2239 | 0.6164 |
| | | 0.0941 | 0.1705 | 0.7355 | 0.1482 | 0.2287 | 0.6231 |
| | | 0.0961 | 0.1741 | 0.7298 | 0.1475 | 0.2281 | 0.6243 |
| | | 0.0946 | 0.1743 | 0.7311 | 0.1445 | 0.2297 | 0.6259 |
| | | 0.0947 | 0.1810 | 0.7243 | 0.1398 | 0.2332 | 0.6270 |
| | | 0.0947 | 0.1781 | 0.7272 | 0.1384 | 0.2318 | 0.6298 |
| | | 0.0948 | 0.1868 | 0.7185 | | | |
| | | 0.0944 | 0.1882 | 0.7174 | 0.1313 | 0.2329 | 0.6359 |
| | | 0.0945 | 0.1905 | 0.7151 | 0.1303 | 0.2335 | 0.6363 |
| | | 0.0956 | 0.1941 | 0.7102 | 0.1286 | 0.2333 | 0.6380 |
| | | 0.0944 | 0.1982 | 0.7075 | 0.1224 | 0.2336 | 0.6440 |
| | | 0.0960 | 0.2008 | 0.7032 | 0.1198 | 0.2313 | 0.6490 |
| | | 0.0986 | 0.2098 | 0.6915 | 0.1152 | 0.2306 | 0.6542 |
| | | 0.1045 | 0.2260 | 0.6696 | 0.1045 | 0.2260 | 0.6696 |
| | 111.46 | 0.1950 | 0.0000 | 0.8050 | 0.4080 | 0.0000 | 0.5920 |
| | | 0.1897 | 0.0415 | 0.7688 | 0.3285 | 0.0594 | 0.6121 |
| | | 0.1898 | 0.0613 | 0.7489 | 0.2929 | 0.0808 | 0.6263 |
| | | 0.1917 | 0.0640 | 0.7443 | 0.2861 | 0.0825 | 0.6314 |
| | | 0.2643 | 0.0921 | 0.6436 | 0.2643 | 0.0921 | 0.6436 |

**COMPONENTS:**

(1) Nitrogen; $N_2$; [7727-37-9]

(2) Santowax R;

**ORIGINAL MEASUREMENTS:**

Grove, N.H.; Whiteley, F.J.
*J. Appl. Chem.* 1960, *10*, 101.

**VARIABLES:**

T/K = 507-682

P/MPa = 0.2-0.5

**PREPARED BY:**

C. L. Young

**EXPERIMENTAL VALUES:**

| T/K | *P*/bar | Solubility* | Ostwald Coefficient |
|---|---|---|---|
| 507 | 3.93 | 13.0 | 0.131 |
| 510 | 2.54 | 7.0 | 0.110 |
| 511 | 1.84 | 5.3 | 0.115 |
| 600 | 2.16 | 9.3 | 0.187 |
| 600 | 2.98 | 12.3 | 0.179 |
| 609 | 4.72 | 22.7 | 0.210 |
| 679 | 2.49 | 12.3 | 0.223 |
| 679 | 3.45 | 16.3 | 0.214 |
| 682 | 5.38 | 29.0 | 0.224 |

*moles of nitrogen per mg of Santowax R.

AUXILIARY INFORMATION

**METHOD /APPARATUS/PROCEDURE:**

Static cell with null pressure transducer. Pressure measured with Bourdon gauge. Temperature measured with thermocouple. Sample placed in cell and gas added at room temperature. Cell then heated to experimental temperature. Pressures on both sides of transducer kept approximately equal. Details in source.

**SOURCE AND PURITY OF MATERIALS:**

(1) No details given.

(2) Analysis by infra-red method showed sample to be 11.8% o-terphenyl, 56.3% m-terphenyl, 29.3% p-terphenyl, 2.6% diphenyl and higher polyphenyls. Obtained from Monsanto Chemicals Ltd.

**ESTIMATED ERROR:**

$\delta T/K = \pm 1$; $\delta P/\text{bar} = \pm 0.01$; $\delta(\text{solubility}) = \pm 10\%$

**REFERENCES:**

| COMPONENTS: | ORIGINAL MEASUREMENTS: |
|---|---|
| (1) Nitrogen; $N_2$; [7727-37-9]<br><br>(2) Hydrocarbon Blend (Heavy naphtha) | Frolich, P. K.; Tauch, E. J.; Hogan, J. J.; Peer, A. A.<br><br>*Ind. Eng. Chem.* <u>1931</u>, *23*, 548-50. |
| VARIABLES: | PREPARED BY: |
| T/K = 298<br>P/MPa = 1.0-14.2 | C. L. Young |

EXPERIMENTAL VALUES:

| T/K | *P*/atm | *P*/MPa | Solubility*, *S* |
|---|---|---|---|
| 298.15 | 10 | 1.0 | 1.05 |
| | 20 | 2.0 | 2.15 |
| | 30 | 3.0 | 3.3 |
| | 40 | 4.1 | 4.5 |
| | 50 | 5.1 | 5.45 |
| | 60 | 6.1 | 6.4 |
| | 70 | 7.1 | 7.2 |
| | 80 | 8.1 | 8.0 |
| | 90 | 9.1 | 8.9 |
| | 100 | 10.1 | 9.85 |
| | 110 | 11.1 | 10.65 |
| | 120 | 12.2 | 11.4 |
| | 130 | 13.2 | 12.75 |
| | 140 | 14.2 | 14.15 |

* Volume of gas measured at 101.325 kPa pressure and 298.15 K dissolved by unit volume of liquid measured under the same conditions.

AUXILIARY INFORMATION

| METHOD/APPARATUS/PROCEDURE: | SOURCE AND PURITY OF MATERIALS: |
|---|---|
| Static equilibrium cell. Liquid saturated with gas and after equilibrium established samples removed and analysed by volumetric method. Allowance was made for vapor pressure of liquid and the solubility of the gas at atmospheric pressure. Details in source. | (1) Nitrogen was of the highest purity available.<br><br>(2) Density 0.8003 g $cm^{-3}$, vapor pressure 80 mmHg at 298.15 K. |
| | ESTIMATED ERROR: |
| | $\delta T/K = \pm 0.1$; $\delta S = \pm 5\%$. |
| | REFERENCES: |

| COMPONENTS: | ORIGINAL MEASUREMENTS: |
|---|---|
| (1) Nitrogen; $N_2$; [7727-37-9]<br><br>(2) Hydrocarbon Blend (Gas oil) | Frolich, P. K.; Tauch, E. J.; Hogan, J. J.; Peer, A. A.<br><br>*Ind. Eng. Chem.* 1931, *23*, 548-50. |
| **VARIABLES:** | **PREPARED BY:** |
| T/K = 298<br>P/MPa = 1.0-14.2 | C. L. Young |

EXPERIMENTAL VALUES:

| T/K | *P*/atm | *P*/MPa | Solubility*, *S* |
|---|---|---|---|
| 298.15 | 10 | 1.0 | 1.1 |
| | 20 | 2.0 | 2.2 |
| | 30 | 3.0 | 3.25 |
| | 40 | 4.1 | 4.25 |
| | 50 | 5.1 | 5.25 |
| | 60 | 6.1 | 6.2 |
| | 70 | 7.1 | 7.15 |
| | 80 | 8.1 | 8.05 |
| | 90 | 9.1 | 8.85 |
| | 100 | 10.1 | 9.65 |
| | 110 | 11.1 | 10.45 |
| | 120 | 12.2 | 11.25 |
| | 130 | 13.2 | 12.05 |
| | 140 | 14.2 | 12.8 |

* Volume of gas measured at 101.325 kPa pressure and 298.15 K dissolved by unit volume of liquid measured under the same conditions.

AUXILIARY INFORMATION

METHOD/APPARATUS/PROCEDURE:

Static equilibrium cell. Liquid saturated with gas and after equilibrium established samples removed and analysed by volumetric method. Allowance was made for vapor pressure of liquid and the solubility of the gas at atmospheric pressure. Details in source.

SOURCE AND PURITY OF MATERIALS:

(1) Nitrogen was of the highest purity available.

(2) Density 0.8319 g $cm^{-3}$, vapor pressure 2 mmHg at 298.15 K.

ESTIMATED ERROR:

$\delta T/K = \pm 0.1$; $\delta S = \pm 5\%$.

REFERENCES:

| COMPONENTS: | ORIGINAL MEASUREMENTS: |
|---|---|
| (1) Nitrogen; $N_2$; [7727-37-9]<br><br>(2) Fuel RJ-1 | Findl, E.; Brande, H.; Edwards, H.<br><br>U. S. Dept. Comm., Office Tech. Ser. Report No. *AD 274 623*, 1960, 216 pp.<br><br>*Chem. Abstr.* 1963, *58*, 6628c. |

| VARIABLES: | PREPARED BY: |
|---|---|
| $T$/K = 310.9 - 533.2<br>$p_1$/kPa = 172.4 - 274.4<br>(25 - 398 psia) | H. L. Clever |

METHOD/APPARATUS/PROCEDURE

The solubility apparatus was constructed of two concentric glass tubes sealed at the end by invar steel caps and silicone O-rings. The caps had attachments for flexible tubes to admit solvent and gas, and to circulate the thermostating fluid.

The inner tube was of *TRUE-BORE* glass tubing. It contained the solvent and a glass ball 0.0083 inch smaller than the tube diameter. The ball was used for both stirring on gas dissolution and viscosity measurements on the gas saturated solution. A steel scale beside the inner tube allowed determination of the solvent volume to within 0.3 per cent.

The thermostating liquid (white mineral oil) circulated in the space between the two tubes.

The solvent was initially saturated with air. The amount of dissolved air was known from a previous Orsat analysis.

The solute gas was contained in a steel bottle at 700 psig maintained at a temperature of 150 ± 1°F. There were two pressure gages. A differential gage between the gas bottle and a reference bottle to measure the pressure drop when gas was admitted to the apparatus, and a Bourdon gage to measure the final total pressure in the cell.

The solvent was added to the cell until the vapor/liquid ratio reached about 0.15. The cell and its contents were heated to the temperature of the measurement. The gas was admitted, and the test cell rocked for 10 minutes more than the time required to reach a stable pressure. The amount of gas dissolved was calculated from the pressure drop in the gas reservoir and the total pressure in the cell. The partial pressure of the gas in the

## AUXILIARY INFORMATION

METHOD/APPARATUS/PROCEDURE: (continued)

cell was calculated taking into account the initial dissolved air, the solvent vapor pressure, and the compressibility of the gas. Both the Ostwald coefficient and the volume of gas at standard conditions per unit weight of solvent were calculated.

The results were presented in large scale graphs. The results on the data sheets are values read from the graphs by the compiler.

Other experimental details and results were in an earlier report of the authors (1).

SOURCE AND PURITY OF MATERIALS:

(1) Nitrogen. Stated to be of high purity.

(2) Fuel. The fuel was specified as MIL-F-25558 Grade RJ-1 formerly designated as Shell UMF, Grade C. A table of fuel characteristics was included in the report.

ESTIMATED ERROR:

$\delta T$/°F = ± 1
$\delta p$/lb in$^{-2}$ = ± 1
$\delta L/L$ = ± 0.02 (minimum), max. ≃ 0.05

REFERENCES:

1. Schlagel, L. A.; Findl, E.; Edwards, H.
Ing Er. Rept. 183, Thompson Products, Inc. Inglewood Lab., Inglewood, CA, USA, Aug 19, 1955.

| COMPONENTS: | ORIGINAL MEASUREMENTS: |
|---|---|
| (1) Nitrogen; $N_2$; [7727-37-9]<br><br>(2) Fuel RJ-1 | Findl, E.; Brande, H.; Edwards, H.<br><br>U. S. Dept. Comm., Office Tech. Ser. Report No. *AD 274 623*, 1960, 216 pp.<br><br>*Chem. Abstr.* 1963, *58*, 6628c. |

EXPERIMENTAL VALUES: continued

| Temperature $t$/°C | $T$/K | Nitrogen Pressure $p_1$/psia | Kuenen Coefficient | Ostwald Coefficient $L/cm^3 cm^{-3}$ |
|---|---|---|---|---|
| 100 | 310.9 | 12.5 | -- | 0.101 |
| | | 12.6 | -- | 0.095 |
| | | 59.0 | -- | 0.085 |
| | | 60.0 | -- | 0.100 |
| | | 110.0 | 0.79 | 0.100 |
| | | 121.0 | 0.77 | 0.090 |
| | | 209.0 | 1.38 | 0.094 |
| | | 222.0 | 1.52 | 0.099 |
| | | 309.0 | 2.27 | 0.104 |
| | | 309.0 | 2.34 | 0.110 |
| | | 408.0 | 2.87 | 0.100 |
| | | 410.0 | 3.18 | 0.110 |
| | | | | (0.100)[1] |
| 200 | 366.5 | 15.0 | -- | 0.152 |
| | | 59.0 | 0.51 | 0.139 |
| | | 108.0 | 0.88 | 0.127 |
| | | 109.0 | 0.94 | 0.139 |
| | | 208.0 | 1.80 | 0.140 |
| | | 209.0 | 1.80 | 0.139 |
| | | 309.0 | 2.55 | 0.131 |
| | | 309.0 | 2.71 | 0.141 |
| | | 406.0 | 3.69 | 0.144 |
| | | | | (0.141)[1] |
| 300 | 422.0 | 56.0 | -- | 0.175 |
| | | 106.0 | 1.12 | 0.185 |
| | | 205.0 | 1.97 | 0.166 |
| | | 205.0 | 2.18 | 0.188 |
| | | 307.0 | 3.08 | 0.175 |
| | | 307.0 | 3.26 | 0.186 |
| | | 404.5 | 4.27 | 0.190 |
| | | 405.0 | 4.19 | 0.180 |
| | | | | (0.182)[1] |
| 400 | 477.6 | 106.0 | 1.12 | -- |
| | | 198.0 | 2.13 | 0.197 |
| | | 200.0 | 2.15 | 0.200 |
| | | 300.0 | 3.30 | 0.208 |
| | | 397.5 | 4.52 | 0.213 |
| | | 400.0 | 4.41 | 0.208 |
| | | | | (0.205)[1] |
| 500 | 533.2 | 19.0 | 0.30 | -- |
| | | 81.0 | 1.04 | 0.247 |
| | | 183.0 | 2.43 | 0.241 |
| | | 184.0 | 2.55 | 0.258 |
| | | 285.0 | 3.70 | 0.245 |
| | | 287.5 | 3.71 | 0.257 |
| | | 380.0 | 5.21 | 0.265 |
| | | 382.5 | 5.05 | 0.266 |
| | | | | (0.258)[1] |

[1] From author's line through the Ostwald coefficient values. psia ≡ pounds per square inch absolute. One lb $in^{-2}$ ≡ 6.89476 kPa.

COMPONENTS:

(1) Nitrogen; $N_2$; [7727-37-9]

(2) Jet A(79) fuel

ORIGINAL MEASUREMENTS:

Rupprecht, S. D.; Faeth, G.M.

NASA CR 1981, NASA-CR-3422, 111 pp.

*Chem. Abstr.* 1981, 95, 153333x.

VARIABLES:

T/K = 298-374
P/MPa = 1-10

PREPARED BY:

R. Battino

EXPERIMENTAL VALUES:

| T/K | P[a]/MPa | $x_1$ |
|---|---|---|
| 298.15 | 1.03 | 0.0094 |
| 373.15 | 1.03 | 0.0116 |
| 298.65 | 2.07 | 0.0196 |
| 373.65 | 2.07 | 0.0240 |
| 298.40 | 4.82 | 0.0458 |
| 373.15 | 4.82 | 0.0558 |
| 296.65 | 10.34 | 0.1006 |
| 373.15 | 10.34 | 0.1186 |

[a] Total pressure.

AUXILIARY INFORMATION

METHOD/APPARATUS/PROCEDURE:

A 1000 $cm^3$ chamber is filled with ca. 500 $cm^3$ of liquid. Gas is added to the desired pressure. The chamber is agitated to attain equilibrium. Temperatures are read via thermocouples and pressures via Heisse bourdon tube gages. A liquid sample is removed and allowed to de-pressurize at atmospheric pressure. The solubility is calculated from the measurement of liquid and gas volumes of the de-pressurized sample. Residual dissolved gas is considered to be negligible.

SOURCE AND PURITY OF MATERIALS:

(1) No details given.

(2) 0.8381 sp. gr. at 60°F. Boiling range is 434.82-600.93K

ESTIMATED ERROR:

$\delta x_1/x_1 = \pm 0.03$, compiler's estimate.

REFERENCES:

**COMPONENTS:**

(1) Nitrogen; $N_2$; [7727-37-9]

(2) Jet A(80) fuel

**ORIGINAL MEASUREMENTS:**

Rupprecht, S.D.; Faeth, G.M.

NASA CR 1981, NASA-CR-3422, 111 pp.

*Chem. Abstr.* 1981, 95, 153333x.

**VARIABLES:**

T/K = 298-373
P/MPa = 1-10

**PREPARED BY:**

R. Battino

**EXPERIMENTAL VALUES:**

| T/K | $P^a$/MPa | $x_1$ |
|---|---|---|
| 299.15 | 1.03 | 0.0121 |
| 373.15 | 1.03 | 0.0135 |
| 298.65 | 2.07 | 0.0251 |
| 373.15 | 2.07 | 0.0286 |
| 298.65 | 4.82 | 0.0594 |
| 373.15 | 4.82 | 0.0680 |
| 299.15 | 10.34 | 0.1242 |
| 373.15 | 10.34 | 0.1381 |

[a] Total pressure.

**AUXILIARY INFORMATION**

**METHOD/APPARATUS/PROCEDURE:**

A 1000 $cm^3$ chamber is filled with ca. 500 $cm^3$ of liquid. Gas is added to the desired pressure. The chamber is agitated to attain equilibrium. Temperatures are read via thermocouples and pressures via Heisse bourdon tube gages. A liquid sample is removed and allowed to de-pressurize at atmospheric pressure. The solubility is calculated from the measurement of liquid and gas volumes of the de-pressurized sample. Residual dissolved gas is considered to be negligible.

**SOURCE AND PURITY OF MATERIALS:**

(1) No details given.

(2) 0.8095 sp.gr. at 60°F.
Boiling range 444.26-547.59K.

**ESTIMATED ERROR:**

$\delta x_1/x_1 = \pm 0.03$ compiler's estimate

**REFERENCES:**

# AIR SOLUBILITIES

**COMPONENTS:**

(1) Air

(2) Water; $H_2O$; [7732-18-5]

**ORIGINAL MEASUREMENTS:**

Baldwin, R. R.; Daniel, S. G.

*J. Appl. Chem.* 1952, *2*, 161-5.

*J. Inst. Petrol., London* 1953, *39*, 105-24.

**VARIABLES:**

$T$/K = 293.15
$p_1$/kPa = 101.325

**PREPARED BY:**

P. L. Long
H. L. Clever

**EXPERIMENTAL VALUES:**

| Temperature $t/^{\circ}C$ | Temperature $T/K$ | Bunsen Coefficient $\alpha/cm^3(STP)cm^{-3}atm^{-1}$ | Ostwald Coefficient $L/cm^3\ cm^{-3}$ |
|---|---|---|---|
| 20.0 | 293.15 | 0.01858 | 0.01994 |
| | | 0.01875 | 0.02012 |

The compiler calculated the Ostwald coefficient values from the author's Bunsen coefficients.

**AUXILIARY INFORMATION**

**METHOD/APPARATUS/PROCEDURE:**

1. Degassing. The liquid is deaerated by passage through a continously evacuated vessel. The liquid is saturated with the gas, then degassed again.

2. Saturation of the liquid with the gas. The gas is passed through a liquid air trap to remove water and carbon dioxide, then through a thermostated coil to bring it to the temperature of the experiment, and then bubbled through the liquid for up to five hours.

3. To determine the amount of gas dissolved in the saturated liquid, the saturated liquid is placed in an evacuated vessel and buret system. The gas released by the liquid is transferred to the buret. Five transfer operations recover all but a trace of the gas. The last trace (2 - 5 %) is recovered by boiling the liquid.

**SOURCE AND PURITY OF MATERIALS:**

(1) Air. From compressed air line. Dried and freed from $CO_2$.

(2) Water. Distilled.

**ESTIMATED ERROR:**

$\delta p/p = \pm\ 0.01$
$\delta\alpha/\alpha = \pm\ 0.01$

**REFERENCES:**

**COMPONENTS:**

(1) Air

(2) Water; $H_2O$; [7732-18-5]

**ORIGINAL MEASUREMENTS:**

McKee, O. L. Jr.

*Ph. D. thesis*, 1953
Purdue University
W. Lafayette, IN USA

**VARIABLES:**

$T$/K = 273.15, 294.8
Total $p$/kPa = 3,450 - 20,680

**PREPARED BY:**

H. L. Clever

**EXPERIMENTAL VALUES:**

| Temperature $t/^0$F | Temperature $T$/K | Total Pressure $p$/lb in$^{-2}$ | Air Solubility $S$/cm$^3$(STP) g$^{-1}$ | | |
|---|---|---|---|---|---|
| 32 | 273.15 | 500 | 0.73 | 0.79 | |
| | | 1000 | 1.52 | 1.72 | 1.75 |
| | | 1500 | 2.29 | 2.44 | 2.46 |
| | | 2000 | 2.95 | 3.10 | 2.97 |
| | | 3000 | 4.15 | 4.19 | |
| 71 | 294.8 | 500 | 0.44 | | |
| | | 1000 | 1.12 | 1.17 | |
| 70 | 294.3 | 1500 | 1.64 | 1.65 | |
| 71 | 294.8 | 2000 | 2.13 | 2.10 | |
| 66 | 292.0 | 3000 | 2.72 | | |
| 67 | 292.6 | | 2.97 | | |
| 71 | 294.8 | | 2.85 | | |

The maximum error is estimated to range from about 5 per cent at 0.75 to about 2 per cent at a solubility value of 4.0.

**AUXILIARY INFORMATION**

**METHOD/APPARATUS/PROCEDURE:**

The apparatus of Zoss (1) was used with modifications to improve the temperature control at 32 F, and to improve the sampling of the liquid phase.

Total Pressure:

| $p$/lb in$^{-2}$ | $p$/atm | $p$/kPa |
|---|---|---|
| 500 | 34.0 | 3450 |
| 1000 | 68.0 | 6895 |
| 1500 | 102.1 | 10340 |
| 2000 | 136.1 | 13790 |
| 3000 | 204.1 | 20680 |

1 atm ≡ 101.325 kPa ≡ 14.6960 lb in$^{-2}$

**SOURCE AND PURITY OF MATERIALS:**

(1) Air. Source not given.

(2) Water. Distilled water. Boiled for three hours before use.

**ESTIMATED ERROR:**

See statement under the table above.

**REFERENCES:**

1. Zoss, L. M.
*Ph. D. Thesis*, 1952.
Purdue University
W. Lafayette, IN USA

**COMPONENTS:**

(1) Air

(2) Water; $H_2O$; [7732-18-5]

**ORIGINAL MEASUREMENTS:**

Eichelberger, W.C.

*Ind. Eng. Chem.* 1955, *47*, 2223-8.

**VARIABLES:**

T/K = 298-338
P/MPa = 6.8-24.2

**PREPARED BY:**

R. Battino

**EXPERIMENTAL VALUES:**

| Temperature | $P_{Air}$/psig | $P_{Air}$[a]/MPa | Solubility[b] | Temperature | $P_{Air}$/psig | $P_{Air}$[a]/MPa | Solubility[b] |
|---|---|---|---|---|---|---|---|
| 65°C | 970 | 6.787 | 0.78 | 45°C | 1010 | 7.063 | 0.91 |
| (338.2 K) | 1475 | 10.27 | 1.14 | (318.2 K) | 1310 | 9.131 | 1.13 |
| | 1645 | 11.44 | 1.27 | | 1950 | 13.54 | 1.61 |
| | 1935 | 13.44 | 1.42 | | 2200 | 15.27 | 1.78 |
| | 2430 | 16.85 | 1.73 | | 2540 | 17.61 | 2.02 |
| | 3430 | 23.74 | 2.27 | | 2750 | 19.06 | 2.09 |
| | | | | | 3440 | 23.81 | 2.54 |
| 25°C | 1050 | 7.339 | 1.15 | | 3500 | 24.23 | 2.54 |
| (298.2 K) | 1850 | 12.85 | 1.83 | | | | |
| | 2775 | 19.23 | 2.58 | | | | |
| | 3420 | 23.68 | 2.98 | | | | |

[a] Calculated by compiler. Partial pressure of air (absolute).

[b] Solubility in units of ml air (STP) per ml water. Gas volumes not corrected for non-ideality of gas at pressure of measurement.

**AUXILIARY INFORMATION**

**METHOD/APPARATUS/PROCEDURE:**

A pressure vessel is charged up to pressure with gas and liquid and rocked for several hours. After equilibration a sample is removed into a buret where the liquid and gas volumes are measured at atmospheric pressure.

**SOURCE AND PURITY OF MATERIALS:**

(1) From standard cylinder.

(2) Distilled.

**ESTIMATED ERROR:**

$\delta T/K = \pm 0.1$
$\delta$Solubility/Solubility = ±0.03, compiler's estimate.

**REFERENCES:**

| COMPONENTS: | ORIGINAL MEASUREMENTS: |
|---|---|
| (1) Air<br>(2) Sodium Chloride; NaCl; [7647-14-5] (brines)<br>(2) Water; $H_2O$; [7732-18-5] | Eichelberger, W. C.<br>*Ind. Eng. Chem.* 1955, *47*, 2223-8. |

EXPERIMENTAL VALUES:

| | | 25°C (298.2 K) | | 45°C (318.2 K) | | 65°C (338.2 K) | |
|---|---|---|---|---|---|---|---|
| $P_{Air}$/psig | $P_{Air}$[a]/MPa | g NaCl/ℓ soln.[b] | Solubility[c] | g NaCl/ℓ soln.[b] | Solubility[c] | g NaCl/ℓ soln.[b] | Solubility[c] |
| 2500 | 17.33 | 0 | 2.36 | 0 | 1.985 | 0 | 1.76 |
| | | 30 | 2.04 | 30 | 1.71 | 30 | 1.51 |
| | | 99 | 1.36 | 98 | 1.20 | 97 | 1.135 |
| | | 203 | 0.77 | 203 | 0.75 | 197 | 0.705 |
| | | 305 | 0.425 | 304 | 0.425 | 250 | 0.55 |
| | | | | | | 309 | 0.43 |
| 3500 | 24.23 | 0 | 3.05 | 0 | 2.55 | 0 | 2.30 |
| | | 30 | 2.63 | 30 | 2.24 | 30 | 1.96 |
| | | 99 | 1.77 | 98 | 1.50 | 97 | 1.43 |
| | | 203 | 0.98 | 203 | 0.945 | 197 | 0.92 |
| | | 305 | 0.55 | 304 | 0.55 | 250 | 0.72 |
| | | | | | | 309 | 0.545 |

[a] Calculated by compiler. Partial pressure of air (absolute).

[b] Concentration of brine solution given as g NaCl/liter of solution.

[c] Solubility in units of ml air (STP) per ml brine solution. Gas volumes not corrected for non-ideality of gas at pressure of measurement.

[d] These are smoothed values read from graphs by the authors.

| COMPONENTS: | ORIGINAL MEASUREMENTS: |
|---|---|
| (1) Air<br>(2) Sodium Chloride; NaCl; [7647-14-5]; (brines)<br>(3) Water; $H_2O$; [7732-18-5] | Eichelberger, W. C.<br>*Ind. Eng. Chem.* 1955, *47*, 2223-8. |
| VARIABLES:<br>T/K = 298-338<br>P/MPa = 7.0-24.2<br>Concentration | PREPARED BY:<br>R. Battino |

EXPERIMENTAL VALUES:

| | | 25°C (298.2 K) | | 45°C (318.2 K) | | 65°C (338.2 K) | |
|---|---|---|---|---|---|---|---|
| $P_{Air}$/psig | $P^a_{Air}$/MPa | g NaCl/ℓ soln.[b] | Solubility[c] | g NaCl/ℓ soln.[b] | Solubility[c] | g NaCl/ℓ soln.[b] | Solubility[c] |
| 1000 | 6.994 | 0 | 1.10 | 0 | 0.91 | 0 | 0.83 |
| | | 30 | 0.93 | 30 | 0.77 | 30 | 0.67 |
| | | 99 | 0.65 | 98 | 0.54 | 97 | 0.50 |
| | | 201 | 0.355 | 203 | 0.34 | 197 | 0.31 |
| | | 305 | 0.21 | 304 | 0.205 | 309 | 0.19 |
| 1500 | 10.44 | 0 | 1.54 | 0 | 1.29 | 0 | 1.15 |
| | | 30 | 1.315 | 30 | 1.105 | 30 | 0.97 |
| | | 99 | 0.89 | 98 | 0.78 | 97 | 0.725 |
| | | 201 | 0.50 | 203 | 0.485 | 197 | 0.45 |
| | | 305 | 0.285 | 304 | 0.28 | 250 | 0.34 |
| | | | | | | 309 | 0.275 |
| 2000 | 13.89 | 0 | 1.96 | 0 | 1.64 | 0 | 1.465 |
| | | 30 | 1.69 | 30 | 1.41 | 30 | 1.25 |
| | | 99 | 1.13 | 98 | 1.00 | 97 | 0.94 |
| | | 203 | 0.64 | 203 | 0.62 | 197 | 0.58 |
| | | 305 | 0.36 | 304 | 0.355 | 250 | 0.45 |
| | | | | | | 309 | 0.36 |

continued on following page

AUXILIARY INFORMATION

METHOD/APPARATUS/PROCEDURE:

A pressure vessel is charged up to pressure with gas and liquid and rocked for several hours. After equilibration a sample is removed into a gas buret where the liquid and gas volumes are measured at atmospheric pressure.

SOURCE AND PURITY OF MATERIALS:

(1) From standard cylinders.

(2) Raw brine from Syracuse plant of Solray Process Division.

(3) Distilled.

ESTIMATED ERROR:

$\delta T/K = \pm 0.1$
$\delta$Solubility/Solubility = ±0.03 compiler's estimate.

REFERENCES:

| COMPONENTS: | ORIGINAL MEASUREMENTS: |
|---|---|
| (1) Air<br>(2) Sulfuric acid; $H_2SO_4$; [7664-93-9]<br>(3) Sodium sulfate; $Na_2SO_4$; [7757-82-6]<br>(4) Water; $H_2O$; [7732-18-5] | Kobe, K. A.; Kenton, F. H.<br>*Ind. Eng. Chem., Anal. Ed.* 1938, *10*, 76-7. |
| **VARIABLES:** | **PREPARED BY:** |
| $T$/K = 298.15<br>$p_1$/kPa = 101.325 | P. L. Long<br>H. L. Clever |

EXPERIMENTAL VALUES:

| Temperature $t$/°C | Temperature $T$/K | Solvent Volume $V$/cm$^3$ | Air Volume Absorbed $v_1$/cm$^3$ | Bunsen Coefficient $\alpha$/cm$^3$ (STP) cm$^{-3}$ atm$^{-1}$ | Ostwald Coefficient $L$/cm$^3$ cm$^{-3}$ |
|---|---|---|---|---|---|
| 25 | 298.15 | 99.54 | 0.51 | 0.0047[a] | 0.0051[a] |
| | | 49.54 | 0.27 | 0.0049 | 0.0053 |

The solvent is a mixture of 800 g $H_2O$

200 g $Na_2SO_4$ (anhydrous)

40 ml $H_2SO_4$ (Conc., 36 normal)

Thus the molality of the solution is

$m_2$/mol kg$^{-1}$ = 0.90 ($H_2SO_4$)

$m_3$/mol kg$^{-1}$ = 1.76 ($Na_2SO_4$)

[a]Calculated by compilers.

AUXILIARY INFORMATION

METHOD/APPARATUS/PROCEDURE:

The apparatus is described in detail in an earlier paper (1). The apparatus consists of a gas buret, a pressure compensator, and a 200 cm$^3$ absorption bulb and mercury leveling bulb. The absorption bulb is attached to a shaking mechanism.

The solvent and the gas are placed in the absorption bulb. The bulb is shaken until equilibrium is reached. The remaining gas is returned to the buret. The difference in the final and initial volumes is taken as the volume of gas absorbed.

SOURCE AND PURITY OF MATERIALS:

(1) Air. Source not given. Purity stated to be 99+ per cent.

(2, 3) Sulfuric acid and sodium sulfate. Sources not given. Analytical grade.

(4) Water. Distilled.

ESTIMATED ERROR:

$\delta\alpha/\alpha$ = ±0.02 (compilers)

REFERENCES:

1. Kobe, K. A.; Williams, J. S. *Ind. Eng. Chem., Anal. Ed.* 1935, *7*, 37.

| COMPONENTS: | ORIGINAL MEASUREMENTS: |
|---|---|
| (1) Air<br>(2) Carbon dioxide; $CO_2$; [124-38-9]<br>(3) Sulfuric acid; $H_2SO_4$; [7664-93-9]<br>(4) Sodium sulfate; $Na_2SO_4$; [7757-82-6]<br>(5) Water; $H_2O$; [7732-18-5] | Kobe, K. A.; Kenton, F. H.<br>*Ind. Eng. Chem., Anal. Ed.* 1938, *10*, 76-7. |
| VARIABLES: | PREPARED BY: |
| $T$/K = 298.15<br>$p_1$/kPa = 101.325 | P. L. Long<br>H. L. Clever |

EXPERIMENTAL VALUES:

| Temperature $t$/°C | $T$/K | Gas Composition volume per cent | Solvent Volume $V$/cm$^3$ | Air and $CO_2$ Volume Absorbed $v_1$/cm$^3$ | Bunsen Coefficient $\alpha$/cm$^3$ (STP) cm$^{-3}$ atm$^{-1}$ | Ostwald Coefficient $L$/cm$^3$ cm$^{-3}$ |
|---|---|---|---|---|---|---|
| 25 | 298.15 | 5% $CO_2$ | 49.54 | 0.67 | 0.0124 | 0.0135 |
| | | 95% Air | 49.54 | 0.67 | | |
| 25 | 298.15 | 10% $CO_2$ | 49.54 | 1.17 | 0.0215 | 0.0235 |
| | | 90% Air | 49.54 | 1.16 | | |
| 25 | 298.15 | 20% $CO_2$ | 49.54 | 2.21 | 0.0410 | 0.0447 |
| | | 80% Air | 49.54 | 2.23 | | |

The solvent is a mixture of 800 g $H_2O$
200 g $Na_2SO_4$ (anhydrous)
40 ml $H_2SO_4$ (Conc., 36 normal)

Thus the molality of the solution is

$$m_3/\text{mol kg}^{-1} = 0.90\ (H_2SO_4)$$

$$m_4/\text{mol kg}^{-1} = 1.76\ (Na_2SO_4)$$

AUXILIARY INFORMATION

METHOD/APPARATUS/PROCEDURE:

The apparatus is described in detail in an earlier paper (1). The apparatus consists of a gas buret, a pressure compensator, and a 200 cm$^3$ absorption bulb and mercury leveling bulb. The absorption bulb is attached to a shaking mechanism.

The solvent and the gas are placed in the absorption bulb. The bulb is shaken until equilibrium is reached. The remaining gas is returned to the buret. The difference in the final and initial volumes is taken as the volume of gas absorbed.

SOURCE AND PURITY OF MATERIALS:

(1,2) Air and carbon dioxide. Source not given. Purity stated to be 99+ per cent.

(3,4) Sulfuric acid and sodium sulfate. Sources not given. Analytical grade.

(5) Water. Distilled.

ESTIMATED ERROR:

$\delta\alpha$/cm$^3$ = ±0.005 (authors)

REFERENCES:

1. Kobe, K. A.; Williams, J. S. *Ind. Eng. Chem., Anal. Ed.* 1935, *7*, 37.

| COMPONENTS: | ORIGINAL MEASUREMENTS: |
|---|---|
| (1) Air<br>(2) Alkanes; $C_7H_{16}$, $C_8H_{18}$, and $C_{10}H_{22}$ | Ijams, C. C.<br>Ph.D. thesis, 1941<br>Vanderbilt University<br>Nashville, TN |
| **VARIABLES:** | **PREPARED BY:** |
| T/K = 298.15<br>P/kPa = 101.325 | H. L. Clever |

EXPERIMENTAL VALUES:

| T/K | Bunsen Coefficient $\alpha$ | Ostwald Coefficient L |
|---|---|---|
| Heptane; $C_7H_{16}$; [142-82-5] | | |
| 298.15 | 0.224 | 0.245 |
| Octane; $C_8H_{18}$; [111-65-9] | | |
| 298.15 | 0.207 | 0.226 |
| 2,2,4-Trimethylpentane or isooctane; $C_8H_{18}$; [540-84-1] | | |
| 298.15 | 0.226 | 0.247 |
| Decane; $C_{10}H_{22}$; [124-18-5] | | |
| 298.15 | 0.165 | 0.180 |

The Ostwald coefficient value was calculated by the compiler.

AUXILIARY INFORMATION

METHOD/APPARATUS/PROCEDURE:

A Van Slyke-Neill Manometric Apparatus manufactured by the Eimer and Amend Co. was used.

The procedure of Van Slyke (1) for pure liquids was modified so that small solvent sample (2 $cm^3$) could be used with almost 100 per cent recovery of the solvent.

An improved temperature control system was used.

SOURCE AND PURITY OF MATERIALS:

(1) Air. Air Reduction Co. Purity 99.5 per cent.

(2) Alkanes. Purified. Vapor pressure and boiling point also reported in the thesis.

ESTIMATED ERROR:

$\delta T/K = 0.05$

REFERENCES:

1. Van Slyke, D. D. *J. Biol. Chem.* 1939, *130*, 545.

COMPONENTS:

(1) Air

(2) 2,2,4-Trimethylpentane or isooctane; $C_8H_{18}$; [540-84-1]

ORIGINAL MEASUREMENTS:

Baldwin, R. R.; Daniel, S. G.

*J. Appl. Chem.* 1952, *2*, 161-5.

*J. Inst. Petrol., London* 1953, *39*, 105-24.

VARIABLES:

$T$/K = 293.15

$p_1$/kPa = 101.325

PREPARED BY:

P. L. Long
H. L. Clever

EXPERIMENTAL VALUES:

| $T$/K | Bunsen Coefficient $\alpha$/cm$^3$(STP) cm$^{-3}$ atm$^{-1}$ | Ostwald Coefficient $L$/cm$^3$ cm$^{-3}$ |
|---|---|---|
| 293.15 | 0.240 | 0.258 |

AUXILIARY INFORMATION

METHOD/APPARATUS/PROCEDURE:

The liquid is deaerated by passing it through a continously evacuated vessel. The gas is passed through a liquid air trap to remove water and carbon dioxide, then through a long thermostated coil to bring the gas to the temperature of the experiment, then through the liquid until saturation is reached.

The air-saturated liquid is degassed, and the gas is collected and measured

SOURCE AND PURITY OF MATERIALS:

(1) Air. From a compressed air line. Dried and freed from $CO_2$.

(2) 2,2,4-Trimethylpentane. Source not given.

ESTIMATED ERROR:

$\delta p/p = \pm\ 0.01$

$\delta\alpha/\alpha = \pm\ 0.01$

REFERENCES:

**COMPONENTS:**

(1) Air

(2) n-Dodecane; $C_{12}H_{26}$; [112-40-3]

**ORIGINAL MEASUREMENTS:**

Rupprecht, S. D.; Faeth, G. M.

NASA CR 1981, NASA-CR-3422, 111 pp.

*Chem. Abstr.* 1981, 95, 153333x.

**VARIABLES:**

T/K = 249 - 373
P/MPa = 1-10

**PREPARED BY:**

R. Battino

**EXPERIMENTAL VALUES:**

| T/K | $P^a$/MPa | $x_1$ |
|---|---|---|
| 248.82 | 1.03 | 0.0142 |
| 373.15 | 1.03 | 0.0161 |
| 297.40 | 2.07 | 0.0312 |
| 373.15 | 2.07 | 0.0330 |
| 297.65 | 4.82 | 0.0722 |
| 373.15 | 4.82 | 0.0768 |
| 297.15 | 10.34 | 0.1491 |
| 373.15 | 10.34 | 0.1579 |

[a] Total pressure.

**AUXILIARY INFORMATION**

**METHOD/APPARATUS/PROCEDURE:**

A 1000 $cm^3$ chamber is filled with ca. 500 $cm^3$ of liquid. Gas is added to the desired pressure. The chamber is agitated to attain equilibrium. Temperatures are read via thermocouples and pressures via Heisse bourdon tube gauges. A liquid sample is removed and allowed to de-pressurize at atmospheric pressure. Solubility is calculated from the measurement of liquid and gas volumes of the de-pressurized sample. Residual dissolved gas is considered to be negligible.

**SOURCE AND PURITY OF MATERIALS:**

No details given.

**ESTIMATED ERROR:**

$\delta x_1/x_1 = \pm 0.03$, compiler's estimate.

**REFERENCES:**

**COMPONENTS:**

(1) Air

(2) Aliphatic alcohols:

1-Heptanol; $C_7H_{16}O$; [111-70-6]

1-Octanol; $C_8H_{18}O$; [111-87-5]

**ORIGINAL MEASUREMENTS:**

Ijams, C. C.

*Ph. D. thesis*, <u>1941</u>
Vanderbilt University
Nashville, TN, USA

**VARIABLES:**

$T$/K = 298.15
$p$/kPa = 101.325

**PREPARED BY:**

H. L. Clever

**EXPERIMENTAL VALUES:**

| $T$/K | Bunsen Coefficient $\alpha$/cm$^3$(STP)cm$^{-1}$atm$^{-1}$ | Ostwald Coefficient $L$/cm$^3$cm$^{-3}$ |
|---|---|---|
| 1-Heptanol | | |
| 298.15 | 0.111 | 0.121 |
| 1-Octanol | | |
| 298.15 | 0.108 | 0.118 |

The Ostwald coefficient values were calculated by the compiler.

**AUXILIARY INFORMATION**

**METHOD/APPARATUS/PROCEDURE:**

A Van Slyke-Neill Manometric Apparatus manufactired by the Eimer and Amend Co. was used.

The procedure of Van Slyke (1) for pure liquids was modified so that the small solvent sample (2 cm$^3$) could be used with almost 100 per cent recovery of the solvent.

An improved temperature control system was used.

**SOURCE AND PURITY OF MATERIALS:**

(1) Air. Air Reduction Co. Purity 99.5 per cent.

(2) Aliphatic alcohols. Purified. Vapor pressure and boiling point also reported in the thesis.

**ESTIMATED ERROR:**

$\delta T$/K = ± 0.05

**REFERENCES:**

1. Van Slyke, D. D.
*J. Biol. Chem.* <u>1939</u>, *130*, 545.

**COMPONENTS:**

(1) Air

(2) 1,1'-Oxybisethane or diethyl ether; $C_4H_{10}O$; [60-29-7]

**ORIGINAL MEASUREMENTS:**

Christoff, A.

*Z. Phys. Chem.* 1912, *79*, 456-60.

**VARIABLES:**

$T$/K = 273.15 - 288.15
$p$/kPa = Atmospheric

**PREPARED BY:**

H. L. Clever

**EXPERIMENTAL VALUES:**

| Temperature $t$/°C | $T$/K | Ostwald Coefficient $L$/cm$^3$ cm$^{-3}$ |
|---|---|---|
| 0 | 273.15 | 0.290 |
| 10 | 283.15 | 0.287 |
| 15 | 288.15 | 0.286 |

AUXILIARY INFORMATION

**METHOD/APPARATUS/PROCEDURE:**

The apparatus is a modified Ostwald apparatus described by Just (1) and modified by Skirrow (2). The apparatus consists of a thermostated gas buret and an absorption flask.

The modification involes the use of vapor free gas in the gas buret. A correction is made for the vapor pressure of the liquid. A steel capillary tube with a stopcock, which prevents the gas and the liquid vapor from mixing in the buret, is used to connect the absorption flask and the gas buret.

**SOURCE AND PURITY OF MATERIALS:**

(1) Air. Stated to be dry air.

(2) Diethtylether. Merck and Co. Stated to be pure and anhydrous.

**ESTIMATED ERROR:**

$\delta L/L = \pm 0.03$

**REFERENCES:**


1. Just, G. *Z. Phys. Chem.* 1901, *37*, 342.

2. Skirrow, F. W. *Z. Phys. Chem.* 1902, *41*, 139.

**COMPONENTS:**

(1) Air

(2) 1-Bromoheptane; $C_7H_{15}Br$; [629-04-9]

**ORIGINAL MEASUREMENTS:**

Ijams, C. C.

Ph.D. thesis, 1941
Vanderbilt University
Nashville, TN

**VARIABLES:**

T/K = 298.15
P/kPa = 101.325

**PREPARED BY:**

H. L. Clever

**EXPERIMENTAL VALUES:**

| T/K | Bunsen Coefficient $\alpha$ | Ostwald Coefficient L |
|---|---|---|
| 298.15 | 0.121 | 0.132 |

The Ostwald coefficient value was calculated by the compiler.

**AUXILIARY INFORMATION**

**METHOD/APPARATUS/PROCEDURE:**

A Van Slyke-Neill Manometric Apparatus manufactured by the Eimer and Amend Co. was used.

The procedure of Van Slyke (1) for pure liquids was modified so that small solvent sample (2 $cm^3$) could be used with almost 100 per cent recovery of the solvent.

An improved temperature control system was used.

**SOURCE AND PURITY OF MATERIALS:**

(1) Air. Air Reduction Co. Purity 99.5 per cent.

(2) 1-Bromoheptane. Purified. Vapor pressure and boiling point also reported in the thesis.

**ESTIMATED ERROR:**

$\delta T/K = 0.05$

**REFERENCES:**

1. Van Slyke, D. D. *J. Biol. Chem.* 1939, *130*, 545.

**COMPONENTS:**

(1) Air
(2) 1-Bromo-1,1,2,2,3,3,4,4,5,6,6,6-dodecafluoro-5-(trifluoromethyl)-hexane; $C_7F_{15}Br$; [50279-29-3]

**ORIGINAL MEASUREMENTS:**

Wesseler, E. P.; Iltis, R.; Clark, L. C. Jr.

*J. Fluorine Chem.*, 1977, *9*, 137-46.

**VARIABLES:**

T/K = 298

**PREPARED BY:**

C. L. Young

**EXPERIMENTAL VALUES:**

| T/K | Solubility† of air $S_{air}$ | Solubility† of oxygen $S_{O_2}$ | Solubility† of nitrogen $S_{N_2}$ |
|---|---|---|---|
| 298.15 | 36.0 | 10.2 | 25.8 |

† volume of gas absorbed (measured at atmospheric pressure) per 100 ml of fluorocarbon. (Total pressure was the barometric pressure, assumed to be 1 atmosphere by the compiler.)

**AUXILIARY INFORMATION**

**METHOD/APPARATUS/PROCEDURE:**

Air bubbled through fluorocarbon in stirred cell. When saturation achieved samples of liquid removed with hypodermic Precision Sampling 50 μℓ Pressure Lok syringe. Five samples taken and chromatographed. Average peak heights compared with those for injections of known amounts of air. Chromatographic column consisted of 13X molecular sieve at 70 °C. Helium was used as carrier.

**SOURCE AND PURITY OF MATERIALS:**

(1) No details given.
(2) Pennwalt Corp. sample, boiling pt. 120 °C.

**ESTIMATED ERROR:**

$\delta T/K = \pm 0.3$; $\delta S = \pm 2\%$.

**REFERENCES:**

| COMPONENTS: | ORIGINAL MEASUREMENTS: |
|---|---|
| (1) Air<br>(2) 1-Chloro-1,1,2,2,3,3,4,4,5,6,6,6-dodecafluoro-5-(trifluoromethyl)-hexane; $C_7F_{15}Cl$; [63243-37-8] | Wesseler, E. P.; Iltis, R.; Clark, L. C. Jr.<br>*J. Fluorine Chem.*, 1977, *9*, 137-46. |
| VARIABLES:<br>T/K = 298 | PREPARED BY:<br>C. L. Young |

EXPERIMENTAL VALUES:

| T/K | Solubility† of air $S_{air}$ | Solubility† of oxygen $S_{O_2}$ | Solubility† of nitrogen $S_{N_2}$ |
|---|---|---|---|
| 298.15 | 37.4 | 10.5 | 26.9 |

† volume of gas absorbed (measured at atmospheric pressure) per 100 ml of fluorocarbon. (Total pressure was the barometric pressure, assumed to be 1 atmosphere by the compiler.)

AUXILIARY INFORMATION

METHOD/APPARATUS/PROCEDURE:

Air bubbled through fluorocarbon in stirred cell. When saturation achieved samples of liquid removed with hypodermic Precision Sampling 50 μℓ Pressure Lok syringe. Five samples taken and chromatographed. Average peak heights compared with those for injections of known amounts of air. Chromatographic column consisted of 13X molecular sieve at 70 °C. Helium was used as carrier.

SOURCE AND PURITY OF MATERIALS:

(1) No details given.

(2) Pennwalt Corp. sample, boiling pt. 108 °C.

ESTIMATED ERROR:

$\delta T/K = \pm 0.3$; $\delta S = \pm 2\%$.

REFERENCES:

| COMPONENTS: | ORIGINAL MEASUREMENTS: |
|---|---|
| (1) Air<br>(2) 1-Bromo-1,1,2,2,3,3,4,4,5,5,6,6,-7,7,8,8,8-heptadecafluorooctane (Perfluorooctylbromide); $C_8F_{17}Br$; [423-55-2] | Wesseler, E. P.; Iltis, R.; Clark, L. C. Jr.<br>*J. Fluorine Chem.*, 1977, *9*, 137-46. |
| VARIABLES: | PREPARED BY: |
| T/K = 298 | C. L. Young |

EXPERIMENTAL VALUES:

| T/K | Solubility† of air $S_{air}$ | Solubility† of oxygen $S_{O_2}$ | Solubility† of nitrogen $S_{N_2}$ |
|---|---|---|---|
| 298.15 | 34.1§ | 9.8 | 24.3 |

§ there is a typographical error in the original paper.

† volume of gas absorbed (measured at atmospheric pressure) per 100 ml of fluorocarbon. (Total pressure was the barometric pressure, assumed to be 1 atmosphere by the compiler.)

AUXILIARY INFORMATION

METHOD/APPARATUS/PROCEDURE:

Air bubbled through fluorocarbon in stirred cell. When saturation achieved samples of liquid removed with hypodermic Precision Sampling 50 μℓ Pressure Lok syringe. Five samples taken and chromatographed. Average peak heights compared with those for injections of known amounts of air. Chromatographic column consisted of 13X molecular sieve at 70 °C. Helium was used as carrier.

SOURCE AND PURITY OF MATERIALS:

(1) No details given.

(2) 3M sample, boiling pt. 140.5 °C.

ESTIMATED ERROR:

$\delta T/K = \pm 0.3$; $\delta S = \pm 2\%$.

REFERENCES:

COMPONENTS:

(1) Air
(2) Octadecafluorooctane (Perfluorooctane); $C_8F_{18}$; [307-34-6]

ORIGINAL MEASUREMENTS:

Wesseler, E. P.; Iltis, R.; Clark, L. C. Jr.

*J. Fluorine Chem.*, 1977, *9*, 137-46.

VARIABLES:

T/K = 298

PREPARED BY:

C. L. Young

EXPERIMENTAL VALUES:

| T/K | Solubility† of air $S_{air}$ | Solubility† of oxygen $S_{O_2}$ | Solubility† of nitrogen $S_{N_2}$ |
|---|---|---|---|
| 298.15 | 37.4 | 10.5 | 26.9 |

† volume of gas absorbed (measured at atmospheric pressure) per 100 ml of fluorocarbon. (Total pressure was the barometric pressure, assumed to be 1 atmosphere by the compiler.)

AUXILIARY INFORMATION

METHOD/APPARATUS/PROCEDURE:

Air bubbled through fluorocarbon in stirred cell. When saturation achieved samples of liquid removed with hypodermic Precision Sampling 50 μℓ Pressure Lok syringe. Five samples taken and chromatographed. Average peak heights compared with those for injections of known amounts of air. Chromatographic column consisted of 13X molecular sieve at 70 °C. Helium was used as carrier.

SOURCE AND PURITY OF MATERIALS:

(1) No details given.
(2) PCR sample, boiling pt. 100-105 °C.

ESTIMATED ERROR:

$\delta T/K = \pm 0.3$; $\delta S = \pm 2\%$.

REFERENCES:

| COMPONENTS: | ORIGINAL MEASUREMENTS: |
|---|---|
| (1) Air<br>(2) Octadecafluorodecahydro-naphthalene (Perfluorodecalin); $C_{10}F_{18}$; [306-94-5] | Wesseler, E. P.; Iltis, R.; Clark, L. C. Jr.<br>*J. Fluorine Chem.*, 1977, *9*, 137-46. |
| VARIABLES: | PREPARED BY: |
| T/K = 298 | C. L. Young |

EXPERIMENTAL VALUES:

| T/K | Solubility† of air $S_{air}$ | Solubility† of oxygen $S_{O_2}$ | Solubility† of nitrogen $S_{N_2}$ |
|---|---|---|---|
| 298.15 | 29.8 | 8.6 | 21.2 |

† volume of gas absorbed (measured at atmospheric pressure) per 100 ml of fluorocarbon. (Total pressure was the barometric pressure, assumed to be 1 atmosphere by the compiler.)

AUXILIARY INFORMATION

METHOD/APPARATUS/PROCEDURE:

Air bubbled through fluorocarbon in stirred cell. When saturation achieved samples of liquid removed with hypodermic Precision Sampling 50 μℓ Pressure Lok syringe. Five samples taken and chromatographed. Average peak heights compared with those for injections of known amounts of air. Chromatographic column consisted of 13X molecular sieve at 70 °C. Helium was used as carrier.

SOURCE AND PURITY OF MATERIALS:

(1) No details given
(2) ISC Chemicals Ltd. sample, boiling pt. 142 °C.

ESTIMATED ERROR:

$\delta T/K = \pm 0.3$; $\delta S = \pm 2\%$.

REFERENCES:

| COMPONENTS: | ORIGINAL MEASUREMENTS: |
|---|---|
| (1) Air<br>(2) 1,1,1,2,4,4,5,7,7,8,10,10,11,13,-13,14,16,16,17,17,18,18,18-Tricosafluoro-5,8,11,14-tetrakis (trifluoromethyl)-3,6,9,12,15-pentaoxaoctadecane; $C_{17}HF_{35}O_5$; [37486-69-4] | Wesseler, E. P.; Iltis, R.; Clark, L. C. Jr.,<br>*J. Fluorine Chem.*, 1977, *9*, 137-46. |
| VARIABLES: | PREPARED BY: |
| T/K = 298 | C. L. Young |

EXPERIMENTAL VALUES:

| T/K | Solubility† of air $S_{air}$ | Solubility† of oxygen $S_{O_2}$ | Solubility† of nitrogen $S_{N_2}$ |
|---|---|---|---|
| 298.15 | 31.8 | 8.6 | 23.2 |

† volume of gas absorbed (measured at atmospheric pressure) per 100 ml of fluorocarbon. (Total pressure was the barometric pressure, assumed to be 1 atmosphere by the compiler.)

AUXILIARY INFORMATION

METHOD/APPARATUS/PROCEDURE:

Air bubbled through fluorocarbon in stirred cell. When saturation achieved samples of liquid removed with hypodermic Precision Sampling 50 μℓ Pressure Lok syringe. Five samples taken and chromatographed. Average peak heights compared with those for injections of known amounts of air. Chromatographic column consisted of 13X molecular sieve at 70 °C. Helium was used as carrier.

SOURCE AND PURITY OF MATERIALS:

(1) No details given.

(2) DuPont sample, boiling pt. 224 °C.

ESTIMATED ERROR:

$\delta T/K = \pm 0.3$; $\delta S = \pm 2\%$.

REFERENCES:

| COMPONENTS: | ORIGINAL MEASUREMENTS: |
|---|---|
| (1) Air<br>(2) L1822 (large mixture of compounds from the electrochemical fluorination of decalin) | Wesseler, E. P.; Iltis, R.; Clark, L. C. Jr.<br>*J. Fluorine Chem.*, 1977, *9*, 137-46. |
| VARIABLES: | PREPARED BY: |
| T/K = 298 | C. L. Young |

EXPERIMENTAL VALUES:

| T/K | Solubility† of air $S_{air}$ | Solubility† of oxygen $S_{O_2}$ | Solubility† of nitrogen $S_{N_2}$ |
|---|---|---|---|
| 298.15 | 32.0 | 9.0 | 23.0 |

† volume of gas absorbed (measured at atmospheric pressure) per 100 ml of fluorocarbon. (Total pressure was the barometric pressure, assumed to be 1 atmosphere by the compiler.)

AUXILIARY INFORMATION

METHOD/APPARATUS/PROCEDURE:

Air bubbled through fluorocarbon in stirred cell. When saturation achieved samples of liquid removed with hypodermic Precision Sampling 50 μℓ Pressure Lok syringe. Five samples taken and chromatographed. Average peak heights compared with those for injections of known amounts of air. Chromatographic column consisted of 13X molecular sieve at 70 °C. Helium was used as carrier.

SOURCE AND PURITY OF MATERIALS:

(1) No details given.
(2) 3M sample.

ESTIMATED ERROR:

$\delta T/K = \pm 0.3$; $\delta S = \pm 2\%$.

REFERENCES:

| COMPONENTS: | ORIGINAL MEASUREMENTS: |
|---|---|
| (1) Air<br>(2) FC75 (mainly perfluorobutyltetrahydrofuran and perfluoropropyltetrahydropyran) | Wesseler, E. P.; Iltis, R.; Clark, L. C. Jr.<br>*J. Fluorine Chem.*, 1977, *9*, 137-46. |
| VARIABLES: | PREPARED BY: |
| T/K = 298 | C. L. Young |

EXPERIMENTAL VALUES:

| T/K | Solubility† of air $S_{air}$ | Solubility† of oxygen $S_{O_2}$ | Solubility† of nitrogen $S_{N_2}$ |
|---|---|---|---|
| 298.15 | 37.7 | 10.5 | 27.2 |

† volume of gas absorbed (measured at atmospheric pressure) per 100 ml of fluorocarbon. (Total pressure was the barometric pressure, assumed to be 1 atmosphere by the compiler.)

AUXILIARY INFORMATION

METHOD/APPARATUS/PROCEDURE:

Air bubbled through fluorocarbon in stirred cell. When saturation achieved samples of liquid removed with hypodermic Precision Sampling 50 μℓ Pressure Lok syringe. Five samples taken and chromatographed. Average peak heights compared with those for injections of known amounts of air. Chromatographic column consisted of 13X molecular sieve at 70 °C. Helium was used as carrier.

SOURCE AND PURITY OF MATERIALS:

(1) No details given.

(2) 3M sample.

ESTIMATED ERROR:

$\delta T/K = \pm 0.3$; $\delta S = \pm 2\%$.

REFERENCES:

**COMPONENTS:**

(1) Air

(2) Nitric acid, propyl ester or 1-propyl nitrate; $C_3H_7NO_3$; [627-13-4]

**ORIGINAL MEASUREMENTS:**

Amster, A. B.; Levy, J. B.

*J. Am. Rocket Soc.* 1959, *29*, 870-1.

**VARIABLES:**

$T/K$ = "Room temperature"
$p/kPa$ = 31 - 88

**PREPARED BY:**

M. E. Derrick
H. L. Clever

**EXPERIMENTAL VALUES:**

| Temperature | Air Pressure[1] | | Solubility[1] |
|---|---|---|---|
| | $p_1$/mmHg | $p_1$/kPa | $10^6 m_1$/mol $g^{-1}$ |
| "Room temperature" | 230 | 31 | 1.6 |
| | 230 | 31 | 1.75 |
| | 240 | 32 | 1.75 |
| | 240 | 32 | 1.6 |
| | 440 | 59 | 3.6 |
| | 440 | 59 | 3.4 |
| | 440 | 59 | 3.5 |
| | 450 | 60 | 3.1 |
| | 450 | 60 | 3.75 |
| | 475 | 63 | 3.75 |
| | 560 | 75 | 3.8 |
| | 620 | 83 | 5.4 |
| | 620 | 83 | 5.5 |
| | 620 | 83 | 5.5 |
| | 650 | 87 | 5.1 |
| | 660 | 88 | 5.6 |
| | 660 | 88 | 5.8 |

[1] The compilers read the values of pressure, mmHg, and solubility, moles air per g of liquid, from a graph in the paper. The air partial pressure was obtained by subtracting the ester vapor pressure from the total pressure.

**AUXILIARY INFORMATION**

**METHOD/APPARATUS/PROCEDURE:**

The apparatus consists of an absorption flask connected by a flexible spiral to a mercury filled capillary (1).

A weighed quantity of liquid is placed in the flask. The gas is shaken with the degassed liquid. The volume of the gas is taken as the volume of the flask minus the volume of the liquid. The solubility is calculated from

$$\text{Solubility/mol g}^{-1} = V\Delta p/WRT$$

$V$ = the gas volume
$\Delta p$ = the pressure change
$T$ = the temperature
$W$ = the weight of the liquid.

The compilers converted the solubility values above to a Bunsen coefficient $\alpha/\text{cm}^3(\text{STP})\text{cm}^{-3}\text{ atm}^{-1} = 0.14_2 \pm 0.01_4$ at a temperature of 293.15 K.

**SOURCE AND PURITY OF MATERIALS:**

(1) Air. Passed through a drying tu tube.

(2) Nitric acid, propyl ester. Source and purity not given.

**ESTIMATED ERROR:**

$\delta p$/mmHg = ± 5
$\delta T$/K = ± 2
$\delta m_1/m_1$ = ± 0.15

**REFERENCES:**

1. Levy, J. B.; Taft, R. W. Jr. Aaron, D.; Hammett, L. P. *J. Am. Chem. Soc.* 1953, *75*, 3955.

| COMPONENTS: | ORIGINAL MEASUREMENTS: |
|---|---|
| (1) Air<br>(2) 1,1,2,2,3,3,4,4,4-Nonafluoro-N,N-bis(nonafluorobutyl)-1-butanamine (Perfluorotributylamine); $C_{12}F_{27}N$; [311-89-7] | Wesseler, E. P.; Iltis, R.; Clark, L. C. Jr.<br>*J. Fluorine Chem.*, 1977, *9*, 137-46. |
| VARIABLES: | PREPARED BY: |
| T/K = 298 | C. L. Young |

EXPERIMENTAL VALUES:

| T/K | Solubility† of air $S_{air}$ | Solubility† of oxygen $S_{O_2}$ | Solubility† of nitrogen $S_{N_2}$ |
|---|---|---|---|
| 298.15 | 30.7 | 8.6 | 22.1 |

† volume of gas absorbed (measured at atmospheric pressure) per 100 ml of fluorocarbon. (Total pressure was the barometric pressure, assumed to be 1 atmosphere by the compiler.)

AUXILIARY INFORMATION

METHOD/APPARATUS/PROCEDURE:

Air bubbled through fluorocarbon in stirred cell. When saturation achieved samples of liquid removed with hypodermic Precision Sampling 50 μℓ Pressure Lok syringe. Five samples taken and chromatographed. Average peak heights compared with those for injections of known amounts of air. Chromatographic column consisted of 13X molecular sieve at 70 °C. Helium was used as carrier.

SOURCE AND PURITY OF MATERIALS:

(1) No details given.
(2) 3M sample, boiling pt. 174 °C.

ESTIMATED ERROR:

$\delta T/K = \pm 0.3$; $\delta S = \pm 2\%$.

REFERENCES:

**COMPONENTS:**

(1) Air

(2) Animal and vegetable oils

**ORIGINAL MEASUREMENTS:**

Vibrans, F. C.

*Oil and Soap* 1935, *12*, 14-5.

**VARIABLES:**

$T$/K = 296-318
$p$/kPa = 101.3

**PREPARED BY:**

P. L. Long
H. L. Clever

**EXPERIMENTAL VALUES:**

| $T$/K | Solubility, $cm^3$ (273.15 K, 1 atm) air per 100 $cm^3$ | | | | | |
|---|---|---|---|---|---|---|
| | Experiments | | | | | Av. |
| Corn oil | | | | | | |
| 296 - 299 | 7.28 | 7.38 | 7.30 | | | 7.32 |
| 318 | 6.98 | 7.14 | | | | 7.06 |
| Cottonseed oil | | | | | | |
| 296 - 299 | 7.18 | 7.12 | 7.23 | 7.24 | 7.22 | 7.20 |
| 318 | 7.12 | 6.93 | | | | 7.02 |
| Hydrogenated Cottonseed oil | | | | | | |
| 318 | 7.20 | 7.36 | | | | 7.28 |
| Steam rendered lard | | | | | | |
| 318 | 7.22 | 7.13 | | | | 7.17 |

**AUXILIARY INFORMATION**

**METHOD/APPARATUS/PROCEDURE:**

Van Slyke manometric method (1).

**SOURCE AND PURITY OF MATERIALS:**

(1) Air. No information.

(2) Corn oil. Commercial sample bought on the market. Representative of its class.

Cottonseed oil. Commercial sample bought on the market. Representative of its class.

Lard. From a mixture of killing and cutting fats. Steam rendered.

**ESTIMATED ERROR:**

$\delta T$/K = ± 3 at room temperature
± 2 at 318 K (45 $^0$C)
$\delta c/c$ = ± 0.04 (Author)

**REFERENCES:**

1. Van Slyke, D. D.
*J. Biol. Chem.* 1924, *61*, 523.

**COMPONENTS:**

(1) Air

(2) Animal and vegetable oils

**ORIGINAL MEASUREMENTS:**

Schaffer, P. S.; Haller, H. S.

*Oil and Soap* 1943, *20*, 161

**VARIABLES:**

$T$/K = 313, 333

$p$/kPa = 101.3

**PREPARED BY:**

P. L. Long

H. L. Clever

**EXPERIMENTAL VALUES:**

| $T$/K | $cm^3$ (273.15 K, 1 atm) Air per 100 $cm^3$ |
|---|---|
| Cottonseed oil | |
| 313 | 8.7 |
| Butter oil | |
| 313 | 10.1 |
| 333 | 9.6 |
| Lard | |
| 313 | 8.8 |

The author's solubility unit is $10^2\alpha$ or 100 x Bunsen coefficient.

**AUXILIARY INFORMATION**

**METHOD/APPARATUS/PROCEDURE:**

The apparatus consists of a reaction flask connected to a water-jacketed gas buret, a mercury leveling bulb, and a differential manometer. The reaction flask is attached to a shaking mechanism in an air bath.

A 50 $cm^3$ sample of oil is placed in the flask. Heat and vacuum are applied to remove dissolved gases. The sample is cooled to the temperature of the measurement. Air is admitted, a zero reading is taken. The sample is shaken until it takes up no more gas at a pressure of one atm. Three determinations are made at each temperature.

**SOURCE AND PURITY OF MATERIALS:**

(1) Air. Source not given. Commercial cylinder.

(2) Cottonseed oil. Good grade of commercial product.

Butter oil. Obtained from butter prepared in the lab from fresh cream.

Lard. Good grade of commercial product.

**ESTIMATED ERROR:**

$\delta T$/K = ± 0.5

$\delta\alpha/\alpha$ = ± 0.003 (Authors, error among three determinations)

**REFERENCES:**

| COMPONENTS: | ORIGINAL MEASUREMENTS: |
|---|---|
| (1) Air<br>(2) Fats (oils) | Vernon, H. M.<br>*Proc. Royal Soc., Ser. B.* 1907, *79*, 366 - 371. |
| VARIABLES:<br>$T$/K = 288, 310, 318<br>$p_1$/kPa = 101.3 (760 mmHg) | PREPARED BY:<br>H. L. Clever |

EXPERIMENTAL VALUES:

| $T$/K | The Air Saturated Fat was Analyzed to Contain | | | | | | |
|---|---|---|---|---|---|---|---|
| | Oxygen, $cm^3$(STP)/100 $cm^3$ | | Nitrogen, $cm^3$(STP)/100 $cm^3$ | | Carbon Dioxide, $cm^3$(STP)/100 $cm^3$ | | Total Bunsen[1] $10^2\alpha$ |
| Olive oil | | | | | | | |
| 288 | 2.20, 2.23, 2.42 | 2.28 Av. | 5.23, 5.30, 5.27 | 5.26 Av. | 0.19, 0.24, 0.16 | 0.20 Av. | 7.74 |
| 310 | 2.33, 2.36, 2.30 | 2.33 Av. | 5.19, 5.23, 5.15 | 5.19 Av. | 0.17, 0.13, 0.18 | 0.16 Av. | 7.68 |
| Cod-liver oil | | | | | | | |
| 288 | 2.34, 2.31, 2.22 | 2.29 Av. | 4.95, 5.15, 5.07 | 5.06 Av. | 0.19, 0.25, 0.19 | 0.21 Av. | 7.56 |
| 310 | 2.21, 2.22, 2.22 | 2.22 Av. | 5.05, 5.10, 5.08 | 5.08 Av. | 0.18, 0.24, 0.20 | 0.21 Av. | 7.51 |
| Lard | | | | | | | |
| 318 | 2.23, 2.40, 2.35 | 2.33 Av. | 5.05, 5.09, 5.18 | 5.11 Av. | 0.12, 0.12, 0.15 | 0.13 Av. | 7.57 |

[1] Air Bunsen Coefficient, $10^2\alpha/cm^3(STP)cm^{-3}atm^{-1}$.

AUXILIARY INFORMATION

METHOD/APPARATUS/PROCEDURE:

In the observations made at room temperature, the oil was shaken violently with air in a bottle for several minutes, and was allowed to stand 1 to 1½ hours till all air bubbles had risen to the surface. It was then weighed, and about 40 - 50 g of it were sucked into a vacuum flask containing 70 - 100 $cm^3$ of 0.5 per cent sulfuric acid, which had been well degassed by boiling. The mixture of oil and dilute acid was boiled for ½ hour. The gas was pumped off and analyzed in a Haldane gas analysis apparatus (1).

At the higher temperatures, the oil presaturated at room temperature was heated to several degrees above the temperature of measurment. It was vigorously shaken with air for 2 minutes, warmed again, and shaken for another 2 minutes. It was held for ½ hour in a water bath and then treated as above.

SOURCE AND PURITY OF MATERIALS:

1. Air.
2. Fats. A table of the oil properties is reproduced below.

| | Cod-liver | Olive | Lard |
|---|---|---|---|
| Solid. Pt. °C | 0 - -10 | -6(?) | 27-30 |
| Sp. Grav. | $0.928^{15}_{15}$ | $0.917^{15}_{15}$ | $0.895^{45}_{15}$ |
| $I_2$ Value | 123-168 % | 79-88 % | 50-70 % |

ESTIMATED ERROR: The author mentions that the unsaturated oils take up some $O_2$, thus the $O_2$ values may be small.

$\delta\alpha/\alpha = \pm 0.10$ (compiler).

REFERENCES:

1. Haldane, J. S. *J. Physiol.* 1898, *22*, 465.

**COMPONENTS:**

(1) Air

(2) Fats (oils)

**ORIGINAL MEASUREMENTS:**

Schmidt-Nielsen, S.

*Ann. Acad. Sci. Fenn., Ser. A.* 1927, *29* (No. 12), 7 pp.

**VARIABLES:**

$T$/K = 293, 323
$p_1$/kPa = 101 (1 atm)

**PREPARED BY:**

H. L. Clever

**EXPERIMENTAL VALUES:**

| $T$/K | Solubility/ $cm^3$(STP) per 100 g | Oxygen Content of the Dissolved Gas Volume % |
|---|---|---|
| Cottonseed oil | | |
| 293 | 5.0 | 32.4 |
| 323 | 6.4 | 16.2 |
| Cod-liver oil | | |
| 293 | 9.6 | 21.8 |
| 323 | 9.0 | 23.5 |
| Herring oil | | |
| 293 | 5.9 | 10.8 |
| 323 | 5.7 | - |
| Linseed oil | | |
| 293 | 10.1 | 19.1 |
| 323 | 10.4 | 10.0 |
| Corn oil | | |
| 293 | 4.7 | 29.6 |
| 323 | 6.4 | 11.6 |
| Olive oil | | |
| 293 | 5.0 | 28.7 |
| 323 | 6.1 | 15.7 |
| Mineral oil | | |
| 293 | 8.4 | 30.6 |
| 323 | 8.1 | 30.5 |

**AUXILIARY INFORMATION**

**METHOD/APPARATUS/PROCEDURE:**

The oil is degassed for two hours in a vacuum.

The oil stands for 24 hours with the dry air.

The dissolved gas is removed from the oil by a Barcroft blood gas pump, and the gas volume is measured.

The oxygen content of the dissolved gas is determined.

---

Except for the mineral oil, these are unsaturated sytems which react with the oxygen. If the oils were allowed to stand for some time with the dissolved air in them, the percentage of $O_2$ fell, and in some cases only pure $N_2$ was left. Less oxygen was recovered at 323 K than at 293 K. This is in accord with ideas of autoxidation.

**SOURCE AND PURITY OF MATERIALS:**

No information.

**ESTIMATED ERROR:**

**ADDITIONAL EXPERIMENTAL VALUES:**

Air solubility in hardened whale fat (m. p. 40 - 42 $^0$C) was:

| | | |
|---|---|---|
| 323 K | 10.3 | with 82.2 % nitrogen |
| 373 K | 6.1 | with 83.1 % nitrogen |
| 378 K | 6.3 | with 91.7 % nitrogen |

after 12, 12, and 120 hours, respec.

| COMPONENTS: | ORIGINAL MEASUREMENTS: |
|---|---|
| (1) Air<br>(2) Oils | Baldwin, R. R.; Daniel, S. G.<br>*J. Appl. Chem.* 1952, *2*, 161-5.<br>*J. Inst. Petrol., London* 1953, *39*, 105-24. |

| VARIABLES: | PREPARED BY: |
|---|---|
| $T$/K = 273.15 - 373.15<br>$p_1$/kPa = 101.325 | P. L. Long<br>H. L. Clever |

EXPERIMENTAL VALUES:

| $T$/K | Bunsen Coefficient $\alpha$/cm$^3$(STP)cm$^{-3}$atm$^{-1}$ | Ostwald Coefficient $L$/cm$^3$ cm$^{-3}$ |
|---|---|---|
| Oil A1, viscosity 615 centistokes at 100 $^0$F, mean mol wt 670 | | |
| 293.15 | 0.0783 | 0.0840 |
| Oil A2, viscosity 268 centistokes at 100 $^0$F, mean molecular weight 610 | | |
| 273.15 | 0.0794 | 0.0794 |
| 293.15 | 0.0792 | 0.0850 |
| 333.15 | 0.0835 | 0.1015 |
| 373.15 | 0.0818 | 0.112 |
| Oil A3, viscosity 181 centistokes at 100 $^0$F, mean molecular weight 570 | | |
| 293.15 | 0.0843 | 0.0926 |
| Oil A4, viscosity 80.3 centistokes at 100 $^0$F, mean molecular weight 530 | | |
| 293.15 | 0.0878 | 0.0942 |
| Oil A5, viscosity 34.9 centistokes at 100 $^0$F, mean molecular weight 400 | | |
| 273.15 | 0.0920 | 0.0920 |
| 293.15 | 0.0903 | 0.0969 |
| 333.15 | 0.0919 | 0.1121 |
| 373.15 | 0.0926 | 0.1265 |
| Oil B, viscosity 260 centistokes at 100 $^0$F, mean molecular weight 630 | | |
| 293.15 | 0.0752 | 0.0807 |

METHOD/APPARATUS/PROCEDURE:

1. Degassing. The liquid is deaerated by passage through a continously evacuated vessel. The liquid is saturated with the gas, then degassed again.
2. Saturation of the liquid with the gas. The gas is passed through a liquid air trap to remove water and carbon dioxide, then through a thermostated coil to bring it to the temperature of the experiment, and then bubbled through the liquid for up to five hours.
3. To determine the amount of gas dissolved in the saturated liquid, the saturated liquid is placed in an evacuated vessel and buret system. The gas released by the liquid is transferred to the buret. Five transfer operations recover all but a trace of the gas. The last trace (2 - 5 %) is removed by boiling the liquid.

SOURCE AND PURITY OF MATERIALS:

(1) Air. From compressed air line. Dried and freed from $CO_2$.

(2) Oils. The oils conform to DTD-472B specifications. The mol wts were determined by f.p. depression in naphthalene to ± 5 %. Other properties are given in 1953 paper.

ESTIMATED ERROR:

$\delta p/p = \pm 0.01$
$\delta\alpha/\alpha = \pm 0.01$

**COMPONENTS:**

(1) Air

(2) Mineral oils

**ORIGINAL MEASUREMENTS:**

Clark, F. M.

*J. Franklin Inst.* 1933, *215*, 39 - 67.

**VARIABLES:**

$T/T$ = 298, 373

$p_1$/kPa = 101.3 (760 mmHg)

**PREPARED BY:**

H. L. Clever

**EXPERIMENTAL VALUES:**

| Temperature $t/^{0}C$ | $T/K$ | Per cent Dissolved by Volume Under Atmospheric Conditions[1] % Nitrogen | % Oxygen | % Air |
|---|---|---|---|---|
| Asphaltic Mineral Oil, Saybolt Viscosity/s is 100 at 37.8 $^{0}C$ and 150 at 100 $^{0}C$. | | | | |
| 25 | 298 | 6.2 | 2.1 | 8.3 |
| 100 | 373 | 4.64 | 0.16 | 4.8 |
| Paraffin Mineral Oil, Saybolt Viscosity/s is 150 at 100 $^{0}C$. | | | | |
| 100 | 373 | 6.4 | 1.2 | 7.6 |
| Mixed Mineral Oil, Saybolt Viscosity/s is 100 at 37.8 $^{0}C$. | | | | |
| 25 | 298 | 4.6 | 0.6 | 5.2 |
| Transformer Type Mineral Oil | | | | |
| 25 | 298 | | | 10.8 |
| 100 | 373 | | | 12.7 |

[1] The per cent volume of gas dissolved is based on gas volumes reduced to 25 $^{0}C$ and 760 mmHg with oil volumes at the temperature of the test.

**AUXILIARY INFORMATION**

**METHOD/APPARATUS/PROCEDURE:**

The gas is removed from the air saturated oil by displacement with carbon dioxide. The gas is collected over concentrated KOH solution which absorbes the carbon dioxide.

In addition to the data above there are figures in the paper which (i) show air solubility in transformer oil as a function of temperature from -15 to 120 $^{0}C$, and (ii) show the rate of air take-up by transformer mineral oil at 80 $^{0}C$.

The paper also contains data on the solubility of oxygen, nitrogen, hydrogen and carbon dioxide in transformer mineral oil.

**SOURCE AND PURITY OF MATERIALS:**

1. Air. No information.
2. Oils. See descriptions in the table above.

**ESTIMATED ERROR:**

The error in the per cent volume solubility of air may be 5 - 10 %. (compiler)

**REFERENCES:**

| COMPONENTS: | ORIGINAL MEASUREMENTS: |
|---|---|
| (1) Air<br>(2) Mineral oils | Gemant, A.<br>*Trans. Faraday Soc.* 1936, *32*, 694 - 701. |
| VARIABLES:<br>$T/K$ = 288 - 293, 353<br>$p_1$/kPa = 101.3 (1 atm) | PREPARED BY:<br>H. L. Clever |

EXPERIMENTAL VALUES:

| $t_1/{}^0C$ | Oil Mass/g | Oil Specific Volume/g cm$^{-1}$ at $t_1$ | Oil Specific Volume/g cm$^{-1}$ at 80$^0$C | Air Absorbed V/cm$^3$(STP) | v[1] | Bunsen[2] Coefficient α at $t_1$ | k[3] |
|---|---|---|---|---|---|---|---|
| Silvertown Lubricants, Russian origin, British Standard Specification A.30. | | | | | | | |
| 16 | 108.95 | 1.164 | 1.225 | 12.48 | 0.64 | 0.092 | -0.001 |
| Silvertown Lubricants, USA Origin (Kansas and Oklahoma), British Standard Specification A.0. | | | | | | | |
| 18 | 113.0 | 1.181 | 1.243 | 12.62 | 0.63 | 0.088 | -0.002 |
| Silvertown Lubricants, Russian Origin, British Standard Specification B.30. | | | | | | | |
| 15 | 111.35 | 1.154 | 1.214 | 12.04 | 0.66 | 0.088 | +0.001 |
| Silvertown Lubricants, USA Origin (Pennsylvania), British Standard Specification B.0. | | | | | | | |
| 16 | 108.60 | 1.176 | 1.239 | 12.00 | 0.55 | 0.088 | -0.005 |
| Silvertown Lubricants, British Standard Specification A.30. | | | | | | | |
| 20 | 107.3 | 1.165 | 1.224 | 12.25 | 0.52 | 0.092 | -0.007 |
| Shell-Mex and B. P., 21 $^0$C viscosity 0.060 poise, 60 $^0$C viscosity 0.022 poise. | | | | | | | |
| 20 | 106.2 | 1.211 | 1.280 | 14.70 | 0.45 | 0.106 | -0.022 |
| Shell-Mex and B. P., 21 $^0$C viscosity 2.0 poise, 60 $^0$C viscosity 0.21 poise. | | | | | | | |
| 20 | 115.3 | 1.083 | 1.135 | 9.80 | 0.66 | 0.073 | +0.020 |
| Vacuum Oil Co., Mobiloil BB viscosity data. | | | | | | | |
| 20 | 111.6 | 1.139 | 1.189 | 10.60 | 0.46 | 0.078 | -0.003 |

AUXILIARY INFORMATION

METHOD/APPARATUS/PROCEDURE:

About 130 cm$^3$ of oil was placed in a glass bulb, weighed, and the oil degassed by vacuum pumping and vigorous shaking for five hours.

The degassed oil was brought into contact with air. The system shaken for 20 minutes and the air uptake measured on a Hg manometer system at the temperature $t_1$.

In a second experiment the difference in air absorbed between $t_1$ and 80 $^0$C was directly measured.

The Bunsen coefficient at $t_1$ and its temperature coefficient over the $t_1$ to 80 $^0$C temperature interval, k, were given. Thus $\alpha_{80} = (1 + k)\alpha_{t_1}$.

[1] v/cm$^3$ is the additional air absorbed between $t_1$ and 80 $^0$C.

[2] Bunsen α/cm$^3$(STP)cm$^{-3}$atm$^{-1}$

[3] k the Bunsen coefficent temperature coefficient for the $t_1$ to 80 $^0$C temperature interval.

SOURCE AND PURITY OF MATERIALS:

(1) Air. No information.

(2) Mineral oils. See entries in the table above.

ESTIMATED ERROR:

REFERENCES:

**COMPONENTS:**

(1) Air

(2) Fuels, see below

**ORIGINAL MEASUREMENTS:**

Schweitzer, P.H.; Szebehely, V.G.

*J. Appl. Phys.* 1950, *21*, 1218-24.

**VARIABLES:**

T/K = 294
P/kPa = 101

**PREPARED BY:**

R. Battino

**EXPERIMENTAL VALUES:**

| Kinetic Viscosity /centistokes | $\rho^a$/g cm$^{-3}$ | $\gamma^b$/g cm$^{-1}$ | Solubility Constant (%) | $10^2L^c$ |
|---|---|---|---|---|
| | | Diesel Fuel | | |
| 4.4 | | | 11.98 | 11.98 |
| | | Aircraft Engine Fuel | | |
| 1.67 | 0.780 | 0.0239 | 17.20 | 17.20 |
| 0.6350 | 0.723 | 0.0191 | 22.80 | 22.80 |
| 0.6350 | 0.692 | 0.0181 | 25.14 | 25.14 |

[a] Density at 294 K.

[b] Surface tension at 294 K. (Authors presumably mean unit to be g $s^{-2}$.)

[c] Calculated by compiler. Ostwald coefficient (= solubility constant/100). Measurements at 70°F (294 K).

AUXILIARY INFORMATION

**METHOD/APPARATUS/PROCEDURE:**

The method consisted of shaking a container partly filled with a known amount of liquid and taking accurate pressure measurements before and after equilibration.

**SOURCE AND PURITY OF MATERIALS:**

No details given.

**ESTIMATED ERROR:**

$\delta L/L = \pm 0.05$, compiler's estimate

**REFERENCES:**

**COMPONENTS:**

(1) Air

(2) Hydrocarbon fuels

Kerosene

**ORIGINAL MEASUREMENTS:**

Baldwin, R. R.; Daniel, S. G.

*J. Appl. Chem.* 1952, *2*, 161-5.

*J. Inst. Petrol., London* 1953, *39*, 105-24.

**VARIABLES:**

$T$/K = 273-293

$p_1$/kPa = 101.325

**PREPARED BY:**

P. L. Long

H. L. Clever

**EXPERIMENTAL VALUES:**

| $T$/K | Bunsen Coefficient $\alpha$/cm$^3$(STP) cm$^{-3}$atm$^{-1}$ | Ostwald Coefficient $L$/cm$^3$ cm$^{-3}$ |
|---|---|---|
| 100 Octane fuel C, density $_4^{20}$ 0.7096 | | |
| 293.15 | 0.243 | 0.261 |
| 100 Octane fuel D, density $_4^{20}$ 0.7188 | | |
| 293.15 | 0.235 | 0.252 |
| Kerosene, density $_4^{20}$ 0.822 | | |
| 273.15 | 0.139 | 0.139 |
| 293.15 | 0.136 | 0.146 |

| | Olefins | Aromatics | Naphthalenes | Paraffins |
|---|---|---|---|---|
| 100 octane fuel C | 2 % | 2-3 % | 22 % | 73-74 % |
| 100 octane fuel D | 1.5 % | 12 % | 14 % | 72.5 % |
| kerosene | 3 % | 22 % | 53 % | 22 % |

In addition the 100 octane fuels C and D contained 4.0 and 4.8 cm$^3$ tetraethyl lead per imperial gallon, respectively.

**AUXILIARY INFORMATION**

**METHOD/APPARATUS/PROCEDURE:**

1. Degassing. The liquid is deaerated by passage through a continously evacuated vessel. The liquid is saturated with the gas, then degassed again.

2. Saturation of liquid with gas. The gas is passed through a liquid air trap to remove water and carbon dioxide, then through a thermostated coil to bring the gas to the temperature of the experiment, and then bubbled through the liquid up to five hours.

3. To determine the amount of gas dissolved, the saturated liquid is placed in a vessel attached to a calibrated buret. The vessel and buret are initially evacuated. The gas released by the liquid is transferred to the buret. Five operations and transfers recover all but a trace of the gas. The last trace (2 - 5 %) is removed by boiling the liquid.

**SOURCE AND PURITY OF MATERIALS:**

(1) Air. From compressed air line. Dried and freed from $CO_2$.

(2) 100 Octane fuels C and D. Molecular weight estimated to be 100. Composition as above.

Kerosene. Mean molecular weight measured, obtained 165. Composition as above.

Other liquid properties in paper.

**ESTIMATED ERROR:**

$\delta p/p$ = ± 0.01

$\delta\alpha/\alpha$ ± 0.01

**REFERENCES:**

**COMPONENTS:**

(1) Air

(2) Aviation fuels, Kerosenes

**ORIGINAL MEASUREMENTS:**

Derry, L. D.; Evans, E. B.; Faulkner, B. A.; Jelfs, E. C. G.

*J. Inst. Petrol.* 1952, *38*, 475-525.

**VARIABLES:**

T/K = 289 K

**PREPARED BY:**

R. Battino

**EXPERIMENTAL VALUES:**

| Fuel No. | Specific Gravity[a] 60/60°F | $\alpha^b$/cm³ (STP) cm⁻³atm⁻¹ |
|---|---|---|
| | 60°F (288.7K) | |
| 14 | 0.7990 | 0.1452 |
| 15 | 0.8120 | 0.1376 |
| 16 | 0.8030 | 0.1322 |
| 17 | 0.7890 | 0.1398 |
| 18 | 0.7820 | 0.1353 |
| 19 | 0.7665 | 0.1528 |
| 20 | 0.8285 | 0.1231 |
| 21 | 0.8080 | 0.1391 |

[a] At 288.7 K.

[b] Bunsen coefficient. Paper reported percentage solubility which they defined as 100$\alpha$.

continued on following page

**AUXILIARY INFORMATION**

**METHOD/APPARATUS/PROCEDURE:**

Used an apparatus similar to that described in CRC Method F-16-245 (1) for the determination of the air content of gasolines.

**SOURCE AND PURITY OF MATERIALS:**

(1) No details given.

(2) The properties of the fuels are given in table II in the original paper.

**ESTIMATED ERROR:**

$\delta\alpha/\alpha$ = ±0.02, compiler's estimate.

**REFERENCES:**

1. CRC Handbook, 1946 ed., p. 222.

| COMPONENTS: | ORIGINAL MEASUREMENTS: |
|---|---|
| (1) Air<br>(2) Aviation fuels, Kerosenes | Derry, L. D.; Evans, E. B.; Faulkner, B. A.; Jelfs, E. C. G.<br>*J. Inst. Petrol.* 1952, *38*, 475-525. |

EXPERIMENTAL VALUES:

continued

| Fuel Number | $\alpha^a$/cm³ (STP) cm⁻³ atm⁻¹<br>0°F (255.4K) | 60°F (288.7K) | 65°F (291.5K) | 108°F (315.4K) | 120°F (322.0K) |
|---|---|---|---|---|---|
| 14 | 0.161 | 0.145 | | | 0.155 |
| 16 | 0.136 | 0.132 | | | 0.143 |
| 19 | | | 0.152 | | |
| 22 | | | 0.147 | | |
| 23 | | | 0.132 | | |
| 21 | | | 0.139 | | |
| 18 | 0.141 | | 0.135 | 0.135 | |
| 20 | | | 0.123 | | |

[a] Bunsen coefficient. Calculated by compiler from a "solubility coefficient", k, where $\alpha = 7.60k$.

**COMPONENTS:**

(1) Air

(2) Hydrocarbon fuels and oil

**ORIGINAL MEASUREMENTS:**

Logvinyuk, V. P.; Makarenkow, V. V.; Malyshev, V. V.; Panchenkov, G. M.

*Khim. Tekhnol. Topl. Masel.* 1970, *15* (5), 27-9.

*Chem. Technol. Fuels Oils (Eng.trans.)* 1970, *15*, 353-5.

**VARIABLES:**

$T$/K = 293.15
$p_1$/kPa = 101.325

**PREPARED BY:**

S. A. Johnson
H. L. Clever

**EXPERIMENTAL VALUES:**

| $T$/K | Bunsen Coefficient $\alpha$/cm$^3$(STP)cm$^{-3}$atm$^{-1}$ | | Ostwald Coefficient $L$/cm$^3$ cm$^{-3}$ |
|---|---|---|---|
| Hydrocarbon fuel, T-1, density $\rho_4^{20}$/g cm$^{-3}$ = 0.816 | | | |
| 293.15 | 0.139 | | 0.149 |
| Hydrocarbon fuel, TS-1, density $\rho_4^{20}$/g cm$^{-3}$ = 0.775 | | | |
| 293.15 | 0.148 | | 0.159 |
| Hydrocarbon fuel, T-6, density $\rho_4^{20}$/g cm$^{-3}$ = 0.84 | | | |
| 293.15 | 0.132 | | 0.142 |
| Hydrogenated fuel, density $\rho_4^{20}$/g cm$^{-3}$ = 0.832 | | | |
| 293.15 | 0.132 | | 0.142 |
| Oil, MK-8, density $\rho_4^{20}$/g cm$^{-3}$ = 0.855 | | | |
| 293.15 | 0.091 | | |
| | 0.093 | | |
| | 0.094 | | |
| | 0.095 | | |
| | 0.093 | | |
| | 0.093 | Av. | 0.100 |

The compilers calculated the Ostwald coefficients.

**AUXILIARY INFORMATION**

**METHOD/APPARATUS/PROCEDURE:**

The details of the apparatus and procedure are described in reference (1) which was not available to the compilers.

The liquid is saturated with air to equilibrium at 20 °C and atmospheric pressure. The gas saturated oil in poured into another apparatus and vacuumized, liberation of the gas being carried out as described in (1) After returning the apparatus to atmospheric pressure, the volume of the liberated gas is measured.

**SOURCE AND PURITY OF MATERIALS:**

No information other than density data in the table above.

**ESTIMATED ERROR:**

$\delta\alpha/\alpha$ = ± 0.06 or less (Authors)

**REFERENCES:**

1. Gogitidze, L. D.; Logvinyuk, V. P. Makarenkov, V. V.; Panchenkov, G.M. Malyshev, V.V.; Yakovlevskii, V. V. *Method of Evaluating the Operating Properties of Jet Fuels and Lubricating Materials (in Russian), Mashinostroenie* [See *Izv. Vyssh. Uchebn. Zaved. Mashinostr.*] 1966

**REFERENCES** (Continued) 2. Borisov, V.D.; Gogitidze, L.D.; Logvinyuk, V.P.; Makarenkov, V.V.; Malyshev, V.V.; Panchenkov, G.M.; Yakovlevskii, V.V. *Apparatus for the determination of the amount of gas dissolved in a fluid,* See *Chem. Abstr.* 1966, *65*, 16041d.

**COMPONENTS:**

(1) Air

(2) Lubricating Oils, see below

**ORIGINAL MEASUREMENTS:**

Schweitzer, P. H.; Szebehely, V. G.

*J. Appl. Phys.* 1950, *21*, 1218-24.

**VARIABLES:**

T/K = 294
P/kPa = 101

**PREPARED BY:**

R. Battino

**EXPERIMENTAL VALUES:**

| Kinetic Viscosity /centistokes | $\rho^a$/g cm$^{-3}$ | $\gamma^b$/g cm$^{-1}$ | Solubility Constant (%) | $10^2L^c$ |
|---|---|---|---|---|
| | | HEAVY LUBRICATING OIL | | |
| 823-974 | 0.882 | 0.0299 | 7.75 | 7.74 |
| 823-974 | | | 8.14 | 8.14 |
| 762-974 | | | 8.15 | 8.15 |
| 762-974 | | | 8.24 | 8.24 |
| 762-974 | | | 8.61 | 8.61 |
| 562-670 | 0.883 | 0.0308 | 8.62 | 8.62 |
| 325-411 | 0.880 | 0.0301 | 9.05 | 9.05 |
| 249-346 | 0.869 | 0.0293 | 9.11 | 9.11 |
| 88.7 | | | 9.18 | 9.18 |
| | | LIGHT LUBRICATING OIL | | |
| 17.5 | 0.870 | 0.0274 | 9.70 | 9.70 |
| 17.5 | 0.870 | 0.0274 | 9.95 | 9.95 |
| 13.5 | 0.938 | 0.0259 | 10.72 | 10.72 |
| 17.5 | 0.867 | 0.0286 | 11.30 | 11.30 |

[a] Density at 294 K.

[b] Surface tension at 294 K. (Presumably the unit is really g s$^{-2}$.)

[c] Calculated by compiler. Ostwald coefficient (= solubility constant/100). Measurements at 70°F (294 K).

**AUXILIARY INFORMATION**

**METHOD/APPARATUS/PROCEDURE:**

The method consisted of shaking a container partly filled with a known amount of liquid and taking accurate pressure measurements before and after equilibration.

**SOURCE AND PURITY OF MATERIALS:**

No details given

**ESTIMATED ERROR:**

$\delta L/L = \pm 0.05$, compiler's estimate

**REFERENCES:**

**COMPONENTS:**

(1) Air

(2) Fuel RJ-1

**ORIGINAL MEASUREMENTS:**

Findl, E.; Brande, H.; Edwards, H.

U. S. Dept. Comm., Office Tech. Ser. Report No. *AD 274 623*, 1960, 216 pp.

*Chem. Abstr.* 1963, *58*, 6628c.

**VARIABLES:**

$T$/K = 310.9 - 533.2

$p_1$/kPa = 172.4 - 274.4 (25 - 398 psia)

**PREPARED BY:**

H. L. Clever

**METHOD/APPARATUS/PROCEDURE:**

The solubility apparatus was constructed of two concentric glass tubes sealed at the end by invar steel caps and silicone O-rings. The caps had attachments for flexible tubes to admit solvent and gas, and to circulate the thermostating fluid.

The inner tube was of *TRUE-BORE* glass tubing. It contained the solvent and a glass ball 0.0083 inch smaller than the tube diameter. The ball was used for both stirring on gas dissolution and viscosity measurements on the gas saturated solution. A steel scale beside the inner tube allowed determination of the solvent volume to within 0.3 per cent.

The thermostating liquid (white minearl oil) circulated in the space between the two tubes.

The solvent was initially saturated with air. The amount of dissolved air was known from a previous Orsat analysis.

The solute gas was contained in a steel bottle at 700 psig maintained at a temperature of 150 ± 1°F. There were two pressure gages. A differential gage between the gas bottle and a reference bottle to measure the pressure drop when gas was admitted to the apparatus, and a Bourdon gage to measure the final total pressure in the cell.

The solvent was added to the cell until the vapor/liquid ratio reached about 0.15. The cell and its contents were heated to the temperature of the measurement. The gas was admitted, and the test cell rocked for 10 minutes more than the time required to reach a stable pressure. The amount of gas dissolved was calculated from the pressure drop in the gas reservoir and the total pressure in the cell. The partial pressure of the gas in the

**AUXILIARY INFORMATION**

**METHOD/APPARATUS/PROCEDURE:** (continued)

cell was calculated taking into account the initial dissolved air, the solvent vapor pressure, and the compressibility of the gas. Both the Ostwald coefficient and the volume of gas at standard conditions per unit weight of solvent were calculated.

The results were presented in large scale graphs. The results on the data sheets are values read from the graphs by the compiler.

Other experimental details and results were in an earlier report of the authors (1).

**SOURCE AND PURITY OF MATERIALS:**

(1) Air. Stated to be of high purity.

(2) Fuel. The fuel was specified as MIL-F-25558 Grade RJ-1 formerly designated as Shell UMF, Grade C. A table of fuel characteristics was included in the report.

**ESTIMATED ERROR:**

$\delta T$/°F = ± 1

$\delta p$/lb in$^{-2}$ = ± 1

$\delta L/L$ = ± 0.02 (minimum), max. ≃ 0.05

**REFERENCES:**

1. Schlagel, L. A.; Findl, E.; Edwards, H. Ing Er. Rept. 183, Thompson Products, Inc. Inglewood Lab., Inglewood, CA, USA, Aug 19, 1955.

| COMPONENTS: | ORIGINAL MEASUREMENTS: |
|---|---|
| (1) Air<br>(2) Fuel RJ-1 | Findl, E.; Brande, H.; Edwards, H.<br>U. S. Dept. Comm., Office Tech. Ser. Report No. *AD 274 623*, 1960, 216 pp.<br>*Chem. Abstr.* 1963, *58*, 6628c. |

EXPERIMENTAL VALUES: continued

| Temperature $t$/°F | $T$/K | Air Pressure $p_1$/psia | Solubility $cm^3$(STP)$g^{-1}$ | Ostwald Coefficient $L/cm^3 cm^{-3}$ |
|---|---|---|---|---|
| 100 | 310.9 | 16 | -- | 0.116 |
| | | 17 | -- | 0.120 |
| | | 63 | 0.40 | 0.119 |
| | | 63 | 0.52 | 0.121 |
| | | 113 | 0.92 | 0.115 |
| | | 113 | 0.95 | 0.121 |
| | | 214 | 1.68 | 0.112 |
| | | 214 | 1.76 | 0.120 |
| | | 313 | 2.57 | 0.121 |
| | | 414 | 3.47 | 0.119,0.126 |
| | | | | (0.119)[1] |
| 200 | 366.5 | 17.5 | -- | 0.153 |
| | | 18 | -- | 0.167 |
| | | 60 | -- | 0.136 |
| | | 62.5 | -- | 0.169 |
| | | 113 | 1.11 | 0.155 |
| | | 113 | 1.14 | 0.163 |
| | | 212 | 2.10 | 0.155 |
| | | 212 | 2.12 | 0.159 |
| | | 311 | 2.99 | 0.149 |
| | | 313 | -- | 0.149 |
| | | 411 | 4.04 | 0.157 |
| | | 412.5 | 4.03 | 0.158 |
| | | | | (0.155)[1] |
| 300 | 422.0 | 17.5 | -- | 0.264 |
| | | 59 | 0.79 | 0.230 |
| | | 110 | 1.61 | 0.263 |
| | | 112 | 1.52 | 0.241 |
| | | 207.5 | -- | 0.235 |
| | | 210 | 3.07 | 0.241 |
| | | 309 | 4.17 | 0.235 |
| | | 310 | 4.37 | 0.246 |
| | | 408 | 5.67 | 0.240 |
| | | 410 | 5.72 | 0.242 |
| | | | | (0.240)[1] |
| 400 | 477.6 | 102.5 | 1.50 | 0.270 |
| | | 203 | 3.00 | 0.275 |
| | | 303 | 4.50 | 0.275 |
| | | 402.5 | 5.99 | 0.280 |
| | | 403 | -- | 0.275 |
| | | | | (0.275)[1] |
| 500 | 533.2 | 87.5 | 1.43 | 0.314 |
| | | 87.5 | 1.46 | 0.316 |
| | | 190 | 3.08 | 0.314 |
| | | 190 | -- | 0.316 |
| | | 287.5 | 4.62 | 0.314 |
| | | 387 | -- | 0.317 |
| | | 390 | -- | 0.316 |
| | | | | (0.315)[1] |

[1] From author's line through the Ostwald coefficient values. psia ≡ pounds per square inch absolute. One lb $in^{-2}$ ≡ 6.89476 kPa.

COMPONENTS:

(1) Air

(2) Jet A(79) fuel

ORIGINAL MEASUREMENTS:

Rupprecht, S. D.; Faeth, G. M.

NASA CR 1981, NASA-CR-3422, 111 pp.

*Chem. Abstr.* 1981, 95, 153333x.

VARIABLES:

T/K = 295-373
P/MPa = 1-10

PREPARED BY:

R. Battino

EXPERIMENTAL VALUES:

| T/K | $P^a$/MPa | $x_1$ |
|---|---|---|
| 295.82 | 1.03 | 0.0118 |
| 373.15 | 1.03 | 0.0132 |
| 296.65 | 2.07 | 0.0239 |
| 373.15 | 2.07 | 0.0278 |
| 298.15 | 4.82 | 0.0571 |
| 373.15 | 4.82 | 0.0635 |
| 297.15 | 10.34 | 0.1184 |
| 373.15 | 10.34 | 0.1306 |

[a] Total pressure.

AUXILIARY INFORMATION

METHOD/APPARATUS/PROCEDURE:

A 1000 $cm^3$ chamber is filled with ca. 500 $cm^3$ of liquid. Gas is added to the desired pressure. The chamber is agitated to attain equilibrium. Temperatures are read via thermocouples and pressures via Heisse bourdon tube gages. A liquid sample is removed and allowed to de-pressurize at atmospheric pressure. The solubility is calculated from the measurement of liquid and gas volumes of the de-pressurized sample. Residual dissolved gas is considered to be negligible.

SOURCE AND PURITY OF MATERIALS:

(1) No details given.

(2) 0.8381 sp. gr. at 60°F.
Boiling range is 434.82-600.93K.

ESTIMATED ERROR:

$\delta x_1/x_1 = \pm 0.03$, compiler's estimate

REFERENCES:

COMPONENTS:

(1) Air

(2) Jet A(80) fuel

ORIGINAL MEASUREMENTS:

Rupprecht, S. D.; Faeth, G. M.

NASA CR 1981, NASA-CR-3422, 111 pp.

*Chem. Abstr.* 1981, 95, 153333x.

VARIABLES:

T/K = 297-373
P/MPa = 1-10

PREPARED BY:

R. Battino

EXPERIMENTAL VALUES:

| T/K | $p^a$/MPa | $x_1$ |
|---|---|---|
| 297.65 | 1.03 | 0.0139 |
| 373.15 | 1.03 | 0.0150 |
| 298.15 | 2.07 | 0.0289 |
| 373.15 | 2.07 | 0.0312 |
| 297.15 | 4.82 | 0.0707 |
| 373.15 | 4.82 | 0.0733 |
| 297.40 | 10.34 | 0.1512 |
| 373.15 | 10.34 | 0.1540 |

[a] Total pressure.

AUXILIARY INFORMATION

METHOD/APPARATUS/PROCEDURE:

A 1000 $cm^3$ chamber is filled with ca. 500 $cm^3$ of liquid. Gas is added to the desired pressure. The chamber is agitated to attain equilibrium. Temperatures are read via thermocouples and pressures via Heisse bourdon tube gages. A liquid sample is removed and allowed to de-pressurize at atmospheric pressure. The solubility is calculated from the measurement of liquid and gas volumes of the de-pressurized sample. Residual dissolved gas is considered to be negligible.

SOURCE AND PURITY OF MATERIALS:

(1) No details given.

(2) 0.8095 sp. gr. 60°F. Boiling range 444.26-547.59K.

ESTIMATED ERROR:

$\delta x_1/x_1 = \pm 0.03$, compiler's estimate

REFERENCES:

## SYSTEM INDEX

Underlined page numbers refer to evaluation text and those not underlined to compiled tables. All compounds are listed as in Chemical Abstracts. For example, toluene is listed as benzene, methyl-, dimethylsulfoxide is listed as methane, sulfinylbis-. Since nitrogen is usually one of the compounds it is omitted in the listing.


A

Acetamide (aqueous) 84, 93
Acetamide, *N*-methyl- 261
Acetic acid 221
Acetic acid (aqueous) 45-47, 54, 77
Acetic acid (ternary and multicomponent) 357 - 360
Acetic acid, cobalt salt (aqueous) 45-47, 54, 77
Acetic acid, copper (2+) salt (aqueous) 45-47, 54, 77
Acetic acid, ethyl ester 224
Acetic acid, ethyl ester (ternary) 105, 107, 108
Acetic acid, manganese (2+) salt (aqueous) 45-47, 54, 77
Acetic acid, mercury (2+) salt (aqueous) 45-47, 54, 77
Acetic acid, methyl ester 223
Acetic acid, 2-methylpropyl ester 225
Acetic acid, nickel (2+) salt (aqueous) 45-47, 54, 77
Acetic acid, pentyl ester 226
Acetic acid, sodium salt (aqueous) 45-47, 54, 77
Acetic acid, zinc (2+) salt (aqueous) 45-47, 54, 77
Acetone, see 2-propanone
*D*-Alanine (aqueous) 308
Aluminium oxide, hydrated 45-48, 60
Ammonia 476, 477 - 483
AMSCO 123-5 330
Amyl acetate, see acetic acid, pentyl ester
Aniline, see benzenamine
Animal and vegetable oils, see under oil
Argon (multicomponent) 482

B

Barium chloride (aqueous) 45-47, 49, 62, 63
Benzenamine 265, 266
Benzene 162, 163 - 168, 252, 443, 444
Benzene (ternary) 104, 111, 112, 300 - 302
Benzene, chloro- 240
Benzene, dimethyl- 171
Benzene, hexafluoro- 242
Benzene, methyl- 162, 169, 170, 445
Benzene, 1,1'-methylenebis- 173
Benzene, nitro- 264, 265
1,1-Bicyclohexyl 161
Blood, cod 292
Blood, dog 464
Blood, eel 292
Blood, human, plasma 284, 285
Blood, human 286, 292
Blood, ox 289, 465
Blood, ox, plasma 289, 295
Blood, rabbit 293
Blood, trout 292
Borane, trimethyl- 497
Brain, rabbit 293
1-Bromoheptane, see heptane, 1-bromo-
1-Bromo-1,1,2,2,3,3,4,4,5,6,6,6-dodecafluoro-5-(trifluoromethyl)-hexane, see hexane, 1-bromo-1,1,2,2,3,3,4,4,5,6,6,6-dodecafluoro-5-(trifluoromethyl)-

(cont.)

System Index (cont.)


1-Bromo-1,1,2,2,3,3,4,4,5,5,6,6,7,7,8,8,8-heptadecafluorooctane see octane, 1-bromo-1,1,2,2,3,3,4,4,5,5,-6,6,7,7,8,8,8-heptadecafluoro-
1,3-Butadiene 418
1-Butanamine, 1,1,2,2,3,3,4,4,4-nonafluoro-N,N-bis(nonafluorobutyl)- + air 532
1-Butanamine, 1,1,2,2,3,3,4,4,4-nonafluoro-N,N-bis(nonafluorobutyl)- 267 - 269
Butane 409, 410 - 413, 418
Butane (ternary) 385, 386, 414, 415
*iso*-Butane, see propane, 2-methyl-
Butane, 1,1,2,2,3,3,4,4-octafluoro-1,4-bis(1,2,2,2-tetrafluoro-1-(trifluoromethyl)-ethoxy)- 250
Butane, 2-methyl- 421
(*R*,*R*)-1,2,3,4-butanetetrol (aqueous) 309
1-Butanol 175, 177, 178, 197 - 200
Butter oil, see oil, butter
*iso*-Butyl acetate, see acetic acid, 2-methylpropyl ester

C

Calcium chloride (aqueous) 346
Carbonic acid, dipotassium salt (aqueous) 45-47, 55, 81, 82
Carbonic acid, disodium salt (aqueous) 71
Carbonic acid, monosodium salt (aqueous) 71
Carbon dioxide 488, 489 - 494
Carbon dioxide (multicomponent) + air 516
Carbon dioxide (multicomponent) 76, 466, 495, 498 - 501
Carbon disulfide 253, 254 - 257
Carbon tetrachloride, see methane, tetrachloro-
Caroxin-D, see butane, 1,1,2,2,3,3,4,4-octafluoro-1,4-bis(1,2,2,2-tetrafluoro-1-(trifluoromethyl)ethoxy)-
Caroxin-F, see hexane, 1,1,1,2,2,3,3,4,4,5,5,6,6-tridecafluoro-6-(1,2,2,2-tetrafluoro-1-(trifluoromethyl)ethoxy)-
Cephalin 302
Chloral hydrate, see 1,1'-ethanediol, 2,2,2-trichloro-
Chloride, sodium, see sodium chloride
Chlorine hexafluoride 474
Chlorine pentafluoride 472, 473
Chlorine trifluoride 471
Chlorobenzene, see benzene, chloro-
Chlorodifluoromethane, see methane, chlorodifluoro-
1-Chloro-1,1,2,2,3,3,4,4,5,6,6,6-dodecafluoro-5-(trifluoromethyl)-hexane, see hexane, 1-chloro-1,1,2,2,3,3,4,4,5,6,6,6-dodecafluoro-5-(trifluoromethyl)-
Chloroform, see methane, trichloro-
1-Chlorohexane, see hexane, 1-chloro-
Cholest-5-en-3β-ol 300
Cobalt acetate, see acetic acid, cobalt (2+) salt
Cod blood, see blood, cod
Copper acetate, see acetic acid, copper (2+) salt
Corn oil, see oil, corn
Cotton seed oil, see oil, cottonseed
Cyclohexane 148, 149 - 154, 439, 440
Cyclohexane, decafluorobis(trifluoromethyl)- 246
Cyclohexane (ternary) 104, 117, 118, 353, 354
Cyclohexane, *cis*-1,2-dimethyl- 157
Cyclohexane, *cis*-1,3-dimethyl- 160
Cyclohexane, *cis*-1,4-dimethyl- 159
Cyclohexane, *trans*-1,2-dimethyl- 158


(cont.)

System Index (cont.)


| | |
|---|---|
| Cyclohexane, *trans*-1,3-dimethyl- | 160 |
| Cyclohexane, *trans*-1,4-dimethyl- | 159 |
| Cyclohexane, methyl- | 155, 441, 442 |
| Cyclohexane, methyl- (ternary) | 355, 356 |
| Cyclohexane, undecafluoro(trifluoromethyl)- | 244 |
| Cyclohexanol | 207 |
| Cyclohexanol (multicomponent) | 357 |
| Cyclohexanone | 220, 449, 450 |
| Cyclohexanone (ternary and multicomponent) | 357 - 360 |
| Cyclohexene | 154 |
| Cyclooctane | 156 |
| Cyclotetrasiloxane, octamethyl- | 275 |

D

| | |
|---|---|
| Decane + air | 517 |
| Decane | 119 - 121, 131, 139, 140, 419, 435, 436 |
| 1-Decanol | 176 - 178, 213, 214, 324 |
| Detergent (aqueous) | 352 |
| Dichlorodifluoromethane, see methane, dichlorodifluoro- | |
| Diborane | 496 |
| Diethyl ether, see ethane, 1,1'-oxybis- | |
| Dihydropyran, see pyran, dihydro- | |
| 1,2-Dimethylcyclohexane, see cyclohexane, 1,2-dimethyl- | |
| 1,3-Dimethylcyclohexane, see cyclohexane, 1,3-dimethyl- | |
| 1,4-Dimethylcyclohexane, see cyclohexane, 1,4-dimethyl- | |
| Dimethylsulfoxide, see methane, sulfinylbis- | |
| Dimethyl glutarate, see pentanedioic acid, dimethyl ester | |
| 1,1-Dimethylhydrazine, see hydrazine, 1,1-dimethyl- | |
| 1,2-Dimethylhydrazine, see hydrazine, 1,2-dimethyl- | |
| 1,5-Dimethyl-2-pyrrolidinone, see 2-pyrrolidinone, 1,5-dimethyl- | |
| Dinitrogen tetroxide, see nitrogen oxide | |
| 1,4-Dioxane | 229 |
| 1,3-Dioxolan-2-one, 4-methyl- | 448 |
| Diphenylmethane, see benzene, 1,1'-methylenebis- | |
| Dipropyl ether, see propane, 1,1'-oxybis- | |
| DMSO, see methane, sulfinylbis- | |
| Dodecane + air | 519 |
| Dodecane | 119, 142, 437 |
| 1-Dodecanol | 177, 178, 215, 324 |
| Dog blood, see blood, dog | |
| Dog fat, see fat, dog | |

E

| | |
|---|---|
| Eel blood, see blood, eel | |
| Erythritol, see (*R*,*R*)-1,2,3,4-butanetetrol | |
| Erythrocytes, see ox erythrocytes | |
| Ethane | 391, 392 - 399 |
| Ethane (ternary) | 376 - 382 |
| 1,2-Ethanediol | 184 |
| 1,1'-Ethanediol, 2,2,2-trichloro- (aqueous) | 84, 92, 237 |
| Ethane, 1,1,2-trichloro-1,2,2-trifluoro- | 239 |
| Ethane, 1,1'-oxybis + air | 521 |
| Ethane, 1,1'-oxybis | 227, 228 |
| Ethane, 1,1'-oxybis (ternary) | 105, 109, 110 |
| Ethane, 1,2-dichloro-1,1,2,2-tetrafluoro- | 238 |
| Ethane, 1,2-epoxy- | 231, 454 |
| Ethanol | 174, 177, 178, 185 - 190, 447 |
| Ethanol (aqueous) | 83, 88 - 91 |
| Ethanol (ternary) | 104, 105, 106 - 114 |
| Ethene | 387 - 390 |
| Ethylene glycol, see 1,2-ethanediol | |


(cont.)

System Index (cont.)


| System | Component | Pages |
|---|---|---|
| Ethylene oxide, see ethane, 1,2-epoxy- | | |
| Ethyl acetate, see acetic acid, ethyl ester | | |
| Ethyl palmitate, see hexadecanoic acid, ethyl ester | | |
| **F** | | |
| Fat, dog | | 281 |
| Fat, human | | 281 |
| FC-75 | + air | 530 |
| FC-80, see furan, heptafluorotetrahydro(nonafluorobutyl)- | | |
| Ferric oxide, see iron (3+) oxide | | |
| Freon-114, see ethane, 1,2-dichloro-1,1,2,2-tetrafluoro- | | |
| Freon-12, see methane, dichlorodifluoro- | | |
| *D*-Fructose (aqueous) | | 311 |
| Fuel | + air | 540 - 543 |
| Fuel (hydrocarbon) | + air | 544 |
| Fuel (hydrocarbon) | + nitrogen | 327, 328 |
| Fuel, jet | + air | 548, 549 |
| Fuel, jet | | 507, 508 |
| Fuel, RJ-1 | + air | 546, 547 |
| Fuel, RJ-1 | | 505, 506 |
| Furan, heptafluorotetrahydro(nonafluorobutyl)- | | 247, 248 |
| Furan, tetrahydro- | | 229 |
| **G** | | |
| Gas oil, see oil, gas | | |
| Gelatin (aqueous) | | 313 |
| Germanium, tetrachloro- | | 252 |
| α-*D*-Glucopyranoside, β-*D*-fructofuranosyl- (aqueous) | | 314 |
| *D*-Glucose | | 312 |
| Glycerol triacetate, see 1,2,3-propanetriol, triacetate | | |
| Glycine (aqueous) | | 307 |
| **H** | | |
| Hemaglobin | | 287, 288, 290, 291, 294, 296 |
| Heptane | + air | 517 |
| Heptane | | <u>119</u>, <u>120</u>, 128 - 131, <u>425</u>, 426 - 430 |
| Heptane (ternary) | | 355, 356 |
| Heptane, 1-bromo- | + air | 522 |
| Heptane, 1-bromo- | | 243 |
| Heptane, hexadecafluoro- | | 245 |
| 1-Heptanol | + air | 520 |
| 1-Heptanol | | <u>177</u>, <u>178</u>, 208 - 210 |
| Hexadecafluoroheptane, see heptane, hexadecafluoro- | | |
| Hexadecane | | <u>119</u>, <u>121</u>, 146, <u>147</u>, <u>419</u>, 438 |
| Hexadecanoic acid, ethyl ester | | 280 |
| Hexafluorobenzene, see benzene, hexafluoro- | | |
| Hexane | | <u>119</u>, <u>120</u>, 124 - 128, <u>419</u>, 422 - 424 |
| Hexane, 1-bromo-1,1,2,2,3,3,4,4,5,6,6,6-dodecafluoro-5-(trifluoromethyl)- | + air | 523 |
| Hexane, 1-chloro- | | 241 |
| Hexane, 1-chloro-1,1,2,2,3,3,4,4,5,6,6,6-dodecafluoro-5-(trifluoromethyl)- | + air | 524 |
| Hexane, 1,1,1,2,2,3,3,4,4,5,5,6,6-tridecafluoro-6-(1,2,2,2-tetrafluoro-1-(trifluoromethyl)ethoxy)- | | 249 |
| 1-Hexanol | | <u>177</u>, <u>178</u>, 205, 207 |
| Human blood, see blood, human | | |
| Hydrated aluminium oxide, see aluminium oxide, hydrated | | |
| Hydrated ferric oxide, see ferric oxide, hydrated | | |


(cont.)

System Index (cont.)


Hydrazine 270, 273
Hydrazine, 1,1-dimethyl- 272, 273
Hydrazine, 1,2-dimethyl- 274
Hydrazine, methyl- 271
Hydrazine, tetrafluoro- 487
Hydrogenated cotton seed oil,
see oil, hydrogenated cottonseed
Hydrogen sulfide 468 - 470
Hydrogen sulfide (multicomponent) 495

I

Iron (3+) oxide, hydrated 45 - 47, 61

J

Jet fuel, see fuel, jet

K

Kerosine + air 541, 542

L

Lard + air 533, 534
Lard 276, 277, 280
Lecithin 301
Lipoidal suspension 303
Lithium chloride (aqueous) 45 - 47, 49, 64
*L*-Arabinose (aqueous) 310
*L*-1822 (mixture of mostly 10-carbon fluorocarbons) + air 529
*L*-1822 (mixture of mostly 10-carbon fluorocarbons) 251

M

Magnesium sulfate, see sulfuric acid, magnesium salt
Manganese acetate, see acetic acid, manganese (2+) salt
Mercury (2+) acetate, see acetic acid, mercury (2+) salt
Methane 361, 362, 363 - 375
Methane (ternary) 376 - 386, 482, 498 - 501
Methane, chlorodifluoro- 455, 456
Methane, chlorotrifluoro- 234, 460, 461
Methane, dichloridifluoro- 235, 457
Methane, nitro- 260
Methane, sulfinylbis- 253, 258
Methane, tetrachloro- 236, 252, 458
Methane, tetrafluoro- 234, 459, 460, 461
Methane, trichloro- 232, 233
Methanol 174, 177, 178, 179 - 183, 446
Methanol (aqueous) 83, 86, 87
Methanol (multicomponent) 495
*N*-Methylacetamide, see acetamide, *N*-methyl
Methyl acetate, see acetic acid, methyl ester
Methylcyclohexane, see cyclohexane, methyl-
2-Methyl heptane, see heptane, 2-methyl-
Methylhydrazine, see hydrazine, methyl-
5-Methyl-*N*-methyl pyrrolidinone,
see 2-pyrrolidinone, 1,5-dimethyl-
1-Methylnaphthalene, see naphthalene, 1-methyl-
2-Methyl propane, see propane, 2-methyl-


(cont.)

System Index (cont.)

*N*-Methylpyrrolidinone, see 2-pyrrolidinone, *N*-methyl-
Mineral oil, white, see oil, mineral

N

Naphtha 503
Naphthalene, octadecafluorodecahydro- + air 527
Naphthalene, 1-methyl- 172
Naphthalene, 1,2,3,4-tetrahydro- 280
Nickel (2+) acetate, see acetic acid, nickel (2+) salt
Nitric acid 59
Nitric acid, propyl ester + air 531
Nitric acid, propyl ester 262
Nitrobenzene, see benzene, nitro-
Nitrogen oxide ($N_2O$) 316
Nitrogen oxide ($N_2O_4$) 317, 318
Nitrogen trifluoride 486
Nitromethane, see methane, nitro-
Nitrous oxide 484
Nitryl fluoride 485
Nonane 119, 120, 137, 138
1-Nonanol 177, 178, 213

O

Octadecafluorodecahydronaphthalene,
see naphthalene, octadecafluorodecahydro-
Octadecafluorooctane, see octane, octadecafluoro-
9-Octadecenoic acid 280
Octamethylcyclotetrasiloxane,
see cyclotetrasiloxane, octamethyl-
Octane + air 517
Octane 119, 120, 131 - 134, 419, 431, 432
Octane, 1-bromo-1,1,2,2,3,3,4,4,5,5,6,6,7,7,-
8,8,8-heptadecafluoro- + air 525
Octane, octadecafluoro- + air 526
1-Octanol + air 520
1-Octanol 176 - 178, 209 - 212
Oil, asphaltic mineral + air 538
Oil, butter + air 534
Oil, butter 277
Oil, cod-liver + air 535, 536
Oil, corn + air 533, 536
Oil, corn 276
Oil, cottonseed + air 533, 534, 536
Oil, cottonseed 276, 277
Oil, gas 504
Oil, herring + air 536
Oil, hydrocarbon + air 544
Oil, hydrocarbon 328
Oil, hydrogenated, cottonseed + air 533
Oil, hydrogenated, cottonseed 276
Oil, linseed + air 536
Oil, lubricating + air 539, 545
Oil, mineral + air 536, 537 - 539
Oil, mineral 319 - 321, 331
Oil, olive + air 535, 536
Oil, olive 279, 280 - 283
Oil, paraffin + air 538
Oil, paraffin 322 - 324
Oil, soybean 278
Oil, transformer type + air 538
Oxygen (multicomponent) 76, 353, 354, 466
Oxygen fluoride 467
Ox blood, see blood, ox
Ox erythrocytes 290, 291, 295

(cont.)

System Index (cont.)

P


| System | | Pages |
|---|---|---|
| Paraffin wax | | 325 |
| Pentadecane | | 119, 145 |
| Pentanedioic acid, dimethyl ester | | 451 |
| Pentane | | 123, 419, 420 |
| Pentane, 2,2,4-trimethyl- | + air | 517, 518 |
| Pentane, 2,2,4-trimethyl- | | 121, 131, 135, 136, 433, 434 |
| Pentane, 2,2,4-trimethyl- (ternary) | | 104, 113, 115 - 118 |
| 1-Pentanol | | 176 - 178, 202 - 205 |
| 3,6,9,12,15-Pentaoxaoctadecane,1,1,1,2,4,4,5,-7,7,8,10,10,11,13,13,14,16,16,17,17,18,-18,18-tricosafluoro-5,8,11,14-tetrakis-(trifluoromethyl)- | + air | 528 |
| 2,5,8,11,14-Pentaoxapentadecane | | 453 |
| Pentyl acetate, see acetic acid, pentyl ether | | |
| Perchloryl fluoride, see chlorine hexafluoride | | |
| Perfluorodimethylcyclohexane, see cyclohexane, decafluorobis(trifluoromethyl)-cyclohexane | | |
| Perfluoro-1,4-di*iso*propoxy butane see butane, 1,1,2,2,3,3,4,4-octafluoro-1,4-bis(1,2,2,2-tetrafluoro-1-(trifluoromethyl)-ethoxy)- | | |
| Perfluorodecalin see naphthalene, octadecafluorodecahydro- | | |
| Perfluoroheptane, see heptane, hexadecafluoro- | | |
| Perfluoromethylcyclohexane, see cyclohexane, undecafluoro(trifluoromethyl)- | | |
| Perfluorooctane, see octane, octadecafluoro- | | |
| Perfluoro-1-*iso*propoxyhexane, see hexane, 1,1,1,2,2,3,3,4,4,5,5,6,6-tridecafluoro-6-(1,2,2,2-tetrafluoro-1-(trifluoromethyl)ethoxy)- | | |
| Perfluorotetrahydrofuran, see furan, heptafluorotetrahydro(nonafluorobutyl)- | | |
| Perfluorotributylamine, see 1-butanamine, 1,1,2,2,3,3,4,4,4-nonafluoro-*N*,*N*-bis(nonafluorobutyl)- | | |
| Petroleum | | 329 |
| Petrowax A | | 326 |
| Piperidine | | 263 |
| Plasma | | 294, 295 |
| Potassium carbonate, see carbonic acid, dipotassium salt | | |
| Potassium chloride (aqueous) | | 45 - 47, 54, 55, 78, 79 |
| Potassium iodide (aqueous) | | 45 - 47, 55, 80 |
| Propane | | 402, 403 - 408 |
| Propane (ternary) | | 383, 384 |
| Propane, 2-methyl- | | 416 - 417 |
| Propane, 2-methyl- (ternary) | | 414, 415 |
| Propane, 1,1'-oxybis- | | 230 |
| 1,2,3-Propanetriol (aqueous) | | 84, 100 - 102 |
| 1,2,3-Propanetriol, triacetate | | 452 |
| 2-Propanol | | 175, 195, 196 |
| 2-Propanol (aqueous) | | 84, 98, 99 |
| 1-Propanol | | 174, 175, 177, 178, 191 - 194 |
| 1-Propanol (aqueous) | | 84, 96, 97 |
| 1-Propanol (ternary) | | 104, 115, 116 |
| 1-Propanol, 2-methyl- | | 175, 176, 201 |
| Propanoic acid (aqueous) | | 84, 94, 95 |
| Propanoic acid, 2-methyl- | | 222 |
| Propanoic acid, 2-methyl- (aqueous) | | 84, 103 |


(cont.)

System Index (cont.)


2-Propanone 216, 217 - 219
2-Propanone (ternary) 105, 106, 114
Propene 400, 401
Propene, 2-methyl- 418
Propionic acid, see propanoic acid
Propylene carbonate, see 1,3-dioxolan-2-one, 4-methyl-
1-Propyl nitrate, see nitric acid, propyl ester
Pyran, 2,3-dihydro- 229
2*H*-Pyran, tetrahydro- 229
Pyridine 263
Pyrrolidine 263
2-Pyrrolidinone, 1,5-dimethyl- 463
2-Pyrrolidinone, 1-methyl- 462

R

Rabbit blood, see blood, rabbit
Rabbit brain, see brain, rabbit
Red cell ghost, human 298

S

Saline 293
Santowax R 502
Seawater 31 - 33, 34 - 44
Seawater + argon + nitrogen + oxygen 37, 38
Serum albumin 297
Silane, tetrachloro- 252
Silane, trichloromethyl- 252
Sodium bicarbonate, see carbonic acid, monosodium salt
Sodium carbonate, see carbonic acid, disodium salt
Sodium chloride (aqueous) + air 513, 514
Sodium chloride (aqueous) 45 - 47, 50 - 52, 67 - 71, 348 - 351
Sodium hydroxide (aqueous) 45 - 47, 49, 50, 65, 66
Sodium sulfate, see sulfuric acid, sodium salt
Sodium sulfite, see sulfurous acid, disodium salt
Steam rendered lard, see lard
Sucrose, see α-*D*-glucopyranoside, β-*D*-fructofuranosyl-
Sulfate, magesium, see sulfuric acid, magnesium salt
Sulfate, sodium, see sulfuric acid, sodium salt
Sulfur hexafluoride 259
Sulfuric acid + air 515, 516
Sulfuric acid + nitrogen 45 - 48, 57, 58, 74, 76
Sulfuric acid, magnesium salt (aqueous) 345
Sulfuric acid, sodium salt (aqueous) + air 345, 515, 516
Sulfuric acid, sodium salt (aqueous) 45 - 47, 52, 53, 73 - 76, 347
Sulfurous acid, disodium salt (aqueous) 45 - 47, 52, 53, 72, 75
Sulfur dioxide 315, 475

T

Tetradecane 119, 144
Tetrahydropyran, see 2*H*-pyran, tetrahydro-
Tetrafluorohydrazine, see hydrazine, tetrafluoro-
Tetralin, see naphthalene, 1,2,3,4-tetrahydro-
Tetramethylene glycol dimethyl ester, see 2,5,8,11,15-pentaoxapentadecane
Toluene, see benzene, methyl-
Triacetin, see 1,2,3-propanetriol, triacetate
1,1,2-Trichloro-1,2,2-trifluoroethane, see ethane, 1,1,2-trichloro-1,2,2-trifluoro-
Trichloromethane, see methane, trichloro-


(cont.)

System Index (cont.)


1,1,1,2,4,4,5,7,7,8,10,10,11,13,13,14,16,16,-17,17,18,18,18-Tricosafluoro-5,8,11,14-tetrakis(trifluoromethyl)-3,6,9,12,15-pentaoxaoctadecane, see 3,6,9,12,15-pentaoxaoctadecane, 1,1,1,2,-4,4,5,7,7,8,10,10,11,13,13,14,16,16,17,17,18,-18,18-tricosafluoro-5,8,11,14-tetrakis-(trifluoromethyl)-

Tridecane 119, 142

Trimethylborane 497

2,2,4-Trimethyl pentane, see pentane, 2,2,4-trimethyl-

Trout blood, see blood, trout

U

Undecafluoro(trifluoromethyl)cyclohexane, see cyclohexane, undecafluoro-(trifluoromethyl)-

Undecane 119, 141

1-Undecanol 177, 178, 215

Urea (aqueous) 304 - 306

Urine, human 299

W

Water + air 510 - 516

Water 1 - 4, 5 - 29, 45 - 56, 57 - 82, 83, 84 - 103, 286, 294, 304 - 314, 333, 334 - 352

Water-$d_2$ 30

Wax, see paraffin wax

X

Xylene, see benzene, dimethyl-

Z

Zinc (2+) acetate, see acetic acid, zinc (2+) salt

## REGISTRY NUMBER INDEX

Underlined page numbers refer to evaluation text and those not underlined to compiled tables.


| Registry No. | Pages |
|---|---|
| 50-99-7 | 312 |
| 56-23-5 | 236, 252, 258 |
| 56-40-6 | 307 |
| 56-81-5 | 84, 100-102 |
| 57-13-6 | 304-306 |
| 57-14-7 | 272, 273 |
| 57-48-7 | 311 |
| 57-50-1 | 314 |
| 57-88-5 | 300 |
| 60-29-7 | 105, 109, 110, 227, 228, 521 |
| 60-34-4 | 271 |
| 60-35-5 | 84, 93 |
| 62-53-3 | 265, 266 |
| 64-17-5 | 83, 88-91, 104, 105, 106-114, 174, 177, 178, 185-190, 447 |
| 64-19-7 | 45-47, 54, 77, 221, 357-360 |
| 67-56-1 | 83, 86, 87, 174, 177, 178, 179-183, 446, 495 |
| 67-63-0 | 83, 98, 99, 175, 195, 196 |
| 67-64-1 | 105, 106, 114, 216, 217-219 |
| 67-66-3 | 232, 233 |
| 67-68-5 | 253, 258 |
| 71-23-8 | 104, 115, 116, 174, 175, 177, 178, 191-194 |
| 71-32-8 | 84, 96, 97 |
| 71-36-3 | 175, 177, 178, 197-200 |
| 71-41-0 | 176-178, 202-205 |
| 71-43-2 | 104, 111, 112, 162, 163-168, 252, 300-302, 443, 444 |
| 71-48-7 | 45-47, 54, 77 |
| 74-82-8 | 361, 362, 363-386, 482, 498-501 |
| 74-84-0 | 376-382, 391, 392-399 |
| 74-85-1 | 387-390 |
| 74-98-6 | 383, 384, 402, 403-408 |
| 75-15-0 | 253, 254-257 |
| 75-21-8 | 231, 454 |
| 75-28-5 | 414-418 |
| 75-45-6 | 455, 456 |
| 75-52-5 | 260 |
| 75-71-8 | 235, 457 |
| 75-72-9 | 234, 460, 461 |
| 75-73-0 | 234, 459-461 |
| 75-79-6 | 252 |
| 76-13-1 | 239 |
| 76-14-2 | 238 |
| 78-78-4 | 421 |
| 78-83-1 | 175, 176, 201 |
| 79-09-4 | 84, 94, 95 |
| 79-16-3 | 261 |
| 79-20-9 | 223 |
| 79-31-2 | 84, 103, 222 |
| 90-12-0 | 172 |
| 92-51-3 | 161 |
| 98-95-3 | 264, 265 |
| 101-81-5 | 173 |
| 102-76-1 | 452 |
| 106-97-8 | 385, 386, 409, 410-415, 418 |
| 106-99-0 | 418 |
| 107-21-1 | 184 |


(cont.)

Registry Number Index


108-32-7 448
108-87-2 155, 355, 356, 441, 442
108-88-3 162, 169, 170, 445
108-90-7 240
108-93-0 207, 357, 358

108-94-1 220, 357-360, 449, 450
109-66-0 123, 419, 420
109-99-9 229
110-19-0 225
110-34-3 119, 120, 124-128

110-54-3 419, 422-424
110-82-7 104, 117, 118, 148, 149-154, 353, 354, 439, 440
110-83-8 154
110-86-1 263
110-89-1 263

111-27-3 177, 178, 205, 206
111-43-3 230
111-65-9 119, 120, 131-134, 419, 431, 432, 517
111-70-6 177, 178, 208-210, 520
111-84-2 119, 120, 137, 138

111-87-5 176-178, 209-212, 520
112-08-8 176-178, 213, 214, 324
112-30-1 176-178, 213, 214, 324
112-40-3 119, 142, 437, 519
112-42-5 177, 178, 215

112-53-8 177, 178, 215, 324
112-80-1 280
115-07-1 400, 401
119-64-2 280
123-75-1 263

123-91-1 229
124-18-5 199-121, 139, 140, 419, 435, 436, 517
124-38-9 76, 466, 488, 489-495, 498-502
127-09-3 45-47, 54, 77
141-78-6 105, 107, 108, 224

142-68-7 229
142-71-2 45-47, 54, 77
142-82-5 119, 120, 128-131, 355, 356, 425, 426-430, 517
143-08-8 177, 178, 213
143-24-8 453

144-55-8 45-47, 53, 71
292-64-8 156
302-01-2 270, 273
302-17-0 84, 92, 237
306-94-5 527

307-34-6 526
311-89-7 267-269, 532
335-02-2 244
335-36-4 247
335-57-9 245

338-69-2 308
373-02-4 45-47, 54, 77
392-56-3 242
423-55-2 525
495-19-8 45-47, 53, 71


(cont.)

Registry Number Index


| | |
|---|---|
| 540-73-8 | 274 |
| 540-84-1 | 104, 113, 115-118, 121, 135, 136, 433, 434, 517, 518 |
| 544-10-5 | 241 |
| 544-76-3 | 119, 121, 146, 147, 419, 438 |
| 556-67-2 | 275 |
| 557-34-6 | 45-47, 54, 77 |
| 584-08-7 | 45-47, 55, 81, 82 |
| 593-90-8 | 497 |
| 624-29-3 | 159 |
| 627-13-4 | 262, 531 |
| 628-63-7 | 226 |
| 628-97-7 | 280 |
| 629-04-9 | 243, 522 |
| 629-50-5 | 119, 143 |
| 629-59-4 | 119, 144 |
| 629-62-9 | 119, 145 |
| 638-04-0 | 160 |
| 638-38-0 | 45-47, 54, 77 |
| 872-50-4 | 462 |
| 1119-40-0 | 451 |
| 1120-21-4 | 119, 141 |
| 1310-73-2 | 45-47, 49, 50, 65, 66 |
| 1330-20-7 | 171 |
| 1333-84-2 | 45-48, 60 |
| 1600-27-7 | 45-48, 54, 77 |
| 2207-01-4 | 157 |
| 2207-03-6 | 160 |
| 2207-04-7 | 159 |
| 2551-62-4 | 259 |
| 5075-92-3 | 463 |
| 5328-37-0 | 310 |
| 6876-23-9 | 158 |
| 7440-37-1 | 37, 38, 482 |
| 7446-09-5 | 315, 475 |
| 7447-40-7 | 45-47, 54, 55, 78, 79 |
| 7447-41-8 | 45-47, 49, 64 |
| 7487-88-9 | 45-48, 345 |
| 7493-90-5 | 309 |
| 7647-14-5 | 45-47, 50-52, 67-71,348-351, 513, 514 |
| 7664-41-7 | 476, 477-483 |
| 7664-93-9 | 45-48, 53, 57, 58, 74, 76, 515, 516 |
| 7681-11-0 | 45-48, 55, 80 |
| 7697-37-2 | 59 |
| 7727-37-9 | 1-4, 5-30, 31-33, 34-44, 45-56, 57-82, 83-85, 86-103, 104, 105, 106-118, 119-122, 123-147, 148, 149-161, 162, 163-173, 174-178, 179-215, 216, 217-252, 253, 254-278, 279, 280-331, 333, 334-360, 361, 362, 363-390, 391, 392-401, 402, 403-408, 409, 410-418, 419, 420-424, 425, 426-475, 476, 477-487, 488, 489-508 |
| 7732-18-5 | 1-4, 5-29, 45-46, 57-82, 83-85, 86-103, 237, 281, 286, 293, 294, 297, 303-314, 333, 334-352, 510-516 |
| 7757-82-6 | 45-47, 52, 53, 73-76, 347, 515, 516 |
| 7782-44-7 | 37, 38, 76, 353, 354, 466 |
| 7783-06-4 | 468-470, 495 |
| 7783-41-7 | 467 |
| 7783-54-2 | 486 |


(cont.)

Registry Number Index


7789-20-0 30
7790-91-2 471
10022-50-1 485
10024-97-2 316, 484
10026-04-7 252

10036-47-2 487
10038-98-9 252
10043-53-4 45-49, 346
10361-37-2 45-47, 49, 62, 63
10544-72-6 317, 318

10579-83-6 45-47, 52, 53, 72, 75
12259-21-1 45-48, 61
13637-63-3 472, 473
19287-45-7 496
23228-90-2 250

25512-65-6 229
37340-18-4 249
37486-69-4 528
40464-54-8 248
41719-16-8 249

50279-29-3 523
56077-96-4 474
63243-37-8 524

AUTHOR INDEX


Adeney, W. E. 11
Adlivankina, M. A. 234
Akers, W. W. 411, 427
Amster, A. B. 262, 531
Arai, Y. 456, 493, 498, 499
Ar, A. 293
Atwell, L. L. 411
Avdeeva, O. I. 70
Azarnoosh, A. 435, 436

Baldwin, R. R. 136, 319, 327, 510, 518, 537, 541
Balog-Megyery, K. see Megyery-Balog, K.
Baranovich, Z. N. 424, 432
Barsuk, S. D. 413
Battino, R. 19, 28, 134, 140, 152, 155-160, 167, 170, 201, 212, 214, 242, 275, 282, 300-302
Becker, H. G. 11
Benson, B. B. 21, 22, 37, 38
Ben'yaminovic, O. A. 413
Berengarten, M. G. 127, 153, 168
Besserer, G. J. 420, 421, 468, 469
Bikov, M. M. 78, 79
Billett, F. 17, 64, 69, 80
Bircher, L. J. 181, 188, 192, 198, 204, 206, 208, 211
Blagoi, Yu. P. 400
Bloomer, O. T. 363, 364, 392
Bohr, C. 9, 58
Boomer, E. H. 426
Boyer, F. L. 181, 188, 192, 198, 204, 206, 208, 211
Bozhovakaya, T. K. 70
Bradstreet, E. D. 24, 352
Brande, H. 505, 506, 546, 547
Brasted, R. C. 77
Braun, L. 62, 63, 67, 68, 94, 95, 304, 305
Brunner, G. 355, 356, 429, 442
Byrne, J. E. 167, 300-302

Cannon, W. A. 387, 394, 406, 467, 471, 473, 474, 481, 485-487, 496, 497
Capell, R. G. 325, 326
Chaikovskii, V. F. 460, 461
Chang, E. T. 270-274, 317, 318
Chang, S. D. 369, 378, 379, 393
Chappelear, P. S. 372, 373, 395, 396
Cheung, H. 368, 403
Christoff, A. 57, 227, 521
Christoforides, C. 286
Cines, M. R. 365-367
Clark, F. M. 538
Clark, L. C. 523-530, 532
Clever, H. L. 19
Conant, J. B. 287, 288
Corinaldi, G. 322
Cosgrove, B. A. 29, 30
Cosway, H. F. 376, 377
Creighton, H. J. M. 289

Dana, L. I. 490
Danforth, W. F. 134, 140, 167, 201, 212, 214, 282, 300-302
Daniel, S. G. 136, 319, 327, 510, 518, 537, 541
Davidson, D. 280
Dean, M. R. 475
Delaney, D. E. 261
Derry, L. D. 542, 543
Dillon, R. T. 15, 71, 290, 291, 295
Dodge, B. F. 443, 479, 480
Dornte, R. W. 315
Douglas, E. 23, 39, 40


(cont.)

Author Index (cont.)


Drucker, K. 100, 103, 222
Drushlyak, O. G. 59
Dymond, J. H. 150, 151, 258
Dymova, R. P. 357, 358, 449, 450

Eakin, B. E. 392
Eckert, C. A. 459
Edwards, A. W. T. 20, 284, 285, 299
Edwards, H. 505, 506, 546, 547
Efremova, G. D. 444
Eggleton, P. 280
Eichelberger, W. C. 341, 512-514
Eicke, H. F. 428, 441
Ellington, R. T. 392
El'natanov, A. I. 445
Emmel, V. M. 36
English, W. D. 387, 394, 406, 467, 471, 473, 474, 481, 485-487, 496, 497
Enns, T. 24, 352
Evans, E. B. 542, 543
Evans, F. D. 201, 242, 282
Ewing, G. J. 296

Faeth, G. M. 437, 507, 508, 519, 548, 549
Farhi, L. E. 20, 284, 285, 293, 299
Faulkner, B. A. 542, 543
Ferguson, C. V. 315
Ferrell, J. K. 495
Ficke, H. F. 434
Field, L. R. 155, 170
Figuiere, P. 430
Findl, E. 505, 506, 546, 547
Findlay, A. 289
Foggie, P. 280
Fox, C. J. J. 7, 8, 34, 35
Fredenslund, A. 388, 397, 401, 408
Friedman, H. L. 260
Frolich, P. K. 12, 410, 447, 458, 503, 504
Fujio, J. 107-112
Furmer, I. E. 127, 153, 168

Gaddy, V. L. 335, 336, 478
Gall, J. F. 259
Gami, D. C. 392
Garanin, V. F. 472
Gardner, G. C. 380-382, 398, 399
Gasem, K. A. M. 389, 390
Geller, E. B. 157-160
Gemant, A. 539
Ginsberg, K. A.
Gjaldbaek, J. C. 124, 129, 132, 137, 149, 165, 184, 191, 203, 207, 244-246, 255
Gniewosz, S. 329
Gokcen, N. A. 270-274, 317, 318
Goodman, J. B. 13, 334
Gorbachev, V. M. 252
Goryunova, N. P. 455, 457
Graham, E. B. 431, 433
Grauso, L. 388, 397, 401, 408
Grensburg, A. G.
Grezin, A. K. 460, 461
Grollman, A. 303
Gross, D. H. 19
Grove, N. H. 502
Guerry, D. Jr. 128, 154, 220, 229, 230, 241, 263
Gupta, M. K. 380-382, 398, 399

Haller, H. S. 277, 534
Hammel, A. Von 102, 237


(cont.)

Author Index (cont.)


Hawkins, J. A. 16, 464, 465
Hedley-Whyte, J. 286
Hegarty, M. J. 398, 399
Heins, C. 14, 335
Heise, F. 483
Hess, L. G. 454
Hiemenz, W. 324
Hildebrand, J. H. 124, 149, 151, 165, 239, 244-246, 255, 256, 267
Hiraoka, H. 239
Hirayama, C. 77
Hiza, M. J. 370, 371, 374, 375, 389, 390
Hogan, J. J. 12, 365-367, 410, 447, 458, 503, 504
Hom, J. F. 430
Homma, T. 20, 284, 285, 299
Horiuti, J. 164, 218, 223, 228, 236, 240
Hüfner, G. 6, 93, 306-312

Ijams, C. C. 131, 209, 243, 517, 520, 522
Ikels, K. G. 281
Iltis, R. 523-530, 532
Ionescu, L. V. 296

Jelfs, E. C. G. 542, 543
Johnson, C. A. 426
Just, G. 88, 163, 169, 171, 179, 185, 202, 217, 221, 224-226, 232, 254, 264, 266

Kalra, H. 416, 417, 420, 421, 470
Kaminishi, G. 491, 493
Kaplan, L. K. 448, 452, 453, 462, 463
Katayama, T. 107-112, 115-118, 125, 182, 189, 193, 196, 199
Katz, D. L. 376, 377
Kelemen, S. 342, 345-348
Kenton, F. H. 74, 76, 515, 516
Khazonova, N. E. 489
Khodeeva, S. M. 353, 354, 357-360, 439, 449, 450
Khuchua, K. S. 414, 415
Kidnay, A. J. 370, 374, 375, 389, 390, 398, 399, 494, 500, 501
Kilgore, C. H. 427

Klots, C. E. 21, 22
Kobatake, Y. 256, 267
Kobayashi, R. 372, 373, 395, 396
Kobe, K. A. 74, 76, 515, 516
Kohn, D. M. 427
Korosy, F. 233
Krase, N. W. 13, 334, 337
Kretschmer, C. B. 89, 106, 113, 114, 135, 180, 187, 195, 197, 219
Krichevskii, I. R. 444, 446, 489
Krishnan, T. 416, 470
Kubie, L. S. 321
Kurata, F. 404, 405
Kusano, K. 278

Laugier, S. 430
Lebedeva, E. S. 446
Leites, I. L. 234
Lesnevskaya, L. S. 489
Levina, M. I. 65, 66
Levy, J. B. 262, 531
Lindroos, A. E. 479, 480
Logvinyuk, V. P. 328, 544

Lu, B. C. Y. 369, 378, 379, 382, 383, 393, 407
Luther, H. 323, 324


(cont.)

Author Index (cont.)


Makarenkow, V. V. 328, 544
Makranczy, J. 123, 126, 127, 130, 133, 138, 139, 141-146, 153, 168, 183, 190, 194, 200, 205, 210, 213, 215
Malyshev, V. V. 328, 544
Manowitz, B. 235, 316, 330
Margaria, R. 15, 71, 290, 291, 295
Maslennikova, V. Ya. 455, 457
Masuoka, H. 492
Matange, J. N. 495
Matous, J. 482
Matyash, Yu. I. 460, 461
Maude, A. H. 320
McHale, J. L. 275
McKee, O. L. Jr. 340, 511
McKetta, J. J. 166, 385, 386, 412, 422, 423, 435, 436
Megyery-Balog, K. 123, 126, 130, 133, 138, 139, 141-146, 183, 190, 194, 200, 205, 210, 213, 215
Metschl, J. 186, 265
Miller, H. C. 259
Miller, R. C. 370, 374, 375
Miller, P. 443
Minnich, B. H. 339
Miranda, R. D. 416, 417
Mishnina, T. A. 70
Modell, J. H. 248-250
Moles, E. 100, 103, 222
Mollerup, J. 388, 397, 401, 408
Morrison, T. J. 17, 64, 69, 80
Muirbrook, N. K. 466
Müller, C. 10, 92, 101, 314
Murray, C. N. 25, 41-44

Nagy, B. 342, 345-348
Namiot, A. Yu. 438
Nasini, A. G. 322
Niemann, H. 184, 191, 203, 207
Nikitina, I. E. 413
Nitta, T. 107-112, 115-118, 125, 182, 189, 193, 196, 199
Nohka, J. 456
Novak, J. P. 482
Nowakowaska, J. 89, 106, 113, 114, 135, 180, 187, 195, 197, 219

Ohta, Y. 293
Olson, J. D. 231
Orobinskii, N. A. 400
O'Sullivan, T. D. 343, 344, 349-351

Panchenkov, G. M. 328, 544
Parent, J. D. 363, 364, 392
Parker, P. D. M. 37, 38
Parrish, W. R. 371, 374, 375
Patyi, L. 123, 126, 127, 130, 133, 138, 139, 141-146, 153, 168
Peer, A. A. 12, 410, 447, 458, 503, 504
Peter, S. 355, 356, 428, 429, 434, 441, 442
Piercey, G. A. 426
Podzolko, L. G. 472
Poon, D. P. L. 383, 384, 407
Poston, R. S. 422, 423
Poston, T. M. 270-272, 274
Potter, O. E. 440
Powell, R. J. 257, 269
Power, G. G. 26, 283, 298
Prausnitz, J. M. 147, 161, 172, 173, 459, 466
Pray, H. A. 339
Pruzansky, J. 235, 316


(cont.)

Author Index (cont.)


Rakestraw, N. W. 36
Renon, H. 430
Richon, D. 430
Ridenour, W. P. 325, 326
Riley, J. P. 25, 41-44
Roach, J. T. 365-367
Roberts, L. R. 385, 386, 412
Robinson, D. B. 416, 417, 420, 421, 468-470
Robinson, J. A. 411
Robson, J. H. 387, 394, 406, 467, 471, 473, 474, 481, 485-487, 496, 497
Rodman, C. J. 320
Roland, C. H. 365-367
Rottger, H. 323
Rousseau, R. W. 495
Rupprecht, S. D. 437, 507, 508, 519, 548, 549
Rusz, L. 123, 126, 130, 133, 138, 139, 141-146, 183, 190, 194, 200, 205, 210, 213, 215
Ryabstev, N. I. 414, 415

Saddington, A. W. 337
Sadilenko, A. S. 127, 153, 168
Safranova, T. P. 331
Saito, S. 456, 493, 498, 499
Sandalova, L. 489
Sarashina, E. 456, 498, 499
Sargent, J. W. 247, 251, 268
Saylor, J. H. 19
Schaffer, P. S. 277, 534
Schindler, D. L. 404, 405
Schmidt-Nielsen, S. 536
Scholander, P. F. 24, 352
Schroder, W. 81, 82
Schweichert, C. E. 339
Schweitzer, P. H. 540, 545
Scott, N. D. 287, 288
Seffl, R. J. 247, 251, 268
Shakhova, S. F. 448, 451-453, 462, 463
Shapka, A. V. 59
Shenderei, L. I. 445
Shilling, C. W. 16, 464, 465
Shkol'nikova, R. I. 60, 61, 297, 313
Sibbitt, W. L. 18
Skripka, V. G. 413, 438
Smirnova, A. M. 424, 432
Smith, N. O. 342-351
Sobr, J. 482
Somait, F. A. 494, 500, 501
Spinks, T. J. 27
Sridhar, T. 440
Steen, J. B. 292
Stegall, H. 26, 283, 298
Steinbach, H. G. 418
Steinberg, M. 235, 316, 330
Steinbrecher, M. 418
Stepanova, Z. G. 127, 153, 168
Stoddard, J. L. 294
Stryjek, R. 372, 373, 395, 396
Stsibarovskaya, N. P. 65, 66
Suciu, S. N. 18, 338
Sukhoverkhov, V. F. 472
Sultanov, R. G. 438
Swift, G. W. 404, 405
Szebehely, V. G. 540, 545
Szwarc, H. 430

Tatsuishi, A. 115-118
Tauch, E. J. 12, 410, 447, 458, 503, 504
Tham, M. K. 248-250


(cont.)

Author Index (cont.)


Thomsen, E. S. 129, 132, 137
Tilton, V. V. 454
Tokunaga, J. 86, 87, 90, 91, 96-99
Tomoto, N. 278
Toriumi, T. 491
Tremearne, T. H. 477
Tremper, K. K. 147, 161, 172
Tret'yakov, G. V. 252
Tsiklis, D. S. 445, 455, 457

Van Slyke, D. D. 15, 71, 290, 291, 295
Verdelli, L. S. 259
Vernon, H. M. 535
Vibrans, F. C. 276, 533

Walfisz, A. 329
Walker, R. D. Jr. 248-250
Walkley, J. 29, 30
Walls, W. S. 475
Wang, D. I. J. 368, 403
Weale, K. E. 431, 433
Weatherford, W. D. 325, 326
Wenzel, H. 355, 356, 429, 442
Wesseler, E. P. 523-530, 532
Whiteley, F. J. 502
Wiebe, R. 14, 89, 106, 113, 114, 135, 180, 187, 195, 197, 219, 335, 336, 477, 478
Wilcock, R. J. 28, 134, 140, 156, 212, 214, 275
Wild, J. D. 440
Wilhelm, E. 134, 140, 152, 155-160, 170, 201, 212, 214, 275
Williams, V. D. 238
Wilson, T. R. S. 25, 41-44
Winkler, L. W. 5
Wood, R. H. 261

Yasunishi, A. 72, 73, 75
Yen, L. C. 166
Yorizane, M. 492
Yoshimura, S. 492
Yudina, L. V. 59

Zakharov, N. D. 460, 461
Zeininger, H. 484
Zenner, G. H. 490
Zhuravleva, A. N. 59
Zhuze, T. P. 331
Zoss, L. M. 18
Zubchenko, Yu. P. 448, 451-453, 462, 463